Seabed Fluid Flow
The Impact on Geology, Biology, and the Marine Environment

Seabed fluid flow, also known as submarine seepages, involves the flow of gases and liquids through the seabed. This includes hot hydrothermal vents, and cold hydrocarbon seeps – the hydrocarbons may be of thermogenic or microbial origin. Such fluids have been found to leak out through the seabed into the marine environment in seas and oceans around the world – from the coasts to deep ocean trenches.

This geological phenomenon has widespread implications for the sub-seabed, seabed and marine environments. Seabed fluid flow affects seabed morphology (including pockmarks and mud volcanoes), mineralisation, and benthic ecology – sustaining unique chemosynthetic biological communities. Natural fluid emissions also have a significant impact on the composition of the oceans and atmosphere; methane emissions have important implications for the global climate. Shallow gas is a significant geohazard to offshore operations, seeps are a petroleum exploration aid, and gas hydrates and hydrothermal minerals are potential future resources.

This book describes seabed fluid flow features and processes, and demonstrates their importance to human activities, natural environments and the workings of our planet. It is targeted at the growing community of research scientists with interests in the marine environment (above and below the seabed), and at those involved in the exploration for, and exploitation of, marine resources. *Seabed Fluid Flow* is also supported by a website (www.cambridge.org/9780521114202) hosting colour versions of many of the illustrations, and additional material – most notably feature location maps.

ALAN JUDD, formerly working at the Universities of Sunderland and Newcastle upon Tyne, is now an independent consultant working mainly for the petroleum and offshore site survey industries. He is a Fellow of the Geological Society of London and a Chartered Geologist.

MARTIN HOVLAND is a Marine Geology Specialist at Statoil ASA, Stavanger, Norway. He is a Fellow of the Geological Society of London, and a member of the American Geophysical Union and the American Association for the Advancement of Science.

Seabed Fluid Flow

The Impact on Geology, Biology, and the Marine Environment

Alan Judd and Martin Hovland

CAMBRIDGE
UNIVERSITY PRESS

CAMBRIDGE UNIVERSITY PRESS
Cambridge, New York, Melbourne, Madrid, Cape Town, Singapore, São Paulo, Delhi

Cambridge University Press
The Edinburgh Building, Cambridge CB2 8RU, UK

Published in the United States of America by Cambridge University Press, New York

www.cambridge.org
Information on this title: www.cambridge.org/9780521114202

First published 2007
This digitally printed version 2009

A catalogue record for this publication is available from the British Library

ISBN 978-0-521-81950-3 hardback
ISBN 978-0-521-11420-2 paperback

Contents

Preface xi
Acknowledgements xii
Note on the accompanying website xiii
List of maps on the accompanying website xiv
List of contributed presentations on the accompanying
website xv

1 **Introduction to seabed fluid flow** 1

2 **Pockmarks, shallow gas, and seeps: an initial
    appraisal** 7
   2.1 The Scotian Shelf: the early years 7
   2.2 North Sea pockmarks 8
      2.2.1 History of discovery 9
      2.2.2 Pockmark distribution 10
      2.2.3 Pockmark size and density 10
      2.2.4 Pockmark morphology 11
      2.2.5 Evidence of gas 15
   2.3 Detailed surveys of North Sea pockmarks and
         seeps 18
      2.3.1 The South Fladen Pockmark Study
              Area 18
      2.3.2 Tommeliten: Norwegian Block 1/9 25
      2.3.3 Norwegian Block 25/7 30
      2.3.4 The Holene: Norwegian Block
              24/9 33
      2.3.5 The Norwegian Trench 36
      2.3.6 Gullfaks 36
      2.3.7 Giant pockmarks: UK Block
              15/25 41
   2.4 Conclusions 44

3 **Seabed fluid flow around the world** 45
   3.1 Introduction 45
   3.2 The eastern Arctic 45
      3.2.1 The Barents Sea 45
      3.2.2 Håkon Mosby Mud Volcano 47
   3.3 Scandinavia 49
      3.3.1 Fjords in northern Norway 49
      3.3.2 The Norwegian Sea 50

      3.3.3 The Skagerrak 52
      3.3.4 The Kattegat 52
   3.4 The Baltic Sea 56
      3.4.1 Eckernförde Bay 56
      3.4.2 Stockholm Archipelago,
              Sweden 58
   3.5 Around the British Isles 59
      3.5.1 Pockmarks, domes, and seeps 60
      3.5.2 'Freak' sandwaves 60
      3.5.3 Methane-derived authigenic
              carbonate 62
      3.5.4 The Atlantic Margin 62
   3.6 Iberia 66
      3.6.1 The Rías of Galicia, northwest
              Spain 66
      3.6.2 Gulf of Cadiz 67
      3.6.3 Ibiza 69
   3.7 Africa 69
      3.7.1 The Niger Delta and Fan 69
      3.7.2 The continental slope of West
              Africa 70
   3.8 The Mid-Atlantic Ridge 72
   3.9 The Adriatic Sea 72
      3.9.1 Seeps and carbonates of the northern
              Adriatic 73
      3.9.2 Pockmarks, seeps, and mud diapirs in
              the central Adriatic 73
   3.10 The eastern Mediterranean 74
      3.10.1 Offshore Greece 75
      3.10.2 Mediterranean Ridge 76
      3.10.3 The Anaximander Mountains 79
      3.10.4 Eratosthenes Seamount 79
      3.10.5 Nile Delta and Fan 79
   3.11 The Black Sea 81
      3.11.1 Turkish Coast 81
      3.11.2 Offshore Bulgaria 81
      3.11.3 Northwestern Black Sea 81
      3.11.4 Central and northern
               Black Sea 83

3.11.5 The 'underwater swamps' of the east
Black Sea abyssal plain 84
3.11.6 Offshore Georgia 85
3.12 Inland seas of Eurasia 85
3.12.1 The Caspian Sea 85
3.12.2 Lake Baikal 86
3.13 The Red Sea 87
3.14 The Arabian Gulf 88
3.14.1 Setting 88
3.14.2 Seabed features 88
3.14.3 Strait of Hormuz 90
3.15 The Indian subcontinent 91
3.15.1 The Makran coast 91
3.15.2 The western coast of India 92
3.15.3 The eastern coast of the
subcontinent 92
3.15.4 Indian Ocean vent fauna 93
3.16 South China Sea 93
3.16.1 Offshore Brunei 93
3.16.2 Offshore Vietnam 93
3.16.3 Hong Kong 93
3.16.4 Taiwan 93
3.17 Australasia 94
3.17.1 Sawu Sea 94
3.17.2 Timor Sea 94
3.17.3 New Britain and the Manus
basins 95
3.17.4 New Zealand 97
3.18 Western Pacific 98
3.18.1 Silicic dome volcanism in the Mariana
Back-arc Basin 98
3.18.2 Serpentine mud volcanoes near the
Mariana Trench 98
3.18.3 The Yellow and East China seas 99
3.18.4 Offshore Korea 99
3.18.5 Japan 99
3.18.6 Sea of Okhotsk 101
3.18.7 Piip Submarine Volcano, east of
Kamchatka 103
3.19 Offshore Alaska 103
3.19.1 Bering Sea 103
3.19.2 Gulf of Alaska 105
3.19.3 The Aleutian Subduction Zone 106
3.20 British Columbia 107
3.20.1 Queen Charlotte Sound 107
3.20.2 The Fraser Delta 107
3.21 Cascadia 109
3.21.1 Hydrate Ridge 109

3.21.2 Axial Seamount 110
3.22 California 111
3.22.1 Northern California 111
3.22.2 Monterey Bay 112
3.22.3 Big Sur 114
3.22.4 Santa Barbara Channel 115
3.22.5 Malibu Point 117
3.23 Ocean spreading centres of the east Pacific 117
3.23.1 Guaymas Basin, Gulf of
California 117
3.24 Central and South America 118
3.24.1 Costa Rica 118
3.24.2 Peru 119
3.24.3 The Argentine Basin 120
3.24.4 The Mouth of the Amazon 120
3.25 The Caribbean 120
3.25.1 Barbados Accretionary Wedge 121
3.25.2 Birth of Chatham Island,
Trinidad 122
3.26 Gulf of Mexico 122
3.27 The eastern seaboard, USA 126
3.27.1 Cape Lookout Bight 126
3.27.2 Atlantic Continental Margin 126
3.27.3 Chesapeake Bay 127
3.27.4 Active pockmarks, Gulf of
Maine 128
3.28 The Great Lakes 129
3.28.1 Ring-shaped depressions, Lake
Superior 129
3.28.2 Pockmark-like depressions, Lake
Michigan 129
3.29 Eastern Canada 129
3.29.1 The Scotian and Labrador shelves,
and the Grand Banks 129
3.29.2 The Laurentian Fan 131
3.29.3 The Baffin Shelf 131
3.30 Finale 132

4 The contexts of seabed fluid flow 134
4.1 Introduction 134
4.2 Oceanographic settings 134
4.2.1 Coastal settings 134
4.2.2 Continental shelves 136
4.2.3 Continental slopes and rises 136
4.2.4 Abyssal plains 136
4.3 Plate tectonics settings 136
4.3.1 Divergent (constructive) plate
boundaries 137

4.3.2 Convergent (destructive) plate
      boundaries 138
4.3.3 Transform plate boundaries 140
4.3.4 Intraplate igneous activity 140
4.3.5 Serpentinite seamounts 142
4.4 Conclusion 143

5 The nature and origins of flowing fluids 144
5.1 Introduction 144
5.2 Hot fluids 144
    5.2.1 Magma and volcanic fluids 144
    5.2.2 Geothermal systems 145
    5.2.3 Hydrothermal circulation systems 145
    5.2.4 Exothermic hydrothermal systems 149
5.3 Water flows 150
    5.3.1 Submarine groundwater discharge
          (SGD) 150
    5.3.2 Expelled pore water 150
5.4 Petroleum fluids 151
    5.4.1 Organic origins 151
    5.4.2 Microbial methane 153
    5.4.3 Thermogenic hydrocarbons 154
    5.4.4 Hydrothermal and abiogenic
          petroleum 157
5.5 Discriminating between the origins 162

6 Shallow gas and gas hydrates 163
6.1 Introduction 163
    6.1.1 The character and formation of gas
          bubbles 163
6.2 Geophysical indicators of shallow gas 165
    6.2.1 The acoustic response of gas
          bubbles 166
    6.2.2 Seismic evidence of gassy
          sediments 167
    6.2.3 Novel gas detection and mapping 178
    6.2.4 Seasonal shallow gas depth
          variations 178
6.3 Gas hydrates – a special type of
    accumulation 178
    6.3.1 Nature and formation 179
    6.3.2 Gas hydrates and fluid flow 182
    6.3.3 The BSR 183
    6.3.4 Other hydrate indicators 186
    6.3.5 Dissociation 187

7 Migration and seabed features 189
7.1 Introduction 189
7.2 Pockmarks and related features 190

7.2.1 Distribution 191
7.2.2 Pockmarks and fluid flow 192
7.2.3 Pockmark activity 194
7.3 Mud volcanoes and mud diapirs 195
    7.3.1 The distribution of mud volcanoes and
          mud diapirs 197
    7.3.2 Mud-volcano morphology 198
    7.3.3 Mud-volcano emission products 201
    7.3.4 Mud-volcano activity 202
7.4 Related features 205
    7.4.1 Seabed doming 206
    7.4.2 Collapse depressions 206
    7.4.3 Freak sandwaves 206
    7.4.4 Shallow mud diapirs and mud
          volcanoes 206
    7.4.5 Red Sea diapirs 207
    7.4.6 Diatremes 208
    7.4.7 Sand intrusions and extrusions 209
    7.4.8 Polygonal faults 211
    7.4.9 Genetic relationships 212
7.5 Movers and shakers: influential factors 213
    7.5.1 The deep environment 214
    7.5.2 Driving forces 215
    7.5.3 Fluid migration 216
    7.5.4 Modelling the processes 226
    7.5.5 Triggering events 228
    7.5.6 Ice-related influences 237
7.6 A unified explanation 239
    7.6.1 Fundamental principles 239
    7.6.2 Explaining seeps 240
    7.6.3 The formation of pockmarks and related
          seabed features 242
    7.6.4 Mud volcanoes and diapirism 245
    7.6.5 Alternative explanations 246
7.7 Fossil features 247
7.8 Related features - looking further afield 247

8 Seabed fluid flow and biology 248
8.1 Seabed fluid flow habitats 248
    8.1.1 Cold seeps on continental
          shelves 248
    8.1.2 Deep-water cold seeps 251
    8.1.3 The link between hydrocarbons and
          cold-seep communities 253
    8.1.4 Shallow groundwater discharge
          sites 253
    8.1.5 Deep-water groundwater discharge
          sites 255

8.1.6   Coral reefs and seabed fluid
        flow 255
8.1.7   Hydrothermal vents 260
8.2   Fauna and seabed fluid flow 262
8.2.1   Microbes – where it all begins 262
8.2.2   Living together: symbiosis and
        seeps 269
8.2.3   Non-symbiotic seep fauna 273
8.3   Seeps and marine ecology 276
8.3.1   Geographical distribution 277
8.3.2   Communities as indicators of seep
        activity and maturity 278
8.3.3   Do shallow-water cold seeps support
        chemosynthetic communities? 280
8.3.4   Do seeps contribute to the marine
        food web? 284
8.3.5   Is fluid flow relevant to global
        biodiversity? 286
8.3.6   The 'deep biosphere' and the origins
        of life on Earth 287
8.4   A glimpse into the past 288
8.4.1   Fossil cold-seep communities 288

9   Seabed fluid flow and mineral
    precipitation 290
9.1   Introduction 290
9.2   Methane-derived authigenic carbonates 290
9.2.1   North Sea 'pockmark
        carbonates' 290
9.2.2   'Bubbling Reefs' in the Kattegat 291
9.2.3   Carbonate mineralogy 291
9.2.4   Other modern authigenic
        carbonates 293
9.2.5   Isotopic indications of origin 295
9.2.6   MDAC formation mechanism 295
9.2.7   Associated minerals 297
9.2.8   MDAC chimneys 299
9.2.9   Self-sealing seeps 301
9.2.10   MDAC: block formation 302
9.2.11   Carbonate mounds 302
9.2.12   Fossil seep carbonates 304
9.2.13   Summary of MDAC occurrences 307
9.3   Other fluid-flow-related carbonates 307
9.3.1   Microbialites and stromatolites 308
9.3.2   Ikaite 311
9.3.3   Whitings 311
9.3.4   Carbonates and serpentinites 313

9.4   Hydrothermal seeps and mineralisation 314
9.4.1   Sediment-filtered hydrothermal fluid
        flow 315
9.4.2   Anhydrite mounds 316
9.4.3   Hydrothermal salt stocks 317
9.5   Other mineral precipitates 318
9.5.1   Iron from submarine groundwater
        discharge 318
9.5.2   Phosphates on seamounts, guyots,
        and atolls 318
9.6   Ferromanganese nodules 319
9.7   Final thoughts 321

10   Impacts on the hydrosphere and
    atmosphere 323
10.1   Introduction 323
10.2   Hydrothermal vents and plumes 323
10.2.1   Plumes 324
10.2.2   Plume composition 325
10.2.3   Plumes and the composition of the
        oceans 326
10.2.4   Heating the oceans 328
10.3   Submarine groundwater discharge 329
10.3.1   Detection and quantification 329
10.3.2   Water quality 330
10.4   Seeps 331
10.4.1   Identifying seeps 331
10.4.2   Eruptions and blowouts 333
10.4.3   Quantifying seeps 335
10.4.4   The fate of the seabed flux 338
10.5   Methane in the 'normal' ocean 341
10.5.1   Rivers, estuaries, and
        lagoons 341
10.5.2   The open ocean 342
10.5.3   The influence of seabed methane
        sources 344
10.6   Emissions to the atmosphere 345
10.6.1   Methane emissions from the
        oceans 345
10.6.2   Seabed sources of atmospheric
        methane 347
10.7   Global carbon cycle 349
10.8   Limiting global climate change 350
10.8.1   Quaternary ice ages 350
10.8.2   Earlier events 353
10.9   Afterword 353

11   Implications for man 355
11.1   Introduction 355
11.2   Seabed slope instability 355
    11.2.1   Gas-related slope failures: case studies 356
    11.2.2   Associated tsunamis 359
    11.2.3   Why do submarine slopes fail? 359
    11.2.4   Predicting slope stability 361
    11.2.5   Impacts of slope failures on offshore operations 362
11.3   Drilling hazards 362
    11.3.1   Blowouts 362
    11.3.2   Hydrogen sulphide 366
    11.3.3   Drilling and gas hydrates 367
11.4   Hazards to seabed installations 369
    11.4.1   Pockmarks as seabed obstacles 369
    11.4.2   Trenching through MDAC 369
    11.4.3   Foundation problems 370
    11.4.4   Effects of gas hydrates 370

11.5   Eruptions and natural blowouts 371
    11.5.1   Gas-induced buoyancy loss 372
11.6   Benefits 374
    11.6.1   Metallic ore deposits 374
    11.6.2   Exploiting gas seeps 374
    11.6.3   Gas hydrates – fuel of the future? 375
    11.6.4   Technological challenge 376
    11.6.5   Benefits to fishing? 383
    11.6.6   Seeps, vents, and biotechnology 383
11.7   Impacts of human activities on seabed fluid flow and associated features 383
    11.7.1   Potential triggers 383
    11.7.2   Environmental protection 384

References 387
Index 442

Preface

'*Seabed fluid flow*' encompasses a wide range of fluids (gases and liquids) that pass from sediments to seawater, involving natural processes that modern science would pigeon-hole into a wide range of disciplines, mainly in the geosciences, biosciences, chemical sciences, environmental sciences, and ocean sciences; they also impinge on (or are affected by) human activities. With our background, it is inevitable that the most prominent fluid in this book is methane. There is a vast literature on hydrothermal vents, and a growing interest in submarine groundwater discharge with which we do not wish to compete. However, we recognise the importance of considering *all* forms of seabed fluid flow so that similarities and differences in the processes may be considered. We have attempted to assimilate all forms, manifestations, and consequences of seabed fluid flow of whatever origin.

It is impossible, in a single volume, to do justice to such a multidisciplinary subject. The pace of research has progressively increased since our own interests in pockmarks and seeps began. Of particular significance is the move of the petroleum industry from the continental shelves into the deeper waters of the continental slope and rise; this has rejuvenated research in deep-seabed processes, and has resulted in rethinking many old ideas not least because of the discovery of many deep-water features associated with seabed fluid flow.

When we wrote our first book (*Seabed Pockmarks and Seepages*; Hovland and Judd, 1988) we believed that seabed features such as pockmarks and seeps deserved more careful scientific study than had, by then, been devoted to them. The blossoming of interest in this specialised topic since then is demonstrated by the number of conferences, workshops, and meetings dedicated to it. Interest has grown not only because more people from an increasing number of countries are involved, but also because of the variety of related phenomena identified and described. Now the tide is in full flow and it is impossible to keep up with new literature coming from all over the world. The task of pulling together all the threads to make a coherent and comprehensive synthesis is impossible, yet we have tried to synthesise current understanding of seabed fluid flow, and to demonstrate the interactions between processes often considered separately. We encourage others to think beyond their own specialism, and accept that 'seabed fluid flow' is far more than a mere geological curiosity.

Acknowledgements

Many individuals, particularly Keith Kvenvolden, Jean Whelan, and the late Gabriel Ginsburg, have encouraged us to complete this new book. Fellow researchers with whom we have worked or met at conferences, etc. have provided invaluable discussions, helping us (perhaps unwittingly) to formulate the ideas we present here, and/or providing us with information, data, figures, comments on sections of text, etc. These include: Adel Aliyev, Alan Williams, Andy Hill, Antje Boetius, Bahman Tohidi, Ben Clennell, Ben de Mol, Beth Orcutt, Bo Barker Jørgensen, Björn Lindberg, Daniel Belknap, Dave Long, Derek Moore, Eric Cauquil, Fritz Abegg, Geoff Lawrence, Geoff O'Brien, Gerhard Bohrmann, Gert Wendt, Giovanni Martinelli, Giuseppe Etiope, Graham Westbrook, Gunay Çifçi, Günther Uher, Helge Løseth, Ian MacDonald, Ibrahim Guliev, Ira Leifer, Irina Popescu, Jean-Paul Foucher, Jeff Ellis, Jens Greinert, John Woodside, Jon Ottar Henden, Lori Bruhwiler, Louise Tizzard, Luis Pinheiro, Lyoubomir Dimitrov, Mandy Joye, Mike Leddra, Mike Sweeney, Nils-Martin Hanken, Peter Croker, Raquel Díez Arenas, Richard Salisbury, Roar Heggland, Rob Sim, Rob Upstill-Goddard, Rolf Birger Pedersen, Ruth Durán Gallego, Soledad García-Gil, Stéphan Hourdez, Steve McGiveron, Tim Francis, Troels Laier, Valery Soloviev, Vas Kitidis, Veronica Jukes, Vitor Magalhães, Wolfgang Bach.

While acknowledging the help of all these people we are responsible for any omissions, oversights, or errors.

We also acknowledge the various publishers, organisations (especially statoil ASA), and individuals who have granted us permission to reproduce figures. The maps presented on the website (http://www.cambridge.org/9780521819501 3) were prepared using the on-line mapping system 'GeoMapApp' (http://www.marine-geo.org/geomapapp/) of the Marine Geoscience Data Management System, Lamont-Doherty Earth Observatory. We are grateful to Zoë Lewin who, with patience and good humour, worked hard to convert our 'copy' into a book.

Finally we thank our wives (Laraine and Målfrid) and families (children and grandchildren) who have patiently put up with us over the years during which we have been working on this book. Laraine has also provided invaluable assistance with the preparation of the text; it could not have been completed without your support.

Alan Judd
High Mickley
Northumberland
UK

Martin Hovland
Sola
Stavanger
Norway

Note on the accompanying website
www.cambridge.org/9780521114202

The figures presented in this book are in black and white. They are all reproduced on the accompanying website where some of them (marked by an asterisk) are in colour. Also on this website are location maps (with an index of place names – see www.cambridge.org/9780521114202), and Powerpoint presentations contributed (by invitation) by various colleagues to supplement the text with additional detail, and more images. We thank these colleagues for their efforts which, we think, are a valuable asset.

Maps on the accompanying website

The maps presented on the website are accompanied by an introductory section, a general key, and an index of place names. Each map is accompanied by notes explaining abbreviations. etc.

1 Index map (see Figure 3.1)
2 Barents Sea
3 Northwest Europe
4 British Isles
5 Iberia
6 West Africa
7 Adriatic
8 East Mediterranean
9 Black Sea
10 Southern Caspian Sea
11 Lake Baikal
12 Red Sea
13 Arabian Gulf
14 Makran coast of Pakistan
15 The Indian subcontinent
16 South China Sea
17 Australasia: Timor Sea
18 New Britain and the Manus Basins
19 New Zealand
20 West Pacific
21 Yellow Sea and Japan
22 Sea of Okhotsk
23 Alaska
24 British Columbia
25 Cascadia Margin
26 California
27 Gulf of California
28 South America
29 Southeast Caribbean
30 Gulf of Mexico
31 Eastern USA
32 Eastern Canada
33 Hydrothermal vents and submarine volcanoes (worldwide)
34 Mud volcanoes (worldwide)
35 Submarine groundwater discharge (worldwide)
36 Gas hydrates (worldwide)

Contributed presentations on the accompanying website

Contribution 1: F. Abegg, Germany. *Structure and distribution of gas hydrates in marine sediments.*

Contribution 2: I.W. Aiello and R.E. Garrison, USA. *Subsurface plumbing and three-dimensional geometry in Miocene fossil cold-seep fields, coastal California.*

Contribution 3: G. Bohrmann, Germany. *Mud volcanoes and gas hydrates in the Black Sea – an important linkage to the methane cycle: the Dvurechenskii Mud Volcano, initial results from M52/1 MARGASCH.*

Contribution 4: L. Dimitrov, Bulgaria. *Black Sea methane hydrate stability zone.*

Contribution 5: S. Garcia-Gil, Spain. *A natural laboratory for shallow gas: the Rías Baixas (Spain).*

Contribution 6: I. Guliyev, Azerbaijan. *South Caspian Basin, seeps, mud volcanoes.*

Contribution 7: R. Heggland, Norway. *Gas-escape features of the Niger Delta.*

Contribution 8: M. Hovland, Norway. *Mystery of mud volcanoes – driven by supercritical water generated deep down?*

Contribution 9: D.C. Kim, Korea. *High resolution profiles of gassy sediment in the southeastern shelf of Korea.*

Contribution 10: H. Løseth, Norway. *Shallow mud mobilisation in the Hordaland Group caused by fluid injection.*

Contribution 11: I.R. MacDonald, USA. *Stability and change in Gulf of Mexico chemospheric communities.*

Contribution 12: P. Meldahl, Norway. *Neural networks used for shallow-gas and chimney detection.*

Contribution 13: G. Papatheodorou *et al.*, Greece. *Gas charged sediments and associated seabed morphological features in the Aegean and Ionian seas, Greece.*

Contribution 14: T. Treude and A. Boetius, Germany. *Anaerobic oxidation of methane (AOM) in marine sediments.*

Contribution 15: UCSB Seep Group. *Characteristic of seepage at the Coal Oil Point seep field, Santa Barbara Channel, California.*

1 · Introduction to seabed fluid flow

Discovery commences with the awareness of
anomaly, i.e. with the recognition that nature has
somehow violated the paradigm-induced
expectations that govern normal science.

Kuhn, 1970

This chapter introduces the concept that seabed fluid
flow is a widespread and important natural process. It
has important consequences for subseabed and seabed
geological features, and also for marine biological processes, and the composition of the oceans. Seabed fluid
flow provides both hazards and benefits for human
activities, and it is recognised that some sites are precious and need protection.

Earth scientists remember the 1960s as the decade of the
plate tectonics 'revolution'. In the same decade, the discovery of two remarkable seabed features; hydrothermal
vents and pockmarks, provided evidence of extensive
emissions of fluids from the seabed. Since then there has
been a growing awareness that dynamic geological processes, driving the exchange of fluids across the seabed–
seawater interface, are of fundamental importance to the
nature and composition of the 'marine system'; not only
to marine geology, but also to the chemical and biological composition of the oceans. Today as in the geological
past, seabed fluid exchange is as important as the interactions between the oceans and atmosphere. Gas bubble
streams and columns of coloured or shimmering water,
mineral crusts and chimneys, and biological communities that thrive without the aid of sunlight are all evidence
of 'seabed fluid flow'.

Spectacular discoveries made during investigations
of ocean spreading centres like the East Pacific Rise
and the Mid-Atlantic Ridge have turned upside-down
concepts of how our planet works. On the continental shelves, and more recently in the deeper waters of
the slope and rise, the oil industry has not only driven

technological advances, but has also been responsible for
an increasing awareness of the fundamental role of fluids in sedimentary processes. Tryon *et al.* (2001) pointed
out that: '*Subsurface fluid flow is a key area of earth science research, because fluids affect almost every physical,
chemical, mechanical, and thermal property of the upper
crust*'. They went on by saying that research in the
deep biosphere, gas hydrates, subduction zone fluxes,
seismogenic zone processes, and hydrothermal systems
all are '*directly impacted by the transport of mass, heat,
nutrients, and other chemical species in hydrogeological
systems*'.

Mankind's activities, particularly during the last
century, have resulted in increasingly serious pollution
of the marine environment. Some of the principal causes
relate to the petroleum industry, yet natural processes
have been responsible for petroleum 'pollution' for a
far greater period of time. In the Bible God instructed
Noah to make an ark and '*coat it inside and out with pitch*'
(Genesis 6:14). Indigenous populations from parts of
the world where seeps occur have made good use of the
special properties of natural petroleum products; Native
Americans in California used 'asphaltum' to caulk their
canoes, hold together hunting weapons and baskets, for
face paints, and even chewing gum (USGS, 1999). The
'eternal flames' of natural gas seeps in Azerbaijan and
elsewhere are central to the Zoroastrian faith.

Such seepages gave the first indications of the presence of petroleum in most of the world's petroleum-
producing regions (Link, 1952). Indeed, Link considered that at least half the reserves proved by 1952 were
discovered by drilling on or near seeps. But petroleum
seeps are not confined to the land. Great lumps of floating tar, such as that illustrated in Fig. 1.1, caused the
Romans to call the Dead Sea *Mare Asphalticum*, and
early navigators of the Gulf of Suez, the Gulf of Mexico,
the Californian coast, and many other parts of the
world's oceans discovered oil slicks and tar-polluted
beaches centuries before the modern oil industry was

Figure 1.1 A giant lump of tar found on the shore of the Dead Sea. Numerous warm groundwater seeps are known to occur in the area. It seems that there are also hydrocarbon seeps. (From Hovland and Judd, 1988; sketched from a photograph in Landes, 1973.)

founded and oil-powered ships and tankers were introduced (Soley, 1910; MacDonald, 1998). Kvenvolden and Cooper (2003) reported that natural seepage introduces between 0.2 and 2.0 × 10⁶ (best estimate 0.6 × 10⁶) tonnes of crude oil per year into the marine environment. This is about 47% of all the crude oil currently entering the marine environment; mankind is responsible for the rest. Hornafius *et al.* (1999) estimated that the present-day natural hydrocarbon seeps in Santa Barbara Channel, California are a significant source of air pollution, the flux being '*twice the emission rate from all the on-road vehicle traffic in Santa Barbara County*'.

Petroleum seeps are not the only form of seabed fluid flow that has been known for thousands of years. Taniguchi *et al.* (2002) identified the following ancient reports of submarine groundwater discharge:

- The Roman geographer, Strabo, who lived from 63 BC to AD 21, mentioned a submarine spring (fresh groundwater) 2.5 miles offshore from Latakia, Syria (Mediterranean), near the island of Aradus. Water from this spring was collected from a boat, utilising a lead funnel and leather tube, and transported to the city as a source of fresh water.
- Pliny the Elder (first century AD) reported submarine '*springs bubbling fresh water as if from pipes*' along the Black Sea coast.
- Pausanius (second century AD) told of Etruscan citizens using coastal springs for 'hot baths'.

Historical accounts tell of water vendors in Bahrain collecting potable water from offshore submarine springs for shipboard and land use.

Considering the long history of knowledge of petroleum and freshwater seeps, it is perhaps remarkable that hydrothermal vents and chemosynthetic biological communities have been discovered so recently. However, they are not the only features hidden in the ocean depths, out of reach of all but the most recent technology. Vogt *et al.* (1999a) made a comparison, highlighting the progress made in one decade between the contents of *The Nordic Seas* (Hurdle, 1986), and current understanding. They noted that: '*Two thirds of that 777-page volume was devoted to topography and geology . . . yet the words "methane", "hydrate", "pockmark", "gas vent or seep", "chemosynthesis", and "mud volcano" do not appear even once in the 42-page subject index*'.

The development in the mid 1960s of the sidecan sonar and towed photographic cameras made widespread high-resolution seabed mapping possible, while the parallel development of high-resolution seismic profilers extended this mapping to include the subseabed sediments and rocks. More recently multibeam echo sounders (MBESs), manned and remotely operated vehicles (ROVs), autonomous underwater vehicles (AUVs), and many more sophisticated instruments have enabled more rapid and detailed inspection of survey areas and individual seabed locations. These developments have enabled the pace of discovery to increase progressively. Features that were only recently regarded as geological curiosities are now known to be widely distributed geographically, from the coasts to the ocean depths, and through geological time. It is amazing how far knowledge of the seabed has advanced in little more than 60 years since the following words were written:

In 1911, Fessenden made the first attempts to determine depths by sonic methods, and from about 1920 sonic depth finders have been in use with which soundings can be taken in a few seconds from a vessel running full speed. This new method has in a few years completely altered our concept of the topography of the ocean bottom. Basins and ridges, troughs and peaks have been discovered, and in many areas a bottom topography has been found as rugged as the topography of any mountain landscape.

Sverdrup *et al.*, 1942

Figure 1.2* Seep bubbles emerging from the seabed. This photograph, taken in 1983, shows the first seeps identified in the North Sea, at Tommeliten (Block NO1/9, North Sea). The location is situated above the Delta salt diapir. Gas bubbles emanate from small, funnel-shaped craters in the sand-covered seabed. (From Hovland and Judd, 1988.)

Today's technology facilitates not only detailed 3D mapping, but also sampling and visual inspection, revealing features Sverdrup could not have dreamed of. This technology now permits an appreciation of how widespread emissions of water, petroleum fluids, and hydrothermal fluids are; it also enables associated features such as mineralised chimneys and chemosynthetic biological communities to be sampled and investigated. Only now is the importance of the natural processes responsible for them being realised in marine science.

It was through our curiosity towards pockmarks that we became aware of the importance of seabed fluid flow. An initial appraisal of pockmarks, in Chapter 2, is an account of the pockmark investigations of the Scotian Shelf and the North Sea that provided us with a preliminary insight into seabed fluid flow. This research, undertaken in the 1970s and 1980s, led us to realise the significance of pockmarks as indicators of fluid flow, and documented North Sea gas seeps (Figure 1.2) and the associated carbonates and benthic fauna for the first time. Chapter 3 (supported by maps provided in the web material) is a review of some key sites around the world that have provided evidence critical to the development of our present understanding of seabed fluid flow. It emphasises the relationship between the natural processes (geological, biological, physical, and chemical) involved, and shows that the study of this topic is not possible without crossing traditional scientific discipline boundaries. It is clear from Chapter 3 that seabed fluid flow is widespread, and that various types of fluid are involved. This book is concerned with three main types of fluid:[1] hydrothermal fluids

---

[1] It is not uncommon in the literature to find 'gas and fluids' mentioned as if they are separate phases. They are not. The Oxford English dictionary defines a fluid as 'a substance that is able to flow freely, not solid or rigid', and specifically states that this includes both liquids and gases. So, throughout this book when we refer to fluids we mean both gases and liquids. However, direct quotes do not necessarily conform to this standard.

generated by the circulation of seawater through the cooling igneous rocks of ocean-spreading centres and submarine volcanoes; gases, particularly methane, generated in marine sediments; and groundwater flowing from catchment areas on land. Perhaps it is normal to deal with each of these fluid types separately. However, although major differences, for example in temperature and chemical composition, result in contrasting behaviour, many processes and associated features are either common, or so closely related that it is hard to consider one without mentioning the others. So, we consider the cycles of generation, migration, and utilisation or escape of these three fluid types, pointing out the similarities and contrasts between them, and the overall significance of seabed fluid flow. Our objective is to be inclusive rather than selective.

It is remarkable how common seabed fluid flow is. As we show in Chapter 4, the examples described in Chapter 3 come from every seabed environment from coastal waters down to the deep ocean trenches. Also, seabed fluid flow is integral to every marine plate-tectonics setting: hydrothermal venting is part of the system that cools igneous rocks at plate boundaries; mud volcanoes and seeps permit the compaction of sediments trapped in the accretionary wedges (prisms) of convergent boundaries; buoyant hydrocarbon fluids escape from intra-plate sedimentary basins through seeps. The nature and origins of the various types of fluid (discussed in Chapter 5) are largely a function of these contexts, and the geological and biological processes operating in them. So, where igneous processes dominate, hydrothermal fluids are formed by the interactions between pore fluids and hot rocks. In sedimentary basins the most significant fluids are hydrocarbons, particularly methane, formed by the degradation of organic matter held within the sediments.

At this point it is appropriate to clarify some terminology. The word 'biogenic' is commonly used, particularly by geoscientists, when referring to methane that has been derived by the activity of micro-organisms, as opposed to 'thermogenic' methane, derived from processes occurring deeper within the sediments. However, in the biological sciences 'thermogenic' methane is also regarded as being 'biogenic' because the source materials are of organic origin; thus 'biogenic' is distinct from 'abiogenic', formed without the involvement of living organisms. We will avoid this confusion by avoiding the word 'biogenic' altogether. Instead we distinguish between 'thermogenic' and 'microbial' methane. This also avoids the use of 'bacterial methane', which is generally incorrect as microbes, 'minute living beings', which generate methane, are actually archaea, not bacteria. However, although these are definitions we stick to, quotations from other authors may imply something different; we do not wish to modify other people's words.

Methane, formed during sediment burial, is buoyant and therefore inclined to migrate towards the surface. Although seepage is a natural result of this migration, geological conditions often result in the formation of accumulations. In deep water, temperature and pressure conditions favour the formation of gas hydrates that also inhibit migration. In order to understand the distribution of seeps in both space and time it is essential to appreciate how and why these accumulations form, and how to identify them. We address these issues in Chapter 6. Diatremes, mud diapirs, gas chimneys, and mud volcanoes form as a result of the pressure that builds up in some subseabed gas accumulations. However, the nature of the migration mechanism is dependent on the stress environment within the sediments. In some places migration is a much more gentle process, and the plumbing system may lead to pockmarks, or to seeps with no associated seabed morphological features at all. As we discuss in Chapter 7, the style of migration and seabed escape is determined by interactions between many factors. Fluid flow is clearly a dynamic process.

Perhaps the most amazing biological discovery of the twentieth century was made in 1977 when deep-ocean chemosynthetic communities were found at the Galapagos Rift. Until then it was inconceivable that life could exist without benefiting from the Sun's energy. Although such communities are probably rare, they are clearly widespread and, as we discuss in Chapter 8, they are not confined to ocean spreading centres or to hydrothermal vents. The principal effect of petroleum seeps, particularly those of the shelf seas, might be expected to be the pollution of the seabed sediments and the overlying waters. This is not the case. Similarities between hot-vent and cold-seep communities are remarkable, as is the suggestion that the first life on Earth may have relied on chemosynthesis. Is photosynthesis a relatively recent adaptation?

Some of the most spectacular seabed scenery is associated with seabed fluid flow. At some locations the

Figure 1.3 Godzilla, a 45 m high sulphide mound with flanges on the Juan de Fuca Ridge. The submersible Alvin is drawn to scale. (Reproduced with permission from Robigou et al., 1993.)

scenery is provided by carbonate chimneys associated with methane seeps, at others by chimneys of hydrothermal metal sulphides. Exceptional examples stand metres tall; *Godzilla*, a structure on the Juan de Fuca Ridge, towers 45 m above the seabed, belching black smoke from its chimneys (Figure 1.3). Mineral precipitation, the subject of Chapter 9, results in changes to the composition of flowing fluids, whether by microbial utilisation, as in the case of methane-derived authigenic carbonate (MDAC), or precipitation as a result of a sudden change in temperature (as at hydrothermal vents). As we see in Chapter 10, the fluids that escape contribute to the composition of the overlying water column, adding heat as well as metals or hydrocarbons; nutrients and substrates that can be oxidised by microbes in the water also contribute to biological productivity. If seabed fluid flow were a rare phenomenon, then these contributions would be of little consequence. However, given the widespread distribution shown in Chapters 3 and 4, perhaps the composition of the oceans has been significantly influenced by geological contributions. Seeps and mud volcanoes may also influence atmospheric concentrations of methane, particularly in shallow water where gas bubbles can survive a journey to the sea surface. Vast volumes of methane are sequestered by seabed gas hydrates during interglacial periods, and may be released during glaciations, so it seems possible that variations in the seabed flux of geological methane moderates the extremes of global climate change.

In the final chapter we discuss both the implications of seabed fluid flow for mankind, and the effects of offshore activities on seabed fluid flow. Marine geohazards include slope failures and drilling hazards associated with shallow gas, and the possible implications of seabed eruptions for seabed installations and shipping. However, seabed fluid flow offers benefits too. The mining of metals from hydrothermal ore deposits on land is a major industry, and active hydrothermal vents provide useful information for mining, as well as having future potential as a resource. The energy potential of gas hydrates has encouraged significant research programmes in several countries, and the oil industry makes use of seeps in petroleum exploration. A more recent concern to marine science is the vulnerability of benthic ecosystems associated with seabed fluid flow. International legislation is now affording some protection; for example, the European Union's *Habitats Directive* has identified '*sub-marine structures made by leaking gas*' as a habitat worth protecting.

In this book we suggest that seabed fluid flow is of fundamental importance to the marine environment and the working of our planet. It is widespread, dynamic, and influential. Although it is essentially a geological process, it affects marine ecology, ocean chemistry, and the composition of the atmosphere. The seabed does not mark the limit of the marine system. Fluids flowing out of the seabed contribute to and, we argue, play a significant role in ocean processes and the global carbon cycle.

2 • Pockmarks, shallow gas, and seeps: an initial appraisal

The North Sea's fattyness is, after its saltiness, a peculiar property, . . . It should be assumed here that in the ocean as on land there exists, here and there, seepages of running oily liquids or streams of petroleum, naptha, sulphur, coal-oils and other bituminous liquids.

Translated from Erich Pontoppidan, 1752

This chapter begins with a review of the pioneering work undertaken on the Scotian Shelf, off eastern Canada, by L. H. King and his colleagues at the Bedford Institute of Oceanography. However, having 'cut our teeth' in the North Sea, the pockmarks and seeps here have become the standard against which we compare those of other areas. Consequently it is appropriate to review our early studies of North Sea pockmarks. This provides a historical perspective on pockmark research, and indicates how this early work led us to the conclusions that pockmarks and seabed seeps are important geological phenomena and indicators of processes associated with seabed fluid flow. In some cases the sites we visited early on have been the subjects of further work. This is also reviewed here.

By the end of this chapter it becomes clear that seeps and pockmarks, along with the associated carbonates and biological communities, are components of the important hydrocarbon cycle.

2.1 THE SCOTIAN SHELF: THE EARLY YEARS

Pockmarks were first described on the continental shelf offshore Nova Scotia, Canada by King and MacLean (1970). Subsequent work in this area was reported by Josenhans et al. (1978). Pockmarks were found to be present over an area of 3000–4000 km^2 in the Roseway,

LaHave and Emerald basins, and two smaller basins. From echo sounder and side-scan sonar records, and from visual observations made from the manned submersible 'Shelfdiver', the features were described as cone-shaped seabed depressions that bottomed at a well-defined point. In plan, most are elongate with a preferred orientation that, on average, is north–south. No raised rims were present, but the pockmark edges were found to be sharply defined, the slope changing from horizontal to an estimated 30° within a distance of only 0.5 m.

The surficial sediments of the Scotian Shelf range in thickness from a few metres to over 200 m. They consist of five formations of which the oldest, the Scotian Shelf Drift, is mainly glacial till. The basins are infilled with Emerald Silt, a fine-grained, muddy sediment, predominantly silt but locally sandy and containing some gravel. This is overlain by the mainly-Holocene LaHave Clay, comprising homogeneous, loosely compacted marine silty clay that locally grades to clayey silt. These three sediment units are illustrated on the seismic profile (Figure 2.1), where it can be seen that the younger two thicken towards the deeper parts of the basin. King and MacLean (1970) found that the pockmark distribution is related to the distribution of the LaHave Clay. However, pockmarks are not found throughout this area, neither are they restricted to this sediment type. Some are found in the Emerald Silt and a few small, isolated pockmarks have been reported in the Sambro Sand (medium- to fine-grained sand, moderate to well sorted, with up to 20% silt and clay-sized material) near the edge of the Emerald Basin (Josenhans et al., 1978). Pockmarks are not found in the intervening Sambro and Roseway banks, and, with the exception of a slight overlap in the Roseway Basin, they overlie the coastal plain sediments. These are a thick sequence of well-stratified, seaward-dipping Tertiary and Cretaceous sediments that wedge out against the basement rocks along a line subparallel to the coast. The basement

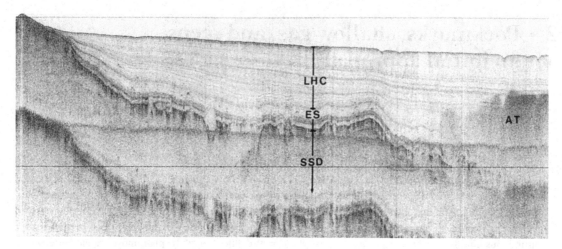

Figure 2.1 A shallow seismic record from the Scotian Shelf. LHC = LaHave Clay, ES = Emerald Silt, SSD = Scotian Shelf Drift, AT = acoustic turbidity. Note how the turbidity decreases below pockmarks. (From Hovland and Judd, 1988; courtesy of L. H. King.)

rocks comprise folded Cambro-Ordovician metasediments and granitic intrusions of Devonian age.

The pockmarks in the three basins are similar, but King and MacLean found those of the Roseway Basin to be more numerous (200 per km^2) and smaller (15–30 m across and 3–6 m deep) than those of the Emerald and LaHave basins (45 per km^2, 30–60 m in diameter and 6–9 m deep). A more detailed study of a small (150 km^2) area in the Emerald Basin (Josenhans et al., 1978) indicated that pockmark density and size were related to the surficial sediment type and thickness, more but smaller pockmarks occurring in the silts, fewer and larger pockmarks in the clays. The largest pockmark they recorded lay in the LaHave Clay and measured 300 m long, 150 m wide and 15 m deep.

King and MacLean (1970) considered that '*the crater-like nature of the pockmarks strongly suggests that they are erosional features*'. After discussing various possible mechanisms, they concluded that the association with the underlying coastal plain sediments suggested a link and surmised that water or gas rising from these sediments (or underlying coal-bearing Upper Carboniferous strata) to the seabed was the most likely cause or agent. Although the currently known petroleum fields on the Scotian Shelf lie further seaward, considerable updip migration cannot be ruled out. It was further envisaged that water currents would disperse suspended sediment, and that the pockmark walls would slump, enlarging the feature, until a stable slope developed. The

preference of pockmarks for fine-grained sediments was considered to reflect the inability of escaping fluids to percolate through such sediments without disturbing them. In contrast, percolation could occur in areas such as the Roseway and Sambro banks where coarse sediments are present. Josenhans et al. (1978) observed that elongate pockmarks are aligned with their long axes parallel to the dominant tidal flow, which has an oscillating tidal component of 10 cm s^{-1} with a major axis oriented north to northwest, and a residual current flow of 3 cm s^{-1} from the north.

Although Josenhans et al. (1978) favoured gas escape as the pockmark-forming process, they could find insufficient evidence to support present-day gas escape from shallow seismic reflection profiles, echo sounder profiles, side-scan sonar records, the analysis of piston core samples (reported by Vilks and Rashid, 1975), or hydrocarbon sniffer data. This led them to conclude that the Scotian Shelf pockmarks are largely relict features.

2.2 NORTH SEA POCKMARKS

The first North Sea pockmarks were discovered in 1970 by Decca Surveys during a rig-site survey in preparation for exploration drilling at BP's Forties field. The following year they were found off the Norwegian coast during a research survey (van Weering et al., 1973). Indications

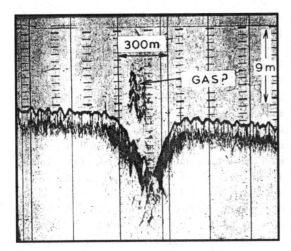

Figure 2.2 Echo sounder profile of a North Sea pockmark. The water-column target was thought to indicate seeping gas. However, more recent data have shown that this is actually a shipwreck, see Section 11.5.1. (From Hovland and Judd, 1988; image courtesy of BP).

of gas seeps from pockmarks were also recorded in the early 1970s (Figure 2.2), but it was not until 1983 that positive proof of gas seepage was obtained (Hovland *et al.*, 1985; 1987).

2.2.1 History of discovery

In 1971 the Netherlands Institute of Sea Research (NIOZ) conducted a survey in the Norwegian Trench between Oslo and Bergen, using a hull-mounted 3.5 kHz sub-bottom profiler. The main objective was to map the thickness of the surficial sediments. Side-scan sonar was not used, but the seabed notches were correctly interpreted as pockmarks by comparing them to the pockmarks of the Scotian Shelf. From this survey it was evident that there are pockmarks along most of the Norwegian Trench, including some parts of the Skagerrak. This conclusion has been confirmed by subsequent work. Indeed, it is now known that pockmarks are present throughout most of the area covered by the youngest sediments, the Kleppe Senior Formation, although they are generally most common along the western slope of the Norwegian Trench. The NIOZ also discovered that pockmarks are extensive in the Witch Ground Basin of the UK sector of the North Sea (Jansen, 1976).

During the period 1974–8 the British Geological Survey (BGS) undertook a research programme to find out more about pockmarks in an area that was then attracting increasing attention from the oil industry. This programme was concentrated in the South Fladen area, northwest of the Forties field in UK blocks 15/28 and 21/3 (i.e. blocks UK15/28 and UK21/3), on the southern side of the Witch Ground Basin. Ten investigations were undertaken. They included gravity and vibrocore sampling, drilling, visual inspection using the unmanned submersible (ROV) Consub, *in situ* geophysical (seismic-velocity and electrical-resistivity) measurements, geophysical surveys (side-scan sonar and seismic profiling) and geochemical studies of core samples and seawater. The results were summarised by McQuillin *et al.* (1979) and referred to by Fannin (1979) and McQuillin and Fannin (1979). Subsequent analyses of the data acquired during some of the surveys were undertaken by Judd (1982). The results of this and some subsequent work (including the regional mapping of the UK continental shelf by the BGS) are reviewed in Section 2.3.1.

These early surveys were intended to obtain basic information to delimit the area in which pockmarks occur, and to give some indication of the mode of formation. In particular it was felt necessary to establish whether or not the process of pockmark formation might be hazardous to offshore installations. During these surveys a range of features were identified, including evidence of the presence and migration of gas. Many of them had not been recognised before, so a terminology was developed to describe them – see Sections 2.2.4 (pockmarks) and 6.2.2 (gas).

Because pockmarks occur over such wide areas of the northern North Sea they have been a source of considerable interest to the oil industry. In the UK sector, many producing petroleum fields (e.g. Balmoral, Britannia, Forties, Ivanhoe, Piper, and Tartan) lie within the Witch Ground Basin, and several pipeline systems cross this area. Several fields (e.g. Troll, Veslefrikk, Snorre, and part of Gullfaks) are located at pockmarked sites in the deeper waters of the Norwegian Trench, and several pipelines (e.g. the Statpipe, Zeepipe, and Europipe systems) also cross the Trench. The work involved in site and route planning surveys for these installations has provided a considerable volume of data about pockmarks. Unfortunately, the operators retain much of this in confidence. Of the oil company data that have been released, Statoil produced the overwhelming majority. These include echo sounder, side-scan sonar records, shallow seismic reflection profiles, and sedimentological data from coring. Also, ROVs have been utilised for seabed inspections.

The vast majority of the survey data acquired prior to 1983 represent a form of remote sensing, using side-scan sonar, shallow seismics etc. As with all remote sensing, a proper interpretation cannot be made without ground-truthing. During two research cruises in 1983 and 1985, Statoil conducted detailed inspections of pockmarks using ROVs. The results of these, and some subsequent surveys, are discussed in Section 2.3.

2.2.2 Pockmark distribution

The North Sea can be subdivided into three bathymetric zones: the southern and northern North Sea (separated by the Dogger Bank), and the Norwegian Trench. For a long time no pockmarks had been located south of the 56° parallel. This was assumed to be a function of the seabed sediment types rather than being due to the absence of gas seepages. However, careful analysis of MBESs and shallow seismic data from the Zeepipe pipeline route data has revealed pockmarks in areas of sandy sediments and sandwaves. These are discussed in Section 3.5.2.

The seabed of the northern North Sea is a gently inclined plateau; water depths gradually increasing from about 60 m in the south to about 250 m in the far north, at the edge of the continental shelf. The largest of several basins within the plateau is the Witch Ground Basin. Here, water depths increase to more than 150 m. Many of the smaller basins are actually channels cut into the plateau sediments during the late Pleistocene and subsequently partially infilled. Sediment types on the plateau vary, but stiff glacial clays covered by varying layers of sand predominate. In contrast, the basins and channels tend to be characterised by soft, muddy sediments (Andrews *et al.*, 1990; Johnson *et al.*, 1993; Gatcliff *et al.*, 1994). The origin of the Norwegian Trench, where waters are as much as 700 m deep (in the Skagerrak), has long been debated. However, it is now generally believed originally to have been cut fluvially in the late Tertiary and subsequently deepened by glaciers and ice-sheets during the Pleistocene. It is an asymmetric trough in form. The western slope is smooth, whereas the landward side is steeper and frequently rugged (Holtedahl, 1993).

Most pockmarks are found in the three muddy sediment formations in the northern North Sea: the Witch Ground Formation, in the Witch Ground Basin, the Flags Formation of the smaller basins further north, and the Kleppe Senior Formation that occupies the floor of the Norwegian Trench. There are also pockmarks in equivalent sediments that infill or partially infill channels cut into the stiffer clays of the plateau. These sediments are all post-glacial and are similar in most respects. Indeed, Hovland *et al.* (1984) noted that both the Witch Ground and Kleppe Senior formations are remarkably similar to the Emerald Silt–LaHave Clay sequence of the Scotian Shelf. This comparison is valid in respect of their lithological characteristics, seismostratigraphic character, and depositional environment. Also, sedimentation has all but ceased in the basins of the Scotian Shelf, as it has in the northern North Sea.

2.2.3 Pockmark size and density

The density of pockmarks varies from area to area both within the North Sea and within the individual pockmarked areas in the North Sea. In the Norwegian Trench the density varies from 0 to about 60 per km^2 (counting only those that are more than 10 m across); the most densely pockmarked area of substantial size lies over the Troll gas field. The sizes of individual pockmarks in any given area are varied, but the only change in the range of sizes within the Trench is associated with the western slope, which is the only area in which elongated pockmarks are found. In contrast, the size and density of the pockmarks in the Witch Ground Basin vary, apparently in response to variations in the thickness and lithology of the seabed sediments (Long, 1986). In general, pockmarks are between 50 and 100 m in diameter with depths in the range 1–3 m. The highest densities (>40 km^{-2}) occur within bathymetric hollows characterised by sandy muds, but here sizes rarely exceed 50 m. In the deepest parts of the basin, where seabed sediments are pure mud, the density is 10–15 per km^2, but sizes are much larger (100–150 m). Both pockmark density and size tend to decrease towards the edges of the basin beyond the outcrop of the Witch Member and particularly where the Fladen Member becomes thinner and coarser. At the basin edge where the underlying Coal Pit and Swatchway formations approach the seabed, there are pockmarks, but they are few and far between and very small in size.

2.2.4 Pockmark morphology

Although early investigations concluded that pockmarks are approximately circular in plan, it is clear that there is a considerable variety of shape. Size also varies considerably both between areas and within an individual area. The following descriptions are based mainly on seismic (mainly deep-towed boomer) profiles and side-scan sonar data from surveys undertaken in the 1970s and 1980s in the South Fladen area (Witch Ground Basin) and the Norwegian Trench, but we have used MBES images as illustrations where appropriate.

Standard circular and elliptical pockmarks are perhaps the most common (Figure 2.3). Length:breadth ratios vary considerably from 1 (circular) to 1.25 or more in the South Fladen area. On the western slope of the Norwegian Trench the axes are generally aligned parallel to the slope suggesting a relationship between the slope and pockmark shape. In the South Fladen area, where seabed slope is less pronounced, there is a preferred orientation that apparently conforms to the dominant tidal current; slope-normal bottom currents may also explain pockmark orientation on the western slope. On detailed inspection individual standard pockmarks are found to be discrete depressions whose perimeters are complicated by indentations and lobes (Figure 2.4). They are not as 'regular' in shape as is commonly supposed. The pockmark floors tend to be undulating rather than smooth.

Composite pockmarks occur where individual standard pockmarks merge with one another. In some instances groups of pockmarks are found together or merging (Figure 2.5), while in others the merger has proceeded to the extent where a single feature with a complex shape has been produced.

Asymmetric pockmarks are best imaged by MBESs (Figure 2.6). On side-scan sonar records they appear to have a distinct and often quite lengthy 'tail' and a strong backwall reflection on one side only. On seismic sections it can be seen that the lack of backwall reflection occurs where the slope up to seabed level is long and gentle. Stoker (1981) suggested that asymmetrical pockmarks are more common than regular pockmarks in the Witch Ground Basin. Surprisingly, the orientation of the asymmetry, although uniform in individual areas, varies considerably from one part of the Witch Ground Basin to another.

Pockmark strings comprise individual pockmarks, commonly symmetrical, shallow and 10–15 m in diameter, arranged in strings or chains (Figures 2.7 and 2.8) that often extend for several hundred metres, but more normally to 100–150 m. A space approximately equivalent to the pockmark diameter occurs between the pockmarks. Many strings end in a single standard pockmark much larger than those forming the string. In some cases the strings radiate out in several directions from a single large pockmark. In the Norwegian Trench the most common orientation (42% of those measured by Hovland, 1981) of these features is northwest–southeast, parallel to the bottom currents. The remainder are aligned north-northwest–south-southeast (30%) or north–south (24%), with only 4% aligned in other directions. Similar alignments were reported from the Troll field (Green *et al.*, 1985). Some long, thin pockmarks may have been formed by the growth and merger of the pockmarks in a string, into a single feature.

Elongated pockmarks and troughs; along some mid-depth sections of the western slope of the Norwegian Trench the pockmarks tend to be elongated to the extent that they resemble gullies or troughs rather than standard (circular or oval) pockmarks. Inside these troughs the topmost sediment layers are absent and older sediments are exposed at the seabed (Hovland, 1983). The Norwegian Geological Survey (Longva and Thorsnes, 1997) mapped similar features in the southern portion of the Norwegian Trench (Bøe *et al.*, 1998).

Near the foot of the western slope, there is a single trough more than 1 km long and about 200 m wide. It is composed of a series of large interlinked pockmarks aligned approximately north–south (Figure 2.9). Shallow seismic sections crossing the channel indicate that it corresponds to a furrow in the surface of the Norwegian Trench Formation, although the axes of the two features do not coincide exactly. The furrow has the appearance of an ice-ploughed mark cut by either the keel of a large iceberg or the irregular underside of an ice-sheet. Although subsequently deposited sediments have smoothed out most irregularities in the surface of the Norwegian Trench Formation and its equivalents, this one has been maintained, probably by a secondary

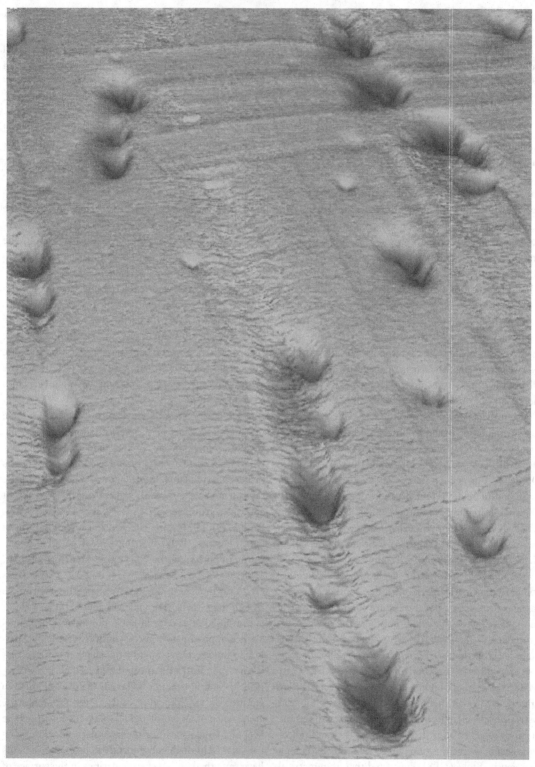

Figure 2.3* Typical North Sea pockmarks, Witch Ground Basin, UK North Sea. Multibeam echo sounder image [Image acquired by the UK government (Department of Trade and Industry) as part of the Strategic Environmental Assessment process.]

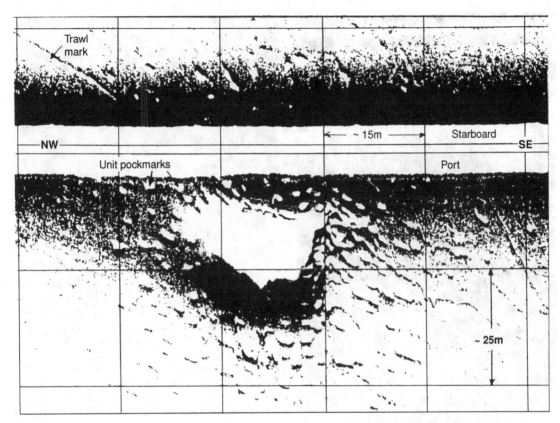

Figure 2.4 A standard pockmark and associated 'unit' pockmarks, central Norwegian Trench. An ROV-mounted side-scan sonar image. Note the trawl mark on the far side of the pockmark wall. (From Hovland and Judd, 1988.)

process during and after the deposition of the Kleppe Senior Formation.

Unit pockmarks are very small (<5 m) seabed depressions found in isolation, in groups and in association with larger pockmarks (Figure 2.4). Harrington (1985) referred to these features as '*pits*' and '*pit clusters*'. He also described areas of '*disturbed seabed*' where individual pits were apparently merging to form a single nascent pockmark. He suggested a genetic relationship between these features, there being a progression from a single unit pockmark or pit to a pockmark proper.

Giant pockmarks are anomalously large when compared to other pockmarks in the vicinity. For example, whereas most pockmarks in the Witch Ground Basin are < 150 m in diameter and < 5 m deep, the well-studied active pockmarks in UK Block 15/25 (see Section 2.3.7) are about 500 m across and > 20 m deep. These, and other giant pockmarks from the central Barents Sea, are discussed in more detail in Section 7.5.6.

Erosive nature

Shallow seismic records from both the Norwegian Trench and the Witch Ground Basin indicate the erosive nature of pockmarks. The internal reflectors of the uppermost sediments are truncated by the V-shaped incisions of the pockmarks, as can be seen in Figure 2.10. However, the pockmarks are not simple inverted cones; some have smooth, relatively flat bottoms while in others the bottom is mounded. Pockmark sides are rarely smooth, but commonly show several breaks in slope or changes in the angle of slope. Stoker (1981) suggested that the combination of shallow tensional depressions close to the edge of these pockmarks and a hummocky floor may be an indication of slumping. In certain cases the profile shape suggests that rotational shear has occurred. Frequently the pockmarks are asymmetric, with one side considerably steeper than the other. In many instances, one side of the pockmark does not regain seabed level for some considerable distance beyond what

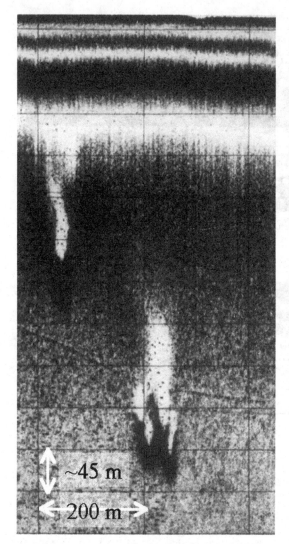

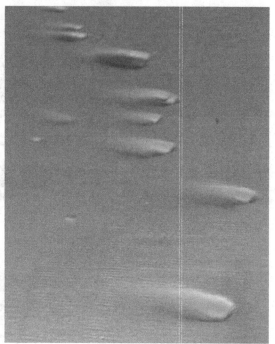

Figure 2.6* Asymmetric pockmarks, Witch Ground Basin, UK North Sea. Multibeam echo sounder image. [Image acquired by the UK government (Department of Trade and Industry) as part of the Strategic Environmental Assessment process.]

Figure 2.5 Composite pockmarks: the feature in the centre of this side-scan sonar (from the Witch Ground Basin, UK North Sea) image comprises at least four craters. (From Hovland and Judd, 1988; reproduced by permission of the British Geological Survey. © NERC. All rights reserved. IPR/67-34C.)

appears at first sight to be the edge of the pockmark. This is probably caused by the restriction of slumping to one side of the feature. These are the same asymmetric features described above from side-scan sonar records.

Partial infilling

British Geological Survey data from the South Fladen area have shown that the pockmarks sampled by gravity core contained 2 m or more silt compared to about 30 cm on the surrounding seabed. Statoil cores from pockmarks in Norwegian blocks 24/9 and 25/7 (i.e. blocks NO24/9 and NO25/7) also showed a greater thickness of the topmost sediment, compared to outside the pockmarks. This 'infill' may have been derived by one or more of three mechanisms: winnowing of the original sediment to leave a coarse 'lag' deposit (the fine sediments having been removed and dispersed), sidewall slumping, or infilling by sedimentation during a period of inactivity since formation.

Particle-size distributions of the BGS samples showed that the sediments inside the pockmarks are identical to those of the Glenn Member outside. Andrews *et al.* (1990) attributed the Glenn Member of the Witch Ground Basin to '*pockmark re-working of underlying sediments*'. This suggests that pockmark activity has played a significant role in the generation of fine sediments in this area, and possibly beyond.

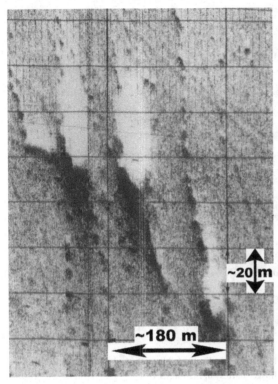

Figure 2.7 Strings of small and large pockmarks consisting of numerous amalgamated unit pockmarks. The strings are connected to the large pockmarks. Note that there are also trawl scars. Side-scan sonar image from the Norwegian Trench. (From Hovland and Judd, 1988.)

Associated features

Close inspection of high-resolution shallow seismic data has resulted in the identification of a range of intra-sedimentary features.

BURIED ('FOSSIL') POCKMARKS

Pockmarks are not found exclusively on the seabed. Buried or 'fossil' pockmarks occur at various horizons within the Kleppe Senior Formation and the Witch Ground Formation (Figure 2.11). Long (1992) described them as '*pockmarks that have ceased venting and have subsequently been covered by sediments*'. When these features are seen only on two-dimensional seismic sections it is possible to confuse them with linear features such as ice-scour marks (as in Figure 2.12). However, in areas for which adequate seismic coverage is available, there can be no such mistake.

DOMES

Localised doming of the seabed (Figure 2.13) has been recorded in a few places in the Norwegian Trench, in Block NO25/7, and the Witch Ground Basin. The domes are isolated circular or elliptical positive topographic features, generally about 100 m in diameter and standing some 1–2 m above the surrounding seabed. They are very unspectacular seabed features with side slopes of <2° (Stoker, 1981). If it were not for the vertical scale enhancement and their rarity on echo sounder and seismic records, they might easily be overlooked.

2.2.5 Evidence of gas

In the original pockmark paper King and MacLean (1970) suggested: '*the main agent responsible for the formation of pockmarks is either ascending gas or water*'. Evidence of gas above and below the seabed takes many forms. The following indicators were found on shallow seismic profiles from North Sea pockmark areas: acoustic turbidity, enhanced reflections, columnar disturbances, and intra-sedimentary doming (Figures 2.10 and 2.13). These terms are explained and illustrated in Section 6.2.2. Various other features not uncommonly found on shallow seismic profiles in association with pockmarks, include hyperbolic reflections and sloping reflections (illustrated in Figure 2.10). These are most probably artefacts produced by the interference of reflections from the pockmark walls beside the path of the seismic equipment, although in some instances hyperbolic reflections are caused by point-source targets, such as dropstones dropped by melting icebergs, and methane-derived authigenic carbonate (described below). Dropstones are not confined to pockmarks.

There have been several reports of shallow cores from pockmark areas in the North Sea expanding on removal from the core barrel, or even extruding themselves. Such cores contain cavities and often they smell of hydrogen sulphide, and analyses have shown elevated concentrations of methane. The expansion and extrusion, and the cavities, are caused by gas expanding with the release of pressure as the cores are brought to the sea surface. Before the removal of the sediment from the seabed the gas may have been in solution or in bubbles. Correlations between this evidence and seismic evidence (acoustic turbidity and bright spots) were first made in the Forties field, where Lucas (1974)

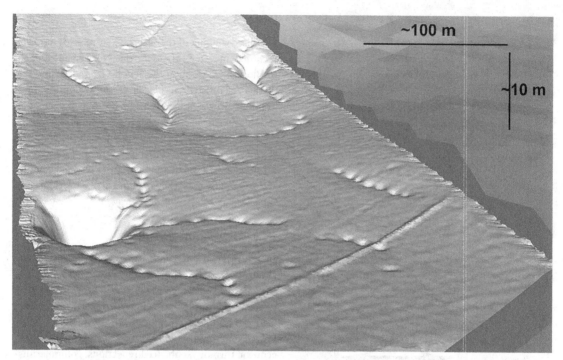

Figure 2.8* An MBES image of pockmark strings in the Norwegian Sea. These strings have no preferred orientation, and some lead to (or from) large standard pockmarks. The 26 inch Haltenpipe pipeline is visible on the lower part of the image. Image courtesy of Statoil ASA.

(a)

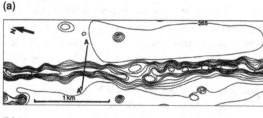

(b)

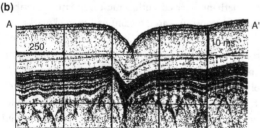

Figure 2.9 The trough shown in (a) and in cross section in (b) is located at the foot of the western slope of the Norwegian Trench. It seems to be composed of a series of large interlinked pockmarks. Note that it overlies a furrow in the Norwegian Trench Formation. One-metre interval contours in (a) show the water depth. (From Hovland and Judd, 1988.)

considered that bright spots represented discontinuous lenses of gas-bearing sand. This interpretation was supported by Caston (1977), who reported that it correlated with high-pressure gas at a depth of 540 m in Well UK21/10-3. The correlation has been demonstrated many times since.

During the many years of North Sea oil and gas exploration every well has been continuously monitored for gas, once the first casing has been set. The results of these investigations are not normally released into the public domain. However, it is certain that there is a considerable body of evidence of gas in the shallow sediments (Tertiary peat, for example) in various parts of the North Sea.

The extent of shallow gas

Seismic evidence of the presence of gas has been recorded from every North Sea area in which pockmarks are known to exist. Gas is not ubiquitous, neither is it always found in close association with pockmarks. Yet, the presence of acoustic turbidity in pockmark areas is sufficiently common to be significant. It is also significant that evidence of gas is not restricted to pockmark

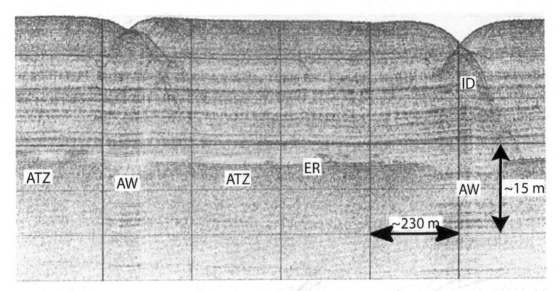

Figure 2.10 Deep-towed boomer profile showing two pockmarks in the Norwegian Trench. Detailed examination of the acoustic layering shows that the pockmarks cut down through the topmost sediments. Deeper layers are obscured by acoustic turbidity zones (ATZs), except in the acoustic windows (AWs). The top of the acoustically turbid zone is marked by an enhanced reflection (ER); this is thought to represent the 'gas front' – the top of the gassy sediments. The acoustic windows (which both lie beneath pockmarks) are presumed to be gas-free zones. An upward inflection of layers beneath the pockmark on the right is termed 'intrasedimentary doming' (ID). Hyperbolic reflections beneath the pockmarks are probably artefacts caused by the pockmark walls. (Adapted from Hovland and Judd, 1988.)

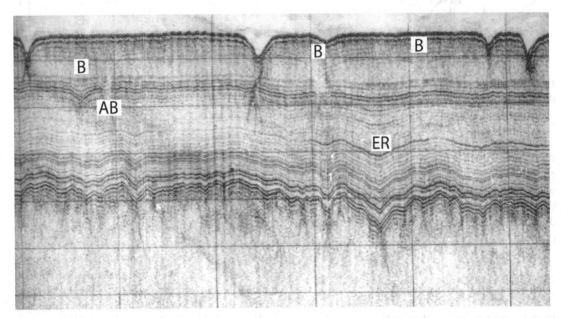

Figure 2.11 Seabed and buried pockmarks (B) in sediments of the Witch Ground Formation; deep-towed boomer profile, UK North Sea. Enhanced reflections (ER) and acoustic blanking (AB) are also present. (From Hovland and Judd, 1988; reproduced by permission of the British Geological Survey. © NERC. All rights reserved. IPR/67–34C.)

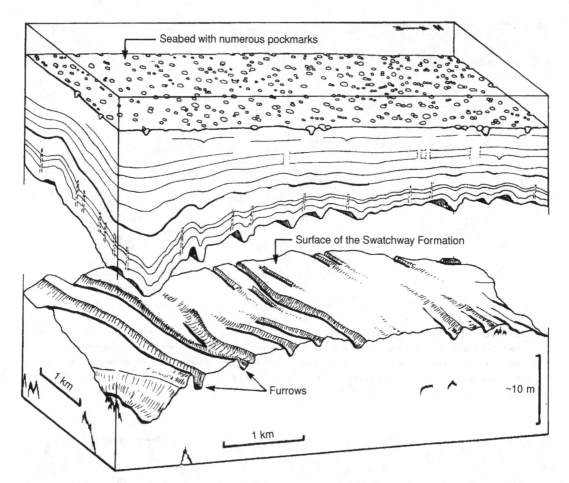

Figure 2.12 Ice-scoured furrows marking the boundary between the Swatchway Formation and the overlying Witch Ground Formation. The distribution of the seabed pockmarks appears to show no correlation with these features. (From Hovland and Judd, 1988.)

areas (see Section 3.5). Regional geochemical studies of the northern North Sea (Faber and Stahl, 1984; Gervitz et al., 1985; Brekke et al., 1997), in which analyses were undertaken on gases from seabed cores, have suggested that gas is quite widespread, although there is some disagreement as to whether the gas is of thermogenic (Gervitz et al.) or microbial (Brekke et al.) origin.

2.3 DETAILED SURVEYS OF NORTH SEA POCKMARKS AND SEEPS

In this section we review the results of some detailed surveys conducted in the North Sea. The locations are shown in Figure 2.14.

2.3.1 The South Fladen Pockmark Study Area

In order to understand more about pockmarks, the BGS undertook detailed studies of an area within the Witch Ground Basin during the 1970s (McQuillin et al., 1979); some of this work was co-funded by British Petroleum (BP). This South Fladen Pockmark Study Area, UK blocks 15/28, 21/3, and 21/4, was selected because the pockmark density is high. Three principal avenues were followed: a study of pockmark morphology and distribution, investigations to see if the geotechnical properties of the sediments varied within and outside pockmarks, and an examination of the possible relationship between pockmarks and gas within the sediments. In both 1977 and 1978 Sonarmarine Ltd provided

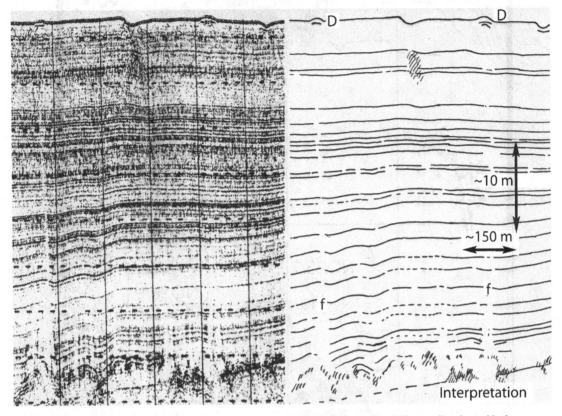

Figure 2.13 Deep-towed boomer profile showing two relatively small seabed domes in the Norwegian Trench. Strong reflections inside the domes (D) might be caused by the presence of gas. Acoustically transparent 'columnar disturbances' in the underlying sediments might indicate vertical fluid migration pathways. From Hovland and Judd, 1988.

complete side-scan sonar coverage of a 3 × 19 km area, and ran shallow seismic (boomer) lines at 150 m intervals. In 1975, the BGS drilled two boreholes using the dynamically positioned drillship Wimpey Sealab (later named the Pholas). Borehole 75/33 was sited within a pockmark and 75/34 was some 200 m away on 'normal' seabed as a control. Borehole 75/33 penetrated to 180 m; however, poor recovery from the top 20 m necessitated the drilling of an additional borehole (75/33a) close by. The following account describes the results of these surveys, and some more recent work.

Pockmark morphology and distribution
Side-scan sonar mapping (Figure 2.15) allowed the sizes, shapes and distribution of the pockmarks to be analysed. However, a clearer image (Figure 2.16) was obtained in 2001 with MBES. Most of the pockmarks are elliptical in shape, rather than circular, and the long axes are generally aligned with the dominant tidal current. They are <130 m long (mean 55 m), <3.5 m deep (mean 1.4 m), <5000 m^2 in area (mean about 1000 m^2), and occupy <5% of the seabed. The overall pockmark density is 33 km^{-2}, but the distribution is clearly not uniform; there are fewer in the south. Furthermore, the pockmarks appear to be preferentially arranged in lineations, or even rings. Statistically the distribution is non-random, and the lineations did not occur by chance (Judd, 1982). It seems probable that there must be some control exercised by the subseabed sediments to account for these lineations.

Evidence of gas
Early surveys produced evidence of gas within the Quaternary sediments of parts of the South Fladen area.

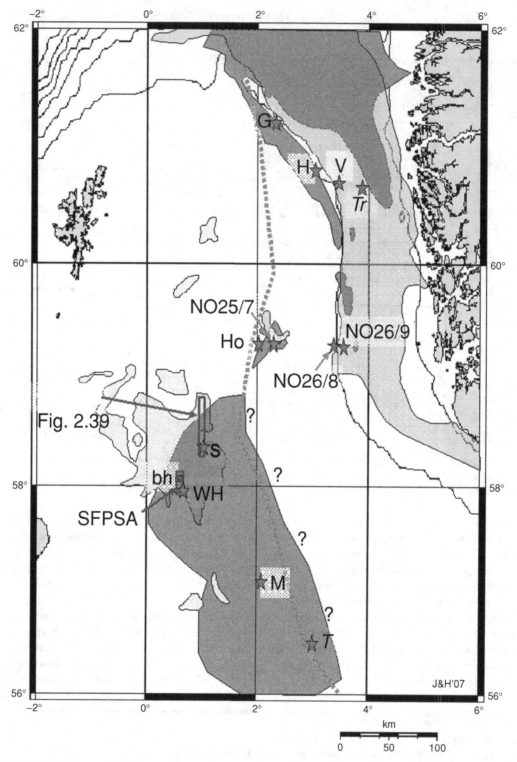

Figure 2.14* North Sea study sites. bh: Borehole 75/33; G: Gullfaks; Ho: The Holene (Norwegian Block 24/9); S: Scanner pockmark (Block UK 15/25); SFPSA: South Fladen Pockmark Study Area; T: Tommeliten; WH: Witch's Hole. Other sites mentioned in the text are H: Huldra; M: Machar; Tr: Troll; V: Veslefrikk. The dark ornament indicates major areas of shallow gas (the ?s indicate an uncertain boundary); light ornament indicates the main pockmark areas. The UK/Norway median line is indicated as a dotted line. (Compiled from various sources.)

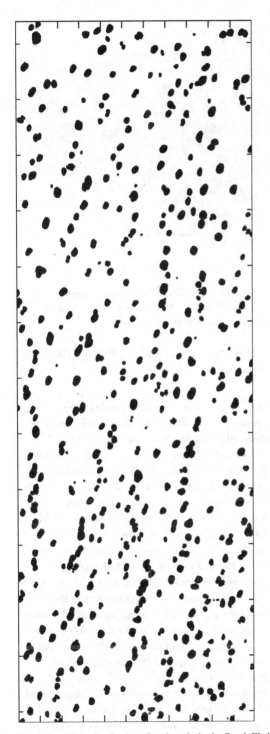

Figure 2.15 The distribution of pockmarks in the South Fladen Pockmark Study Area; derived from manually digitised side-scan sonar records acquired in 1977. The area shown is approximately 3 km wide (east–west) by 9 km long (north–south).

Bright spots recorded during a 1975 site survey for Total Oil Marine were thought to indicate gas at a depth of 320 to 380 m (this was confirmed by data acquired in 1987; see Judd, 1990), and Holmes (1977) reported bright spots covering about 46% of a 44 km^2 area centred on BGS borehole 75/33. The sediments of the Ling Bank Formation at a depth of 59 m in this borehole were described as *'stiff sandy clay: honeycombed with ovoid gas cavities'* Holmes (1977 – see Figure 2.17); gas voids were also found within stiff, silty clays of the Aberdeen Ground Formation at a depth of 160 m in the same borehole. However, no obvious differences between any of the measured geotechnical properties of the sediments beneath and away from the pockmark were found – at least not in the Witch Ground Formation and the underlying Swatchway Formation (borehole 75/34 went no deeper). Holmes also reported that acoustic turbidity within the Swatchway Formation occupied 10% of the same area. Acoustic turbidity within the Witch Ground Formation is confined to the southeast corner of the area, around the unusually large pockmark known as the Witch's Hole. Columnar disturbances and enhanced reflections are present within the Witch Ground Formation throughout the area, indicating the presence of gas or its previous passage through the sediments.

The presence of hydrocarbon gases within the water column was demonstrated by a survey undertaken in 1975 as a collaborative venture between BGS and BP. This survey, one of the first such surveys to be performed in the North Sea, was done with the Interocean Sniffer; which continuously pumped water from close to the seabed for gas chromatographic analysis onboard the ship. The main survey covered an area of 340 km^2. A total of 700 line kilometres were run at approximately 0.5 km intervals. Three detailed study areas were also surveyed, two within the main area and the third further south, to provide coverage of the location of the (then proposed) borehole, 75/33, and two places where BP had previously reported gas bubbles coming from the seabed. The average total hydrocarbon concentration was approximately 0.5 ppm, but values of up to 0.9 ppm were recorded. The majority (approximately 98%) of detected hydrocarbon was methane, but the higher hydrocarbons were all found to be present. C_1: $(C_2 + C_3)$ ratios were generally in the range 80 to 110.

These reports confirm the presence of gas, and it is tempting to suggest that they represent stages in the

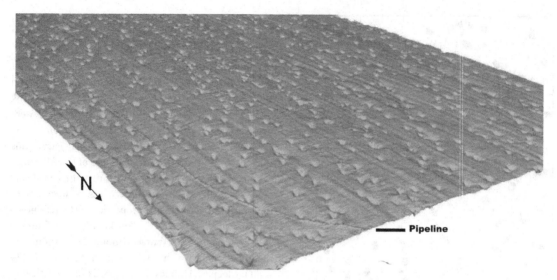

Figure 2.16* Pockmarks in the northern part of the South Fladen Pockmark Study Area; MBES survey, 2001. The pipeline (Scott-Forties Unity pipeline; 24 inch, 61 cm, diameter) gives an idea of the scale. [Image acquired by the UK government (Department of Trade and Industry) as part of the Strategic Environmental Assessment process.]

vertical migration of gas from underlying Tertiary and Mesozoic sediments to the seabed sediments, and then into the water column.

Evidence of pockmark growth and activity
In order to see if new pockmarks were formed, or existing pockmarks grew during the period between the 1977 and 1978 surveys, detailed comparisons were made of pockmarks in part of the South Fladen study area. This was done by first identifying the best possible side-scan record of each individual pockmark from each year, and then digitising the pockmark outlines. No new pockmarks were found, but the results suggested a slight increase in the total area occupied by pockmarks. Comparisons between the new (2001) and old (1978) maps were attempted, but proved not as straightforward as might be expected because of discrepancies between position-fixing systems, acoustic geometry etc. Although no quantitative comparisons could be made, it seems that there were no major changes to the pockmark distribution over the 23 years between surveys. However, there is a hint that gas beneath the seabed is mobile. Judd (1990) found minor discrepancies between the 1975 and 1978 extents of the acoustic turbidity near the Witch's Hole.

A second kind of evidence of pockmark activity was reported by McQuillin *et al.* (1979). The following quotation describes what occurred on 30 July 1978 during one of the geophysical surveys:

At 07:36 hours survey commenced of a line across the main survey area. Large dark clouds were observed on the sonar record which were interpreted as being due to suspended material in the water mass. These clouds are not similar to any previously detected fish shoal. The boomer records showed evidence of a thin layer of recently deposited material at seabed. Over the period 07:36 hours to 15:30 hours, this cloud was seen on a number of records to disperse within the eastern part of the area being studied. A closely spaced group of E–W lines were being surveyed. At 13:00 hours a new event occurred to the west of the earlier event. On the line being surveyed, the sonar record showed an even stronger reflection from a concentrated cloud of material in the water mass [shown here as Figure 2.18b] and the boomer record showed plumes of material in the water elevated to about 10 m above seabed [Figure 2.18a]. This same event was again detected on a parallel line 150 m to the south at 14:40 hours though now smaller and less elevated.

The sonar record on this line indicated some dispersion of the main cloud though the boomer

Figure 2.17 Photograph of a core sample showing gas voids in slightly muddy sand of the Aberdeen Ground Formation, 59 m below seabed in borehole 75/33, South Fladen area, North Sea. (From Hovland and Judd, 1988; reproduced by permission of the British Geological Survey. © NERC. All rights reserved. IPR/67–34C.)

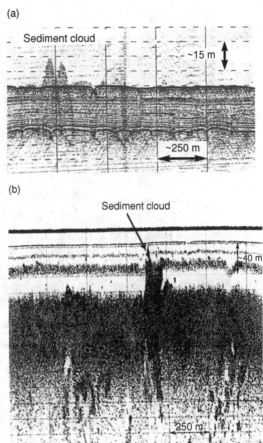

Figure 2.18 Suspended sediment cloud in the South Fladen area: (a) on a boomer record; (b) on a side-scan sonar record. (From Hovland and Judd, 1988; reproduced by permission of the British Geological Survey. © NERC. All rights reserved. IPR/67–34C.)

record indicated close association between the smaller plume and a particular pockmark. This association, however, may not be significant. At 15:44 hours, a line was surveyed across the same course as that which had first detected the plume (at 13:00 hours) and crossed the plume site at 15:54 hours. By this time all evidence of the plume had disappeared and there was little evidence of suspended sediment on the sonar record.

McQuillin *et al.*, 1979

It was concluded that neither gas nor fish produced these features. Fish shoals were frequently observed

during this survey on side-scan sonar (40 kHz) and echo sounder (100 kHz) records, yet they are seldom recorded by boomer (1–6 kHz). The clouds seen on sonar records are larger and denser than those normally associated with fish shoals. Had the clouds been caused by gas they would have dispersed rapidly and upwards, but the recordings showed that the clouds dispersed over a period of a few hours, and settled gradually to the seabed. The conclusion drawn was that two separate disturbances occurred at or near the seabed, probably over a relatively long time period, resulting in large masses of fine sediment being lifted into suspension to a height of at least 10 m above the seabed, and that the sediment gradually settled back to the seabed. These

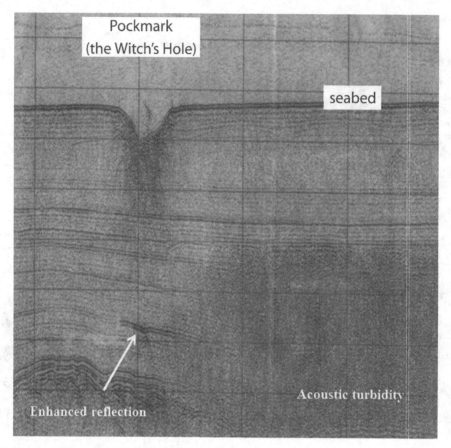

Figure 2.19* Boomer profile across the Witch's Hole, an unusual pockmark in the South Fladen area. (Reproduced by permission of the British Geological Survey. © NERC. All rights reserved. IPR/67–34C.)

events occurred on a calm day, although there had been gales two days previously. A similar cloud was recorded on a single boomer record during the 1977 survey. McQuillin *et al.* surmised that *'the disturbances recorded on 30th July were associated with pockmark growth though not necessarily with the initial development of any new pockmarks'*.

This evidence suggests that the pockmarks in this study area include some features that are at least periodically active.

Post-glacial pockmark activity

The persistence of pockmark formation throughout the post-glacial period represented by the Witch Ground Formation is indicated by the presence of buried pockmarks at several layers. Long (1992) reported that there have been several periods of non-deposition over this time, each represented by a pockmarked former-seabed surface. He suggested that the pockmark density was generally proportional to the time period over which an individual seabed surface had been preserved. However, he concluded that a high density of buried pockmarks at the Witch/Fladen Member boundary about 13 000 years before present (BP) was consistent with an increase in gas-escape activity. This coincided with a period of rapid climatic amelioration and *'the degradation of subseabed permafrost ice lenses'*, which had previously trapped gas.

The Witch's Hole

Figure 2.19, a seismic profile across the Witch's Hole, not only indicates that gas lies close beneath the seabed, but also that there is a 'plume' rising from the pockmark itself (the shadow of this 'plume' can also be seen on Figure 2.20). The acoustic turbidity, the unusual ('fresh') appearance of the pockmark, the Sniffer

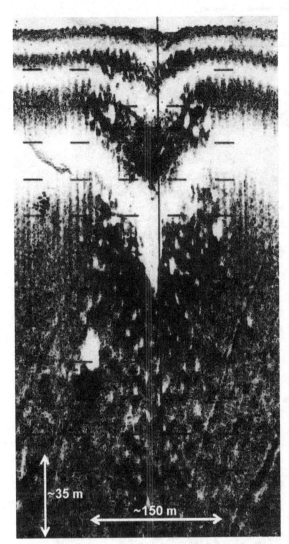

Figure 2.20 Side-scan sonar record showing the Witch's Hole. This was previously (Hovland and Judd, 1988) interpreted as a 'fresh' pockmark. However, more recent data have shown that the shadow in the centre of the pockmark is in fact a shipwreck – see Section 11.5.1 for further details. (From Hovland and Judd, 1988; reproduced by permission of the British Geological Survey. © NERC. All rights reserved. IPR/67–34C.)

evidence of methane within the seawater, and this plume were all taken as evidence that this pockmark was actively seeping gas. However, a side-scan sonar survey undertaken in 1987 by Total Oil Marine produced a surprising result. The 'plume' is in fact a shipwreck (discussed further in Section 11.5.1)!

2.3.2 Tommeliten: Norwegian Block 1/9

This location lies in the Greater Ekofisk area over the Tommeliten Delta salt diapir. Tommeliten was a small gas-condensate field associated with three salt diapirs: Alpha, Gamma, and Delta. The total net hydrocarbon pore volume for the Alpha and Gamma reservoirs was estimated as between 120 and 148×10^6 m^3 (D'Heur and Pekot, 1987), but the Delta structure, which has domed and pierced the enclosing sedimentary rocks, lacks a proper seal. Well number 1/9-5, which was drilled on this structure, turned out to be dry.

There are seismic 'chimneys' (chaotic or turbid zones) above all three Tommeliten salt diapir structures (D'Heur and Pekot, 1987). The strongest may be seen above the Delta structure (Figure 2.21), where gas seeps were recorded during a routine site survey in 1978. The seabed at this site appears flat and featureless on echo sounder records. Water depths range between 74 and 75.5 m, with a very slight slope towards the southwest.

Geophysical reconnaissance and sediment sampling
Shallow seismic (3.5 kHz pinger) profiles indicate that acoustic turbidity (indicating gas) is present in the surficial sediments over an area of approximately 120 000 m^2. Water-column targets visible on pinger and echo sounder records were located in various parts of this area, but most were concentrated in the 6500 m^2 area indicated on Figure 2.22.

Side-scan sonar records showed that the seabed was essentially flat and featureless, but a few areas of high reflectivity were recorded. The sizes of the patches ranged from 100 m^2 upwards; the larger ones are shown in Figure 2.22. During a subsequent site survey, small (about 5 m diameter) oval to circular shallow depressions noted on short-range side-scan sonar records were found to be distributed with a spacing of between 50 and 500 m. Each depression was found to include a patch of highly reflective material, so they were called 'eyed pockmarks'.

Piston core samples demonstrated that the seabed sediments consist of fine- to medium-grained sands and silty sands that frequently contain an abundance of shell material. In some cores the lower sections of this sand were partially cemented by a calcareous cement. This topmost sediment was found to be about 30 cm thick in most places, but near the centre of the area the underlying stiff clay was exposed at the surface. Interstitial

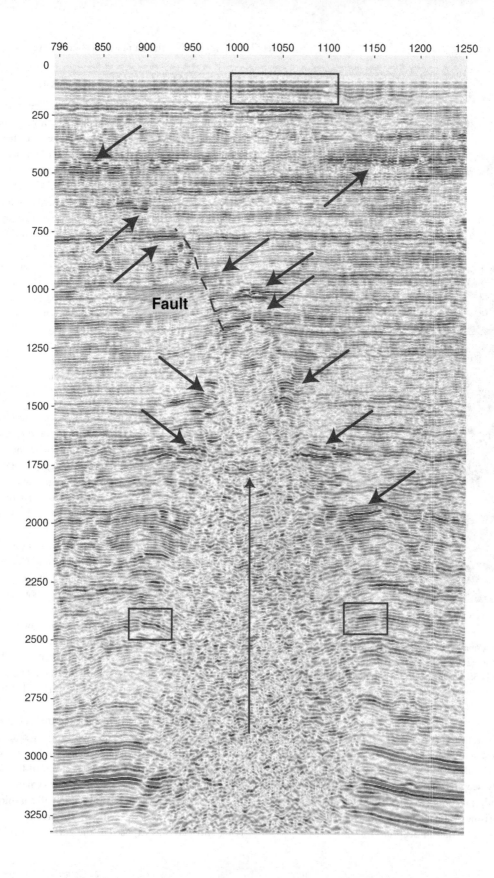

Fault

sediment gases were taken from the piston cores and analysed for hydrocarbons (see below).

ROV surveys

The first objective of the ROV surveys was to locate and identify the water-column targets. It was found that these were gas seeps. It was estimated that there were about 120 individual seeps, all but a few of which are located within the area of gas-charged sediments. Most commonly each seep occurred as a single stream of bubbles, about 1 cm in diameter, rising from a discrete hole of the same size within the sandy seabed. These holes were generally located within a shallow funnel-shaped depression approximately 20 cm in diameter (Figure 1.2). As an experiment one of these small craters was filled with sand. After about 1.5 min the bubble stream was re-established from the same vent, eroding a similar small crater with an initial burst of a few large bubbles. It was concluded that the gas was emanating from a single conduit from the underlying clay and that new escape routes were not easily formed, it was estimated that, on average, each seep produced one bubble every 6 s, and that the total gas production from the main seepage area was about 24 m^3 per day (at ambient pressure, 75 m water depth). Samples of gas were collected from two vents by filling partially evacuated flasks via a funnel.

All the seeps observed from the ROV were located on a sandy seabed. Apart from small ripples, the seabed was found to be featureless with only occasional signs of biological activity. However, this desert-like appearance changed abruptly where the highly reflective seabed, mapped by side-scan sonar, was encountered. These

Figure 2.21* Seismic section across the Tommeliten Delta structure, a salt piercement diapir. The noisy zone is interpreted as a gas chimney through which gas rises vertically (as indicated by the large arrow). Some gas escapes laterally to produce brightening of adjacent reflectors, and reducing the acoustic velocity (v_p) to produce 'pulldown' (examples are in the rectangles). A fault rising from the top of the gas chimney, from about 1200 ms to about 750 ms, provides a gas migration pathway; bright spots adjacent to this fault and elsewhere (examples are arrowed) indicate shallow gas accumulations. Reflection enhancement near the seabed (in the rectangle) occurs where the near-seabed sediments are slightly domed upwards; it is above this feature that gas seeps into the water column. (Image courtesy of Helge Løseth, Statoil ASA.)

zones proved to be 'oases' of biological activity containing a rich fauna. They are located on a slight mound (about 0.5 m high) littered with disarticulated shells and skeletal debris. The dramatic change from barren seabed to 'bioherm' is illustrated in Figure 2.23. The benthic organisms include sea anemones, sea pens, bivalves, gastropods, crabs, starfish plus fish, shrimp, and various planktonic species and some bacterial mats. In 1997, during a subsequent ROV survey, it was discovered that the bioherms are not based on sandy mounds, but on hard, methane-derived authigenic carbonate (MDAC) crusts up to at least 20 cm thick.

The 'eyed pockmarks' proved to be 3–5 m wide. Their origin is uncertain. It could be argued that they developed as a result of current scour around a group of sessile animals; however, there are no cometary markings typical of current erosion features, so they are interpreted as small pockmarks. Inspection by ROV (illustrated in Figure 2.24) revealed that here, too, the reflectivity of the seabed is caused by a concentration of debris, fauna, and carbonate. Although numerous gas seeps, and also bacterial mats, were found close to the bioherms, no signs of gas seepage were seen either in the bioherms or the eyed pockmarks. However, they are all situated above or adjacent to gas-charged sediment. The fauna, and the link between seeps, bacterial mats, and the carbonates are discussed in Section 8.1.1.

Gas

Gas stripped from the sediment core samples, and seep gases collected by the ROV were analysed by gas chromatography for their hydrocarbon content. In the sediments the dominant gas was methane 72%, but ethane, propane, butane, pentane, and hexane were also present. In contrast, the seep gases proved to be almost pure methane. This suggests that mineral grains of the clayey sediment adsorb the higher hydrocarbons, while methane passes through. Carbon-isotope ratios (δ^{13}C) of these gases were found to be in the range −26.7 to −47.7%. When combined with the gas ratios (C_1:$C_2 =$ 0.75 to 0.95), these results indicate a thermogenic origin.

The water-column targets seen on the pinger and echo sounder records extended several metres up into the water. However, data acquired in 2002 show distinct plumes extended all the way to the sea surface. These, and their significance, are discussed in Section 10.4.1.

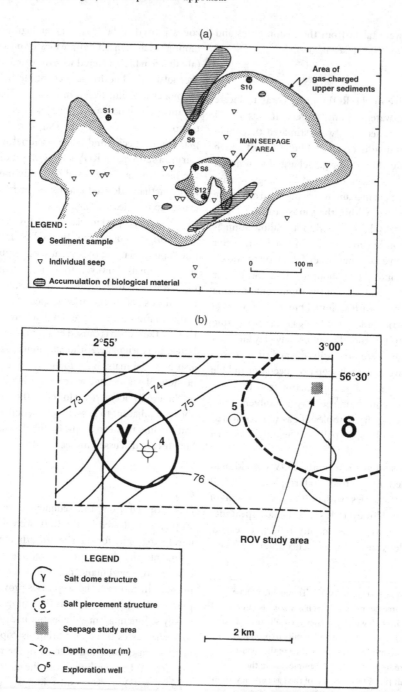

Figure 2.22 The Tommeliten seep study area, Block NO1/9, North Sea. (a) The area surveyed in detail by ROV. The main seep area is located in the centre of the area underlain by gas–charged sediments. Note the locations of relatively prolific benthic flora and fauna. (b) Location map showing that the area shown in (a) is located centrally above the Delta diapir. (From Hovland and Judd, 1988.)

Figure 2.23* A relatively large 'bioherm' as seen from a 'barren' seabed near the gas seeps at the Tommeliten field (Block NO1/9, North Sea). The distance from the ROV-mounted camera to the 'bioherm' is about 2 m, and it is about 0.5 m high. (From Hovland and Judd, 1988.)

Figure 2.24* An 'eyed pockmark' at the Tommeliten field (Block NO1/9, North Sea). The 'eye' consists of a small 'bioherm' (with a large variety of shell debris and other fauna) located within a 'unit' pockmark 3–5 m wide. (From Hovland and Judd, 1988.)

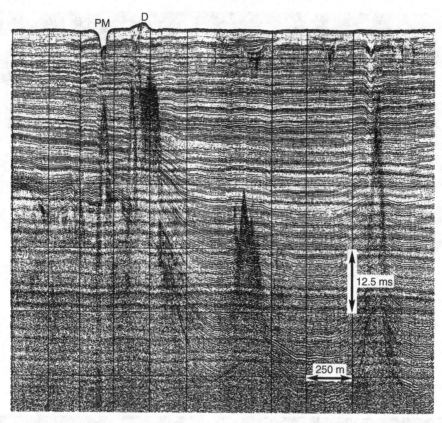

Figure 2.25 A shallow seismic profile across the study site in Block NO25/7, North Sea. Note the pockmark (PM), seabed dome (D), and chimney-like zones of acoustic turbidity. (From Hovland and Judd, 1988).

2.3.3 Norwegian Block 25/7

The study area in Block NO25/7 is located on the plateau (Fig. 2.14) where five pockmarks and two seabed domes were recorded during an earlier survey. Pockmarks are not characteristic of this part of the plateau and the presence of the five pockmarks was itself noteworthy. The seabed, about 125 m deep, was found to be generally flat and very gently sloping towards the east. A thin (< 50 cm) veneer of silty sand covers the seabed and is underlain by 25–60 m of thick multi-layered marine clayey sandy silt. Acoustic turbidity is seen on the seismic profiles (Figure 2.25) and occurs directly beneath the pockmarks and the domes. Zones of columnar disturbance were recorded in several places. Some stem from zones of acoustic turbidity and one extends from the acoustic turbidity to one of the seabed domes. These domes are both approximately 100 m in diameter and about 1 m higher than the surrounding seabed.

Two varieties of pockmark were recorded by side-scan sonar. The first are less than 10 m in diameter and occur singly or in clusters of four or five throughout the area. These were not seen on either echo sounder or pinger profiles. There are five much larger (50–200 m across) pockmarks that present a strong, dark reflection with little acoustic shadow on the side-scan records (Figure 2.26). On pinger profiles they appear as steep-sided features with a flat but rugged floor. Small (unit) pockmarks of the first variety are located around several of these larger pockmarks.

ROV surveys

Away from the pockmarks and domes the seabed proved to be typically covered with grey cohesionless sediment containing a number of small burrows, presumably made by crustaceans (Figure 2.27). There was no change in the appearance of the seabed on the one dome that was

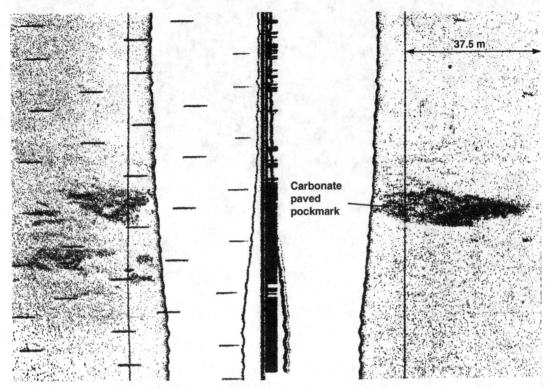

Carbonate
paved
pockmark

37.5 m

Figure 2.26 Side–scan sonar record from the study site in Block NO25/7, North Sea. The pockmark is the one shown in

Figure 2.25. The high acoustic reflectivity is caused by MDAC. (From Hovland and Judd, 1988.)

inspected; however, a single small pockmark, not noted on the side-scan records, was located within the area of the dome. The pockmark sides were found to be about half a metre high. Within the pockmark a considerable accumulation of broken shell (mainly bivalve) material was found. On video it was noted that many crustaceans (?shrimps) were swimming just above the seabed, and that the water immediately above the seabed shimmered. Although no gas bubbles were seen, this seems to indicate some form of seabed fluid flow.

Between the dome and one of the large pockmarks the ROV again passed over an area of unspectacular seabed, but when entering the pockmark a marked change immediately became obvious. The seabed inside the pockmark was found to be rugged and littered with light grey rocks. These rocks were not ice-dropped erratics, which would be smooth. They were too rough in appearance, yet no other type of rock was expected in the middle of the North Sea. The size and nature of this material is shown in Figures 2.28 and 2.29; it appears

that there is an almost continuous crust, covered or partially covered with fine-grained sediment (Figure 2.30). One piece of this crust was grabbed with the manipulator and recovered with the ROV. Once on deck the specimen was measured (30 cm long), weighed (about 10 kg), photographed (Figure 2.31) and doused in liquid nitrogen before being stored in the freezer. Torleiv Brattegard (University of Bergen) identified 38 species attached to this specimen. Although this number seems large, none of them would be unexpected on a stony substrate in the North Sea. What was unexpected was that the nearest stony substrate should have been near the Norwegian coast or the Shetland coast, not midway between the two in an area of unlithified soft Quaternary sediments. This was our first encounter with what we now know to be MDAC. The sample was described by Hovland et al. (1985, 1987), see Section 9.2.1.

Not only was the rock sample teeming with life, but also a considerable diversity of fauna was seen within the pockmark both on and off the carbonate material. There

Figure 2.27* The general sandy seabed scenery at 126 m in Block NO25/7, North Sea. The fish are an argentine (*Argentina silus*), left, and a haddock (*Melanogrammus aeglefinus*), right. (From Hovland and Judd, 1988.)

Figure 2.28* This methane-derived authigenic carbonate (MDAC) was located on the slope of the pockmark shown in Figures 2.25 and 2.26. The fish is a torsk (*Brosme brosme*). The beer can is an example of seabed litter, and provides a scale. (From Hovland and Judd, 1988.)

Figure 2.29* The centre of the pockmark shown in Figures 2.25 and 2.26. Note the torsk fish (*Brosme brosme*), hiding beneath the slab of MDAC. This slab was recovered by the ROV. In the background is a sponge and a feather star. (From Hovland and Judd, 1988.)

was an abundance of shells and shell debris as well as sessile, planktonic, and nektonic fauna; in addition, bacterial mats, fluffy and white in appearance, were found on the pockmark floor (Figure 2.32).

2.3.4 The Holene: Norwegian Block 24/9

The Holene (The Holes) is an open channel about 1 km wide and several kilometres long cut into the plateau (Fig. 2.14), and partially filled with soft, muddy sediments. The general level of the seabed is about 120 to 125 m below sea level, but a maximum water depth of 148 m was recorded during the 1983 survey of an area in the southern part of the Holene. The surficial sediments are not dissimilar to those of the adjacent plateau and the area described in the previous section, except that they are thicker, softer, and finer grained. The layer immediately beneath the surface veneer of sand, which is less than 2 m thick on the plateau, attains thicknesses of up to 20 m in the Holene. Acoustic turbidity is common within the topmost 30 m of sediment in that part on the plateau adjacent and to the south and southeast of the channel, while in the channel itself, or at least that part covered by this survey, it is ubiquitous, generally at a depth of 10–20 m. There are two places in the channel where the acoustic turbidity extends upwards to the seabed. Both are pockmark locations.

Two varieties of pockmark observed in this area are essentially the same as those described from the previous area (Section 2.3.3), which lies some 20 km further east. The majority are small (< 10 m diameter) in size and occur either singly or in small groups within the channel and on the adjacent plateau, but mainly along the eastern margin where they attain a maximum density of about 30 per km^2. Within the Holene the pockmarks are larger, mainly up to 40 m across, but a

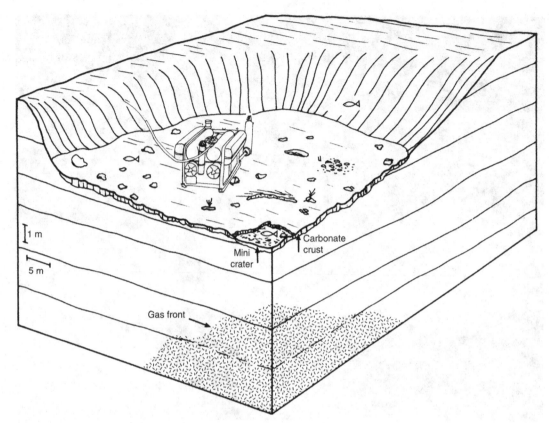

Figure 2.30 Artist's impression of pockmark paved with MDAC, such as those seen in Blocks NO24/9 and NO25/7, North Sea. The ROV provides an approximate scale. (From Hovland and Judd, 1988.)

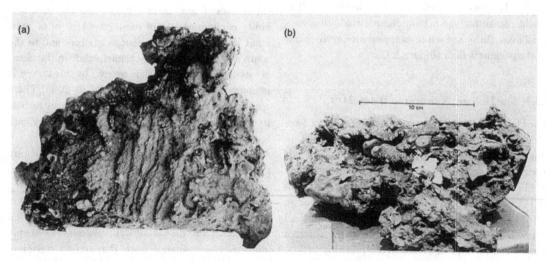

Figure 2.31 An MDAC sample 25/7–83 after recovery; (a) shows the ribs characteristic of the underside, (b) side view. Note the abundant fauna, and the shell debris held by the cement. (From Hovland and Judd, 1988.)

Figure 2.32* Bacterial mats (probably *Beggiatoa spp.*), MDAC, shell debris, and a variety of fauna inside the pockmark shown in Figures 2.25 and 2.26. The bacterial mats are the white patches in the foreground; they indicate the seepage of sulphidic fluids. (From Hovland and Judd, 1988.)

few attain diameters of 100 m or more. These are not well-defined features; rather they appear to be areas in which a considerable number of individual features of varying size have encroached upon one another. They are characteristically only 1 or 2 m deep with flat-bottomed, acoustically reflective floors. As in Block NO25/7, this high reflectivity is caused by MDAC on the pockmark floors. This was demonstrated during an ROV inspection. The MDAC is usually found where acoustic turbidity (gas blanking) is present and

most commonly where it reaches or nearly reaches the seabed.

During an ROV dive on a 40 m diameter pockmark within the Holene it was found that the floor was partly covered with convex-up shells of the bivalve *Arctica islandica*. Where the shells are most densely congregated, at the foot of the steep sidewall, there are about 12 or 13 per m^2 (Figure 2.33). In one part of this pockmark the seabed was completely covered by shell material, most of it disarticulated bivalves.

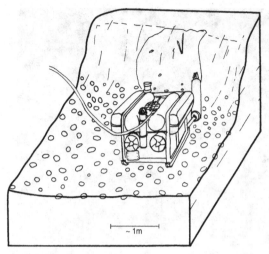

Figure 2.33 Artist's impression of the shell-floored pockmark studied in Block NO24/9. The sidewall is approximately 1.5 m high. (From Hovland and Judd, 1988.)

2.3.5 The Norwegian Trench

Short investigations were undertaken at two sites approximately 17 km apart, one at the top of the western slope of the Norwegian Trench (water depths of 148–167 m), the other due east of the first, near the foot of the western slope (water depths of 200–245 m).

Top of the western slope: Norwegian Block 26/8
An ROV inspection of the first site showed the seabed to be flat. There was a sparse benthic fauna comprising mainly bivalves, scaphopods, crustaceans (principally *Nephrops norvegicus*, the Norwegian lobster) and occasional fish and shrimp. The single pockmark investigated proved to be steep-sided, with sidewall slopes of at least 12° and a depth of approximately 3 m. The seabed within the pockmark was hummocky; this was attributed to slumping. Unlike the pockmarks of the plateau, there was no evidence of MDAC, and the benthic fauna was not abundant in the pockmark. However, it was noted that the benthic and nektonic crustacean (scampi and shrimp) populations were somewhat higher inside the pockmark than outside. The seabed sediments in the pockmark were soft but cohesive and apparently capped by a thin, weak crust (Figure 2.34). This crust did not resemble the carbonate cement described in previous sections. Nevertheless, it was found that, when pushed by the ROV's manipulator, parts of the seabed would move as a single block.

A glacial dropstone, a rounded, crystalline boulder about 50 cm in diameter, was found embedded in the sidewall of this pockmark. Crabs, sea anemones, and other creatures that favour a hard substrate had colonised this stone.

Bottom of the western slope: Norwegian Block 26/9
Near the bottom of the western slope the appearance of the seabed was similar to that at the top. One difference was that the sediments were finer; as a consequence it took much longer for the 'dust' to settle after the ROV had been parked on the seabed. The sediment inside the pockmarks investigated was cohesive, balling up in front of the ROV's skids. This did not happen outside on the normal seabed.

The two pockmarks investigated were both 30–40 m in diameter and attained depths of about 6 m. Smaller (unit) pockmarks were found around the main features. The main pockmarks comprised a broad, shallow depression with slopes of about 5° or less, with a deeper central feature inside. Slopes of up to 19° were recorded in these central parts, and even steeper slopes were thought to exist; at one time the ROV was parked with one skid suspended over a very steep drop into the deep, central part of a pockmark. Again, there was evidence of sidewall slumping; however, this was not as pronounced as at the previous site (in Block NO26/8).

The only evidence of an increase in faunal activity in these pockmarks was a dramatic increase in the density of shrimp and other nekton (Figure 2.35). They were present outside the pockmarks, but not in anything like the same concentration as inside.

2.3.6 Gullfaks

Gullfaks is one of three giant oil fields (Snorre, Statfjord and Gullfaks) located on the edge of the North Sea Plateau in the western limb of Viking Graben. The reservoirs are charged by hydrocarbons generated in the Oseberg Kitchen to the south, filling and spilling from one field to another. Gullfaks is the last in this succession, so it had been expected to be fully charged. However, in 1978 when it was drilled there was some disappointment because the seal was leaky, restricting the amount of petroleum in the reservoir (Karlsson, 1986). Figure 2.36, a shallow seismic profile across the field, shows that there is shallow gas within several horizons

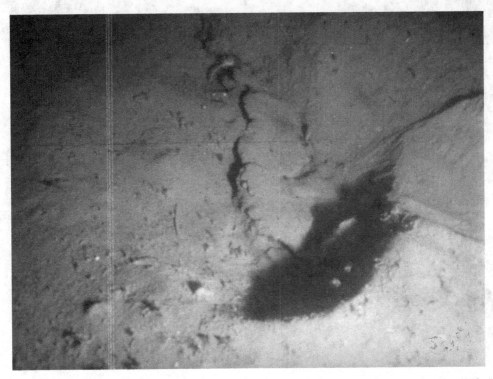

Figure 2.34* A small pockmark, about 3 m long and 1 m wide, with some dish-shaped sediment crusts. When sampled these crusts broke and crumbled. It is assumed that they represent an early stage of MDAC formation. The location is on the upper part of the western slope of the Norwegian Trench, Block NO26/8, North Sea. (From Hovland and Judd, 1988.)

right across the area. The gas accumulations lying 300–450 m below seabed presented serious problems for exploration and production drilling, as we explain in Section 11.3.1.

Gullfaks lies partly under the North Sea Plateau, and partly under the western slope of the Norwegian Trench. The soft, plastic, silty clays of the Kleppe Senior Formation on the western slope (see Section 2.2.2) coarsen upslope, and partially grade into the coarse sands and gravels of a submerged beach zone. Above the beach zone are the stiff clays of the plateau.

Geophysical reconnaissance
Pockmarks are abundant on the western slope of the Norwegian Trench. Those of the Gullfaks C area are elongate, up to 250 m long, 125 m wide and generally 5 or 6 m, but up to 10 m, deep. The long axes show a remarkably constant preferred orientation that is approximately parallel to the seabed contours. It is possible that this orientation is affected not only by the slope, but also by the dominant southward flow of bottom waters (the Atlantic Inflow Water) that prevails in this part of the Norwegian Trench (Eide and Andersen, 1984). Unit pockmarks are also found. They are most densely concentrated at the southern ends of larger pockmarks.

The pockmarks were found to be steep-sided, and they all extend down through the topmost sediments (fine, silty, clayey sand) into the underlying slightly overconsolidated silty, sandy clays, and sometimes into the clayey, silty sand lying beneath them. All were found to be associated with a columnar zone of disturbance, and these zones were found to be elongate like the pockmarks. Some zones terminated beneath the seabed, while others extended to the seabed where there was no pockmark. The zones originate from a deeper layer of acoustic turbidity found to be ubiquitous in this area. Analyses of gases taken from borehole samples demonstrated that this acoustic turbidity is caused by gas, principally methane. The top surface of the acoustic turbidity (the 'gas front') has a marked topography. The depressions in the gas front normally coincide with

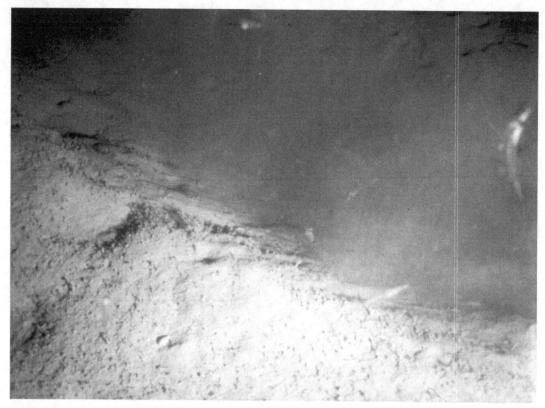

Figure 2.35* Turbid water and numerous shrimp seen inside a small pockmark at the bottom western slope of the Norwegian Trench, Block NO26/8, North Sea. Also seen are a squat lobster (*Munida sarsi*), krill (*Meganyctiphanes norvegicus*), and arrow worms (cheatognaths). (From Hovland and Judd, 1988.)

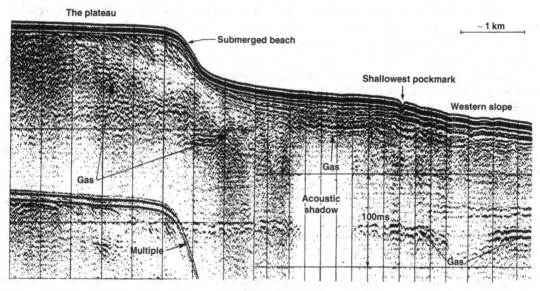

Figure 2.36 Shallow seismic (sparker) profile across the Gullfaks field, Norwegian North Sea. Note the gas accumulations in sand lenses at 410 ms (about 375 m). The acoustic shadow is caused by acoustic energy absorption by overlying gassy sediments. The 'shallowest pockmark' is at about the position of PM204. (From Hovland and Judd, 1988; record courtesy of Institutt for Kontinentalsokkelundersøkelser, IKU.)

Figure 2.37* View into the 'cavern' inside PM204, Gullfaks, Norwegian North Sea. The cavern is about 1.5 m wide and 1 m high. Clay/silt blocks have evidently fallen from the roof. It seems that they are partially cemented, and have sufficient strength to bridge the mouth of the cavern. The fish are torsk (*Brosme brosme*) and ling (*Molva molva*). (From Hovland and Judd, 1988.)

zones of columnar disturbance, which lead to seabed pockmarks.

Evidence from this survey indicates that gas, originating from depth, migrates upslope through the surficial sediments of the western slope, periodically escaping to the seabed pockmarks through vertical zones of columnar disturbance. However, near the very top of the slope the gas apparently migrates updip through the relatively coarse (and therefore more permeable) sediments deposited near the old beach zone. Gas also migrates laterally to the beach zone from beneath the plateau to the west where there are no pockmarks. Having reached the beach zone the gas can relatively easily migrate to the seabed because of the coarseness of the sediments, but there are no pockmarks, presumably because the gas can percolate through the large pore spaces. On one occasion, when a cable was being buried across a section of the beach, thick (< 0.5 m) layers of MDAC were encountered, covered by a thin layer of loose sand and gravel.

ROV surveys

Two pockmarks, referred to as pockmarks PM204 and PM249 because of the depths of water around them, were inspected during the 1985 survey. The seabed outside the pockmarks was found to be generally smooth with a few ripples and undulations. As the pockmarks were approached a gradual increase in the frequency of shells, trails, and tracks of benthic animals was noted. Unit pockmarks, ranging in size from approximately 0.25 to 5.0 m in length, also increase in frequency towards the pockmarks. Some unit pockmarks are deeper and more sharply defined than others, their appearance suggesting that they are relatively fresh; others seem to have been smoothed and infilled with time. Shells, sea anemones, and occasionally starfish were observed within most unit pockmarks. Fish were observed occupying several of them, and shrimps were seen in and around them. Only one, relatively large (40 × 30 × 30 cm), round block of MDAC, was seen within PM204. A few bacterial mats were also noted inside this pockmark, but the most interesting features were 'caverns' in the southern sidewall of the pockmark. One of them was occupied by a group of fish (Figure 2.37). It seems that the roof of the cavern has partially collapsed. This roof material is full of holes, yet it is evidently strong enough to remain intact when roof collapses occur. At first, it was thought that it might consist of MDAC. However, it was easily crushed by the manipulator, which left a smooth groove in the sediment, demonstrating its cohesive nature. We think the caverns, and the perforated nature of the cohesive clay in the southern, steep end of the pockmark result from

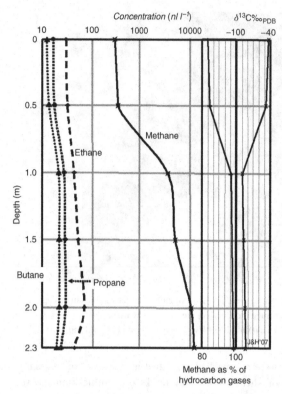

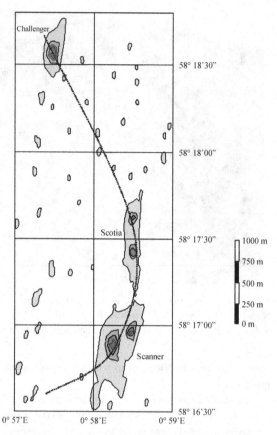

Figure 2.38 Occluded hydrocarbon gases in sediments beneath PM204, Gullfaks, Norwegian North Sea. Left: concentrations of methane, ethane, propane, and butane (*iso-* and *n-*); next panel: methane as a percentage of these hydrocarbon gases; right: methane carbon isotope signature.

Figure 2.39 The giant pockmarks of Block UK15/25. Pockmarks > 1 m deep are shown. The three giant pockmarks, Challenger, Scotia, and Scanner, are significantly larger and deeper than the numerous 'normal' pockmarks; they are 15, 14, and 22 m deep, respectively, whereas the normal pockmarks extend no more than 3 m below the general seabed depth of 150 m. Figures 2.41 and 2.42 are boomer and side-scan sonar sections across Scanner; the dotted line shows the approximate location of the seismic section shown in Figure 2.44. Compiled from MBES data acquired by the UK government (Department of Trade and Industry) as part of the Strategic Environmental Assessment process. This location is shown on Figure 2.14.

repeated expulsions of fluid. It seems that the surface crust, undermined by the erosion of soft clay from the caverns during these events, periodically collapses. So, this pockmark progressively grows at the southern end, and may fill from the north; it migrates southwards along the slope.

No gas bubbles were seen during the exploration of PM204 and PM249. However, MDAC seen at several places on the plateau and in the beach zone indicate seepage, and prolific gas bubbling from a seabed covered in bacterial mats was found during ROV inspection of the seabed immediately downslope of the beach.

Sediment gases

A 2.3 m gravity core was taken in PM204. Analysis of the interstitial gases showed that methane was dominant, but the higher hydrocarbon gases (ethane, propane, butane, and pentane) were also present.

Figure 2.38 shows that the concentration of methane decreases towards the seabed, but the relative increase in the higher hydrocarbon gases indicates that methane is either escaping or the most quickly utilised. This depletion is also reflected in the carbon isotope data; ratios ranged between −44.3% (a thermogenic signature) and the very low value of −90.6% (microbial). The adsorbed methane at 1.5 and 2.3 m has a 'thermogenic' isotope signature compared to the occluded methane (see

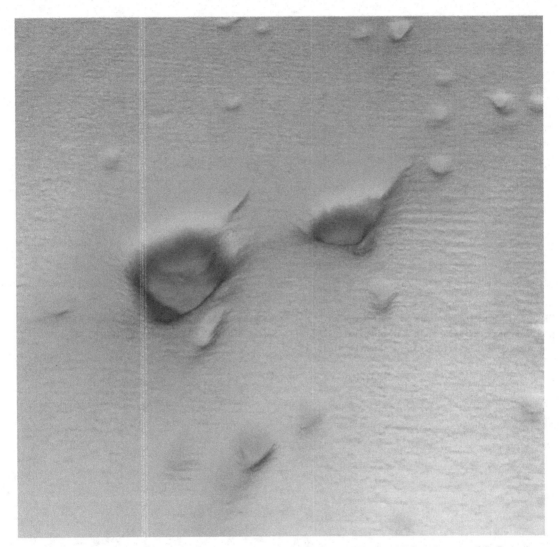

Figure 2.40* An MBES image of the Scanner pockmark, Block UK15/25, North Sea. [Image acquired by the UK government (Department of Trade and Industry) as part of the Strategic Environmental Assessment process.]

Figure 2.38), which shows a typical 'microbial' signature. This could either be due to migration of thermogenic methane and *in situ* production of microbial methane, or due to fractionation via the oxidation of thermogenic methane near the surface (Whiticar *et al.*, 1986).

2.3.7 Giant pockmarks: UK Block 15/25

During a routine site survey by Geoteam UK Ltd for Conoco UK Ltd in 1983, an unusually large pockmark was found to occupy the site of a proposed exploration well (Hovland and Sommerville, 1985). This pockmark, and a few similar features that have since been found nearby (see Figure 2.39), have been studied in great detail: shallow seismic and side-scan sonar surveys (1983, 1991, 1992, 2001, and 2002), seabed sediment sampling (1988, 1989, 1990, 1991, and 2002), ROV inspection (1985) and manned submersible survey (1990). Studies of the geological history, benthic ecology, and water chemistry have been undertaken. Together, these surveys have provided a wealth of detailed

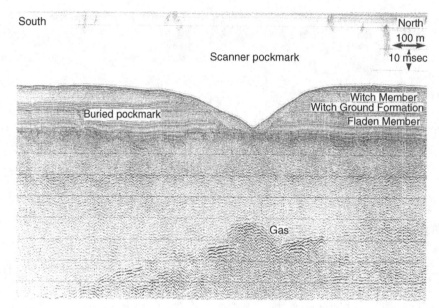

Figure 2.41 Digital boomer profile across the Scanner pockmark, Block UK15/25, North Sea. Shallow gas is indicated approximately 50 m below the pockmark. (From Hovland and Judd, 1988; reproduced by permission of the British Geological Survey. © NERC. All rights reserved. IPR/67–34C.)

Figure 2.42 Uncorrected side-scan sonar image of the Scanner pockmark, Block UK15/25, North Sea, showing discrete water-column targets related to individual gas-bubble plumes [Image acquired by the UK government (Department of Trade and Industry) as part of the Strategic Environmental Assessment process.]

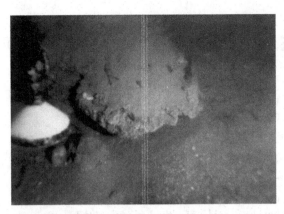

Figure 2.43* A large slab of MDAC found near the centre of the Scanner pockmark, Block UK15/25, North Sea. (From Hovland and Judd, 1988.)

information which has been presented in various publications and unpublished reports, most notably: Hovland and Sommerville (1985), Dando *et al.* (1991), Judd *et al.* (1994), and Dando (2001). The findings are briefly summarised here, but more detailed discussions of some aspects appear later in this book.

To avoid confusion, Judd *et al.* (1994) named the three large pockmarks in this area after three of the survey ships used here: Scanner, Scotia, and Challenger. Figure 2.39 shows that two of these pockmarks (Scanner and Scotia) are in fact composite features, each having two individual deep areas. Overall Scanner (also shown in Figure 2.40) is roughly oval in shape, 900 m long, 450 m wide, and ~22 m deep. Data from an ROV-mounted side-scan sonar survey of the Scanner pockmark showed, as in the Gullfaks pockmarks, that unit pockmarks increase in density towards the main feature, and that they are particularly concentrated at one end. The volumes of these pockmarks (Scanner: approximately 1 million m^3) are considerably greater than the normal pockmarks in the area. They cut down right through the Witch Ground Formation sediments to reveal the stiffer clays of the underlying Coal Pit Formation (Figure 2.41). Detailed analysis of this section (Judd *et al.*, 1994) showed that this pockmark was formed about 13 000 years ago (described in more detail in Section 7.5.6).

Shallow gas

Evidence of shallow gas has been recorded on airgun, sparker and boomer profiles from depths ranging from > 1000 m to < 60 m below the seabed. Two possible sources are present: the Upper Jurassic Kimmeridge Clay, the primary petroleum source rock in the northern North Sea, and peats in Tertiary (Hordaland and Nordland groups) sediments. The shallowest gas, which is trapped beneath the Coal Pit Formation, covers an area of at least 11 km^2, and its depth varies from > 75 m to < 55 m; the three giant pockmarks overlie places where the gas is shallowest (Judd *et al.*, 1994).

Gas seeps

Echo sounder, side-scan sonar and boomer profiles have all shown water-column targets emanating from the deepest parts of all three pockmarks (see Figure 2.42) indicating that they are all actively seeping; this seepage has been recorded on nine separate surveys between 1983 and 2002. Inspections by the ROV Solo (in 1985) and the submarine Jago (in 1990) provided confirmation that gas is bubbling from the seabed. In 1985 only three gas seeps were found during a close visual search grid. They were all located at the southwestern end of the pockmark, clustered beside a large slab of MDAC. Two of the bubble streams were continuous. Small bubbles, 5 mm in diameter, emanated from a mound of loosely packed sediment adjacent to the hollow under the MDAC (Figure 2.43). The third seep was different. Here, trains of five or six large (30–40 mm diameter) bubbles intermittently burst out of the loose sediments in the slope between the mound and hollow under the slab. This suggests that gas builds up in a subsurface (probably very shallow) reservoir before the intermittent release. This gas was sampled. Bubble streams were also recorded on video and sampled in 1990 when gas-flux measurements showed that individual bubble streams produced between 0.14 and 0.6 litres of gas per hour; about 5.74 l per hour at surface temperature and pressure (Clayton and Dando, 1996).

Although bubbles were seen on each visual survey, the number of bubble plumes was smaller than indicated by some of the acoustic records (such as Figure 2.42). On some other records (e.g. Figure 2.44) single large plumes were recorded above each pockmark. The significance of this is discussed in Section 10.4.2.

Gas analyses

Gas samples were taken from seabed sediments (1985, 1990), by sampling the bubbles streams (1985 and 1990), and from the overlying water (1990). Surprisingly,

Figure 2.44* Seep plumes from the Scanner (left), Scotia (centre), and Challenger (right) pockmarks, Block UK15/25, North Sea acquired during the Heincke 180 cruise, October 2002 (Alfred Wegener Institute) using the parametric sediment echosounder system (SES-2000DS) developed at Rostock University, Germany; this scan shows depths from 140 to 190 m. (Courtesy of Gerdt Wendt, University of Rostock.)

although the seeping gas is almost pure (> 99%) methane, sediment interstitial gases contained only 82 to 87% methane, with 6.4 to 7.4% ethane, plus minor amounts of propane, butane, and pentane. Carbon isotope ratios also suggest a difference between the sediment gases (−39.3 to −40.0%) and the seep gas: −71.0% in one 1985 sample; −73 to −79% in 1990 samples (Clayton and Dando, 1996). These results suggest that the seep gas comes from a microbial source (Clayton and Dando, 1996, suggested the Tertiary peats), whilst the sediment gas is of thermogenic origin (possibly the Kimmeridge Clay). This explanation may need further consideration. Methane concentrations in the water were significantly higher (< 60 nM) above the pockmarks than away from the pockmarks (< 20 nM).

Carbonates

Visual inspections of the Scanner pockmark revealed the presence of large slabs of MDAC (Figure 2.43); the largest measured approximately 2×1 m and was 20–50 cm thick. These slabs were generally oval discs that, in at least some cases, appeared to be supported centrally by a pillar or pedestal (i.e. mushroom-shaped). One small nodule was collected by the ROV, another by the submarine. Like the carbonates from the other sites, it was cemented with ^{13}C-depleted aragonite and calcite (δ^{13}C −52.0%).

Benthic ecology

The benthic ecology of these pockmarks, particularly Scanner, has been studied in detail. As in the other sites described above, a greater diversity and concentration of fauna was found inside the pockmark than outside it. However, the implications of this are debateable (see Section 8.1.1).

2.4 CONCLUSIONS

This initial phase of investigations proved the importance of pockmarks not just as morphological features, but also as evidence of fluid migration, and various processes associated with fluid migration. The primary conclusions reached as a result of the research reviewed above are the following.

- Pockmarks are indicators of gas-escape events, but not all pockmarks are actively seeping gas.
- Gas seepage may occur with or without the formation of pockmarks.
- Gas seeps/pockmarks are frequently associated with carbonate precipitates.
- Gas seeps/pockmarks are frequently associated with unusual biological activity.

We do not claim to be the first or the only people to have worked on pockmarks and seeps. We worked along with many others; not least the group led by Lewis King at the Bedford Institute of Oceanography (Dartmouth, Nova Scotia), Bob McQuillin, Nigel Fannin, David Long, and others at the BGS, Paul Dando and colleagues at the Marine Biological Association of Great Britain and more recently at the University of Wales, Bangor, and several oil-industry groups. However, our work on the compilation of our 1988 book helped us recognise the threads that nature has tangled together in pockmarks and seeps. The door was now opening to a whole new area of scientific research, discovery, and understanding.

In the years that have passed since we arrived at these conclusions there has been an explosion of interest in features associated with seabed fluid flow. A new scientific paradigm shift is clearly underway. In the following chapters we attempt to provide a synthesis of the present understanding of the features associated with seabed fluid flow, and the processes responsible for them.

3 · Seabed fluid flow around the world

The development of an observational science like geology depends upon the personal experience of individual workers. I suggest that, with certain reservations, the best geologist is he who has seen the most rocks.

H. H. Read

It is impossible to review all the data, reports, and publications of examples of seabed fluid flow. This chapter, organised as a round-the-world tour, comprises descriptions of examples of features associated with seabed fluid flow: pockmarks, seeps, mud volcanoes, and gas hydrates. The chapter provides the basic information required to undertake more analytical studies of these features, the processes that are responsible for them, and their implications.

3.1 INTRODUCTION

To understand the nature and variability of seabed fluid flow, and to realise the extent of its implications, it is necessary to review as much evidence as possible. We do not claim to be the ones who have seen the most, but we have tried to review as many data and publications as possible in developing our ideas about seabed fluid flow features. Seabed mapping has progressed at such an incredible rate that it is impossible to include every example. Here we describe manifestations of seabed fluid flow from many parts of the world. We have selected excellent examples, and features that have provided critical evidence for the development of our present understanding of seabed fluid flow. This chapter therefore provides the compendium of 'observations' necessary as a foundation for the more analytical approach of subsequent chapters.

Compilations of observations must be organised in some logical way. In this case we have chosen to embark on a 'world tour'. Our route starts in the Arctic, north of Scandinavia, and ends in eastern Canada. We visit pockmarks, seeps, vents, and other features formed by or associated with fluid flow: mud volcanoes, gas hydrates, carbonates, chemosynthetic biological communities, etc. We focus mainly on 'cold' seeps as the literature on hydrothermal vents, particularly those of the ocean spreading centres, is more extensive. In many cases, more detailed descriptions and discussions in subsequent chapters follow the observations reported in this chapter. Location maps showing the places mentioned in the text appear on the accompanying website; Figure 3.1 (Map 1) indicates the coverage of these maps.

3.2 THE EASTERN ARCTIC

3.2.1 The Barents Sea

Pockmarks and iceberg scours
Pockmarks 50 km south east of Hopen Island (Map 2) occur in about 200 m of water in an area where Mesozoic bedrock is covered only by a thin unit (seldom more than 8 m) of clayey till and a thin cover of Holocene mud. The pockmarks are small, 10 to 20 m in diameter and < 1 m deep, but in places they occupy about 25% of the seabed. The low organic content in these thin Quaternary sediments persuaded Solheim and Elverhøi (1985) that these pockmarks were caused by seepage of thermogenic gas. Pockmarks occur elsewhere in the Barents Sea; most of those over the Snøhvit petroleum field are about 15 m wide and < 3 m deep.

In much of the Barents Sea icebergs heavily scoured the seabed during the late Weichselian, as the ice retreated. The iceberg ploughmarks tend to be more pockmarked than the seabed outside the ploughmarks (see Figure 3.2), suggesting some relationship. Possible explanations are that the icebergs affect the character of the sediments, enabling fluids to migrate

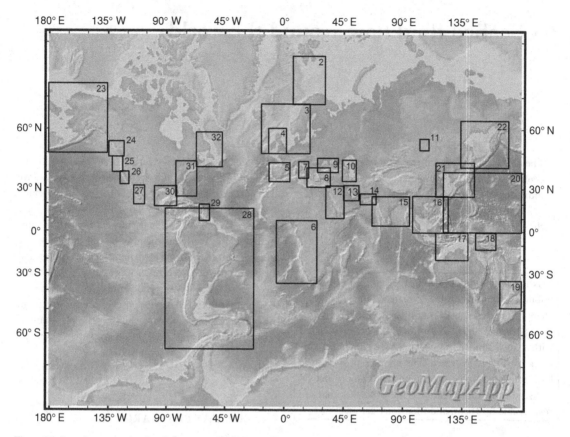

Figure 3.1 Location maps: showing the areas covered by location maps presented on the website. [Prepared using the on-line mapping system 'GeoMapApp' (www.marine-geo.org/geomapapp/) of the Marine Geoscience Data Management System, Lamont-Doherty Earth Observatory.]

through them more easily, or that the 'pockmarks' are actually indentations made by the ice, and not true pockmarks.

Giant blowout craters

In the central Barents Sea Solheim and Elverhøi (1993) mapped a cluster of 15 to 20 large craters. They are 10 to 30 m deep and generally 300 to 400 m wide, but the largest is 600 by 1000 m across (Figure 3.3). Because of their steep (< 50°) walls and large size, they are distinctly different from normal pockmarks (Figure 3.4). Indeed, seismic profiles show that they have been cut down right through the unlithified sediments into the Triassic sedimentary bedrock. Video surveying and photographs showed that angular slabs of mudstone and siltstone lie on the surface of seabed mounds inside the craters. Long *et al.* (1998) attributed

these mounds to the presence of gas hydrates. A large plume of methane-charged seawater over, and adjacent to, the craters indicates that they are in hydraulic contact either with a subsurface gas hydrate deposit or a seeping reservoir of methane gas (Lammers *et al.*, 1995a). Acoustic profiling, coring, and visual inspection with an ROV failed to pinpoint the source of this plume. However, the fauna may provide a clue: Long *et al.* (1998) reported that there were suspension feeders (sponges, soft corals and anemones) attached to hard substrates. The sponges were tentatively identified to include '*typical colonists of hard substrate in deep waters*' (*Phakettia* sp.; *Chonelasma* sp.; *Polymastia* sp.; and *Geodia* sp.), plus a member of the genus *Cladorhiza* (see Section 8.2.2). They also reported dense swarms of krill, occasional shrimps; fish, particularly cod, were found to be numerous in the deepest parts of the depressions. The fact that

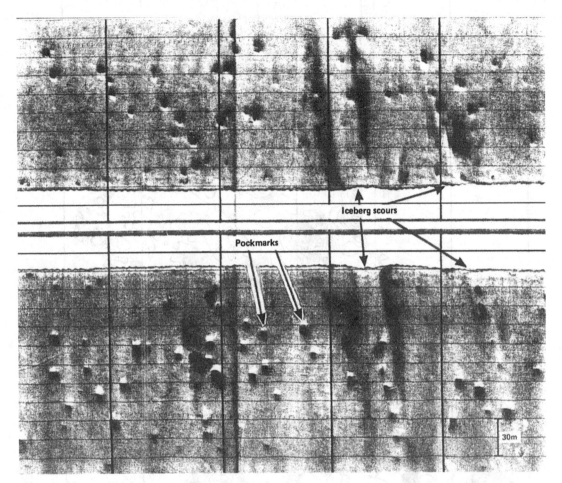

Figure 3.2 Uncorrected side-scan sonar image showing seabed pockmarks and iceberg scours over the Snøhvit petroleum field. A thin cover of fine-grained sediments drape over the stiffer iceberg-scoured underlying till. Note that many of the pockmarks formed in lines at the bottom of the scour marks. (From Hovland and Judd, 1988.)

crustacean burrows were uncommon suggested to them that there was only a thin sediment cover.

Although Andreassen *et al.* (1990) found geophysical evidence for gas hydrates in the southern Barents Sea, there is no such evidence from the central region where the craters are located. Even so, the steep-sided craters are features one would expect to result from a violent natural blowout event, through competent and over-consolidated sediments (see Section 7.5.6).

3.2.2 Håkon Mosby Mud Volcano

Since its discovery in 1989–90 the Håkon Mosby Mud Volcano (HMMV), in the north eastern Atlantic, has attracted considerable attention from American,

European, and Russian researchers. It is one of the most intensively studied marine mud volcanoes; a special issue of *Geo–Marine Letters* has been devoted to some of this work (see Vogt *et al.*, 1999a for a summary – although more work has been done subsequently). Håkon Mosby Mud Volcano is located at 1270 m water depth, on the Bear Island Fan Slide complex (Laberg and Vorren, 1993; see Map 2). It is about 1 km in diameter, but has a height (above normal seabed) of only 5–10 m (Figure 3.5), and was discovered by chance as a slight feature on side-scan sonar; had the side-scan not crossed directly over it, it may never have been found (Vogt *et al.*, 1999a).

On a subsequent HMMV survey the team measured very high temperature gradients, cored methane

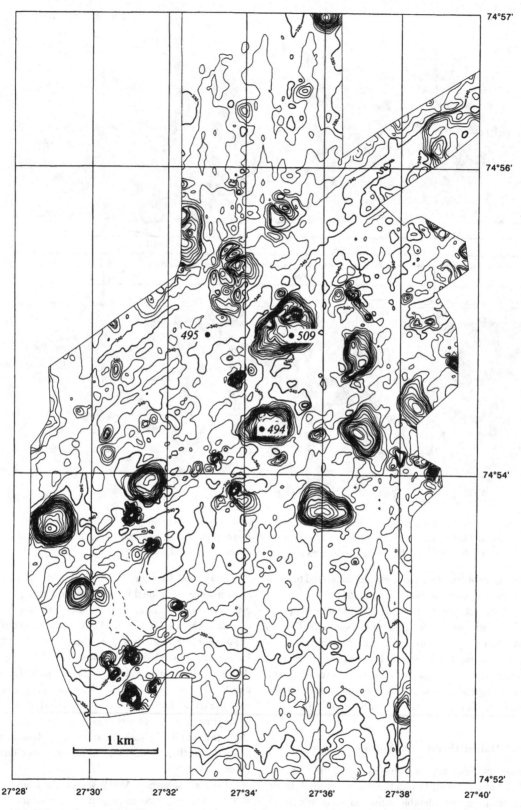

Figure 3.3 Bathymetric map of giant blowout craters in the Barents Sea. The large numbers indicate sediment sampling locations. (Reproduced from Lammers *et al.*, 1995a with kind permission from Springer Science and Business Media.)

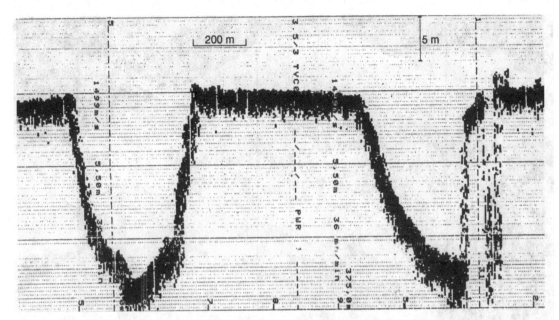

Figure 3.4 Parasound profiler section across two of the Barents Sea giant blowout craters; numbered 494 (left) and 509 (right) on Figure 3.3. Crater 494 is 26 m deep and about 600 m across with sidewall slopes of < 50°. (Reproduced from Lammers *et al.*, 1995a with kind permission from Springer Science and Business Media.)

hydrate at 2 m sub-bottom, and recovered chemosynthetic tubeworms (the fauna of HMMV is discussed in Section 8.1.2). Detailed video mapping (Jerosch *et al.*, 2004) showed the extent of the various seabed 'habitats': the outer rim (occupied by tubeworms), sediments covered by bacterial mats, and sediments covered by benthic organisms.

In 1996 the temperature in the centre of the mud volcano was found to be 15.8 °C, the temperature gradient an astounding 10 °C m^{-1}, and the heat flow was 7500 mK m^{-1}; this is one of the highest heat flows measured in the ocean, away from plate boundaries. Egorov *et al.* (1999) modelled the stability of the HMMV gas hydrates, providing valuable clues about the migration of fluids to the seabed, gas hydrate production, and a distinct methane water-column plume. Because the concentration of dissolved methane is about 10^3 ml l^{-1}, and the normal content of methane in seawater is about seven orders of magnitude less, a high diffusional dispersion of methane into seawater normally hinders the formation of methane hydrates near the seabed (Egorov and Trotsuik, 1990). Hydrates occur as nodules and thin layers at a distance of 200–830 m from the centre of the volcano. A methane concentration of 100 μl l^{-1} was found in the water column 60 m above the mud volcano, whereas the residual gas content in sediments recovered from the central zone was as high as 100–200 ml l^{-1} of wet sediment.

The HMMV was formed by the rise of deep-rooted methane-saturated pore waters along a fault; it lies near the Senja Fracture Zone. Because this upwelling transports fine-grained sediments from deep within the sediments, the process ultimately formed the mud volcano. Whereas the estimated methane flux rate is in the order of 30 m^3 m^{-2} y^{-1}, the total annual flux of methane into the water column from associated seabed gas hydrates may be as much as 50 m^3 m^{-2} (Egorov *et al.*, 1999).

3.3 SCANDINAVIA

3.3.1 Fjords in northern Norway

A marine geophysical survey of two fjords in northern Norway undertaken in 1982 located pockmarks, deepwater corals, and various associated features in an area of crystalline bedrock (Map 2).

Lyngenfjord: pockmarks, doming and high reflectivity

A series of normal pockmarks was found on the gently sloping seabed in Lyngenfjord, spaced at intervals of

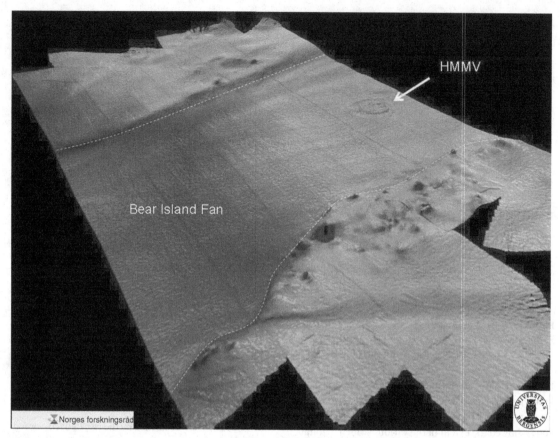

Figure 3.5* An MBES image of Håkon Mosby Mud Volcano (HMMV) showing its location on the Bear Island Fan; sediments derived from the Barents Sea shelf (located to the right of the image) occupying a wide channel (the edges of which are indicated by the dotted lines) cut into continental slope. (Image courtesy of Rolf Birger Pedersen.)

150–400 m. The largest, a composite feature, is about 100 m wide and 5 m deep; this represents an eroded volume of about 4×10^4 m^3. From Figure 3.6 it may be seen that none of these pockmarks penetrate through the whole of the acoustically transparent (muddy?) upper sediment layer (Hovland and Judd, 1988). However, the strong reflection, R1, is breached below most of the pockmarks and is domed below some of them. The deepest reflection, R3, is partly obscured by an overlying patch of acoustic turbidity at the southern end of the profile. This is thought to represent gas charging of the sediment, but recently acquired geochemical core analyses have shown no indication of methane (Björn Linberg, 2002, personal communication).

At one location a seabed dome, which stands only about 0.6 m above the otherwise flat seabed, is situated immediately over a zone of strong reflectivity or acoustic turbidity within the upper transparent sediment layer.

A few hundred metres further north, numerous small (5–10 m wide) pockmarks occur, together with an area of flat seabed that appears from side–scan sonar records to be highly reflective (Figure 3.7). This 'dark' patch may be caused by a highly disturbed seabed or by the presence of coarse sediment (gravel, stones, or shells). Alternatively, it may be caused by the presence of methane-derived authigenic carbonate (MDAC).

3.3.2 The Norwegian Sea

Since 1990, many new petroleum fields (Åsgard, Draugen, Heidrun, Kristin, Midgard, Mikkel, Njord, and Norne) have been found off mid Norway (Map 3). To the east of the Draugen field, Tertiary and Mesozoic sedimentary rocks subcrop on the seabed. The sequence dips seaward, with the Jurassic subcropping closest to shore (Bugge et al., 1984). Because the Cretaceous rocks

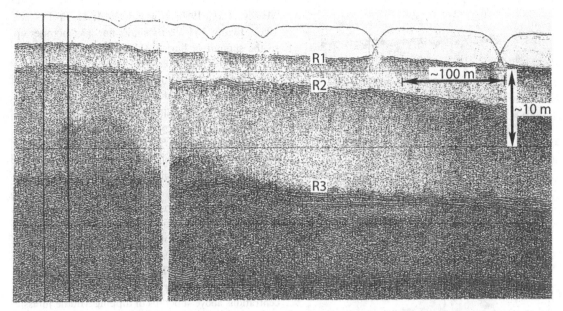

Figure 3.6 Shallow seismic reflection (boomer) profile showing a series of pockmarks in the southern part of Lyngenfjord. Pockmarks cut down through the topmost acoustically transparent sediments. Reflector R1 is breached under most of them. Note the acoustic turbidity at about 15 m below seabed on the left of the section. (From Hovland and Judd, 1988.)

are less competent (weaker) than the adjacent Palaeocene rocks, the Cretaceous subcrops correspond with large-scale seabed depressions or basins that have been infilled with ponded, stratified glaciomarine and marine sediments. In contrast the Palaeocene subcrops represent topographic highs, capped with stiffer Quaternary clay.

Pockmarks have been found at many locations off mid Norway. In one of the basins of subcropping Cretaceous strata, there are up to 10 m deep pockmarks of about 150 m diameter. There are also kilometre long strings of small pockmarks reaching out from some of the large pockmarks, and also strings of such pockmarks not associated with other 'normal' pockmarks. Also, there are pockmarks over the Åsgard and Kriston fields, where the seabed topography is very complex due to relict iceberg ploughmarks (Vasshus, 1998; Hovland et al., 1998a).

Living deep-water reefs constructed by the stony coral *Lophelia pertusa* (see Figures 3.8 and 3.9) occur where Palaeocene, Oligocene, and Jurassic rocks subcrop under a thin (< 10 m) cover of Quaternary clay (Hovland et al., 1998a; Hovland and Mortensen, 1999). These features are up to 30 m high and several hundred metres in width. Hovland et al. (1998a) suggested that they are supported by pore water laden with light hydrocarbons, migrating from the Mesozoic sediments and seeping through the seabed (see Section 8.1.6).

Large parts of the shelf off mid Norway are characterised by a seabed with a rugged topography, and composed of plastic clays. There is abundant seismic evidence of gas within the sediments, and some evidence of gas bubble plumes in the water column (Figure 3.10). Although relict iceberg ploughmarks are common, some of the seabed ridges are thought to be formed by the plastic deformation of the clay due to their high gas content (Hovland 1990a), as discussed in Section 7.4.4.

Pockmarks and other features indicating vertical fluid flow, such as mounds (possibly mud volcanoes and diapirs), and subsurface columnar disturbances (acoustic voids) have also been found further west, on the continental slope, at water depths between 1020 and 960 m on the Vøring Plateau (Vogt et al., 1999b; Bouriak et al., 2000), and > 1000 m over the Vema Dome (Hovland et al., 1998b). These features are especially common in areas underlain by a bottom-simulating reflector (BSR) that suggests the presence of gas hydrates. These areas lie immediately to the north of the Storegga Slide, one of the largest seabed slides known; the volume of the displaced sediment is estimated to be

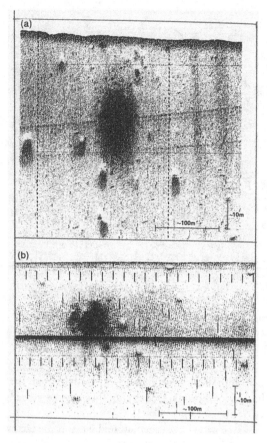

Figure 3.7(a) Small pockmarks and an area of high seabed reflectivity (the dark area) in Lyngenfjord. The high reflectivity might be caused by exposed coarse sediments, shell accumulations, or a cemented seabed sediment (e.g. MDAC). Above some petroleum fields drill cuttings might also produce such reflectivity. (b) For comparison, similar features above a petroleum field in the Barents Sea. Both (a) and (b) are side-scan sonar images. (From Hovland and Judd, 1988.)

5500 km^3, and the slide scar covers an area of about 34 000 km^2 (Bugge *et al.*, 1987; 1988). Bouriak *et al.* (2000) reported pockmarks, mud diapirs, and evidence of gas and gas hydrates within the sediments near the Storrega headwall. The possible association between slides and gas hydrates and/or gassy sediments is discussed in Section 11.2.1.

3.3.3 The Skagerrak

Several people have reported pockmarks and gas-charged sediments at the bottom of the Norwegian Trench in the Skagerrak (van Weering *et al.*, 1973;

Maisey *et al.*, 1980; Hovland, 1991; Hempel *et al.*, 1994). Figure 3.11 clearly shows how gas migrates up to the summit of buried sedimentary rocks before it breaks through the overlying 'impermeable' silts and clays. Hovland (1991) noted the presence of clay diapirs, some of which reach the seabed (Figure 3.12). However, the true distribution (Map 3) and density of seabed fluid flow features was only fully appreciated when the Norwegian Geological Survey (Longva and Thorsnes, 1997) and others conducted a multibeam echo sounder seabed mapping campaign in the mid 1990s (Longva and Thorsnes, 1997).

Pockmarks were found to be especially dense and abundant in certain regions. Rise *et al.* (1999) reported that more than 90% of them occur above Mesozoic rocks, whereas densities are low in areas with a thick Quaternary sequence. This suggests that they are not related to a Quaternary microbial gas source. Concentrations above subcrops of dipping Middle Jurassic sandstones, and a linear concentration above a sub-Quaternary fault suggest that their locations are controlled by fluid migration pathways: updip along the sandstone, and along the fault plane. However some enormous (< 400 m wide, < 2 km long, and < 45 m deep) elongated pockmarks in the mid-southern slope of the Norwegian Trench were attributed by Bøe *et al.* (1998) to the action of strong bottom currents that sweep away particles lifted by long-term continuous fluid flow in an area with a high sedimentation rate.

Dando *et al.* (1994a) reported a detailed study of the ecology of a seep site. In the immediate vicinity of the seep they found a benthic community, which included tubeworms and other species known to be tolerant of the anoxic conditions found at methane seeps. Further discussions of this site and other studies of the tubeworms from the Skagerrak are presented in Sections 8.2.2 and 8.3.2

3.3.4 The Kattegat

Since 1865 it has been known that there are small pockets of shallow gas near the town of Frederikshavn on the eastern coast of the northern Jutland peninsula, Denmark. During the 1930s and 1940s $4 \times 10^7 \, \text{m}^3$ of methane was extracted for domestic use; they even powered some local buses with the gas! A visit to the city of Frederikshavn during rainy weather may reveal gas bubbles

Figure 3.8* A healthy *Lophelia* colony of about 1 m height located inside pockmark KA2, at the Kristin field off mid Norway. The pockmark is about 8 m deep and 120 m in diameter. The coral reef is located at the bottom of this pockmark. A typical characteristic of pockmark ecology is the abundance of *Acesta excavata* bivalves living in some sort of symbiosis with the corals. They are clearly seen in this picture, which was taken by an ROV at a water depth of 326 m.

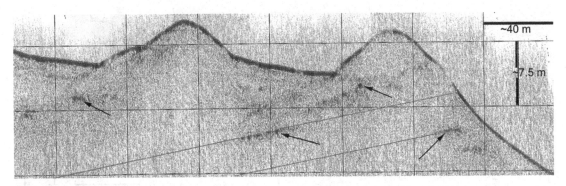

Figure 3.9* A shallow seismic record (Chirp, 4 kHz) run over the Haltenpipe Reef Cluster with an ROV. Two *Lophelia* reefs are crossed. Although the dipping 'coherent' reflections are not evident on this type of record (shown by two continuous interpretation lines dipping towards the left, northwest), the anomalous reflections are (arrowed). Because higher-than-background values of light hydrocarbon concentrations have been documented in the sediments adjacent to these reefs, the anomalous reflections are inferred to represent gas-enhanced reflections.

coming out of cracks in roads and pavements. There are still abandoned wellheads in several back yards. Gas seeps can be seen along some beaches just below the water line, south of Frederikshavn (Dando *et al.*, 1994b and c). There are many more seeps offshore and the gassy area extends northwards into the Skagerrak (see Map 3).

It seems that there is gas of two different origins in this area. Acoustic turbidity indicates gas in post-glacial silty clays. Methane extracted from these sediments is

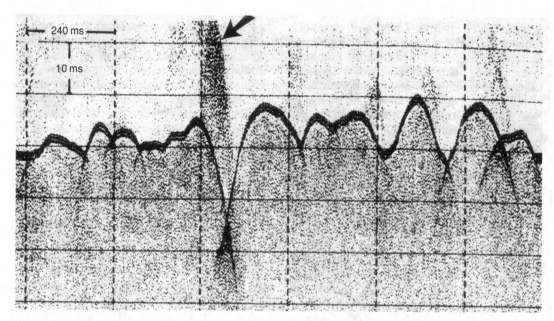

Figure 3.10 Rough seabed topography in the Haltenbanken area, offshore mid Norway seen on a 3.5 kHz seismic profiler record. The ridges are thought to be caused by clay diapirism; note the (gas?) plume (arrowed). (Reproduced from Judd and Hovland, 1992 with permission from Elsevier.)

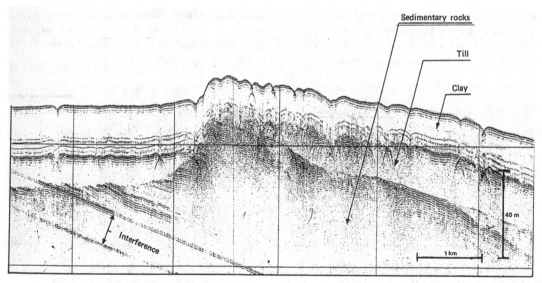

Figure 3.11 A shallow seismic (boomer) profile from the Norwegian Trench in the Skagerrak. The acoustic turbidity and the pockmarks are above the highest points of the inclined sedimentary rocks suggesting that gas is migrating through them, then escaping to the younger sediments and the seabed. (From Hovland and Judd, 1988; after Maisey et al., 1980.)

very young, dating from 2600 years BP (Laier et al., 1992). In contrast, gas seeping from closer to the coast, which is also of microbial origin, is thought to be from an older, Eemian – Early Weichselian source (Laier et al., 1996).

Much of the pioneering work on MDAC was undertaken in this area by Niels Oluf Jørgensen of Copenhagen University (Jørgensen, 1976, 1979, 1989, 1992a and b). Fishermen have long known about tall concrete-like structures protruding out of the seabed off the

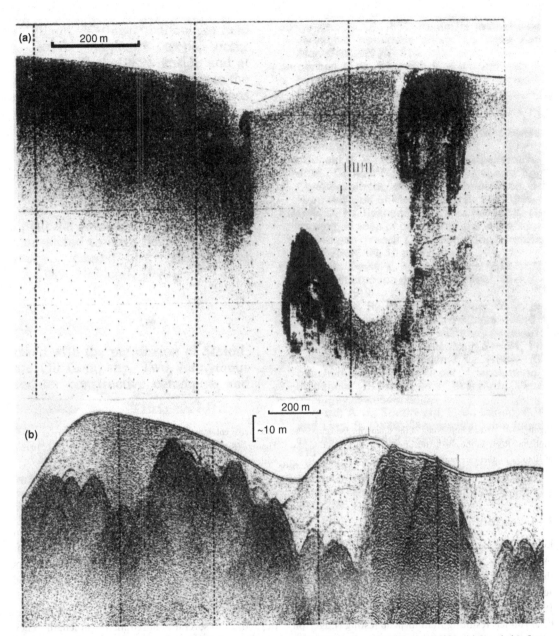

Figure 3.12 Clay diapirs exposed on the seabed in the Skagerrak; (a) side-scan sonar, (b) corresponding shallow seismic profile. (Reproduced from Hovland, 1991 with permission from Elsevier.)

notrh-east coast of Denmark (Figure 3.13). On inspection by local scuba divers it was noted that they were associated with vigorously bubbling water and therefore probably not of anthropogenic origin (Hansen, 1988). Marine biologists and geologists were attracted to the location following a feature film presented on Danish television in the late 1980s. Since then, these seeps, the 'Bubbling Reefs', and the associated ecology have been studied in detail by various groups (Jørgensen et al., 1990; Laier et al., 1992; Jensen et al., 1992; Dando et al., 1994b and c; O'Hara et al., 1995 etc.). They are discussed in more detail in Sections 8.1.1 and 9.2.2).

Figure 3.13* Artist's impression of the submarine landscape, the 'Bubbling Reefs' in the northern Kattegat. Here, in water depths of 10 to 12 m, there are sandstone formations, gas seeps, and spectacular columns of MDAC. Artwork by Chr. Würgler Hansen. (Reproduced with permission from Jensen *et al.*, 1992 with permission from Inter-Research.)

3.4 THE BALTIC SEA

3.4.1 Eckernförde Bay

Eckernförde Bay, on the Baltic Sea coast of Germany (Map 3), was where, in 1952, Schüler first described the 'Becken Effekt' (the basin effect); now recognised as acoustic turbidity. Pockmarks were recorded here before 1966, although they were interpreted at the time as artificial runnels resulting from torpedo testing (Edgerton *et al.*, 1966). During the late 1970s and early 1980s Friedrich Werner and Michael Whiticar studied the pockmarks, and recognised the abundance of methane in the sediments and waters of the bay, and the relationship between acoustic turbidity and gas (Werner, 1978; Whiticar, 1978; Whiticar and Werner, 1981). Whiticar demonstrated that the upper limit of the acoustic turbidity represents the maximum concentration of methane in the sediments. This methane was found to result

from the decomposition of the organic content of the Holocene sediments. Relatively high fluxes of methane were measured at the base of the sulphate reduction zone in parts of the bay, but above this level (about 150 cm below seabed) bacterial consumption results in the decrease and eventual disappearance of methane (Whiticar, 1978).

The pockmarks are elongate, oval, and crescentic, and up to several hundred metres in length. They are found in soft Holocene muds on both sides of the bay, and adjacent to the Mittelgrund, a morainic ridge (Figure 3.14). The pockmarks lie above sub-seabed ridges in glacial tills and outwash sands. The location of pockmarks not far from the Schwedeneck oil field, which lies on the margin of a salt dome, suggests the possibility that petroleum migration was responsible for their formation. However, despite a high general methane concentration in and around

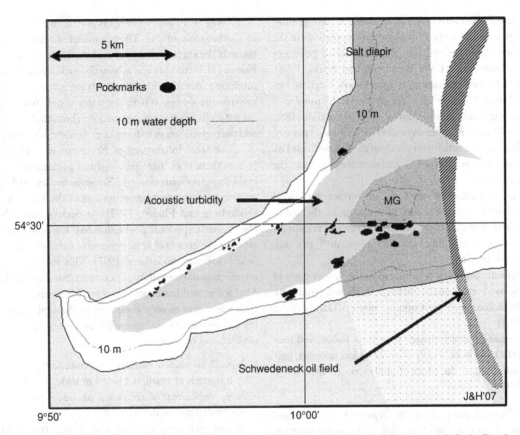

Figure 3.14* Eckernförde Bay: showing the relationship between acoustic turbidity (shallow gas) and pockmarks, and the locations of the Mittelgrund (MG) – a moraine ridge, and the Schwedeneck oil field lying adjacent to a salt diapir. (Based on Hovland and Judd, 1988; Wever *et al.*, 1998; and Whiticar, 2002.)

the pockmarks, Whiticar and Werner (1981) found the following.

- Methane concentrations were substantially lower in the pockmark than in the unpockmarked areas.
- $C_1:C_2$ ratios were considerably higher in the pockmark than in samples from the oil field.
- Carbon isotope ratios ($\delta^{13}C$) of methane from in and around the pockmarks were significantly more negative ($-80‰$) than those from the oil field ($-60‰$).
- The salinity of pore waters dramatically decreased with depth in and around the pockmarks ('*the interstitial fluids are practically freshwater at the depth where the underlying glacial till and sand lithologies are encountered*') whereas it was constant at the other locations.

They concluded that the pockmarks were not formed by gas of either thermogenic or microbial origin, but by groundwater flowing through the glacial out-

wash sands that extend beyond the shoreline on either side of the bay (see Section 7.2.2). This water rises into the sand ridges and breaks through the Holocene muds where they are thinnest.

The gassy sediments of Eckernförde Bay were the focus of a major research campaign led by US and German groups (Richardson and Davis, 1998). Results included the following.

- Intrasedimentary bubbles were imaged with X-ray computer-aided tomography (CT) scans, and bubbles were found to occur as coin-shaped, up to 10 mm in diameter 'flat' bubbles in the sediments (Abegg and Anderson, 1994; Anderson *et al.*, 1998).
- Groundwater from the large underlying aquifer, the *Untere Braunkohlesande*, discharges through the pockmark and dilutes bottom water salinities to values as low as 2.9‰.

- Experiments with a bottom-mounted scanning sonar, and measurements of methane concentrations in the water demonstrated that there are active gas seeps (Jackson *et al.*, 1998; Bussmann and Suess, 1998). The flux of methane at the pockmark location varied from 0.95 μmol m^{-2} per day to 412.62 μmol m^{-2} per day, and at the nearby control location from -0.42 μmol m^{-2} per day to 343.29 μmol m^{-2} per day. Actually, the entire water body of the Eckernförde Bay appears to be affected by methane seeping from the sediments.
- Concentrations of methane in surface waters of < 2800% indicate a flux of methane to the atmosphere; however there are strong seasonal variations. The maximum flux, 200–400 μmol m^{-2} per day, occurs in winter.
- Acoustic turbidity, widespread in the deeper parts of the bay, is attributed to methane bubbles derived from the decomposition of organic matter (Dando *et al.*, 2001).
- Methane bubble release, 'gas-escape holes', and bacterial mats were '*widely observed*' in late summer, particularly near the edges of pockmarks (Dando *et al.*, 2001).

These results reveal '*the important role of coastal oceans in the global methane cycle, as an intense but variable source of methane of largely unknown magnitude*' (Bussmann and Suess, 1998).

3.4.2 Stockholm Archipelago, Sweden

In 1992 Tom Flodén and Per Söderberg of Stockholm University reported that pockmarks and gas-charged sediments had been recorded in the Baltic Sea '*for more than 20 years*', and that pockmarks and gas seepages are '*common features*' in glacial and post-glacial clays '*from sources in crystalline bedrock*' along the Swedish coasts and in the Baltic Sea (Söderberg and Flodén, 1992). The Stockholm Archipelago (Map 3) is a complex group of islands and coastal inlets on the southeast coast of Sweden. Here, glacial and post-glacial clays that range in thickness between zero and 100 m overlie the metasedimentary crystalline basement of the Fennoscandian shield. The basement is cut by deep-seated tectonic lineaments. Gas seeps and large fields of pockmarks have been found along these lineaments (Flodén and Söderberg, 1988, 1994; Söderberg and Flodén, 1991).

Söderberg and Flodén (1992) reported that the gas is of thermogenic origin. They considered that it seeps towards the surface along parts of the Furusundsleden Fairway at Vettershaga, reaching the rockhead and then migrating laterally within the overlying bottom moraine towards the shores. Where the water is only 6 to 16 m deep it seeps '*more or less continuously*' through the glacial and post-glacial clays to form large fields of pockmarks.

The local inhabitants of Stavnsnäs and Tranarö (where there is an '*impressive*' field of pockmarks), several kilometres apart along the Strömma lineament, have reported intermittent gas eruptions at the sea surface (Söderberg and Flodén, 1992). At Södergårdsfjärden Bay, Löparö, the local population has, for generations, avoided the area '*due to its mysterious bubbling and diabolic behaviour*' (Söderberg, 1997). This site lies at the intersection of two tectonic lineaments, has shallow (3 to 5 m) water, and has experienced recurrent earthquakes with epicentres of only about 900 m. Several interesting features were found on side-scan sonar, sub-bottom profiler, and by divers:

- A patch of seabed more or less completely covered by hundreds of small, 0.2 to 0.7 m wide, 0.1 to 0.2 m deep, 'bubblemarks' (recorded on side-scan sonar); see Flodén and Söderberg, 1994.
- Numerous interspersed acoustically light (low reflectivity) circular mud mounds, typically 1.5 m wide and 0.3 m high.
- Several fluid flow conduits (seemingly inactive at the time of observation) were found on the seabed at the mounds and pockmarks. The conduits are 3 to 7 cm in diameter and consist of vertical holes into the sediments (Söderberg and Flodén, 1991).
- Alga (*Vaucheria dichterma*), which normally forms thin (0.5 to 2 cm) layers near the surface, is found here to be more than 3 m thick. The decomposition of this organic matter is thought to be an important local source of microbial gas (Söderberg, 1997).

The 'bubblemarks' are not permanent features. They are created intermittently by '*explosive eruptions*' as pockmarks '*up to several metres across*', that are partially filled with sediment to form smaller (0.3 to 1.5 m) pockmarks.

To find out more about the gas and where it came from, steel pipes were driven up to 10 m into the sediment from the winter ice. After passing through 3.5 m of algae, they encountered thin mud and clay over the glacial clays. A clever arrangement of perforations of the steel pipe, internal pipes, and seals enabled them to take uncontaminated samples of gas. This proved to be 97% methane, but did not provide the information they wanted. To sample the gas coming from beneath the surficial sediment they drilled a 412 m borehole to intersect fractures within the basement rocks.

> To get a passage between the drill hole and the natural local fracture 72 kg of dynamite was lowered into the hole. The 30 m long string of explosives was lowered down to around 300 m depth, the level where drilling fluids earlier were lost in some minor fractures. The string of explosives was fired at both ends to get maximal force at the level of the minor fracture. Just after the explosion all the water in the hole came up. After some minutes a major out blow of gas came. The hole was hermetically sealed by a valve arrangement and the extracted gas was set on fire. Furthermore, immediately after the hole had been sealed, samples of free gas were taken . . .
>
> Söderberg, 1997

Gas collected by this unique experiment included helium, apparently of crustal origin, and methane of probable thermogenic origin. Analyses of the gases in the seabed sediments suggest that both microbial and thermogenic gases are present. Microbial gas may be generated within the post-glacial clays; these have total organic carbon contents in excess of 2%. Söderberg (1997) suggested that the thermogenic methane was generated from organic material in continental sediments that were subducted. This methane mixed with helium (produced by the radioactive decay of thorium and uranium in the crust) and rose through overlying crystalline bedrock. Because the thermogenic gas is more highly concentrated along the lineaments, Söderberg inferred that they act as gas conduits, particularly where the cap rock has been fractured either by earthquake activity or as a result of isostatic uplift.

3.5 AROUND THE BRITISH ISLES

The regional mapping programmes of the national geological surveys have shown that most of the North Sea, and much of the rest of the continental shelf around the British Isles, is underlain by a viable source of gas of one geological age or another (from Devonian to Holocene). Mapping results, combined with information from commercial site investigation surveys, also suggest that shallow gas and seeps are widespread (Map 4); Taylor (1992), for example, described 'a wealth of evidence' of shallow gas in near-shore marine sediments around Great Britain.

There are thermogenic, ancient microbial and modern microbial sources of gas, and evidence that each of these is associated with shallow gas accumulations and seeps. In some places more than one viable gas source is present. Thermogenic sources include both the coal-bearing Carboniferous and the oil-prone late Jurassic sediments of the North Sea that are responsible for the major petroleum provinces (dry gas in the south, oil and wet gas in the north), as well as the less prolific petroleum fields of the Irish Sea.

- Gas from intertidal seeps at Torry Bay, Firth of Forth comes from coal-bearing Carboniferous rocks (Judd et al., 2002b).
- The seeps at Tommeliten produce thermogenic gas (Section 2.3.2), and shallow gas and seeps are known to be associated with other salt diapirs (e.g. at the Machar oil field; Salisbury, 1990; Thrasher et al., 1996). These are linked to a Jurassic (Kimmeridge Clay) source.
- Gas seeps occurring very close to the cliffs at Anvil Point, Dorset (on the southern coast of England) were first reported by subaqua divers (Hinchliffe, 1978), but have since been studied in some more detail by the oil industry. They are also associated with a Jurassic source.

Ancient microbial sources include extensive peats and lignites, and organic-rich clays of Tertiary and Quaternary age.

- The seeps in Block UK15/25 have been attributed to Tertiary peat (see Section 2.3.7).
- Peats and organic-rich clays of Quaternary age are widespread in areas that were above sea level during glacial periods. These include pro-deltaic and pro-grading Pleistocene deposits of the southern North Sea and outer Thames Estuary (Cameron et al., 1984; C. Laban, 1986, personal communication; B. D'Olier, 1986, personal communication; Missaien et al., 2002; Schroot and Schüttenhelm, 2003).
- In the fill of Late Weichselian glacial channels of the southern North Sea (Schroot and Schüttenhelm, 2003).

• Late to post-glacial organic-rich clays and peats of Scottish sea lochs (Firth of Forth) and between the islands of the Inner Hebrides (Farrow, 1978; R. J. Whittington, 1986, personal communication; Evans, 1987).

According to Taylor (1992) and other sources, shallow gas of (probable) modern microbial origin is present in major estuaries (Mersey, Tees, and Thames), bays (Bridgwater, Cardiff, and Tremadog), rias (drowned river valleys: Plym and Tamar estuaries, and Bantry Bay; García-Gil and Croker, 2001, personal communication), and fjords (Loch Eribol; Loch Etive – see Howe, 2002). Offshore, there are some extensive, continuous areas of gassy sediments, for example: in the Norwegian Trench and the Witch Ground Basin (discussed in Chapter 2), the Flemish Bight of the southern North Sea (Cameron et al., 1984), and in the Inner Hebrides (Evans, 1987). In the Irish Sea shallow gas is widespread; in areas of muddy sediments there are pockmarks and seabed domes (Yuan et al., 1992; Jackson et al., 1995), and there is evidence of seabed fluid flow (seabed doming, anomalous sandwaves, seeps, and a small mud volcano) in areas of coarser seabed sediments.

Even where there is a source of gas, seabed fluid flow only occurs where there is also a migration pathway. In the North Sea, for example, the thickness of the Tertiary and Quaternary sediments hinders migration. However, some faulting associated with salt diapirs allows gas to reach the seabed (for example at Tommeliten and Machar), and seeps in places with no salt diapirs show that migration is still possible. In the Irish Sea, carbonate mounds have been identified along the line of the Codling Fault (see Figure 3.15), indicating that it is a fluid migration pathway (Croker and García-Gil, 2002). Also, the association of evidence of seeps with seabed trenches, such as the Lambay Deep (6 km by > 1 km, and > 40 m deep), has led Croker to suggest several such features in the Irish Sea might have been formed by gas escaping from subcropping Carboniferous strata (Peter Croker, 2004, personal communication).

Because of the hazard potential of shallow gas (see Section 11.3.1), the offshore industry has identified circumstances in which accumulations are likely to occur. Fyfe et al. (2003) identified the following as being particularly gas-prone: Tertiary and Lower Pleistocene sands, sediments on the flanks and within

large buried channels, the fill of buried iceberg plough marks.

3.5.1 Pockmarks, domes, and seeps

In the North Sea, pockmarks are present in a few places outside the Witch Ground Basin and the Norwegian Trench (discussed in Chapter 2), mainly in sediments of similar age, nature and origin. Occasional pockmarks are also found in other contexts. For example, Schroot and Schüttenhelm (2003), reported isolated pockmarks (one about 140 m in diameter) in the northern Dutch sector. In the Irish Sea, an area of pockmarks lies within, and extends beyond, the extensive area of shallow gas in the 'western mud belt' (Map 4). Here and elsewhere gas plumes rise from the acoustic turbidity to seabed domes, suggesting vertical gas migration (Figure 3.16), but we are not aware of any evidence to confirm seepage at these locations. Boulton et al. (1981) reported doming ('gas domes') associated with acoustic turbidity in the Inner Hebrides, between the Island of Muck and the Scottish mainland.

Although extensive surveys have been undertaken in and around the British Isles, the number of confirmed seeps is very small. The majority of them have already been mentioned in this section, or in Chapter 2; some others reported in various publications have not been verified by 'ground truthing'. However, we are sure that there are many more seeps. In a review of archived shallow seismic profiles from the UK sector of the North Sea, Judd et al. (1997) identified similar water-column targets they interpreted as being evidence of gas bubbles. The validity of this interpretation is not accepted by all, as is discussed in Section 10.4.1, but, by extrapolation from the number of targets (1707) on the seismic profiles reviewed, they estimated that there are about 173 000 seeps on the UK continental shelf, of which most are in the North Sea (Figure 3.17).

3.5.2 'Freak' sandwaves

In the southern North Sea, off the coast of the Netherlands, there are wide expanses of sand that form large sandwaves on the seabed. They rise to heights of up to 10 m, in water depths of only 40 m, and form as a consequence of strong, shifting tidal currents. When Statoil mapped a 3 km wide corridor for its Zeepipe gas trunk pipeline in the early 1990s, they found some strangely

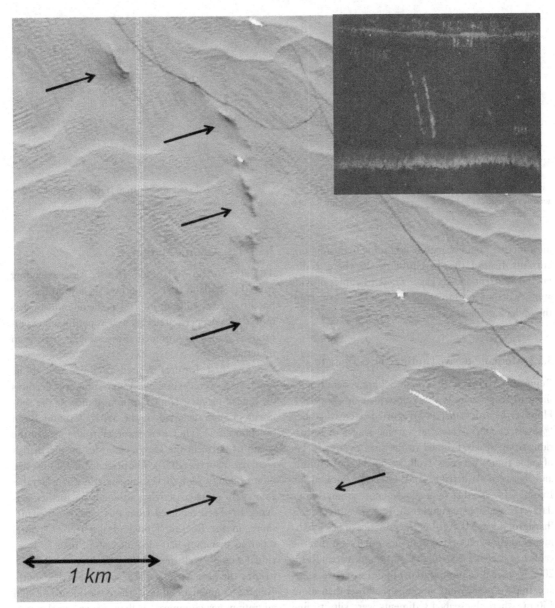

Figure 3.15* An MBES image of carbonate mounds marking the line of the strike-slip Codling Fault in the western Irish Sea; the inset photograph of a fish-finding echo sounder screen shows gas seeping from one of the mounds. (Images courtesy of Peter Croker.)

formed sandwaves, which seemed to be totally out of step with the regional, regular sandwave pattern. Characteristic deep depressions that consistently occurred near these 'freak' sandwaves (Figure 3.18) aroused the suspicion that there was an active agent hindering normal behaviour. Close inspection of the original side-scan sonar, echo sounder, and sub-bottom profiler records provided the independent evidence of seeping gas as the responsible agent (Figure 3.19).

The explanation for the formation and survival of these freak sandwaves is that the seepage activity prevents the migrating sand particles from settling at the

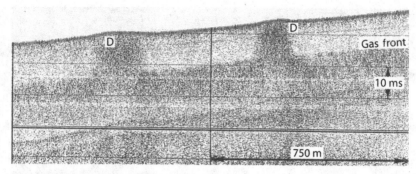

Figure 3.16 Seabed domes (D) associated with plumes of acoustic turbidity rising above a generally level gas front; 3.5 kHz shallow seismic reflection profile (from Hovland and Judd, 1988; courtesy of Robert Whittington).

location that would be natural according to the reigning currents. Instead, they have to settle elsewhere, as dictated by current and seepage. These examples provide another demonstration of the ability of seeps to sculpt the underwater landscape.

3.5.3 Methane-derived authigenic carbonate

Reports of MDAC in the North Sea are generally confined to seeps such as those described in Chapter 2. However, Comrie *et al.* (2002) reported extensive seabed '*gas-cemented hardground*' at varying stages of formation, from weakly lithified to massive, over > 30 km of cable routes between Troll, Veslefrikk and Huldra, and between Veslefrikk, and Oseberg (located on Figure 2.14). Neither mineralogical nor carbon isotope data were available to support the interpretation of this 'hardground' as MDAC, but the presence of black sediments and bacterial mats within the trench soon after the cable had been buried (Robin Comrie and Adrian Read, 2002, personal communication) indicates active seepage. In parts of the route in waters shallower than about 150 m, where the seabed sediments were silty to fine sands, there were no visible indications that carbonate might be present just beneath the surface. However, in the greater water depths of the section between Veslefrikk and Troll, carbonates were found to be concentrated in pockmarks. This carbonate proved problematic for trenching operations, as shown in Section 11.4.2.

Scoffin (1988) reported '*one or two occurrences*' of localised hardgrounds to the west of Scotland; a sample from the Passage of Tiree was confirmed as MDAC ($\delta^{13}C$ −36.5‰).

3.5.4 The Atlantic Margin

The area to the west of the British Isles, extending from the Shetland Islands southwards beyond Ireland, has attracted significant interest as the oil industry has moved into deeper waters. As well as petroleum exploration, detailed environmental assessments have been undertaken.

The first features related to fluid flow to be found here were hyperbolic water-column reflections on pinger profiles, similar in appearance to the sediment plumes recorded in the Witch Ground Basin (Section 2.3.1). They were recorded to the west of the island of Barra in the Outer Hebrides by the British Geological Survey (BGS). The presence of nine discrete plumes along a 7 km survey line suggests a considerable amount of seepage activity, although no further evidence has been reported from this area. Further offshore, in the Rockall Trough, Isaksen *et al.* (2001) reported that they had identified more than 50 individual seep areas during regional exploration, demonstrating the presence of a viable source. Further south, over the Connemara oil field, there is evidence of fluid migration: gas chimneys, shallow gas, and pockmarks, some of which appear to contain carbonate concretions (Games, 2001). However, the most interesting features in this area are mounds.

Mounds of various descriptions have been reported from both the Rockall Trough and the Porcupine Basin (Map 3). Work on them has been undertaken by the oil industry, the TTR (Training Through Research) cruises of the United Nations Educational, Scientific and Cultural Organization (UNESCO)-funded 'Floating University', and several large,

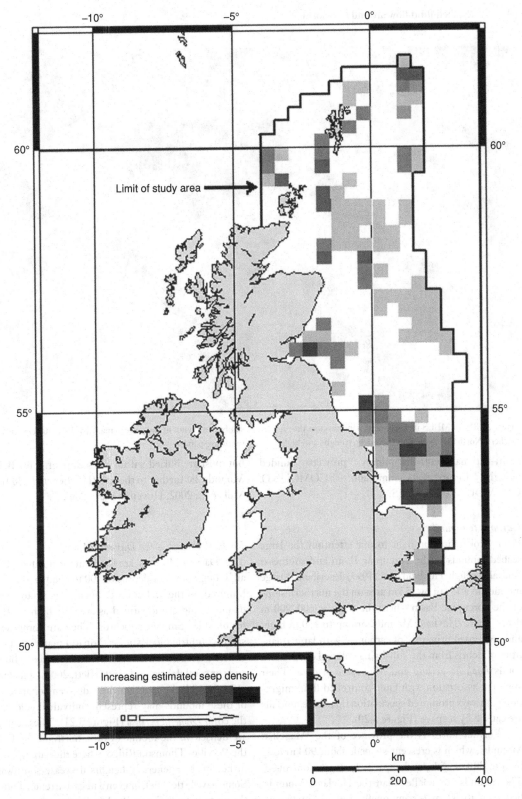

Figure 3.17 Estimated density of seeps in the UK sector of the North Sea; based on an analysis of water-column targets on 3.5 kHz seismic profiles (adapted from Judd *et al.*, 1997).

Figure 3.18* An MBES image of freak sandwaves in the southern North Sea. Note how their deep troughs and high sandwaves do not fit the pattern made by the regular, background sandwave pattern.

multi-national/multi-disciplinary projects funded by the European Commission: ECOMOUND, GEOMOUND, and ACES.

Carbonate mounds

Peter Croker first brought to our attention the large seabed mounds in the Porcupine Basin and southeast Rockall Trough. Hovland *et al.* (1994) described 31 large mounds in a 15 km by 25 km area on the northern slope of the Porcupine Basin where the water is 600–900 m deep. These 'Hovland Mounds' are up to 150 m high, and composed primarily of carbonate with large quantities of debris from the still-living colonial deep-water corals *Lophelia pertusa* and *Madrepora oculata*. Their apparent association with fault-controlled fluid migration pathways prompted speculation that these mounds are caused by seepage (Figure 3.20).

Further north is the province of the 'Magellan Mounds', which is crescent-shaped, about 90 km long by 8 to 20 km wide, and apparently depth controlled. The mounds are smaller than the Hovland Mounds, being 60 to 90 m (exceptionally 130 m) in height,

but most are buried within the sediment. The 'Belgica Mounds' lie further to the east (Henriet *et al.*, 2001; De Mol *et al.*, 2002; Huvenne *et al.*, 2002).

Rockall Trough – the Darwin Mounds

The 'Darwin Mounds' are found in the northern Rockall Trough in water depths of 900 to 1060 m. There are hundreds of them. They are subcircular, up to 75 m in diameter and are identified as areas of high backscatter on side-scan sonar records. They are composed of sand with little bioclastic (carbonate) material, 'blocky rubble', which may be cemented carbonate sediments and/or coral debris, and sands (Bett, 2001). The highly reflective (black) spots on the side-scan sonar records of these mounds may represent individual colonies of the coral *Lophelia pertusa* (Figure 3.21). These colonies are denser on the mounds to the north, at the foot of the Wyville–Thomson Ridge, where they are up to 5 m in height, but generally heights decrease southwards. Some have little relief, or even a negative relief. Those in the north have distinctive 'tails'; elongate to oval patches

(a)

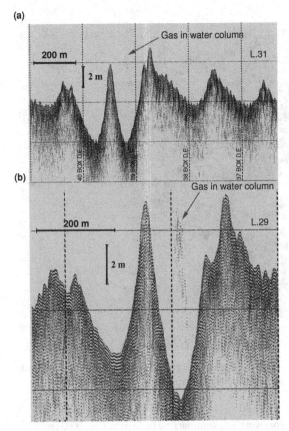

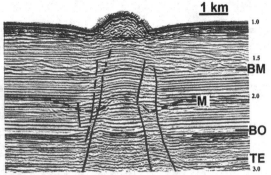

Figure 3.20 Seismic section across a 'Hovland Mound' in the Porcupine Basin, offshore Ireland. The annotations indicate inferred faults/suspected fluid migration pathways, base Miocene (BM), base Oligocene (BO), top Eocene (TE), and the first seabed multiple (M). (Reproduced from Hovland *et al.*, 1994 with permission from Elsevier.)

Figure 3.19 (a) Echo sounder and (b) shallow seismic profiles of freak sandwaves in the southern North Sea showing evidence of gas seepage (acoustic plumes in the water column).

of moderate to high backscatter. These are up to 500 m long, aligned southwest–northeast, and they all lie to the southwest of the mounds, suggesting that they are oriented in the direction of the prevailing bottom current. These tails have no topographic signature, and their sediments appear, from cores and photographs, to be the same as the surrounding seabed (Bett *et al.*, 1999). However, they have an unusual fauna characterised by xenophyophores, the 'giant protozoa'.

South of the mounds, where the seabed sediments are muddier and the water depth is 1000–1200 m, lies a > 3000 km² area of pockmarks. These are typically circular, 50 m in diameter and have a low relief. Unlike the mounds, the pockmarks appear to have the same fauna as the 'normal' seabed (Masson *et al.*, 2003).

It seems that there is a north–south, depth-related trend in the nature of the seabed features associated

with the fining of the sediments. They gradually change from positive (in the north) to negative (in the south) relief. According to Masson *et al.* (2003) this suggests variations in a single geological process. The consensus view is that all these features are actually sand volcanoes formed by the expulsion of fluids (probably pore water) from the sediments. Where the sediments are coarse, subsurface sands are brought to the seabed and accumulate. The escaping fluids are generally thought to be sourced by dewatering of slumping on the southwest side of the Wyville-Thomson Ridge (W-TR). However, Bett (2001) asked why *Lophelia* grows on the mounds. He noted that they may benefit from the elevated position, but also recognised that they need a hard substrate. He suggested that this might indicate the presence of cemented sediments, which may further suggest seabed fluid flow. If there is such a cement, and this results from fluid escape, what is the nature of the escaping fluids? Escaping pore water may be able to explain the topographic features, but perhaps the escape of methane might be indicated if there are carbonates.

We will not be convinced by either interpretation until either seeps are documented, or MDAC is found. Instead we favour a less conventional idea. Consider that the W-TR acts as a massive dam that prevents the very cold water (about −1 °C at 1000 m depth) in the Norwegian Sea Basin from entering the Rockall Trough. On the south side of the W-TR the temperature is about 7 °C. The density contrast between the two sides of

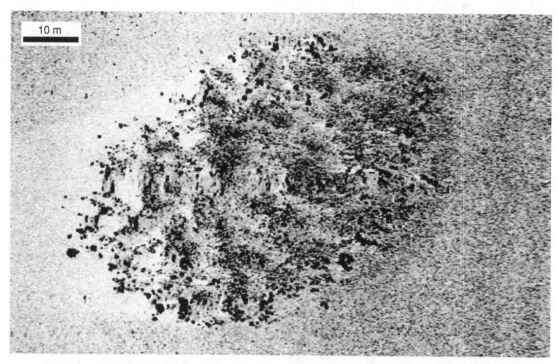

Figure 3.21 High-resolution 410 kHz side-scan sonar image of one of the Darwin Mounds. Dark (high-backscatter) areas (<1 m across) are thought to correspond to individual colonies of corals and other organisms. (Reproduced from Masson *et al.*, 2003 with permission from Elsevier.)

the ridge is therefore large, and should be more than enough to drive water through inhomogeneities (cracks and fissures) in the ridge. We therefore suggest that the cause of both the pockmarks and the mounds on the south side of the ridge is pycnoclinal, that is, a hydraulic head caused by a horizontal density contrast.

3.6 IBERIA

3.6.1 The Rías of Galicia, northwest Spain

'Ria', the internationally accepted geomorphological term for a submerged coastal valley, is a Galician word originally applied to the 'Rías Baixas' of Galicia, northwest Spain (Map 5). Granites and metamorphic rocks of palaeozoic age typify this area, but the rias are partially infilled with Tertiary and Quaternary sediments of fluvial, deltaic, and marine origins. Acosta (1984) reported gassy sediments in Ría de Muros, and Soledad García-Gil and her research students at the University of Vigo have shown that gas is present in all the four main rias of this coast: Vigo, Pontevedra, Arousa, and Muros y Noia;

this work was summarised by García-Gil, 2003 (see also García-Gil, website).

The presence of gas is indicated by extensive acoustic turbidity on shallow seismic profiles. In Ría de Vigo, for example, about 9% of the 136 km^2 covered by seismic surveys is affected, and seeps (water-column targets) are found over about 54% of this survey area. The interpretation of this acoustic turbidity as gassy sediment is supported by the appearance of gas bubbles on X-rays of gravity cores. In view of the age of the bedrock in this region, the gas is most probably generated microbially within the sediments that infill the rias. The sediments of the central and inner parts of Ría de Vigo are dominated by clay and silt which contain up to 10% organic matter. Local fishermen have reported that the waters of the innermost part of the Ría, Ensenada San Simón, sometimes '*boil*', and active intertidal seeps have been sampled. Methane is the dominant gas, and is associated with a peat deposit (García-Gil, 2003; website).

Morphological features associated with the gas include pockmarks, seabed domes, and intrasedimentary collapse structures. Pockmarks are common in the rias. Many of them are small (< 10 m in diameter), but most are about 50 m across, and some are much larger (up to 200 m wide). Depths are generally less than 3 m, but some are a little deeper. In Ría de Arousa there are 50–100 m wide domes, up to 3 m high, on the seabed. They are believed to be associated with the gas-charged sediments, possibly due to locally overpressured pore water. Within the younger sediments there are structures, 70 to 130 m in diameter and up to 8 m deep, thought to have been caused by sediment collapse following dewatering or degassing about 3000 years before present (BP).

The distributions of acoustic turbidity, seep plumes, pockmarks, and collapse structures do not coincide. García-Gil (2003) suggested that mode of gas escape is controlled by the sediment type.

1. Seeps occur where the seabed sediment has a high porosity.
2. Where the seabed sediments are impermeable gas accumulates until a catastrophic gas release is triggered by gas overpressure, and a pockmark is formed.
3. Where the seabed sediment is both semi-porous and semi-cohesive gas release would occur at a lower gas overpressure, and the loss of volume would result in the formation of a collapse structure.

Although the occurrences of acoustic turbidity and pockmarks are very interesting, perhaps the most exciting features seen in the Rías Baixas are the water-column targets. They are prolific, and targets are found over large parts of the rias. Their persistence has been proved by repeated transects of Ría de Vigo at different times of year over a period of several years. These targets are interpreted as plumes of gas bubbles, and suggest that the sediments of the rias are actively degassing. This is significant because, in the shallow (< 10 to 55 m) waters of the rias, it is probable that a significant proportion of this gas escapes to the atmosphere (Vasilis Kitidis, 2004, personal communication). The gas may have an impact on the ecology of the rias; does it also explain why mussel farming is so successful in the Rías Baixas?

3.6.2 Gulf of Cadiz

The tectonic framework of the Gulf of Cadiz reflects the complex boundary between the Iberian, African, and Eurasian plates. This area has experienced several phases of rifting, convergence, and strike-slip faulting as the North Atlantic opened, Tethys closed, and then the basins of the western Mediterranean were formed (Maldonado *et al.*, 1999).

The upper slope – gassy sediments and pockmarks
The location of this gulf between Portugal, Spain, and Morocco, close to the Straits of Gibraltar, means that the Mediterranean Outflow Water (MOW) may dominate sedimentation. However, because the Gibraltar sill is so shallow, sedimentation is dominated by the input from the local rivers. Pliocene and Quaternary low-stand deltas formed under these conditions are thought to have been rich in organic matter. Baraza and Ercilla (1996) reported shallow gas, pockmarks, and gas seeps in these sediments; the gas is thought to be of microbial origin, although Casas *et al.* (2003) suggested that thermogenic methane is also present.

Curiously, pockmarks do not occur in the area of acoustic turbidity, but further down the slope where the seabed sediments are coarser. Where the strong MOW current impinges on the seabed, the sediments are typically silty sands; these are pockmarked. Any sediment lifted by escaping gas will be swept away by 80 to 180 cm s^{-1} currents, leaving only the coarser material behind. In this area, pockmarks are oval or near-circular, and typically 300–500 m across (ranging from 125–850 m) and up to 16 m (generally 3–12 m) deep. Water-column targets recorded on a 3.5 kHz profile suggest active gas seepage in the pockmark area. Baraza *et al.* (1999) thought that shallow gas might be affecting the slope stability as there is evidence of slumping nearby.

The lower slope – mud diapirs, mud volcanoes, and gas hydrates
West of the Strait of Gibraltar, there have been several periods of extension and compression since the Triassic (Gardner, 2001) associated with plate-boundary movements. According to Baraza *et al.* (1999), more than 1000 km^2 of the seabed is affected by mud diapirism, mud volcanoes, and other features associated with seabed fluid flow resulting (according to Somoza, 2001; Somoza *et al.*, 2003) from a combination of tectonic compression and along-slope gravitational sliding. They grouped the features into four 'fields' (Map 5).

- *The Spanish–Moroccan field* (Adamastor, Baraza, Ginsburg, Kidd, Rabat, Student, TTR, Yuma mud

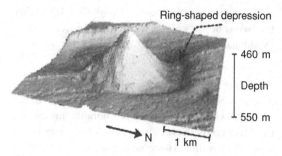

Figure 3.22* Three-dimensional (3D) MBES image showing an example of one of several mud volcano morphologies in the Gulf of Cadiz: the circular mud volcano, Anastasya, which is surrounded by a collapse depression. (Reproduced from Somoza et al., 2003 with permission from Elsevier.)

volcanoes) where '*fluid venting seems to be related to compression of the Gibraltar Arc*'.

- *The Gualdalquivir Diapiric Ridge field* where mud volcanoes (Anastasya, Gades, Ibérico, Pipoca, and Tarsis) are found in front of an accretionary wedge at the boundary between Triassic salt and Early Miocene marls. Here, fluid and sediment mobilisation is thought to be triggered by compression associated with the convergence of the African and Eurasian plates.
- *The TASYO field* (named after a Spanish-Portuguese research project), located at the boundary between the MOW and Atlantic waters where there are large pockmarks, slump scars, and mud volcanoes (Almazán, Aveiro, Cibeles, Faro, Hespérides, St Petersburg, and TASYO).
- *The Deep Portuguese Margin field* where three mud volcanoes (Bonjardim, Olenin, and Ribeiro) are associated with '*contractional structures*' at the toe of a gravitational salt nappe.

Fluid-flow features in these fields include mud volcanoes, pockmarks, carbonate crusts and dolomite chimneys, gas hydrates, and cold-seep communities. There is a variety of mud-volcano morphologies. For example, Anastasya is a simple cone surrounded by a collapse depression (Figure 3.22); in contrast the 'Hespérides mud volcano complex' in the TASYO field which has a diameter of > 5 km and a height of 80 m, is complex, with at least five conical structures and steep-sloped depressions. These depressions '*are interpreted as formed by gas venting and brine seepage (brine pools?)*' (León et al., 2001). The cones are mud-volcano vents

from which dolomite (ankerite) slabs, chimneys, pyrite, and siboglinids have been collected by trawling.

León et al. noted that the depth of the gas hydrate stability zone (GHSZ) varies between areas affected by the MOW (top at 850 m) and the Atlantic waters (top at 680 m). So, any changes in the respective influence of these waters, for example during times of sea-level rise or fall associated with global warming or cooling, would affect the distribution of gas hydrates. During the last glacial maximum, when sea level was below the Gibraltar Sill, there would have been no MOW, so the influence of the Atlantic waters would have prevailed throughout the area. As sea level rose towards its present height the top of the GHSZ would have risen too. However, when the Gibraltar Sill was breached, the influence of the MOW must have depressed the GHSZ, destabilising gas hydrates and, possibly, producing some of the vent structures visible today.

A geochemical study of cores from several of the Gulf of Cadiz mud volcanoes allowed Blinova and Stadnitskaya (2001) to conclude that two mud volcanoes, Ginsburg and Bonjardim, are '*very active*' because '*their deposits are characterised by extremely high concentrations of hydrocarbon gases and the presence of gas hydrate*'. Gas hydrates also occur on Yuma, Ribeiro, and Olenin (Mazurenko et al., 2001). In contrast, the Rabat, Baraza, TASYO, and Ribeiro were found to be less active, although hydrocarbon concentrations were still above background. In fact Blinova and Stadnitskaya considered that the whole of their study area, which covered seven mud volcanoes, showed evidence of a flux of deeply-sourced fluids as even hemi-pelagic sediments contained 'enhanced' concentrations of hydrocarbon gases. Gas (iso-C_4/n-C_4) ratios suggested that the gases were of mixed thermogenic and microbial origin, and Stadnitskaya et al. (2000) deduced that there were two mechanisms of upward fluid flow: dispersed upward migration and focussed fluid flow, of which the latter is more important. They considered that the almost complete absence of the unsaturated gases, ethene, and propene, may imply that microbiological processes are slow, and hydrocarbon flux rates are high.

During the Anastasya 2000 cruise of the TASYO project, some impressive chimneys were dredged from Ibérico, a 120 m high feature in 870 m water depth, described by Diaz-del-Rio et al. (2001) as a 'carbonate mound'. The chimneys were found to be composed of mineral aggregates dominated by ankerite (iron-enriched dolomite), and with minor amounts of quartz, pyrite, rutile, and zircon. As the dolomite was 'remarkably depleted' in ^{13}C (-35 to -56‰), the chimneys were judged to be methane-derived. This conclusion was supported by the identification of abundant microcrystalline framboids comprising aggregates of sulphate-reducing bacteria replaced by haematite. Individual framboids were about 60 μm in diameter, and contained about 60 individual bacteria. Some of these framboids were located within the chambers of the abundant foraminifera (globigerinoids and milioids), possibly suggesting a symbiotic association between the forams and the bacteria (Diaz-del-Rio et al., 2001).

Preliminary work by Cunha et al. (2001) on samples from the top 30 cm of gravity cores from eight mud volcanoes identified siboglinids and thyasirid bivalves (which are known to be seep-associated), and the deepwater corals, Lophelia and Madrepora, amongst other macrofaunal species. Siboglinids were particularly common in the samples from the Bonjardim mud volcano (Deep Portuguese Margin field) and mud volcanoes of the Spanish–Moroccan field.

Work in the Gulf of Cadiz is still producing exciting results (Pinheiro et al., 2003; Somoza et al., 2003). Also, there is evidence that the mud volcano area extends eastwards; Sautkin et al. (2003) reported several mud volcanoes located in a diapiric belt in the west Alboran Sea, to the east of the Strait of Gibraltar.

3.6.3 Ibiza

Two pockmark areas, one either side of the island of Ibiza in the western Mediterranean Sea (Map 5), were reported by Acosta et al. (2001). The first area is in the channel separating Ibiza from the Spanish mainland. In water depths of 600 or 800 m, there are circular pockmarks ranging from 80 to 700 m in diameter, and 2 to 55 m deep, and smaller ones < 10 m across and about 5 m deep. The pockmarks found east of Ibiza are very different. Although they lie at considerable water depths (500 to 700 m), they are on the periphery of the Monte Norte seamount. These pockmarks are more numerous and more varied in morphology; they include circular features, 150 to 500 m across and 10 to 35 m deep, some of which form chains (fault-related?). Some of the chains are merged to form gullies.

Various possible formation mechanisms of these features were investigated. However, Acosta et al. con-cluded that they were probably formed by the expulsion, in some cases catastrophically, of hydrothermal gases and waters. The Balearic promontory, on which these sites (and the Balearic Islands themselves) lie, has seen extensive volcanic activity during the last 30 million years. Seabed slope failures, found in association with some of the pockmarks, may also be related to hydrothermal fluid releases and, like the formation of pockmarks, may have been triggered by active faulting and associated earthquake activity.

3.7 AFRICA

Our knowledge of seabed fluid flow around Africa is restricted to a relatively small number of occurrences off the west coast; no doubt this is, at least in part, a function of a lack of data from much of the coast. The following examples are identified on Map 6.

3.7.1 The Niger Delta and Fan

During the exploration mapping of the middle and outer Niger Delta, at water depths between 500 and 1500 m, many seep-related seabed features have been found (Hovland et al., 1997a; Cauquil et al., 2003, etc.). These include pockmarks and mud volcanoes; gas, oil, and gas hydrates have been identified in seabed cores (Bernard and Brooks, 2000). The pockmarks generally range in width between about 20 m and 100 m, and up to about 5 m in depth. They are typically aligned along flexures and surfacing faults (Heggland et al., 1996; Graue, 2000), and in association with shallow buried palaeochannels (Cauquil et al., 2003).

Statoil found about 100 seabed craters clustered in a 65 km² part of their survey area. These are substantial features; up to 500 m across and 30 m deep, although most are 15–25 m deep. They lie at the top of pipes that extend vertically down to a reservoir about 1000 m below the seabed. These pipes are visible on seismic profiles and can be seen as circular structures on time slices. Løseth et al. (2001) considered that the pipes, which cut through sediment layers and faults, were formed by 'a high pressure, violent, explosive gas expanding on the way from the reservoir to the seabed'; they are discussed in more detail in Section 7.4.6.

The westward slope of the Niger Delta front, down to a depth of over 2500 m is topographically complex. It is largely controlled by compressional ramp movement,

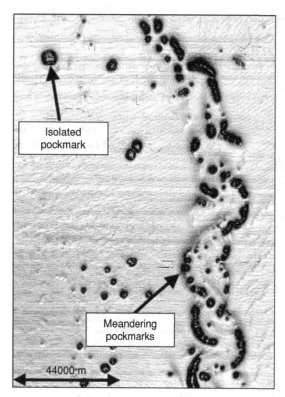

Figure 3.23 Dip map from 3D-seismic data showing meandering pockmarks related to a palaeochannel on the Niger Fan. (Reproduced with permission from Cauquil *et al.*, 2003.)

with fold crests forming the highs. Bottom-simulating reflectors occur in patches that tend to follow the up-dipping strata to straddle, or occur very close to, the crests of anticlinal structures. The patches vary from small circular 'blobs' about 2 km in diameter, to 70 km long and 5 km wide patches. They cover a total area of 568 km^2, 8.5% of the area studied by Hovland *et al.* (1997a). Mud volcanoes are present in several of the patches, mainly at the very summit of the ramps. Gas-hydrate nodules have been sampled in 2 m long seabed cores from mud volcanoes (Brooks *et al.*, 1995; Hovland *et al.*, 1997a; Graue, 2000).

Cauquil *et al.* (2003) described pockmarks associated with palaeochannels (Figure 3.23) and others away from the channels. They conducted a detailed autonomous underwater vehicle (AUV) survey of one of these, lying in about 1265 m of water (Figure 3.24). It was 650 m in diameter and 64 m deep, contained 'carbonate hardgrounds', and fragments of shell material of the

lucinid bivalves. These bivalves are known to be associated with methane seeps (see Section 8.2.2). As there were no indications of vertical fluid flow beneath this pockmark, Cauquil *et al.* suggested that it was formed by the lateral migration of fluids.

3.7.2 The continental slope of West Africa

The West African continental shelf and slope have attracted considerable attention from petroleum exploration companies in recent years. Evidence of shallow gas and related features, including gas hydrates and mud volcanoes, has been found in many areas from the Niger Delta to South Africa. The slope is cut by deep canyons that channel (or have channelled during past climatic and sea-level regimes) sediment from the continental shelf down to the abyssal plain. Although many are still active, particularly those which are adjacent to the major rivers, there is clear evidence that some have become disused, and others have changed course. Some have been fully or partially infilled, and many are marked by a sinuous string of seabed pockmarks. The origin of these pockmarks is discussed in Section 7.6.5.

Both microbial and thermogenic gas have been produced in the sediments of the Congo Fan, and there are gas hydrates and pockmarks (Uenzelmann-Neben *et al.*, 1997; Kasten *et al.*, 2001). Further offshore, in the deep (> 3000 m) waters of the lower Congo Basin, there are gas-charged sediments and '*giant ultra-deep pockmarks*', which are up to 40 m deep and 2 km in diameter (Volkhardt Spiess, 2002, personal communication). Gas bubbles have been seen escaping from the seabed even at this great depth (Sibuet and Olu-Le Roy, 2002), and during the Biozaire cruise, shrimps, bivalves (*Calyptogena* sp.), mussels, and small tubeworms (*Escarpia* sp.) were found in one of the pockmarks. The giant (800 m across and 20 m deep) pockmark studied by Charlou *et al.* (2004) emitted a heterogeneous plume rich in methane, iron and manganese, and particulate matter (as measured by nephelometer). Small pieces of gas hydrate were observed rising in this plume, which extended about 200 m above the 3100 m deep seabed.

Further south, the Orange River Fan (Map 6) has extensive gas hydrates and some mud volcanoes (Ben-Avraham *et al.*, 2002). In relatively shallow water off Namibia, large rafts of gas-charged sediments periodically 'float'. They ground on the seabed in shallow water, and some become stranded along the beaches; a

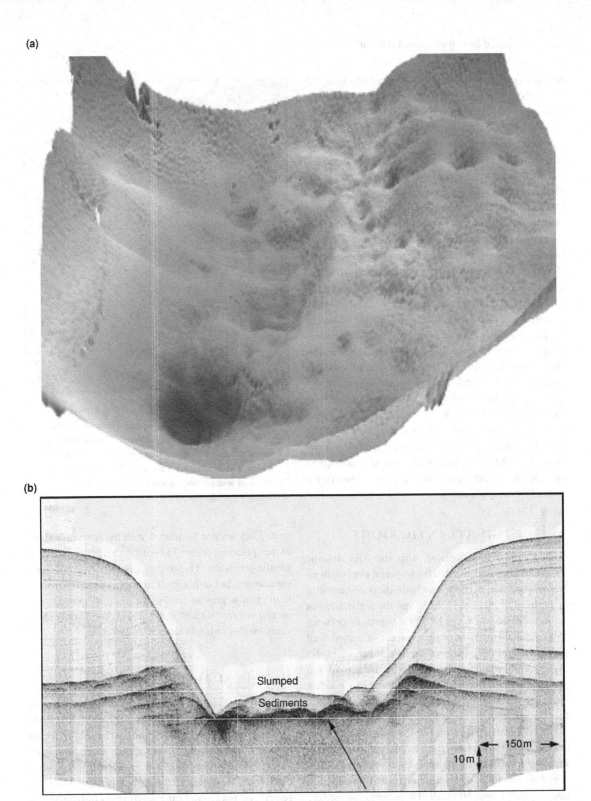

(a)

(b)

Slumped

Sediments

150 m

10 m

Figure 3.24* (a) Three-dimensional bathymetric image from
AUV-mounted MBES data showing the isolated pockmark
indicated on Figure 3.23; (b) AUV-mounted sub-bottom profiler
section of the same pockmark. This feature is 650 m across and
64 m deep. (Reproduced with permission from Cauquil *et al.*,
2003.)

newspaper photograph from 1918 showed blocks of sediment about 8 m high and 20 m long on the beach near Walvis Bay (Bo Barker Jørgensen, 2004, personal communication). Gassy sediments (organic-rich diatomaceous ooze) are seen on high-resolution shallow seismic records in near-shore regions of intense upwelling and high organic productivity; over an area of about 1500 km^2 the gas lies within about 1 m of the seabed (Emeis *et al.*, 2004). Periodically, during the summer months, large volumes of methane and H$_2$S erupt from the seabed killing fish and other fauna in vast numbers. Lobsters detecting the danger walk away; about 3000 t of lobster were found walking up the beach (Brüchert *et al.*, 2004). These events are large enough to have a significant effect on the economy of Namibia whose third largest source of income is fishing. 'Whitings' caused by the precipitation of elemental sulphur in the water column, are extensive enough (< 20 000 km^2) to be visible from space (Figure 3.25). The triggering of sediment disruption and gas release may be the lowered atmospheric pressure as storms pass, or seasonal variations in the microbial oxidation rates (Emeis *et al.*, 2004). However it seems more likely that periodical rainfall along coastal mountains recharges aquifers which extend offshore beneath the Holocene sediments; the subsequent increase in sediment pore fluid pressure triggers the events (Weeks *et al.*, 2004).

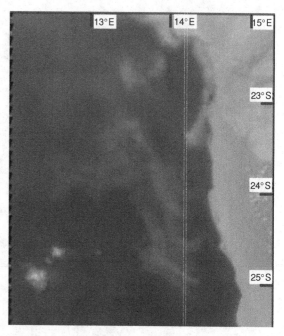

Figure 3.25* Quasi-true colour satellite (SEAWiFS) image showing high concentrations of suspended sulphur granules (lighter colour) in surface waters offshore Namibia. This sulphur is believed to have precipitated following an outburst of gas from the seabed. The image was recorded on 29 March 2001. (Reproduced with permission from Weeks *et al.*, 2004 with permission from Elsevier.)

3.8 THE MID-ATLANTIC RIDGE

Seabed fluid flow associated with the Mid-Atlantic Ridge has been documented from several sites offshore Iceland, but most of the known hydrothermal-vent sites lie between 14 and 38° N. Although the Icelandic sites are in quite shallow water (< 500 m), most lie between 1500 and 4000 m. Most of the evidence of seabed fluid flow at the sites visited during various campaigns involving numerous countries show hydrothermal activity and vent communities typical of ocean spreading centres; these are discussed in subsequent chapters. However, the 'Lost City' vents are different.

The Lost City lies on the 'Atlantis Massif', not on, but about 20 km from the Mid-Atlantic Ridge. It is made up of steep-sided, actively venting chimneys composed of carbonate and magnesium hydroxide minerals, one of which is over 60 m tall. These vents are exciting because they emit warm (40 to 75 °C), rather than hot, fluids with a very high (9.0 to 9.8) pH and calcium con-

tent. They are not associated with the igneous activity of the spreading centre, but with the serpentinisation of mantle peridotite. The vents support a dense microbial community, but unlike hydrothermal vents, the macrofauna that is present (crabs, sea urchins, sponges, and corals) is rare (Kelley *et al.*, 2001). We discuss these vents further in later chapters.

3.9 THE ADRIATIC SEA

According to Conti *et al.* (2002), the first report of a gas seep in the Adriatic Sea was in 1940 when Morgante described continuous gas bubbling at the sea surface off Rovinj, Croatia (Map 7). Since then there have been numerous reports of features associated with seabed fluid flow. Most relate to gas, but perhaps the most remarkable is a groundwater seep in the Gulf of Kastela. Here, water 'boiling' is so intense that it poses a hazard to the navigation of small vessels.

3.9.1 Seeps and carbonates of the northern Adriatic

After the report of the Rovinj seeps, the next report of a gas seep in the northern Adriatic was associated with an oceanographic platform placed about 13 km off Venice in 1970. Gas was reported bubbling continuously a few metres from one of the legs, and other vents in the area were detected acoustically. Then, in 1978, locals in Venice lagoon heard an explosion. It was attributed to a 'methane burst' in the Cavallino area. Two weeks later when the site was visited a 'smoking spot' was found where the water temperature was about 43 °C (Conti et al., 2002). These authors also reported three types of carbonates in the northern Adriatic: beachrock, algal reefs (three types), and calcareous sandstones. They thought that most of the calcareous sandstones were potentially related to methane seeps.

3.9.2 Pockmarks, seeps, and mud diapirs in the central Adriatic

Fluid-flow features have been reported from three areas in the central Adriatic Sea (Map 7).

Bonaccia gas field

In the northern region, which overlies the Bonaccia gas field, shallow seismic and side-scan sonar surveys undertaken by the Marine Biological Institute of Venice located large, actively seeping pockmarks in a water depth of 80–90 m. The flows were emanating from 'erosional depressions up to a few hundred metres wide and a few metres deep' (Stefanon, 1981). The pockmarks are broad and irregular features that seem to be related to large gas-charged structures at a depth of about 30 m below seabed. Stefanon (1981) ascribed them to violent gas eruptions rather than the continuous winnowing of sediments by continuous seepage. Curzi (1998) and Colantoni et al. (1998) reported that small mud volcanoes and carbonate-cemented sediments are also present here. Colantoni et al. reported carbon isotope ratios ($\delta^{13}C$) of between -31 and $-27‰$ for the carbonate cements, yet Curzi et al. considered that the gas was of microbial origin.

Jabuka Trough

Both seabed and buried pockmarks were reported in the Jabuka Trough (Map 7) by Curzi and Veggiani (1985),

and by Mazzotti et al., (1987). The seabed pockmarks vary between 30 and 500 m across and 1 to 6 m in depth. There are up to 10 per km^2; they occur in a water depth of 180–250 m (Mazzotti et al., 1987) where soft silty clays and clays 'provide an excellent medium for recording these features' (Curzi and Veggiani, 1985). Some of the seabed pockmarks overlie deep successions of buried pockmarks connected by vertical columns of disturbed sediment. Columns at least 80 m deep are seen in Figure 3.26, but Mazzotti et al. reported a maximum extension of 96 m below seabed, the deepest being in deep-water areas. Curzi and Veggiani concluded that the pockmarks at the present seabed were formed by gas, probably microbial, whose migration was made relatively easy by a preexisting pathway made available by the differential subsidence in the Jabuka Trough. It was suggested that once the gas had escaped the seabed sediments collapsed into a void caused by the deflation of the sediments in the column. Each column has evidently been continually active throughout the period of deposition of the surficial sediments, and the progressive enhancement of some of the reflections seen on the profiles towards the columns suggests that the gas is still being channelled towards them. Mazzotti et al. noticed 'reflectors curved upwards' (doming) before the lower point of spreading under some of the pockmark stacks. Such doming has also been found beneath pockmarks in the North Sea. We ascribe these features to high pore pressure in the sediments, a suggestion supported by reports of violent blowouts during exploration drilling (Colantoni, et al., 1998).

Reservoirs of oil and gas lying at depths of less than 2000 m have been found during exploration drilling in the central Adriatic (Pieri and Mattavelli, 1986). Near the pockmark areas in the Jabuka Trough lie the Emma and Vest gas fields (Map 7). As there is petroleum generated beneath the pockmark areas, the possibility that the gas responsible for their formation is of thermogenic origin should not be dismissed.

Stefanon (1981) pointed out that this region is seismically active, and that most of the earthquake epicentres lie in the central Adriatic. It is envisaged that once gases have accumulated in the shallow sediment an earthquake would trigger an event during which large volumes of gas, pore water, and sediment would be expelled into the water column. Curzi (1998) reported that fishermen working in this area observed strange phenomena, which may be associated with fault

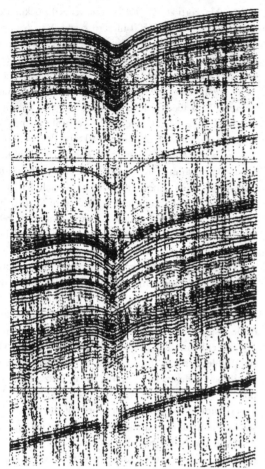

Figure 3.26 Shallow seismic profile (and interpretation) across a typical pockmark in the Jabuka Trough, central Adriatic Sea. It is about 500 m wide and 5 m deep and lies above a stack of buried pockmarks which extends vertically for at least 80 m; Dz marks a columnar acoustic void underlying the deepest (from Hovland and Judd, 1988; courtesy of Mazzotti *et al.*).

movements and earthquake events, over a period of a few months in 1978. These included: 50 m high columns rising from the sea surface, anomalous waves, red lights rising 200 to 300 m into the sky from the sea surface before disappearing, anomalous radar signals, and the sinking of a fishing boat.

Offshore Ortona

Hovland and Curzi (1989) described mud diapirs, about 2 to 3 m high and 50 m in diameter, from an area between the coast off Ortona and the Central Adriatic Depression. There is evidence of gas within the sediments beneath them, and of gas seeping from some of them. However, seismic records indicate that these features are confined to the surficial sediments. Their mode of formation is discussed in Section 7.4.4.

3.10 THE EASTERN MEDITERRANEAN

The structure and topography of the eastern Mediterranean is dominated by the collision between the African and Eurasian Plates, the African Plate being subducted beneath the Eurasian Plate. However, the situation is complicated by the presence of two microplates, the Anatolian Microplate, which is moving westwards, forcing the Aegean Microplate to the southwest, towards the African Plate (McClusky *et al.*, 2000). The consequence of this plate convergence, increased by back-arc extension in the Aegean Microplate, is the formation of the Mediterranean Ridge, the accretionary wedge of the subduction zone, and hydrothermal activity associated with back-arc volcanism in the Aegean Sea (Map 8). This area has played an important role in the development of ideas about seabed fluid flow in the last decade. A key factor has been the pioneering work undertaken by the TTR cruises of the Floating University Programme. Training Through Research has worked in the Mediterranean since 1991 and has identified some of the key sites, and made some very significant discoveries. Further work has been undertaken by various collaborative international and national projects, including several projects funded by the European Commission; MEDINAUT and MEDINETH are examples.

3.10.1 Offshore Greece

The geological location of Greece, in a seismically active area, provides particular interest. Some of the sites described in this section are considered in more detail in later chapters because of the association of seabed fluid flow with earthquake events.

Prinos Bay

Gas is 'quite extensive' in the sediments of the northern Aegean Sea and the Ionian Sea (G. Ferentinos, 1986, personal communication). In Prinos Bay, on the north side of Thasos Island (Map 8), a 4 ha man-made island was planned. During the site survey, pockmarks < 25 m in diameter and < 3 m deep were found in water depths ranging from 6 to 25 m (Newton *et al.*, 1980). The seabed consists of silty sand covered by '*a nearly ubiquitous blanket of seagrass (Posidonia)*'. The area was mapped with echo sounder, shallow seismics and side-scan sonar. Four of the pockmarks were inspected visually by divers who reported that '*either no seagrass was present in the depressions or that seagrass within the depressions was dead*' (Newton *et al.*, 1980). They occur over the southwestern flank of a structural high that lies about 25 m below the seabed. Newton *et al.* (1980) suggested the pockmarks may have formed by submarine groundwater springs flushing out the fine fraction of the seabed sediment. Alternatively, thermogenic gas found during exploration drilling further offshore in the Gulf of Kavala may have migrated laterally and updip to the pockmark site. Whatever the source and nature of the

fluid, the interesting thing about this site is that conditions seem to be detrimental rather than beneficial to the seabed flora.

Aetoliko Lagoon

For a long time the local fishermen and inhabitants have known that Aetoliko Lagoon, from time to time, experiences catastrophic events of anoxia. After periods of strong southerly winds, mass mortalities of fish occur, together with a strong smell of hydrogen sulphide. The Aetoliko Lagoon (Map 8) forms the northern extent of the Messolongi Lagoon, in western Greece, a wetland of great ecological importance. Whereas the Messolongi Lagoon is only 2 m deep, the northwest part of Aetoliko Lagoon has a maximum depth of 30 m; it is anoxic with hydrogen sulphide present in the bottom waters.

Papatheodorou *et al.* (2001) investigated the seabed of the lagoon with side-scan sonar and coring. They distinguished between two areas, the southeast and northwest of the lagoon, on the basis of both sediment type and seabed features. The southeast was characterised by gravelly and sandy sediments and pockmarks, some of which occur in strings consistently aligned in a uniform direction. In contrast, the muddy sediments of the northwest were characterised by highly reflective patches, and acoustic targets were identified in the water column. The escape of microbial gas from the sediments is the most probable cause for the seabed and water-column features, and the high reflectivity may be caused by cemented sediments. Papatheodorou *et al.* (2001) thought that the detection of fluid-flow features may provide a new insight for the identification of the causes of the catastrophic events that take place in the area.

Active pockmarks and earthquakes

An association between pockmarks and earthquakes has occurred at two places in Greece: in the Gulf of Patras, off Patras, and in the Gulf of Corinth, off Aigion in northern Peleponnesos (Map 8). Essentially, evidence suggests that at both places (Patras in 1993, Aigion in 1995) pockmarks became active by emitting gas and/or warm water a few hours before major earthquakes occurred. This coincidence is clearly more than just 'interesting'. Firstly, it sheds some light on pockmark formation processes. Secondly, and more important, it may demonstrate that pockmark activity in seismically active areas such as this could be used as evidence that an earthquake event is imminent. This fascinating topic is discussed in more detail in Section 7.5.5. *In situ* measurements made by Christodoulou et al. (2003) showed elevated methane concentrations in the water in and above these pockmarks, but in Elaiona Bay (Gulf of Corinth) freshwater was found seeping through the seabed.

Hydrothermal seeps

The back-arc volcanism of the Hellenic volcanic arc is responsible for the hydrothermal systems on and around many of the Greek islands: Kos, Lesbos, Methana, Milos, Nisiros, Santorini, etc. (Dando et al., 1999; Map 8). Investigations of offshore systems have concentrated on those in water depths of < 300 m, but most attention has been paid to those of Milos (summarised by Dando et al., 2000).

Large volumes of gas and water are emitted at these hydrothermal vents, generally at temperatures below 150 °C. Gas bubbles have been seen at the sea surface in water as deep as 110 m. Carbon dioxide is the dominant gas, but compositions vary both between sites and over time. The hydrothermal origin of these gases was confirmed by studies of carbon, helium, and hydrogen isotopes described by Botz et al. (1996). Gas and water flow rates of 0.1 to 56.6 $1\,h^{-1}$ (gas) and 0.7 to 124.3 $1\,h^{-1}$ (water) were measured at Milos, and 1.7 to 65.2 $1\,h^{-1}$ (gas) and 4.8 to 96.0 $1\,h^{-1}$ (water) at Kos. Dando et al. (1995a) showed that gas venting occurs at several places (a total area of about 34 km^2) around Milos; Dando et al. (1995b) estimated the total flux as about 2×10^{11} $1\,y^{-1}$. Vent areas are generally located along fault lines, the most prolific spots being where faults intersect, and where the rocks are highly fractured; south of Milos they extend to water depths of at least 110 m.

According to Dando et al., 1995b the emission of gas is affected by seismic activity. Two earthquakes (magnitude M5.0 and M4.4) occurred in March 1992, whilst they were working in Paleohori Bay in southeast Milos, causing the surface water in parts of the bay to 'boil'. Within an hour they conducted an echo sounder transect along the 10 m depth contour. They recorded 99 gas plumes (Figure 3.27). Five hours later a repeat survey showed only 67, and another hour later there were only 60. Three months later the number recorded was again 60 (Dando et al., 1995a).

Seawater near the vents is enriched in lithium, rubidium, strontium, barium, and silicon; white and yellow fluffy mats composed mainly of sulphur (50 to 80%) and silica are found around the vents. The acidity and chemical composition of emitted fluids indicates three sources: a magmatic source, fluids derived from seawater–rock interactions, and fluids associated with bacterial activity close to the seabed. The effects on the local ecology are discussed in Section 8.1.6.

3.10.2 Mediterranean Ridge

The Mediterranean Ridge is a Neogene to Recent accretionary wedge formed as a result of plate convergence and subduction, as described at the start of this section. It is a major topographic feature about 1500 km long and 150 to 300 km wide, and rises to 1200 m from 3000 m below sea level. This is 1 to 2 km above the surrounding seabed: the narrow, deep-water trenches of the Hellenic Trench system to the north, and abyssal plains to the south. Mud volcanoes were first discovered here in the late 1970s. It has become the practice to name individual mud volcanoes after the hometowns of researchers on surveys. This explains the otherwise mystifying appearance of names, such as Gelendzhik, Leipzig, Milford Haven, Moscow, and Stoke on Trent, in the middle of the Mediterranean Sea (Map 8). The initial discoveries were identified on seabed mosaics produced by long-range side-scan sonars, first GLORIA (Kenyon et al., 1982; later revised by Fusi and Kenyon, 1996), and later OKEAN (Limonov et al., 1996; and 1998, etc.), as well as new multibeam data (Mascle et al., 1999). On analysing some of these images, Limonov et al. (1996) realised that mud volcanism does not occur in random, discrete areas along the Ridge, but as a contiguous belt: the 'Mediterranean Ridge mud diapiric belt'. However, Limonov et al. were able to discern patterns within the belt. Assuming that highly reflective patches on the sonographs (apart from trench areas) are mud volcanoes, then:

- the mud volcanoes occur mainly along the Ridge crest;
- they generally form lineations associated with the main thrust planes;
- many individuals are elongated parallel to the thrust trends;
- diameters decrease away from the Ridge crest.

It is thought that mud-volcano activity has been going on for at least one million years. Evidence of present day seabed fluid flow includes gas seeps, gas

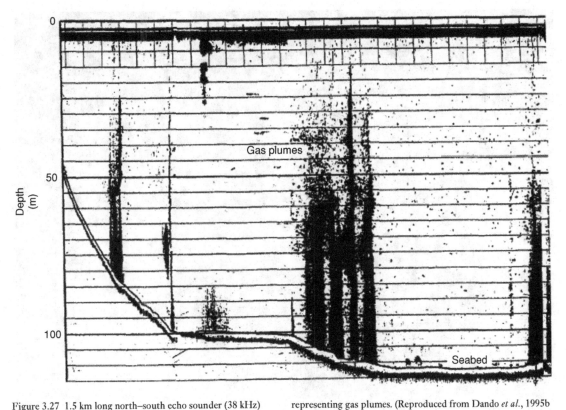

Figure 3.27 1.5 km long north–south echo sounder (38 kHz) profile from the south of Milos, showing water-column targets representing gas plumes. (Reproduced from Dando *et al.*, 1995b with permission from Elsevier.)

hydrates, chemosynthetic communities, and MDAC. It is now clear that there is a wide range of mud extrusion features here: mud diapirs, mud volcanoes, and elongate mud ridges, some of which are impressive topographic features.

- Napoli is believed to be in an active phase. It has a diameter of about 4 km, is circular in shape, and has a flat ('mud pie') top which stands 200 m above the surrounding seabed.
- Toronto is 4.5 km across and 120 m high; its steep-sloped (5–6°) cone has a central 200 m wide crater from which flows of liquid mud have emanated (Ivanov *et al.*, 1996).
- Maidstone, which is more than 7 km across and probably at least 200 m high, may be either a single structure or a complex feature. Certainly there are several individual vents.
- Stvor is described as a clay diapir rather than a mud volcano. It is 240 m high and > 1 km in diameter,

with a very irregular relief and steep (< 10°) smooth flanks.

The morphology is thought to reflect the age and duration of development, the style of activity, and the source of the emitted material. Some (such as Moscow) clearly show multiple phases of activity, with overlapping mudflows, whilst others (such as Jaén) represent single episodes of activity. From seismic sections and core samples (Ocean Drilling Project, ODP boreholes in the cases of Napoli and Milano) it is clear that they are composed of mud breccia containing clasts that have been brought to the surface from considerable depth – possibly as much as 7.5 km (Schulz *et al.*, 1997). Some (such as Napoli) are active, with gas seeps and related features in their craters, gassy mud breccia (which has the appearance of 'mousse'), and gas hydrates, so fluid flow is evidently implicated in their formation. However, gas venting is not restricted to these features; it also occurs along related fault systems, as was discovered during submersible dives here

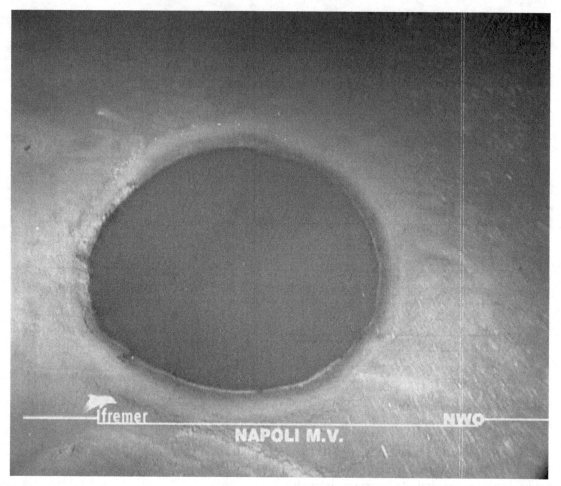

Figure 3.28* Brine lake on the Napoli mud volcano; the white rim around the brine is caused by bacteria which have colonised the brine/seawater interface. (Reproduced from MEDINAUT/ MEDINETH Shipboard Scientifc Parties, 2000; image courtesy of Jean-Paul Foucher, IFREMER.)

(MEDINAUT/MEDINETH Shipboard Scientific Parties, 2000; Woodside *et al.*, 2001). Evidence of prolonged and widespread fluid flow includes carbonate pavements up to several tens of centimetres thick and a chemosynthetic cold-vent-type animal community. At all mud volcanoes visited with the French submarine, Nautile, abundant organisms were observed, including siboglinid tubeworms, bivalves, gastropods, crabs, urchins, sponges, shrimp, and fish (including rays). One report noted widespread deposits of mollusc shells: *'the impression was of a catastrophe that had killed off a far more extensive molluskan community than currently exists there'*. It was suggested that these molluscs had thrived whilst recent mud flows were degassing, but

had died off as the gas supply diminished (MEDINAUT/MEDINETH Shipboard Scientific Parties, 2000).

The Nadir brine lakes are depressions filled with hypersaline fluids that have migrated up along deep faults. These brines are rich in methane (133 ml l^{-1} in one example), and provide a pathway by which dissolved methane can be brought to the seabed. Some of the mud volcanoes also have brine lakes (Figure 3.28). For example, the summit of Napoli appeared from the submersible, *'to form a vast salt marsh where brines, emitted by numerous small vents, have accumulated in shallow depressions on the seafloor'*. The landscape also included *'large crevasses and graben structures several metres wide'* where mud had flowed from the summit of the mud volcanoes (Woodside *et al.*, 2001).

Pockmarks on the Mediterranean Ridge vary in size, the largest being about 250 m across and more than 10 m deep. Some are filled with methane-rich brines, and others are associated with active gas seeps, but some are partially infilled with sediment and appear to be inactive. They mainly occur on mud volcanoes or are associated with faults (Dimitrov and Woodside, 2003). Those found around mud volcanoes are probably the marine equivalents of gryphons and salses of terrestrial mud volcanoes (see Section 7.3.2).

Authigenic carbonate slabs, pavements, and mounds have been seen in active central parts of at least five mud volcanoes, and in brine lakes. Aloisi *et al.* (2000) reported isotopic and mineralogical evidence that these are methane-derived. Analyses of water samples collected above the Napoli and Milano mud volcanoes during the MEDINAUT programme showed '*methane concentrations of 1 to 20 µl l* $^{-1}$, *up to one thousand times the background value*' (Charlou *et al.*, 2000). Together these reports demonstrate that considerable volumes of fluid are flowing to the seabed in this area.

Although many features on the Mediterranean Ridge have been investigated, even more have not. Clearly it is an important area and we have not been able to do justice to it here. Limonov *et al.* (1996) provided a comprehensive review of the investigations here during the 30 years prior to 1996, and there is now an extensive literature about them, most notably special issues of *Marine Geology* (volume 132 issue 1–4, and volume 186 issue 1–2) and *Geo-Marine Letters* (volume 18 issue 2).

3.10.3 The Anaximander Mountains

The Anaximander Mountains, which lie at the junction between the Hellenic and Cyprus arcs (Map 8), are crustal blocks that probably rifted from southwestern Turkey (Woodside *et al.*, 1998). The three main mountain groups rise to depths of 700 to 1200 m from the abyssal plain (> 2000 m). Mud volcanoes and seeps have been identified here (Woodside *et al.*, 1998, 1997; MEDINAUT/MEDINETH Shipboard Scientific Parties, 2000). Also, between two of the peaks there is a large mass of gassy sediment; this is believed to be a large slide (Woodside *et al.*, 1997, 1998; Dimitrov and Woodside, 2003). Large, circular, V-shaped seabed depressions occur within this mass. Although most of

these are about 30 m across and aligned with faults, some are much larger, with diameters of up to 1 km and depths of 300 m. Their origin is uncertain. Dimitrov and Woodside (2003) considered that they may have been formed by seabed collapse following the escape of fluids from the failed gassy sediment.

3.10.4 Eratosthenes Seamount

The elliptical plateau of Eratosthenes Seamount rises 1300 m above the surrounding abyssal plain (Map 8; Figure 3.29). Its flat top, which is inclined somewhat to the north, is about 120 km by 80 km across, and cut by a series of normal faults (Galindo–Zaldivar *et al.*, 2001; Robertson *et al.*, 1998). Dimitrov and Woodside (2003) reported that pockmarks are abundant on this seamount. Sizes range from about 40 m to 200 m in diameter with depths of probably no more than 5 m. There are thought to be several generations present, the older ones being smooth and partially infilled, whilst the younger ones, some of which may be active, have sharper rims and are deeper. Some of the pockmarks seem to be associated with fault systems. The migration of fluids along faults is also thought to have been responsible for the excavation of large cracks.

3.10.5 Nile Delta and Fan

In July 1999 the Institute for Exploration undertook an archaeological survey on the continental slope off Ashkelon, Israel. They located two Phoenician shipwrecks, the oldest ever discovered in the deep sea, loaded with amphorae and other artefacts. Then, during a sidescan sonar survey, they located some additional features that they thought must be more wrecks. However, dives with the ROV Jason showed that they were actually carbonate mounds. One mound was about 4 m across and stood about 1.5 m above the floor of an oval-shaped depression (pockmark?) about 10 m wide and 3 m deep. Three carbonate (aragonitic) structures, each about 0.5 m across at the base, rise from the flat, cylindrical-shaped mound. A cold-seep community of clams, mussels, and polychaetes was found here, and gas bubbles were seen rising from the mound. Profiler data showed that the mound lies on the top of Holocene sediments of the Nile Fan.

Although they inspected only two of the structures with the ROV, the one described above, at a water depth

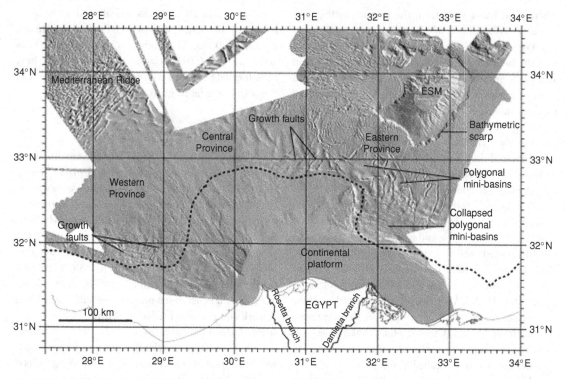

Figure 3.29 An MBES image of the Nile Fan and Eratosthenese Seamount (ESM). The Damietta Province described by Bellaiche *et al.* (2001) lies south of the Eastern Province.

(Reproduced with some minor modifications from Loncke *et al.*, 2004 with permission from Elsevier.)

of 790 m, the other at 340 m; they identified 141 other *'potential gas seeps'* during the side-scan survey (Coleman and Ballard, 2001). This detailed survey is placed into a regional context by subsequent surveys, for example those of the French Prismed II and Fanil projects (see Figure 3.29). Bellaiche *et al.* (2001) identified four different provinces on the Nile Delta Fan:

1. Western province: growth faults in the upper part are responsible for the failure of the young (Pliocene–Quaternary) sediments that have moved downslope. Meandering leveed channels cut the slope. Mud volcanoes, generally 300 to 1000 m in diameter, are numerous in the deeper part of this area where they are associated with growth faults. There are also *'huge sub-circular depressions'*; these are 5 to 10 km across, and look like calderas. The presence of mud volcanoes lead Loncke *et al.* (2001) to suggest that these are areas of *'intense fluid and mud escape'*, probably associated with faults rooted in the underlying Messinian salt.

2. Central province: there are huge debris flows in the upper part of the fan, large growth faults in the eastern area, and salt ridges further offshore. Pockmarks are present in the debris flows.

3. Eastern province: fault-bounded, sediment-filled graben in this area are thought to be associated with salt movement. Mud volcanoes and gas seeps are found in association with the faults, and where faults intersect there are thought to be brine pools. Loncke *et al.* (2001) reported 'gas chimneys' in this area beneath flat, subcircular areas of seabed which are up to 4.5 km in diameter.

4. Damietta Province: where there are long, sinuous channels originating from the Damietta branch of the Nile. No fluid-flow-related features were reported from this area.

Although these are clearly preliminary results, it seems that Loncke *et al.* were justified in concluding that the deep-sea fan of the Nile is *'releasing huge quantities of fluids (maybe chiefly gas)'*. More recent work (e.g. Loncke *et al.*, 2004a and b) has provided exciting images of active mud volcanoes, abundant cold-seep biological communities, brine lakes, and much more.

3.11 THE BLACK SEA

The Black Sea is the largest euxinic basin in the world so, compared to other seas and oceans, the role of methane is different here. Kruglyakova *et al.* (2002) reported that the background concentration of methane is quite low in the surface waters ($< 3.7 \times 10^{-4}$ ml l^{-1}), but is much higher within the anoxic waters at a depth of 200 m ($< 61.4 \times 10^{-4}$ ml l^{-1}), and higher still in the bottom waters ($2300–2730 \times 10^{-4}$ ml l^{-1} at 2000–3000 m water depth). Inevitably this affects the significance and fate of any seabed methane emissions, and it seems that there are many of these. As shown on Map 9, seeps and mud volcanoes are found in many parts of the continental shelf and slope, and in deeper waters of the basins; gas hydrates have been found in a few places, but Vassilev and Dimitrov (2000) thought that they could be present over large areas (see Dimitrov, website).

3.11.1 Turkish Coast

The narrow shelf on the Turkish Black Sea coast shelves rapidly down to the abyssal plain, except where the Yesilırmak River enters the sea with its load of organic-rich sediment. In this area there is evidence of considerable volumes of gas-charged sediments and, in deeper waters, gas hydrates. Çifçi *et al.* (2001, 2003) described pockmarks up to 10 m deep and 150 m wide, and some anomalous 'pockmark'-like elongated depressions 200–300 m wide and 2–3 km long (Figure 3.30). These features were surveyed prior to construction of the Blue Stream pipeline from Georgia to Turkey, Çifçi *et al.* speculated that they are associated with the formation of hydrogen sulphide-enriched methane hydrates associated with a strong acoustic reflector occurring about 40 m below seabed (Gunay Çifçi, 2002, personal communication).

3.11.2 Offshore Bulgaria

Pockmarks were reported off the Bulgarian coast in 1988 (Dimitrov and Dontcheva, 1994). They occur in a long (41 km) narrow (2 to 5 km) zone above the Balkanides, a belt of rocks subjected to Alpine folding. However, shallow gas, pockmarks, and natural gas seeps are now known to be more widespread (Dimitrov, 2002a).

Dimitrov explained that much of the Bulgarian continental shelf and slope are underlain by at least one of several potential methane sources. These include Holocene organic-rich mud swept into the area from the Danube by the south Black Sea current, Quaternary peat, petroleum source rocks of Devonian to Palaeocene age, and Cretaceous coals. To establish the number and distribution of active gas seeps, Dimitrov reviewed archived shallow seismic, side-scan sonar, echo sounder records, and other relevant data, and questioned local fishermen and divers. The results showed that seeps are not uncommon. He estimated that there are more than 6000 in coastal waters (< 20 m water depth), about a third of which are focussed in three prolific areas. The number of seeps in the deeper waters of the Bulgarian shelf was '*estimated to be at least 19 735 individuals*' (Dimitrov, 2002a).

3.11.3 Northwestern Black Sea

The northwestern Black Sea is dominated by the sediments from the present-day River Danube and its predecessors in the modern delta, on the continental shelf and in the deep-sea fan. According to Kutas *et al.* (2002), the first reported gas-escape activity in this area occurred during the Crimea earthquake of September 1927 when flames 500 m high and 30 m wide were noticed about 55 km offshore. They were attributed to explosive methane releases from submarine mud volcanoes. Studies of seeps in this part of the Black Sea started in the 1980s, and have continued ever since. Although much of the literature about them is written in Russian, some publications (e.g. Luth *et al.*, 1999; Peckmann *et al.*, 2001; Kutas *et al.*, 2002; Kruglyakova *et al.*, 2002) are in English.

The Danube pro-delta is obscured by '*a complete acoustic mask*' within a metre of the seabed, and there is a similar complete acoustic mask on the shelf break (Irina Popescu, 2004, personal communication). More than 200 seeps have been identified on the shelf and slope in water depths between 35 and 785 m; these are found along the oxic/anoxic boundary at 130 to 180 m (Luth *et al.*, 1999; Peckmann *et al.*, 2001) and in the vicinity of

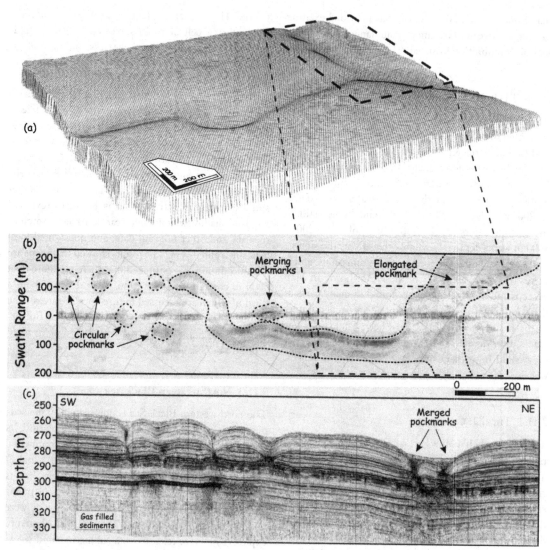

Figure 3.30 Merged pockmarks in the Turkish Black Sea; (a) MBES bathymetry; (b) side-scan sonar; (c) 5 kHz sub-bottom profiler. Pockmarks above gassy sediments seem to be merging to form elongate seabed depressions as a result of repeated periods of gas expulsion. (Reproduced from Çifçi et al., 2003 with permission from Elsevier.)

the palaeo Dnepr Canyon (see Figure 3.31). The relationship between the canyon, its gassy sediments, and seeps is uncertain (Popescu et al., 2004); are the seeps there because of the canyon, or was the canyon formed as a result of slumping induced by the gas? Actually both are associated with a major deep fault, suggesting the hypothesis that both fluid migration and the alignment of the canyon are fault controlled – Irina Popescu, 2004, personal communication. Some seeps are associated with gas hydrates; it seems that free gas migrates

up the slope beneath the gas hydrate stability zone as far as its upslope limit, where it rises to the surface (Irina Popescu, 2004, personal communication). The gas seeps comprise mainly methane (80% according to Ivanov et al., 1991 – cited by Peckmann et al., 2001). They are easily seen on echo sounder records, some occurring as 'flares' (see Figure 3.32). Strong seeps have resulted in 'many visual observations from research vessels' of bubbling at the sea surface in calm conditions (Kruglyakova et al., 2002).

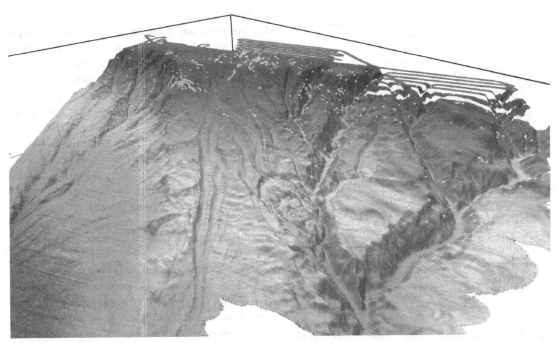

Figure 3.31* An MBES bathymetric image of the palaeo Dnepr Canyon area, west of the Crimea Peninsula, Black Sea recorded during the EC-funded CRIMEA project in 2003 and 2004. Water depths range from 75 to 1200 m. Bubble-releasing seeps detected by ship-based single-beam echo sounders (Y. Artemov, Institute of Biology of the Southern Seas, Sevastopol, Ukraine) are shown as spots. [Image courtesy of Jens Greinert, IFM-GEOMAR (the Leibniz-Insitit fur Meeresuissen schaflen), Kiel, Germany.]

Detailed surveys by ROV and the submarine Jago have shown that the seeps are associated with carbonates and bacterial mats. The nature of the carbonates changes at the oxic/anoxic boundary. In oxic waters, where the boundary is beneath the seabed, they are flat. In the suboxic zone they are thick porous plates, and 'coral-like structures' standing on plates in the lower interface zone. In the anoxic zone they form chimneys, up to 1 m tall, standing on thick platforms. Treude et al. (2002) reported carbonate chimneys standing 4 m above the seabed in an area of vigorous gas seepage in the anoxic zone (see Treude and Boetius, website). Luth et al. (1999) explained that 'the carbonate structures start to rise from the sediment into the near-bottom water, following the gradual elevation of anoxia from the sediment to the water column'. The bacterial mats change in a similar way: from small inclusions inside the carbonates of the oxic zone, to layers several centimetres thick, jelly-like, and pink-brown in colour in the anoxic zone. Pimenov et al. (1997; cited by Luth et al., 1999) reported that they are methanogenic bacteria similar to genus Methanotrix.

3.11.4 Central and northern Black Sea

The presence of mud volcanoes on the Kerch and Taman peninsulas of the Ukraine and Russia, respectively, has been well documented. Further south, mud volcanoes have been reported from two areas: in part of the central Black Sea and in the Sorokin Trough (Map 9). In both cases the sediments and fluids are thought to be from the Maikopian Clay (Oligocene–Miocene).

Central Black Sea

The mud volcanoes of the central Black Sea are randomly distributed over an area of 6500 km^2 in water depths of > 2000 m (Ivanov et al., 1996; Limonov et al., 1997). Three types have been distinguished. The largest, for example Yuzhmorgeologiya and MSU (Moscow State University), are > 2.5 km in diameter with > 100 m tall complex crater-like summits. There is evidence of at least three phases of catastrophic activity, and massive mud flows, but they are considered to be old features, now 'mature' and passive. The second type (e.g. Kornev and Malyshev) are considered less

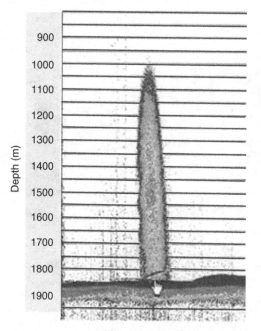

Figure 3.32* Parametric echo sounder image of a 'flare' (intense water-column target caused by vigorous gas seepage) rising 850 m from the seabed in the northwestern Black Sea. (Image acquired in May 2003 during the 58th cruise of the RV 'Professor Vodyanitskiy' during the CRIMEA project. Reproduced with the permission of the Institute of Biology of the Southern Seas, Sevastopol, Ukraine, from Egorov *et al.*, 2003.)

mature and smaller (1–1.5 km across, 60–80 m tall) than the first type. Craters are not well defined, yet these mud volcanoes are considered to be younger and still active. The third type (e.g. Tredmar) is believed to be the youngest and most active. Tredmar is 1.8 km in diameter, rising only 20 m above the seabed, but with a 40 m deep central crater from which gas hydrates have been recovered. A 3–4 m thick liquid mud flow extends 300–400 m from one side of the crater, adjacent to a coarser, blocky flow that stands 30 m above the seabed.

The breccias of all the mud volcanoes studied are characteristically gas charged, and there is evidence of gas seeps and gas hydrates in most of those studied. The gas is dominated (90–99%) by methane, but higher hydrocarbon gases (up to C_7) are also present. This composition and carbon isotope data indicate that at least a component of the gas is of thermogenic origin. This seems to have migrated from depth along deep feeder channels and faults (Ivanov *et al.*, 1996).

Sorokin Trough

Mud volcanoes are abundant in the Sorokin Trough (Map 9), a foredeep of the Crimea Mountains where water depths range from 800 to > 2000 m (Blinova *et al.*, 2003; Bohrmann *et al.*, 2003; Krastel *et al.*, 2003; Bohrmann, website). They are in three shapes: cones (most common), collapse structures, and flat-topped (one example: Dvurechenskii mud volcano, DMV). All but one, Kazakov (the largest) overlie diapirs or diapiric ridges that rise to about 400 m beneath the seabed (Figure 3.33). The diapirs are linked to the mud volcanoes by conduits, many of which are fault-related. Some diapirs have no associated mud volcano. Although there is faulting in the area, some with a seabed expression, not all the diapirs are associated with faults (Krastel *et al.*, 2003).

Gas hydrates have been recovered from several of the mud volcanoes, but DMV seems to be particularly active. This activity is indicated by gas hydrates, MDAC, bacterial mats, and chloride, temperature, and hydrocarbon anomalies (Bohrmann *et al.*, 2003; Bohrmann, website). The hydrocarbons are dominated by methane (about 99.5%), but minor concentrations of ethane, propane, and butane have been measured in sediment samples. Blinova *et al.* concluded that the hydrocarbon gas ratios, carbon and deuterium ratios, and high sediment organic-carbon contents indicated the presence of both microbial and deep-sourced thermogenic gases; Spiess *et al.* (2004) explained that this is because gases are contributed to the feeder channels by young sediments overlying the tops of the diapirs.

3.11.5 The 'underwater swamps' of the east Black Sea abyssal plain

In the central part of the east Black Sea basin, where the flat seabed lies at a depth of about 2200 m, there is an area (shown on Map 9) with 'hills and hummocks' up to 2.5 m high and 5 m in diameter. Here the muddy seabed sediments hold between 63.4 and 169.1 ml kg^{-1} of gas; the dominant gas is methane, but carbon dioxide and nitrogen are also present. Kruglyakova *et al.* (2002) thought that these seabed features were formed by the '*generation and escape to the water column*' of microbial gas; they considered the seabed to be like '*boiling mud*', and the areas to be '*underwater analogies of swamps*'.

Amazingly, even in these great water depths, the presence of gassy sediments seems to be contributing

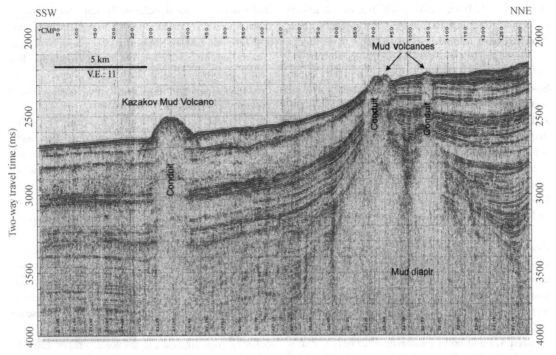

Figure 3.33 Seismic profile (CMP stack) of mud volcanoes in the Sorokin Trough, Black Sea. Kazakov, 2.5 km in diameter and 120 m high, is the largest in this area; those on the right are more typical. (Reproduced from Krastel *et al.*, 2003 with permission from Elsevier.)

methane not only to the water column, but also to the atmosphere. Kruglyakova *et al.* (2002) analysed gases in the sea surface water (6 m depth) in the area of the '*underwater swamps*'. Methane concentration, for example, was found to be 28.4 to 920 × 10^{-4} ml l^{-1}, > 300 times more concentrated than in 'background' areas, and significantly oversaturated relative to the air.

3.11.6 Offshore Georgia

Tkeshelashvili *et al.* (1997) studied a small (16 km^2) area off Georgia where a canyon, cut by the Supsa River when sea level was lower than it is now, cuts across the shelf. Here, in water depths ranging from about 25 to 150 m, they found gas seeps, and inspected a small, shallow water area in detail. The seeps were unevenly distributed, but in places the density of individual seeps was 10 to 25 per m^2. Gas bubbled continuously in bubbles 2–8 mm in diameter, but every 3–5 s larger (> 10 mm) bubbles appeared. The 'high-rate' gas seeps occupied an area of about 0.1 km^2, and here an estimation made from ROV-mounted video recordings suggested flux

rates of 0.02–0.08 $1s^{-1}$. They concluded that the area yields about 172 800 m^3 of gas per day. The composition was dominated by methane (94.7%), ethane (4.7%), and propane (0.6%). This seems to be an exceptionally prolific methane source!

3.12 INLAND SEAS OF EURASIA

3.12.1 The Caspian Sea

The Caspian Sea is actually a freshwater lake without a drainage system except for evaporation. Water input is from the Volga River and other smaller rivers. Natural oil seepage can be described as prolific, and oil slicks are abundant. The Caspian coast of Azerbaijan and Turkmenistan, and the neighbouring offshore region together form the Southern Caspian Basin, considered the most active area of mud volcanoes in the world. An Atlas of 220 large terrestrial mud volcanoes in this area was issued by the Azerbaijan Academy of Science in 1971 (Jakubov *et al.*, 1971). There are more than 90 mud volcanoes in the Southern Caspian Basin

(Jakubov *et al.*, 1983; see Map 10). The largest under-water mud volcano found in the Caspian Sea is about 7 km across according to Guliev (2002).

In the mid 1990s western oil companies were invited by the Azeri government to develop fields in the Caspian Sea. British Petroleum (BP), Statoil, and other compa-nies acquired the three linked fields Guneshli, Chirag, and Azeri in the middle of the Caspian Sea. The seabed mapping revealed three large (about 1 km wide) mud volcanoes on each individual field. Their roots go much deeper than the reservoired oil in the fields. Mud vol-canoes, including those of Azerbaijan and the Southern Caspian Basin, are discussed in more detail in Chapter 7, and images are presented by Guliyev (website).

Extensive seabed and shallow seismic mapping has shown that, as well as mud volcanoes, there are various geohazards, many of them related to fluid flow (gas-charged sediments, seeps, pockmarks, and slope insta-bility), in the Caspian Sea. Ginsburg and Soloviev (1994) reported gas hydrates in pockmark fields on the sum-mit of two large mud volcanoes, Buzdag and Elm, at between 475 and 660 m water depth. Here, hydrates were observed in 19 out of 20 gravity cores, each of 1.2 m length, from Buzdag, and 5 out of 7 cores from the Elm mud volcanoes. At both places the gas hydrates formed semi-transparent inclusions of various shapes and sizes within clay breccia. In fact, gas hydrates are probably quite widespread in the Southern Caspian Basin. Muradov and Javadova (2002) identified 11 locations where they had been found, and esti-mated that they could be present over an area of some 122 000 km^2.

3.12.2 Lake Baikal

In eastern Siberia, far from the nearest sea, lies Lake Baikal. Perhaps there should be no place for this fresh-water lake in a book about seabed features, but some lakebed features in Lake Baikal are far too interesting to leave out. Lake Baikal is arcuate in shape, 620 km long, but only 14 to 80 km wide, and covers a total area of about 31 500 km^2; the third largest lake in Asia and, at > 1600 m, the deepest in the world. It lies in an active continental rift zone, so the presence of hydrother-mal vents is not unexpected. However, the presence of gas hydrates is unusual, both because this is an exten-sional tectonic environment, and because its water is fresh. The review presented here is mainly based on the

preliminary results of Belgian–Russian studies under-taken since 1997. Earlier reports, summarised by Granin and Granina (2002), show that gas seeps have been known here for a very long time. Apparently the first record was in 1775, and the Siberian Department of the Imperial Geographical Society undertook a special expedition to study them in 1868.

The lake is covered by ice during the winter, but in many places the ice cover forms later and melts earlier because of gas bubbles breaking the water sur-face. In a few places bubbling is so strong that ice never forms. Granin and Granina call these places 'ice streamthroughs'. They recounted several tales that described evidence of gas seepage activity. For example: '*sometimes the gas is seeping from the bottom so intensively that the ice streamthrough looks like* [a] *huge boiling kettle*'. During the first part of winter the gases escaping from the lakebed accumulate under the ice, or as flattened, egg-shaped bubbles included in the ice. Some accumu-lations inflate the ice to form hillocks. When these are broken with a crowbar, the gases escape '*very loudly, and with great strength*'; if the ice is broken carefully escaping jets of gas can be lit, demonstrating that the main constituent is methane. These streamthroughs are quite numerous, and their distribution (Map 11) is sim-ilar to that of more recently documented evidence of gas seeps.

Granin and Granina (2002) also reported what may be the earliest descriptions of natural gas hydrates. According to one report a borehole drilled through the ice in the winter of 1931–2 encountered '*permafrost*' at a depth of 30 m below the lakebed that was so strong that it froze the sandy sediments. Further evidence has come in the form of '*kolobovnik*'; ball-like friable lumps of autumnal ice, sometimes up to 80 cm in diameter, that become trapped in smooth ice making the surface irregular and difficult to drive on. Sometimes they have sediment inclusions in them. Granin and Granina sug-gested that these might be lumps of gas hydrate that have floated to the surface.

Bottom-simulating reflectors are widespread in two of the three deep-water basins, and on the Selenga delta (de Batist *et al.*, 2000); the Selenga River drains a consid-erable part of central Mongolia, and deposits sediments rich in organic matter in the lake. In 1997 hydrate was recovered from between 120 and 161 m below lakebed (water depth 1428 m) in a borehole sited in the south-ern basin. It occupied 10% of the pore volume of coarse sandy turbidites, and contained gas with a microbial isotopic signature (δ^{13}C -58 to -68‰); other samples have been collected since (see Matveeva *et al.*, 2000, for example). Vanneste *et al.* (2001) estimated that these gas hydrates represent a reservoir of about 9×10^{12} m^3 of methane at standard temperature and pressure (4.6 Gigatonnes of carbon).

In the southern basin, the BSR generally lies parallel to the seabed, but it undulates near faults, and, at one location, a 'vertical chimney' interrupts it. The faulting, which in some places seems to be quite recent, displaces the BSR; in one place the lakebed is itself displaced by about 20 m (van Rensbergen *et al.*, 2000). Disruption of the BSR is thought to indicate the flow of warm fluids along the fault (Golmshtok, 2000). Close to this fault, the Posolskiy Fault, there are at least five lakebed features, one of which (Bolshoi) is the crater of a conical mud volcano about 800 m by 500 m across and 10 m deep. Gas hydrates, shallow gas, and seeps are associated with these features.

At Frolikha Bay in northern Lake Baikal hydrothermal vents at depths from 32 to 415 m emit low-temperature waters. The vents are marked by sulphur encrustations, mats of the sulphur bacterium *Thioploca*, and macrofaunal communities which include amphipods, oligochaetes, chironomidae, and sponges. Namsaraev *et al.* (2002) found that the benthic macrofauna utilises both carbon derived from planktonic carbon and carbon from microbial methane rising from within the sediments. They concluded that the abundant methane-oxidising bacteria '*play a pivotal role*' in the benthic community.

Preliminary studies of the gas hydrate sites in the southern basin indicate another flourishing microbial community and a diverse fauna that includes chironomidae, oligochaetes, ostracods, nematodes, and a cyclops in greater abundance than is typical for this part of the lake (Namsaraev *et al.*, 2000).

3.13 THE RED SEA

The Red Sea is a proto-ocean. The northern Red Sea is currently in a phase of continental rifting, whilst the central axis of the southern Red Sea is a young ocean spreading centre. High heat-flow and hydrothermal activity typify this setting. Active oil seepages in the northern Red Sea and abnormally high temperatures and pressures in oil exploration wells in the southern

Red Sea (Ahmed, 1972) are considered to be associated with hydrothermal activity.

Perhaps the most spectacular features of the Red Sea are the axial deeps, located in the transition zone between continental rifting and ocean spreading (Map 12). According to Schmidt *et al.* (2003), about 25 of these are brine filled, and some are hydrothermally active. Individual depressions are up to 70 km^2 in area, and several hundred metres deep, extending to depths of 2200 m below sea surface. The brines, some of which are > 150 m thick, are warm (20 to > 65 °C), dense, and very saline (Cl = 0 to > 158‰). According to conventional theories, they probably formed as a consequence of interactions between hydrothermal fluids and evaporites deposited in the rift valley during the Miocene; however, according to Hovland *et al.* (2005a) this exemplifies the hydrothermal formation of large salt deposits (as explained in Section 9.4.3). Active hydrothermal venting is indicated by dramatic variations in temperature; the relatively long time period since the first temperature measurements were made in the 1960s makes this a unique area for monitoring hydrothermal activity. According to Hartmann *et al.* (1998) the temperature in the Atlantis II Deep rose from 55.9 °C in 1965 to 67.2 °C in 1995. The level of the brine in the neighbouring Discovery Deep rose by about 17 m between 1977 and 1995. It seems that each deep has its own hydrothermal regime.

Sediments in the deeps are impregnated with hydrocarbons; mainly but not exclusively methane. Michaelis *et al.* (1990) suggested that sediment 'maturity' increased rapidly, even within the top 3 m of the sediments. However, Schmidt *et al.* used isotope studies to show that thermogenic hydrocarbons are introduced to the brine pools at depth, and that at least some of the methane is of abiogenic origin. Although there is a significant flux of methane across the brine–seawater interface, it seems that the brines effectively store methane at concentrations of > 4.8×10^5 nmol l^{-1} (Schmidt *et al.*, 2003).

The brines attracted attention in the 1960s when it was realised that they contained high concentrations of heavy metals. Sediment cores were brightly coloured because of the concentrations of copper, iron, manganese, zinc, and other metals (Degens and Ross, 1969). Although these represent a significant resource (the Atlantis II Deep contains an estimated 94×10^6 t of 'ore') the conclusion of 'mining' trials conducted from a modified drill ship proved uneconomic (Scott, 2001).

3.14 THE ARABIAN GULF

First reports of seeps in the Arabian Gulf related to artesian groundwater springs. Williams (1946) and Fromant (1965) reported freshwater discharges from narrow crevices in the seabed in the shallow waters of the Bahrain Archipelago, and Chapman (1981) reported that pearl fishermen could stay at sea for extended periods because they collected freshwater from submarine springs; *'they would dive down with a jar, or skin, invert it over the spring, then close it'*. However, the water demands of Bahrain have since depleted the aquifers and stopped the springs.

According to Emery (1956), natural submarine oil seepages have also been known since 1860. Reports of pockmarks and related features, most of which lie over major petroleum fields, came in the late 1970s from contracting surveyors who had worked in the area. The first publication about them was by Jeffrey Ellis and William McGuinness of the Hydrographic Survey Unit, Arabian American Oil Company (Aramco; Ellis and McGuinness, 1986). The Aramco operational area covers 1300 km^2 of the northwestern Gulf, an area with several petroleum fields. The following section has been compiled mainly from the above-mentioned paper, Ellis (2000, unpublished poster presentation), and other detailed information provided by Jeff Ellis; the context is provided by Uchupi *et al.* (1996) who mapped the distribution of active pockmarks, seeps, and water-column targets over a large part of the Arabian Gulf (Map 13).

3.14.1 Setting

Water depths in the Gulf range from 5 to 50 m, but are mainly less than 30 m. Because of the shallow water and relatively good survey conditions, side-scan sonar images are clear and detailed. The surficial sediments, which are mainly soft, include both cohesive and cohesionless sediments of various grain sizes, often with a high carbonate content. In addition to normal sediments, 'submarine hardground' ('faroush') and coral are found on the seabed. Faroush is a crust of sand, shells, and lime muds with a carbonate cement. Although it resembles MDAC, carbon and oxygen isotope analyses of a sample from the Berri oil field demonstrated that it was a tropical marine carbonate (this material is discussed further in Section 9.2.4). Acoustic turbidity is common in the shallow sediments of this area; it has been mapped over an area of about 149 km^2, but may also be present further north and east of the Berri field. It never reaches the seabed; rather there is always 2.5–5 m of 'clear' (transparent) sediment above this gas. Reflection enhancement is displayed on some shallow seismic profiles.

The surficial sediments are of late-Holocene age and overlie an undulating regional unconformity, which is generally at a depth of about 10 m and truncates the underlying beds. These are semi-consolidated to consolidated rocks such as hard clay, limestone, and sandstone.

3.14.2 Seabed features

There is a wide range (in terms of size and form) of pockmarks throughout the region. Most commonly the larger features (< 500 m across) occur above high points in the underlying rocks. There is a range of sizes. In the Berri field, where the water is about 30 m deep and the sediments are soft silts, pockmarks are 80 m across and 1–5 m deep, and the density is about 20 per km^2. In the silty to sandy sediments of shallower areas, pockmarks are only 12 m across and 0.5 m deep, with a deeper section about 4 m wide in the middle, but densities are much higher: about 240 per km^2. Pockmarks tend to have lips or, in some cases (see, for example, the pockmarks from the Zuluf field, Figure 3.34), raised rims, and a dark/reflective surrounding seabed, indicating that coarser sediment has been expelled from the pockmarks.

Although pockmarks are common, not all seabed depressions are pockmarks. Over the Zuluf field there is a 6 m deep depression 335 m long and 235 m wide. Seismic profiles show that, originally, the depression was about 12 m deep, and cut through the surficial sediment apparently exposing a gypsum layer. Subsequently it has been partially infilled. Mapping has shown that the topographic highs on either side of the depression do not extend all the way round it. They are natural features, possibly composed of material from inside the depression. There is no clear explanation for this feature. It may have been formed by the solution of the gypsum, but it may be a true pockmark.

Small (1 m or less) seabed craters are widespread throughout the deeper parts of the Aramco operating area, where the seabed is smooth, flat, and of low to

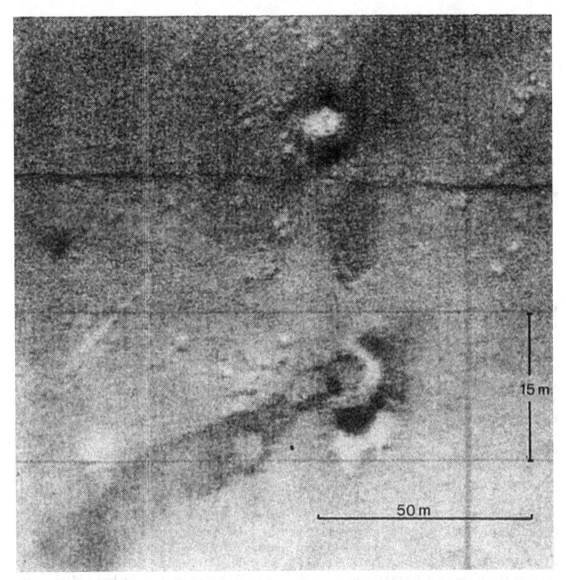

Figure 3.34 Side-scan sonar image of two pockmarks over the Zuluf oil field, Arabian Gulf; the lower one has a clear rim, and a seep plume (extending to the bottom-left of the figure). (From Hovland and Judd, 1988; image courtesy of Jeff Ellis.)

moderate reflectivity. In our 1988 book we reported that Ellis and McGuinness (1986) considered that these 'bubblemarks' were gas-escape features. However, fish and shrimp have since been seen to disturb the seabed sediment whilst feeding, and sediment samples are riddled with burrows. This, and the absence of gas bubbles, suggested to Ellis that these features are caused by bioturbation, not gas escape.

Over the Berri field the hard, grey clays, which underlie the regional unconformity, frequently appear as peaks protruding through the younger sediments.

They are exposed at the seabed within round or elongate pockmarks (Figure 3.35). Some of these features are associated with active gas venting, and it is probable that gas escape has played an important role in their formation, particularly in keeping depressions open. It is thought that gas migrates along the interface above the unconformity until it finds an escape route to the surface.

There is ample evidence in the Arabian Gulf that fluid seeps from pockmarks. About 340 active pockmarks have been identified by Aramco; this is about 5%

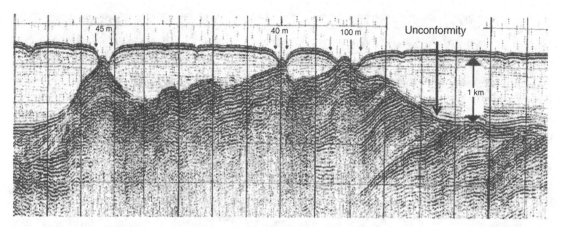

Figure 3.35 Sub-bottom profile of pockmarks over the Berri field, Arabian Gulf, located around peaks of hard, grey clay protruding through the regional unconformity. It seems that gas migrates along the unconformity to escape around these peaks. The horizontal scale is indicated by the down arrows which mark pockmark widths. [Seismic (boomer) profile courtesy of Jeff Ellis.]

of the total number recorded. Active pockmarks with plumes rising from them are seen in Figure 3.36. In the Safaniya field several widely scattered pockmarks are present on a highly mobile (rippled) sandy seabed. If these had not recently been active, they would have been filled in by water-transported sediments. Ellis and McGuinness (1986) reported that at one location '*a pockmark appears to be turned off and on by the tide*': no doubt the variation in hydrostatic pressure is critical to the escape from the seabed sediments of pore fluids. At another location a pre-construction site survey for a structure showed that there were no pockmarks in the area; yet one year later, after installation of a platform, a survey located seven pockmarks of which five were active. During routine ROV pipeline inspections in late 1999 and early 2000, active gas seeps were observed at a place where seeps had previously (1977) been seen by divers; evidence of seeps was also seen on geophysical data recorded in 1984–5. Clearly seepage is ongoing although there are no obvious pockmarks at this location. The seabed sediment is sandy clay.

Summary
The association between pockmark activity and thermogenic hydrocarbons in this area is confidently acknowledged, even though no samples of the seeping fluid have been collected. In one borehole, oil-saturated sands were recovered from a subseabed depth of only 25–28 m, and oil-saturated limestones were recovered from 35 m in the same hole; yet a hole drilled to a depth of

45 m some 3 km away located no oil shows. This suggests, firstly, that hydrocarbons are migrating from the petroleum-producing strata at depth, and secondly, that such near-surface accumulations are localised. Shallow seismic data support this statement. No doubt leakage from these shallow reservoirs is responsible for the seepage and for the formation of pockmarks. An example of this comes from the Berri field where 25 pockmarks occur in a 4 km arc at the contact between silty sediments and a layer of faroush. Ellis and McGuinness (1986) concluded that the hardground acted as a cap which trapped ascending petroleum over a wide area, and that the arc represented a joint in the cap through which the petroleum can leak to the seabed.

Pockmarks and associated features in the northwestern Arabian Gulf are in most cases located over oil and gas fields. The association between petroleum and pockmarks is so strong that it has been said that '*where there are hydrocarbon deposits there will be pockmarks*' (J. P. Ellis, 1986, personal communication) – but please note, the reverse statement is not necessarily valid! It is expected that the gas seen on shallow seismic profiles is mainly hydrocarbon gas (methane). The seep gas or liquid emanating from various seabed features is probably hydrocarbon with a thermogenic isotope signature.

3.14.3 Strait of Hormuz

Preliminary interpretation of side-scan sonar profiles, seismic profiles, and bottom samples (Kenyon *et al.*,

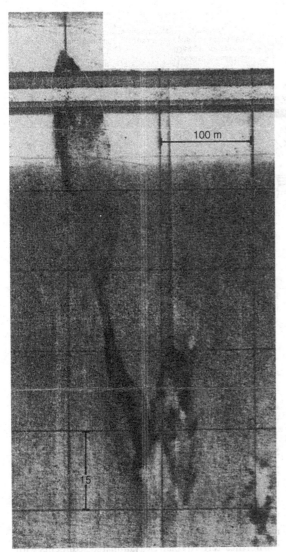

Figure 3.36 Seeping pockmark, Arabian Gulf, on a 100 kHz side-scan sonar record. (From Hovland and Judd, 1988; image courtesy of Jeff Ellis.)

1997) showed an area dominated by strong currents with coarse-grained sediments and mobile bedforms. Another area, where currents are weak, has fine-grained material with numerous pockmarks. They occur in water depths of between 80 and 100 m, are < 10 to > 40 m across and < 7.5 m deep, and are found in densities that range from a few per km² to 330 per km². Individuals are circular to elliptical in shape, the elliptical ones being aligned parallel to the direction from which the bottom current is thought to flow.

3.15 THE INDIAN SUBCONTINENT

3.15.1 The Makran coast

On 15 March 1999 a new island, Malan Island, appeared in shallow (<10 m) water close to the Makran coast. It proved to be short-lived. By November the same year it had already disappeared, although Delisle *et al.* (2002) reported that the site was identified by the last remnants of a mud volcano and '*vigourous gas venting*'. Apparently this was a reincarnation of Hingol Island, which had temporarily existed at the same spot in 1945. Clearly this mud volcano experiences brief periods of activity, in which large volumes of gas and mud are expelled before deflation occurs and the monsoon storms reclaim the spot for the sea. This is just one example of the many seabed-fluid-flow features associated with the accretionary wedge caused by the subduction of the Arabian Plate beneath the Eurasian Plate (see Map 14).

Studies undertaken in this area in the 1990s (von Rad *et al.*, 1996, 2000; Kukowski *et al.*, 2000, 2001; Wiedicke *et al.*, 2001, amongst others) have identified mud diapirs, mud volcanoes, pockmarks, seeps, cold-seep communities, and MDAC. A distinct and widespread BSR indicates that gas hydrates are present where the water is more than 350 m deep. Sain *et al.* (2000) reported that there might be a gas column 200 to 350 m high trapped beneath gas hydrates. Fluid flow from the sediments is concentrated in relatively shallow (300–800 m) waters of the continental slope, and is probably focussed through faults. However, at 'Calyptogena Canyon', seepage occurs where a > 100 m deep canyon cuts into the gas hydrate-bearing zone at a water depth of 2100–2500 m. At shallower seep sites there are bacterial mats (*Beggiatoa* and *Thioploca*), and carbon isotope studies indicate that carbonates ($\delta^{13}C < -40‰$) at vent sites are derived from microbial methane ($\delta^{13}C -77.8‰$) oxidised under anaerobic conditions. Macrofaunal assemblages (*Calyptogena*, *Acharax*, and siboglinids) were only found in Calyptogena Canyon (Wiedicke *et al.*, 2001).

Further east, the Indus Fan extends southwards into the Indian Ocean, but its northwest flank is bounded by the Murray Ridge, which marks the plate boundary. The sediments of the fan are gassy, as was demonstrated when a Deep Sea Drilling Programme (DSDP) borehole was drilled here in the 1970s, and by later surveys (Collier and White, 1990). There are also mud diapirs associated with both the fan and the ridge. These are unusual features in that they originate at depths of less

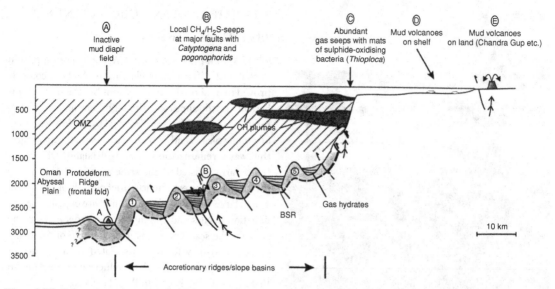

Figure 3.37* Schematic section across the Makran accretionary margin showing locations where mud diapirs/volcanoes, gas seeps, gas hydrates, and BSRs occur. Note the methane-rich plumes within the oxygen minimum zone (OMZ). (Reproduced from von Rad *et al.*, 2000 with kind permission of Springer Science and Business Media.)

than 500 m below the seabed, significantly shallower than those of most areas. It seems that they formed because of the gravitational instability of overpressured mud within the rapidly deposited sediments of the fan (Collier and White, 1990).

Methane is being injected into the waters of the Arabian Sea by the mud volcanoes and seeps described here. Water profiles show high methane concentrations (> 400 nl l^{-1}) at depths of about 750 m, as well as in near-surface waters (Figure 3.37; Delisle and Berner, 2002). However, these profiles do not extend far enough south or west to overlap with those taken during biogeochemical investigations (see Section 10.6).

3.15.2 The western coast of India

The Indian National Institute of Oceanography has surveyed the western continental margin of India (Map 15) from the southern side of the Gulf of Kutch, in the north, to Cape Comorin, in the south; an area of about 0.13×10^6 km^2. The continental shelf is more than 300 km wide in places, and the shelf and slope are characterised by sediments with a high organic carbon content. Gassy sediments have been mapped on several parts of this coast: off Tarapur, Mumbai, Karwar, Coondapur, and Mangalore (Karisiddaiah *et al.*, 1993). They cover

a total area of about 6500 km^2 (Karisiddaiah and Veerayya, 1994). The interpretation of geophysical evidence (acoustic turbidity, etc.) of gas has been verified by sampling. Siddiquie *et al.* (1985) reported that seabed sediment cores from the inner shelf off Mumbai contained 10% by volume of gas, the dominant gas being methane. Some of the gassy areas seem to be associated with petroliferous basins (e.g. off Ratnagiri and Mumbai), but there are two other possible sources. Some may be coming from the decomposition of vegetation derived from the drowning of coastal forests during the post-glacial marine transgression (about 9000 to 10 000 years BP). In the northeastern Arabian Sea laminated sediments with organic-rich bands representing blooms of organic productivity are associated with monsoon seasons. Karisiddaiah and Veerayya (2002) reported organic carbon concentrations of 6–7% in parts of the upper slope.

In places the gassy sediments are associated with pockmarks, mud volcanoes, mud diapirs, and gas seeps. Rao *et al.* (2001) showed that, in the deeper water areas, BSRs indicate the widespread occurrence of gas hydrates.

3.15.3 The eastern coast of the subcontinent

Although the eastern coast of India has a narrow continental shelf, there is evidence of shallow gas and gas seeps on the shelf and slope close to the mouths of some of the major rivers (Map 15). Murthy and Rao (1990) identified 'acoustic wipeouts' associated with the Krishna, Godavari, and Mahanadi rivers.

3.15.4 Indian Ocean vent fauna

So far, two locations of hydrothermal vent fauna have been discovered in the Indian Ocean: Kairei and Edmond vent fields. These are important discoveries as they are isolated from the known hydrothermal vents in the Pacific and Atlantic Oceans, as we discuss in Section 8.1.6. The primary producers at the two vent locations include hydrogen-oxidising, sulphate-reducing, and methanogenic bacteria (*Methanococcus* sp.). Two hydrogen-oxidising isolates from Edmond sulphides were grown at 90 °C and represent the first *Aquifex* spp. ever recovered from deep-sea hydrothermal vents (Van Dover *et al.*, 2001). The invertebrate fauna at the two locations is dominated by dense swarms of shrimp and peripheral beds of anemones.

3.16 SOUTH CHINA SEA

The earliest reports of pockmarks and seeps in the South China Sea came from Platt (1977). He indicated that pockmarks occur widely along the east coast of Malaysia, the east coast of Thailand, and the east and northwest coasts of Borneo, commonly associated with seeps. Since then there have been numerous reports of seabed fluid flow in this area, but the significance of the 1977 report was that this was the first report of pockmarks in an unglaciated part of the world. Map 16 shows the locations mentioned in the following text.

3.16.1 Offshore Brunei

The first publishable evidence of seabed fluid flow offshore Brunei was made available to us by the Brunei Shell Petroleum Company Sendirian Berhad many years ago. Their data showed pockmarks of various sizes in several parts of the shelf, some associated with seeps (Hovland and Judd, 1988). However, it is known that seeps occur over a wider area, including onshore. According to van Rensbergen and Morley (2001) some of the seeps occur in conjunction with mud volcanoes.

3.16.2 Offshore Vietnam

Gas and oil seeps, and also pockmarks, have been reported off the coast of Vietnam. The large Nam Con Son sedimentary basin is located south of the Mekong Delta, in southern Vietnam. According to Traynor and Sladen (1997) there is a variety of seep indicators here: gas anomalies, seabed mounds, pockmarks, gas in the water column, and sea-surface oil slicks.

Further north, exploration of the 10 000 m thick Yinggehai Basin, between Vietnam and Hainan, has identified Tertiary gas reservoirs that contain microbial and thermogenic hydrocarbons, nitrogen, and carbon dioxide; in some of the reservoirs carbon dioxide is the dominant gas. Huang *et al.* (2002) described a complex history, with gas coming from separate hydrocarbon source rocks, and from a deeper carbon dioxide source, during four migration episodes. High sedimentation rates and high geothermal gradients contributed to overpressuring, resulting in the formation of diapirs. Faulting and fractures associated with these diapirs have provided migration pathways, allowing gas to migrate to the shallowest reservoir, which is of Pliocene age.

3.16.3 Hong Kong

Areas of gassy sediment were found in the waters of Hong Kong during a survey for sand resources, and subsequently during surveys for Chek Lap Kok airport. Extensive areas, > 50 km^2 in total, of acoustic turbidity, were found in muddy Holocene marine sediments, and this was confirmed to represent gassy sediments during an intensive sampling and *in situ* testing programme (Premchitt *et al.*, 1992). It seems that the acoustic turbidity is confined to basins and channels that were infilled during the late Pleistocene–Holocene. The presence of organic matter in the infilling sediments, and the fact that the surficial sediments overlie igneous bedrock, suggests a microbial origin for the gas. Samples yielded gases dominated by nitrogen, carbon dioxide, and methane, which are consistent with the origin.

3.16.4 Taiwan

Taiwan lies over the subducting Luzon arc, so earthquakes are commonplace. Tectonic compression is responsible for the rugged mountains on the eastern side of the island, and mud volcanoes on the coastal

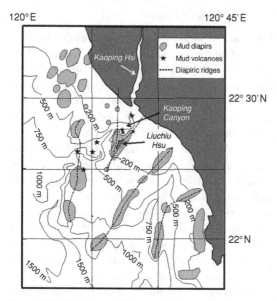

Figure 3.38* Mud diapirs offshore southwest Taiwan. (Adapted from Sun and Liu, 1993, and Chow *et al.*, 2001.)

plain of southwest Taiwan. The mud-volcano belt continues offshore into a foreland basin, the Tainan Basin, formed during the Pliocene–Quaternary. The sediments in the basin are 6 km thick, but in palaeochannels the thickness is up to 10 km. Mud diapirs that have developed in these channels are elongate, the larger ones being ridges about 15 km long. Sun and Liu (1993) explained that these diapirs *'were induced and developed along lines of anticlinal folding axes during the Pleistocene time by unbalanced loading forces under thick channel deposits'*. However, Chow *et al.* (2001) found that within the present day Kaoping submarine canyon, which is incised > 200 m into the slope, the diapirs have broken through to the seabed to become mud volcanoes (see Figure 3.38).

Extensive BSRs indicate the presence of gas hydrates in the sediments, and shallow gas, seeps, and pockmarks are also present in this area (Chow *et al.*, 2001).

3.17 AUSTRALASIA

3.17.1 Sawu Sea

Between the islands of Flores and Timor, eastern Indonesia (Map 17), the Australian continent is colliding with the Eurasian Plate, causing continental-shelf sedi-

ments to be accreted in a complex subduction melange or accretionary wedge; the Banda Arc (Barber *et al.*, 1986). Breen *et al.* (1986) found that mud volcanoes characterised one of three structurally distinct areas of the accretionary wedge. Acoustic void structures reported from this area by van Weering *et al.* (1987, personal communication) are probably the result of the migration of buoyant gases (or liquids), and may indicate that seabed fluid flow has occurred.

3.17.2 Timor Sea

The petroleum potential of the Timor Sea/northwest Australian shelf (Map 17) is indicated by seeps, and other features associated with seabed fluid flow. These have been described in various papers by Geoff O'Brien and his colleagues. Seeps have been detected by side-scan sonar and sniffer surveys, airborne laser fluorosensor (ALF), and satellite-mounted SAR (synthetic aperture radar) in four areas: Yampi Shelf, Heywood Shoals, Sahul Shoals, and Karmt Shoals. They are associated with leakage from petroleum reservoirs in sedimentary basins (the Bonaparte, Browse, and Vulcan basins) of the Timor Trough. This is a foreland basin system formed since the mid Tertiary by the collision between the Australian and Eurasian plates.

Bathymetric maps made in the 1960s showed the presence of mounds or knolls in the Vulcan sub-basin. Some are more than 1 km wide at the base, and stand > 200 m above the seabed (Hovland *et al.*, 1994). Dredging has shown that these features have reef-forming algae (*Halimeda* sp.) on their summits (where the water is < 40 m deep), as well as large (< 2 cm diameter) foraminifera, and bryozoa. Seismic data show that these features are associated with faults that are thought to be migration pathways for hydrocarbon-bearing fluids (Figure 3.39). Work summarised by O'Brien *et al.* (2002) has linked other modern reefs and carbonate banks in the Bonaparte and Browse basins to hydrocarbon seeps; this is discussed in more detail in Section 9.2.12. O'Brien *et al.* (1999) showed that within Eocene sediments there are carbonate cements with carbon isotopes (δ^{13}C about −25‰) consistent with a thermogenic methane source. These 'hydrocarbon-related diagenetic zones' (HRDZs) typically occupy areas 3–5 km long by 0.5–1 km across, and they are attributed to leakage of hydrocarbons through the fault zones from breached Mesozoic petroleum reservoirs. The cementation

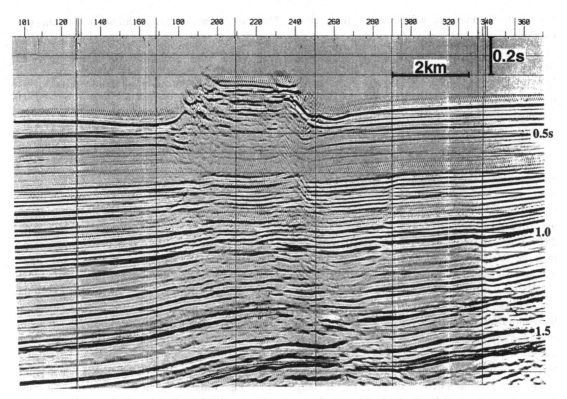

Figure 3.39 Seismic profile of the Pee Shoal, a carbonate knoll in the Timor Sea. Note the faults – suspected fluid migration pathways. (Reproduced from Hovland *et al.*, 1994 with permission from Elsevier.)

is sufficiently intense to increase the acoustic impedance contrast with adjacent, uncemented sands, and to cause significant signal starvation, giving the appearance of gas chimneys underneath; these features allow the cemented zones to be imaged clearly on seismic data. They have been found over the Challi, Cornea, Jabiru, and Skua fields. Clearly there is a long history of hydrocarbon leakage in this area, and it seems likely that even those modern seabed carbonate knolls and reefs with no active seepage originated as methane-derived carbonates.

3.17.3 New Britain and the Manus basins

Active and historically active volcanoes are characteristic of New Britain Island (Papua New Guinea), but in the last 20 years exploration has shown that the surrounding seas are also geologically active (Map 18). There is an active back-arc spreading centre in the Manus Basin to the northwest, and to the northeast, in the east Manus Basin there is pull-apart rifting between trans-form faults. To the southeast the Solomon Sea plate is being subducted down the New Britain trench; further to the southeast the Woodlark Basin is an active ocean spreading centre that merges into a continental rifting zone on the Papuan Peninsula. Both *et al.* (1986) reported that a bottom-towed camera had photographed collapsed and standing hydrothermal chimneys on a small axial volcano inside a rift valley in the Manus Basin to the north of the island. Photographs provided evidence of recent hydrothermal mineralisation, and fauna including bacterial mats and tubeworms. Since then a considerable amount of work has been done in this area by multi-national teams including a joint USSR–Australia–USA–PNG (Papua New Guinea) campaign (reported in a special issue of *Marine Geology*: Crook *et al.*, 1997) and drilling on ODP Leg 193.

There is ample evidence of hydrothermal activity in these waters. Venting, mineral-rich chimneys, and chemosynthetic communities have been reported from the spreading centre in Woodlark Basin, and venting and chimneys are present in the Franklin

Seamount. However, we will focus on three other locations.

Matupi Harbour

In Matupi Harbour, on New Britain Island, hydrothermal vents are related to the local volcanic activity; the most recent eruption occurred in 1943. The vents described by Tarasov *et al.* (1999) are under the water, in depths of up to 27 m. Volcanic gases are emitted in two areas, one of 500 m^2, the other of 1000 m^2. The high temperatures of the emitted fluids, 50–60 °C and 80–90 °C, respectively, clearly affected the ecology, as did the nature of the fluids. These were found to be dominated by carbon dioxide (62 to 94%), but nitrogen (4.9 to 28.8%), oxygen (1.4 to 2.9%), hydrogen (0.006 to 1.0%), methane (0.003 to 0.045%), ethane (<0.00007%), and hydrogen sulphide (< 14.2%) were also present.

Bacterial mats were found to be abundant at only one of these vent sites. Here the sulphate-reducing bacterium *Desulfovibrio* sp. and its syntrophic partners *Spirochaeta* sp., *Thiovulum* sp., and *Beggiatoa* sp. were abundant. Where temperatures exceeded 60 °C various thermophilic microbes, including the archaea *Thermococcus*, were present. The vent areas were '*practically devoid*' of macrofauna and had only a limited meiofauna, dominated by polychaetes or a single temperature-resistant nematode. In contrast, '*the periphery of the vent fields was characterised by the richest benthic communities found in Matupi Harbour*' (Tarasov *et al.*, 1999). Chemosynthesis was found to account for between 15 and 60% of the bacterial production. This detailed report concluded that semi-isolation from the open sea has enabled the hydrothermal fluids to have a '*pronounced influence*' on the ecology and biological processes here.

Manus Basin

Lisitzin *et al.* (1997) reported water-column plumes, typified by anomalous turbidity, manganese concentrations, temperature, and salinity, in the central Manus Basin. Venting seems to be concentrated in the spreading centre graben where three sites were reported on by Galkin (1997): HF1 (Vienna Woods), HF2, and HF3. Of these only one has high-temperature (<276 °C) black smoker vents and well-developed (> 20 m tall) chimneys; the others are much cooler (<30 °C), but have noticeable '*shimmering water*'. All three sites have rich and well-developed vent fauna communities, but curiously, only one (the cool-water HF3) has tubeworms.

However, tubeworms, along with bresiliid shrimp, gastropods, pink anemones, and crabs, plus white to yellow highly flocculent bacteria, were reported from another site, Worm Garden, by Sinton (1997).

Lein *et al.* (1997) recognised that the hydrothermal vents had a notable effect on the organic-carbon content in the water column. A daily production of 2.7 kg organic matter was estimated for a 32 400 m^2 area of seabed around the Vienna Woods. We discuss the significance of this compared to photosynthetic productivity in the surface waters in Section 8.3.4.

Eastern Manus Basin

Three hydrothermal sites have been investigated: PACMANUS, Susu Knolls, and DESMOS as part of the ODP studies. Together these represent a hydrothermally active area of some 4000 km^2 (Binns *et al.*, 1995). There are three hydrothermally active volcanoes in the Susu Knoll area, with small vents emitting fluids at temperatures ranging from 50 to 280 °C, whilst at DESMOS (a 250 m deep caldera, > 1.5 km across) white smokers vent very acidic (pH 2), sulphate-rich fluids at 88–120 ° C. The PACMANUS hydrothermal field lies on the Pual Ridge, which is composed of dacitic and andesitic volcanic rocks. Hydrothermal vents and mineral deposits occur along the crestal zone over a distance of nearly 10 km. Within this distance there is a 2.5 km area with significant mineral deposits, whilst active hydrothermal venting, indicated by faunal abundances, occurs in a 1.2 km zone. Vent emissions range from high-temperature (< 300 °C) metal-bearing fluids that build chimneys < 30 m tall, to diffusive, low-temperature discharge.

The ODP targeted two hot-vent sites (Satanic Mills and Roman Ruins), and one low-temperature site (Snowcap), at PACMANUS. This, the first subseafloor examination of an active hydrothermal system hosted by felsic volcanics, produced rather disappointing results for those who expected to find ore minerals. Contrary to expectations, they showed that the rocks beneath the seabed had been heavily altered by hydrothermal activity, producing surprising porosities (25%) and permeabilities. Anhydrite ($CaSO_4$) was found in veins, breccia matrices, within vessicles, and disseminated within the altered wallrock of zones bleached by hydrothermal alteration. Binns *et al.* (2001) suggested that this implies '*a major role for seawater infiltrated or swept into the rising column of hot hydrothermal fluid*', made possible by the high porosity. The abundance of clay minerals (assemblages of chlorite, chlorite–vermiculite, chlorite–vermiculite–smectite, and illite–smectite, etc.) in altered rocks throughout all levels was also a surprise, providing *'insights into the chemical changes due to hydrothermal alteration'* (Lackschewitz *et al.*, 2001). Is this a clue to where some of the clay in the deep ocean originates from?

Luckily, however, some of the mineral deposits at PACMANUS are *'unusually rich'* in zinc, copper, gold, and silver, as was a chimney sampled on Susu Knoll. Consequently, this area is considered to be particularly important as a guide to understanding economic metal sulphide deposits on land.

3.17.4 New Zealand

New Zealand's location on a plate margin makes for exciting geology and, it seems, some interesting examples of seabed fluid flow associated with subduction (on the Hikurangi active margin, offshore eastern New Zealand), petroleum generation (in the Taranaki Basin), and active volcanism (in Poverty Bay and in the Taupo Volcanic Zone).

The Hikurangi active margin has developed an accretionary wedge with a narrow fore-arc basin. Gas hydrates seem to be *'ubiquitous'* on this margin, according to Pecher *et al.* (2003). In the Taranaki Basin, off the west coast, shallow gas, gas chimneys, pockmarks, and seeps have been reported during site surveys (Falconer, 1991). It is probable that this gas is thermogenic in origin, and active faults provide migration pathways towards the seabed. Off the east coast Lewis and Marshall (1996) reported seep sites between East Cape in the northeast to Otago and Puysegur Ridge in the southwest (the sites are indicated on Map 19). Seven of these were found to have distinctive faunas; others had carbonate chimneys, carbonate crusts, or seep plumes on sounder records. Some seep plumes extended over 250 m from the seabed in 900 m of water. Lewis and Marshall identified seeps in three distinct geological environments.

- Where permeable layers outcrop, or over fault zones on mid to upper slope ridges of subduction margins, seep plumes were recorded, and MDAC crusts and chimneys were found in association with a fauna that included *Calyptogena* sp., *Bathymodiolus*-like mussels, and siboglinid tubeworms.

- At shelf edges and canyons heads (often away from subduction margins) MDAC crusts and chimneys were found, along with communities including the thyasirid bivalve *Maorithyas* sp.
- At the edges of large slumps that had collapsed into sedimentary basins 'diapirs' and *Calyptogena*-based communities were found.

Poverty Bay

During a side-scan sonar survey of Poverty Bay, North Island, Nelson and Healy (1984) discovered pockmark-like features, < 20–40 m across and only 0.3–0.5 m deep. With a density of up to 30 per 100 m², they *'impart a distinctly pitted or cratered appearance to the seafloor and locally produce a submarine topography not unlike a lunar landscape in miniature'*.

In this area tight compressional folding and faulting is associated with the under-thrusting of the Pacific Plate beneath the Australian Plate. Under-compacted calcareous and bentonitic mudstones of Palaeogene age are driven upwards into thick Neogene flysch and mudstones by a combination of abnormally high fluid and gas pressures. In the adjacent onshore area, mud diapirism is not uncommon and has occurred intermittently throughout the Quaternary. Nelson and Healy described quiescent mud volcanoes that emit saline water and gas (mainly methane). One onshore site, about 24 km from the pockmarked area, *'bears a remarkable topographic resemblance to the cratered seafloor'*, encouraging them to *'tentatively postulate'* that the Poverty Bay pockmarks are offshore equivalents.

It seems that this area is subject to periodic mud-volcano activity, associated with seismic events. A few days after a catastrophic earthquake in 1931 the seabed in part of the bay (Sponge Bay) *'rose 2.1 m to become dry land, apparently as a result of extrusion of muddy breccia'* (Strong, 1933). Local doming of the seabed, charted in 1954, was found to be absent during Nelson and Healy's 1981 survey (Nelson and Healy, 1984). It would appear that this dome had deflated after a period of diapiric activity; the pockmarks may result from this deflation. This suggestion is supported by a report of *'violent foaming of the sea'* in the area.

Taupo Volcanic Zone

The Taupo Volcanic Zone, a product of back-arc extension, is a 300 km long depression that extends 50 km offshore from North Island to White Island, an active fumarolic volcano. The onshore area is a well-known geothermal province; less is known about the offshore area. However, early indications (Duncan and Pantin, 1969; Glasby, 1971) of seep plumes and bubbles breaking the sea surface indicated seabed activity. White Island is one of several submarine volcanoes and seamounts between the coast and the edge of the continental crust. Seabed venting of hot water and gas is focussed along faults that mark the bounds of the Whakatane Graben. Whitfield (1999) noted that vents emit fluids at a temperature of < 200 °C. The vents are inhabited by bacterial mats, and many samples *'came up coated in an oily film that smelt of petroleum'*, suggesting that thermogenic petroleum generation is being induced by the high temperature gradient. However, the most interesting fact is that hydrothermal minerals precipitated at the vents include an arsenic–antimony–mercury association, as well as native sulphur, barytes, and anhydrite. Some samples produced silver droplets of mercury when recovered (Whitfield, 1999).

3.18 WESTERN PACIFIC

Some of the most noteworthy points about the southern and western Pacific are the geological complexity, the number of islands, and the seamounts (see Map 20). Merle *et al.* (2003) indicated that there are about 20 000 km of volcanic arcs in the western Pacific, but only about 13% of them have been systematically surveyed. However, knowledge is growing rapidly. Although Ishibashi and Urabe (1995) identified only seven submarine arc volcanoes with active hydrothermal activity in the Pacific, de Ronde *et al.* (2001) found that seven of thirteen submarine volcanoes along a 260 km section of the Kermadec Arc (effectively a continuation of the Taupo Volcanic Zone; Map 19) had active hydrothermal vents. More recently, active venting was found on 11 out of 56 submarine volcanoes identified on the Mariana Arc (Merle *et al.*, 2003).

It seems that about 25% of the submarine volcanoes that have been identified are actively venting. If this proportion and the density per kilometre of arc are representative, there must be almost 1000 submarine volcanoes of which about 250 have hydrothermal vents. We will return to the significance of this, and the likely number of seamounts, in later chapters.

3.18.1 Silicic dome volcanism in the Mariana Back-arc Basin

During a detailed geophysical and photographic survey of the Mariana Trough back-arc basin in the western Pacific, Lonsdale and Hawkins (1985) found clusters of 50–1000 m diameter domes or mounds. They recovered rhyodacite pumice that they thought had erupted at a depth of 2600–3600 m, and suggested that it was the result of partial melting of sediments. Apparently this is a deep-ocean equivalent of the silicic dome volcanism, a form of quiescent or exhalative (non-explosive) volcanism previously described from the Obsidian Dome volcano in eastern California. Eichelberger *et al.* (1986) found that this type of eruption results from rapid, subsurface gas release from magma ascending as a permeable foam. The degassed foam then collapses during extrusion. If the Mariana Trough domes are silicic volcanoes, then significant fluid flow can be expected. This may explain the large methane plumes and temperature anomalies reported by Horibe *et al.* (1983) at 3000 m depth, some 600 m above seabed.

3.18.2 Serpentine mud volcanoes near the Mariana Trench

On the southeastern Mariana Fore-arc there are several prominent seamounts. Sampling, dredging, and visual inspection has shown that these are actually serpentine mud volcanoes (Fryer and Fryer, 1987; Fryer *et al.*, 1990; Mottl *et al.*, 2004). Conical Seamount, located about 90 km landward of the trench, was inspected using Alvin, as part of the planning for an ODP drilling project. Fryer *et al.* (1990) reported that 'mud' flows extended up to 18 km from the 1300 m–high summit of the seamount, and covered an area of about 550 km^2. Although the flows were initially thought to be of mobilised fore-arc sediments, this proved not to be the case. The flows presumed to be the youngest are *'light green'* in colour, and have *'contorted surfaces'*. Like the breccia from some mud volcanoes, they contain cobbles and boulders; but these are composed mainly of serpentinised peridotites and dunites, along with some metabasalts and metagabbros. These are rock types typical of the mantle.

Although serpentinite samples had been collected in grab samples from other mud volcanoes in the area, it must have been quite a surprise to see 'sedimentary serpentinite' flowing from the summit! On the summit of the seamount, at a depth of about 3150 m, Alvin located a 100 m^2 field of carbonate chimneys. They were composed of varying proportions of calcite and aragonite. At this depth, below the 'carbonate compensation depth' (CCD; the depth beyond which calcium carbonate becomes soluble in seawater), these minerals are not stable, and the chimneys showed signs of corrosion. Silicate chimneys were also found, and these too were encrusted with black ferromanganese oxides. Although no active venting was found, these chimneys clearly indicated that venting has occurred, and fluids sampled from within one of the silicate chimneys proved to be slightly cooler than ambient seawater, very alkaline (pH 9.3), and enriched in methane, silicate, sulphate, and hydrogen sulphide.

Subsequently, during drilling, the pore water fluids were found to be fresher than seawater, highly alkaline (pH of 12.6), and enriched in hydrogen sulphide and the light hydrocarbons: methane to propane (Fryer and Mottl, 2000). The hydrocarbons were considered to have originated from sediments on the subducted slab.

Another of these structures, the Chamorro Seamount, rises from a general water depth of just over 4000 m to 2930 m, and measures 20 to 25 km in diameter at its base. It lies 85 km from the Mariana Trench where the subducting slab is at a depth of about 26 km (Isacks and Barazangi, 1977). Visual inspection at the summit of the seamount revealed seeps and seep-associated biota. These are found in fissures, some a metre or so deep, that cut across the cemented blue-green serpentinite mud on the summit knoll. Fryer and Mottl (2000) found that these seeps support a 'vigorous biological community' including mussels (probably *Bathymodiolus* sp.), gastropods, tubeworms, and galatheid crabs. Mottl *et al.* (2004) sampled several similar subduction-related serpentinite mud volcanoes near the Mariana Trench (Map 20).

3.18.3 The Yellow and East China seas

The Yellow and East China seas (Map 21) represent one of the world's widest continental shelves, into which flow two of the world's great rivers, the Huang Ho and the Yangtse (Changjiang). According to Milliman *et al.* (1985), the majority of this shelf has less than 100 m of water, and much of it, particularly off the Yangtze River mouth, has more than 10 m of Holocene sediments.

There are shallow gas, gas seeps, and pockmarks in these areas (Butenko *et al.*, 1985; Milliman *et al.*, 1985).

3.18.4 Offshore Korea

Along the southwest and southeast coast of Korea, there are thick deposits of Holocene mud (Map 21). Off Pohang, these muds are charged with gas. Kim *et al.* (2002) surveyed this gas with sub-bottom profiler (Chirp) and by downcore electric resistivity and acoustic-velocity logging. The sub-bottom profiler records show '*a lot of gas seepage, pock marks and other typical features of shallow gas*' (Kim *et al.*, 2002). The spectacular occurrences of acoustic turbidity correlate with higher resistivity and dramatically reduced velocity (from 1500 m s^{-1} to 1300 m s^{-1}); further details of the features in this area are presented by Kim (see website). Shallow gas and pockmarks are also present in > 1800 m of water in the Ulleung Basin, Sea of Japan (Lee and Chough, 2003).

3.18.5 Japan

The Japanese islands are classically divided into two segments, southwest Japan and northeast Japan, by the Tanakura Fault (Faure *et al.*, 1986). The opening of the Sea of Japan and the bending of the Japanese islands are closely related to complex and tectonic interactions of the east Asian–Pacific margin. A combination of differential rotation, flexural-slip folding, and low-angle thrusting (subduction) are important components. Active volcanism and associated hydrothermal activity, and the subduction of the Pacific Plate beneath Japan result in fluid flow in Japan and the surrounding offshore areas (see Map 21).

Sagami Bay

The Hatsushima cold-seep site, in Sagami Bay, central Japan, occurs in surprisingly deep water (1100 to 1200 m) considering how close it is to the coast. The extensive biological community associated with these seeps was found in an area with coarse, black, sulphide-rich sediments. The macrofaunal species here includes dense beds of *Calyptogena soyoae*, a giant clam. The meiofauna were found to be 1.6 times more abundant than at a nearby 'control' site, and its composition was '*distinct*'; for example the diversity of nematode species was lower than at the control site (Shirayama and Ohta, 1990).

Kagoshima Bay

There is a great contrast between the deep-water Sagami Bay cold seeps and shallow (80 to 200 m) water hydrothermal venting in Kagoshima Bay. This bay comprises two volcanic calderas separated by an active volcano, Mount Sakurajima. Active venting in the inner part of the bay causes bubbling ('*Tagiri*') that can be seen at the sea surface. Sampling of the venting fluids showed that they are hot (28 to 30 °C), warmer than the normal bottom water (16 °C), acidic (pH 5.1–6.9), and dominated by carbon dioxide (80.6–84.3%), but methane (12.5–15.3%) and hydrogen sulphide (0.07–0.43%) were also present (Hashimoto *et al.*, 1993). However, the most interesting thing about this site is the fauna. Within a small (150 m by 200 m) area on the eastern side of the summit of a small knoll adjacent to the caldera dense clusters of siboglinid tubeworms were found. One cluster had a density of about 3000 individuals per m^2; that is a net biomass weight of > 9 kg m^{-2}. Remarkably, the water depth was only 82–110 m. These sibogglinids, along with solemyid bivalves, polychaetes, palaemonid shrimps, and galatheid crabs, demonstrate that typical hydrothermal vent communities are not restricted to the deep oceans.

Yuigahama Beach

Of a completely different nature are some small pits that form after stormy weather on Yuigahama Beach in the southern end of the Izu Peninsula. The pits vary from 0.7 to 1.5 m in diameter and from 0.4 to 0.5 m in depth. They form recurrently on the seabed at a water depth of only 1–6 m, about 20–50 m offshore. On 5 July 1980 three people drowned by falling into pits of this kind. Subsequently, the local community demanded research into the formation mechanism of the pits. It seems that the beach is underlain by a confined aquifer. Although the water table fluctuates in response to the tides (with a phase lag of about 3.5 h), the pore water pressure in the aquifer remains above hydrostatic. During severe weather the waves cause a pumping effect, which is transmitted down to the aquifer. The sandy seabed becomes disturbed, and the sealing layer that confines the aquifer is broken down allowing the groundwater to flow out, resulting in localised liquefac-

tion of the sands and pit formation (Takumi Sakamoto and Shigeyasu Okusa, 1985, personal communication).

Subduction zones

To the west of the Japanese islands lie the Ryuku and Japan trenches, which meet at a triple junction with the Bonin Trench. They are formed by the subduction of the Pacific and Philippine Sea plates beneath the Eurasian Plate. This simplistic description belies the complex and dramatic bathymetry of this region. In 1985 and 1989 the Franco–Japanese Kaïko programmes identified evidence of seabed fluid flow in various geological settings (subduction–erosion, accretionary-wedge formation, and the subduction of a seamount beneath the continental margin) during dives with the French submersible Nautile in the Nankai Trough, the Japan Trench, and the Kuril Trench. The results of these expeditions are reported in collections of papers in the journal *Earth and Planetary Science Letters*, volumes 83 (1987) and 109 (1992).

Nautile, working in water depths of 2000 to nearly 6000 m, located more than 50 seeps, but subsequent work has found seeps at 7326 m (Fujikura *et al.*, 1999). Apparently, at these great depths, the easiest way to locate seeps is to look for the 'oases' of biological communities that depend upon them. Confirmation is given by temperature anomalies and elevated concentrations of methane in the seawater. The biological communities are found where pore water emerges from fault planes that act as migration pathways. The main macrofaunal genus present has been found to be the vesicomyid clam *Calyptogena*. It sometimes occurs in colonies of more than 100 densely packed individuals, but is accompanied by other species, particularly holothurians, galatheid crabs, actinarians, polychaetes, sibogglinids, and amphipods (Sibuet *et al.*, 1988; Le Pichon *et al.*, 1992). In the seep areas the biological colonies are quite common. At one site the submersible was used to estimate this density. Le Pichon *et al.* explained that their visual search of a 312 500 m^2 area found 445 colonies covering a total area of 94 m^2; but they actually only saw 4% of the seabed, implying that the true number is > 10 000! One of the significant conclusions of these expeditions was that different types of cold-seep communities colonise seeps according to the strength of the seepage. This is discussed further in Section 8.3.2.

Fluid sampling at one site during the 1989 cruise (described by Le Pichon *et al.*, 1992) identified a methane concentration more than 10 000 times higher than the 'normal' (9×10^4 nl kg^{-1} compared to 5 nl kg^{-1}). The carbon isotope ratios (δ^{13}C values of -35 to -58‰) of carbonate crusts, concretions, and chimneys sampled from some of the sites confirmed that the carbon was derived from methane.

One of the objectives of the 1989 expedition was to estimate the fluid flux. This was achieved in the toe area of the accretionary wedge (Nankai Trough) where the flow: '*is estimated to be about 200 m³ per metre of the width of the wedge (m^2. a^{-1})*'. This value is so high that Le Pichon *et al.* concluded the emerging fluids represent not only dewatering caused by the compaction of the sediments, but also some recirculation of seawater. This suggestion of dilution is supported by chemical evidence. Yet, not all the active faults studied were associated with seeps. Some must intersect aquifers, whilst others do not. Attempts made to determine whether or not there were temporal fluctuations in the fluid flow were not entirely conclusive. However, the duration of venting could be estimated at one of the sites (Site 2, at 2000 m in the Nankai Trough) by dating shells locked within the carbonates. Carbon-14 and uranium/thorium dating provided an age range of 20 000 to 130 000 years, suggesting continuous or sporadic activity throughout this time period.

Elsewhere in this region there are mud diapirs, including some that breach the seabed to form mud volcanoes. These were found on the fore-arc slope of the Ryukyu Trench, between the trench and the southern islands of Japan, in water depths of 1500 to 2400 m. Here the accretionary wedge is between 8 and 12 km thick. Individual mud volcanoes are 4–5 km across and < 270 m high. The description of these features, by Ujiié (2000) focussed mainly on the nature and origin of the sediments emitted by the mud volcanoes. However, a BSR seen on seismic records between the diapirs suggests that they are associated with dewatering and degassing.

Gas hydrate resources
Gas hydrates are regarded by many as the 'fuel of the twenty-first century', so the proximity of gas hydrates to Japan provides an important opportunity to satisfy future energy demands. According to Uchida (2002), seismic surveys have indicated BSRs covering an area of 33 000 km^2 in the relatively shallow (< 1000 m) waters between the Nankai Trough and the coast of southern Japan. Subsequent drilling (ODP Site 808, and wells sponsored by METI, the Japanese Ministry of Economy, Trade, and Industry) has proved the presence of gas hydrate, and that it is composed of almost pure methane of microbial origin. The results from METI's 'Nankai Trough' wells, announced in January 2000, were summarised by Max (2000a). Hydrates were proved in three sandy formations with a total thickness of about 16 m between 207 and 265 m below seabed (water depth 945 m). It was estimated that the hydrate occupies about 20% of the bulk sediment volume (80% of the pore volume), indicating that the volume of gas (at standard temperature and pressure) is about 525×10^6 m^3 km^{-2}. Overall, the Nankai Trough area has an estimated 1.6–2.7×10^{12} m^3 of gas in hydrate and associated free-gas accumulations.

3.18.6 Sea of Okhotsk

The first reports of gas venting in the Sea of Okhotsk came from fishermen working to the west of Paramushir Island in the spring of 1982. Researchers from the Institute of Volcanology, Russian Academy of Sciences soon followed up these reports, and discovered methane-rich waters rising from a gas hydrate area on the slope. During several research cruises on the RV (research vessel) Vulkanolog (1982–5) and later with the manned submersibles, Pisces VII and Pisces XI (1986) it was found that the vigour of the gas escape varied over time. However, the cigar-shaped 'flare' (similar to Figure 3.32) seen on echo sounder records generally rose 500 m above the 800 m deep seabed. It is now clear that this is just one vent in a vent field considered (by Gaedicke *et al.*, 1997) to be '*one of the largest sources* [of methane] *ever described*'. This field lies in a back-arc basin (see Map 22), where the sediments are at least 1500 m thick, and sedimentation is rapid (Zonenshayn *et al.*, 1988; Kamenev *et al.*, 2001).

Since this site was first discovered a great deal of research has been done in the Sea of Okhotsk. Although many reports have been published, the majority are (understandably) in Russian; English language reports are relatively scarce. Kamenev *et al.* (2001) identified 12 individual '*cold-water methane-rich seep zones*' in the Sea of Okhotsk, some on the east side on the slope adjacent to the Kamchatka Peninsula, and most of the others close to Sakhalin Island, to the west.

Northeast of Sakhalin Island

Cranston *et al.* (1994) described ten active vent areas, with footprints typically 100–300 m in diameter. Gas hydrates, recovered from cores, occurred in thin layers '*millimetres to centimetres thick*' in the upper two metres of the sediment. Evolved gases were dominated by methane (97%), but ethane (30 ppm), propane (15 ppm), and butane (7 ppm) were also measurable. Nevertheless carbon isotope ratios ($\delta^{13}C = -60$ to -67) brought Cranston *et al.* to the conclusion that the hydrate methane is microbial in origin.

A 20 kHz fish-finding sonar showed gas bubble plumes rising '*hundreds of metres above the sea floor*' from these vents, and when the sea was extremely calm, '*small gas bubbles (estimated to be 1 mm in radius) were observed at the sea surface 700 m above one vent*' (Cranston *et al.*, 1994). This bubble escape clearly indicates the ability of some gas to escape to the atmosphere even from these great depths. Subsequent mapping has revealed extensive BSRs, and several spectacular acoustic 'flares' in this area (Lüdman and Wong, 2003; Shakirov *et al.*, 2004).

West of the Kuril Islands

The seabed slopes gently from Paramushir Island to a depth of about 1000 m in a back-arc basin. Here, volcaniclastic sediments accumulate rapidly, currently reaching a thickness of about 1500 m. On the slope there are some topographic highs underlain by strong seismic reflections. Associated magnetic anomalies caused Zonenshayn *et al.* (1988) to interpret these as igneous bodies, and this idea was supported when andesitic rocks were found on one of the rises. It seems likely that these 'lava domes' affected the surrounding sediments, no doubt raising their temperature. This is significant because there are gas hydrates in them, and gas accumulations deeper in the sediment sequence. Zonenshayn *et al.* concluded: '*the mobilization of the gas phase is most logically attributed to the heat attending the intrusion of volcanic domes*'. Similar underwater gas sources are '*apparently widespread*' in the Sea of Okhotsk. Indeed, mapping reported by Gaedicke *et al.* (1997) identified six individual seep areas, and extensive BSRs (see Map 22). The seeps are generally found close to the edge of the gas hydrate area. In most of these areas plumes on echo sounder records indicate seep locations, and methane concentrations in bottom waters exceed 1000 nl l^{-1} (compared to a background of 30–40 nl l^{-1}). However, in the main vent area, west of Paramushir Island, which Gaedicke *et al.* indicated to be very large, gas is emitted '*in a regional and diffuse manner over the whole area*'. Nevertheless, the flare reported by Zonenshayn *et al.* lies within this area. Analyses of the gases showed that they are > 98% methane, with ethane, propane, and hydrogen also present. Gas bubbles coming from the seabed are up to 1 cm in diameter.

The source of the Zonenshayn Flares was found to be at a water depth of 768 m on a swell 400 m wide, 800 m long, and about 15 m higher than the surrounding seabed. On this swell an area about 50 m^2 marks the gas source. Gas hydrates were found in cores about 1.8 m below the seabed, and the seabed sediments are covered by near-continuous bacterial mats (identified as *Beggiatoa*). Zonenshayn *et al.* described in detail '*subsidence funnels*'. These are pits < 2 m across and at least 1 m deep and narrowing into a funnel shape where they extend laterally into the seabed. There is clearly a complex of these structures honeycombing the sediments, with water freely circulating between those that are interconnected. '*Grottoes and caverns*' are large enough for large (0.5 m) fish to inhabit them. The seabed sediment is '*strengthened by calcareous crusts*' which '*often overhang the pits like cornices*'. These crusts are '*riddled*' with empty tubes (0.5 to 1.5 cm across), which Zonenshayn *et al.* attributed to escaping gas. These features sound remarkably like the 'fish caverns' of Gullfaks (see Section 2.3.6).

The bottom fauna includes '*abundant populations*' of *Conchocele*, and also populations of polychaetes (from the Ampharetidae family). In the Sea of Okhotsk, Kamenev *et al.* (2001) reported seven occurrences of *Conchocele bisecta*; a thyasirid bivalve widely distributed in the northern Pacific. For example, dense congregations of *Conchocele bisecta* were found off Paramushir Island: up to 15 or 20 individuals per m^2, and no less than 2.5–3.0 kg biomass per m^2. It seems that this species is found only at seep sites where the methane concentration in the bottom waters and sediments is at least ten times the background. Like other thyasirids, this species is host to sulphur- and methane-oxidising symbiotic bacteria.

The significance of the Sea of Okhotsk

Results of a regional survey of hydrocarbon concentrations in bottom sediments (reported by Geodekian, 1976 – referred to by Lammers *et al.*, 1995b) indicated two areas where concentrations of methane and the higher hydrocarbon gases were anomalously high: the Tinrio Basin (northeast Sea of Okhotsk) and the entire shelf and slope off eastern Sakhalin Island. To these must be added the gas-prone area west of the Kurils. Lammers *et al.* said that '*more than 20% of the area of the Sea of Okhotsk might be underlain by CH_4 accumulations both of biogenic* [microbial] *and thermogenic origin*'. They considered that it '*is one of the marginal seas with the highest potentials for a large marine contribution to the atmospheric methane in the northern hemisphere*'.

Work, undertaken as part of a Russian–German co-operative programme, has studied flares in both the areas described above. Obzhirov *et al.* (2000) reported that the number of flares varies from year to year; 1988 was a good year for flares, but between 1998 and 2002 another 200 were discovered (Obzhirov *et al.*, 2004). Apparently seismic activity and fault movements affect this variability. Although most of the flares rose to depths of 300–500 m below sea level, some penetrate to the surface waters. Methane concentrations in the waters near the flares were 1000–3000 nl l^{-1}, but those inside the flares reached 10 000–20 000 nl l^{-1} (compared to a regional background of < 90–100 nl l^{-1}). In these areas even surface waters had concentrations of greater than 1000 nl l^{-1} in spring and autumn, but concentrations were found to vary with water temperature and salinity. The sea is entirely covered by ice several metres thick during the winter months (Lammers *et al.*, 1995b), so the emission of methane to the atmosphere is clearly seasonal. This is discussed in more detail in Section 10.6.1.

3.18.7 Piip Submarine Volcano, east of Kamchatka

Seepage associated with the Piip (or Peepa) Submarine Volcano (Map 22) was first discovered in 1986. Subsequent surveys on the RV Vulkanolog, and investigations using the Mir submersible, proved that there are seeps in water depths from 40 to 500 m (Seliverstov *et al.*, 1994). They include both high- and low-temperature fluids. Piip, lying in a rift basin associated with the strike-slip boundary between the Pacific and North American plates, is several kilometres in diameter and stands about 3000 metres tall. There are hydrothermal vents on both the northern and southern peaks, in water depths of 450 to 750 m. Fluids are thought to be emitted at near boiling point (about 250 °C), through vents and chimneys composed of sulphates (anhydrite, gypsum, and barytes), and carbonates (aragonite and calcite), with associated pyrite (Torokhov and Taran, 1994). Torokhov and Taran (1994) found that, for hydrothermal fluids, these

fluids seem to be unusually rich in methane (81% by volume) and other hydrocarbon gases (ethane, propane, and butane), as well as CO_2 (1.35%), nitrogen (16.75%), oxygen (0.73%), argon (0.33%), helium (0.04%), and hydrogen (0.001%). Also, the high-temperature fluids (from which the anhydrite and gypsum precipitate), are unusually depleted in sulphur (<1%), and do not form the black or white smokers commonly seen at hydrothermal vents. Large areas of the seabed are covered in bacterial mats: ('*there are areas here as much as 100 m^2 entirely covered with bacterial mats*'; Torokhov and Taran, 1994). The bivalve *Calyptogena* has also been found, but generally the bivalves are '*absolutely absent*' from the chimneys.

Torokhov and Taran explained that seawater circulating through the volcano '*draws up pore waters rich in hydrocarbon gases from the surrounding sedimentary strata*', and this mixes with rising hydrothermal fluids. So, thermogenic methane is derived from the underlying strata. At the shallower sites MDAC cements, crusts, and concretions comprise calcite and Mg-calcite with carbon isotope ratios of −24.7 to −62‰. Seliverstov *et al.* (1994) concluded that these were derived from microbial methane formed where conditions are suitable for rapid sedimentation.

3.19 OFFSHORE ALASKA

Extensive surveys of various parts of the seas off the northern, western, and southern coasts of Alaska have been undertaken by a number of organisations, most notably the United States Geological Survey (USGS). Gas seeps are not widespread, but shallow gas accumulations and gas hydrates are not uncommon, and pockmarks are present in several areas. Investigations by Sandstrom *et al.* (1983), and Venkatesan *et al.* (1983), showed that hydrocarbon gases are '*ubiquitous*' in sediment samples collected in the Beaufort Sea, Norton Sound, southeast Bering Sea, Kodiak Shelf, Navarin Basin, Gulf of Alaska, and Lower Cook Inlet (Map 23). However, concentrations are generally very low (methane < 75 ppb of dry sediment).

3.19.1 Bering Sea

Chirikov Basin and Norton Sound

The shelf adjacent to the west coast of Alaska is quite shallow with water depths less than 50 m in the Chirikov Basin and less than 20 m in Norton Sound. This, and the exposed northerly location, makes the seabed susceptible to a range of dynamic processes – strong bottom currents, storm waves and storm surges, seiches, ice-gouging etc. – which might be expected to prevent the preservation of seabed features. However, pockmarks are known to be present over an area of 5000–7000 km^2 and may be even more widespread (Nelson *et al.*, 1979; Thor and Nelson, 1979). In other parts, especially in the southern part of the sound, intense ice-gouging may have obliterated all other surface features including pockmarks (C. H. Nelson, 1979, personal communication). The pockmarks are characteristically broad and flat-floored, and tend to be circular rather than elongate in plan. Although they range in size from 1 to 10 m, the average diameter is 2 m; they are thought to be no more than 0.5 m deep. Pockmark densities generally vary from 200 to 1000 per km^2 with a maximum of 1340 per km^2. Despite this apparently extreme density, the pockmarks occupy less than 2% of the seabed.

The pockmarks occur in Holocene marine fine-grained sandy silts derived from the Yukon River. They range in thickness between 0 and 10 m, but are generally less than 2 m thick in the pockmarked area. Sediments containing enough gas to appear as acoustic turbidity on boomer records underlie an extensive part of the Chirikov Basin and Norton Sound, including the whole pockmark area. This occurs beneath the parallel horizontal reflectors of the topmost sediments and coincides with late-Pleistocene peaty muds (< 1–5 m thick), which were formed above sea level under tundra conditions. The peat contains 2–8% organic carbon, which is subject to microbial decomposition. The resultant gas, which has been sampled from a few vibrocores, is principally methane, but Kvenvolden and Redden (1980) reported the presence of minor quantities of higher hydrocarbons; C_1:($C_2 + C_3$) ratios and methane $\delta^{13}C$ values were considered by Nelson *et al.* (1979) to indicate a microbial origin. It is possible that it is the escape of this gas through the seabed that is responsible for the formation of the pockmarks.

The gas escape may be continuous, but Larsen *et al.* (1980) considered the absence of methane in the bottom waters during their survey, suggesting that this is not the case. They think it more probable that the gas is retained within the peaty mud and only escapes during winter storms when rapid changes in pore water pressure occur. This may be caused by sea-level set-up, seiches, the erosional unloading of the overlying sediments and the liquefaction (by storm-wave or earthquake loading) of the

seabed sediments. Certainly gas escape and pockmark formation are present-day occurrences because pockmarks occur within ice-gouge grooves cut by the pack ice which covers the surface of Norton Sound each winter. The gouges are caused when collisions between floes push ridges of ice to the seabed (Thor and Nelson, 1980).

Pockmarks are absent from a large part of the Chirikov Basin, which is also underlain by the peaty muds. In these areas the seabed sediments consist of cohesionless fine to medium sand and gravel. It is probable that gas from the peat layer diffuses through the porous seabed without disturbing it. The pockmarks near Port Clarence occur in mud-covered swales between sand ridges in an area underlain by organic-rich muds (Nelson *et al.*, 1979).

To the south of Nome, where the seabed sediments are very fine sand and coarse silt, acoustic turbidity on boomer records and the total disruption of reflections on sparker records over an area of about 50 km^2 seabed suggest that the sediments contain gas. In this area gas bubble plumes on echo sounder records indicate the presence of seeps (Nelson *et al.*, 1979; Kvenvolden *et al.*, 1979). Video surveys showed that the seep sites are located within deep, conical pits 10–40 cm in diameter that differ from pits produced by benthic organisms in that these tend to be smaller and unfilled, whereas some of the gas-escape pits '*contain vertical holes descending to unknown depths*'. These pits were observed along a 1 km video traverse. They are too small to be visible on side-scan sonar records, unlike the pockmarks described above. Gas samples from vibrocores were, unusually, dominated by carbon dioxide (98%). The methane (362 ppm) is of thermogenic origin, whereas that of other locations is apparently microbial. The seismic data from this site indicate the presence of folded and faulted Tertiary rocks unconformably overlain by horizontally bedded Quaternary sediments. At a depth of 100 m an extensive zone of disturbed seismic reflections is interpreted (Nelson *et al.*, 1979; Kvenvolden *et al.*, 1979) as an accumulation of gas extending over an area some 9 km wide. This gas is apparently migrating to the seabed through a 12 km long fault zone.

Navarin Basin

Carlson *et al.* (1985) identified a total of 280 gas-related acoustic anomalies on 18 000 km of seismic lines shot over the Navarin continental margin, to the west of St Matthew Island. Acoustic turbidity on uniboom and mini-sparker profiles, and bright spots on airgun records indicate gas between 15 and 250 m below seabed, but mainly between 25 and 100 m. These occur on the continental shelf in water depths of less than 400 m. In the deeper waters of the slope they found two BSRs. The upper one is of limited extent, about 5 km; but the lower one: '*extends intermittently beneath the base of the slope from Navarinsky Canyon to Zhemchung Canyon*' and is '*traceable on one seismic line for a distance of about 450 km*' (Carlson *et al.*, 1985). The upper BSR was attributed to gas hydrate, however the authors thought that the more extensive feature was caused by '*diagenetic change*'. Carlson *et al.* (1982) and Carlson and Marlow (1984) described a plume of gas that extended about 55 m above the seabed (water depth 115 m) on 3.5 kHz mini-sparker records. There is evidence of gas within the sediments beneath the seep and airgun records indicate the presence of an eroded anticline whose crest lies about 350 m beneath the seep. It may be that this seep was the only one detected because of the wide spacing (15–30 km) of the survey lines rather than the rarity of seeps.

The collection (by gravity core) and analysis of 141 seabed sediments complemented the seismic surveys described above. Headspace gases were analysed by gas chromatography for the hydrocarbon gases. Concentrations were generally low (methane $< 200\,\mu l\,l^{-1}$ of wet sediment), and ratios of methane to the higher hydrocarbons ($C_1/[C_2 + C_3)]$) generally suggested a microbial origin. There were some anomalies. In the Navarinsky Canyon six cores showed methane concentrations that exceeded $1000\,\mu l\,l^{-1}$, again thought to be of microbial origin. Seventeen cores, mainly from the slope, were considered to show gas of thermogenic origin, and cores taken near the gas seep held principally methane, but the presence of some ethane and propane suggest that this gas is of thermogenic origin.

Apart from the BSRs, almost all the seismic anomalies lie over the Navarin Basin, a sedimentary basin with up to 12 km of Cenozoic sediments considered in 1985 to be a promising hydrocarbon prospect. This conclusion is perhaps supported by the fact that the methane concentrations tended to be higher over the basin.

Bering Sea pockmarks made by whales?

Nelson and Johnson (1987) reported an intriguing relationship between biological activity and pockmarks. Eighteen per cent of a 22 000 km² area in the central Chirikov Basin and a small area south of St Lawrence Island (Map 23) are occupied by shallow pits. These features, 1–10 m long, 0.5–7 m wide, are visible on side-scan sonar records and at first sight appear to be pockmarks. However, they are only 10–40 cm deep. They occur in water depths of 30–50 m in an area characterised by a sandy seabed that is considered to be too porous for the pits to have been formed by fluid escape. Research has demonstrated that the pits are formed by California grey whales, which migrate north from off Baja California to this area in large numbers (about 16 000) from May to November. It seems that the whales filter the seabed sediments through the baleen, the fibrous plates that hang from each side of the upper jaws, in order to feed on ampeliscid amphipods. These shrimp-like crustaceans occur in an area that closely matches the pitted area. Their mucus-lined tubular burrows bind the loose sediments into an erosion-resistant mat that is destroyed by the whales' feeding habits. Once this mat has been removed erosion may occur; indeed, the pits are frequently aligned parallel to the dominant bottom current. But this apparently destructive process, disturbing perhaps 5–6% of the seabed each year, actually enables the amphipod population to survive. Juveniles not trapped by the baleen thrive on the plankton, whose growth is stimulated by the nutrients released from the disturbed sediment, and survive to regenerate the sediment mat.

In neighbouring areas of the seabed, where the sediments are relatively coarse, long, thin furrows averaging 47 m long by 40 cm wide have been recorded. They are shallower and less angular than ice gouges, and are not straight. It appears that walruses hunting for clams make them. However, there is no proof that there is not a link between seepage and the mussels harvested by the whales and walruses.

3.19.2 Gulf of Alaska

Shelikoff Strait

Pockmarks have been reported from a 1500 km² area in the Shelikoff strait (Hampton and Winters, 1981; see Map 23). The strait is underlain by a thick sequence of Tertiary and Mesozoic sediments unconformably overlain by a blanket of Quaternary sediments which are about 100 m thick in the northeast of the area and more than 500 m thick in the southwest. These overlie Mesozoic sediments of the fore-arc basin associated with the boundary between the Pacific and North American plates. There are active volcanoes on the Alaska Peninsula on the northern shore of the Strait, and this is recognised as one of the most seismically active regions on Earth.

The seabed sediments grade from muddy sand in the northeast to slightly sandy mud in the southwest. A decrease in undrained shear strength reflects this change, but most samples are classified as 'very soft' (undrained shear strength, S_u < 12 kPa). At all but two locations, undrained shear strength increases with depth. Hampton and Winters surmised that the two exceptions, which occur in relatively sandy sediments, may have acquired additional strength from cementation or an '*internal fabric effect*'. Shallow seismic profiles, particularly those from the northeast of the strait, contain reflection enhancements and areas of acoustic blanking. The absence of direct evidence of gas prevented Hampton and Winters from concluding that this was the cause of the 'anomalies'; however, it is probable that they represent gas. Certainly hydrocarbon gases were found to be present in the shallow sediments. The distribution of pockmarks also varies with the character of the seabed sediments, there being more in the finer-grained, softer sediments of the southwest. They are typically circular, 50 m in diameter and less than 5 m deep. Some are reported to have low (< 1 m) rims. In addition to the seabed pockmarks, there are a few buried pockmarks within the topmost 25 m of sediment.

Although both pockmarks and shallow gas are present in this area, it seems that the two are rarely seen in association with one another. This, and the apparent lack of reflection disturbance beneath the pockmarks, inclined Hampton and Winters toward an origin unrelated to gas escape, namely sediment liquefaction. Having seen relatively little acoustic turbidity in some of the pockmark areas of the North Sea, we consider that a gas-escape formation should not be discounted. It is tempting to suggest that earthquake activity plays a role in triggering fluid escape, but there is no evidence of this.

Alsek River Sediment Instability Area

Various papers by Molnia *et al.* (Molnia *et al.*, 1978; Molnia, 1979; Molnia and Rappeport, 1984) reported the presence of several areas of shallow gas on the shelf of southern Alaska (Map 23). Of these, the largest, which covers an area of 200 km^2, contains 'pockmarks' which range in size from less than 2 m to nearly 400 m across and as much as 15 m deep. It is said that they are being formed at the present day and that they are commonly the sites of gas seepages into the water column. The gas is principally methane and is believed to have a microbial origin. The maximum-recorded concentration, 3×10^7 nl l^{-1} of wet sediment, is about three to four times the background level.

This is an area of rapid sedimentation. One detailed study of an area in the Alsek Delta demonstrated that Holocene sediments are between 40 and 120 m in thickness. There is apparently no visible evidence on high-resolution seismic profiles of gas at, or migrating from below, these sediments. It is presumed that the gas is generated by the decomposition of the organic matter, which represents as much as 2% of the sediment. The topmost metre of sediment is sand; this covers 2–4 m of under-consolidated clayey silt, with a high water content and containing thin sand layers; this in turn overlies a much thicker layer of dewatered clayey silt. Various features characteristic of an unstable seabed of rapid sedimentation are found in this area. As in the Mississippi Delta, there are slumps and flows and bottleneck depressions. Those who have worked in the area attribute these features to a combination of degassing and dewatering, which may be triggered by wave-loading, seismic activity, or overpressuring of the gas. The similarity between this area and the Mississippi Delta inclines us to consider that the 'pockmarks' from this area are in fact collapse depressions.

3.19.3 The Aleutian Subduction Zone

Where the northward-moving Pacific Plate collides with that part of the North American Plate occupied by Alaska and the Aleutian Archipelago, there is a subduction zone, the Aleutian Subduction Zone (ASZ). Suess *et al.* (1998) summarised the results of three cruises on the RV Sonne by the Research Centre for Marine Geosciences, GEOMAR. They looked at four sections of the ASZ (Map 23). Geophysical evidence suggests that venting occurs right along the 850 km covered by these study areas, and another 250 km. Researchers from GEOMAR made 25 transects with their towed TV sled, EXPLOS (Ocean Floor Exploration System), undertook water sampling and analysis, and CTD (conductivity–temperative–depth) casts. Results indicated that fluid flow is quite widespread. They found '*distinctive faunas, mineral precipitates, methane anomalies, a temperature anomaly*', also the chemistry of pore waters contrasted between seep and non-seep locations.

Detailed investigations of targetted sites used a TV-guided grab, and another TV-guided instrument, VESP (VEnt SamPler) made *in situ* fluid flux measurements. These studies, undertaken in water depths of > 4000 m, demonstrated that fluid is venting by identifying water-column plumes rich in methane. These are confined to the lower part of the water column, and methane concentrations (50 to 150 nl l^{-1}) are '*significantly higher than the oceanographic background*' (Suess *et al.*, 1998). Seepage is focussed on discrete locations near the deformation front where tensional faults intersect ridges. Fluid-flow sites are most often at the base of steps, or in depressions. They are typified by cold-seep communities with bacterial mats, siboglinid tubeworms and '*omnipresent large colonies of bivalves*'; these included vesicomyid clams and solmyid protobranchs.

On one of the cruises an ROV (ROPOS) was deployed for detailed video and still photography. Altogether they documented 56 m^2 of active seeps, the average clam field covering about 0.7 m^2. Population densities were reported to be between 159 and 200 individuals per m^2. With carbon isotope ratios of the seep fauna lying in the range -57.1 to -64.3‰, it was concluded that these communities were methanotrophic. Because of the great water depths here, it is relatively easy to distinguish between active and passive seeps. If there are any organisms with a calcium carbonate shell, such as the vesicomyid clams described above, then the seep must be active. Once the seep activity stops and they find themselves without a source of energy, the clams die. When they die the $CaCO_3$ dissolves because these sites are beneath the CCD (Gerhard Bohrmann, 2002, personal communication). Both barium sulphate and calcium carbonate precipitates were found to be present as disseminated microcrystals, '*fragile linings of fluid channels*', and dense, impermeable precipitates and crusts.

The venting fluids are waters rich in methane rather than gas bubbles. Water is derived by the compaction of the sediments during compression, but bivalves living at the vents add to the fluid flow by pumping seawater through the sediment. This 'bioirrigation' makes the measurement of the natural (tectonically driven)

fluid flow more complicated. However, a biogeochemical approach based on oxygen-flux measurements and vent fluid analysis (described by Wallmann *et al.*, 1997) enabled Suess *et al.* (1998) to estimate the tectonically driven flow; on average individual vents in the Aleutian subduction zone produce 5.5 ± 0.7 l m^{-2} d^{-1}.

Water-temperature readings taken by the EXPLOS sled provided indications of heat input into the ocean from the seeps. However, this '*temperature anomaly*' was very small: 0.010 ± 0.002 °C (Suess *et al.*, 1998). Although they found higher temperatures over the crests of all the ridges, the GEOMAR team acknowledged that naturally their data were not unambiguous. Nevertheless, they considered that this '*small, yet significant*' temperature anomaly '*documents an extra heat source to the oceanic bottom waters from the shallow seafloor, thus giving a new meaning to the concept of cold seeps*'.

3.20 BRITISH COLUMBIA

3.20.1 Queen Charlotte Sound

Shallow gas and gas seeps were reported (Hamilton and Cameron, 1989) from Hecate Strait (between the Queen Charlotte Islands and the mainland) and Queen Charlotte Sound (between the Queen Charlotte Islands and Vancouver Island; see Map 24). These waters are underlain by a sedimentary basin which may be as much as 10 km thick, with potential petroleum sources in Upper Triassic to Lower Jurassic, Cretaceous and possibly Tertiary sediments. Petroleum exploration has not been allowed here because of the area's beauty and environmental sensitivity. However, if any drilling were to be done, Hamilton and Cameron suggest that shallow gas, trapped beneath Quaternary tills, may be a drilling hazard.

3.20.2 The Fraser Delta

Visitors to Vancouver fly in over the Strait of Georgia to land at the international airport, close to the shoreline of the Fraser Delta. The delta itself is still being built from the sediments brought down from the Rocky and Coastal mountains by the Fraser River. This is the largest fluvial system on the Pacific coast of Canada, draining an area of approximately 250 000 km^2, and discharging about 17.3 million tonnes of sediment into the Georgia Strait every year (Luternauer, 1994). Above sea level it covers an area of about 1000 km^2, but it extends a further 9 km offshore into waters more than 300 m deep. Some parts of the delta are susceptible to slope failure (as discussed in Section 11.2.1), a worrying prospect as so much infrastructure (a large ferry terminal and a coal exporting port) has been built across the tidal flats, particularly as this area is periodically affected by subduction-related earthquakes.

A review of the seismic data held at the Geological Survey of Canada's Pacific Geoscience Centre (Judd, 1995, unpublished report to the Geological Survey of Canada) showed that there are many features associated with gas. The first indications of the presence of pockmarks were identified on a single sparker profile recorded on 1 April 1966 as eight indentations in the seabed reflection, with 'bow-tie' side reflections. A pencil annotation on the record, presumably dating from the time of the survey, indicates the approximate shape of the features – four years before the first published description of pockmarks by King and MacLean, 1970.

Evidence of gas
ACOUSTIC TURBIDITY AND ENHANCED REFLECTIONS

Hart and Hamilton (1993) estimated that the acoustic turbidity is nearly continuous over a 530 km^2 area seaward of the tidal flats, and out into the strait (Map 24). However, there are gas-free zones around topographic highs in the pre-postglacial sediments. The gas generally continues up the delta slope as far as the survey lines extend. One line extended onto the delta front within the dredged basin of the Westshore Terminal, a coal-exporting pier. Acoustic turbidity is present beneath this basin, showing that gas is present in the delta front sediments; gas flows reported from boreholes at the terminal confirm this (Christian et al., 1995).

Over much of the delta foreslope the top of the 'gas front' lies less than 3 m (< 4 ms) below seabed, but the depth generally increases away from the delta. The areas with the deepest acoustic turbidity are the deepwater (> 200 m) basins, and the edges of the affected area where the gas front generally deepens to 45–55 ms (26–42 m) below seabed. The thickness of the gassy sediments can be estimated where enhanced reflections extend laterally beyond the gas front. Vertical separations of 20 ms suggest that the gassy sediments are at least 15 m thick. On many seismic profiles the acoustic

turbidity is 'capped' by enhanced reflections. It seems that gas is generally rising towards the seabed, but its vertical progress is halted at the horizons that are brightened. These may be relatively coarse sediment layers in which gas is effectively trapped by the overlying fine sediments. However, when able to penetrate the overlying sediments, the gas seems to rise in plumes of acoustic turbidity, some of which lie beneath pockmarks.

GAS IN CORES

Indications of gas were recorded in several seabed sediment cores collected between 1983 and 1993. Generally the indications were in the form of 'gas cracking', and this was most commonly recorded in clays. These clays were frequently described as being black in colour, which would be expected in the reducing environment in which methane occurs. Evidence of gas was never reported from coarse sediments, or from fine sediments above coarser layers. However it was reported more than once from fine-grained sediments immediately beneath a coarse horizon. Gas has been noted in the logs of numerous gravity and piston cores collected on the delta foreslope. Christian et al. (1995) reported that gas contents exceeded 3% in distal parts of the deltaic sediment. Gas blowouts have occurred during site investigation drilling on the tidal flats (Christian et al., 1995). Onshore a few gas seeps have been reported (Pat Monahan, 1995, personal communication), and gas has been encountered in site investigation and other boreholes (Dallimore et al., 1995; Pat Monahan, 1995, personal communication).

Hart and Hamilton (1993) suggested that most of the gas in this area is microbial, derived from the organic matter in the sediments of the Fraser River. They remarked: '*the gas is generally closest to the sediment–water interface in areas proximal to the mouths of the Fraser River*'. The higher sedimentation rates close to the river mouths cause rapid burial, and inhibit oxidation, permitting an early onset of microbial degradation.

Gas-related features
POCKMARKS

Pockmarks are present in four discrete areas: the two bays (Burrard Inlet and Boundary Bay) on either side of the delta, a large (at least 240 km^2) area in deep (> 200 m) water towards the foot of the delta, and a small area in deeper (> 300 m) water further offshore. Sizes and densities vary considerably; those in the bays are generally larger (< 200 m across and < 8 m deep) than elsewhere.

SEABED DOMES

There is a small number of seabed domes. The largest is about 150 m across with a very small (< 1.5 m) topographic expression. In most cases plumes of acoustic turbidity extend to the seabed immediately beneath them. This suggests that their formation is related to the presence of the gas. The majority of these domes occur in a restricted area: a corridor extending down the delta slope to a water depth of about 200 m. This corridor seems to correlate with a restricted zone of silty sands and sandy muds.

3.21 CASCADIA

Off the coasts of British Columbia, Washington, and Oregon lies the small Juan de Fuca Plate, and two even smaller plates: the Explorer Plate to the north, and the Gorda Plate to the south (see Map 25). These microplates are sandwiched between the Pacific and North American plates. The Cascadia Margin marks the convergent boundary of the Oregon Subduction Zone where the North American Plate overrides the Juan de Fuca Plate at a rate of about 4 cm y^{-1} (Wells et al., 1984). Here, cold seeps are associated with an accretionary wedge. In contrast, active volcanism and hydrothermal venting characterise the opposite side of these microplates. Hydrothermal activity on the Juan de Fuca Ridge is described in detail elsewhere, and we refer to it in later chapters.

3.21.1 Hydrate Ridge

Kulm et al. (1986) found widespread seepage of cool fluids associated with communities of bivalves, tubeworms, and carbonate mineral precipitates on a fault-bend anticline at a water depth of 2036 m. Following the discovery of extensive BSRs, four boreholes were drilled in the area in 1992 on ODP Leg 146 to investigate areas of diffusive and focussed fluid flow (Westbrook et al., 1994). Site 892, located close to a fault, encountered gas hydrates (nodules up to 4 cm long and white; 5 mm spherical 'hailstones') intercalated and evenly dispersed in the soft mud between 2 and 19 m below the seabed

(Kastner et al., 1995; Hovland et al., 1995). The chlorinity of the pore water suggested that hydrates probably were present to about 68 m, the depth of the BSR, a supposition confirmed by subsequent drilling on ODP Leg 204. Beneath this level acoustic measurements indicated 'free gas': vertical seismic profiles showed an acoustic velocity of < 1400 m s^{-1} between 68 and 71.5 m (Ginsburg and Soloviev, 1998).

Since ODP Leg 146, in 1992, American, Canadian, and German teams have done more work on the seabed ridge, with peaks at 600 and 840 m water depth, located just to the south of Site 892; this ridge is now, officially, known as Hydrate Ridge (see Fig. 3.40). This site covers an area of 250 km^2, of which about 5% is occupied by fluid-flow-related features (exposed gas hydrates, MDAC, cold-seep communities, bacterial mats, and seeps), as has been shown by detailed side-scan sonar and video mapping (Klaucke et al., 2004). Active venting associated with faults at the northern summit is indicated by vent communities, MDAC, and bubble streams support a 'huge methane plume' (see Section 10.4).

Samples of gas hydrate, including one of nearly 50 kg, have been collected. The hydrates are dominantly composed of methane (> 96%) and hydrogen sulphide (<3.0%), but ethane, propane, and carbon dioxide are also present. Carbon isotope ($\delta^{13}C$) values of between −71.5 and −62.4‰, and the ratio of methane to the higher hydrocarbon gases, suggested a microbial origin to Suess et al. (1999b). The samples comprised muddy host sediment with layers, some 10 cm thick, of pure white hydrate parallel to the bedding, but overall the samples were described as having a 'globular fabric', with hydrate occurring only on cavity surfaces. According to Bohrmann et al. (1998), gas bubbles trapped beneath impermeable sediment layers created cavities, then the subsequent addition lead to globular hydrate formation. They said: 'the globular fabric of the gas hydrate constitutes definitive evidence for the existence of free gas prior to hydrate precipitation'.

There are two distinct types of carbonates here. One is a high-magnesium calcite precipitated as submicrometre crystals within the pore space of the terrigenous sediment, with cemented sediment clasts showing various degrees of brecciation, often present within a matrix of pure hydrate. Oxygen isotope (^{18}O) enrichment suggested that the high-magnesium calcite precipitated in water released during hydrate destabilisation. The other

Figure 3.40* Bathymetry of Hydrate Ridge, viewed from the north; image generated using Fledermaus™ from data recorded during the LOTUS project in 2002. At the northern (590 m) and southern (800 m) summits bubbling seeps were detected. The spatial distribution of acoustic backscatter caused by bubbles in the water column (as seen on single-beam echo sounder) are shown here as flare-like features. (Image courtesy of Jens Greinert, IFM-GEOMAR, Kiel, Germany.)

carbonate is pure aragonite, finely disseminated and as aggregates of 2–10 μm thick needles forming botryoids or thin layers. Aragonite from one sample had a surface morphology that partly imaged that of the gas hydrate bubble fabric. Bohrmann *et al.* thought that the aragonite had precipitated in cavities within the hydrate.

3.21.2 Axial Seamount

In August 1983 the Canadian American Seamount Expedition discovered hydrothermal seepage and a hydrothermal vent community in the summit caldera of Axial Seamount, a submarine volcano located on the central segment of the Juan de Fuca Ridge (see Map 25). Since then three other vent fields have been described: The ASHES vent field (Axial Seamount Hydrothermal Emissions Study), and the East and South vent fields. The ASHES vent field emits clear, brown, and black fluids at high temperature (up to 328 °C, according to Massoth *et al.*, 1989); the other three vent fields emit low-temperature fluids. They are all located in the summit caldera at about 1500 m depth (Rogers, 1994). There is a biological community associated with each of them. Tunnicliffe *et al.* (1985) reported siboglinids, limpets,

and alvinellid polychaetes around these vents, and two biological indications of the proximity of the vents: an increased abundance of white bacterial mats in crevices within 100 m of the vents, and an increase in the number of large crabs.

The true nature of Axial Seamount as a 'sleeping giant' became evident after an eruption in January 1998. *In situ* sensors detected an earthquake 'swarm' first, followed about three hours later by caldera subsidence, lateral contraction, and a temperature anomaly (Davis, 2000). Thermal profiling during a cruise soon afterwards showed a large 'lens' of warm water with an anomalously low ^{3}He/heat ratio to the southwest of the summit. It seems that this lens had been produced by rapid heat exchange during the ten-day eruptive phase, and was subsequently swept some 20 km from its source by ocean currents. It was estimated that as much as 200 GW of heat energy was injected into the ocean during this event; that is more than two orders of magnitude greater than the typical heat output from the caldera during quiescent periods.

One intriguing finding was the increase, relative to typical values in hydrothermal vent water, of methane concentrations in the plume created by the volcanic activity. Although the source is unknown, Davis (2000) suggested the possibility that microbiologically 'mature' water in the crust, cool enough to support life, was suddenly heated by the volcanism and buoyantly ejected. This latter comment on the source of the methane shows that 'primordial' methane from the upper mantle of the Earth is still not considered seriously by most academic Earth scientists (see our discussion on this topic in Section 5.4.4).

A set of near real-time monitoring instruments deployed inside the caldera of the Axial Seamount, at about 1600 m water depth, included a camera, temperature sensor, and an acoustic modem. This NeMO (New Millennium Observatory Network) station was set up to monitor a hot spring of the volcano every three days (Weiland *et al.*, 2000), the latest 'news' being published on the Internet.

3.22 CALIFORNIA

There is a series of petroleum-bearing sedimentary basins along the coast of California. The basins between Redondo Beach and the Oregon border all contain at least some evidence of seabed seepages, and one of them, the Santa Barbara Basin, contains what may be the most prolific offshore petroleum seeps in the world. Richmond and Burdick (1981) claimed that gas seeps occur where bedrock outcrops at the seabed or is covered by a thin veneer of surficial sediments, and that near-surface structures act as conduits for any gas that is present. A considerable volume of research and exploration has been undertaken here, and it would be neither possible nor appropriate for us to present more than a brief review; the areas mentioned are indicated on Maps 25 and 26.

3.22.1 Northern California

There is an extensive area of northern California underlain by gas hydrates, shallow gas, gas seeps, and associated features. The Eel River Basin, a fore-arc sedimentary basin with Miocene to Recent sediments, underlies this area. It is tectonically active because it lies at the southern end of the Cascadia Subduction Zone, where the Gorda Plate and North American Plate converge. Also, the shelf here receives large volumes of sediment, particularly from the Eel and Klamath rivers. This combination of rapid sedimentation, tectonic compression and the associated earthquake activity, and a sedimen-

tary basin combine to make the sediment pore fluid regime here particularly interesting.

In a review of this area, based on data from several sources (some of it unpublished), Ginsburg and Soloviev (1998) noted that the extent of the area affected by BSRs is '*at least 3000 km²*'. It lies on the upper continental slope (water depths 725–2000 m), and occurs between 135 and 315 m below seabed. The presence of gas hydrates has been confirmed by coring in water depths as shallow as 510 m (Brooks *et al.*, 1991). Shallow gas, gas seeps, and associated features have also been reported from the continental shelf and slope.

The Eel Shelf and Slope
There have been several studies of the continental shelf and slope between Cape Mendocino and Point St George. The most comprehensive was the STRATAFORM programme, funded by the US Naval Research Laboratory (Nittrouer, 1999, introduced a special issue of the journal *Marine Geology* devoted to this programme). Although stratigraphy and sedimentation were the principal foci, the role of gas featured quite strongly; it was summarised by Yun *et al.* (1999).

A large proportion of their study area, from about 30 m down to > 700 m, is underlain by evidence of shallow gas (acoustic turbidity, bright spots, BSRs, and gas chimneys). The gas is at various depths between seabed and 700 m below seabed, but generally seems to be shallower (< 20 m) on the shelf.

- Numerous water-column acoustic targets were observed on 3.5 kHz profiles on the shelf (< 200 m water depth).
- Nearly 4000 pockmarks were mapped. These generally range in size from 10 to over 25 m in diameter, but there are a few much larger (< 200 m diameter) ones associated with the major slope instability feature, the Humboldt Slide (see Section 11.2.1), that occurs within the study area. In some areas pockmarks are distributed at random, but in others they are '*clustered or occur along linear trends*'.
- Side-scan sonar surveys were examined for high reflectivity anomalies that might indicate MDAC, but none were found. However, a previous study (Kennicut *et al.*, 1989) described seep organisms dredged from the slide area.

Yun *et al.* remarked on how the nature and distribution of the gas, seeps, and pockmarks varied across the area, with the greatest concentrations of pockmarks where there was little evidence of gas, but no detectable pockmarks where the sediments are gassy. They suggested that this is because diffuse flow and high sedimentation rates prevent the formation or preservation on the shelf. In contrast, on the slope gas is expelled '*episodically and catastrophically, creating pockmarks and redistributing sediment*'; the pockmarks are preserved because of low sedimentation rates. The comment by Yun *et al.* about the redistribution of sediment is interesting as it indicates that pockmark activity is a significant factor in sediment redistribution. They estimated that, over an area of about 2100 km^2, over 6.6×10^5 m^3 of sediment had been '*excavated and redistributed by pockmark-facilitated gas expulsion*'.

Within this generalised pattern there are certain unusual gas-related features.

1. Gullies that run down the slope, cutting into the topmost sediments. Yun *et al.* noted that some pockmarks occur within these gullies, but Orange *et al.* (2002) found no visual evidence of seepage.
2. The Little Salmon Fault, an active thrust fault, is associated with a seabed topographic high, which seismic profiles show to be either a breached anticline or a mud diapir. The seabed on this prominent feature has a high reflectivity, and there is a '*crater-like depression*' on the top (Orange *et al.*, 2002). Remotely operated vehicle surveys showed that the top of this feature was covered by '*extensive authigenic carbonate*' in the form of slabs, blocks, and chimneys, which explained the high reflectivity. They also found numerous active cold-seep communities and one bubbling seep.
3. On a separate cruise, scientists from Scripps Institute of Oceanography also studied cold-seep communities in this area. Rathburn *et al.* (2000) described active methane venting, some of it '*rapid*', at a site in about 520 m of water. Here, in an area of at least 1 km^2, they videoed many '*clam beds*' with dense clusters of the bivalve *Calyptogena pacifica*, and MDAC. Some of the clam beds covered an area of more than 15 m^2. Actually, the first signs of cold-seep communities in this area came about ten years before the dives described above when some shells thought to be of vesicomyids were found in core samples. Subsequently a trawl, deployed for an hour in water depths of > 400 m, retrieved 1000 kg of material which included samples of *Calyptogena pacifica* and gastropods from the genus *Neptunea* (Kennicut *et al.*, 1989). It was thought that,

like *Calyptogena pacifica*, this species has chemosynthetic symbionts.

4. In shallow (40 m) water not far north of the mouth of Eel River, there is an extensive bright spot at a sub-seabed depth of about 700 m (~750 ms two-way time) indicating that gas is accumulating at the crest of an anticline, the Table Buff Anticline. The gas seems to have mobilised the sediment to form a mud diapir that has penetrated more or less to the seabed. Orange *et al.* (2002) reported that gas had been seen bubbling from the seabed, and local fishermen told them of '*acoustic anomalies and bubbling at the sea surface*'. During ROV dives they saw several seeps of bubbling gas, but found no evidence of MDAC or macrofauna, but bacterial mats that, in places, '*covered the seafloor for tens of meters*'. Where the mats were less dense, they were confined to the ripple troughs.

A detailed study of the fauna of seep sites on the slope (500–525 m water depth) and on the shelf (31–53 m water depth) was described by Levin *et al.* (2000). They specifically addressed the question: '*do methane seeps support distinct macrofaunal assemblages?*'. We consider their findings in more detail in Section 8.3.3.

Klamath Delta

A major, magnitude 7, earthquake occurred off the Klamath Delta on 8 November 1980. This area had been covered by high-resolution geophysical surveys before the earthquake and was surveyed again afterwards. Comparative studies showed that a series of ridges, mounds, and pockmarks had formed, as had a large number of gas vents. The incidence of shallow gas was significantly increased. Whereas pockmarks had been produced by single-point gas venting, we believe that the ridges and mounds may represent mud diapirism associated with gas release. Field and Jennings (1987) concluded that a 20 km^2 area on the Klamath Delta had failed partly by liquefaction caused by sediment degassing triggered by the earthquake.

3.22.2 Monterey Bay

In Monterey Bay the Packard family (of Hewlett Packard fame) have built an impressive marine research facility at Moss Landing with a fleet of surface and underwater vehicles. Over recent years the Monterey Bay Aquarium Research Institute (MBARI) has, in collaboration with other organisations, surveyed and mapped large expanses of ocean floor, including 'the other Grand Canyon' – Monterey Canyon (which is over 3,000 m deep), in the Monterey Bay National Marine Sanctuary. Many seep features have been encountered (most of them by chance!) and studied. This work has resulted in numerous publications; for this review we have made particular use of papers by James Barry and Dan Orange and their colleagues (Barry et al., 1996; Orange et al., 1999, 2002).

Monterey Bay straddles the western part of the active transform margin of central California where the Pacific Plate has been sliding northwards past the North American Plate for the last 27 million years or so (Map 26). The well-known San Andreas Fault is only one of the fault zones here. Offshore, the Ascension, San Gregorio, and Monterey Bay fault zones run approximately north-northwest to south-southwest. The Monterey Canyon that has been cut down through Neogene sediments into the underlying Cretaceous granitic basement crosses all these faults.

The numerous individual sites identified and studied in this area (some with interesting names: Mt Crushmore, Tubeworm City, Clam Field, Chimney Field, Camp Cup, etc.) are each characterised by evidence of one or other of the following situations in which seabed fluid flow occurs.

- Artesian flow from heavily faulted exposures of an aquifer (the porous Purisma Formation) that is recharged in the Santa Cruz Mountains.
- Aquifer-related flow at heavily faulted exposures of the hydrocarbon-laden Monterey Formation.
- Dewatering of tectonically-compressed (transpressional) hemipelagic sediments.
- Or expulsion of hydrocarbons derived from depth (or both).
- Focussing of pore fluid flow by seabed geomorphological features.
- Gas esape from relict organic debris deposits.

The following are descriptions of example sites that demonstrate these situations.

Mt Crushmore: artesian flow from heavily faulted exposures of the porous Purisma Formation, an aquifer that is recharged in the Santa Cruz Mountains.

There are small (< 3 m diameter) patches of cold-seep fauna in a belt about 1 km wide between 580 and 700 m below sea level. Fluid flow is not obvious, but is identified by whitish-grey bacterial mats, MDAC, and seep-associated fauna. The vesicomyid clam *Calyptogena pacifica* and the bacteria *Beggiatoa* were the most conspicuous fauna here, but a variety of seep-related fauna has been reported: other vesicomyids, solemyid clams, columbellid gastropods, patellacean limpets, chitons, galatheid crabs, and polychaete worms (details are presented by Barry et al., 1996). The seeping fluid is sulphide-rich pore water; a sulphide concentration of only 0.09 mM hydrogen sulphide was measured in the top 10 cm of sediment in a clam bed, but in nearby sediments it was not detectable. Methane was not detected here.

Clam Field: aquifer-related flow at heavily faulted exposures of the hydrocarbon-laden Monterey Formation.

Clam Field occupies a 500–1000 m wide swathe at 875 to 920 m water depth where the highly fractured hydrocarbon-bearing Monterey Formation is exposed in the western wall of Monterey Canyon. Fluid flow may be artesian, as at Mt Crushmore, as the Monterey Formation is also exposed in the Santa Cruz Mountains, but tectonic compression may be the cause. Although vesicomyid clams and bacterial (*Beggiatoa* sp.) mats dominated the fauna here, unlike Mt Crushmore, *Calyptogena kilmeri* was the most common clam. Columbellid gastropods were abundant, whilst solemyid clams and siboglinid worms were in smaller numbers. In clam groups sulphide levels were higher than at Mt Crushmore (10.9 mM hydrogen sulphide in the top 10 cm), and methane was found in concentrations 10.6 μM methane. Barry et al. described carbonate deposits as '*sparse*'.

Smooth Ridge/Clam Flats: tectonically-compressed (transpressional) hemipelagic sediments.

The Smooth Ridge area was uplifted and compressed by oblique movements associated with the San Gregorio Fault. The seabed here is relatively smooth, but there are some features identified by a high acoustic backscatter. Inspection of one of these areas, Clam Flats, showed that the high reflectivity is caused by widespread authigenic carbonate and modern shells. The carbonate occurs as slabs lying on the seabed, and as high-angle veins jutting out of the seabed. Cold-seep communities (dominated by vesicomyid clams, especially *Calyptogena kilmeri*) occur in two patterns: circular (1–2 m diameter) clusters which occur within an area of 100–400 m²; and linear communities (1–3 m wide by 10–30 m long) arranged along the bases of small (50–80 cm high) escarpments. Bacterial mats are common, and there are also some small pockmarks (1 m in diameter and 50 cm deep) that Orange *et al.* (1999) described as having '*small (< 10 cm) blowout rims above the seafloor*'. Push cores collected within the clam fields demonstrated the presence of high concentrations of both hydrogen sulphide (< 20 000 μM) and methane (< 7000 μM).

Cabrillo Canyon Head/Chimney field: focussing of fluid flow by seabed geomorphology.

Seeps here are confined to the steep, canyon walls where it sharply changes direction. The scarps are sometimes near vertical, individual sections being up to 8 m high, but the total scarp height being 20–30 m. Extensive carbonate slabs and chimneys (< 2 m high and < 2 m in diameter) were scattered at the base of scarp walls, and on the seabed up to 60 m away. With individual samples weighing as much as 80 kg, it is clear that these carbonates are substantial, suggesting fluid flow over a considerable time period. Carbon isotope data ($\delta^{13}C$ −55.1 to −11.8‰) indicated a mixture of microbial and thermogenic methane, plus some normal marine carbonate. Oxygen isotope values ($\delta^{18}O$ 3.21 to 6.82 ‰) indicated formation at temperatures close to that of the present seabed. However, the absence of seep fauna or any indicators of fluid flow suggested to Orange *et al.* (1999) that there was no active seepage at the time of the survey, or that seeping fluids contained too little hydrogen sulphide or methane to support a chemosynthetic community.

The descriptions of these sites, and others, show that seepage is demonstrated by MDAC, cold-seep communities, bacterial mats, and small pockmarks. However, no bubbling gas seeps have been found. Seeping fluids are either methane-rich or hydrogen sulphide-rich. Lorenson *et al.* (2002) reported that the background hydrocarbon gases in this area are generally of microbial origin, to which varying amounts of thermogenic hydrocarbons have been added; the Monterey Bay Formation, which outcrops in part of the area, is rich in methane and higher hydrocarbons. Methane-derived authigenic carbonate is more common than cold-seep communities. It seems that in many cases the carbonates indicate locations where either seepage occurred in the past but has since stopped, or there is insufficient methane

or hydrogen sulphide to support a community (Stakes *et al.*, 1999). The carbonates exhibit a broad range of stable isotope and mineral composition, which reflects local differences in sources and flux of carbon.

The widespread occurrence of cold-seep communities in a variety of settings and variable supply of methane-rich fluids suggested to Lorenson *et al.* (2002) that '*a direct, copious supply of thermogenic hydrocarbons is not necessary for their existence*'. But, the predominance of vesicomyid clams, which depend on endosymbiotic sulphide-utilising bacteria, suggested to them that '*the distribution and maintenance of H_2S in pore water is likely the most important variable in the location of cold-seep communities*'.

3.22.3 Big Sur

A large field of pockmarks was discovered off the Big Sur coast in 1998. Paull *et al.* (2002) estimated that there are about 1500 pockmarks within a 560 km² area (see Map 26) where water depths are 900–1200 m, but their surveys did not reach the southern edge of the field. Individual pockmarks are 130–260 m in diameter, have steep sides (5–6°), and are typically 8–12 m deep. They overlie the Sur Basin, another of the sedimentary basins along the California coast in which the petroleum-bearing Monterey Formation is thought to be present. The pockmarks lie on the updip end of the stratigraphic sequence, where fluids migrating through the sequence are likely to be released into the surficial sediments, but they are confined to Quaternary continental-margin sediments. Paull *et al.* noted that previous work (McCulloch, 1989) had identified seismic anomalies likely to indicate shallow gas in this area. However, the presence of shallow gas is unlikely because of the water depth; at these depths gas would be sequestered by gas hydrates, yet there are no indications of them in the form of BSRs.

In 1999 and 2000 the MBARI undertook a detailed campaign of investigations of this pockmark field. This involved geophysical surveys, coring, water analyses, use of the METS™ methane sensor, and ROV inspections (with the ROV Tiburon). They expected the results of these investigations to show the 'usual' evidence of seepage. '*However, no visual evidence of active fluid venting was seen during 19 Tiburon transects across pockmarks in five different areas within the pockmark field*' (Paull, *et al.*, 2002). Furthermore, neither samples nor the METS provided any evidence of methane in the water, and pore waters from piston cores and ROV-deployed push cores contained less than 0.10 mM methane. Analyses of sediment samples indicated no difference between the seabed inside and outside pockmarks, even ^{14}C ages of carbon contained in disseminated organic matter of sediment samples were the same. The data indicated no change in the depositional regime within the 45 000 y time period covered by the piston core samples, not even between glacial and interglacial periods. There is no indication from the lithologies or carbon isotope compositions of sampled sediments that authigenic carbonate or organic matter has contributed to the sediments. As Paull *et al.* said: '*there is no evidence that these pockmarks formed or have been active within the last 45 000 yr*'.

These results are fascinating, and the conclusion of this exhaustive and comprehensive study, that pockmarks are not necessarily formed by fluid escape, must be considered seriously. This is done in Section 7.2.2.

3.22.4 Santa Barbara Channel

The presence of natural petroleum seeps in the Santa Barbara area was first recorded in 1775 or before, when Spanish explorers noted that the local Native Americans near the site of present-day Carpinteria used tar to caulk their boats, to seal their water pitchers, and in tool and weapon making. Tar on the beaches of the Goleta area was reported in 1776 by Father Pedro Font: '*much tar which the sea throws up is found on the shores*', and Captain Cook's navigator, Vancouver, made the following observations when passing through the Santa Barbara Channel in 1792:

> The surface of the sea, which was perfectly smooth and tranquil, was covered with a thick, slimy substance, which when separated or disturbed by a little agitation, became very luminous, whilst the light breeze, which came principally from the shore, brought with it a strong smell of tar, or some such resinous substance. The next morning the sea had the appearance of dissolved tar floating on its surface, which covered the sea in all directions within the limits of our view . . .
>
> Imray, 1868

Commercial exploitation of the petroleum began in or around 1857, when illuminating oil was dis-

tilled. Most of the tar came from onshore quarries, but some was gathered from the beaches. Drilling for oil commenced onshore in 1866 and extended offshore in 1896. Gas and oil production is mainly from sandstones and interbedded siltstones of Tertiary (principally Pliocene) age that occur in faulted anticlines. The reservoir rocks are generally very porous and permeable; they are also heavily fractured and faulted. Surficial sediments (Holocene), including sand, silt, and mud, cover much of the shelf, except in nearshore areas and the crests of some anticlines. Generally they get finer to seaward.

Seabed seeps have been reported from many places in the channel, but are most prolific off Point Conception, Coal Oil Point, and Carpinteria. They have attracted particular attention not only because they may be the most vigorous seeps anywhere, nor because some have been exploited by capturing the escaping methane in 'seep tents' (see Section 11.6.2), but because of concerns that the offshore oil industry here may be affecting the marine environment. In fact, several detailed studies of the ecology of these seeps have suggested that they benefit the marine biology (see Sections 8.1.1 and 8.3.3).

The seeps are obviously related to the occurrence of petroliferous sedimentary rocks. Although some are associated with slump scarps, the majority are located where the Holocene sediments are thin or absent, especially over the crests of anticlines and along fault lines. The seeping material includes tar, heavy and light oils, and gases which, according to Eichhubl *et al.* (2000), come from shales of the Monterey Formation (Miocene). Fluid migration pathways are provided by conductive fractures and faults. There are several discrete seepage areas, three of which cover areas of approximately 1000 m^2. In these areas oil is exuded as globules weighing up to 4.0 g from 0.5 cm openings in the superficial sediments. In some places there are as many as 100 such seeps per m^2, but in others they are few and far between.

Sniffer data from this area (e.g. Figure 3.41) indicate the massive seepage rates; propane concentrations of 200 nl l^{-1} and methane concentrations of 20 000 nl l^{-1} (J. J. Sigalove, 1986, personal communication). In comparison most Gulf of Mexico seeps have a propane concentration of 10 nl l^{-1}. Measurements made by Allen and Schleuter (1970) indicated that oil flow rates in the Coal Oil Point area vary; they found that 50–70

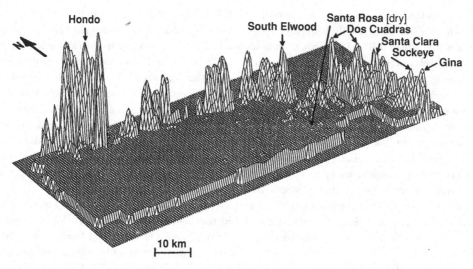

Figure 3.41 Propane concentrations (log scale) in the waters of Santa Barbara Channel; results of a Sniffer survey. Note the high concentrations above known oil and gas fields. (From Hovland and Judd, 1988; courtesy of Inter Ocean Systems, Inc.)

barrels per day flowed on average, but as much as 100 barrels per day occurs at times. Early estimates of the total rates for the area (including the Santa Monica Basin and south to Point Fermin, an area of some 2600 km²) range between 100 and over 900 barrels per day (Wilson *et al.*, 1974). Allen and Schleuter considered that tar deposition on the beach at Coal Oil Point was more than a hundred times that found at any other location between Point Conception and the Mexican border.

Seabed mounds were first described from the area in the mid 1950s by scuba-diving geologists; they were reported by Vernon and Slater (1963), who referred to them as tar mounds. Certainly, they have tar seeping from them, but some investigators described them as calcareous reefs (R. L. Kolpack, 1987, personal communication); more recent work (by Eichhubl *et al.*, 2000) described the widespread occurrence of authigenic carbonates associated with not only tar mounds, but also gas seeps. The mounds (referred to by some as mud volcanoes) occur as roughly circular mounds off Point Conception, Coal Oil Point, and Carpinteria. Off Point Conception the mounds cover an area of at least 0.6 km² and individual mounds are up to 30 m in diameter, 2.5 m high in the centre and have a scarp 3–3.5 m at the seaward edge. Tar is exuded from a central vent as whip–like strands 3.5 m or more in length (Figure 3.42). The strands may break away and float; but if flow rates are too low, gas and light petroleum fractions are lost,

they become denser than seawater and sink to become part of the mounds. Vernon and Slater (1963) described core samples whose fractures were filled with tar, and they estimated that the top 3 m of the shales were tar-filled. Individual tar mounds are located over major fractures. The tar is accompanied by oil and gas, but where the fractures have been sealed by tar the lighter fractions are barred from escaping, at least until they are under sufficient pressure to clear the blockage (see Section 10.4.2).

In a paper entitled '*The world's most spectacular marine hydrocarbon seeps*', Hornafius *et al.* (1999) described results of sonar estimates of the natural hydrocarbon emission rates from the Coal Oil Point seep field. Off Coal Oil Point there are more than 1000 individual vents emitting methane, ethane, and higher hydrocarbon gases, as well as 40 to 200 bbl oil per day. The contribution of seep gas to the atmosphere is thought to be about 100 000 m³ d⁻¹, and a similar volume enters the hydrosphere. By the time this 'seep gas' reaches the sea surface it is composed of about 60% methane, 30% air, and 10% higher hydrocarbons (Clark, 2002). However, studies by Fischer and Stevenson (1973) and Quigley *et al.* (1999) showed that there have been significant reductions in these natural emissions due to the exploitation of petroleum reservoirs (as we discuss in Section 11.6.2), so Hornafius' estimates significantly understate the 'natural' flux rate from this area. (The Coal Oil Point seeps are described in more detail by the

Figure 3.42 Tar extrusion on a 'mud volcano', Molino anticline, south of Gaviota, Santa Barbara Channel, California.   (Reproduced from Eichhubl *et al.*, 2000, with permission from Elsevier.)

University of California, Santa Barbara seeps group on the web material.)

3.22.5 Malibu Point

Earthquakes, whatever the causes, are major candidates for triggering the release through the seabed of gas trapped in shallow sediments; indeed, the triggering of a gas seep by an earthquake has been documented at Malibu Point on the coast of California, USA (Map 26). Nardin and Henyey (1978) presented evidence of gas seeps here, and Clifton *et al.* (1971) reported that, on 9 February 1971, an earthquake (Richter magnitude 6.6) occurred at San Fernando some 50 km to the northeast. At this time a 'large volume' of gas was observed bubbling from the sea surface about 500 m offshore. On the following day, when the seepage site was inspected by scuba divers, bubbles 0.5–2.0 cm across were still appearing at the sea surface. The gas was seen emanating from small craters (40 cm across and 10–15 cm deep) in a rippled, fine to very fine sand with an abundance of organic debris. A better-sorted, medium sand was present at a depth of 10 cm; it was found that bubble size was related to crater size and that the craters were sometimes several metres apart. They were arranged in a linear zone 12 m wide and 120 m long, aligned parallel to a geological structure; however, there was no evidence of faulting on the seabed.

Unlike the seeps of the Santa Barbara Channel and further north, the gas at Malibu Point was probably from a shallow zone as no hydrocarbons other than methane were detected.

3.23 OCEAN SPREADING CENTRES OF THE EAST PACIFIC

As with ocean spreading centres elsewhere, we do not propose to devote space to describing those of the east and southeast Pacific. However, research undertaken on the East Pacific Rise and the Galapagos Rift is referred to in some detail in later chapters.

3.23.1 Guaymas Basin, Gulf of California

One of the most interesting seep sites associated with 'spreading centres' is perhaps the Guaymas Basin in the

Gulf of California (Map 27). Large seep plumes were detected in the water column in a water depth of about 2000 m by towing a deep-towed acoustic sensor carrier fitted with an upward-looking 23.5 kHz echo sounder, a side-scan sonar, and a 4 kHz sub-bottom profiler. The plumes were found to have relatively slow rise rates of 12–18 cm s^{-1}. Merewether et al. (1985) explained this by assuming the rising methane to be contaminated with heavier hydrocarbons. Subsequently, Welhan and Lupton (1987) reported that the seeping fluids were light hydrocarbon gases (methane to pentane), but in the Southern Trough seeps emit petroleum condensate (C_1 to C_{40}) from a young vent mound at about 2000 m (Simoneit et al., 1990).

Simoneit et al. (1990) reported that the methane seeps are associated with pockmarks situated on eroding crests of steep anticlines, at a water depth of about 1600 m. Methane-derived authigenic carbonate and dense communities of bivalves (Calyptogena) and tubeworms (Lamellibrachia) have colonised the carbonates, which are mainly aragonite in composition. Simoneit et al. commented that this fauna is very similar to those of cold seeps in Oregon and Japan, despite being only a few kilometres from high temperature vents that lie further south. The presence of hydrothermal vents discharging fluids at < 350 °C shows that this spreading centre is active, and indicates a high geothermal gradient in the sediments. The sediments covering the new ocean crust are only 300 to 500 m thick, and the hydrothermal fluids migrating through them cause a host of reactions. About 130 vent sites have been mapped with side-scan sonar, and about 20 have been visited by the submersible Alvin (Lonsdale and Becker, 1985). Hydrothermal fluids discharge at high velocities and high temperatures (up to 314 °C) through chimneys, or diffuse at much lower temperatures (up to 100 °C) through mounds. The mineral precipitates include sulphide (e.g. pyrrhotite), sulphate (e.g. anhydrite), carbonate, and silicate minerals, but petroleum is also present in some mounds. Welhan and Lupton (1987) reported that the petroleum was derived predominantly thermogenically from organic carbon in sediments intruded by volcanic rocks of the ocean spreading centre.

3.24 CENTRAL AND SOUTH AMERICA

There is probably much more evidence for seabed fluid flow around South America than might be suggested by the number of English-language publications we have identified. Butenko and Barbot (1979) and Mahmood et al. (1980) both described seabed depressions from the Orinoco Delta, Venezuela, and we are aware of seabed fluid flow indicators off the Colombian coast. There is extensive evidence of seabed fluid flow in the Magellan Strait, between Tierra del Fuego and the Chilean mainland (A. Beckett, 2002, personal communication), but we found no papers about it. Also, boundaries between the Antarctic, Scotia, and South American plates, and the Phoenix and South Shetland microplates in the Drake Passage and the Scotia Sea (southwest South Atlantic) have attracted interest. For example, Tinivella et al. (1998) reported that there is a BSR indicating the presence of 50 m of gas-bearing sediments beneath gas hydrates in the accretionary wedge at the convergent margin between the Phoenix and South Shetland microplates, between the Antarctic Peninsula and the tip of South America. Dählmann et al. (2001) and others have reported on hydrothermal venting on the Hook Ridge, Bransfield Strait; gas hydrates or indications of them, have been reported from several locations off the Pacific coast of Central America, and a few places off South America (Kvenvolden, 2002).

In this section we focus on just a few of the relevant South American sites (indicated on Map 28).

3.24.1 Costa Rica

Off the Pacific coast of Costa Rica, the Cocos Plate is being subducted beneath the Caribbean Plate. Three major investigations of this plate margin have been undertaken. Silver (1996) introduced a collection of papers reporting on work associated with a series of dives with the submersible Alvin in 1994, in 1996 ODP Leg 170 undertook drilling here, and GEOMAR led a campaign here which was summarised by Bohrmann et al. (2002). Initially it was thought that the sediments here were accumulated by the accretion of sediments scraped off the subducting Cocos Plate, but the ODP drilling showed that the sediments of the marginal wedge are of continental, not marine, origin. Bohrmann et al. reported that a more recent interpretation is that sediments are being 'tectonically eroded' off the upper (Caribbean) plate (see Ranero and von Huene, 2000). This situation is further complicated by major slumps that have occurred on the steep seabed slope between the coastal margin and the deep (> 3500 m) ocean floor.

The largest of these, the Nicoya Slump, is about 50 km across.

There is no evidence of seabed fluid flow near the toe of the marginal wedge, but venting does occur in five different situations on the wedge.

1. Close to the toe, there is a zone of active normal faulting (McAdoo *et al.*, 1996). Seep communities (clams, bacterial mats, plus galatheid crabs) here show that either the faults are acting as fluid migration pathways, or they are barriers, preventing further fluid migration, and diverting the fluid to the seabed.

2. Mid-slope mud diapirs act as conduits for fluidised sediments. One of these, with a seabed area of 0.5 by 2 km, was found to have active venting. Pockmarks (1.5 to 4 m diameter and 1 to 1.5 m deep), abundant carbonates, and seep communities (bacterial mats, clams and tubeworms with associated egg mats, limpets, gastropods, and galatheid crabs) indicate active venting in at least one of these. The carbonates occur as < 3 m massive, yellowish grey blocks, slabs, and crusts that cover 10–50% of the seabed. Some are encrusted with corals and sponges (Kahn *et al.*, 1996).

3. Another feature, described as a mud volcano (it stands about 100 m above the surrounding seabed) showed signs of active venting with bacterial mats, vesicomyid and solemyid clams, and '*huge bunches*' of siboglinid, as well as massive carbonates, at the centre of the crest. High (94 nM l^{-1}) methane concentrations in the water above the mud volcano also indicated active venting (Bohrmann *et al.*, 2002).

4. In the upper slope, vent communities were found on the walls of canyons. As there is no indication of fractures or faulting, Kahn *et al.* thought the fluids were emerging from permeable horizons in the canyon walls. The communities comprised lamellibrachiid tubeworms, vesicomyid and mytilid clams, and bacterial mats. Where subducting seamounts have deformed the wedge sediments, for example at the Jaco and Parrita scarps, seabed fluid flow results from the deformation of the sediments. The GEOMAR team found carbonates associated with radial, 40 m high scarps above the Jaco Seamount. These were considered to indicate former long-term seepage activity.

5. Downslope, the 280 m high headwall of the Jaco Scarp exposes sedimentary rocks. These rocks are correlated with probable gas hydrate-bearing sediments identified by BSRs on seismic surveys. These outcrops were covered with bacterial mats. Further down the scarp slope seep communites were found. These comprised bacterial mats, clams, and siboglinids. Hydrocasts showed that the bottom waters contained high concentrations (up to 237 nM l^{-1}) of methane.

Major slumps and landslides cross the slope. The GEOMAR team traversed both the Nicoya Slump and the Quepos and Cabos Blanco Slides with their TV-sled (OFOS). They reported carbonates, bacterial mats, and seep fauna, and elevated methane concentrations in the bottom waters. Where the slide plane of the Quepos Slide outcrops, the bacterial mats were described by Bohrmann *et al.* as '*spectacular*', with white, blue/grey, and orange mats extended for about 220 m along the traverse. It was noted that the slide plane outcrop (at about 400 m water depth), lies above the top of the gas hydrate stability zone (at about 580–600 m). It was conjectured that decomposing gas hydrates might have been implicated in the failure of the slopes.

3.24.2 Peru

The Andean continental margin, where the Nazca Plate is subducting beneath the South American Plate, extends along most of the west coast of the South American continent. A characteristic of this margin is that there is no accretionary wedge. A small part of this margin, between 5 and 7° S (offshore Peru), was explored during the French NAUTIPERC cruise in 1991. Using the submersible Nautile, fluid-flow sites in water depths down to > 5400 m were inspected; here we focus on just one area, off the coast of Paita and Chiclayo, northern Peru. Within this zone, the seabed sloping from the continent into the Peru Trench is broken by major escarpments. These mark a major detachment fault (the Upper Slope Scarp), the headscarp of a major debris avalanche (the Middle Slope Scarp), and the subduction zone (the Trench Scarp). Fluids venting in these three zones are associated with faults. Venting fluids are rich in barium, and there are some thick barytes deposits, as well as cold-seep communities.

The strontium isotopes of fluid samples collected by Nautile were analysed in order to investigate the origins of the fluids. The results suggested that they were derived, at least in part, from the metamorphic rocks of the onshore Andean basement. However, it was not clear whether they migrated to the seabed sites through aquifers, or as a result of subduction erosion (Dia *et al.*, 1993).

The vesicomyid bivalves, *Calyptogena* and *Vesicomya*, dominate the seep communities, but there are also '*thickets*' of serpulid worms. Numerous other species were also reported by Olu *et al.* (1996). One of the vent sites on the Middle Slope Scarp covers an area of about 4000 m² (Dia *et al.*, 1993) including a clam field that covers about 1000 m². Within this clam field the mean density of bivalves is 40 individuals per m², but locally this rises to 1000 individuals per m². This represents a biomass of 30 kg m⁻² – that's about two tonnes of bivalves! (Olu *et al.*, 1996).

During a survey off Peru, Bialas and Kukowski (2001) found large plateaux of authigenic carbonate, up to 15 m high and 500 m wide at several locations in the Yaquina Basin, which contains a strong BSR. They also found living specimens of *Calyptogena* and siboglinids, indicating hydrogen sulphide occurrence, associated with methane escape. They took these indicators, and associated heat-flow measurements, as confirmation of the presence of gas hydrates.

3.24.3 The Argentine Basin

In the deep (> 5000 m) waters off the coast of Argentina (Map 28) there are large, apparently mobile, sediment waves (mudwaves) on the abyssal seabed. Piston coring and sub-bottom profiling undertaken in 1987 (reported by Manley and Flood, 1989) showed that the seabed sediments in one of these areas, the Zapiola Drift, are predominantly olive-green muds with thin black mud laminae. The lower parts of the cores commonly smell of hydrogen sulphide. Organic carbon (< 1.0%) is thought to be of terrigenous origin because $\delta^{13}C$ values lie in the range −21.9 to −23.3‰.

Seismic profiles showed indications of BSRs, and gas voids were found at depths of 10–17 m below seabed in the cores. Analyses showed that the gas is mainly methane, with traces of ethane and air. Manley and Flood thought this to be of microbial origin, generated within the shallow, organic-rich sediments. Calcite nodules found in some of the cores ($\delta^{13}C$ −14.9 to −19.4‰) were thought to be authigenic.

At about 700 m depth on the continental slope off the southeast coast of Brazil there are some large (1 km diameter and 100 m deep) pockmarks; mounds about 20 m high and 180 to 360 m wide stand on the edges of these pockmarks. Sumida *et al.* (2004) reported that deep water corals live on these mounds.

3.24.4 The Mouth of the Amazon

At present the sediments transported by the River Amazon are deposited in a submarine delta on the continental shelf. However, since the Miocene, sediments also have been carried across the continental shelf to be deposited as a deep-sea fan, particularly during glacial periods when the sea level has been lower than it is now. The river sediments are rich in organic matter and nutrients, and the biological productivity is high in the waters of the shelf. In these conditions it is not surprising that seismic and coring surveys have identified a large area (31 000 km²) of gassy sediments in the submarine delta. Near the river mouth gas causes the brightening of reflectors in sandy sediments. In the distal portions of the delta, where the sediments are finer-grained, acoustic turbidity is more common. Gas was also seen as bubbles in piston cores and '*degassing fractures*' on X-ray photographs of cores (Figueiredo *et al.*, 1996). These authors showed that gas generation and accumulation occurs in areas with the highest sediment accumulation rates. They reported that methane concentrations are < 100 μM l⁻¹ of sediment beneath the sulphate reduction zone, except where it is trapped beneath a near-seabed over-consolidated muddy layer. In such cases concentrations of up to 1000 μM l⁻¹ were measured.

Mapping by the Lamont-Doherty Geological Observatory in the mid 1980s (Manley and Flood, 1988) identified extensive BSRs in the sediments of the deep-sea fan, and identified '*small diapirlike structures*' that '*protrude upwards from the BSR region*'. Some of these are domes standing 10–20 m above the surrounding seabed. The interpretation of the BSR as evidence of gas hydrates was supported by the recovery of gassy cores.

3.25 THE CARIBBEAN

The Caribbean (Map 29) hosts two environments for seabed fluid flow: there are active submarine volcanoes such as Kick 'em Jenny, and mud volcanoes, cold seeps etc., some, such as the tar lakes of Trinidad, associated with petroleum. These are both associated with the convergence of the South American and Caribbean plates. We focus on just two 'cold' sites.

3.25.1 Barbados Accretionary Wedge

In front of Barbados Ridge the South American Plate is subducting beneath the Caribbean Plate, and the sediments overlying the Atlantic Plate are being scraped off to accumulate as an accretionary wedge. A considerable amount of research has been conducted in this area. This has resulted in a large number of publications, for example: Biju-Duval et al., 1982; Zhao et al., 1986; Brown and Westbrook, 1987, 1988; Moore et al., 1987; Le Pichon et al., 1990; Henry et al., 1990, 1996; Martin et al., 1996; Faugères et al., 1997; Olu et al., 1997; Lance et al., 1998; Aloisi et al., 2002.

In 1987 Brown and Westbrook reported a patch of 'rough seafloor and disturbed reflectors (water or gas escape features)' on the seabed here. The following year the same authors reported that a total of about 450 mud diapirs and volcanoes had been mapped here. The GLORIA (long range side-scan sonar) data showed they were up to 17 km long and 1 km wide, and they were concentrated along the crests of ridges – or maybe the ridges themselves were diapiric? Brown and Westbrook (1987) assumed that mud diapirism is one of the processes that allows the accretionary sediment complex to dewater, but methane expulsion also plays a part; it seems that the process of defluidising is still in progress. Leg 110 of the ODP penetrated the decollement of the Barbados Accretionary Wedge, and reported that, even after a complex history of deformation, the sediments retained a high water content (20–40%). Moore et al. (1987) also reported that evidence for fluid migration along fault zones included high temperatures, low chloride concentrations in pore water, and high methane concentrations. The origin of the methane was assumed to be thermogenic, produced at a depth below seabed of over 2 km. It was found that methane-bearing fluids at and below the decollement travel more than 25 km beneath the shallowly inclined decollement beds before they are expelled somewhere beyond the deformation front of the accretionary wedge.

The mud volcanoes are on a 1 km high basement scarp that extends away from the deformation front. The most distant, Mt Manon, is 25 km from the front. Lance et al. (1998) noted that the mud volcanoes expel mud containing fauna of Oligocene to Quaternary age.

They estimated the age of the features by identifying the ages of the surrounding and underlying sediments; ages vary between 40 000 y (Volcano A) and about 750 000 y (Volcanoes F and J). Atalante is at least 65 000 years old. These ages allow for some interesting speculation about the relationship between deformation and mud volcanoes:

> The deformation front is expected to advance about 15 km eastward in 750 000 years (Le Pichon et al., 1990). Because 15 km is the length of the zone between the deformation front and Atalante where most mud volcanoes are found, the mud volcano field may well be a permanent feature.
>
> Lance et al., 1998

The morphology of the mud volcanoes varies between 'mud pies' (e.g. Atalante and Cyclops) and conical mounds (e.g. Mt Manon). The mud pies are flat-topped circular to elliptical structures. Atalante covers an area of about 505 000 m^2, with a central depression (or 'eye') 2 m deep and 180 m in diameter. The majority (60%) of the feature's surface is essentially flat, but there are concentric ridges in the outer portion, and it is surrounded by a 17 m deep depression. Cyclops is a little smaller (600 m across), but has a similar concentric zonation. In contrast, Mt Manon is a conical mound about 1000 m in diameter and standing 180 m above the surrounding seabed, with steep (25°) sides; it also has a surrounding depression (15 m deep). Fluid venting has been observed from the 'eyes' of Atalante and Cyclops, and from the central flat top of Mt Manon. It is thought that these central areas are also the sites of active mud expulsion (Lance et al., 1998). It seems that there is a correlation between these facts.

- The convection of mud results in high temperature gradients in the centre of the mud volcanoes; temperatures of 21 °C 1.8 m below seabed, and a gradient of $> 1 °C \, m^{-1}$ was recorded in the centre of Atalante (Le Pichon et al., 1990; Olu et al., 1997).
- The high heat flow results in the dissociation of gas hydrates, which in turn accounts for the low chlorinity of the pore fluids (for example, in Atalante the chloride content of pore fluids is 220 mM, compared to 560 mM in the seawater).
- The highest fluid fluxes were observed in these central areas.

Henry et al. (1996) estimated that Atalante emits water at a rate of at least 50 000 m^3 y^{-1}, and methane at 3×10^6 kg y^{-1}.

The cold-seep communities from several of the mud volcanoes were investigated by Olu et al. (1997). They reported that these communities are dominated by vesicomyid bivalves (Calyptogena), by large bushes of a sponge (genus Cladorhiza), and by long, tubicolous polychaetes. In addition there were several other macrofaunal species, and high densities of meiofaunal species, especially nematodes, copepods, polychaetes, and kinorhyncha. The vesicomyids were found in small (<2 m diameter), dense (<150 individuals m^{-2}) clusters, and in dispersed (<10 individuals m^{-2}), but widespread populations. The presence of large 'bushes' of sponges is perhaps even more interesting because of their 'exceptional size': 1.5–2 m diameter and 40 cm tall, and 'several hundreds to thousands of individuals' (Vacelet et al., 1995, 1996). This is because they host methanotropic endosymbionts, as we explain in Sections 8.1.2 and 8.2.2.

The distribution of dense, and dispersed clam beds, and sponge bushes must reflect the availability of reduced sulphur and methane, respectively. For example, on Atalante Olu et al. (1997) reported that the central 'eye' is colonised by methanotrophic sponges, indicating the upwelling of methane-rich fluids. Outside this central area, sulphur-oxidising clams occur in dense beds where there is a focused flow of sulphur-rich fluids, probably associated with pore fluid convection. Elsewhere within the confines of the feature, a more dispersed flow of sulphur-rich fluids supports only dispersed clam beds. Differences between the distributions and proliferation of cold-seep communities between the various mud volcanoes studied by Olu et al. led them to suggest that the 'mud pies' and the 'conical mounds' were 'at different stages of activity'.

'Diagenetic carbonates' were found on all the mud volcanoes inspected by Nautile. On some, the carbonate blocks ranged in size from a few centimetres to a few decimetres across, but on El Pilar 'the expelled sediments have been entirely lithified from a thickness of several metres' (Aloisi et al., 2002). Single blocks up to 10 m wide in the slope debris attest to the extent of MDAC formation. These carbonates are most abundant where seabed fluid flow is focused. The majority have δ^{13}C values lower than −30‰, indicating a methane carbon source. However, some (e.g. those from El Pilar) have values of between −10 and 10‰ which suggested to

Aloisi et al. (2002) that they were derived from the mixing of seawater-derived carbon and carbon derived from the oxidation of organic matter. Some of the carbonates they sampled were also enriched in ^{18}O, possibly due to their association with decomposing gas hydrates.

Lance et al. (1998) suggested that the gas hydrate dissociation influenced the morphology of the mud extrusion features. Once the water derived from dissociating hydrates raises the water content to about 70% 'the mud behaves as a liquid and mud volcanoes of the "mud pie" type are obtained'; episodic mud flows make the central vent area unsuitable for chemosynthetic communities, which live on the outer areas. However, if the water content is lower, then 'the mud behaves plastically and forms mounds', providing a more suitable place for the seep communities to grow.

3.25.2 Birth of Chatham Island, Trinidad

An unusual natural event, demonstrating the dynamics of mud diapirism, occurred in Erin Bay, Trinidad, in August 1964. A new island appeared above the sea about 2.5 km from Chatham town on the south coast. It rose to a maximum height of 8 m above sea level and had an original area of about 4.25 ha at low tide (Higgins and Saunders, 1967). Extrusion of plastic mud took place during the first two days, beginning from the east and extending along the axis of the southern anticline, a tectonic feature that shows mud volcanic activity on the adjacent land. Although gas was given off in quantity, Higgins and Saunders believed that tectonic (thrust compression) movement was the driving force causing the event. From samples taken on the island, the gas was found to be methane. Like the Apsheron Peninsula (Azerbaijan), the Chatham area of Trinidad is well known as a prolific hydrocarbon-producing area. Chatham is only 20 km from Pitch Lake, a natural lake of heavy crude oil and asphalt.

The island disappeared below sea level eight months after its birth. Chatham Island comes and goes. It is a mud volcano that is normally submerged, but appears every few decades during, and for a short time after, eruptions. Further details of these are dicussed in Section 7.3.4.

3.26 GULF OF MEXICO

When the Spanish explorers reached the coasts of the Gulf of Mexico they found that the Native Americans in some areas used natural tar or pitch to waterproof their canoes and pottery. During the fifteenth and sixteenth centuries, the Spaniards used tar from beaches of Texas and Louisiana to caulk their ships. Throughout the subsequent period of time, there have been reports of oil slicks on the waters of the Gulf. In the 1900s all ships crossing the Gulf were asked by the New Orleans Hydrographic Office to complete oceanographic reporting forms. Many reported patches of oil, some measuring more than 150 km in length and several kilometres in width. A map produced by Captain Soley in 1910 (and subsequently republished by Texas A & M University), showed the distribution of pelagic tar lumps and drifting oil on the sea surface. A survey conducted by Hildebrand in 1954, before appreciable quantities of oil were produced in the area, concluded: '*petroleum residues occur on all beaches in the northern Gulf from Washing Beach, Tamaulipas, Mexico to Fort Walton Beach, Florida*' (quoted in Geyer, 1979). These occurrences correlate well with oil-seepage locations along the deep-water portion of the continental slope (MacDonald, 1998). Oil pollution resulting from man's activities, from shipping and from petroleum exploration and exploitation (including the Ixtoc blowout in the Gulf of Campeche, Mexico, in 1979) must be seen against this background of natural oil 'pollution'.

The Gulf of Mexico opened during rifting as Pangea, the supercontinent, started to break up in the Triassic. The basin contains deeply buried marine carbonate source rocks and extensive salt deposits, both of Mesozoic age. Rapid deposition during the Tertiary led to salt movements. There are salt diapirs and domes both onshore (in Texas, Louisiana, Mississippi, and Alabama) and on the adjacent continental shelf. Beyond the edge of the shelf and out to a depth of about 3000 m, the edge of the Sigsbee Escarpment, there are extensive thrusted salt sheets (Sassen *et al.*, 1994). Faulting associated with these salt structures, and with other movements within the thick sedimentary sequence, has provided migration pathways for hydrocarbons to move upwards into younger (Miocene to Pleistocene) reservoirs. These pathways also enable hydrocarbons to reach the seabed, which is why there is such extensive seepage.

There has been so much exploration of the seeps in the Gulf of Mexico over such a long time that it is impossible to provide a comprehensive review here. Additional information is provided in the presentation by McDonald, see website. Research, conducted by various research groups (perhaps most notably from Texas A & M University) and oil companies, has benefited from extensive use of every type of vehicle from manned submarines (Alvin, Johnson Sea Link, and NR-1) to the Space Shuttle. Seeps have been documented from the deep waters (> 2000 m) at the base of the continental slope, across the continental shelf and into the coastal waters of the Mississippi Delta. See, for example, the following.

- In a review of features associated with seeps recorded offshore Louisiana, Sieck (1973) reported that '*circular seabottom depressions*' were present in several licence blocks representative of a shallow marine environment overlain by '*soft sediments*'. They are up to 46 m in diameter and 1.5 to 3.0 m deep. Sieck also reported seeps ('*full column bubble clusters*') and mud volcanoes.

- Watkins and Worzel (1978) estimated the presence of more than 19 000 individual seeps in the 'Serendipity Gas Seep Area', offshore southern Texas. They reported two types of evidence of gas within the shallow sediments, as seen on 3.5 kHz seismic profiles: 'smearing' (acoustic turbidity) and 'wipeout' (acoustic blanking). Of these they found 'smearing' to be associated with gas seeps whereas 'wipeouts' were generally not. They considered that the seeps were fault controlled and reported that 'mud volcanoes', some with gas seeping from the top, were common in the area. The pore fluids of eight sediment cores collected from this area did not contain anomalous concentrations of hydrocarbons (Brooks *et al.*, 1987). However, anomalous concentrations were found at three locations in an area further south. Methane, ethane, and propane concentrations in both seawater and pore waters indicate that this gas is thermogenic in origin, although molecular fractionation in the sediments is reducing the relative concentration of the higher hydrocarbons in the seawater. Watkins and Worzel could not provide positive evidence of the origin of the gas in the Serendipity area. However, they considered that the seeps originate from faults; consequently it is probable that the gas is of thermogenic origin.

- Geyer (1979) demonstrated that tar from seeps off the Yucatan Peninsula and Laguna Taruahua can drift northwards to appear on the beaches of Texas and Louisiana.

- Two hundred and forty seeps were recorded in an area off Panama City, Florida by Addy and Worzel (1979), who estimated that these were representative of some 8600 seeps in their 2350 km² study area. The area contains regions of acoustic blanking and acoustic turbidity; mud volcanoes, which according to Addy and Worzel are 'common features in swampy oil fields' in this area, are also present. The gas seeps are associated with extensional faults at the crest of a dome, which was possibly caused by salt tectonics. No analyses were reported, so the nature of the gases is unknown. Addy and Worzel argued that for the production of microbial methane one would expect a reducing, anoxic environment; however, in this area there is an abundance of infauna. Reef growth and a highly diversified shell fauna indicate aerobic conditions, so they consider it 'reasonable to believe' that the gas is thermogenic.
- Bryant and Roemer (1983) stated that sediments of the continental shelf and slope of the northern Gulf, particularly near the Mississippi Delta, are 'extremely gassy'. They presented a series of seismic profiles illustrating acoustic turbidity, acoustic masking, enhanced reflections, and bright-spots. Their evidence came from 41 licence blocks in the following areas: Eugene Island, East Cameron, West Cameron, Ship Shoal, South Marsh Island, Vermilion, Grand Isle, and South Timbalier; gas seeps are shown from 15 blocks, and 'mud lumps' or 'mud mounds' in at least 10 of these. No doubt their evidence is no more than representative of the areas covered, at least some of which may contain large numbers of seeps. South Marsh Island block 142, for example, contains 'literally hundreds' of gas seeps.
- Some seeps and related features have been located in very deep waters. Two sediment cores from the Nowlin Knolls area, in about 3000 m of water, were found to contain tar (Bouma and Rezak, 1969; Foote et al., 1983). O'Connell (1985) reported pockmarks from a water depth of about 3000 m at the foot of the Mississippi Fan.

Map 30 gives an indication of the extent of seabed fluid flow and associated features. This map has been compiled from various sources in addition to those mentioned above, including:

- a distribution map of seeps, gas hydrates, and cold-seep communities, presented by Milkov and Sassen, 2001;

- an inventory of cold-seep communities in the northern Gulf, compiled by MacDonald et al. (1996), who identified 43 individual sites with cold-seep communities.

MacDonald and Leifer (2002) reported that 'comprehensive remote sensing surveys indicate there are about 350 constant seeps that produce perennial slicks of floating oil in consistent locations' which 'must contain a sizable fraction of gas' (see also MacDonald et al., 2002).

Migration and seabed features
Sassen et al. (1993a) saw clear links between petroleum reservoirs and seeps evidenced by 'geophysically obvious migration conduits' extending from depth to the seabed, and by the presence of oil seeps and chemosynthetic communities over shallow salt structures and faults close to many oil fields, e.g.: Auger, Bullwinkle, Cooper, Joillet, Mars, Popeye, Vancouver, and Ram-Powell. Roberts and Carney (1997) emphasised the importance of the close relationship between faults and seeps. They noted that seabed features associated with seabed fluid flow include a spectrum ranging from 'features that result from rapid expulsion of large volumes of fluid much frequently containing crude oil and hydrocarbon gases' to 'features related to very slow seepage of hydrocarbons or hydrocarbon-charged (gas and/or crude oil) waters'. They considered that these might result from single or multiple venting episodes, or continuous venting, and that some individual sites may change in venting style over time. These features include pockmarks and mud volcanoes of various sizes and a range of water depths. Some of the most spectacular are the 'giant mud mounds' (which we would now call mud volcanoes). They are large gas-exhaling features; some measure 2 km across and rise over 100 m above mean seabed level. It was on one of these that deepwater hydrocarbon-related cold-seep chemosynthetic communities were first found in the mid 1980s (see Section 8.1.2).

Some of the mud volcanoes discharge brine as well as hydrocarbons. Because the brines are denser than normal seawater, they collect in summit craters, but may overflow to run down the flanks of the mud volcano. There are numerous pockmarks associated with seep areas, sometimes close to mud volcanoes or aligned along faults. Prior et al. (1989) described one spectacular example as a 'large explosion crater'; this feature is discussed in more detail in Section 7.3.4.

Cold-seep biological communities

Seeps in the shallower waters of the shelf are characterised by bacterial (*Beggiatoa*) mats, described as white or orange in colour. However, cold-seep communities that include macrofauna associated with chemosynthesis seem to be confined to waters more than 280 m deep. There are known to be at least 43 deep-water sites in the northern Gulf (MacDonald *et al.*, 1996). Seep fauna include siboglinid tubeworms, bathymodiolid mussels, and lucinid and vesicomyid clams. Other species may be associated with specific environments, for example, a polychaete referred to by Fisher *et al.* (1998) as '*the ice worm*', seems to inhabit gas hydrate. The composition of each community is dependent upon the availability of the energy sources for chemosynthesis: hydrogen sulphide and methane. Similarly the fluid flux determines the success of each community: in some places, such as at Bush Hill, tubeworms grow in dense, bush-like communities, with individuals growing to nearly 2 m in length.

Fluids and hydrates

In seep areas, seabed piston cores have yielded as much as 15% oil by weight, and gases with thermogenic carbon isotope values. Gas hydrates have also been recovered in piston cores. Some contain ethane, pentane, and butane as well as thermogenic methane ($\delta^{13}C$ −43 to −56‰), but others contain gas with $\delta^{13}C$ values of −66 to −71‰, indicating a microbial origin (Sassen *et al.*, 1993b). Visual inspections by submarine have also shown that gas hydrates are exposed on the seabed in some places, including on some mud volcanoes. The hydrates take the form of nodules, lumps, and mounds. At Bush Hill, for example, hydrates: '*form sediment-draped mounds 30–50 cm high and up to several metres in width, with exposed hydrate visible at the edges of the mounds*' (Sassen and MacDonald, 1997). MacDonald *et al.* (1994) reported a: '*steady stream of tiny bubbles and occasional bursts of larger bubbles escaped along the edges*' of the hydrate, and that '*small pieces of hydrate broke free and floated away*' when they probed around the mound with the submarine's manipulator.

This hydrate exposure was visited in both July 1991 and August 1992. MacDonald *et al.* (1994) noticed that between these dates the shape of the mound changed, indicating that '*the hydrate was decomposing*'. By deploying a 'bubbleometer', they measured the rate of gas release over a 44 day period. They found that gas releases only occurred when the seawater temperature increased to 8 °C, and stopped again when it fell back below this temperature. They concluded that temperature variations like this are probably related to warm-water eddies that move along the shelf edge a few times every year. These findings have important implications as they demonstrate the ability of seawater temperature increases to cause gas hydrate dissociation.

Milkov and Sassen (2001) presented results of a detailed study of gas hydrate distribution in the northwestern Gulf. Their study covered an area of about 59 000 km². It extends from the 440 m bathymetric contour, the shallowest depth from which gas hydrates have been reported, downwards. They considered this area to be the most suitable for gas hydrate formation; apparently there is no definite evidence of hydrates elsewhere in the US EEZ (Exclusive Economic Zone) of the Gulf. Within this area they concluded that gas hydrates occur:

1. on the rims of salt-flanked mini-basins, where thermogenic and microbial gas hydrates occur in '*structurally focussed*' accumulations, that is salt deformation structures and faults;
2. as disseminated microbial hydrates within mini-basins.

The volumes of gas hydrates in these two contexts were estimated as about $8–10 \times 10^{12}$ m³ C_1 to C_5 hydrocarbon gas (at STP), and about $2–3 \times 10^{12}$ m³ of microbial methane, respectively.

Asphalt mud volcanoes

In a 5000 km² area of the abyssal plain of the southern Gulf of Mexico there is a group of 22 topographic features known as the Campeche Knolls. They rise 300 to 400 m above the seabed in about 3200 m of water. MacDonald *et al.* (2004) reported that light oils coming from eight of these knolls produce oil slicks on the sea surface, and sediment lobes extend up to 4 km from nine of them; in one case this lobe consists of asphalt. High concentrations of hydrocarbon gases and isotopically heavy carbon dioxide were detected in the sediments on the knolls; a $\delta^{13}C$ ratio of −7.5‰ suggests that the carbon dioxide has a magmatic source.

The Florida Escarpment groundwater seeps

Submarine groundwater discharge occurs on the Florida Escarpment (see Map 30) at the contact between the Cretaceous limestones of the Florida Platform and the hemipelagic sediments of the abyssal plain in the Gulf of Mexico at a depth of about 3270 m (Commeau et al., 1987). The water is sulphidic and hypersaline, but the temperature does not contrast with that of the sea bottom water, and there is no significant hydrocarbon content. However, the sulphide content is sufficient to attract chemosynthetic biological communities (described in Section 8.1.5), and to cause the precipitation of pyrite (see Section 9.5.1).

3.27 THE EASTERN SEABOARD, USA

3.27.1 Cape Lookout Bight

Cape Lookout Bight (shown on Map 31) is a 1 km^2 coastal marine basin with water depths of 2–9 m. It has been semi-enclosed by the Outer Banks of North Carolina, and is a trap for fine-grained sediments leaving the lagoons landward of the Outer Banks, which accumulate at a rate of > 10 cm y^{-1}. There are extensive seagrass beds in the lagoons; productivity is high, so the sediments are rich in organic matter. Studies of the biogeochemistry of the Bight have continued for over 20 years. The following brief summary (based on Kipphut and Martens, 1982; Martens and Klump, 1984; and Martens et al., 1998) is inadequate to do justice to the pioneering work done here, but we will refer to it again in later chapters.

Methane is generated in the sediments of the Bight, production rates being significantly higher in summer than winter. At this time sediment pore waters are saturated with methane to within 10 cm of the seabed. Methane escape to the water column is achieved by a combination of diffusion and bubbling, but bubbling only occurs in summer months, and is triggered by the falling tide. The bubbles rise to the seabed through 'bubble tubes'. In the winter, methane concentrations increase linearly with depth, indicating diffusion to the water column. It is significant and unusual that methane oxidation does not occur in the summer as production rates are matched by the flux into the seawater.

3.27.2 Atlantic Continental Margin

Geophysical mapping by the USGS and others has proved that gas hydrates are present over an enormous area (shown on Map 31) on the continental rise between 31 (offshore Georgia) and 39° N (offshore New Jersey) – a distance of about 1000 km. The hydrates are concentrated in areas of rapid sediment deposition, and in association with a line of salt diapirs, in water depths from 2000 to > 5000 m. The potential of gas hydrates as a future energy resource has encouraged research in this area, and so have the major slope failures (e.g. the Cape Fear and Cape Hatteras slides).

Blake Ridge

Blake Ridge, off the coast of South Carolina, is a broad ridge of deep-sea contourite sediments that have accumulated since the Oligocene because of the interaction of ocean currents. One of the earliest reports of pockmarks in the deep sea resulted from a detailed side-scan sonar survey at a depth of 4400–4800 m on Blake Ridge. During the survey, described by Flood (1981), 40 pockmarks with diameters of 30–40 m and depths of about 5 m were found within a 100 km^2 area in fine-grained cohesive sediments; they also found some long seabed furrows. However, because the Ridge is underlain by gas hydrates it has attracted a great deal of research attention. This includes drilling campaigns by DSDP (Deep Sea Drilling Programme) and, more recently, ODP Leg 164. The hydrates vary in thickness, in places being > 45 m thick. In the ODP boreholes finely disseminated hydrates occupied between 1% and 4% of the sediments between 200 and 450 m below seabed; but there were denser concentrations, the largest being a nodule layer > 30 cm thick. From seismic data Dillon et al. (1993) estimated that there is an area of about 3000 km^2 where the total thickness of hydrates is at least 30 m. The volume of hydrate in this area alone is about 0.113 $\times$ 10^{12} m^3. This is equivalent to about 18 $\times 10^{12}$ m^3 of methane at surface temperature and pressure. However, this is only part of the story.

During ODP Leg 164 three boreholes were drilled, two of them penetrating the BSR, but all three encountered disseminated gas hydrates at subseabed depths of between 200 m and 450 m (the depth of the BSR). Seismic velocities as low as 1400 m s^{-1} were measured below the BSR in one hole, and methane was collected with a pressurised core sampler from below the depth at which gas hydrates are stable. Not only did this work demonstrate that free gas exists beneath the hydrates but, according to Paull et al. (1998), it also indicated that '*the amount of free gas in the sediments beneath the base of hydrate stability may exceed the volume stored in the hydrates above'.* Several estimates of the total methane resource have been made; these vary between 18 and 80×10^{12} m^3 (the lower estimates including only gas hydrates, the higher ones also including underlying free gas). Dillon and Max (2000) remarked: *'with our present state of knowledge, the fact that these numbers are all within a factor of 10 is actually rather encouraging'.*

The Blake Ridge Diapir

A large salt diapir, the southern-most of a line of 25 diapirs that extends along the coast, lies beneath the landward end of the ridge. This was where Site 996 of ODP Leg 164 was drilled. Within an area of about 7 km^2 above the diapir the sediments are domed and faulted. The faults extend right up to the seabed, despite active sedimentation, so the diapir must still be moving. In this area the BSR is elevated towards the seabed. According to Taylor *et al.* (2000) this may be because of a higher heat flow associated with the high thermal conductivity of the salt, with fluid expulsion or with an increase of pore water salinity. The seabed above the diapir is pockmarked, and seabed photographs and dives in Alvin have shown that there are authigenic carbonates and cold-seep communities here (Van Dover *et al.*, 2003). Seep plumes rise about 320 m from the seabed. The seeps are associated with the faults that must be acting as gas migration pathways (Paull *et al.*, 1998; Taylor *et al.*, 2000). Clearly the gas from deep beneath the seabed is escaping today, but Naehr *et al.* (2000) suggested that seeps have been active for *'at least 600 000 years'*. They argued that MDAC is formed near the seabed, so the depth distribution of MDAC indicates the history of formation, and hence seepage.

The Blake Ridge Depression

On Blake Ridge there is a 25 km wide depression that, unlike the surrounding area, is not underlain by a BSR. Dillon *et al.* (1998) concluded that it was caused by a structural collapse. They suggested that overpressured pore fluid and gassy muds just beneath the GHSZ escaped, causing a massive loss of sediment volume (> 13 km^3), along with an enormous volume of gas which, according to Dillon and Max (2000), *'might have increased the methane content of the atmosphere by 4% on the basis of gas concentrations sampled by drilling in*

the area'. However, a detailed seismic study reported by Holbrook *et al.* (2002) demonstrated that the feature is actually a large field of sediment waves characterised by a complex pattern of sedimentation and erosion. According to this new interpretation the absence of gas is explained by the intersection of the free gas zone at the base of the GHSZ by permeable sediments of the sandwave field. They provided a gas-escape pathway for enormous volumes of gas, estimated by Holbrook *et al.* to exceed *'by an order of magnitude'* the volume suggested by Dillon *et al.* Holbrook *et al.* concluded that gas escape occurred over the last two million years; they said that escape was not continuous, but they could not tell if it had been rapid, gradual, or episodic.

Cape Fear Slide

Schmuck and Paull (1993) described the Cape Fear Slide as *'the largest mass movement feature that has been mapped on the US Atlantic Margin'.* It was initiated over one of the salt diapirs, but it extends for over 300 km down the continental slope. It is just one of several overlapping seabed slides and slump scars located over the gas hydrate area described above. This is significant, and not just a coincidence. At the head of the Cape Fear Slide, as at the Blake Ridge Diapir, the BSR rises towards the seabed over the diapir. This feature is discussed in Section 11.2.1.

3.27.3 Chesapeake Bay

Chesapeake Bay was formed by the drowning of the lower reaches of a river system, the Susquehanna River and some of its tributaries, incised during the last glaciation. It stretches more than 260 km from the open sea. As early as 1969 it was known that there was gas in the sediments of the bay (Reeburgh, 1969), but detailed geophysical mapping and seabed sediment coring showed that about 30% of the bay is underlain by gas (Hill *et al.*, 1992). In the inner reaches of the bay, north of 39° N the gas reaches up to the seabed. Hill and his colleagues attributed this to *'low salinity modern diagenetic processes'.* Closer to the open sea, the distribution of acoustic turbidity shows a dendritic pattern that clearly suggests that gas is confined to the drowned river channels (Figure 3.43).

On X-ray images of core samples collected from the gassy areas, Hill *et al.* noted elongate gas voids. These

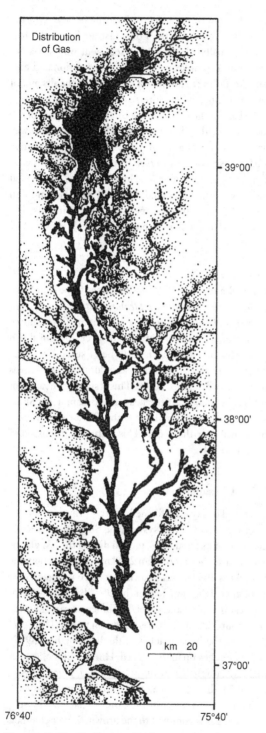

Figure 3.43 The distribution of shallow gas (dark shaded areas) in Chesapeake Bay, eastern USA. (Reproduced from Hill *et al.*, 1992 with permission from Elsevier.)

were 1–2 mm across and 3–5 mm long, the long axes standing nearly vertical in the cores. They reported that these voids often occupied as much as 15 to 20% of the cross–sectional area of the cores. In these highly gas-charged sediments the voids generally occupied about 6% of the total sediment volume. Gas analyses showed that methane was the dominant gas. One 4.5 m long core contained gas voids from top to bottom, and they estimated that gas extended right up to the seabed in 60% of the area in which gas was present. Hagen and Vogt (1999) suggested that the gas was no more than 2 m thick, and showed that the top of the gassy layer varied in depth according to the seasons. The gas only reached the seabed during the summer, when the increased water temperature affected methane solubility. During the winter months the gas front sank by about 0.5 m. When the gas reached the seabed Hagen and Vogt detected high reflectivity on side-scan sonar data. They attributed this either to the side-scan 'seeing' gas bubbles, or to surface roughness caused by escaping bubbles.

Hill *et al.* suggested that, as sea level rose after the last glaciation, sediments with an organic carbon content '*probably exceeding 5%*' were rapidly deposited. As this occurred in a freshwater environment, sulphate was absent, so methanogenesis was rapid. However, Hagen and Vogt suggested that the rapid genesis could be accounted for by the influx from below of fresh groundwater.

3.27.4 Active pockmarks, Gulf of Maine

Pockmarks were discovered in Penobscott and Belfast bays, Gulf of Maine during a regional geological survey by the USGS and detailed surveys for a power cable route by the Nova Scotia Power Corporation (Scanlon and Knebel, 1985, 1989; McClennen, 1989). They occur where acoustically transparent Holocene marine muds unconformably overlie coarse Upper Weichselian (Wisconsinan) glaciomarine sands and gravels and Palaeozoic bedrock. The Holocene sediments were deposited in an enclosed bay behind a bedrock sill that restricted water circulation and encouraged rapid sedimentation and the preservation of organic matter. Acoustic turbidity, which is widespread, may be attributable to gas formed by the decomposition of this organic matter. The pockmarks, which are entirely restricted to the area of muddy seabed, were described by Scanlon and Knebel (1985) as nearly circular in plan, normally 20–60 m in diameter and 5–10 m deep (although diameters up to 200 m and depths of 1–30 m have been recorded). The bottoms of some extend down to, but not across, the unconformity marking the boundary with the underlying coarse sand and gravel. The association of pockmarks with shallow gas suggests that gas escape is responsible for these pockmarks. However, Scanlon and Knebel, noting the presence of acoustic turbidity beyond the pockmark area, consider that this may not be the case. A possible alternative explanation, suggested by Kathy Scanlon (1987, personal communication), is that they were formed by the flow of freshwater from the Holocene–Pleistocene unconformity, as there is an aquifer below it in nearby locations onshore. Whatever their origin, the features are fresh in appearance and unfilled by recent sedimentation, which suggests present-day activity.

During a later survey reported by Kelley *et al.* (1994), pockmarks were discovered in Belfast Bay; '*they are found in densities up to 160/km², are up to 350 m in diameter and 35 m in relief, and are among the largest and deepest known*'. The largest pockmark occurs at a general water depth of only 30 m (which is deep for the Gulf of Maine). Kelley *et al.* also recorded a pockmark (on side-scan sonar) that emitted bubbles and sediments, and recounted fishermen's reports of '*enormous numbers of gas bubbles*' erupting from the bay in 1992. Repeat surveys in 1998 and 2000 provided more evidence that the pockmarks in Belfast Bay are active. Gontz *et al.* (2001) stated: '*Comparison of the 2000 survey with the 1998 survey revealed significant changes in the seafloor over the 2-year period. Creation and destruction of pockmarks and fishing drag marks was clearly evident*'. Another indication of activity is the steepness of pockmark sidewalls. Slopes of 30° or more have been measured, but were shown to be unstable; apparently slope failure could be induced easily with a little encouragement from a passing submersible! It is very unlikely that such slopes could remain intact for long (Daniel Belknap, 2004, personal communication).

3.28 THE GREAT LAKES

3.28.1 Ring-shaped depressions, Lake Superior

Ring-shaped lakebed depressions were described from western Lake Superior by Berkson and Clay (1973)

and Flood and Johnson (1984). They are 100–300 m in diameter, and the depressions forming the outsides of the rings are generally 2–5 m deep, the lakebed inside and outside being at the same level (Figure 3.44). They are cut into the 4–5 m thick, acoustically transparent, varved post-glacial lacustrine clays. The origin of these curious features is uncertain. Possible formation mechanisms include syneresis (proposed by Berkson and Clay) and sediment dewatering (Johnson, 1980; Flood and Johnson, 1984). However, a survey using pseudo-three-dimensional high resolution (28 kHz) and multibeam bathymetry provided new evidence. This included evidence of small normal faults, with throws of '*usually less than 50 cm*', and led Wattrus *et al.* (2001) to suggest that the features represent a polygonal fault system similar to those described from Tertiary sediments of the North Sea by Cartwright and Dewhurst (1998). If this is correct, this system may still be active, enabling sediment dewatering.

3.28.2 Pockmark-like depressions, Lake Michigan

Lake Michigan 'pockmarks' were first reported by Berkson and Clay (1973). The lake floor was resurveyed with side-scan sonar, single-beam echo sounder, and 3.5 kHz sub-bottom profiler by Colman *et al.* (1992). They were not convinced that the subcircular depressions they found were true pockmarks, and concluded that their mode of formation was '*problematic*'. Their evidence brought them to the conclusion that many of the possible mechanisms, with the exception of '*compaction dewatering*', could be ruled out.

3.29 EASTERN CANADA

3.29.1 The Scotian and Labrador shelves, and the Grand Banks

The original discovery of pockmarks was made by Lew King and Brian MacLean offshore Nova Scotia in the late 1960s. Their pioneering work was discussed in Section 2.1 of this book. Subsequently Gordon Fader, also from the Geological Survey of Canada's Bedford Institute of Oceanography, made a very thorough review of the pockmarks and other gas-related features off the coast and on the continental shelf of eastern Canada (Fader, 1991). He reported that pockmarks occur

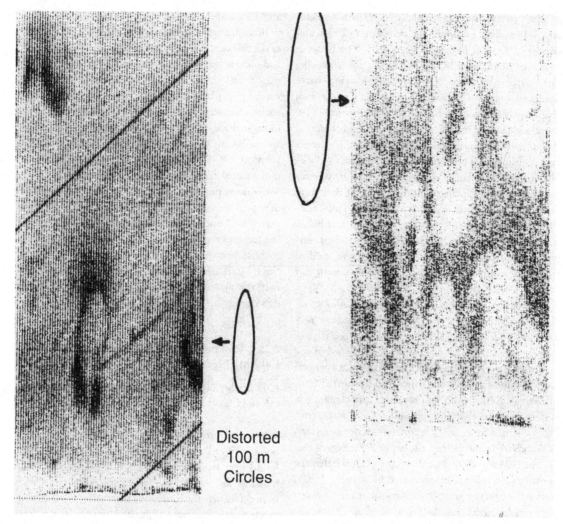

Distorted
100 m
Circles

Figure 3.44 Two side-scan sonar records showing three ring depressions found on the bed of Lake Superior. In some areas the rings are indicated by weak reflections (light rings), which suggests that they are infilled by fine-grained sediment; in other areas the reflections are strong (dark), suggesting infilling by coarse sediments. The two ellipses represent 100 m diameter circles distorted in the same way as the side-scan image. (From Hovland and Judd, 1988; courtesy of R. D. Flood.)

in numerous areas: Passamaquoddy Bay, the eastern Scotian Shelf, the Scotian Slope, the Laurentian Channel, the Gulf of St Lawrence, the Halibut Channel, and on the Labrador Shelf (see Map 32). He noted that several of these occurrences are associated with either microbial or thermogenic gas. In the Laurentian Channel and in the Karlsefni Trough (Labrador Shelf) pockmarks are associated with iceberg furrows and iceberg pits.

Josenhans and Zevenhuizen (1983) reported that the pockmarks off Labrador range in size up to 100 m in width and 7 m in depth, and that gas is frequently found close to, and in one example directly beneath, a pockmark. They are most frequently found immediately adjacent to the rims of the iceberg furrows. Each year icebergs driven south by the Labrador Current actively plough the seabed. Josenhans and Zevenhuizen suggested that there may be a causal relationship between the iceberg furrows and the pockmarks. Possibly, grounded icebergs trigger the release of the gas. Alternatively, the compaction of the sediments caused by the grounding of an iceberg may create excessive pressures in the pore waters, which escape to the seabed through the pockmarks.

On the Newfoundland Grand Banks acoustic turbidity is widespread, but it is most frequently found in sediments that overlie buried channels infilled with glacial and post-glacial sediments. Fader (1985) reported that in one area of acoustic turbidity in Downing Basin, numerous point-source reflectors were recorded within the water column. The appearance of bubbles at the sea surface around the survey ship ruled out the possibility that they were caused by fish. Bottom photographs taken at the site show numerous small bubbles, small mounds and cracking of the seabed sediments, and a great deal of particulate matter in the water. The density of particulate matter seen on these photographs is regarded as highly unusual as '*only a few of the frames of most stations ever show any evidence of (sediment) disturbance*' (G. B. J. Fader, 1986, personal communication). Despite the obvious intensity of this gas seep, it proved impossible to relocate it on a later occasion. This suggests that it may be intermittent. The analysis of gases extracted from piston cores at this site demonstrated the dominance of methane, but ethane and propane were present in minor quantities (G. B. J. Fader, 1986, personal communication). This suggests a thermogenic origin for the gas. Despite the presence of gas seeps, no pockmarks have been recorded in this area during investigations over several years, probably because the sandy texture of the seabed sediments (sandy muds) makes them unsuitable for pockmark formation (Fader and Miller, 1986).

Fader (1985; 1986, personal communication) has reported the presence of bacterial mats in two areas of the Grand Banks. The first observation was made from the submersible Pisces IV when areas of the seabed were found to be covered with a '*white filamentous material*'. Localised patches of dense shells have been located in the southeast Shoal area of the Grand Banks and on Sable Island Bank. They appear as dark patches on side-scan sonar records, and occur in a wide variety of shapes and relationships: as isolated features, in small groups, in large groups, and as continuous or discontinuous (beaded) patches (Fader and Miller,

1986). Samples collected by epibenthic sled from the Sable Island Bank shellbeds produced a large sample of diverse fauna; however, most of the specimens collected were dead. Fader (1985) reported that, in the Tail of the Banks, shell beds are found in the same area as large acoustic anomalies, at a depth of 300 m. These anomalies have been interpreted as gas-charged sediments within the Tertiary. It seems likely that gas at this depth has migrated from a deeper thermogenic source. Consequently the possibility of a relationship between this and other similar occurrences of gas and the shell beds cannot be ruled out.

3.29.2 The Laurentian Fan

The 1929 Grand Banks earthquake (magnitude 7.2) induced sediment failure, and a turbidity current carried some 100 km^3 of sediment down the continental slope and onto the Laurentian Fan, a feature which extends onto the Sohm Abyssal Plain (Map 32). In 1986 the submersible Alvin was used to study the sediments of the Laurentian Fan, off the eastern coast of Canada. During a dive onto the floor of the Eastern Valley, in water 3840–3890 m deep, dense biological communities were discovered; they were described by Mayer et al. (1988), and are discussed in Section 8.1.2.

3.29.3 The Baffin Shelf

Scott Inlet and Buchan Gulf

Oil and gas seeps have been reported from the continental shelf off Scott Inlet and Buchan Gulf on the northeast coast of Baffin Island (Map 1). Oil slicks fed by the seeps were first observed in 1976 and subsequent investigations were carried out in 1977, 1978, 1981, and 1985 (MacLean et al., 1981; Grant et al., 1986).

The site of the oil slick off Scott Inlet in 1976 was revisited and relocated in 1977. During calm weather oil droplets and gas bubbles were seen erupting at the sea surface, a small patch of 'chocolate mousse' was observed and the slick itself was thick enough in places for the display of interference colours. Water samples from the surface microlayer were found to contain up to 1726 µg l^{-1} of petroleum residues containing a broad range of saturated and aromatic hydrocarbons typical of a partially weathered crude oil. A more extensive survey area was covered in 1978, when petroleum residues of up to 1200 µg l^{-1} were found in a slick off the Buchan Gulf. Analyses of water samples collected throughout the water column, and of seabed sediment samples revealed anomalous concentrations of hydrocarbons. Of these, the sediment samples were considered to be of greater interest. To some extent, high concentrations of petroleum residue are correlated with the deep-water areas, as the fine-grained sediments of these areas are better able to absorb organic materials. In both the Scott and Buchan troughs there are anomalously high concentrations in areas unrelated to changes in grain size or water depth, which suggest a local source for the hydrocarbons. This conclusion was supported by the results of an experiment conducted in 1985 when samples of dyed diesel fuel were released into the water at a depth of 400 m. The samples surfaced at, or within, 400 m of a point directly above the release point.

Seismic, gravity, and magnetic surveys and bedrock sampling (using an underwater rock drill) indicated that the area is underlain by a structural depression – a thick (at least 4 km) sequence of Tertiary, Upper Cretaceous, and possibly older sediments. These sediments were considered to be suitable in age and origin to be the source of the seeping oil. As the most persistent slick was located over a basement high, it would seem that oil is migrating updip through these sediments to escape through the seabed where they have been truncated during the erosion of the trough.

During 1981 and 1985, the supposed seepage site in Scott Trough was investigated from the submersible Pisces IV (Grant et al., 1986). The seabed sediments were found to include mud, gravel, and boulders arranged into low boulder ridges with intervening linear gravel-free muds, probably formed by iceberg scouring. Bacterial mats (probably Beggiatoa) were found within 'a roughly circular, saucer-like depression, estimated to be about 30 m in diameter and 2 to 3 m deep'. The floor of this feature was fissured, and there was a carbonate crust around the fissures. Samples retrieved and examined onboard ship were described as indurated bottom sediment cemented in a fine carbonate matrix; it seems that this was MDAC. When disturbed on the seabed one of the samples exuded droplets of oil that floated upwards in the water. Oil also dripped from the sample once it was onboard the survey ship; several cubic centimetres of a dark-brown, medium viscosity oil was collected from it. No gas bubbles were observed during the seabed surveys, but small amounts of methane, ethane, and hydrogen sulphide were trapped in open-bottomed aluminium pyramids fitted with collection chambers deployed on the seabed.

Fresh water seeps, Cambridge Fjord

A freshwater spring in Cambridge Fjord, Baffin Island (Canada) was identified because it results in the formation of a polynya, a patch of ice-free water (Sadler and Serson, 1981). The 50 m diameter polynya is situated 300 m from the prodelta at the head of the fjord. It was reported that upwelling water was obvious to the naked eye and that the water was fresh to the taste. The feature was considered to be the result of a freshwater spring in the bed of the fjord, the water rising to the surface in a plume under the influence of buoyancy forces. Alex Hay from Memorial University of Newfoundland subsequently monitored this plume during September 1982 and 1983. He found that a submarine spring discharges from a depth of 47 m, and that maximum vertical plume velocities are 37 cm s^{-1}. The discharge volume was estimated to be 0.14 m^3 s^{-1} and the vent diameter 1.1 m. The apparatus used by Hay to monitor the buoyant plume was a launch-mounted high-frequency (200 kHz) echo sounder. The acoustic backscatter images of the buoyant plume appear as a 37 m high 'acoustic plume' (Hay, 1984). This shows that even freshwater seepages can appear as acoustic reflections in seawater. It is a well-known fact that buoyant plumes of gases or liquids can be used to keep areas ice-free. Perhaps searching for strange polynyas could be another way of finding submarine seeps in arctic areas.

3.30 FINALE

For one or more of the following reasons, our rapid journey around the world has by-passed many places where there is seabed fluid flow.

- There is an absence of data. There are still many parts of the world where there has been no exploration and investigation with modern equipment.
- There is an absence of publications. The petroleum industry has acquired data, much of them of the highest quality, from many areas, but the data have not been released to the public domain. The priorities of industry do not always coincide with those of science!

- There is the inaccessibility of publications. It is inevitable that interesting work, relevant to this chapter (and the remainder of this book) has been undertaken and reported in publications and languages unavailable to us.

There has been an explosion of interest and publications about seabed fluid flow in recent years; this has continued whilst we have been writing this book. So, we are guilty of missing out some places because we have not been able to read everything available on this fascinating subject. In the interests of space, many descriptions have been shortened, and examples have been left out; Fleischer *et al.* (2001) and Mazurenko and Soloviev (2003) provide global-distribution reviews.

4 · The contexts of seabed fluid flow

... geology concerns fluids every bit as much as it does solids. That is, geology usually conjures the idea of just the rock the earth is made of; but this rock is a locus of major interactions between fluids and solids. Fluids firstly concern the environments of sedimentary deposit, which differ according to location: salt ocean water, continental fresh water, atmospheric air, to name a few. Then the movements of the water tables in the beds running along accidents, evidenced by water springs, for example, are important phenomena. The oil and gas fields that petroleum explorers look for are yet another illustration of how fluids participate in geological processes.

Biju-Duval, 2002

> Seabed fluid flow occurs in a wide range of oceanographic environments (coastal, continental shelf, slope and rise, and the deep ocean) and geological (plate tectonics) contexts (convergent, divergent, and transform plate boundaries, and intraplate settings). In this chapter each situation in which seabed fluid flow occurs is described. In some cases only a brief description is necessary. However, in others, particularly those that are less well known, more detail is appropriate.

4.1 INTRODUCTION

Understanding the contexts in which it occurs may lead to an appreciation of the true geographical distribution of seabed fluid flow, and the features associated with seabed fluid flow. The purpose of this chapter is to consider two types of context: oceanographic and plate tectonics. These are discussed separately.

4.2 OCEANOGRAPHIC SETTINGS

The seas and oceans occupy about 67% of our planet. For simplicity, we will consider the following settings:

coastal, continental shelf, continental slope and rise, and the abyssal plains (Figure 4.1). The importance of the oceanographic setting must be considered in terms of not only the present sea level, but also the sea level during former times. For example, global sea level fell about 120 m during the last glacial period (about 23 000 y before present, BP). Sediment regimes changed considerably between these high and low sea-level stands, and during the intervening transition periods. Sediments being carried into the sea by rivers are now being deposited near the coast or on the continental shelf, but during glacial periods they may have been swept off the shelf to accumulate at the base of the continental slope. In the following subsections we briefly describe each setting.

4.2.1 Coastal settings

In many coastal settings sediment accumulation has been affected by local or eustatic changes in sea level. Locations of present-day (modern) seabed fluid flow are not necessarily 'permanent', and sediments that accumulated under previous conditions may continue to affect modern fluid flow.

The flow of groundwater into the sea has been detected '*on all the continental shelves of the world*' (Faure *et al.*, 2002). It seeps through the seabed from unconfined aquifers in near-shore areas, and as artesian flow from outcropping (or subcropping) confined freshwater aquifers that extend further offshore beneath the continental shelf (Church, 1996).

The relatively quiet conditions found in embayments and coastal indentations with restricted water movements may also be suitable for the accumulation of fine-grained sediments and for microbial methane generation. The following settings are particularly suitable for modern methanogenesis.

Estuaries are dominated by the delivery of sediment from present-day rivers, and/or the trapping of

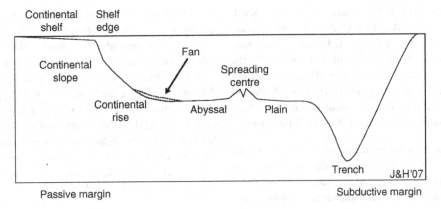

Figure 4.1* The principal global oceanographic and plate
tectonics settings.

sediments moved from offshore or along the coast by
water currents and long-shore drift.

Rias are valleys that have been drowned by a rise in
relative sea level. Generally, they were partially infilled
by rapid sedimentation during the marine transgression following the last glaciation. The waters of the Rías
Baixas (northwest Spain), for example, are marine rather
than estuarine in terms of their salinity.

'*Fjord*' is generally applied to a '*deep, high-latitude
estuary which has been (or is presently being) excavated
or modified by land-based ice*' (Syvitksi *et al.*, 1987). The
'typical configuration' is: '*a long, narrow, deep, and steep-
sided inlet, which is frequently branched and sinuous, but
may be remarkably straight in whole or in part where the
ice has followed major fault zones*' (Syvitksi *et al.*, 1987).
Fjords, the deepest type of estuary, are predominantly
features of mountainous, high-latitude regions; there
are extensive fjord coasts in both hemispheres.

Sediment inputs from inflowing rivers are generally high, and many fjords contain modern river deltas,
and have high sedimentation rates. Most, but not all,
fjords contain one or more shallow water 'sills' which
restrict the exchange of water between the open sea
and the deep-water basins. As a consequence many
fjords, although not even a majority, contain anoxic
waters. Rapid sedimentation and anoxic conditions are
favourable for the generation of microbial methane.

Deltas have formed by the rapid accumulation of
terrestrial sediments. Stanley and Warne (1994) identified 36 deltas that started to form since the last glacial
maximum, generally between 8500 and 6500 years ago.
This is true of all the major deltas in the world. During

glacial periods, the lowered sea level caused the rivers
to flow across the continental shelves, eroding all or
part of preexisting deltas. However, some, such as the
Mississippi and Nile deltas, have been built on ancestral
deltas formed during previous glacial periods (Davis,
1994). Deltas are characterised by sediments rich in
organic matter, so they are commonly associated with
methane generation and seepage. This is true of modern deltas, such as the Niger Delta, where microbial
methane dominates, and also of ancient deltas (such as
the Brent Delta, a Jurassic feature of the northern North
Sea) whose deeply buried sediments now produce thermogenic petroleum.

Drowned coastal plains and lowland areas
drowned during the Holocene transgression include
areas with extensive and prolific vegetation. Under
modern conditions this vegetation may have been covered in sediment and is now converting to peat. The
North Sea provides three generations of examples.

- As sea level rose following the last glaciation, coastal
  peat was formed in estuaries and close to the modern
  coastline. The outer Thames estuary is thought to
  contain extensive Holocene peat.
- During the Lower and Middle Pleistocene a large
  delta complex, fed by the River Thames and River
  Rhine, occupied the area south of about 55° N. The
  delta plain, known as Ur-Frisia, was characterised by
  wetlands, since submerged. The resultant sediments,
  the Yarmouth Roads Formation, contain abundant
  plant debris, peat, and wood clasts (Cameron *et al.*,
  1992) from which methane is being generated.

• During the Tertiary the western coast of the North Sea lay further to the east. Deltaic lignite (brown coal) is found in sediments deposited close to this coastline, between about 57° N and about 59°15′ N: the Upper Palaeocene–Lower Eocene (Beauly Member of the Moray Group), Oligocene to Miocene (Hordaland Group), and the Pliocene (Nordland Group). These are all thought to be sources of microbial methane.

Coastal lagoons, protected from the open sea by barrier reefs or islands, enjoy quiet conditions that may be suitable for the accumulation of sediments rich in organic matter, and therefore the production of microbial methane. The most intensively studied is Cape Lookout Bight on the eastern coast of the USA, where methane is being generated from organic-rich sediments.

4.2.2 Continental shelves

The continental shelves are drowned areas of the continents. As such, they share the geological variability of the terrestrial continents, so some areas are more likely than others to be associated with seabed fluid flow. In the context of this book, the most significant areas are sedimentary basins where organic matter in the sediments is degraded by microbial and thermogenic processes. Whether or not petroleum has been generated in a sedimentary basin, it is highly likely that some sediments contain sufficient organic carbon for microbial methanogenesis to occur.

4.2.3 Continental slopes and rises

Sediment transported across the continental shelves is transported by slope failures, debris flows, turbidity currents, etc. to accumulate on the continental slope or the abyssal plain. The principal transport pathways down the slope are marked by canyons, but like rivers, these may change course over time. Abandoned canyons may become filled with sediment, and pockmarks in this fill indicate fluid emission. However, the seabed fluid flow is most commonly associated with the sediment accumulations on the continental rises, particularly deep-sea fans.

The sediments accumulating in deep-sea fans are mainly derived from major river systems. During times of lowered sea level (glacial periods), when more of the continental shelves are exposed and sediments may be eroded from them and swept onto the slope, considerable thicknesses accumulate. Deep-sea fans associated with major rivers and deltas (such as those of the West African margin) are attractive targets to the petroleum industry; many have evidence of seabed fluid flow. However, the fans of the Atlantic margin of northwest Europe (and probably other high-latitude continental margins) are dominated by glacigenic sediments containing less organic carbon, and there seems to be less evidence of seabed fluid flow associated with them.

4.2.4 Abyssal plains

The abyssal plains, at water depths of 3000 to 6000 m, are generally flat and featureless areas of little relevance to seabed fluid flow, but in places there are troughs, hills, and craters, which may be associated with fluid flow. Deep-water currents form accumulations of sediment called contourites. These currents are driven by density contrast, and follow the seabed topography. 'Contour currents' may carry a sediment load, lifted from the soft, fine detritus of the abyssal plain. Deposition of these sediments occurs, presumably, when current strength is reduced, and where different current systems interfere with one another. The resultant deposits, contourites, may be up to a kilometre thick and several tens of kilometres (< 200 km) long. This thickness may be explained by the fact that deep-water currents have persisted for very considerable periods of time. For example, the construction of contourite drift deposits such as the Feni Drift, off the west coast of the British Isles, may have been continuous since the Middle Miocene (Stoker *et al.*, 1993). There is evidence of seabed fluid flow in at least two such deposits: Argentine Basin (Section 3.24.3), and Blake Ridge (Section 3.27.2).

4.3 PLATE TECTONICS SETTINGS

The Earth's crust (the lithosphere) is composed of ocean plates (composed primarily of basalt), and continental plates (composed of a variety of rock types, but generally considered to have an 'average' composition like granite). Together these plates cover the entire surface of the Earth, but forces acting on them from underneath

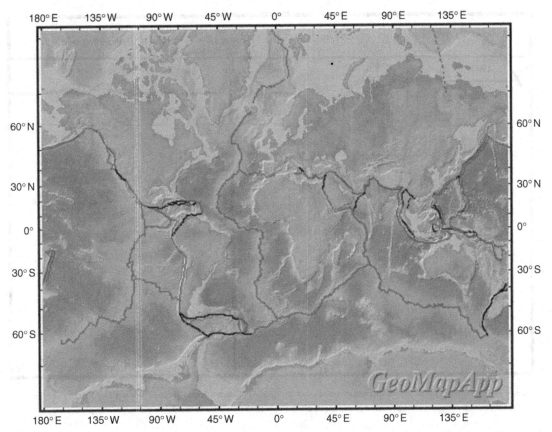

Figure 4.2* The main plate boundaries; light grey: subduction zones; dark grey: spreading centres; black: transform boundaries. (Plate boundaries obtained from the Institute for Geophysics, University of Texas: http://www.ig.utexas.edu/research/projects/plates/– accessed June 2005.)

(in the mantle) move each plate independently. The plates are in contact with each other, but the nature of the boundaries varies according to whether adjacent plates are moving towards each other, away from each other, or side-by-side. In the following sections we will consider each type of plate boundary, and intraplate volcanic settings associated with mantle hot spots, and the types of seabed fluid flow associated with them.

4.3.1 Divergent (constructive) plate boundaries

Divergent plate boundaries occur where adjacent plates move away from each other, and new crust is formed by igneous (magmatic) activity along the boundary. This most commonly occurs at ocean spreading centres

(Figure 4.2). These form a more or less continuous seabed mountain chain with a total length of some 55 000 km. Although commonly referred to as mid ocean ridges, this term can be confusing; the Mid-Atlantic Ridge may lie in the mid Atlantic, but part of the East Pacific Ridge lies in the Gulf of California.

At active ocean spreading centres (although the mechanisms of spreading are debated), tectonic plates are believed to grow by the steady accretion of solidifying basaltic magma generated by the partial melting of the material forming the upper mantle at depths of > 100 km (Figure 4.3). Seismic imaging of a crustal magma chamber along the East Pacific Ridge has shown that the molten magma below the spreading centre lies only 1.2–2.4 km below the seabed (Detrick *et al.*, 1987). This magma chamber is probably only 4–6 km wide, but

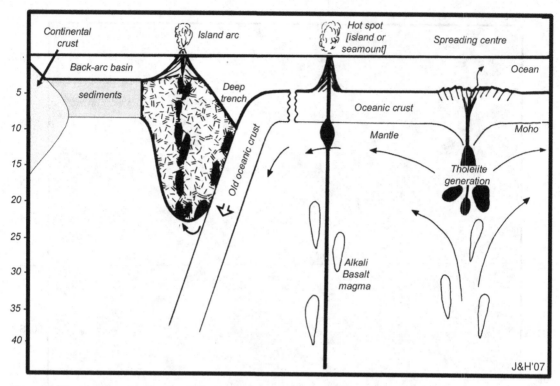

Figure 4.3 Plate tectonic settings: ocean spreading centres, hot spots, and subduction zones involving oceanic plate–oceanic plate collisions. Island arcs are produced by volcanism associated with the melting of the down-going oceanic plate.

it was traced for tens of kilometres along the rise axis (Detrick *et al.*, 1987). Basaltic magma from such chambers is then injected towards the seabed as dykes, which feed lava piles on the seabed. Although volcanism seems to be focussed mainly at the spreading centres, Alexander and Macdonald (1996) showed that there are significant, if relatively small, off-axis volcanic centres (Map 33).

Seabed fluid flow occurs at divergent boundaries primarily because of hydrothermal systems, and because of the degassing of magmas. Although many people think of these as places where recently formed igneous rock is exposed at the seabed, this is not always the case. On the northern Juan de Fuca Ridge (eastern Pacific Ocean), for example, a substantial sediment cover has buried the spreading axis at three locations: Escanaba Trough (Gorda Ridge), the Dellwood Knolls, and the Middle Valley (Davis *et al.*, 1987); at some locations the sediment thickness exceeds 1.5 km. The temperatures at the base of the sediments exceed 300 °C in some places, the highest temperatures (> 300 °C) being found

near the centre of the valley and the lowest (about 100 °C) near the faults bounding the valley (Davis and Lister, 1977).

4.3.2 Convergent (destructive) plate boundaries

Convergent boundaries occur where two plates move towards each other, and one sinks beneath the other.

Subduction zones and accretionary wedges
Where plates converge, one is subducted beneath the other (Figures 4.3 and 4.4). Most subduction zones occur where a dense oceanic crust is subducted beneath less dense continental crust (as on the west coast of South America), but in some cases two ocean plates, or two continental plates, converge. The subducted material is dragged downwards, heats, and eventually partially melts, the molten material rising to feed volcanoes. There are significant differences between different margins. For example, on the Peru Margin sediments lying on the down-going plate are subducted, the continental

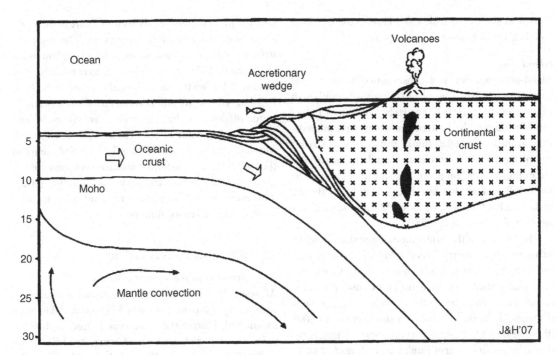

Figure 4.4 Plate tectonic settings: oceanic plate–continental plate collisions. The subduction zone is where the oceanic plate dives beneath the continental plate; accretionary wedges form where sediments accumulate at the plate boundary. (From Hovland and Judd, 1988.)

plate is steep and the fluids are expelled along extensive tectonic structures. In contrast, at the Barbados Margin most of the sediments are scraped off and accreted to the edge of the upper plate, forming an accretionary wedge (prism) (Figure 4.4) in which the sediments are squeezed and contorted, and fluids are expelled. The fluids include dissolved methane, radon, and slightly elevated concentrations of ammonia, iron, and manganese, which are all indicative of active porewater expulsion (Suess and Massoth, 1984). According to Spivack et al. (2002): 'the entire ocean volume is recycled through accretionary prisms in 300 million years'. However, fluid-flow budget measurements suggest that the amount of fluid derived from compaction is insufficient to explain all the fluid produced (Olu et al., 1997).

Fluid transport in accretionary wedges has been thermohydraulically modelled. Kukowski and Pecher (1999) modelled the Peruvian accretionary wedge at 12° S. They found that fluid expulsion at the seabed gradually decreases with distance from the deformation front, and that faults act as fluid migration pathways. The efficiency with which water is squeezed from the sediments is clearly significant. Saffer and Bekins (1998)

considered that 71% of the original Nankai Trough sediment porewater is expelled as diffuse flow through the seabed, < 5% escapes by focussed flow along the decollement and about 1% is subducted.

Seabed fluid flow, with associated cold-seep communities, has been described from numerous subduction zones/accretionary wedges. Mud volcanoes are a common feature of accretionary wedges.

Fore-arc basins

Fore-arc basins are depressions that lie between the island arc and the subduction zone of an ocean–ocean convergent margin. The basin traps sediments eroded from the island arc and from the accretionary wedge (if it is lifted above sea level). In Shelikoff Strait, Alaska (Section 3.19.2) and the Eel River Basin, California (Section 3.22.1) thick sediment sequences have been deposited in such basins, and shallow gas, pockmarks, plus evidence of seeps and gas hydrates in the latter example, are evidence of seabed fluid flow.

These basins are not normally associated with volcanic or hydrothermal activity, although there are exceptions such as the New Ireland and Woodlark spreading centres in the west Pacific (Van Dover, 2000). These arise because of tensions within a plate.

Island arcs

Island-arc volcanism is a consequence of the partial melting of a subducting plate (Figure 4.3). Although volcanic islands such as Grenada, St Vincent, St Lucia, and Martinique (the Eastern Caribbean Arc) immediately spring to mind when thinking of examples of island arc volcanoes, not all arc volcanoes reach the sea surface. For example, Kick 'em Jenny, 8 km north of Grenada, is about 1800 m high, but lies about 150 m below the sea surface; it has been active at least 12 times since it was discovered in 1939 (SRU, 2003).

By the turn of the Millennium, more than 35 active submarine hydrothermal vent fields had been located and sampled for vent fluids, minerals, and fauna, and more than 100 instances of hydrothermal mineral occurrences had been documented at seabed volcanic sites (Butterfield, 2000), but during a single cruise in 2003 (the National Oceanic and Atmospheric Administration's 'Submarine Ring of Fire'; see Merle *et al.*, 2003) a further 11 were identified in the Mariana and Kermadec arcs of the west Pacific; such is the pace of discovery. The total length of the island arc systems around the world is about 23 000 km (de Ronde, 2001). If the Mariana and Kermadec arcs are typical, and 20% of submarine volcanoes are active on all island arcs, then the total number of hydrothermal vents must be very large!

Back-arc basins

Back-arc basins are depressions lying behind volcanic island arcs; the island arc lies between the back-arc basin and the subduction zone (Figure 4.3). In some cases, such as in the Manus Back-arc Basin (Section 3.17.3) the down-going plate pulls on the edge of the overlying plate causing it to split open along the line of greatest weakness; normally the volcanic arc (Perfit and Davidson, 2000). Evidence of volcanic activity in the Mariana Trough Back-arc Basin (Section 3.18.1) seems to be indicative of the partial melting of the sediments. Very old back-arc basins in the Black and Caspian seas are currently significant sedimentary basins.

4.3.3 Transform plate boundaries

A transform boundary is where two plates, approximately travelling in parallel but opposite directions, slide laterally past one another. Crustal rocks are neither created nor destroyed at these boundaries. The boundary itself may comprise a zone of parallel faults and intensely shattered rock through which fluids may migrate. The most well-known transform boundary, the San Andreas Fault, cuts through California and then continues northwards offshore as the Ascension, San Gregorio, and Monterey Bay fault zones. These faults cut through Monterey Canyon (see Section 3.22.2). Sediments in the fault zones at transform boundaries experience compression if the plate movements are not exactly opposed. This compression may squeeze the fluids from the sediments, causing seabed fluid flow.

4.3.4 Intraplate igneous activity

Sill intrusions in sedimentary basins

At some 'passive' continental margins sedimentary basins may develop, into which igneous dykes may be injected. During the Palaeocene to Eocene, basaltic dykes were injected into Upper Cretaceous shales in the Vøring Basin, and in a > 400 km belt extending from the Møre Basin, off mid Norway, into the Faroe–Shetland Basin (Stoker *et al.*, 1993). According to Jamtveit *et al.* (2004) who studied ancient analogies in the Karoo Basin in South Africa, subvolcanic intrusions in sedimentary basins cause strong thermal perturbations and frequently cause extensive hydrothermal activity: '*Hydrothermal vent complexes emanating from the tips of transgressive sills are frequently observed*'. Distinct features include: '*Inward dipping sedimentary strata surrounding a central vent complex comprised of multiple sandstone dykes and pipes, potholes, and hydrothermal breccias*'. Over 600 such vent complexes have been found to date offshore mid Norway. The investigators reckon the complexes consist of large mud volcanoes (now buried), but which still influence the fluid plumbing system of the basins, including the migration of hydrocarbons.

It is probable that such intrusions affect fluid migration in similar geological settings elsewhere.

Large sedimentary abyssal craters

Large craters have been found in abyssal carbonate-dominated sediments, at amongst other places on the super-fast spreading Cocos Plate off Panama and Ecuador. These structures were first mentioned by

Mayer (1981) who described them as '*erosionally exca-vated troughs in an area of relatively rapid accumulation of carbonate*', possibly explained by '*differential accumula-tion and compaction*'. Multichannel seismic lines shot in connection with the planning of ODP Leg 206, together with swathe bathymetry, show that these depressions are actually more like craters formed in the seabed sed-iments. They occur on top of high reflective underlying crust, as if there are volcanic sills beneath them. The craters are up to 2.5 km long and 90 m deep. We suggest that they form as a consequence of sill injection and the escape of supercritical or superheated fluids through the sediments above the injection.

Hot spots and seamounts

Isolated mid-ocean volcanic islands like Pitcairn in the Pacific Ocean and Tristan da Cunha in the South Atlantic would be called seamounts had they been sub-merged beneath sea level. Seamounts, guyots, and atolls are generally circular or conical in shape, and are, per-haps, some of the most spectacular topographic features on the surface of our planet (Figure 4.3). They are genet-ically related, with volcanic origins.

Seamounts are isolated submarine mountains, some rising more than 4000 m from the abyssal depths. The term seamount was broadly defined by Menard (1964) as any isolated >1000 m elevation of the ocean floor with a circular or elliptical horizontal cross section and a slope ranging between 5 and 35 degrees of arc with respect to the surrounding terrain (Hekinian, 1984).

Guyots were discovered by Harry Hess during World War II when he continued his oceanographic investigations whilst captain of a US Navy ship (Hess, 1946). They are flat-topped seamounts, commonly found far from the ocean spreading centres.

Atolls are submerged seamounts upon which a coral reef has grown, extending to the surface to form a ring-shaped island. Charles Darwin explained their formation in a presentation to the Geological Society of London in May 1837, and in his journal of the voyage of the Beagle (Darwin, 1839). Eventually their volcanic origin was demonstrated by drilling.

These three types of feature are important if for no other reason than that there are so many of them; they may also be related to seabed fluid flow. In 1980 a group of British oceanographers led by Roger Searle, making only two passes with the long-range side-scan

sonar, 'GLORIA' in the southeastern Pacific between the East Pacific Rise and the coast of South America, identified about 200 volcanoes with a diameter of >1 km at the base (Hekinian, 1984). Searle estimated that the average density of volcanoes (including seamounts) in the South Pacific is about eight per 10 000 km^2. Work by Mukhopadhyay *et al.* (2002) included smaller (> 50 m high) volcanic features as seamounts, but the densi-ties they reported suggest enormous numbers: 6000 per 10^6 km^2 in the area they studied in the Indian Ocean, and 7000 and 9000 per 10^6 km^2, respectively, in the Atlantic and Pacific oceans (data from Batiza *et al.*, 1989 and Fornari *et al.*, 1987). Assuming an average of only 5000 per 10^6 km^2 would suggest a total of about 1.8 mil-lion seamounts in the world's oceans (total area about 360 × 10^6 km^2). Although this figure is much too high for seamounts standing >1000 m high, it clearly indi-cates the enormous scale of volcanic activity on oceanic plates.

According to the 'hot-spot theory', the volcanic activity occurs when a particular seamount occupies a 'hot spot', which is believed to overlie upwelling con-vection cells in the mantle. However, the seamount does not remain stationary over the hot spot, but rides the ocean crust at the speed of the spreading ocean floor. As the lithospheric plate moves over the hot spot, it is punctuated with volcanoes. In this way, long chains of seamounts form; perhaps the most spectacular is the Hawaiian-Emperor chain. The Hawaiian Islands are the only active subaerial volcanoes in the chain (and the tallest mountains on Earth; Mount Kea stands more than 10 000 m above the abyssal plain – Everest stands only 8848 m above sea level), and the age of the volcanic rocks progressively increases away from Hawaii. Loihi Seamount is the newest and southern-most part of this chain. It is active, and no doubt will form a new island in due course. However, not all seamounts are formed at hot spots. Mukhopadhyay *et al.* (2002) considered that most are formed at the flanks of spreading centres, whilst others are formed at ridge-transform fault junctions and where spreading centres overlap.

Seamounts represent great volumes of rock occu-pying a relatively small local area. Numerous reports have noted the relatively low density of igneous rock samples from seamounts; in many cases this is attributed, at least in part, to the presence of vesicles. Menard (1964) reported two examples: the first was a fragment of relatively unaltered vesicular pahoehoe lava dredged from Jasper Seamount, west of Baja California with a density of *'only slightly more than 2 g/cm³'*. The second was highly vesicular basalt from 3500 m on the steep-sided Popcorn Ridge, west of Baja California. This was decrepitating when brought on deck, and *'breaking with sharp snapping sounds because of the expansion of gas in the vesicles'*. The low density and highly porous nature of these rocks suggests that they came from magmas containing large amounts of gas. Clearly, this activity will result in the flux of large volumes of fluids into the ocean water, as is illustrated by the following dramatic account:

> As researchers on a ship above MacDonald Seamount (south-central Pacific Ocean) dredged rock samples from the summit, large bubbles suddenly broke the sea surface and turned the water dark brown. Gas bubbles shook the ship, causing loud 'clangs and clamours' and a sulphur smell. The top of 1 large bubble reached 2 m above the ocean surface and burst, forcefully ejecting gas and exposing 20–30 volcanic clasts in its core. The rocks floated briefly, and one piece of dark volcanic glass was recovered while still hot.
>
> Anon., 1988

It seems probable that hydrothermal activity continues for some time after volcanic activity has ceased. Thießen *et al.* (2004) reported very high methane values (> 30000 nl l⁻¹ in one sample) in fluids venting from the MacDonald Seamount in 1989, a year after the eruption. They found that concentrations from waters above other seamounts were not so high (< 5600 and < 2300 nl l⁻¹ in waters above the Teahitia and Bounty seamounts, respectively), but significantly above background (< 50 nl l⁻¹). In a review of the biology of seamounts, Rogers (1994) noted that, because most of them are of volcanic origin *'it is of no surprise that some are associated with hydrothermal venting'*. He identified only five with hydrothermal activity: Loihi (off Hawaii), Axial Seamount (see Section 3.21.2), Red Volcano (one of the Larson Seamounts, 30 km east of the East Pacific Rise), Piip Submarine Volcano (in the Bering Sea, see Section 3.18.7), and Marsili Seamount (in the Tyrrhenian Basin, north of Sicily in the Mediterranean Sea). However, he clearly thought that this is only the tip of the iceberg: *'The presence of hydrothermal vents on mid-oceanic seamounts and off-axis seamounts may indicate that hydrothermal emissions are more common in the deep sea than previously thought'*.

More recently, it has been observed that some seamounts are recharge areas, where seawater flows into the seabed. This may account for the fluid plumes, indicated on acoustic transects, rising into the water high above the Hancock Seamount, one of the Hawaiian chain (Hovland, 1988). Harris *et al.* (2002) simulated buoyancy driven flow, then estimated that there are 15000 seamounts of up to 2 km height, and that the mass water flux through them is 10¹⁴ kg per year. This is similar to estimates of the mass flux at ridge axes and through ridge flanks.

So, it seems that seamounts are important sites for different types of seabed fluid flow at different stages in their evolution: volcanic activity during, and for some time after formation, then hydrothermal activity, and finally the cycling of seawater in through the base and out into the relatively porous oceanic crust.

Subducting seamounts
Seamounts are carried like passengers on top of moving tectonic plates, eventually sliding into subduction zones and pulled down beneath the over-riding plate. As they are subducted, they cut a furrow in the sediments of the overlying wedge from the deformation front back to wherever their present position is. For example, off Costa Rica, seamounts on the subducting Cocos Plate are indicated by uplifted domes. Associated scarps and landslide features represent the furrows (Bohrmann *et al.*, 2002). Because they deform the sediments of the overlying wedge, fluid-flow pathways are affected, and seabed fluid flow may result; such as at the Jaco and Parrita scarps, off Costa Rica (Section 3.24.1). Information from Lonsdale (1986) suggests that the subduction of the Louiseville guyot chain into the Tonga Trench is causing the formation of *'a small accretionary prism'*. Surely these contorting processes must also lead to widespread seepage of compressed fluids.

4.3.5 Serpentinite seamounts

The Lost City, near the Mid-Atlantic Ridge (Section 3.8), and Conical and other seamounts near the Mariana Trench in the Pacific (Section 3.18.2; Map 33) are examples of hydrothermal chimneys and mud volcanoes with a non-igneous origin. They are formed as a result of the serpentinisation of mantle peridotites, and may be quite common. Kelley *et al.* (2001) commented that hydrothermal systems of this exothermic type may be common in ocean ridge systems because of the number of peridotite/serpentinite occurrences, and the numerous methane/hydrogen anomalies that have been reported.

4.4 CONCLUSION

In this chapter, we have shown that examples of seabed fluid flow occur in a wide variety of oceanographic and geological contexts. In order to ascertain the true distribution and significance of seabed fluid flow it is necessary to establish the distribution of each of these contexts – and any others that we have inadvertently overlooked, or that have yet to be identified.

5 · The nature and origins of flowing fluids

Given the immense amount of organic matter in the sediments of the Earth's crust, it is evident that the geological consequences of this subterranean generation of gas must also be immense.

H. Hedberg, 1974

> The fluids available to flow through the seabed include various types (liquid and gaseous, organic and inorganic) derived from microbiological and geological processes – some close to the seabed, some distant both in space and time. To provide a foundation for subsequent chapters in which we look at the fate of these fluids, this chapter provides a brief review of the nature and origins of the fluids, and identifies some areas where controversy still persists.

5.1 INTRODUCTION

Although methane gas is the dominant fluid escaping from the seabed in the majority of the examples described in the previous chapters, it is clearly not the only fluid. In this section we provide a brief summary of the nature and origins of the various fluids observed flowing through the seabed. Following geological tradition, we start with fluids associated with igneous activity before considering those associated primarily with sediments.

5.2 HOT FLUIDS

5.2.1 Magma and volcanic fluids

The scale of volcanic activity in the oceans is enormous, as was made clear in the previous chapter. Whether or not magma is expelled at the surface, fluids, being much more mobile, are more likely to escape, and the escape

of gases may continue long after other indicators of volcanic activity have ceased. That is, dormant volcanoes may still produce gas. The presence of gas in magma is clearly demonstrated by the explosion of bubbles from the magma surface as it comes from volcanic vents, and by the presence of gas bubbles or vesicles 'frozen' in cooled and solidified magma. Although the composition and concentration of gas varies according to the nature of the magma (i.e. whether the magma is 'acidic' or 'basic' in composition), all magmas contain some gas. Indeed, the expansion of gases as they rise towards the surface provides much of the energy required to lift the magma towards the surface The following, in order of abundance, are the most common: CO_2, H_2O, CO, SO_2, S_2, and H_2S (Apps and van de Kamp, 1993). There are a few others, such as nitrogen, methane, chlorine, and hydrogen.

As magmas cool, the chemical composition of the melt changes as a sequence of individual minerals crystallises progressively. Volatiles concentrated in the residual melt include compounds produced by reactions between the melt and minerals that crystallised earlier in the cooling process. Evidence for these reactions comes mainly in the form of fluid inclusions, minute pockets of fluids trapped within crystals of the rock-forming minerals. Methane is generally a minor component relative to water and carbon dioxide, but Apps and van de Kamp (1993) identified the following situations in which it is synthesised.

- Degassing and/or cooling of mafic magmas at spreading centres. Gases are dominated by carbon dioxide and hydrogen sulphide, but methane is also present. These gases contribute to the fluids emitted by hydrothermal vents.
- Serpentinisation of ultramafic rocks (peridotite). This occurs near spreading centres and near subduction zones where diapiric serpentinite bodies have been

identified, and methane-rich plumes rise into the water column.

- The cooling of layered mafic intrusions. High-pressure gas escapes have occurred during the mining of platinum-group metals from the Bushveld (South Africa) and Stillwater (Montana) complexes. The gas is composed of methane and hydrogen or hydrogen sulphide. Their presence within the still-cooling melt is shown by the presence of 'potholes', thought to be pockmark-like evidence of the rise of fluids through the crystals settling at the bottom of the magma chamber. This type of intrusion is thought to form during early stages of continental rifting, and may be found on continental margins.

Steaming intrusions

During the late Palaeocene/early Eocene igneous intrusions were injected into sedimentary basins off mid Norway, causing strong thermal perturbations and extensive hydrothermal activity (Planke *et al.*, 2002; Svensen *et al.*, 2003).

Planke *et al.* (2002) envisaged that the intrusions would cause heating and local boiling of the sediment porewater, and that hot water and/or steam would rise through the overlying sediments to erupt through the seabed. They suggested that the migration pathways formed by such events could later be used by hydrocarbon fluids. This process and the resultant features are discussed in more detail in Section 7.5.5.

5.2.2 Geothermal systems

As well as the heat generated when the Earth was accreted some five billion years ago, 'geothermal' heat is derived from the natural decay of the radioactive isotopes of elements such as potassium, uranium, and thorium in the rocks of the Earth's crust and mantle. A consequence of this heat at depth is the geothermal gradient – the increase in temperature with depth below the surface. Anomalous geothermal gradients occur where the crust is unusually thin and where igneous intrusions are still dissipating heat as they cool down. Crustal thinning occurs at plate boundaries, for example where plates are diverging above sites of mantle upwelling. Geysers, fumaroles, and hot springs, are commonly found in such places on land. In contrast, submarine geothermal springs seem to be rare. This apparent rarity is probably because they are difficult to detect, visually or acoustically, and are easily mistaken for hydrothermal (mineral-rich) or cold freshwater seeps.

5.2.3 Hydrothermal circulation systems

Hydrothermal activity results from the heating, or superheating, of water as it comes into contact with magma and hot volcanic rocks. Wherever there is submarine volcanism, there is also hydrothermal activity due to the active role that circulating seawater plays in cooling, altering, and cracking newly emplaced magma. At spreading ridges, back-arc rifting systems, and in various seamount environments, hot igneous rocks are permeable due to wrenching, faulting, and fissuring. Dense, cold (typically 2–4 °C) seawater flows through this 'stockwork' system, penetrating to various depths. It is heated near shallow magmatic chambers, sills, and dykes which, in the case of basalts, are typically at temperatures up to 1200 °C. Leaching of the surrounding rocks chemically alters the heated seawater. This heated, buoyant, and chemically altered fluid then rises to the seabed where it is expelled through hydrothermal vents. Here, the sudden change in ambient conditions causes the precipitation of minerals, but plumes of the remaining hydrothermal fluid rise into the water column.

The potential for hydrothermal systems in the ocean is enormous. The majority of the Earth's crust is oceanic, and is exposed to seawater or to water-saturated sediments. Several authors have commented on the remarkable porosity and permeability of ocean crust near hydrothermal vents. Binns *et al.* (2001), for example, reported *average* porosities of nearly 25% and intense fracturing in hydrothermally altered rocks encountered during ODP Leg 193 drilling in the Manus Basin. If this is generally true of hydrothermally altered rocks below and adjacent to spreading zones and other hydrothermal localities, it means that the crust is highly porous and permeable in places, just like aquifers and 'reservoirs' in sedimentary basins. According to Fyfe (1992), the total mass of water contained at any one time in the ocean crust is actually the same as that in the ocean. He then estimated that the entire ocean mass is recycled through seabed hydrothermal systems in about 10 million years (Fyfe, 1994).

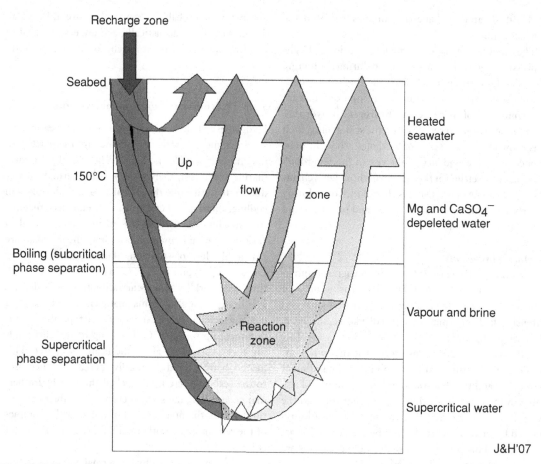

Figure 5.1* Cartoon showing the principal zones of hydrothermal circulation systems. Threshold depths depend on water depth, geothermal gradients, and water salinity.

Hydrothermal systems are comprised of three zones (Figure 5.1):

- a recharge zone, where seawater percolates down through the crust and is slowly heated and chemically modified by low-temperature reactions;
- a high-temperature reaction zone, near the heat source where fluid equilibrates with hot rock;
- an upflow zone where heated, buoyant fluids ascend rapidly to the seabed (Butterfield, 2000).

The recharge zone

Whereas it has proved to be relatively easy to find the vent sites for hydrothermal flow, it is much more difficult to find the recharge zones where seawater enters the oceanic crust. However, this was achieved by the Deep-Sea Drilling Project (DSDP) and the Ocean Drilling Project (ODP) at Site 395 in the Atlantic Ocean. Hole 395A was completed in 1976. It is located 70 km west of the Mid-Atlantic Ridge crest, at an isolated sediment pond, the 'North Pond', which measures 8 km by 15 km. The hole is located in a water depth of 4440 m and penetrates 92 m of sediments and over 500 m into the basaltic oceanic crust. This hole has been re-entered four times, and each time temperature logs, fluid samples, and flow meters *'clearly demonstrated that ocean bottom water was flowing down the hole at consistent rates of about 1000 liters/hr'* (Becker, 1998). Results from a similar setting on the flank of the Endeavour Ridge (ODP Leg 168) also show that there are huge fluxes of low-temperature seawater through very transmissive upper basement in thinly sedimented, young, oceanic crust. This flow seems to occur regardless of whether the sediment cover is continuous or patchy, and regardless of spreading rate.

The reaction zone – introducing supercritical water

As seawater circulates through the oceanic crust its behaviour and composition are affected by both temperature and pressure. The first important changes occur at temperatures of about 130–150 °C. Above this temperature anhydrite ($CaSO_4$) becomes saturated and precipitates. Butterfield (2000) explained: 'As sulfate is the second most abundant anion in seawater, this has a major impact on the fluid chemistry and can produce substantial volumes of anhydrite in pore spaces'. At about the same temperature magnesium is removed from the seawater, and hydrated minerals such as smectite and chlorite are formed. The removal of magnesium is important. It means that remaining fluid is more acidic, and that salts produced by processes in hotter parts of the system will be depleted in magnesium.

The boiling point of seawater rises with increased pressure, and is significantly higher than 'boiling point' (100 °C) under the great pressures experienced at and beneath the ocean bed. When seawater boils 'fluid phase separation' occurs. We are familiar with the formation of vapour bubbles within boiling water, but when seawater boils most of it (about 95% by volume) becomes vapour; the rest (about 5% by volume) condenses to brine. The vapour phase holds significantly less salt, but the salinity of the brine progressively increases as phase separation continues (Dählmann et al., 2001). However, there is a change in the very nature of water at a pressure of 298 bar (about 2800 m water depth) and a temperature of about 407 °C. At this critical point (Figure 5.2) standard seawater does not boil, but becomes 'supercritical' (Lowell et al., 1995).[1] Butterfield (2000) stated that brine produced from seawater at 450 °C would be five times more saline than seawater; above 700 °C the salinity would be more than ten times that of seawater. Fluid inclusion studies show that salinities and temperatures increase with depth. These brines are too dense to flow to the seabed, except when diluted by circulating seawater. The average ocean depth is > 3000 m, and the temperature of basaltic magma may exceed 1100 °C, so

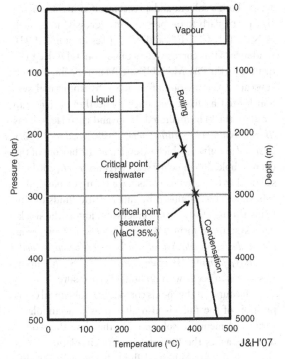

Figure 5.2 The critical point at which seawater becomes 'supercritical'. (Adapted from Charlou et al., 2000; Dählmann et al., 2001.)

there must be a fantastic amount of supercritical water in the ocean crust.

Liquid water may be the most abundant solvent on Earth, but not much is known about water under extreme conditions. Beyond its critical point, even freshwater exhibits many peculiar properties:

- it acquires a bulk density of only 0.32 g cm^{-3}, and a specific volume of 0.31;
- viscosity is reduced significantly;
- dissociation increases, affecting both pH and Eh;
- the dipolar character is strongly reduced;
- it is a non-polar solvent, that is, organic materials and gases are soluble but salts are more-or-less insoluble.

This peculiar behaviour results from a change in intermolecular structure, particularly the hydrogen-bond network (Bellissent-Funel, 2001). Supercritical water

---

[1] For those who thought that 'supercritical' and 'superheated' liquids are the same thing: a superheated liquid is: 'a metastable thermodynamic state of a liquid resulting from rapid heating to a temperature well above the boiling point' (Morrissey and Mastin, 2000) – who said that thermodynamics was easy?

acts as both a solvent and a reactant; it is highly corrosive, particularly in the presence of aggressive ions, such as Na^+, H^+, Cl^-, F^-, and NH_4^+ (Tester *et al.*, 1993). Evidence of this corrosiveness came from ODP Leg 193 drilling on the PACMANUS site, Manus Basin, the first subseabed examination of an active hydrothermal system hosted by felsic volcanics. Beneath a surface cap (< 40 m thick) of unaltered dacite and rhyodacite lavas *'there was a rapid transition to pervasively and thoroughly altered derivatives'*, which extended to the bottom of the deepest hole, 400 m below seabed (Binns *et al.*, 2001). It seems that the igneous rocks have been corroded, dissolved, and recrystallised by supercritical fluids percolating through the network cracks and joints (the stockwork system) to form *'a soft mud composed of anhydrite, clay and silica in which a sub-seafloor massive sulfide body is actively forming'* (Barriga *et al.*, 2001). Surprisingly, this seems to occur with no fluid overpressure.

Of course, as the rock is corroded the chemical composition of the fluid is also altered as it accumulates various elements in solution. By the time the seawater approaches the final stage of its circulation it has changed to a hydrothermal fluid that is radically different in both nature and composition. According to Chevaldonné (1997), the undiluted hydrothermal fluid is typically between 350 and 404 °C, anoxic, acidic (pH < 3), has variable salinity (*'from a tenth to many times that of seawater'*), and is enriched with reduced compounds such as hydrogen sulphide (up to 110 mmol kg^{-1}), methane, carbon dioxide, helium-3, hydrogen, and with trace elements (lithium, manganese, iron, barium, copper, zinc, cadmium, and lead). It is also depleted in magnesium, sulphates, nitrates, and phosphates. Usually the magnesium concentration is used as a guide to the extent to which the hydrothermal fluid has mixed with seawater on its way to the seabed. This is used to calculate the composition of the undiluted, magnesium-free, fluid (although Dählmann *et al.*, 2001 argued that a zero-sulphate was a more realistic representation of the undiluted fluid at sedimented sites). Thus, most reports of hydrothermal fluid compositions quote the composition of the 'end-member fluid', rather than the measured compositions.

Submarine volcanism and associated hydrothermal systems are not confined to ocean spreading centres, but also occur in island arcs, fore-arc and back-arc basins, and intraplate hot spots (as explained in Chapter 4). The circulation systems are essentially similar, but the rock types may differ significantly from the typical mid-ocean ridge basalt (MORB), and the depth of the interaction between water and magma or hot rock may be too shallow for the formation of supercritical water. So, the composition of the resultant fluid may differ considerably.

ROCK TYPE

Several different igneous rock types come into contact with hydrothermal systems. Each has its own chemical composition and degree of stability in the new environment into which it has been intruded. The most extensive rock type in oceanic crust is basalt, which is formed at ocean spreading centres (commonly termed MORB) and intraplate hot spots such as the Hawaiian Islands (and associated submarine volcanoes such as Loihi). Basaltic magmas are formed by the partial melting of ultramafic (periditic) material in the upper mantle. At convergent margins, crustal material is progressively heated as it descends in subduction zones. Partial melting results in the formation of magma more felsic than the oceanic crust from which it is derived. This is partly because of the incorporation of subducting sediments, which also make this magma water-rich compared to basalt. The resultant magma, of intermediate (andesite) or acidic (rhyolitic) composition is more viscous than basaltic magma. This is reflected in the type of volcanic eruption, which tends to be more violent, and sometimes explosive.

THE DEPTH OF THE REACTION ZONE

The nature of the interaction between the circulating seawater and the magma or hot rock differs according to the pressure at which it occurs. Water circulating freely through the rock can be assumed to be at hydrostatic pressure; consequently pressure is a function of depth below sea level, rather than seabed. In shallow waters, even at high temperatures there is insufficient pressure in the reaction zone for supercritical seawater to form. So, although boiling may cause the separation of a vapour phase, it will not have the corrosive power of the supercritical fluids described above. According to Dählmann *et al.* (2001), who worked on the hydrothermal vents on Hook Ridge, Bransfield Strait (Antarctica), boiling and phase separation at depths too shallow for the formation of supercritical water produces a vapour phase with extremely low concentrations of sulphide-forming metals. This may be due to the high degree of boiling, in which case *'the metals are effectively partitioned into the remaining liquid phase'*. Alternatively, this may be because of a low temperature, in which case precipitation of sulphide minerals will occur beneath the surface. In either case, the fluids escaping through the seabed *'do not build visible edifices and diffuse low-temperature emanations dominate the hydrothermal flux even at bare rock sites'*.

Upflow and venting

The rising fluids are cooled in the upflow zone by adiabatic decompression, conductive heat loss, and mixing with ambient fluids around the conduits (Butterfield, 2000). The extent of cooling depends on the nature of the pathway to the seabed, how well it is insulated, and the extent to which it mixes with unheated pore water. Cooling results in the precipitation of solutes, but also causes a change to the nature of the fluid. For example, when metal sulphides precipitate, acid is released. Substantial cooling and precipitation may result in sufficient acidification for other minerals to be dissolved as the fluid migrates towards the seabed. Butterfield (2000) explained that cooling a 400 °C fluid to below 350 °C results in precipitation of much of the copper as chalcopyrite and lowers the pH, so if the resulting fluid comes into contact with zinc sulphides (such as sphalerite) they may be dissolved, causing a significant increase in the zinc concentration in the fluid. This process accounts for the concentrations of different minerals in separate zones in metallic ore deposits (discussed further in Section 9.4).

When, finally, the hydrothermal fluids emerge through the seabed, commonly at temperatures of 300 to 400 °C, they encounter seawater at a temperature of about 2 °C. Rapid cooling results in the instantaneous precipitation of minerals (see Section 9.4). The venting fluid may represent either the vapour phase, or the brine phase of a fluid that has undergone phase separation – or a mixture of the two. We discuss the composition of hydrothermal plumes in Section 10.2.2.

SEDIMENT CAPPING

In some cases, hydrothermal vents are situated where the seabed is made of solid rock, so the venting fluids pass straight into the water column. In other places the fluids have to pass through a layer of sediment, where the pore waters cool them; this cooling is increased by water-circulation cells whereby pore water entrained by the venting fluid is replaced by seawater drawn in close to the vents. The result is not only that the venting fluids are lower in temperature than they would be had there been no sediment cap, but also the cooling causes the precipitation of some minerals within the sediment, so the composition of the venting fluid is also altered.

The effects of the rising fluids on the sediment may also be significant. For example, increased temperatures may accelerate the decomposition of organic matter, and acidity may result in the leaching of certain elements. These processes may explain the presence of thermogenic methane, higher hydrocarbon gases, ammonia, and magnesium in fluids vented through sediments.

5.2.4 Exothermic hydrothermal systems

Low-temperature hydrothermal fluids venting from spectacular carbonate chimneys such as those of the Lost City (Section 3.8) and Conical Seamount (Section 3.18.2) owe their origin to a completely different process to those described in the previous subsection. Kelly *et al.* (2001) found that fluids at the Lost City were warm (40 to 75 °C), rich in calcium (21 to 23 mM kg^{-1}), depleted in magnesium (9 to 19 mM kg^{-1}), and very alkaline (pH 9–9.8); pore water encountered during drilling on Conical Seamount had a pH of 12.6. Peridotite-related hydrothermal systems at the Rainbow and Logatchev sites in the Atlantic have higher-temperature fluids showing indications of reactions involving gabbroic and basaltic rocks. Unlike other divergent-margin hydrothermal fluids, no magma is involved in these reactions. Stable carbon- and oxygen-isotope analyses support the conclusion that these fluids are heated by exothermic reactions between seawater and mantle-derived peridotite during serpentinisation. These reactions are summarised in Equations (5.1a) and (5.1b).

$$5Mg_2SiO_4 + Fe_2SiO_4 + 9H_2O \rightarrow 3Mg_3Si_2O_5(OH)_4$$
$$\text{Forsterite} \quad \text{Fayalite} \qquad \qquad \text{Serpentinite}$$
$$+ Mg(OH)_2 + 2Fe(OH)_2$$
$$\text{Brucite} \qquad (5.1a)$$

$$\rightarrow 3Fe(OH)_2 \rightarrow Fe^{2+}Fe_2^{3+}O_4 + H_2 + 2H_2O$$
$$\text{Magnetite} \qquad (5.1b)$$

It should be noted that the serpentinisation of peridotite causes a volumetric expansion of about 40% because of the uptake of water. Forsterite and fayalite are end-members of the olivine series of minerals. So, such reactions occur where olivine-rich rocks such as peridotites encounter seawater percolating from the seabed. Carbon dioxide-rich fluids and the excess hydrogen may then combine to form methane and some higher hydrocarbons in a Fischer–Tropsch reaction, as has been demonstrated by the industrial production of synthetic hydrocarbons (Selley, 1998). Several individual reactions using certain metallic (iron, nickel, etc.) minerals and/or graphite as catalysts are thought to be involved; these are summarised as in Equations (5.2)–(5.5).

$$C + Fe + H + OH \rightarrow CH_4 + C_2H_6 + C_3H_8$$
$$+ C_4H_8 + FeO \qquad (5.2)$$

$$CO_2 + 2HOH + 4Fe \rightarrow CH_4 + 4FeO \qquad (5.3)$$

$$C + 4HOH + 2Fe \rightarrow CH_4 + O_2$$
$$+ 2H_2 + 2FeO \qquad (5.4)$$

$$CO_2 + 4H_2 \rightarrow CH_4 + 2H_2O \qquad (5.5)$$

Experiments performed by Horita and Berndt (1999) showed 'that abiogenic methane forms rapidly in the presence of even small amounts of hydrothermally generated Ni–Fe alloys under reducing conditions'. They reported that nickel–iron alloys are common products of serpentinisation of olivine-rich mantle rocks, and that awaruite (Ni_2Fe to Ni_3Fe), nickel–iron alloys, and nickel–iron sulphides have been reported from ultramafic rocks in both continental and oceanic settings, and from meteorites; 'Indeed, the oceanic crust is largely composed of ultramafic rocks, and serpentinite containing awaruite and Ni–Fe alloy could make up a large fraction of the oceanic crust'. So, they concluded that abiogenic methane formed by this pathway 'may occur more commonly in Earth's crust than is currently recognized'. This conclusion sits nicely with the discovery of methane-rich fluids from the Lost City, Conical Seamount, and similar sites.

5.3 WATER FLOWS

Water in the pore spaces of rocks includes 'meteoric' water, derived from precipitation and surface water on land (commonly known as 'groundwater'), 'connate' water (fossil water retained in sediments during burial), and 'juvenile' water (from magmatic sources). The first of these is important in the context of seabed fluid flow

because, outside enclosed basins, groundwater flows towards the sea, entering the marine environment as the base flow of rivers and as submarine groundwater discharge (Figure 5.3).

5.3.1 Submarine groundwater discharge (SGD)

Church (1996) pointed out that 97% of the Earth's liquid freshwater reservoir is groundwater. Groundwater occupies pore spaces, joints, and fissures in soils and rocks (aquifers) on land. The most important source of groundwater on land is 'meteoric' water. This is constantly replenished by precipitation and surface water flows in recharge areas. Apart from the presence of rivers, and aquifers that outcrop or subcrop beneath the seabed, the parameters that determine whether or not SGD is likely in an area are: precipitation, the permeability of the rocks (thus sandstones and, particularly, limestones are favourable), vegetation (natural or agricultural), topography, and groundwater abstraction rates (Burnett et al., 2003).

We discuss SGD, and its importance to the hydrosphere in Section 10.3.

5.3.2 Expelled pore water

Pore water, the water retained within the pore spaces of sediments, is expelled during sediment compaction. Sandy sediments experience very little compaction. However, mud, which can be composed of 80% water when it is deposited, may be reduced in volume by as much as 40% during burial and compaction. This volumetric reduction is achieved by the expulsion of water (and other pore fluids), initially as the plate-like clay minerals are rearranged to occupy the available space more efficiently. Deep within a sedimentary column, temperature (and pressure) increases caused by burial bring about a change to certain clay minerals. The smectites include a layer of water molecules within the crystal lattice. The majority of this water is expelled when the temperature rises to between 100 and 200 °C, causing a volume reduction of about 15–30%, and the alteration of the smectite to illite (see Selley, 1998).

Normally the expelled water is retained within the overlying or adjacent sediments. However, under some circumstances, particularly during periods of compaction caused by rapid sedimentation, or in areas undergoing tectonic compression, the expelled pore

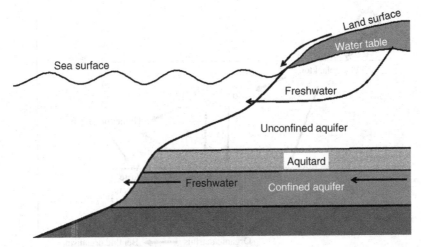

Figure 5.3 Submarine groundwater discharge (SGD). Arrows indicate fluid movement. (Simplified from Burnett *et al.*, 2003.)

water may be forced through the seabed. An example of this occurs offshore New Jersey, where rapid sedimentation on the upper continental slope during the Pleistocene contributed to elevated pore fluid pressures. Dugan and Flemings (2000) explained that spatial variations in sedimentation rates created a flow field that drove fluids laterally towards the middle and lower slope. Near the toe of the slope where sediments are thinner, the increased pore fluid pressure decreased the effective stress, promoted slope failure, and drove groundwater seeps (SGD). Dugan and Flemings considered that other passive margins might be affected in the same way.

5.4 PETROLEUM FLUIDS

Numerous geochemical surveys in the world's oceans, including ODP drilling results, show that methane is the most common gas in marine sediments. When considering seabed fluid flow, it is clear from Chapters 2 and 3 that methane is the most important fluid. In this section we investigate the various origins of methane in the marine environment. These fall into two main categories: biogenic (Sections 5.4.1–5.4.3) and abiogenic (Section 5.4.4).

5.4.1 Organic origins

Living organisms make a fundamental contribution to the marine environment and how it works – both above and below the seabed, in life and after death. In the context of seabed fluid flow, microbes play a critical role in the generation of petroleum fluids, and their utilisation. Without a grasp of some of the microbial processes it is impossible to appreciate just how critical their roles are.

We must start at the beginning, with primary production by photosynthesis. This process fuels the marine biosphere, supplies to the seabed the organic matter that is degraded, first through microbial and then thermogenic degradation, to form methane and other petroleum fluids that return to the seabed through seeps. However, this is not intended to be a detailed analysis of these processes; that is left to the many specialist texts: Schoell, 1980; Rice and Claypool, 1981; MacDonald, 1983; Whiticar *et al.*, 1986; Floodgate and Judd, 1992; Marty, 1992; Martens *et al.*, 1998; Chester, 2000, etc., and numerous papers in Howell, 1993. The biology of methane seeps is discussed in Chapter 8.

Assembling complex hydrocarbons

It is in the photic zone, the top few tens of metres of the water column, that photosynthetic primary production occurs. Here phytoplankton and, in shallow water, marine plants absorb light energy in order to manufacture organic material from CO_2. This process can be summarised by Equation (5.6),

$$6CO_2 + 6H_2O \rightarrow C_6H_{12}O_6 + 6O_2 \qquad (5.6)$$

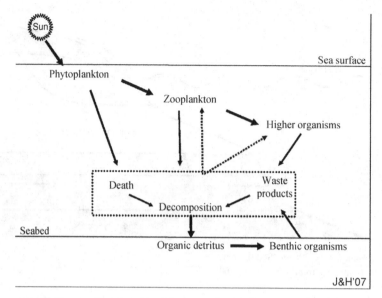

Figure 5.4* The 'conventional' (photosynthesis-based) marine food web.

where $C_6H_{12}O_6$ is taken to represent sugars, which are water-soluble carbohydrates.

To keep things simple, we will only say that the carbohydrates (cellulose, starch, etc.), along with the proteins (complex polymers of amino acids), lipids (waxes, fats, etc.) and lignins (a constituent of woody material), are the principal constituents of organic matter. They are all composed of complex hydrocarbon molecules containing repeated units, so they are sometimes known as 'biopolymers'. They can be divided into two groups: those that can, and those that cannot be converted by hydrolysis into water-soluble material.

Phytoplankton are at the base of the marine food web (summarised in Figure 5.4). They are consumed by zooplankton, which, in turn, are eaten by more sophisticated organisms, and so on. All living things in the sea, it is said, are dependent upon the photosynthesisers and, ultimately, sunlight. Without phytoplankton there would be no organic matter in the sea, apart from what is washed in from land – but of course, land plants evolved from aquatic plants, so without phytoplankton there would be no organic matter! At least, that is what some textbooks would have us believe. Even the organic detritus, eggs, faeces, dead remains (corpses, cast-off exoskeletons, etc.), of all groups of marine organisms is recycled, mainly by oxidation within the water column. The small proportion that remains sinks to the seabed.

Disassembling complex hydrocarbons

In a 'normal' seabed environment, the conventional wisdom says that the benthic ecosystem is essentially dependent upon the rain of organic detritus ('marine snow') that falls from above. Any of the material not utilised (eaten by detrivores such as nematodes) is included in the sediment, where it may be broken down. During decomposition the complex biopolymers constructed during growth are disassembled by microbial processes, the most easily broken down compounds going first. Carbohydrates go to sugars (polysaccharides then monosaccharides), proteins to amino acids, and lipids to fatty acids. In turn these will be further broken down to methanol, glycerol, then to alcohols and acids: ethanol, pyruvate, butyrate, succinate, formate, and hydrogen (Floodgate and Judd, 1992). Water-soluble by-products enter the sediment pore water and are available as substrates for microbial activity.

In sediments at, and close to the seabed, the microbial breakdown ('remineralisation') of this organic matter is done in the presence of oxygen. Oxygen is provided by seawater circulating through the sediment, often with the aid of 'bioturbation', the disturbance of the surface sediment layer by burrowing organisms, etc. The rate of oxygen supply is critical. Organic-rich sediments encourage rapid microbial activity, but this demands oxygen (the 'biological oxygen demand', BOD). When oxygen is no longer available, various oxygen-bearing compounds (oxidants) are utilised to fuel microbial metabolism and the breakdown of carbon compounds; these are (in order): SO_4^{2-}, NO_3^-, NO_2^-, and iron and manganese oxides. With depth in the sediment, microorganisms progressively use up the oxidant they require. This leads to a succession of horizons, each dominated by the utilisation of a particular oxidant. Aerobic organisms give way to sulphate-reducing bacteria (the SRBs). Equation (5.7) summarises their activity (the first term, the 'Redfield composition', represents organic matter) – see Curtis, 1983.

$$(CH_2O)_{106} (NH_3)_{16} (H_3PO_4) + 53SO_4^{2-}$$
$$\rightarrow 106CO_2 + 16NH_3 + 53S^{2-}$$
$$+ H_3PO_4 + 106H_2O \qquad (5.7)$$

Once sulphate utilisation exceeds supply, the next group, the methanogens (methane-producing microorganisms), become dominant. So, a vertical profile through the seabed normally shows a downward decline in the concentrations of first oxygen and then sulphate as the aerobic organisms and SRBs, respectively, deplete the supplies from seawater. The depletion in sulphate is accompanied by a rise in methane concentration (Figure 5.5).

5.4.2 Microbial methane

Methanogenesis is undertaken by methanogenic archaea. They are strict anaerobes, just a trace of oxygen is too much for them, but within anoxic environments with an adequate supply of organic matter (previously broken down by SRBs, etc.); they are very widespread. They can operate in a wide range of temperatures, generally 4–55 °C, but at least one species can exist at 97 °C (Wiese and Kvenvolden, 1993). As well as in marine sediments, they are found in freshwater sediments, soils, sewage sludge, land-fill sites, and in the intestines of animals – even termites – even humans! They are less thermodynamically efficient than the SRBs, and can form hydrocarbons from a limited number of the substrates available in marine sediments. The most important is carbon dioxide reduction Equation (5.8).

$$CO_2 + 4H_2 \rightarrow CH_4 + 2H_2O \qquad (5.8)$$

The hydrogen and carbon dioxide for this pathway are derived from fermentation reactions. Other substrates

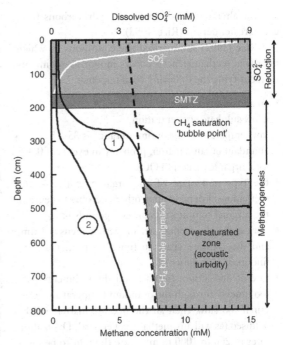

Figure 5.5 Methane and sulphate concentration profiles in seabed sediments. Notice how sulphate is depleted with depth to the sulphate–methane transition zone (SMTZ). Beneath the SMTZ methane concentrations increase with depth; lines 1 and 2 indicate high and moderate formation / accumulation rates, respectively. When methanogenesis results in concentrations exceeding methane saturation free methane bubbles form; they rise (due to buoyancy) until methane oxidation brings concentrations below saturation. At this point bubbles dissolve; this equates to the 'gas front', the top of acoustic turbidity on seismic profiles. (Adapted from Whiticar, 2002; this example is representative of sediments in Eckernförde Bay, Germany.)

are: acetate (CH_3COO^-), formate ($HCOO^-$), methanol (CH_3OH), methylamine (CH_3NH_2), and dimethylsulphide [DMS – $(CH_3)_2S$]. Each of these is derived during earlier stages of the breakdown of bioploymers.

Whiticar *et al.* (1986) suggested that carbon dioxide reduction is the dominant process in marine sediments, and fermentation in freshwater sediments. In marine sediments, the dominance of carbon dioxide reduction over the other pathways is partly explained because it provides the greatest energy yield, but also because of the availability of carbon dioxide from sulphate reduction, Equation (5.7). With increasing depth carbon dioxide may also be made available from the thermal decarboxylation (fermentation) of organic matter and the alteration of thermogenic hydrocarbons from even greater depths (Rice, 1993).

Rice (1992) summarised the conditions in which microbial methanogenesis occurs in marine sediments by identifying the following requirements:

- an anoxic environment;
- a low sulphate concentration;
- low temperature: the optimum range is 35–45 °C;
- abundant organic matter, probably in excess of 0.5% total organic carbon (TOC);
- the type and state of the organic matter – the methanogens depend upon other microbes to degrade the original complex organic compounds first;
- pore space: the average size of methanogens is 1 μm, and they require space to function, particularly in fine-grained sediments;
- rate of deposition: too fast and the sediments pass too quickly through the optimum temperature zone, but if too slow, then all the organic matter will be oxidised (as will any methane generated). Deposition rates of 200 to 1000 m my^{-1} are thought to be most favourable, depending upon the geothermal gradient (Clayton, 1992).

The shallowest conditions suitable for methanogenesis are usually found within two metres of the seabed; the temperature determines the lower limit. Exploration of the 'deep biosphere' has pushed the depth limits lower and lower in recent years (see Section 8.3.6), but it is probably realistic to say that methanogenesis continues to depths of >1 km in most areas. According to Clayton (1992) about 10% of the TOC can be converted to methane under optimum conditions. This is equivalent to 4.9 m^3 per m^3 of source rock per % TOC, or 5 × 10^7 m^3 per km^2 for a 10 m thick section of 1% TOC source. Clayton also said that pore waters at typical generation depths would become saturated with methane even if less than 0.02% of the carbon is converted to methane, so, 10% conversion would lead to the formation of a free microbial gas phase.

What we have described above is the 'normal' sub-seabed environment. Whenever there is an excess or a deficiency in one or more of the normal constituents of seawater and/or sediment pore water, or the amount of organic matter in the sediment, the balance between the various processes described will be changed. Seabed fluid flow may introduce 'abnormal' concentrations of fluids such as methane or freshwater, thereby favouring one or other of the competing microbe groups. Many separate and interacting processes operate on the sediments (including the pore fluids) during burial, including the expulsion of pore fluids as the mass of the overlying sediments causes compaction. Overall, these processes are generally referred to as 'diagenesis', and they culminate in the transformation of sediments into sedimentary rocks.

Burying the remains

As burial causes the temperature to rise, the rate of microbial methanogenesis declines and eventually stops. Despite the best efforts of the microbial communities contributing to diagenesis, some of the organic matter reaching the seabed may remain unused. This is particularly true where the supply of organic matter is high, sedimentation is rapid, and fine-grained sediments dominate (restricting the circulation of oxygen through the sediments). The unused remains are mainly composed of the least-soluble organic compounds such as lipids and lignins, but also more-easily degraded components that are protected by a more resistant covering (for example proteins held within the shell material of invertebrates). With burial, the proportion of unaltered organic matter is progressively reduced, and the water-soluble and gassy by-products of microbial activity are utilised, or squeezed out as the sediments compact.

The remaining material, insoluble and resistant to degradation, forms condensed polymers ('geopolymers'), and eventually, kerogen. A small proportion of the original material may remain, having suffered only minor alteration. Some of the original lipids retain their original carbon skeleton, and are preserved as geochemical fossils or 'biomarkers'.

5.4.3 Thermogenic hydrocarbons

Many texts discuss in great detail the formation of petroleum from organic matter buried in sedimentary rocks, so it is not necessary for us to provide more than a very brief summary.

Sedimentary rocks retaining a significant proportion of organic matter (normally > 2% TOC), in the form of kerogen, are called 'source rocks'. Kerogen in source rocks buried beyond the zone of diagenesis has characteristics dependent on the nature of the original organic matter. In general terms it can be described as insoluble amorphous organic remains.

Three kerogen types are usually distinguished on the basis of hydrogen:carbon and oxygen:carbon ratios:

- Type I kerogen (high H:C ratio, low O:C ratio) is composed mainly of lipids derived from algae and microbes from freshwater, lacustrine environments. During catagenesis, this kerogen produces oil.
- Type II kerogen (intermediate H:C and O:C ratios) contains material derived from algae, phytoplankton, and zooplankton from anoxic marine environments. This kerogen produces oil, condensate, and 'wet' gas during catagenesis.
- Type III kerogen (low H:C ratio, high O:C ratio) is derived primarily from the lignin and cellulose of vegetation from onshore environments. This kerogen type produces virtually no oil or 'wet' gas, but mainly 'dry' gas. This is produced mainly in late catagenesis to early metagenesis.

A fourth kerogen type (Type IV or IIB) may also be distinguished. Like Type III kerogen, it is derived from onshore vegetation, but it has been oxidised, probably in the soil, or during transport. Consequently it has negligible hydrocarbon potential.

The production of hydrocarbon from kerogen is achieved by the thermal 'cracking' of the kerogen molecules to form relatively simple hydrocarbon molecules. A commonly adopted shorthand for describing the formulae of the individual hydrocarbon compounds is to report only the number of carbon atoms present in the molecule: thus methane (CH_4) becomes C_1.

At this point we need to make an important distinction between different types of hydrocarbon: petroleum and coal. Both are hydrocarbons. It is surprising that many petroleum textbooks make little reference to coal. Like petroleum, it is formed from kerogens under the influence of temperature and pressure. Also, as we explain below, the formation of coal involves the expulsion of methane. Indeed, methane can be produced by simply breaking up pieces of coal; laboratory experiments have shown that when coal is ground to progressively smaller (even sub-millimetre) particle sizes, methane escapes at each stage (G. Gray, 2003, personal communication).

Petroleum

The term 'petroleum' covers natural gas, crude oil, and bitumens. These fall into four categories, described

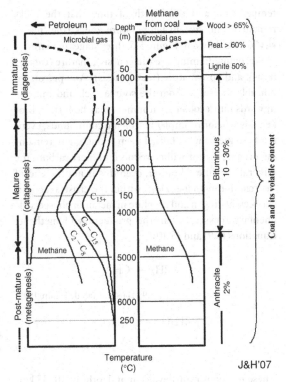

Figure 5.6 The depth of formation of hydrocarbons: microbial and thermogenic petroleum and coal.

here in the order of formation as maturity increases (Figure 5.6).

- 'Crude oil' (C_{15+}): a wide range (> 260) of relatively complex petroleum compounds most of which are in liquid form at surface temperature and pressure; the exceptions are the bitumens (tar).
- 'Condensate' (C_6–C_{15}): petroleum which occurs as gas at the elevated temperatures and pressures deep within the sedimentary rocks, but which condense to liquid at surface temperature and pressure.
- 'Wet gas' (or 'associated gas') (C_2–C_5): the wet gases are produced in association with methane, and may also be associated with condensate or oil.
- 'Dry gas' (or 'non-associated gas'); C_1: methane.

The complexity of petroleum is not indicated by this list, and it is beyond the scope of this book to go into details. We will say only that the main groups are the paraffins (or alkanes), the napthenes (cycloalkanes), and the aromatics. The type(s) of petroleum formed during continued burial depends on the kerogen type, the temperature, and the length of time that the rocks remain within certain temperature zones (the 'maturity'). As is shown in Figure 5.6, three stages of maturity are recognised: immature (diagenesis), mature (catagenesis), and post-mature (metagenesis). Any petroleum not able to escape from the source rock and migrate upwards into a cooler environment may be further broken down to form simpler petroleum compounds; eventually to methane. If buried deep enough, any remaining organic matter will be reduced to inert carbon; the mineral graphite. However, even this is not the end of the road for methane. Deep within the crust, reactions between hydrogen and graphite in metamorphosed sedimentary rocks result in the production of methane, Equations (5.9) and (5.10):

$$C + 2H_2 \rightarrow CH_4$$

(5.9; Apps and van de Kamp, 1993)

$$2C + 2H_2O \rightarrow CO_2 + CH_4$$

(5.10; Burruss, 1993)

These reactions probably occur at depths of 10–35 km, deeper than most sedimentary basins, but Burruss (1993) argued that this methane is of biogenic (microbial) origin.

Areas where source rocks are present and there is prolific petroleum generation are known as 'kitchens'. Of course, there are rocks with some organic carbon, but not enough to qualify as 'source rocks'. These would be overlooked by the petroleum industry as having no commercial significance, but they are affected by the same processes, and will also generate some petroleum.

The production of thermogenic petroleum may continue to quite considerable depths (>10 km) within a sedimentary basin. However, this depends upon the geothermal gradient; the necessary temperature range is about 60–260 °C, depending on the nature of the original organic matter. The influence of time is also important. Just like baking a cake, putting the ingredients into the oven at the right temperature is not enough. They must be left there to cook for a suitable time period. However, similar results can be achieved by slow cooking at a low temperature, or raising the temperature and removing the cake quickly, before it gets burned. An example of the latter is described in Section 5.4.4.

Coal

The heating and burial of vegetation (Type III kerogen) results in the more or less sequential formation of the following 'coals', although different plant types, and environmental and burial conditions, affect the nature of these products:

- Peat: dark brown or black fibrous plant debris, about 55% carbon;
- lignite: soft, brown material with still-recognisable vegetation, 70 to 75% carbon, and a high moisture content;
- bituminous coal: this category includes a wide variety of types according to the rank (sub-bituminous, bituminous, semi-bituminous coals) and the nature of the original material (humic and sapropelic coals); about 85% carbon and variable volatile content;
- anthracite: hard, and black; 92 to 98% carbon.

The main difference between these products is the amount of moisture and volatiles. With burial, the material is compressed and heated, and volatiles are driven off, increasing the concentration of carbon (the 'rank'); this increase in rank can be compared with the increase in the maturity of petroleum (Figure 5.6). Microbial activity is important in the degeneration of vegetation at relatively shallow depths. This is indicated by the generation of methane from peat, and the appearance of methane in drilling mud when lignite is penetrated during petroleum drilling. At greater depths and higher temperatures, thermogenic reactions take over, converting the organic matter to kerogen, and then to carbonaceous mineraloids: fusain, durain, clarain, and vitrain. This process entails the expulsion of water and volatiles, principally methane. Further heating leads to the complete expulsion of all volatiles and the formation of graphite, pure carbon.

Although there are clear similarities between the origins and formation of coal and petroleum, one significant difference is that the organic matter from which coal is derived is terrestrial rather than marine or lacustrine.

Escape from the kitchens

Petroleum, including dry gas derived during coal formation, wet gas, crude oil, and microbial gas, is generally mobile, and less dense than the waters that normally occupy the pore spaces of sedimentary rocks. Because of the unremitting pressure below the seabed, these fluids are buoyant, and tend to migrate towards the surface. However, they must first escape from the source rock. Because source rocks are generally fine grained, their permeability is low, and it is not entirely clear how this 'primary migration' is achieved. There is an extensive literature on primary migration; Selley (1998) provides a good review.

Once out of the source rock, buoyant fluids may make good their escape by migrating upwards or laterally towards the seabed (or the land) surface. How this is achieved is discussed in Section 7.5.3.

5.4.4 Hydrothermal and abiogenic petroleum

It seems that Western geologists in the oil industry are now preparing to open their minds to the possibility that oil may be generated by abiogenic processes also. This is demonstrated by the appearance of conferences on this subject. The following is an excerpt from the invitation to contribute to the London American Association of Petroleum Geologists (AAPG) Hedberg Research Conference in June 2003 (later cancelled):

> For half a century, scientists from the former Soviet Union (FSU) have recognized that the petroleum produced from fields in the FSU have been generated by abiogenic processes. This is not a new concept being reported in 1951. The Russians have used this concept as an exploration strategy and have successfully discovered petroleum fields of which a number of these fields produce either partly or entirely from crystalline basement. Is this exploration strategy limited to the petroleum provinces in Russia or does such a strategy have application to other petroleum provinces like the Gulf of Mexico or the Middle East? Some believe this is a possibility for fields in the Gulf of Mexico, and others argue for application to fields in the Middle East.
>
> AAPG, 2003

The world is certainly changing; it is only three decades since the following 'warning' was printed in front of an article published in the Bulletin of the AAPG:

> Many persons may wonder why this article is published in the AAPG Bulletin. AAPG editorial policy always has been that there are at least two sides to every question and that all sides must be

heard. Dr. Porfir'ev is one of a large group of petroleum geologists and geochemists in USSR and eastern Europe who are convinced that the organic origin of petroleum is unproved and unsound.

> Editor's footnote, Porfir'ev, 1974

By including this section, we do not want to 'kick life into a dead horse', but we do want to keep all options open; although we would like to believe that we know all of the natural processes possible, we must admit that the natural sciences have thrown a few surprises at us during the last 30 years! Not least has been the increasing awareness that petroleum reservoirs can be associated with igneous and metamorphic rocks. Petford and McCaffery (2003), in their book on hydrocarbons in crystalline rocks, cited the example of the Bach Ho field in Vietnam, which has produced oil at a rate of 130 000 barrels per day; our preferred example is the methane-rich Stockholm Archipelago (Section 3.4.2). Sugisaki and Mimura (1994) looked at 227 rocks from 50 places around the world for evidence of abiogenic hydrocarbons. They did find hydrocarbons (n-alkanes) in mantle-derived rocks: in peridotites from ophiolite suites and peridotite xenoliths in alkali basalts. However, they did not find them in gabbros and granites. They identified the following three possible origins.

1. Inorganic synthesis by Fischer–Tropsch-type reactions in the mantle.
2. Derivation from comets and meteors early in the Earth's history.
3. Recycling through subduction.

Their conclusion was that hydrocarbons can survive the high temperatures and pressures found in the mantle, but that they are decomposed into light hydrocarbon gases such as methane at the lower pressures of magmas injected into the crust. In this way they explained that peridotites do not contain heavier hydrocarbons, only the gases up to C_4H_{10} (butane).

In their review of abiogenic hydocarbons and mantle-derived helium, Jenden et al. (1993) identified five arguments supporting the idea that oil and natural gas could have an abiogenic origin.

1. The abundance of methane in the outer planets of the solar system.
2. The abundance and variety of organic compounds identified in some meteorites that may have been incorporated into the primitive Earth.

3. Occurrences of bitumen and hydrocarbon gases in igneous rocks.
4. The association between prolific hydrocarbon accumulations and major faults that could act as migration pathways for deep-seated fluids.
5. The fact that mantle-derived helium is found together with methane and higher-molecular-weight hydrocarbons in some geothermal and commercial natural gases.

However, there are several ways in which petroleum might come to be associated with crystalline rocks, only some of which require what might be termed an 'unconventional' explanation. The first consideration is whether the origin is associated with the crystalline rocks, or whether they have migrated into fractures etc. from neighbouring sediments.

Accelerated sediment maturity

Conventional petroleum generation is a slow geological process during which source rocks mature as a sedimentary basin subsides and fills. In contrast, in places with exceptionally high heat flows, such as active spreading centres, the process is accelerated. In the Guaymas Basin, for example, the sediments overlying the oceanic crust are only 300 to 500 m thick (Curray et al., 1982), yet thermogenic petroleum is being generated. This is because of the high heat flow associated with active hydrothermal vents that penetrate the sediments to discharge fluids at < 350 °C. This early petroleum generation also has been described from other locations: on the Mid-Atlantic Ridge, north of Iceland, on the East Pacific Rise, in the Escanaba Trough (Juan de Fuca Ridge), where seeps of live oil have been reported (Koski et al., 2002), in the Red Sea; the gases coming from the Piip Volcano, off Kamchatka (Section 3.18.7) are dominated by hydrocarbons (methane, 80.58%; ethane, 0.27%; propane, 0.26%; iso-butane, 0.0024%; and n-butane, 0.0013%); other constituents were: carbon dioxide, 1.35%; nitrogen, 16.75%; oxygen, 0.73%; hydrogen, 0.001%; argon, 0.33%; and helium, 0.04% (Torokov and Taran, 1994). At 13° N on the East Pacific Rise, although there is no sediment there is petroleum. This has been correlated with bacterial lipid residues and low-molecular-weight cyclic hydrocarbons, indicating a thermogenic (rather than abiogenic) origin (Brault et al., 1988; Simoneit, 1993).

Post-magmatic inorganic synthesis

According to Mahfoud and Beck (1995), the location and orientation of large petroleum fields in areas like the Middle East are controlled by and related to subduction and rifting activities. They suggested that the necessary ingredients for hydrocarbons, carbon and hydrogen, might originate from either organic compounds located in subducted sediments, or 'from the dissociation of carbonates ($CaCO_3$), and the reduction of carbon dioxide (CO_2) and water (H_2O) that seeps into subduction zones, or deep into rifts and fractures'. They went further by suggesting that CO_2 might be derived from mafic minerals (olivine and pyroxene) in basalts and other rocks in the lithosphere (as discussed in Section 5.2.4). At temperatures of 300–500 °C carbon and hydrogen might then combine to form paraffins and napthenes, such as those present in the oils of the Middle East.

> The continuous formation of hydrocarbons by this process, and the field locations along, near, or above subduction/rift zones, would account for the continuous increase in oil reserves, would explain why hydrocarbons are found close to those zones, and why the reserves are modest in Syria, Turkey, and Oman, relatively to the huge oil reserves found in the countries along the Gulf.
>
> Mahfoud and Beck, 1995

Mantle-derived methane

It is well known that igneous activity is a very effective release mechanism for mantle volatiles. The presence of mantle helium has nearly always been detected when gases from fumaroles, volcanic vents, or hot springs associated with volcanic activity have been measured. Hydrogen, nitrogen, methane, and carbon dioxide are also regularly found at hydrothermal vents and in volcanic eruptions. Some of these gases are the products of serpentinisation and Fischer–Tropsch reactions (as discussed in Section 5.2.4). However, although Potter and Konnerup-Madison (2003) and others think it unlikely, some may be derived from the mantle. Mantle-derived 'juvenile' helium can be distinguished from crustal helium of purely radiogenic derivation as it contains more ^3He. Concentrations of juvenile methane have been measured at exposed ocean spreading centre sites. Abnormally high concentrations of methane, ^3He (helium), and manganese exist in the effluent plumes above hydrothermal vents along the entire length of the East Pacific Rise (Lupton *et al.*, 1980; Craig, 1981; Kim *et al.*, 1983).

It is therefore likely that mantle-derived fluids are also present in other, non-igneous, areas of the world's seas and oceans. Oxburgh *et al.* (1986) investigated the helium content and the helium isotope composition of groundwaters and natural gases of some European and North American sedimentary basins. They found that the migration of volatiles from the mantle into the overlying sediments was most effective in extensional basins. The strongest mantle signature in Europe was found in the sediments of the Rhine Graben, which has been a locus of Tertiary igneous activity as well as extension. In the North Sea, radiogenic crustal helium was found to be dominant, but they found a *'weak signature'* of mantle helium where the Rhine Graben system extends into the North Sea. This could have been acquired either by petroleum migration or by leakage from the mantle along major faults. The main rifting phase in the North Sea was in the Triassic to Jurassic, but there was a late rifting phase during the late Cretaceous–early Tertiary; during the Tertiary and Quaternary there was regional subsidence centred on the existing graben system (Glennie and Underhill, 1998). So, the introduction of mantle fluids during the late rifting phase would have coincided with the onset of maturity. In places the Moho rises to depths of less than 25 km, 5 km less than normal, beneath the North Sea. It also rises beneath the petroliferous areas of the northern Gulf of Mexico (Lehner, 1969). It is often said that Moho highs account for the high heat flows in these areas (and therefore the maturity of the petroleum source rocks), but they could also be candidate locations for the flow of mantle volatiles into the overlying sedimentary strata.

If a deep convection system existed beneath thin-crusted sedimentary basins, the basins would show steeper than normal geothermal gradients. It is also tempting to suggest that a system in which mantle volatiles are driven upwards into organic-rich sediments by excess heat would also support petroleum generation. Is this the case for prolific petroleum-producing basins, such as the Arabian, North Sea, Gulf of Mexico, and Caspian Sea basins? Although this may be coincidental, these relatively high methane values in plumes above rift zones are in full accordance with theory of deep Earth outgassing (discussed in the following section).

The Deep-Earth Gas hypothesis

Thomas Gold has become the advocate of the 'Deep-Earth Gas Hypothesis' (Gold and Soter, 1980; Gold, 1999). He is not alone in thinking that hydrocarbons of the crust are derived from deep within the Earth. However, most of the other proponents are from the former Soviet Union. In the West Gold's ideas tend to attract little support, as is implied by the second quotation in this section (relating to Porfir'ev, 1974).

According to Gold, methane and other hydrocarbons were trapped during the accretion of Earth materials 4.5 billion years ago, since when the Earth has been degassing. On the basis of laboratory experiments Gold *et al.* (1986) proposed that the surfaces of clay minerals encountered by rising abiogenic methane would act as catalysts for a variety of reactions. The methane dissociates in the clay to produce unsaturated hydrocarbons such as ethylene and acetylene, which, in turn, may form aromatic hydrocarbons. Gold *et al.* demonstrated that complex hydrocarbons might form from methane at high pressure and temperature. In his 1999 book he even suggests that the hydrocarbons in coal are formed from deep-Earth gas.

Although some aspects of Gold's hypotheses have merit, and although several very deep wells have been drilled into crystalline rocks, no known commercial amounts of mantle-derived hydrocarbons have yet been found. The most detailed test of the hypothesis was the drilling of a 5000 m deep well into the largest meteorite impact crater in Europe, the Siljan Ring in central Sweden. Light hydrocarbons were found, but these were of no commercial significance.

A thorough review of 'energy gases' of abiogenic origin was reported by Apps and van de Kamp (1993). This included deep-Earth gases as, they said, the economic implications of Gold's arguments *'could be profound and merit a thorough review of all evidence that might clarify this issue'*. Apps and van de Kamp's review did not provide much comfort to Gold. This is how it was summarised:

> In this paper, we explore the likelihood that methane might be migrating from the Earth's interior. We show that the Earth's upper mantle is too highly oxidised to support the presence of other than minor concentrations of methane or hydrogen. Gases from the mantle are transported to the Earth's surface mainly dissolved in magmas. Gas

released from these magmas are composed of CO_2 (+ CO), H_2O, SO_2 or H_2S. The concentrations of methane in these gases is insignificant, and they could not be the source of commercial gas fields. Although gas including methane could migrate advectively from the mantle by flow through intergranular pores, there is no evidence that methane is transported this way. Gases migrating out of basement rocks into sedimentary basins containing natural-gas reservoirs are primarily of crustal origin and contain no detectable methane.

Apps and van de Kamp, 1993

Sugisaki and Mimura (1994), as a result of their study of abiogenic hydrocarbons (referred to earlier in this section), closed their paper with the following comments about Gold's ideas:

> We would withhold a comment on these hypotheses for the present, but these hypotheses are not inconsistent with our results. The concentrations of mantle hydrocarbons in mantle-derived rocks, however, are extremely low, and the hypothetical deep Earth hydrocarbons should be examined particularly in terms of their volume estimation.

Sugisaki and Mimura, 1994

One of the reviewers of our previous book 'accused' us of being too supportive of the deep-Earth (mantle) gas hypothesis of Thomas Gold. In this book we will go even further, even if the late Thomas Gold was an *'unconventional scientist'* (Bourque, 2002). Whilst we do not accept all his ideas, like Sugisaki and Mimura, we do not think they should be dismissed out of hand because some have considerable merit. It is sad that the scientific community seems too eager to dismiss all of them, just because some of them seem too far-fetched; this is too much like throwing out the baby with the bathwater! We have both been inspired by his book, *The Deep Hot Biosphere* (Gold, 1999), and strongly recommend it to all open-minded scientists.

Biogenic vs abiogenic

Jenden *et al.* (1993) compared the production of biogenic and abiogenic gases, and concluded that *'the potential contribution of abiogenic methane to world gas reserves could be large'*. They said that, if Tissot's and Welte's (1984) figure for the net accumulation rate of organic

carbon in sediments (about 3.2×10^9 kg y^{-1}) is correct, and if the thermal alteration took place as in Equation (5.11)

$$CH_{1.0} \text{ (kerogen)} \rightarrow 0.75C \text{ (graphite)} + 0.25CH_4$$
(5.11)

then, the maximum rate of microbial methane production can be no more than 1.5×10^9 m^3 y^{-1} (0.8×10^9 kg C y^{-1}). They compared this to the rate of mantle carbon being emitted from ocean spreading centres, 26×10^9 kg y^{-1} (48×10^9 m^3 y^{-1} CO_2) according to Des Marais (1985); with further contributions coming from volcanism on land, and geothermal activity. They then concluded that *'if methane composed only 1 percent of the total flux of carbon out of the mantle, the biogenic and abiogenic methane production rates would be of comparable magnitude'*. However, their detailed study of the carbon isotope ratios of methane and the higher hydrocarbon gases (ethane to *n*-butane), and helium isotope ratios of 1700 commercial natural gases concluded that *'far less than 1 percent of the methane in most oil and gas fields is of abiogenic origin'*.

Apps and van de Kamp (1993) evaluated the evidence of abiogenic methane in a wide range of geological contexts, and considered the likelihood of commercial accumulations. Firstly, they dismissed the mantle-derived methane accumulations, stating that mantle gases are dissolved in magma until they reach the surface, and that the gas released from the magmas comprise carbon dioxide (plus carbon monoxide), water, and sulphur dioxide or hydrogen sulphide. They thought that *'the concentrations of methane in these gases are insignificant, and they could not be the source of commercial gas fields'*. Then they considered accumulations arising from the hydrolysis of ultramafic and mafic igneous rocks in the crust, finding that *'the quantities of gas that could form from bodies of such rocks can be of the same order of magnitude as those of typical gas fields'*. They explained the apparent absence of commercially viable accumulation of abiogenic hydrocarbon gases to an absence of structural traps in the places where the gases can be generated, the limited access of water, or the presence of other substances that might inhibit hydrolysis. They finally concluded *'that resources of abiogenic energy gases in the Earth's crust are probably small and of little or no commercial interest'*. Lollar *et al.* (2002) came to a similar conclusion. Although these conclusions are generally discouraging for exploration, this does not mean that

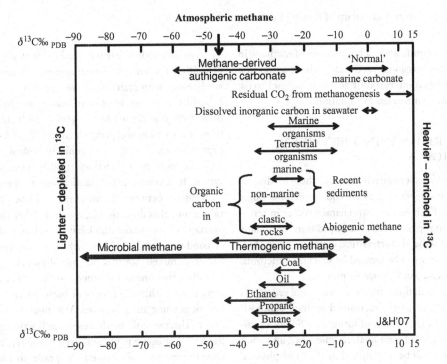

Figure 5.7* The use of carbon isotope ratios to discriminate
between methane of different origins.

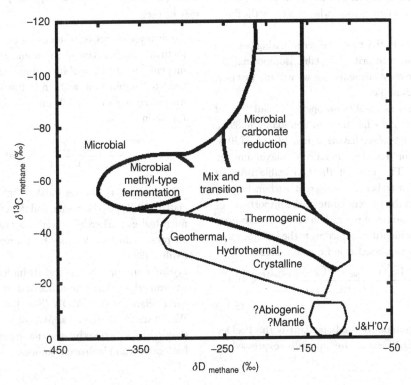

Figure 5.8 The use of carbon and hydrogen isotopes to
discriminate between methane of different origins.

they are of no significance. The presence of methane in hydrothermal emissions, particularly those associated with serpentinisation, certainly seems significant to us, even if not for commercial exploitation.

5.5 DISCRIMINATING BETWEEN THE ORIGINS

From the previous section it is clear that there is a diverse range of fluids generated by an equally diverse range of processes. Each has its own characteristics in terms of temperature, pH, chemical composition, etc. Many share at least some characteristics; methane is the obvious example as it can be formed by biogenic (microbial and thermogenic) and abiogenic processes.

Once formed, fluids in seabed rocks and sediments behave and suffer a fate determined by their composition, and not by their mode of formation. So, for example, the migration of microbial, thermogenic, and abiogenic methane will be determined by the same physical principles; methane-consuming microorganisms are not fussy about where the methane has come from. Nevertheless, it is important to know where and how the fluids originated.

The most effective tool for discriminating between elements of different origins is stable isotope analysis. This applies to most elements, but we will use just one, carbon, as an example.

Carbon has two stable isotopes: ^{13}C and ^{12}C, of which the latter is by far the most abundant: 98.89%. Carbon-14 is unstable, having a half-life of 5730 y, and is used for age determinations (maximum age 30 000 years). The ratio of the two stable isotopes varies significantly between different carbon-bearing compounds. In the present context, the importance of this is that the origin of the carbon atoms in carbonate minerals can be identified according to the isotopic ratio. This is usually expressed as in Equation (5.12).

$$\delta^{13}C\text{\textperthousand} = \frac{(^{13}C/^{12}C_{sample} - {}^{13}C/^{12}C_{standard})}{^{13}C/^{12}C_{standard}} \times 100$$

$$(5.12)$$

As the most commonly used standard is the PeeDee Belemnite (PDB), ratios are normally expressed as $\delta^{13}C\text{\textperthousand}_{PDB}$.

We have found that publications providing figures delimiting the ranges of methane of different origins rarely agree with each other! We regard $\delta^{13}C$ data with a 'healthy' degree of scepticism, and choose to interpret them with due regard to the geological 'field associations' rather than accepting them blindly at face value. Figure 5.7 suggests that abiogenic, microbial, and thermogenic methane are characterised by ranges of $\delta^{13}C$ values. It is common for publications to quote a specific cut-off between thermogenic and microbial $\delta^{13}C$ values, but this diagram clearly shows that the ranges overlap. This is understandable considering the range of conditions under which each may be formed. Perhaps the 'mixed' sources that are often referred to are actually either one or the other, but the $\delta^{13}C$ value happens to lie within the ranges of both. Of course, there may be genuine mixed sources. Also, microbial degradation of thermogenic methane causes fractionation (the lighter ^{12}C isotope is utilised preferentially), causing a clear thermogenic signature to 'degrade' to a microbial signature. Carbon isotopes alone are unable unambiguously to distinguish between origins. Two other clues may be used.

1. To distinguish between microbial and thermogenic methane, and between oil-prone and gas-prone thermogenic methane, carbon-isotope ratios may be used in conjunction with the ratio of the higher hydrocarbon gases. Gas 'wetness' is defined as in Equation (5.13).

$$\frac{C_2 + C_3 + C_4}{C_1 + C_2 + C_3 + C_4} \times 100 \qquad (5.13)$$

This value is very low (e.g. 0.01%) for 'dry' gas (associated with Type III kerogen, and therefore coals) and microbial gas, whereas a 'wet' gas (associated with oil-prone source rocks) may have a wetness of 90% (Miles, 1994).

2. Hydrogen isotope abundances are indicative of maturity and origin. The abundance of deuterium is measured relative to the SMOW (Standard Mean Ocean Water) standard, quoted as $\delta D\text{\textperthousand}_{SMOW}$. Figure 5.8 shows the distinctions between methane origins using both carbon and hydrogen isotopes.

6 · Shallow gas and gas hydrates

The importance of gas in transmitting, marking and altering sediments and of its traces as clues to depositional, paleo-ecological and diagenetic history is not generally appreciated.

Cloud, 1960

Gas generated beneath the seabed is buoyant, and tends to migrate towards the surface. Geological conditions may impede its progress, so shallow gas accumulations are formed. The exact nature of these accumulations depends on the type of sediment they are held in. Also, in certain pressure and temperature conditions, migrating gas may be sequestered by the formation of gas hydrates. Evidence is provided by various forms of acoustic signal recorded on seismic profiles, and is supported by the results of drilling, as well as the occurrence of natural gas seeps at the seabed.

Gas beneath the seabed is not a geological curiosity, but is common-place and widespread.

6.1 INTRODUCTION

After formation, gas migration and accumulation processes apply to any gas, regardless of its origin. Gas may be present in solution in the pore water, or as a free gas phase (bubbles); both are buoyant. Water saturated with dissolved methane is lighter than normal pore water (Park *et al.*, 1990), so will tend to rise. The pressure decrease during upward migration may cause gas to come out of solution to form free gas bubbles. Free gas is strongly buoyant; even at a depth of 3–4 km, the density of methane is only 200–300 kg m^{-3} compared with 1024 kg m^{-3} for seawater (Clennell *et al.*, 2000). Also, over-pressure at depth may provide an additional driver for migration. In certain conditions, found in water depths of > 350 m, some gases may be trapped in gas hydrates (discussed in Section 6.3).

Although some gas may escape through the seabed as gas seeps, the majority will become trapped on its way up to form the natural gas reservoirs exploited by the petroleum industry, and in shallow gas accumulations. In many areas such accumulations occur at more than one level below the seabed, suggesting that accumulations are not permanent features but staging posts. The term 'shallow gas' means different things to different people. Many choose an arbitrary depth below seabed (1000 m is common), but the offshore industry tends to be more pragmatic calling a gas accumulation 'shallow' if it lies above the depth of the first casing point (i.e. the depth to which a well must be drilled before the blowout preventer is installed). Anderson and Hampton (1980a) suggested that the gas fraction of a sediment normally varies between zero and 10% by volume, the maximum probably being about 16%.

6.1.1 The character and formation of gas bubbles

It might be expected that gas bubbles would simply occupy sediment pore spaces. However, Anderson *et al.* (1998) recognised three types of gas bubble in shallow sediments (illustrated in Figure 6.1); the term 'gas voids' comes from Wheeler (1988):

- **interstitial bubbles** – *'very small bubbles within undistorted interstitial pore spaces of the sediment'*;
- **reservoir bubbles** – *'a reservoir of gas occupying a region of undistorted sediment solid framework larger than the normal pore space (the bubble is a gas-filled, liquid-free region of the sediment solid framework)'*;
- **gas voids** – *'a cavity that is larger than the normal sediment interstitial space, contains only gas and is surrounded*

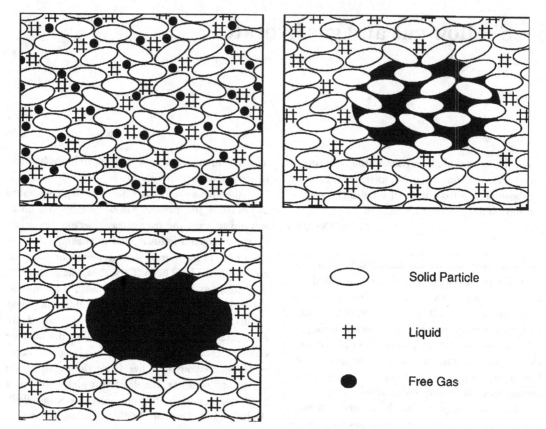

Figure 6.1 The three different types of gas bubble in a sediment: Type I (top left) – interstitial bubbles; Type II (top right) – reservoir bubbles; Type III (bottom left) – sediment-displacing bubbles–gas voids. (Reproduced with permission from Anderson *et al.*, 1998.)

by sediment that is either undisturbed or slightly distorted by cavity expansion due to formation of the bubble of free gas'.

In fine-grained sediments, gas is expected to occur in gas voids. Gas voids have been modelled in the laboratory, and explained in theoretical terms (Wheeler *et al.*, 1990; Sills *et al.*, 1991; Sills and Wheeler, 1992). Anderson and Hampton (1980a) observed them in shallow lacustrine, estuarine, and marine mud; they may be seen in core samples (Figures 2.17 and 6.2) and scanning electron micrographs (Figure 6.3). Gas voids are thought to vary between 0.5 and 50 mm, those less than about 5 mm in diameter being spherical or approximately spherical, whilst larger voids tend to be elongate (Anderson and Hampton, 1980a). In coarse-grained sediments, free gas

most likely occurs initially as interstitial bubbles. It is envisaged that microbubbles will move with the flow of pore water, or under the influence of buoyancy. When they accumulate below an impermeable sediment they coalesce to form larger bubbles. The continued accumulation of gas causes bubble growth, however, growth will be constrained by the physical size of the pore space; consequently pressure within the bubbles must rise.

Direct observations of bubbles are obtained only with difficulty. Fritz Abegg and Aubrey Anderson used computerised-tomography (CT) scan technology to provide the first images of natural bubbles in marine sediments. Divers collected samples of seabed sediment from Eckernförde Bay in pressurised core barrels, keeping them under pressure until they had been scanned. They provided the first images (see Figure 6.4) of

Figure 6.2* A gassy core, recently arrived on deck. Gas bubbles grow as a consequence of the reduction in pressure as a core is recovered from the seabed. (Photo: Alan Judd.)

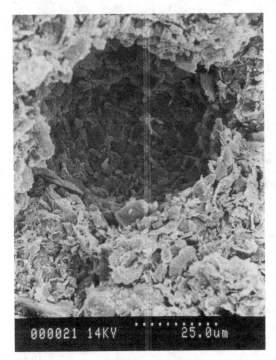

Figure 6.3 Photomicrograph of a gas void. (Reproduced from Yuan *et al.*, 1992 with permission from Elsevier.)

bubbles in their *in situ* size and shape (Abegg *et al.*, 1994; Abegg and Anderson, 1997; Anderson *et al.*, 1998). In these sediments most bubbles are gas voids, ranging in size from '*the resolution limit of 0.42 mm equivalent radius (of a sphere with volume equal to that of the bubble) up to about 5 mm equivalent radius*' (Anderson *et al.*, 1998). Most are not spherical, but 'coin shaped' (or cornflake shaped – see Boudreau *et al.*, 2005) oblate spheroids. Their orientation might be expected to be horizontal (that is parallel to the bedding), but the 'coins' were '*standing in vertical or near-vertical orientation*'; this suggests vertical gas migration.

Without a pressurised corer, the reduction in pressure when sediment samples are recovered from the seabed results in the expansion of any free gas that may be present, and the exsolution of gas dissolved in pore waters; many people have experience of gassy sediments extruding themselves from a core barrel as it is recovered to the deck – like toothpaste coming out of a tube. This makes it difficult to be sure how much, or indeed if any, free gas is present under *in situ* conditions. Beneath the reach of such corers, seismic techniques provide the best indications of the presence of gas.

6.2 GEOPHYSICAL INDICATORS OF SHALLOW GAS

In 1952 Schüler described seismic profiles on which the sediment layering was obscured by what we now know as 'acoustic turbidity'. He noted that this was generally found in sediment-filled basins, so he called it the 'Becken Effekt'. Since then, the conclusion that gas is present in sediments is, in most cases, founded on the interpretation of evidence from seismic surveys. Several people, notably Anderson and Hampton (e.g. Anderson, 1974; Hampton and Anderson, 1974; Anderson and Hampton 1980a and b; Abegg and Anderson, 1997; Anderson *et al.*, 1998) have devoted considerable efforts to the study of the acoustic effects of gas in marine sediments. A detailed example was part of the international study of the gassy sediments of Eckernförde Bay, Germany; particularly the work by Anderson *et al.*, Stoll and Bautista, and Wilkens and Richardson reported in the special edition of *Continental Shelf Research* (Richardson and Davis, 1998). Readers are referred to these papers for an in-depth discussion of this topic.

Figure 6.4* A CT scan of gas voids (white) at *in situ* pressure in a core recovered from Eckernförde Bay, Germany. The image shows a section of the core lying in a plastic core liner (the inner ring) and an aluminium chamber (outer ring). (Reproduced from Richardson and Davis, 1998 with permission from Elsevier; image from A. L. Anderson.)

6.2.1 The acoustic response of gas bubbles

Bubbles, whether in sediment pore water or in the seawater (or in the swim bladders of fish), affect the acoustic and mechanical properties of the medium: sound attenuation increases, sonic energy becomes scattered, the speed of sound propagation changes, and the tensile strength of both seawater and sediment is grossly reduced (Hampton and Anderson, 1974; D'Arrigo, 1986). The amount of attenuation depends primarily on the size of the bubbles relative to the acoustic wavelength.

$$\text{wavelength}\,(\lambda) = \text{acoustic velocity}\,(v_{\mathrm{p}})/\text{frequency}\,(f) \tag{6.1}$$

Attenuation is greatest when the acoustic frequency matches the resonant frequency of the bubbles.

1. When the bubble sizes are approximately the same as, or greater than the wavelength the seismic response is essentially that of the surrounding medium, the bubbles scatter the sound.
2. When the acoustic frequency is intermediate (see below).
3. When the wavelength is considerably greater than the bubble size, the acoustic response is that of the bulk medium (bubbles + sediment or water). In this case the acoustic velocity (v_p) may be dramatically reduced by even small (approximately 1% by volume) quantities of gas. Attenuation is increasingly affected by increasing quantities of gas.

In case (2), where bubble sizes are a little smaller than, but of the same order of magnitude as the wavelength, attenuation is significantly greater than in either (1) or (3). In this category it is the bubble size, rather than the volume of gas, that is most critical. Of course, most seismic sources emit signals with a broad frequency band, and there is likely to be a range of bubble sizes, so it is to be expected that the categories described above merge into one another rather than being distinct. In soft, fine-grained, seabed sediments with an acoustic velocity of about 1500 m s^{-1}, a seismic wavelength of 0.5 to 50 mm (the likely range of bubble sizes) would come from a system with a frequency of 30 to 3000 kHz. However, systems (echo sounders and sonars) with a source frequency in this range do not penetrate the seabed. So, in practice category (1) only applies in the water column. In contrast, the wavelength of low-frequency systems (e.g. airguns) is so great (a 250 Hz signal in seawater with an acoustic velocity of 1500 m s^{-1} would have a wavelength of 6 m) that only category (3) will apply. For these systems gas in the water column and near-seabed sediments will have little effect other than to reduce the acoustic velocity and the amplitude of the reflected signal; these systems will only detect larger gas accumulations at depth. The most suitable systems for detecting gas in near-seabed sediments are those with a high-frequency source (echo sounders, 3.5 kHz profilers, etc.; see Figure 6.5). As the penetration of these systems is restricted to a few tens of metres, deeper accumulations of gas can only be recognised on lower frequency (poorer resolution) systems.

Other factors which should be taken into account when considering gassy sediments are the following.

- Whilst bubble resonance effects are possible in all gassy sediments, they will be greatest in those with a low shear strength, such as soft, fine-grained sediments. So, acoustic turbidity is more likely to occur in silts and clays than in sands or gravels.
- A reflected signal may be detectable even after attenuation by gas if the initial signal is of sufficient amplitude.
- Small concentrations of gas in sediment, producing only small changes in acoustic velocity, can produce large attenuations.
- Whereas gas bubbles are flexible in shallow water, in water more than a few hundred metres deep they are stiff, so the acoustic properties are different; gas in seabed sediments causes a strong reflection (Mayer, 2004).

The acoustic response of gas makes it impossible to quantify the amount in a sediment. The problem is that even a small concentration of gas will cause a marked reduction in acoustic velocity (v_p), as can be seen on Figure 6.6. Effectively acoustic turbidity is either switched off or on at < 2% gas by volume. Wang *et al.* (1998) suggested an alternative way of estimating the amount of gas. They noted that pore fluid compressibility is *'sensitive'* to gas concentration, so concentrations could be calculated by measuring tidally-induced pore fluid pressure variations.

6.2.2 Seismic evidence of gassy sediments

The principal types of evidence of gassy sediments are reviewed below, starting with features seen on shallow, high-frequency seismic profiles.

Acoustic turbidity

The most commonly cited evidence of the presence of gas in shallow marine sediments is acoustic turbidity (also described as 'acoustic masking', 'acoustic blanking', and 'smears') on shallow seismic reflection profiles. It is easily recognised and has been reported from countless areas. Acoustic energy is attenuated (absorbed and scattered) by gas bubbles causing chaotic reflections. These reflections resemble a dark smear, obliterating any reflections that would otherwise originate from affected sediments, and obscuring reflections from deeper sediments (Figure 2.10). Various authors have presented

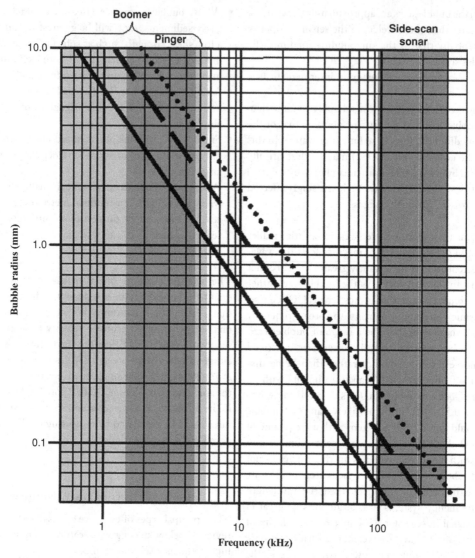

Figure 6.5* Bubble resonance frequency as a function of gas-bubble radius. The three diagonal lines show typical resonance frequencies in typical shallow-water sand (dotted) and mud (dashed), and seawater (solid). The shaded areas show typical operating frequencies for boomers, pingers, and side-scan sonar. (Adapted from Wilkens and Richardson, 1998.)

descriptive terms (blankets, curtains, etc.) for areas of acoustic turbidity with different shapes. We do not see the need for these terms, preferring to accept that shallow gas may occur in areas of any size and shape. However, we do accept and use the terms 'plume', and 'gas chimney' to describe vertical extensions, rising above the general top surface of the acoustic turbidity towards the seabed, and we refer to the limit of acoustic turbidity as the 'gas front' (Figure 2.10).

In 1986, Jones *et al.* reported the results of coring into areas of acoustic turbidity in the Irish Sea. They found that the acoustically transparent upper sediments were virtually devoid of gases. In one core the methane concentration in the upper 1.6 m of the sediments was < 10 nmol ml^{-1}. In contrast the 'gas plumes' indicated local patches of high concentration (> 100 nmol ml^{-1}). Many similar examples of acoustic turbidity associated with high methane concentrations (and with gas seeps)

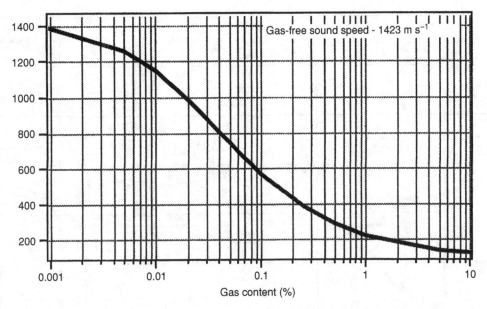

Figure 6.6 The effect of gas on acoustic velocity (v_p). Note the logarithmic x-axis scale; y-axis: v_p in m s^{-1}. The influence of even small concentrations of gas is substantial. (Reproduced from Wilkens and Richardson, 1998 with permission from Elsevier.)

have been published so the relationship with seabed fluid flow now seems unchallengeable. Yet, it is reassuring that samples from a zone of acoustic turbidity in Eckernförde Bay, collected in a pressurised core barrel, were shown by CT scan (Figure 6.4) to hold bubbles; they were between 1 and 100 mm equivalent diameter (Abegg and Anderson, 1997).

Acoustic turbidity is most common in soft, fine-grained sediments, but similar effects may also be caused by gravel beds, shell beds, and peat layers; in the first case gravel scatters the acoustic energy, in the others energy is absorbed by gases retained within the shells or peat.

Enhanced reflections

Enhanced reflections (also known as '*gas brightening*') are seen on high-resolution shallow seismic systems (e.g. boomer, 3.5 kHz pinger, and frequency modulated, 'Chirp', systems). They are coherent reflections that are markedly higher in amplitude (darker) over some of their length (Figure 2.11). They are interpreted as minor accumulations of gas, probably within thin, relatively porous, sediment layers. The gas causes a large 'negative' impedance contrast that results in large-amplitude, phase-reversed reflections (i.e. these are the shallow equivalents of 'bright spots' seen on digital seismic profiles). Sometimes (Figure 2.19), enhanced reflections

stick out from areas of acoustic turbidity, suggesting that gas is migrating laterally away from the acoustic turbidity. We suspect that this occurs where fine-grained sediment includes layers of coarser and/or more permeable sediment. Gas voids in the fine sediment cause acoustic turbidity, but gas migrating into the coarser sediment probably forms 'reservoir'-type bubbles which scatter the sound, producing enhanced reflections.

The impenetrable nature of acoustic turbidity means that it is not possible to tell how thick gassy sediment is, but enhanced reflections emerging at different depths from zones of acoustic turbidity may provide clues.

Intrasedimentary doming

Beneath some pockmarks individual reflectors are domed upwards (Figure 2.10). It is often concluded that these features are artefacts caused by the difference in the seismic velocity of water and the sediments overlying the features. However, since the seismic velocity of water is normally lower than that of the sediments, the expected effect would be a pulldown of reflections. Most commonly reflections are unaffected by the occurrence of a pockmark above them, so the seismic velocities must be closely matched. By the same token, the updoming must be a genuine feature. As they are often found in

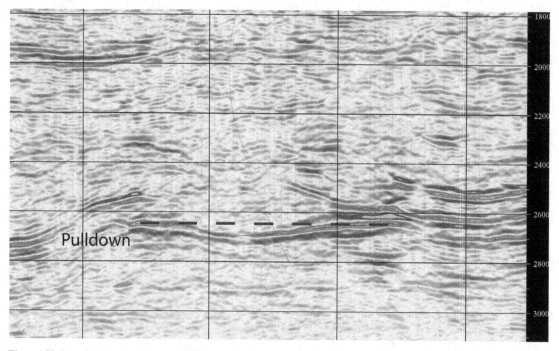

Figure 6.7* Part of a seismic section from the Statfjord field (Norwegian North Sea) illustrating pulldown. The dashed line (at 2600 ms) is horizontal. The concave reflection beneath it is being pulled down by the low acoustic velocity (v_p) of the gas held within the reservoir; it actually represents the horizontal contact between the gas (above) and oil (below) in the reservoir (i.e. a 'flat spot'). The vertical scale is in milliseconds. (Image courtesy of Jon Ottar Henden, Statoil.)

association with columnar disturbances, they may be related to vertical gas migration.

Pulldown

Pulldown is the effect seen on coherent reflections located below units of gas-charged sediments (Figure 2.21). If the gas-charged zone is thick enough to make a significant difference, this effect gives the impression that the sediments are sagging (Figure 6.7), but this is just an artefact. The presence of gas in the overlying sediments reduces the acoustic velocity (v_p), as described in the previous section. Consequently the two-way travel time of the acoustic pulse is increased, even though the sediments are horizontal. The opposite effect, 'pullup', would be caused by the presence of a high-velocity zone.

Gas chimneys

Gas chimneys (Figure 2.21) are vertical zones detected on two-dimensional (2D) and three-dimensional (3D) seismic data sets, which in some way or other have been 'disturbed' by previous or on-going gas migration. Exactly what has caused this acoustically-detected disturbance is still unknown, although it is believed that small (metre-sized) parcels of trapped gas and slightly displaced sediments may be involved. In many cases, rather than a distinct chimney, gas may be present as an amorphous cloud, such as in Figure 6.8.

A 3D-seismic method has been developed (Meldahl et al., 1999) for the detection and mapping of gas chimneys (Heggland et al., 2000; Løseth et al., 2003; Heggland, website). The method uses a standard 3D-cube, applies a set of seismic attributes, and then uses neural network software to classify ('automatically interpret') the features. This method makes chimneys and clusters of chimneys, some related to faults, visible to the interpreter.

Bright spots

These are discontinuous darkened sections of individual reflections caused by high-amplitude, negative-phase reflections (Figure 6.9, see also Figure 2.21), seen on

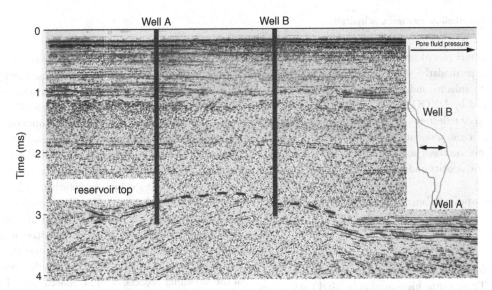

Figure 6.8* Seismic section across the Gullfaks South field (Norwegian North Sea) showing a gas cloud (a zone in which the acoustic layering is obscured by the presence of gas) above the petroleum reservoir. Well A lies outside the gas cloud, Well B passes through it. In Well B mud gas readings were found to be much higher; there were more higher-hydrocarbon gas components (C_2–C_5), more gas shows, and higher pore fluid pressure (see inset) than Well A, outside the chimney; there was a 10.2 MPa pore fluid pressure difference between the two wells at the depth indicated by the two-ended arrow on the inset diagram. NB The vertical scale of the seismic section (in seconds, two-way time) does not match the vertical scale (depth in km) of the pressure diagram. (Courtesy of Helge Løseth, Statoil.)

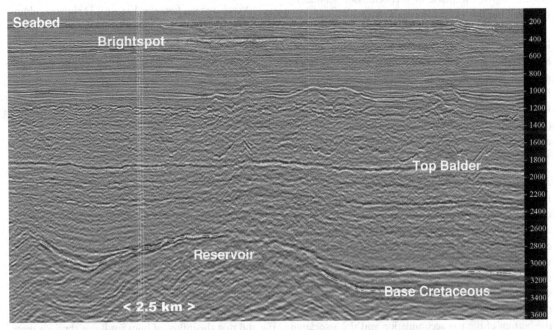

Figure 6.9* Seismic section across the Gullfaks South field (Norwegian North Sea) showing an amplitude anomaly (bright spot) 400 ms below the seabed, and vertically above the petroleum reservoir (2600 ms depth). This indicates the presence of a shallow gas accumulation. The dim zone beneath the bright spot may be partially an artefact caused by signal starvation, however this section is from approximately the same profile as shown on Figure 6.8, so it is confirmed that it is caused by a gas cloud – gas escaping from the rseservoir and rising to the shallow gas accumulation. The vertical scale is in ms of two-way travel time. (Image courtesy of Jon Ottar Henden, Statoil.)

multichannel (deep) seismic (e.g. airgun and sparker) profiles, particularly when the data have been processed to enhance indications of shallow gas (e.g. as prescribed by UKOOA, 1997). They are comparable to enhanced reflections on shallow seismic profiles (see above). Bright spots are caused by a strong acoustic impedance contrast; acoustic impedance is a function of acoustic velocity and density.

$$\text{acoustic impedance}\,(Z) = \text{acoustic velocity}\,(v_\mathrm{p})$$
$$\times\,\text{density}\,(\rho) \qquad (6.2)$$

There is a large acoustic impedance contrast when unlithified sediments overlie harder (usually older) rocks. These cause high-amplitude (dark) reflections on seismic profiles, aiding the interpretation of the stratigraphy. The reduced acoustic velocities of gas-charged sediments contrast with those of 'faster' overlying water-saturated sediments, so there is a marked impedance contrast and a high-amplitude reflection. Because the 'low-velocity' layer is beneath the 'high-velocity' layer, the phase of this reflection should be reversed, but this is not always apparent.

The recognition of bright spots is of particular importance to pre-drilling site investigation surveys because they are regarded as primary characteristics of gas pockets, which may be under significant pressure. Bright spots can be detected easily on 3D seismic data, which therefore are used for drilling-hazard evaluation (Sharp and Samuel, 2004). However, similar reflections are caused by lignite, and by channel-base gravels.

Flat spots

Flat-spot reflection is the term used for the coherent reflection occurring at the gas–water interface in a hydrocarbon reservoir (Figure 6.10). It is caused by contrast in acoustic impedance between gas-filled and water-filled sediment, and is most often horizontal, unless 'pulled down' by the overlying gas. Where flat spots are visible it is possible to estimate the height of the gas column: the separation between the flat spot at the bottom of the gas accumulation and the associated overlying bright spot marking the top. A simple calculation (discussed in Section 11.3.1) allows the pressure of the gas to be estimated, and if the area of closure of the structure can be measured, then the volume of gas can also be estimated.

Acoustic blanking and columnar disturbances

A feature commonly associated with those mentioned above (also known as 'acoustic voids', 'acoustic transparency', 'amplitude blanking', 'blank zones', 'wipeouts', and 'white zones'), but not, in our opinion, indicating the presence of free gas, is a zone devoid of reflections. This acoustic transparency, which is often close to the seabed, is generally vertical in orientation; in some cases as a thin vertical 'columnar disturbance'. It is quite commonplace on high-resolution shallow seismic profiles in areas where gas is present. It may also be seen with subtle features such as intrasedimentary doming. In our experience such features are frequently overlooked because they are presumed to be either artefacts of the recording process, or of no consequence. In fact they may provide important evidence of the presence, or former presence, of fluid within the sediments. However, there is some doubt about what they are; indeed it is probable that similar effects can result from several different causes. We need to look at a few examples to investigate their significance.

North Sea pockmarks: Figure 6.11 shows a columnar zone of acoustic blanking directly beneath a North Sea pockmark. The acoustic layering, representing sediment layers, is completely absent within the area of blanking.

Central Adriatic Sea: Hovland and Curzi (1989) described numerous features associated with seabed fluid flow in the Italian central Adriatic Sea (see Section 3.9.2). These include small columnar acoustic voids or transparent zones, and acoustic voids in deformed sediments at the root of shallow diapiric structures (Figure 6.12). Slight pore fluid overpressure is thought to be responsible for these voids (see below).

Sumba Basin, Indonesia: van Weering (1987, personal communication) reported some remarkable acoustic void structures from the Sumba Basin, west of Timor, Indonesia (Figure 6.13). Here we assume that either salt or mud diapirism is responsible for upward block faulting and that the inflow of buoyant fluids through these zones has caused local overpressure of the pore fluid and the loss of the seismic reflections (see below).

Pagoda structures, offshore West Africa: Emery (1974) drew attention to 'pagoda structures' found in

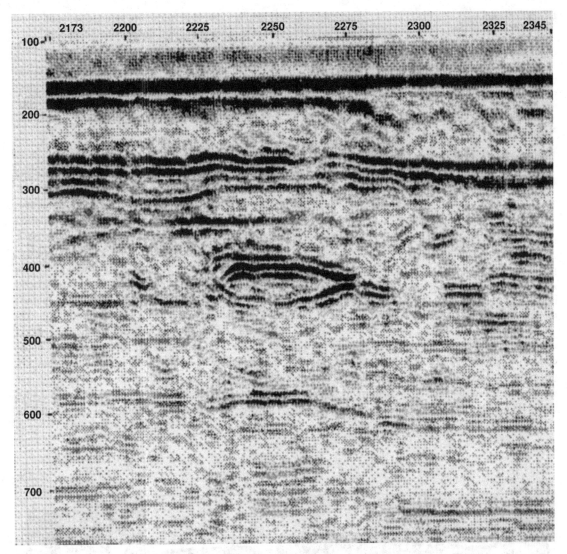

Figure 6.10* A 'flat spot', located on this 3D seismic section beneath shot points 2225 to 2275 at 400 ms two-way time. Notice how the convex-up bright spot is almost mirrored by a convex-down reflection. The intersection between these reflections marks the horizontal gas–water interface. Notice also how the reflections beneath the gas pocket are less pronounced; this is the result of signal starvation. (Adapted from Salisbury *et al.*, 1996.)

deep-water sediments. When these features were discovered they were thought to be artefacts; on one occasion the survey vessel propulsion was turned off and the vessel drifted slowly across one of the features. Because the feature still remained it was concluded that it represented a physical property of the seabed sediments. They have a widespread distribution and occur in a great range of water depths (2000–5000 m). Various kinds of pagoda structure are shown in Figure 6.14. Emery

discussed possible modes of formation and arrived at the conclusion that the acoustically dark areas may be capped by cementation by gas hydrates. This could be the case, but another interpretation was offered by Hovland *et al.* (1999a): on close examination of the features it can be seen that they represent slight seabed doming. We interpret this as a consequence of a slight elevation of pore fluid pressure. Clay/silt cores were taken from the Nares Abyssal Plain where acoustic features

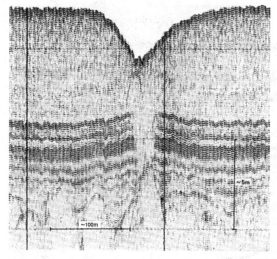

Figure 6.11 An acoustically transparent columnar disturbance (a zone without coherent reflectors) beneath a North Sea pockmark; the adjacent reflectors are updomed (intrasedimentary doming). (From Hovland and Judd, 1988.)

resembling pagoda structures occur. The porosity of the clay varies between 75 and 80%, with water-content values in the clay layers of 140%. In adjacent silt layers the water content was less than 30% (Schüttenhelm *et al.*, 1985). The sediments are therefore very soft, but have

vertically variable permeability values that could allow the migration of buoyant fluids.

Offshore Nigeria: Figure 6.15 shows acoustically transparent sediments interpreted as overpressured aquifers. Here amplitude anomalies, bright spots, and enhanced reflections occur adjacent to acoustically transparent sediments.

Examples associated with bottom-simulating reflectors (BSRs): Figure 6.16 shows examples of vertical blank zones from the Sea of Okhotsk (from Lüdman and Wong, 2003); similar features have been reported from offshore Vancouver Island (Riedel *et al.*, 2002; Wood *et al.*, 2002), near the Storegga Slide (Gravdal, 1999; and Gravdal *et al.*, 2003), and the Niger Delta (Løseth *et al.*, 2001). In each case the features extend from the seabed to a depth of several hundred metres; they are from about fifty to several hundred metres across. The various gas and hydrate indicators associated with them (bright spots, BSRs, pulldown, pockmarks, and carbonates) clearly suggest that they are gas-escape features; Løseth *et al.* described the Niger Delta examples as 'blowout pipes' (see Section 7.4.6). However, care must be taken to avoid jumping to

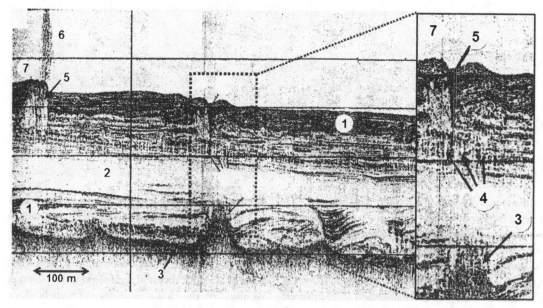

Figure 6.12 Seismic (3.5 kHz pinger) profile from the Adriatic Sea showing: 1: finely layered sediment; 2: acoustically transparent sediments; 3: acoustic turbidity; 4: columnar zones of acoustic transparency; 5: acoustic transparency associated with shallow diapirs; 6: gas seepage; 7: diapiric structures at the seabed. (Adapted from Hovland and Curzi, 1989.)

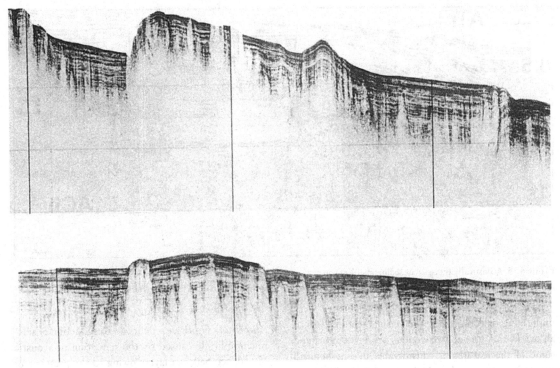

Figure 6.13 Acoustic voids in the Sumba Basin between the Indonesian islands of Flores and Timor. The features are related to seabed domes, and may be related to Neogene faulting possibly associated with fluid migration. (From Hovland and Judd, 1988; courtesy of Tjeerd van Weering.)

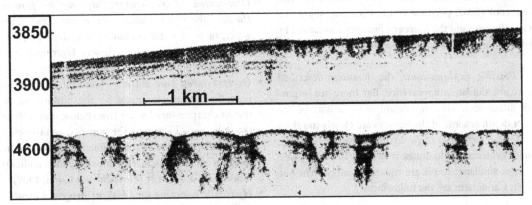

Figure 6.14 'Pagoda' structures seen on early single-channel seismic profiles from the deep ocean and discussed by Emery (1974). Both the upper and lower images show seismic profiles. Emery concluded that they result from the formation of gas hydrate at the seabed; Hovland et al., 1999b, suggested that they result from increases in pore fluid pressure beneath the seabed. (From Hovland and Judd, 1988; courtesy of K. O. Emery.)

conclusions. Figure 6.16 shows indications of pullup, normally associated with sediments with a high acoustic velocity (v_p). Riedel et al. reported a '*mud/carbonate mound*', about 6 m high, at the top of one of their features (with gas hydrates present close to the seabed), and suggested the pullup would be consistent with the presence of seabed carbonate (methane–derived authigenic carbonate, MDAC – as we suggest in the caption to Figure 6.16). A hard seabed reflects acoustic energy, causing signal starvation, manifested by weak

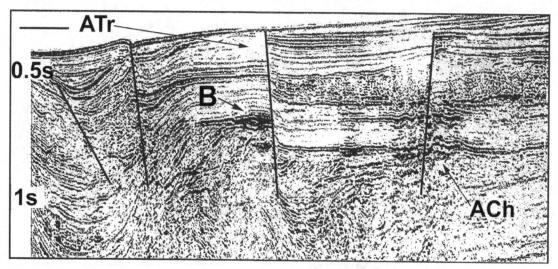

Figure 6.15 Acoustically transparent sediments (ATr) on a 2D seismic record from offshore Nigeria: ACh marks an acoustic chimney; the bright spot (B) is bounded by a possible flat spot (gas–water contact). The horizontal scale bar (top left) is 1 km.

underlying reflections. Are the vertical features real or an artefact? Upturned reflections are evident around some of the features at considerable depths beneath the seabed. Previously (Hovland and Judd, 1988) we suggested that such evidence (from the Gulf of Mexico) indicated physical disturbance of sediment layering by the upward movement of gas; however, Graham Westbrook (2002, personal communication) suggested that pullup around the Storegga features was caused by hydrates forming around the perimeter of the conduit.

Possible explanations: the features described above are similar in appearance, but there are important differences. Most important of these may be the water depth as some of the features are clearly too shallow for the formation of gas hydrates (see Section 6.3), whilst evidence for hydrates is present in other cases. Perhaps similar features are caused by more than one effect. Candidates are the following.

- *Signal starvation* – relatively dense material (e.g. glacial dropstones, carbonate, or gas hydrate) at or near the seabed reflects a higher proportion of the acoustic energy than the 'normal' seabed sediment of the surrounding area. The reduced amplitude of the acoustic blanking is therefore a product of a reduction in the strength of the acoustic signal reaching the layers beneath this barrier.

- *Amplitude blanking* – defined as a reduction of seismic amplitude caused by the reduction of acoustic impedance between layers owing to the presence of gas hydrate in sediments (Dillon *et al.*, 1993). A discussion of this effect and examples are provided by Lee and Dillon (2001).

- *Destruction of sedimentary layering by flowing fluids* – the upward migration of fluid (gas, porewater, or pore water containing dissolved gas) may mechanically disturb the sediment layering so that only incoherent (scattered) reflection occurs.

- *Overpressured pore water* – localised increases in pore fluid pressure, caused by the upward migration of overpressured water from below, may 'inflate' the sediments, increasing the porosity, and therefore reducing the rigidity, of the sediments (because grain-to-grain contact is reduced), impairing their ability to reflect acoustic energy (Hovland and Curzi, 1989).

- *Hydraulic pumping by adjacent gassy sediments* – each minute gas bubble within the gas-charged sediment acts as an effective 'actuator' as it reacts to pressure fluctuations by expanding and contracting. At high tide the increased hydrostatic pressure causes bubble contraction; expansion occurs at low tide. As all the bubbles within a sediment volume expand and contract together, the whole sediment body contracts and expands in harmony with the pressure fluctuation (Hovland *et al.*, 1999a). In impermeable

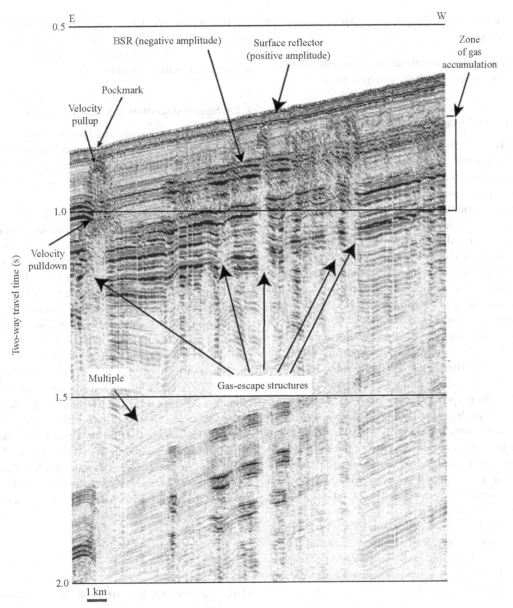

Figure 6.16 Seismic profile from the Sea of Okhotsk showing vertical blank zones (marked by Lüdman and Wong as gas-escape structures); notice also the bright spots (within the 'zone of gas accumulation'), velocity pulldown (adjacent to the vertical blank zones), velocity pullup (beneath seabed pockmarks – possibly caused by the presence of MDAC), and a BSR. (Reproduced from Lüdman and Wong, 2003 with permission from Elsevier.)

sediments adjacent gas-free sediments are affected by this 'hydraulic pumping' because their pore water is incompressible. The result is overpressuring of the gas-free sediments, with the consequence described above.

- *Acoustic scattering by gas* – Wood *et al.* (2002) suggested that columnar zones of acoustic blanking associated with gas hydrates offshore Vancouver are caused by there being sufficient gas in the sediments of these zones to scatter the acoustic energy, preventing the reflection of sufficient energy to produce coherent reflections.

We think that these explanations are possible, but none of them can apply to all the examples given.

6.2.3 Novel gas detection and mapping

3D-seismic techniques used for detection of hazardous gas

It is fair to say that Statoil pioneered the use of the 3D-seismic technique to detect shallow-gas pockets, dangerous for exploration drilling. After the West Vanguard blowout in 1985, several internal projects to improve gas detection and drilling technology in gas-prone areas were initiated. During one it was found, by time-slicing the 3D-seismic cube at various depths, that the cause of the blowout was a large and very complex shallow gas reservoir. The level where the gas accumulated proved to have been an iceberg-gouged clayey palaeo-seabed, reminiscent of the current seabed in that area of the Norwegian Sea. The iceberg ploughmarks, which extend for tens of kilometres in a criss-cross manner, had subsequently been infilled with silty sand (Gallagher *et al.*, 1989). The gas pockets were actually visible on the 2D-seismic records acquired before the well was drilled, but they were small. It was only on the 3D-seismic image that the true nature and the relief of the extensive and complex sandy gas reservoir were recognised. Using this technology, new wells were safely drilled, one only 90 m away from the blowout well, without sand even showing up at the respective depth.

The 3D-technology has now been refined further by, amongst others, Roar Heggland of Statoil (Heggland, 1997, 1998), and has been developed via the use of neural networks for chimney (volume) and fault mapping (Meldahl *et al.*, 1999).

High-resolution gravity

Marine gravity surveys may be used to provide a guide to regional geology, but they are rarely considered to be useful in petroleum exploration, other than to identify large-scale features such as salt structures, igneous intrusions, and basement highs. However, Bauer and Fichler (2002) explained that modern high-resolution gravity techniques are suitable for detecting features in the shallow sediments, such as shallow gas accumulations and gas chimneys. This is possible because the gas that replaces the pore water lowers the density of the bulk sediment. The technique is, apparently, particularly powerful when used in conjunction with 3D seismics and accurate bathymetric data.

6.2.4 Seasonal shallow gas depth variations

In Section 5.4.2 we explained how methane can be formed in shallow sediments by microbial activity. If methanogenesis results in the elevation of methane concentrations beyond the solubility limit (when the partial pressure of methane equals the hydrostatic pressure), any excess production will lead to the formation of methane gas bubbles. These bubbles are buoyant, and therefore, if conditions (principally permeability) in the sediment permit, they will rise. Any bubbles migrating towards the surface will normally be re-dissolved once they enter a zone where the pore water is undersaturated (Whiticar, 2002). For this reason methane bubbles will not normally survive to escape from the seabed. The gas front, the top of the acoustic turbidity, coincides with the depth of bubble solution.

In coastal settings it has been found that the depth of supersaturation, which is temperature dependent, migrates up and down according to the seasons. This can be seen on seismic profiles that show that the gas front migrates seasonally by 0.5 m or so. This effect has been described from Eckernförde Bay (Wever and Fiedler, 1995), and Chesapeake Bay in the eastern USA (Hagen and Vogt, 1999). This seasonal variation is caused by the activity of the methane-generating microbes, which are affected by the temperature of the bottom water.

6.3 GAS HYDRATES – A SPECIAL TYPE OF ACCUMULATION

Marine gas hydrates have attracted massive scientific and, not least, political attention over the last decade or so because of their duality as being a potential blessing (resource) and a threat (geohazard and potential climate mediator). This interest is not only reflected in the number of scientific projects aimed at studying and sampling marine hydrates, but also by a virtual flood of books, articles, and websites dedicated to this mysterious substance, for example: Ginsburg and Soloviev, 1998; Henriet and Mienert, 1998; Sloan, 1998; Vogt *et al.*, 1999a; Max, 2000b; Kukowski and Pecher, 2000 and associated papers; Paull and Dillon, 2001.

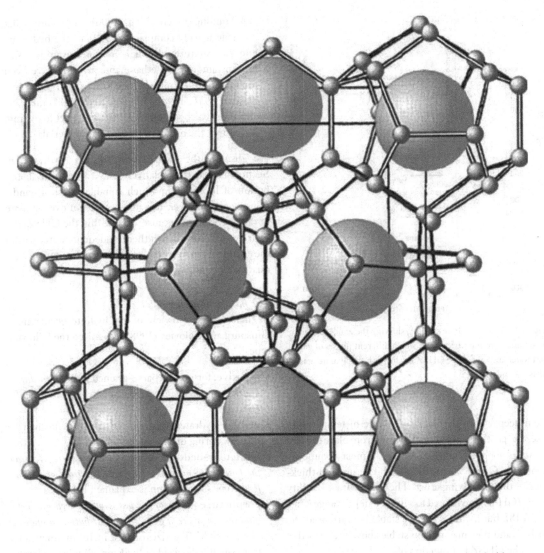

Figure 6.17* Diagramatic representation of the structure of gas hydrates. Water molecules form cage-like lattices around 'guest' molecules of methane or other gases. (Reproduced with permission from Hovland, 2005.)

6.3.1 Nature and formation

Gas hydrates are crystalline, ice-like compounds composed of water and gas; gas molecules are physically trapped within a cage-like framework of hydrogen-bonded water molecules (Figure 6.17). The guest molecule can be any constituent of natural gas. There are three hydrate structure types in nature: structures I, II, and H. They belong to a group of substances called 'clathrates' (Sloan, 1998). Gas hydrates are stable only in very specific temperature and pressure conditions in an environment with adequate water and gas. These conditions can be found on land in polar regions where surface temperatures are very cold, and in the marine environment where there is a combination of low temperature and high pressure.

In the marine environment gas hydrates only exist in the pressures found at water depths greater than 300–500 m, and the temperature is very low. Water circulation keeps ocean bottom waters at low temperatures (generally < 10 °C) all over the world. Beneath the seabed the temperature rises because of heat emanating

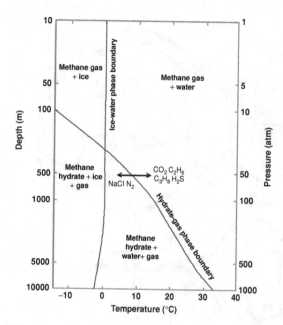

Figure 6.18* Gas hydrate stability, showing the zones (defined by temperature and pressure) in which different phases of gas and water are stable. The GHSZ lies beneath the hydrate–gas phase boundary.

gen bonding. Kvenvolden (2000) commented that methane usually comprises > 99% of the hydrocarbon gas mixture, although in some examples significant amounts of ethane and propane have been found; a sample from Hydrate Ridge (ODP Leg 146 – Kastner et al., 1998) contained about 90% methane and 10% hydrogen sulphide. With respect to hydrate formation, the origin of the gas is immaterial.

We discuss the distribution of gas hydrates in Section 11.6.3. It is relatively easy to map the distribution of locations in which suitable physical conditions occur, however, this may lead to an overestimate of the distribution because, even within the GHSZ, gas hydrates will only form if there is adequate water within the sediments to form the 'host' molecules, and an adequate supply of gas.

The nature of occurrences

In their excellent review of gas hydrate occurrences, Ginsburg and Soloviev (1998) recognised the following types:

- **massive:** formed in coarse-grained sediments 'during hydrate precipitation by fluid soaking through a granular reservoir';
- **veined:** hydrates occurring in fractures formed by the dewatering of the host sediment;
- **lenticular-bedded:** formed as 'water segregates from the host sediments and migrates to the front of hydrate formation' (segregation is explained below);
- **porphyraceous:** 'more or less isometric hydrate inclusions showing more or less regular distribution through the sediment body'. The sizes of these inclusions may range from small (pea-sized or 'hailstones') to considerably larger lumps;
- **hydrate-rock:** in some sediments the hydrates actually occupy a greater proportion of the sediment than the sediment grains: a 'body of hydrate of this kind, 4 m thick, was penetrated at DSDP Site 570 in the Middle America Trench; the concentration of terrigenous particles in this hydrate was 5 to 7%';
- **brecciated:** cataclastic hydrate aggregates, common in fracture zones.

This variety of hydrate forms shows that the intuitive idea that hydrates completely fill the pore spaces of sediments, cementing the sediment column within the GHSZ, is wrong (except, perhaps, in porous

from deep within the Earth. The rate of temperature rise, the 'geothermal gradient', varies from place to place between about 15 and 75 °C km^{-1} (most commonly about 30 °C km^{-1}) according to the nature and thickness of the underlying crust. The 'gas hydrate stability zone' (GHSZ) is confined to a limited depth range (Figure 6.18), but hydrates may be stable beneath or at the seabed, and they may remain stable above the seabed.

The GHSZ is a function of:

- water temperature;
- geothermal gradient (the rate of temperature rise with depth within the seabed sediments);
- depth below the sea surface (which determines the fluid pressure within the sediments);
- pore water salinity and the concentrations of other dissolved chemicals (inhibitors);
- the composition of guest gas(es) – these might be light hydrocarbons (methane, CH_4; ethane, C_2H_6; propane, C_3H_8; butane, C_4H_{10}), and a few non-hydrocarbon gases (including carbon dioxide, nitrogen, and hydrogen sulphide) which stabilise the water molecule lattice thermodynamically by hydro-

coarse-grained sediments). Hydrates may relatively small proportions of the pore volume if there are not enough of the 'ingredients' (water and gas) to make more hydrate. Clearly, if all the pore spaces are not filled with hydrate, then it is not true to say that a hydrate layer must act as an impermeable barrier to fluid migration. Instead, gas will be able to migrate freely through the GHSZ if the available water has already been taken up by the existing pieces of hydrate. This provides clues to the process of formation.

Gas hydrate formation processes
During ice formation at high pressure, large complex molecules enclosing large cavities form; these collapse to form ice unless gas molecules occupy the spaces. In this case gas hydrates are formed. Ginsburg and Soloviev (1998) considered various possible formation models (cryogenetic, transgression, fault, sedimentation, and filtration), but concluded that there was no evidence to suggest that any of these, except the filtration models, were responsible for the natural gas hydrate occurrences reported when they wrote their book. According to the filtration models, the hydrate-forming fluid is supplied to the formation zone by a flowing fluid. The flow may be generated by compaction, or buoyancy. Ginsburg and Soloviev accepted the suggestion of Hyndman and Davis (1992) that there are three important types of geological condition favourable for the migration of fluids towards the seabed, and they identified an example of each.

1. In accretionary wedges where sediments are squeezed by tectonic movements – for example in the Cascadia Margin (Section 3.21.1).
2. In subduction zones with no accretionary wedges, but where sediments are loaded and compressed by underthrusting – for example in the Middle America Trench (e.g. offshore Costa Rica – Section 3.24.1).
3. Where sedimentation is rapid and underconsolidated sediments are compacted by the expulsion of fluids – for example in the Blake Ridge (Section 3.27.2).

We consider that there is a fourth type.

4. Gas hydrates associated with deep-water mud volcanoes such as Håkon Mosby, and those in the Gulf of Cadiz (described in Sections 3.2.2 and 3.6.2).

It has been demonstrated that gas hydrate formation requires the migration of fluids, and Ginsburg and Soloviev (1998) found that the shape of hydrate bodies (whether stratified or veined) is determined by the fluid pathways. But, how do hydrates actually form? Ginsburg and Soloviev identified two mechanisms.

- *Precipitation* from pore waters oversaturated with gas which results in the formation of veins and massive hydrates in fractured and porous sediments, respectively. This mechanism is similar to the precipitation of minerals from hydrothermal fluids because the gas-saturated water is migrating through the sediment.
- In contrast, if the migrating fluid is free gas, then the water required for hydrate formation must be extracted from the sediment; this is termed '*segregation*'. Continued hydrate formation will therefore cause water to migrate towards the 'freezing front' as happens in the formation of permafrost (Clennell *et al.*, 1999). The source of the water may be surrounding sediments or, if the freezing front is close to the seabed, seawater. Segregation also occurs near the interface between ascending gas-saturated water and 'normal' pore water, and where there are contrasts in sediment permeability and pore water salinity. Segregation may be responsible for the formation of lenticular and porphyraceous hydrate formations.

Experiments performed under a microscope at Heriot-Watt University, UK demonstrated that hydrate formation removes methane from solution, increasing the solubility gradient in the pore water. This gradient encourages the diffusion of methane to the site of hydrate formation at the expense of free gas; gas goes into solution, and bubble sizes are reduced, to feed the diffusive migration. So, hydrates may form in the absence of free gas (Bahman Tohidi, 2004, personal communication).

Gas hydrate formation at Håkon Mosby Mud Volcano
At the Håkon Mosby Mud Volcano (HMMV) in the Norwegian Sea (Section 3.2.2) there are gas hydrates at the seabed, despite the very high heat flow from the mud volcano (7500 mK m^{-1}). The bottom-water temperature is very low (~-1 °C), but because of the high geothermal gradient, the lower boundary of the GHSZ is only about 3 m below the seabed (Egorov *et al.*, 1999). For hydrates to form and be stable at the seabed, there must be a continuous supply of methane at a rate sufficient to maintain a concentration in excess of the solubility concentration. Calculations suggest a steady-state

methane flux of 30 m^3 m^{-2} per year. This represents a very interesting exercise in the thermodynamics of gas hydrates and seepage. Pore water at a temperature of 16 °C with a methane content of 3.2 m^3 m^{-3} enters the hydrate stability zone from below. The sudden drop in temperature to about zero causes the methane solubility to decrease, and methane is precipitated in the form of gas hydrates. This process was demonstrated during video surveys when seep vents were found to be acting like snow blowers, creating a surreal scene in which snow falls upwards, accumulating on the underside of the remotely operated vehicle's manipulator arm (Michael Schleuter, 2004, personal communication). The residual upwelling pore water retains the residual methane (about 1.2 m^3 m^{-3}) and rises as a methane-rich plume over the mud volcano (Egorov et al., 1999; Damm and Budéus, 2003 – see Section 10.4.4).

The consequences of hydrate formation

The formation of hydrate by segregation may result in significant dewatering of sediments in the vicinity of the freezing front (Clennell et al., 1999). In extreme cases sediments may be desiccated and show signs of shrinkage. As vertical gas migration pathways become progressively blocked by the formation of hydrates, the consequent segregation of water from a halo of surrounding sediments may eventually produce a pathway in which hydrates cannot form because there is no water available. In these circumstances we can understand how a gas chimney may form, allowing bubbles to pass through the GHSZ to escape through the seabed as seeps.

As gas hydrates form, salt is excluded from the crystals so the residual pore water is saltier. This excess salinity diffuses away with time, but if hydrate formation is rapid the pore water may end up as brine (White, 2002). This allows the amount of hydrate formed to be calculated. Also, it probably means that gas hydrate will be inhibited in future, another factor that causes sedimentary inhomogeneities. In contrast, when hydrates decompose the water that is released is 'fresh', so a negative salinity anomaly will form – if diffusion is inhibited by sediment impermeability.

6.3.2 Gas hydrates and fluid flow

Because gas hydrates tend to block pipes and flowlines, well-controlled laboratory work on gas hydrates in pipes has been done for many years. Experiments described by

Austvik et al. (2000) clearly demonstrated how dynamic gas hydrates can be. They found that a plug of hydrate in a pipe continued to be porous when gas was bubbled through the system. It seems that the bubbles passed through channels in the plug, but when more hydrates formed and blocked these channels, new channels were formed. As successive channels were blocked, the hydrate plug became progressively less permeable.

Although natural gas hydrates are most common where there is active seabed fluid flow, laboratory experiments have shown that a flux of bubbles is not absolutely necessary for their formation and maintenance (Buffett and Zatsepina, 2000) and that the gas is actually removed from solution, not from bubbles (see previous section). The following examples illustrate the relationship between gas hydrates and active seabed fluid flow.

In regions where there is a very high fluid flux through the seabed, masses of gas hydrates may form within the sediments. At Hydrate Ridge it was found that gas hydrates formed in secondary pore spaces (fractures and joints) rather than in the pore spaces between mineral grains. This suggests that the hydrate formed from fluids flowing through the discontinuities. However, it even created its own space by fracturing or pushing apart the sediment framework as crystals grew to form: 'a peculiar structure with large filled pores not unlike bubble wrap material' (Suess et al., 1999b). Gas-rich plumes in the water column rise from pockmarks above the Cape Fear Diapir (Blake Ridge), where chemosynthetic biological communities are evidence of active seepage. This is associated with a small fault that extends down toward a dome in the underlying BSR, suggesting that the seeping fluid comes from the hydrate-bearing sediment (Paull et al., 1995). Doming has also been reported in the BSR in Lake Baikal (see Section 3.12.2). This is also associated with faulting, and it is thought that warm fluids migrating up the fault are responsible for the doming.

In the eastern part of the Sea of Okhotsk, there are scattered occurrences of BSRs associated with venting methane. According to Basov et al. (1996), localised increases in heat flow are associated with 'upward migration' of the GHSZ, and decomposition of gas hydrate, resulting in gas seeping from the seabed.

These examples and others reviewed in Chapter 3 (the Mediterranean Ridge, Section 3.10.2; the Black Sea, Section 3.11; the Makran Accretionary Wedge, Section 3.15.1; the Barbados Accretionary Wedge, Section 3.25.1; and the Gulf of Mexico, Section 3.26)

provide plenty of empirical evidence that a continuous upward stream of gas molecules is responsible for gas hydrate formation in many oceanic environments. These observations are in full agreement with the model of Hyndman and Davis (1992) and also with conclusions made by Trofimuk et al. (1973), Soloviev and Ginsburg (1994), Hovland et al. (1997a), Ginsburg and Soloviev (1998), and Clennell et al. (1999). We conclude that the migration of fluids (mainly methane) through the sediments is necessary for the formation of natural gas hydrates.

Gas hydrates are present in areas with different gas fluxes and there is evidence that the seabed topography in hydrate-prone regions is directly linked to the hydrocarbon flux, and also with tectonic activity (Zhou et al., 1999; Torres et al., 1999).

- In regions with a 'very low hydrocarbon flux', such as on parts of the Blake Ridge, there are well-developed BSRs, but few surface topographic features.
- In regions with 'low to moderate hydrocarbon flux', such as off Nigeria (Hovland et al., 1997), the eastern Black Sea (Ergun and Çifçi, 1999), and offshore Norway (Bouriak et al., 2000) there are not only BSRs, but also surface features such as pockmarks, mud diapirs, and mud vents (Brown, 1990).
- In regions with 'high tectonic activity and a moderate flux' of hydrocarbons, such as Hydrate Ridge (Torres et al., 1999; Suess et al., 1999b), there is a wide range of topographic features: pockmarks, mud vents, large authigenic carbonate structures, and large-scale spring sapping and possibly slide structures (Orange et al., 1999).
- In regions with 'high hydrocarbon flux' and low to moderate tectonic activity, such as the Gulf of Mexico, there are pockmarks, mud vents, and large complex gas hydrate mounds (MacDonald et al., 1994; etc.), but normally no BSRs.

6.3.3 The BSR

Evidence for gas hydrate in ocean-floor sediments first came from anomalous reflections, BSRs, observed on seismic profiles recorded during the DSDP (Deep-Sea Drilling Project; Leg 11) on the Blake Outer Ridge (Lancelot and Ewing, 1973). The BSR is anomalous firstly due to its highly reflective appearance, representing an abrupt change in acoustic impedance; this reflection is a result of a relatively dense/rigid (high acoustic velocity) hydrate-bearing layer overlying gassy (low acoustic velocity) sediment, and is normally of reverse polarity (negative phase). Secondly, the BSR lies at a constant depth beneath the seabed, mimicking seabed topography (hence the name BSR). Where coherent reflections indicate that sediments do not lie parallel to the seabed, the BSR cuts across these normal reflections producing a seismic pattern that cannot be explained in terms of sedimentary processes (Figure 6.19). Seismic reflections below the BSR are usually more strongly defined (i.e. higher amplitude) than those above it, where sediment layers tend to generate only weak reflections. In some cases, this upper zone is 'seismically transparent', possibly as a result of being cemented by hydrate and relatively high acoustic velocity (Paull and Dillon, 1981; Dillon and Paull, 1983; Paull et al., 2000).

Although BSRs normally appear to be smooth on high-frequency shallow seismic profiles, they may be irregular and discontinuous. Woodside and Ivanov (2002) attributed this to the variability of sediment properties such as porosity and grain size, and to a variable gas supply.

The true nature and development of BSRs are not yet fully understood, but there is some consensus that they develop mainly as a consequence of the acoustic contrast between relatively high-velocity, gas hydrate-charged sediments above low-velocity sediments (Paull, 1997; Xu and Ruppel, 1999). The low velocity is attributed to free gas beneath a hydrate layer (Figure 6.20); the negative phase of the reflection is consistent with this. Sain et al. (2000) concluded that these data indicate a 200–350 m thick layer of gas-charged sediments beneath the BSR.

A BSR may therefore be indicative of both gas hydrates above the BSR, and free gas beneath the BSR; it also may indicate the diffusive migration of gas through the whole sediment package, from below the GHSZ to the seabed (MacKay et al., 1995; Soloviev and Ginsburg, 1997; Clennell et al., 1999). However, results of drilling through BSRs have shown that things are not necessarily this simple.

ODP pioneers drilling through the BSR
During ODP Leg 141 a BSR was penetrated for the first time at the Chile Triple Junction. The IODP (Integrated ODP as of 2003) employs a board of academic and industry specialists called the Environmental Protection and Safety Panel (EPSP) to consider and overlook the

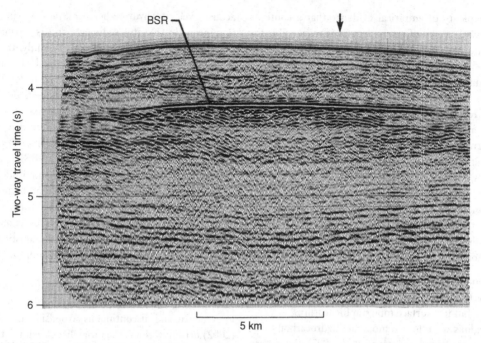

Figure 6.19 A BSR caused by gas hydrates (in the light zone beneath 4 s two-way time) lying above gas-saturated sediments (the dark zone beneath the BSR); seismic (multichannel airgun) section from the Blake Outer Ridge. (From Hovland and Judd, 1988; courtsey of William P. Dillon.)

safety implications of drilling at locations proposed by scientists. One of the main concerns of the EPSP is to avoid gas blowouts – so far with over 1000 holes drilled since 1986, ODP and IODP have experienced neither blowouts nor other dangerous hydrocarbon-related encounters. Because the first BSR penetration (in deep water) was uneventful, the EPSP approved drilling of BSRs in shallower water. The new interpretation of the BSR was that it was formed by as little as 1–2% free gas below sediments that hardly contained any gas hydrates at all (Hyndman and Davis, 1992). In 1992 the safety panel approved penetration and sampling for Leg 146, scheduled to drill on the Cascadia Accretionary Wedge (including Hydrate Ridge) where prominent and relatively shallow BSRs occur. Thereafter, Leg 164 was drilled on Blake Outer Ridge, later followed by ODP Leg 204, at Hydrate Ridge, in 2002. The following is a brief summary of the findings from these legs.

ODP LEG 146, CASCADIA ACCRETIONARY WEDGE

Two sites with relatively strong BSRs were targeted for drilling on Leg 146: one in deep water off Vancouver

Island, and a second, later to be named 'Hydrate Ridge', in shallower water (about 700 m) off Oregon. The first visible evidence of gas hydrates came in the form of 5 mm white pellets and up to 3 cm long clear nodules of gas hydrates located in soft clay within 19 m of the seabed of the shallow water Site 892 (Kastner *et al.*, 1995; Hovland *et al.*, 1995). In addition, this leg provided the following facts related to the BSR and associated natural gas hydrates.

- The BSR is not necessarily located at the base of the GHSZ, but may be shifted by up to 50 m to shallower (cooler) sediment depths (Westbrook *et al.*, 1994).
- The sediments above the BSR contain gas hydrates in various forms, but mostly as small, finely disseminated aggregates concentrated in layers and lenses up to several metres thick (Kastner *et al.*, 1995).
- There are only small amounts of free gas immediately below the BSR, which occur in a relatively dry (actually 'freeze dried'), overconsolidated, and fractured sediment layer (Hovland *et al.*, 1995; Clennell *et al.*, 1999).
- For the first time, it was discovered that the former presence of disseminated gas hydrates in sediment

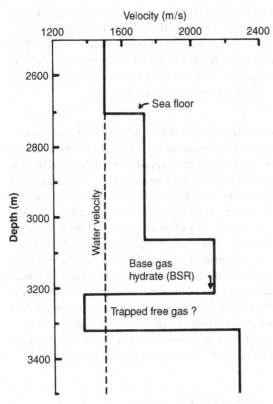

Figure 6.20 The vertical velocity profile at the location arrowed on Figure 6.19. The abrupt change from a sound velocity of > 2000 m s^{-1} in the gas hydrate zone to 1274 m s^{-1} in the underlying gassy zone gives rise to a strong negative-phase acoustic reflection. (From Hovland and Judd, 1988.)

layers might be detected by low core-temperature anomalies caused by endothermal heat loss during gas hydrate dissociation.

- Chloride depletion was detected in the pore waters of the sediments containing gas hydrates.

One of the main results of this leg was that only very small amounts of gas hydrates were found directly and indirectly above the BSR, and only up to an estimated 5% by volume free gas below the BSR (MacKay et al., 1995).

ODP LEG 164, BLAKE OUTER RIDGE

At the Blake Outer Ridge the EPSP strategy was to '*creep up on the BSR*', by first penetrating sediments without a BSR, before moving to neighbouring areas with varying strengths of BSRs. The results proved rather disappointing as no thick, massive gas hydrate layers were found. However, gas hydrates were visible in small amounts (up to 15 cm thick) in all three sites, including the site where there was no BSR. The maximum amount of free gas below the BSR was found to be 12% of the pore volume (i.e. up to 6% by sediment volume – if the pore volume is 50%). The other basic results from this leg, summarised from Matsumoto *et al.* (1996) and Holbrook *et al.* (1996), are as follows.

- Gas hydrates occur within a 200–250 m thick zone that is very homogeneous; they occupy 1–8% of the sediment volume (mainly calculated on the basis of chloride concentrations).
- Temperature measurements, made after cores had been recovered to deck, were found to be the most promising method of estimating the concentration of *in situ* gas hydrate. Even pea-size gas hydrate pellets inside the retrieved cores could be detected with suitable temperature recording devices.
- Resistivity logs showed spikes indicating the presence of hydrates between 200 and 450 m below seabed, but VSP (vertical seismic profiling) did not prove diagnostic of gas hydrates.
- Sampling gas hydrate-bearing sediments was very difficult; the core recovery rate matched the spiked chloride pattern, i.e. where chloride goes down, core recovery was poor.

ODP LEG 204, HYDRATE RIDGE

Hydrate Ridge is located only a few kilometres south of where Leg 146 was drilled (Site 892). The objective of Leg 204 was to find out more about fluxes of gas and amounts of hydrates in sediments, using all available tools. The results from the leg proved that there are varying amounts of finely disseminated and layered gas hydrates from about 45 m to the BSR at about 124 m below seabed. Based on chlorinity measurements, the hydrate concentrations ranged from 2 to 8% of the pore space. One novel aspect of this leg was the use of both hand-held and track-mounted infrared (IR) thermal imaging camera on all the cores obtained. The IR camera on the 'catwalk' indicated numerous occurrences of nodular and disseminated hydrates as the cores came fresh out of the drill barrel. This result also corresponded well with zones of strong and highly variable resistivity in the logging while drilling (LWD) data, which turned out to be a valuable tool for predicting hydrates in the sediment. Also, a new innovation was the use of a pressurised sampler developed by the HYACINTH consortium; this enabled the measurement of the physical properties of sediments at *in situ* pressure (Francis, 2004, unpublished work).

What is the BSR?

It had long been suspected that a BSR indicated the presence of gas hydrates, and that the absence of a BSR indicated that none were present. However, ODP results showed that this assumption was not reliable (Paull, 1997). In some places drilling through BSRs has failed to find gas hydrates, and in other cases gas hydrates have been found where there is no BSR. More light was shed on this problem in 1997 during geotechnical investigations for the Ormen Lange petroleum field, offshore Norway. This investigation involved drilling a BSR adjacent to the Storegga Slide scarp. However, no gas hydrates or temperature anomalies were found in the sampled sediments. Indeed, very little methane was found in the pore waters below the BSR. It was concluded that the feature perhaps represents a palaeo-BSR, and that although the gas and gas hydrates may once have been present, they are no longer there in significant quantities. These 'negative' results indicate that the existence of a BSR is no guarantee that there are detectable amounts of gas hydrates or free gas in the sediments. The question 'what actually produces the seismically detectable BSR' is still not fully answered.

One of the most surprising things found during ODP Leg 146 was that microbial activity was 'stimulated' in a zone about 10 m above the BSR, 225 m below the seabed. The methane oxidation rate was about nine times higher than the average rate at other depths, and the microbial population was considerably increased (Cragg *et al.*, 1996). Similarly, on ODP Leg 164, carbon cycling by methane, acetate, and carbon dioxide occurred '*near and below the BSR*' (Wellsbury *et al.*, 2000). At 691 m below seabed at Site 995 acetate was about a thousand times more concentrated than in typical seabed sediments (~15 mM compared to 2–20 μM). Wellsbury *et al.* said this was explained by upward migration of dissolved organic carbon into these sediments.

This evidence suggests that the BSR is a biogeochemical 'reaction zone' where advecting light hydrocarbons are utilised by microbial activity. We conclude that this activity is possible because free gas is able to accumulate below the GHSZ, presumably because migration is impeded by hydrates formed above it. This conclusion is consistent with the geophysical evidence; nevertheless the complete relationship between BSRs and gas hydrates remains enigmatic.

Gas trapped at the BSR – where does it come from?

At first it may seem that there are two possible sources for the gas (apparently) trapped at the BSR: microbial gas generated *in situ*, or thermogenic gas originating from depth. However, Pecher *et al.* (1998), amongst others, suggested a third possibility: that the free gas comes from the melting of preexisting gas hydrate. According to Clennell *et al.* (2000), on the Central and South American margins at least, BSRs are confined to areas of tectonic downwarping and continued sedimentation and subsidence. This may sound possible, but the gas must have originated from somewhere, so that riddle is not solved. It may be that the accumulated gas represents the results of successive generations of gas hydrate melting, each associated with an interglacial period.

6.3.4 Other hydrate indicators

Although BSRs may be the main indicator of the presence of gas hydrates, we have explained that these are unreliable. Fortunately there are other indicators. These include water-column plumes, which have been used to guide investigators to locations where gas hydrates are present at the seabed (see Section 10.4.1), and velocity amplitude features, or 'VAMPs'. These are similar in appearance to bright spots (described in Section 6.2.2). Gas hydrates may provide extra rigidity to the sediments in which they occur, causing an increase in the seismic velocity (v_p). This may be indicated by updoming of seismic reflections. However, whereas free gas in the sediments beneath the GHSZ will cause a reduction in v_p (as explained in Section 6.2.1), so beneath the updoming there may be downwarping (Figure 6.21). This contrast between 'fast' and 'slow' sediments is also a useful indicator of the presence of gas hydrates in boreholes, and can be detected on sonic logs. Various other well logging tools provide useful indicators, for example resistivity logs will respond to the variations in pore water salinity resulting from gas hydrate formation by segregation.

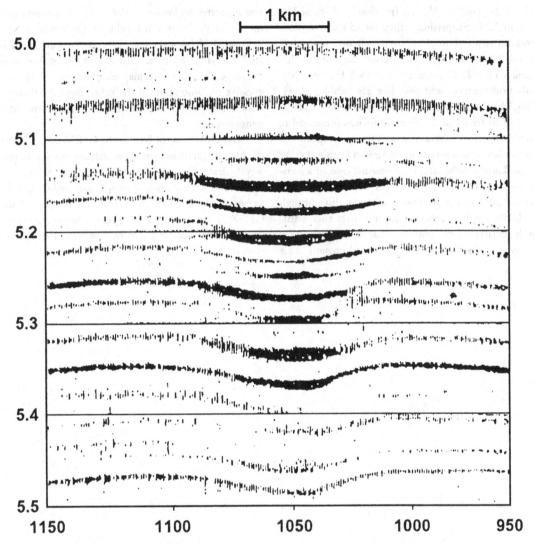

1 km

Figure 6.21 Velocity amplitude feature (VAMP) from a seismic profile taken in the Makran Accretionary Wedge; see Section 6.3.4 for an explanation. (Reproduced with permission from Minshull and White, 1989.)

6.3.5 Dissociation

Gas hydrates dissociate (decompose or melt) rapidly when heated or depressurised, releasing large volumes of gas and *fresh* water. At standard temperature and pressure the volume of gas is estimated to occupy 172 times the volume of the original hydrate (Sloan, 1998). The release of these fluids may totally change the physical character of sediment, effectively turning it to 'mush' or 'soup'. This may have consequences for seabed topog-

raphy and sediment strength and stability. We will discuss the implications of gas hydrates for offshore operations, seabed slope stability, and atmospheric methane concentrations in Chapter 11. However, it is not only human intervention that can result in gas hydrate dissociation. Natural dissociation over short time-scales occurs when gas hydrates exposed on the seabed are affected by changes in seawater temperature. Roberts *et al.* (1999a) continuously recorded bottom-water temperatures at about 550 m water depth at Bush

Hill in the Gulf of Mexico (previously discussed in Section 3.26). Surprisingly, they found that temperatures varied by about 3 °C over their one-year study period. Over a shorter time period they found that changes of < 2 °C could cause seabed hydrates partially to decompose and emit free gas bubbles. Previously, MacDonald *et al.* (1994) had reported that a large piece of hydrate had disappeared (it was presumed to have '*broken free and floated away*') from an exposure at this site between observations separated by months. We do not know whether this is an unusual case, or a common occurrence when hydrate is exposed at the seabed, but considering the likelihood of a gas flux through the GHSZ, it does seem logical to expect such natural dissociation to be common. Indeed, in some places it may be a significant agent for seabed erosion. This

was suggested by Pecher *et al.* (2004) who thought that the top of the Southern Ritchie Ridge, offshore New Zealand, had been truncated as gas hydrate dissociated, carrying sediment with it as it broke free from the seabed. This process may have been caused by progressive tectonic uplift of the ridge, but they thought it more likely to be caused by fluctuating bottom-water temperatures.

Gas evolved at the base of the GHSZ may re-enter the hydrate phase as it rises towards the seabed. In this way hydrate dissociated as a result of the change in pressure as sea level falls during global cooling may be 'recycled'. In contrast gas evolved at the top of the GHSZ as a result of seawater temperature excursions during global warming will be lost from the hydrate reservoir.

7 · Migration and seabed features

Migration is like Einstein's watch. Observations concerning its operation can be made, but since opening the system is not permitted, only hypotheses about its operation, consistent with these operations, can be made. The movement of hydrocarbons in the deep and shallow subsurface is a complex balance of processes. We can draw conclusions based only on our observations. We may never know if these conclusions are correct for any given situation.

Martin D. Matthews, 1996

Wide varieties of seabed features are formed as fluids migrate towards and emerge from the seabed; these include pockmarks and mud volcanoes. The nature and distribution of these features are dependent upon the supply of fluids, and the formations (rock and/or sediment) through which they migrate. To understand the formation of seabed fluid flow features and processes, various physical forces, including external triggers, must be considered.

7.1 INTRODUCTION

Geological processes are not often thought of as spectacular or rapid, and the features they produce do not often excite the public's interest. Earthquakes, tsunamis, and volcanoes are exceptions – but they are not the only exceptions. Mud volcanoes can be violently explosive, producing flames hundreds of metres high. Not only do they arouse the interest of the populations of the areas in which they occur, but, in Iran and Azerbaijan, the 'eternal fires' associated with gas seeps and mud volcanoes have religious significance for Zoroastrians.

In contrast to mud volcanoes, pockmarks may seem much less exciting. However, before they are dismissed as simple seabed depressions, consider their size and the energy required to form them. In the South Fladen area of the North Sea (described in Chapter 2) the 'average' pockmark is 55 metres in diameter and 1 or 2 metres deep. However, some are much bigger. The much-studied Scanner pockmark in Block UK15/25 is about 900 m long, 450 m wide, and 22 m deep; bigger than a large football stadium! That means that more than $1\,000\,000\ m^3$ (approximately 1.8 million t) of sediment has been removed from the seabed.

Seeps, mud volcanoes, and pockmarks are just three of the features associated with seabed fluid flow; the spectrum features:

- fluid escape (vents, seeps, and flows) which has no effect on the sediment;
- fluid escape which disturbs or erodes the sediment;
- fluid escape which mobilises sediments and/or the seabed;
- sediments mobilised as a result of the presence of fluids;
- mobile sediments with no significant fluid component.

These categories do not describe separate processes or phenomena, rather they represent a continuum. In fact, the continuum is multifaceted:

- processes may/may not have a surface expression;
- the features may be found at any water depth: at shallow coastal sites, on continental shelves, in the deep oceans – and in freshwater lakes;
- the grain size of affected sediments may be coarse, permeable (gravelly) sediments, or fine, impermeable muds – or anything in between;
- the sediments affected may be soft, unconsolidated sediments, compacted sediments, or lithified rocks;
- processes may occur at the seabed, at shallow depths, or deep within a sediment sequence;

- gas or water, singly or in combination, or gas dissolved in water may be involved;
- the fluids may be of any origin: microbial, thermogenic, abiogenic, geothermal, hydrothermal, meteoric, etc.;
- the components (gas, liquid, sediments) may be of deep or shallow origin;
- activity may be gentle, vigorous, or explosive;
- periods of activity may be short lived, or enduring; features may be active, dormant, or extinct;
- driving forces may be internal (buoyancy) or external (e.g. tectonic, associated with fault movement, etc.).

Of course, the continuum indicates that, unlike scientists, Nature does not feel the need to identify pigeonholes, classifying anything that has variety. An additional complication is that, whilst we are primarily considering marine features and processes, there are onshore (and even extraterrestrial) equivalents, as well as 'fossil' examples. Morphologically similar features, totally unrelated to fluid flow, also exist in the geological record. Stewart (1999) noted '*at least 10 different geological processes can result in seismically resolvable "circular" structures in sedimentary basins*'. In this chapter we will discuss the principal features directly associated with seabed fluid flow (many of which are 'circular': pockmarks, mud volcanoes, diatremes, etc.), and acknowledge that between them is an infinite variety of features. We also recognise that:

> The sea floor is a complex, generally poorly understood environment that differs fundamentally from terrestrial areas, where a longer history of engineering practice and experience exists.
>
> Prior and Hooper, 1999

Generally, hydrothermal and other hot fluids are considered to emanate from 'vents', cold fluids such as petroleum (oil and gas) come from 'seeps', and groundwater comes from 'springs'. However, there are various other definitions. For example, Clarke and Cleverly (1991) defined a petroleum seep as: '*the surface expression of a leakage pathway along which petroleum is currently flowing, driven by buoyancy, from a subsurface source*'. Link (1952) distinguished between 'seeps' (that can be seen) and 'microseeps' (that can only be detected by other means).

In this chapter we are concerned with the seabed features formed as a result of seabed fluid flow, although in many cases there is no seabed expression at all. The principal seabed features are pockmarks (Section 7.2) and mud volcanoes (Section 7.3). These, and other less common features (Section 7.4), are 'genetically related', and their formation is influenced by a number of common factors (Section 7.5). Once these are understood it is possible to see how this wide variety of features are formed (Section 7.6). In the final two sections we consider various fossil features and extraterrestrial features. 'Secondary' features, associated with seeps but produced as a result of biological activity or mineral precipitation, are dealt with in later chapters, as is the fate of the fluids once they have escaped from the seabed.

7.2 POCKMARKS AND RELATED FEATURES

Descriptions of pockmarks off the coast of Nova Scotia and in the North Sea were presented in Chapter 2. Many other examples were described in Chapter 3, but it is unnecessary to repeat these descriptions here. It is enough to summarise by saying that pockmarks are shallow seabed depressions, typically several tens of metres across and a few metres deep that are generally formed in soft, fine-grained seabed sediments. However, this can be accepted only as a general description. Pockmarks are not as 'regular' in shape as is commonly supposed, as shown in Section 2.2.4. Examples of the variety of forms come from many areas (see Chapter 3). Individual pockmarks have perimeters complicated by indentations and lobes, and floors tend to be undulating rather than smooth. They come in a considerable range of sizes, particularly in terms of the volume of sediment removed during their formation. One-metre diameter unit pockmarks may represent a volume of $0.5 \, \text{m}^3$; very little compared to the $1 \times 10^6 \, \text{m}^3$ of the Scanner pockmark (Figure 2.40). In some areas pockmarks occupy a significant proportion of the seabed. In the North Sea densities vary up to about 60 pockmarks per km^2, and up to 30% of the seabed is occupied by pockmarks (see Section 2.2.3). In the Gulf of Maine (Section 3.27.4) they are > 350 m wide, 35 m deep, and seem to occupy a considerable proportion of the seabed (Figure 7.1). However, in other areas there are only a few isolated individuals.

The key factors that distinguish pockmarks from other, morphologically similar features are that they

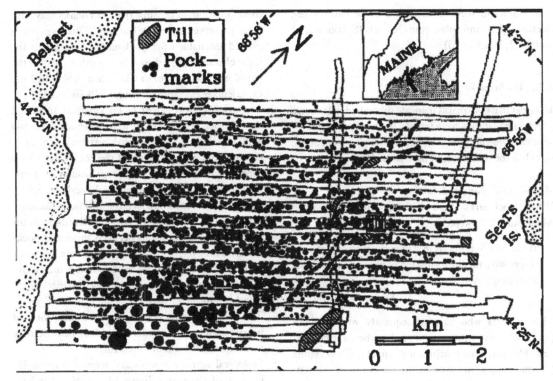

Figure 7.1 Distribution of pockmarks in Belfast Bay, Maine.
(Reproduced with permission from Kelley *et al.*, 1994.)

are erosive, and that the eroding agent comes from beneath the seabed. The truncation of sediment layers seen on seismic sections (Figures 2.11 and 3.24b are examples) clearly demonstrates that pockmarks are erosive features; sediment material has been removed from the seabed during their formation. However, some features described as pockmarks are, strictly, not pockmarks at all. These include 'collapse depressions' (described in Section 7.4.2), and depressions found within the summit craters of mud volcanoes. On land these depressions would be termed 'salses' (see Section 7.3.2).

7.2.1 Distribution

Pockmarks have been reported from every sea and ocean, and some lakes. They occur in shallow, coastal waters, and in the deep oceans; they have been reported from many different areas, a few of which are discussed in Chapter 3. The factors that determine the distribution of pockmarks are:

- a formation mechanism;
- a stable seabed, with no large-scale active erosion or deposition;
- a sediment suitable for their formation.

Sediment type influences pockmark size: there is a trend of decreasing size with increasing grain size. This explains why the sandy sediments of Tommeliten (Section 2.3.2) and the sands and gravels of the beach sediments in the Gullfaks field (Section 2.3.6) have no true pockmarks, only small depressions (unit pockmarks) about 1 m in diameter and a few tens of centimetres deep. Data from other areas show similar trends. Soft silty clays seem to provide the ideal sediments for pockmark formation, but examples have been reported from coarser and finer sediment types. However, additional factors are needed to explain the presence of the 'giant' pockmarks of Block UK15/25 (see Section 7.5.6).

Because of the importance of the sediment type, pockmarks are often restricted to areas with relatively fine sediments, but they have also been found in association with iceberg scours (e.g. in the Barents Sea, Section 3.2.1), and submarine channels (Sections 3.7.1 and 3.7.2; Figure 3.23).

7.2.2 Pockmarks and fluid flow

It is clear that pockmarks truncate subseabed reflections, indicating that they were formed by the removal or erosion of sediment. However, agents working from above or below the seabed could have caused this.

Action from above or below?

When pockmarks were first reported, King and MacLean (1970) wrote:

> We favor a hypothesis in which the main agent responsible for the formation of pockmarks is either ascending gas or water.
>
> King and MacLean, 1970

Most authors who have subsequently written about pockmarks have come to more or less the same conclusion. However, this conclusion was not readily accepted by some people involved in the operation of oil platforms and pipelines in the North Sea. During the 1970s, at both 'official' and 'unofficial' discussions between representatives of various oil companies, institutions, and university groups, several other possibilities for pockmark formation were suggested. These included: meteor impacts, scour hollows around boulders, iceberg dropstones, man-made artefacts such as wrecks, bombs, etc., biological activity (e.g. the action of sub-bottom or bottom-dwelling creatures), and mechanisms essentially related to the last ice age (e.g. strudel scour: depressions caused by the drainage of meltwater through holes in sea-ice; kettlemarks: which imply the previous existence of large bodies of ice trapped within the sediment; collapsed submarine pingos: conical mounds with an ice core that would collapse on the melting of the ice leaving a relatively easily erodable sediment and subsequently a depression).

Of these 'possible' mechanisms, the investigations described in Chapter 2 demonstrated that, in the North Sea at least, some are not viable.

1. The various glacial mechanisms were considered because the seabed in the North Sea has remained essentially unchanged, with no significant erosion or deposition since the end of the last ice age. It was therefore thought possible that pockmarks may have been preserved as relict features. This hypothesis proved untenable when pockmarks of essentially the same character were discovered in other parts of the world, such as the South China Sea, which had not experienced glaciation during the Pleistocene.

2. The geographical extent of pockmarked areas and the number of pockmarks were considered far too great for man-made features to have been responsible, either as the primary cause of the pockmarks or for them to have been caused by scour around artefacts.

3. Glacial dropstones, identified as hyperbolic reflections on high-resolution shallow seismic profiles, are sufficiently common in the late-glacial sediments for there to be a relationship between them and pockmarks. However, there is little evidence of scour around the dropstones we have seen; it is inconceivable that even the average North Sea pockmark could have been formed by scour.

4. The possibility that pockmarks were some form of impact crater was considered unlikely because they rarely have any form of raised rim.

5. Biological activity has always seemed improbable because of the vast volumes of material that have been removed during the formation of pockmarks. It is recognised that a slow process may produce a large feature, over a long period of time (about 8000 years is available in the North Sea), and there are some possible biological agents, for example the following.
   - Various organisms live in the seabed, some making large burrows. De Haas et al. (1997) calculated a 'sedimentation rate' of 29 cm 100 y^{-1} for the activities of the burrowing shrimp, Callianassa subterranea, in the Oyster Ground (southeastern North Sea, off the Netherlands) where this species is particularly abundant. This suggests that the species could 'turn over' sufficient sediment. However, it seems that the consequence of this activity is usually regarded as being the mixing of the sediment ('bioturbation') rather than erosion.
   - Whales are known to make seabed pits as they feed on burrowing amphipods in parts of the Bering Sea (Nelson and Johnson, 1987).

None of these modes of formation is operative over a sufficiently wide geographical area to have been responsible for the pockmarks described in Chapters 2 and 3.

We confidently accept in principle the hypothesis that pockmarks are erosive features formed by fluid escape.

Evidence of fluid escape?

It is dangerous to accept paradigms blindly. They must be continually challenged, and revised or discarded as the evidence dictates. This applies as much to the general acceptance that pockmarks are formed by fluid escape as to any other scientific paradigm, so it is appropriate to consider the evidence, or its absence.

> While pockmarks are commonly inferred to be associated with gas venting, there have been few attempts to verify this assumption. The Sur Pockmark field is now one of the few areas where systematic geochemical investigations have been conducted. The failure to find evidence for active gas or other fluid venting within the Sur Pockmark field opens the question as to how valid the assumption is that the existence of a well-defined field of pockmarks on the seafloor implies active fluid or gas venting. Perhaps other and perhaps purely sedimentological mechanisms should be considered.
>
> Paull *et al.*, 2002

During our North Sea work (described in Chapter 2) we have found plenty of evidence of a link between pockmarks and gas escape:

- evidence (acoustic turbidity and bright spots) of gas at various depths beneath the seabed;
- indicators of the microbial utilisation of methane: bacterial (*Beggiatoa*) mats, macrofauna with chemosynthesising endosymbionts, MDAC – see Chapters 8 and 9 for further details;
- high methane concentrations in seabed cores;
- gas bubbling from pockmarks (acoustic plumes and visual evidence);
- methane anomalies in the near-seabed water;
- sediment plumes in the water column.

It is fair to say that this evidence is not widespread. Examples of further evidence of active pockmarks has come from several other places, for example Stockholm Archipelago (Section 3.4.2), the Arabian Gulf (Section 3.14), and the Gulf of Maine (Section 3.27.4). Gas is thought to be the escaping fluid in most of these cases, but this is not necessarily always the case.

ECKERNFÖRDE BAY

The pockmarks of Eckernförde Bay, Germany, previously described in Section 3.4.1, have been studied in detail over a number of years. Whiticar (2002) discussed their origin, pointing out that although acoustic turbidity indicates the presence of shallow gas, pockmarks are located adjacent to, but not in these areas. Methane concentrations in pockmark sediments were, he thought, too low to support gas escape as a pockmark-forming mechanism. However, very low chlorinity indicated the presence of freshwater. Monitoring over a three-year period showed that temperature, salinity, and methane concentrations in the water column immediately above the pockmark varied according to rainfall in the recharge area of an aquifer that underlies the bay (Bussmann and Suess, 1998). Whiticar therefore concluded that pockmark formation should be attributed to groundwater flow.

Without arguing that, in suitable locations, this mechanism of pockmark formation is viable, we note that gas is clearly seeping from at least one of the pockmarks (Figure 7.2). It seems that the absence of sulphate in the seeping groundwater encourages microbial methane production. Whichever seeping fluid is responsible for the pockmark formation, both groundwater and methane-rich gas are emitted from them.

Absences of evidence

We do not presume that the presence of pockmarks necessarily implies *active* fluid flow. In the absence of gas it could be argued that those pockmarks were formed by some mechanism other than gas escape. However, it is also possible to argue that, if gas was responsible, then that gas has already gone! It is clear from our North Sea work that the majority of pockmarks are *not* active at the present day, and the vast majority of the extensively pockmarked Witch Ground Basin is gas-free. The seabed in the Witch Ground Basin of the North Sea has been unaffected by erosion or sedimentation since the end of the Flandrian Transgression (i.e. the start of the Holocene) about 8000 years ago. As there is no supply of sediment to fill them in, it is quite possible that pockmarks that have been formed by gas remain 'relict' features, or that they are currently 'dormant' waiting for another supply of gas. The present-day pockmark distribution reflects the cumulative pockmark activity over that time period, some of it ancient, some modern.

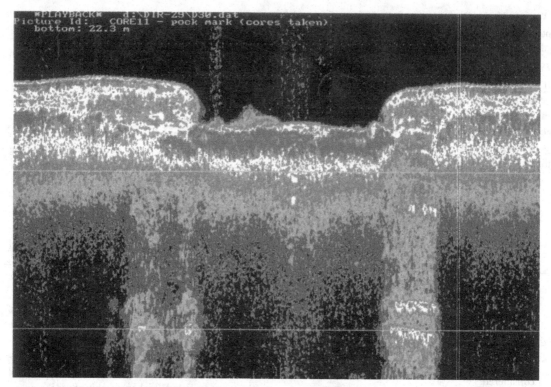

Figure 7.2* Acoustic sediment classification system (ASCS) profile showing a gas plume rising from a pockmark in Eckernförde Bay, Germany. Horizontal lines are 10ms TWT apart. (Reproduced from Richardson and Davis, 1998 with permission from Elsevier; image from D. Lambert, US Naval Research Laboratory.)

The length of time that inactive pockmarks can be preserved on the seabed is determined by the sedimentation rate. In the absence of sedimentation, features such as pockmarks will be preserved indefinitely. We see no problem in applying this conclusion to any pockmark field, including Big Sur (Section 3.22.3), even though Paull *et al.* (2002) concluded that these features are at least 45 000 years old. The absence of evidence of gas in the sediments of a pockmark today is not necessarily evidence that gas has not been present in the past.

7.2.3 Pockmark activity

A relatively small number of the reports of pockmark occurrences mention fluid-escape activity. Those that do, generally describe gentle emanations of gas. Does this mean that pockmark activity is always like this? We think not. Apart from reports that activity varies with the tides in the Arabian Gulf (Ellis and McGuinness, 1986; see Section 3.14.2), we have seen evidence of occasional more vigorous activity, including the following.

- The sediment plumes reported from the South Fladen area of the North Sea (McQuillin *et al.*, 1979; Section 2.3.1) were interpreted as evidence of gas-escape events.
- On the side-scan sonar record shown as Figure 7.3 there are numerous water-column targets which are interpreted as gas bubble plumes coming from the seeps that have been observed visually (Section 2.3.7). There is also a larger target that seems to represent a burst of gas bubbles.
- The side-scan sonar targets in pockmarks of Patras Bay, after an earthquake (Section 7.5.5).
- A natural 'blowout' event has been recorded at the Shane Seep in Santa Barbara Channel; although transient, this event seems to have produced new seabed morphological features (Leifer *et al.*, 2006; I. Leifer, 2002 personal communication).

Figure 7.3 Unprocessed side-scan sonar record showing gas bubble plumes and a larger burst of gas bubbles, from an intermittent gas escape, rising from the Scanner pockmark, Block UK15/25, North Sea. [Image acquired by the UK government (Department of Trade and Industry) as part of the Strategic Environmental Assessment process.]

These observations seem to indicate that 'normal' activity may be interrupted by occasional gas-escape events. Probably most pockmarks in the Witch Ground Basin of the North Sea are inactive, some experience occasional periods of activity, and a few (e.g. those described from Block UK15/25) are continuously active, although the rate of activity may vary. Accounts of '*enormous numbers of gas bubbles*' erupting from Belfast Bay, Maine (Kelley *et al.*, 1994; see Section 3.27.4) are not unique. However, pockmark activity was further demonstrated here when repeat surveys revealed '*significant changes*' in the pockmark distribution over a two-year period (Gontz *et al.*, 2001; also reviewed in Section 3.27.4). It is probable that the relative proportion of active, dormant, and inactive pockmarks varies from area to area. An example of a transient pockmark in the Stockholm Archipelago (Figure 7.4) suggests that pockmarks may be infilled or destroyed once the supply of gas has been depleted, but pockmark activity associated with the triggering of events by other forces must also be considered.

We will discuss formation mechanism in detail in Section 7.5. From the above, and from details presented in Chapters 2 and 3, we can summarise by stating that any formation mechanism(s) must account for the size, shape, and density of pockmarks, and the erosion of seabed sediments.

We have provided evidence of the involvement of seabed fluid flow, particularly, but not always gas: evidence of gas beneath the seabed; indications of active seepage: methane-derived authigenic carbonate (MDAC), cold-seep communities, water-column targets, high methane concentrations, gas bubbling from pockmarks. However, more often than not, these are *not* present, so their absence must also be explained.

7.3 MUD VOLCANOES AND MUD DIAPIRS

Mud volcanoes are some of the world's most dynamic and unstable sedimentary structures. A mud volcano is a positive feature constructed mainly of mud and other sedimentary constituents, which periodically or continuously vent liquid mud, including water, oil, and gas. Features range in size from just a few centimetres to > 1 km across, and tens or hundreds of metres high.

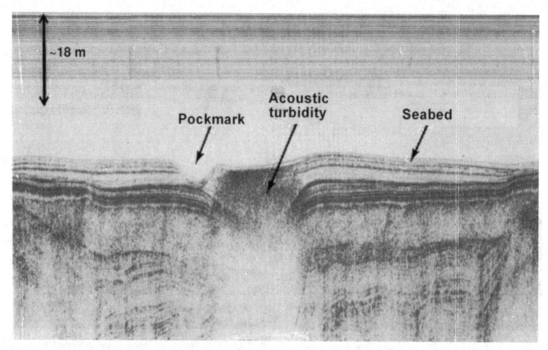

Figure 7.4 Transient pockmark (2.9 m deep, 35 m diameter) and shallow gas in an enclosed bay in the southern Stockholm Archipelago. This mud penetrator profile was recorded in 1985; five years later when the site was resurveyed there were no traces of either the pockmark or the acoustic turbidity. (From Söderberg, 1993.)

There is a confusing array of terms used to describe positive seabed features, on land and the seabed, associated with fluid emission. Perhaps the most commonly used term is 'mud volcano', but Dionne (1973) identified in the literature the following: mud volcano, sand volcano, mud cone, mud mound, mud conelet, sand blow, sand mound, mud and gas blowout, mud and gas vent, gas blister in mudpot, gas dome, gas pit, gas ring, gas volcano, spring dome, spring pit, and air-heave structure. We can extend this list with: mud lump, mud pie, mud spine, mud diapir, mud ridge, giant gas mound, shale diapir, and diatreme. Are these terms interchangeable?

Kevin Brown (1990), in a study of the formation of 'mud intrusions', compared:

- mud diapirs: '*bodies of muddy sediments driven upwards by buoyancy forces arising from the bulk density contrast between an overpressured muddy mass and an overburden of greater density*'

with

- diatremes: '*structures formed by the fluidization and entrainment of unlithified sediment by flowing liquids or gases*'

Brown concluded that the wide variety of surface features associated with 'mud intrusions' indicates that diatremes and mud diapirs are 'intimately related'. In some cases, the processes (and the resultant morphological features) are dominated by the solid (mud) phase, but in others, either water or gas (most commonly methane) is the dominant mobile element.

In Azerbaijan it is generally considered that an essential feature of mud volcanoes is the presence of breccia including clasts of rocks derived from deep below the surface. Features that emit only gas, water, and mud, particularly if they do not experience vigorous eruptions, are not entitled to this term, but are '*manifestations of mud volcanism*' (Adil Aliyev, 2003, personal communication). According to this classification many of the mud volcanoes in Italy and Taiwan (for example) are not mud volcanoes at all!

We prefer to keep the number of terms to a minimum, and suggest that the following definitions (loosely following those of Brown, 1990, and Milkov, 2000) are sufficient to cover all these features:

Mud volcano: a topographically expressed surface edifice from which solid material (at least mud, but generally also breccia comprising clasts of solid rock in a mud matrix) and fluid (water, brine, gas, oil) flow or erupt.

Mud diapir: a sediment (generally muddy or shaley) structure that has risen through a sediment sequence, piercing or deforming younger sediments. Diapirs may, but generally do not, reach the seabed. If they do, and they display evidence of extruding sediment and/or fluid, then they would be termed a 'mud volcano'.

These descriptions are generally applied to features with roots deep (sometimes several kilometres) within a sediment sequence, but morphologically and genetically similar features may also form within near-seabed features. In such cases we would use the same descriptive term, prefixed with the word 'shallow' ('shallow mud diapir', etc.).

The term mud diapir seems to be commonly used to describe large seabed features with a considerable topographic relief, but no obvious vent. Brown (1990) used the term to describe both intrusive *and* extrusive mud bodies rising from depth, but we prefer to distinguish between diapirs that remain beneath the seabed (i.e. they are intrusive) and volcanoes that are extrusive. Although mud volcanoes are morphologically similar to igneous volcanoes, they are not genetically related. They are formed by the emission of a mixture of solid, liquid, and gaseous materials that move up from depth within the underlying sediments, to be emitted at the surface.

Mud diapirs are seen on seismic sections as vertically extensive features, generally with a seismically transparent or chaotic internal character, which disrupt or pierce the normal sediment layering (see Figure 3.33). Collier and White (1990) suggested two possible explanations for the seismic transparency: the physical disruption of the sediment layers, or 'overwriting' of the layering by concentrations of gas. A third possibility is acoustic signal starvation caused by the presence of gas, gas hydrates, or MDAC at or near the seabed.

In many respects mud diapirs are similar to the more familiar salt diapirs (which are also frequently associated with gas migration and seepage – but for different reasons). Hedberg (1980), Brown (1990), Milkov (2000), Kholodov (2002a and b), and Dimitrov (2002b) all provided detailed analyses of mud volcanoes and mud

diapirs. Kopf (2002) provided a succinct explanation of the controlling processes, summarising earlier works (particularly Bishop, 1978; Yassir, 1989; Brown, 1990; Clennell, 1992); he also provided a review of mud volcanoes in 44 areas (onshore and offshore) around the world.

Not all seabed features with positive relief are mud volcanoes or mud diapirs. For example, outcrops of solid rock (particularly in coastal areas) might be mistaken for mud features. Ellis described some 'mud volcano pseudomorphs' in the Arabian Gulf which turned out to be mounds of oyster shells (Jeff Ellis, 2001, personal communication). Equally, not all mud volcanoes have an obvious topographic expression, as can be seen from the account of the discovery of Håkon Mosby Mud Volcano (Section 3.2.2).

7.3.1 The distribution of mud volcanoes and mud diapirs

Mud diapirs and volcanoes have been described from many parts of the world, on land as well as offshore – in the shallow waters of the continental shelves and the deeper waters of the continental slopes and rises (Milkov, 2000; Dimitrov, 2002b; Kopf, 2002). They occur in three main situations, but a characteristic of mud volcano/mud diapir areas is the thickness, commonly several kilometres, of the sedimentary sequence. The Southern Caspian Basin has about 20 km of post-Jurassic sediments of which about 10 km are of Pliocene age or younger. These great piles of sediments owe their origins to:

- rapid sedimentation in subsiding basins;
- rapid sedimentation in deltas and deep water fans; or
- accumulations of sediment in accretionary wedges.

Each of these geological environments is suitable for the formation of organic-rich sediments.

Another characteristic is the presence of soft, plastic, fine-grained sediments of low density in the lower portion of the sediment sequence. In most cases gas (and oil) is generated at depth. Most mud volcanoes are in mobile compression belts and accretionary wedges associated with convergent plate margins where the sediments are affected by lateral compression. Within mud-volcano 'provinces', individuals tend to be associated with faulting; they are often located at fault intersections. Submarine mud volcanoes have been reported from many areas, some of which are described in Chapter 3. Their distribution is indicated in Map 34, but the rate of discovery of offshore (especially deep water) examples is such that this map is inevitably out of date already! Dimitrov (2002b), Milkov *et al.* (2003), and Etiope and Milkov (2004) have estimated the numbers of mud volcanoes. A 'consensus' approximation is that there are about 1000 onshore, 500 on the continental shelves, and a further 5000 in deep waters, but the offshore estimates are very speculative.

In addition to the above there are mud volcanoes and diapirs in some places where shallow fine-grained sediments are overlain by denser, coarser sediments. In many cases (e.g. the Mississippi Delta) these have been deposited rapidly. These are described in Section 7.4.4.

Conical Seamount in the west Pacific, and similar features on this and other convergent plate boundaries, form where the mobile 'sediments' are not sediments at all, but serpentinite. This rock type is derived by the breakdown of ultra-basic peridotite, the igneous rock characteristic of the upper mantle. Although the rock types are so different from those of 'normal' mud volcanoes, the flows of low-viscosity, green serpentinitic mud described from Conical Seamount (Section 3.18.2) seem to behave just like typical mud-volcano sediments.

7.3.2 Mud-volcano morphology

Mud volcanoes found on the seabed are essentially the same as those found on land. Consequently it is convenient to use both on- and offshore examples to describe them.

The classic mud volcanoes are 'volcano-shaped' conical hills with a summit crater. Their size and shape vary in relation to the nature (viscosity, density, grain size) of the emission products, the nature and frequency (slow, rapid, or even explosive) of the emissions, and the volumes of the material and fluids produced. Other forms include distinct cones, domes, flat plateaus, shallow depressions, and calderas. Large mud volcanoes may be complex structures with numerous craters and cones, whereas smaller mud volcanoes may comprise only a single vent. The tallest examples are 300–400 m, or even more, in height. The widest may measure up to 3 or 4 km across, circular or elongate, and cover an area of < 100 km^2 (Figure 7.5). Smaller ones are less than 1 m high; some examples in central Italy are obliter-

ated during heavy rain or when the fields are ploughed (Martinelli and Judd, 2004).

Simple mud volcanoes consist of a cone that surrounds a central feeder channel. The summit of the volcano may be a flat or bulging plateau, normally flanked by a rim. Within this area there may be one or more active vents, each surrounded by a small cone or 'gryphon'. A large summit crater, or caldera, may form as a result of either explosive venting or summit collapse. On land, parts of these may fill with water (such pools are called 'salses'; offshore summit craters may be occupied by brine pools.

AZERI EXAMPLES

In order to discover more about how large mud volcanoes work, we have visited mud volcanoes in eastern Azerbaijan; detailed work was reported in Hovland *et al.* (1997b) and Planke *et al.* (2003).

One of the mud volcanoes we visited was Dashgil, a major structure which rises 262 m above Caspian sea level. According to Jakubov *et al.* (1971) it had major eruptions during 1882, 1902, 1908, 1926, 1958, and 2001 (Aliyev *et al.*, 2002). During our visits the oval-shaped summit-plain, which probably represents a caldera, slopes gently towards the east at an inclination of 3–5° (Hovland *et al.*, 1997b). On this caldera-plain there are four features that tell us something about the internal plumbing of the mud volcano: a cluster of gryphons, central salses, an elevated salse, and a string of sinter mounds (Figure 7.6). A cluster of gryphons was located near the highest point of the caldera. There were about 20 cones of differing sizes: from 1 m to 3 m in height, and 1 m to 10 m across (at their base). Each was partly filled with cool liquid and bubbling clay of varying viscosity. The internal mud levels differed from one cone to another, and about half of them were filled to spill-point and expelled water and mud down their sides. Small gas bubbles, with diameters of 0.5 to 3 cm, were being emitted in the cones containing low-viscosity mud. In the cones with more viscous mud, the bubble sizes were >1 cm, the largest being about 40 cm diameter bubbles.

Near the centre of the caldera-plain there were two distinct active salses. The largest (Figure 7.6, photo C) was 75 m long and 60 m wide and skirted by wet clay stained with oil and salts giving a fascinating halo of colours, ranging from white to yellow, brown, and grey. Although the pool did not have a rim, it was sharply

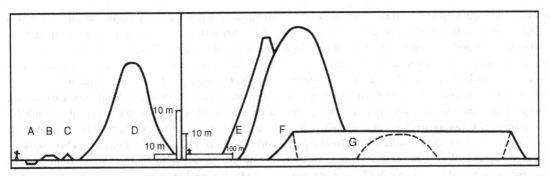

Figure 7.5 Sizes and shapes of various terrestrial mud volcanoes (note the figure of a man in both frames). Locations: A = Maghaehu Stream, New Zealand; B = Volcanito, near Cartegena, Colombia; C = Moruga Bouff, Trinidad; D = El Totumo, near Cartegena, Colombia; E = Chandragup, Makran Coast, Pakistan; F = Napag, Makran Coast, Pakistan; G = Gharniarigh-Tapeh, Gorgan region, northern Iran. (Adapted from Hovland and Judd, 1988.)

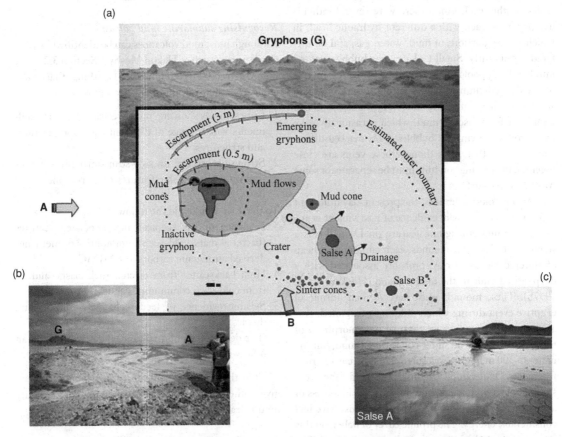

Figure 7.6* Dashgil Mud Volcano, Azerbaijan. The map (adapted from Planke *et al.*, 2003) shows the principal features of the central cone. The arrows show the approximate locations from which the photos were taken: (a) shows the main group of gryphons; (b) shows (in the foreground) one of the sinter mounds; (c) is a close up of the main salse (Salse A). (Photos: Alan Judd.)

defined, and had at least four centres of bubbling activity. Surplus water spilling from the pool drained into cracks in the clay surrounding it, and finally drained away in a narrow stream leading eastwards. The elevated salse lay about 60 m north of the first, inside a 5 m high, steep-sided (45–60°) mud mound. It was about 10 m long and 5 m wide, contained low-viscosity mud, and there were two centres of bubbling activity. Because mud, water, and gas were being emitted simultaneously from different gryphons and salses at different elevations, it seems that each is supplied by a hydraulically separate conduit, each with its own internal pore fluid pressure. This complex plumbing was also seen at the Bakhar Mud Volcano.

At Bakhar there were three groups of gryphons a few tens of metres apart within a single mud-filled crater. Within each group there were several individual gryphons, each with a different hydraulic head, in which the proportions of mud, water, gas, and oil differed significantly. Small (<1 cm) gas bubbles vented into low-lying pools of clear water within a few metres of a dark mud-rimmed pool emitting watery mud, oil, and gas bubbles, and steep-sided gryphons up to 4 m high built from viscous mud, which occasionally gave off large (10–20 cm) gas bubbles. This marked difference suggests that, even though these vents are close together, the feeding conduits must be separate for some way below the surface.

At both mud volcanoes, the present-day quiescent activity does not explain evidence of past violent eruptions. The most intriguing feature on Dashgil was a string of 10–12 sinter mounds, each containing a heap of sintered blocks of clay, probably associated with a deep-seated fault in the mud volcano (Hovland et al., 1997b). These mounds must have formed during an eruptive event during which escaping gas ignited, creating enough heat to cause thermal metamorphism of the clay. At Bakhar no sinter cones were found, but the crater, >100 m across, was surrounded by ejected rock fragments. One example (Figure 7.7), a sandstone, measured about $1 \times 1 \times 2$ m and must have had a mass of about 2 t. The eruption that expelled it must have had considerable energy. The prolific cover of lichen on this stone indicated that it was ejected a long time ago. The last recorded eruption of Bakhar was in October 1992 (Aliyev et al., 2002).

Both these mud volcanoes seem to be in a quiescent phase following a major eruption. Presumably pressure is building at depth prior to another eruption. Perhaps the materials that we saw being vented are draining from the material forming the plug.

These descriptions may be helpful in understanding the morphology of submarine mud volcanoes that are more difficult to observe in detail. Also, gas escaping from the already emitted mud (we saw minute bubbles breaking from the surface of the mud flowing down the sides of the gryphons, and measurements have shown a significant gas flux from mud flows away from the centres of other mud volcanoes – Giuseppe Etiope, 2003, personal communication) may explain the widespread occurrence of Beggiatoa mats on submarine mud volcanoes such as Håkon Mosby (Figure 8.1). Further descriptions of the mud volcanoes of Azerbaijan are presented by Guliyev, web material.

Recognising submarine mud volcanoes

Although most mud volcanoes can be identified by their shape, some (such as Håkon Mosby – Section 3.2.2) are not. Milkov (2000) recommended two things that enable submarine mud volcanoes to be recognised.

1. Core samples showing 'mud breccia' containing sediments with a range of different ages, compositions and structures.
2. Strong backscatter on side-scan sonar records representing topographic features (craters, cones, mud flows, etc.).

To these we would add the following.

3. Evidence of gas seepage and associated features (bacterial mats, cold-seep communities or methane-derived authigenic carbonate – MDAC).
4. High backscatter from ejected rock clasts, and/or from cold-seep communities and MDAC.
5. Seismic evidence of feeder channels and/or mud diapirs.
6. (For deep-water mud volcanoes) gas hydrates in an otherwise hydrate-free area.

On these criteria some features which have been given other names (e.g. 'giant gas mounds') are really mud volcanoes.

Internal structure – the plumbing

Besides the visible surface feature (cone and crater), mud volcanoes have a subsurface feeder pipe (or channel) that may have subsidiary feeder pipes. Numerous oil exploration and production wells have penetrated

Figure 7.7* Boulder ejected by the Bakhar Mud Volcano, Azerbaijan; the baseball cap provides a scale. In the background is a group of gryphons. (Photo: Alan Judd.)

mud volcanoes, shedding light on the internal structure (Bitterli, 1958; Kerr *et al.*, 1970; Jakubov *et al.*, 1971; Hovland *et al.*, 1997b). However, three-dimensional (3D) seismic imaging (Heggland *et al.*, 1996; Graue, 2000) provides more detail. Seabed cores containing oil, gas, gas hydrate clasts, and sand/shale clasts originate from deep below the seabed demonstrate vertical transport through the feeder pipe of material from several kilometres down.

Generally, the impression given is that diapirs and feeder pipes are single entities feeding an individual mud volcano. However, this does not explain the differences between the emission products of individual gryphons and salses. Davies and Stewart (2004) showed that by reprocessing seismic data from the Chirag Mud Volcano (Caspian Sea) they could resolve reflections within the diapir demonstrating that it is not a single mass of mobile sediment. They identified at least eight clastic dykes

rising like the petals of a rose from a common origin about 2 km below the surface. Their interpretation was that these dykes fed small mud volcanoes (about 600 m across) which eventually grew and amalgamated to form the present structure (about 20 km in diameter).

7.3.3 Mud-volcano emission products

Mud volcanoes typically emit a mixture of gas, water, and solid sediment derived from deep within the underlying sediments. Those of Azerbaijan produce sediments derived from up to about 20 km beneath the surface (Sokolov *et al.*, 1969; Jakubov *et al.*, 1971; Davies and Stewart, 2004). The nature and relative proportions of these products, the volumes in which they are produced, and the rate and frequency of their emission all affect the shape of the surface feature.

Sediments

The solid component of the emission products of a mud volcano often takes the form of a liquid mud containing clasts (broken fragments of lithified rocks). This is generally known as a mud breccia, olistostrome, or diapiric mélange. Kopf (2002) noted that the ratio of clasts to matrix '*covers the entire spectrum from clast-supported deposits to virtually clast-free homogeneous muds*'. Individual fragments, which may be angular or rounded, range in diameter from a few millimetres to very large boulders. This may include material derived from different stratigraphic layers, sometimes at considerable depths below the surface (their ages can be identified by micropalaeontological analysis). Mud volcanoes thus provide a useful insight into the underlying geology; in 1923 Golubyatnicov, a famous Russian geologist, called mud volcanoes '*natural survey boreholes – free of charge*' (Dimitrov, 2002b). During explosive activity, rock fragments can be ejected to considerable distances from the mud crater. According to Kopf (2002), clast-free muds often are associated with the final phase of eruption events, but they also seem to come from non-eruptive mud volcanoes.

Mud flowing away from the vents forms gryphons and cones as it dries out. Considerable volumes of very liquid mud may be emitted, and mudflows may extend hundreds of metres or even kilometres from the vent. The fluid behaviour of the mud is attributed to the high water content, which means that the muds may be unstable even on gentle slopes; failures (slides and slumps) are quite common both on land and on the seabed. Some of the large mudflows in Azerbaijan actually resemble glaciers and volcanic lava flows. Where mud volcanoes lie close to each other these flows may coalesce to cover large areas of the land surface (e.g. in Irian Jaya, New Guinea; Williams *et al.*, 1984) or the seabed (e.g. Gelendzhik Mud Plateau on the Mediterranean Ridge; Limonov *et al.*, 1996).

Waters

Chemical and isotopic analyses enable the source of water to be identified; they may be connate or meteoric waters, and individual mud volcanoes may emit mixed waters of deep and shallow origins. Martinelli and Judd (2004) found that the mud volcanoes of Italy emit brackish connate waters; Aliyev *et al.* (2002) reported that those of Azerbaijan fell into four categories, according to the dominant dissolved components: (i) hydrocar-

bons and sodium (the dominant type), (ii) chlorine and calcium, (iii) chlorine and magnesium, and (iv) sulphur and sodium. They pointed out that some mud volcanoes (such as Khamamdag) emitted all four types from different gryphons demonstrating the independence of the individual feeder pipes rising from different depths.

In some parts of the Gulf of Mexico and of the Mediterranean Ridge the waters include dense brines derived from evaporites within the sediment sequence. These brines accumulate as pools within salses (Figure 3.28).

Gases

The gases emitted by the majority of mud volcanoes are dominated by hydrocarbons (principally methane), but in some cases carbon dioxide or nitrogen dominate. However, Shakirov *et al.* (2004) found that the dominance of carbon dioxide, nitrogen, and methane emitted by mud volcanoes on Sakhakin Island varied considerably over a 35 year period. Where hydrocarbon gases are present, they may be of microbial or thermogenic origin. In petroleum provinces (such as Azerbaijan, Romania, Mexico, Nigeria, and Columbia) gases have a relatively high concentration of the higher hydrocarbon gases (ethane, propane, etc.) and may be accompanied by liquid hydrocarbon (crude oil). These emissions, and the accompanying mud and rock clasts are of considerable value in petroleum prospecting as they indicate the nature and maturity of the source rock.

Because mud volcano gases are dominated by methane, an important 'greenhouse' gas, they are potentially significant in an environmental context. Gas composition, emission estimates, and the fates of these gases are discussed in more detail in Section 10.4.3.

7.3.4 Mud-volcano activity

The morphology of mud volcanoes and their activity, which varies between continuous gentle emissions and spasmodic violent eruption, were used by Kalinko (1964; referred to by Dimitrov, 2002b) as the basis for a classification. Three types were described, each being exemplified by mud volcanoes in the former Soviet Union.

I *Lokbatan type:* mud volcanoes characterised by short periods of explosive activity separated by long periods of dormancy. The mud breccia is viscous, so

Figure 7.8* The eruption of Lokbatan Mud Volcano, Azerbaijan; October 2001. The flame was estimated to be about 100 m in diameter and about 400 m high, and burned for about 24 hours (Mukhtarov *et al.*, 2003). The dense black smoke indicates that higher hydrocarbons are burning. (Photo: Phil Hardy.)

the mud volcanoes are steep-sided cones. At the end of a phase of activity, this mud breccia blocks the feeder channel. Subsequent phases of activity occur when there is sufficient gas pressure beneath this 'cork' to blow it out. During explosive events the emitted gas may spontaneously ignite (Figure 7.8; see also Guliyev, website). Lokbatan, a mud volcano in Azerbaijan, has erupted 23 times since 1829, most recently in October 2001. Some of these events have been very violent, with flames shooting up to 500 m into the air and the expulsion of a large volume of mud (Aliyev *et al.*, 2002). The main eruptive phases tend to be short lived, generally lasting two to four hours, but when we visited Lokbatan 12 months after the 2001 eruption there was still a small flame burning.

This type of eruptive behaviour is not unusual. Approximately 25% of the mud volcanoes in Azerbaijan are eruptive, and about a third of these ignite; Bagirov *et al.* (1996) estimated that, on average, there are 9.7 eruptions per year in Azerbaijan, 3.4 of which are major events. It is believed that spontaneous ignition occurs because of the supersonic velocity at which the erupting gas jet escapes. Even the jets from submarine mud

volcanoes ignite. For example, on 12 September 1927 a large mud volcano and a high flame (about 500 m) was observed in the Black Sea approximately five nautical miles from the Navy Observation Station near Feodosia on the Kerch Peninsula (Ukraine). A similar event on Makarov Bank in the Caspian Sea is described in Section 10.4.3 (Figure 10.5).

Dimitrov (2002b) identified reports of 11 eruptive mud volcano events around the world in the year 2001, eight of them in Azerbaijan, two in Barbados, and one in New Zealand. Eyewitness accounts illustrate the dramatic nature of these events:

> The eruption of Irantekjan mud volcano took place at night on 6th October 1965. The observers state that they first heard the underground rumbling and after that the explosion took place; the column of flame rose to a height of 100 to 150 m over the mountain. The vicinity of the volcano for 10–15 km around was brightly lightened. The temperature of the atmosphere rose so much, that it was hard to breath near the borehole, which was located within 1 km of the center of eruption. The height of the

flame fell to 90–100 m in 20–30 min after the beginning of the eruption and the high temperature decreased near the volcano. Mud breccia had been expelled from the volcano along with the burning gas. By 4–5 o'clock in the morning on the 7th October the intensity of the mud volcano eruption had slightly decreased. The flame rose only to a height of 10–20 m. The mud volcano threw out vast volumes of mud breccia (about 1.5 million t) which filled the crater and moved down the flank of the volcano. The extent of the largest flow of breccia is suggested to be 1600–1800 m.

Sokolov et al., 1969

The following two reports describe eruptions of the same submarine mud volcano just off Trinidad (previously mentioned in Section 3.25.2).

An exceptionally violent eruption occurred off Erin on the southern coast of Trinidad on November 4th, 1911. Masses of material were observed being ejected from the sea, accompanied by enormous volumes of gas which became ignited, and by considerable noise. The flames could be seen many miles away, as they rose to a height of 100 ft (30 m), or more. An island was noticed to be steadily rising out of the sea as the material increased in quantity, until after several hours of activity it had attained an area of $2^{1}/_{2}$ acres (1 ha) and was 14 ft (4.3 m) above the sea level at its highest point. Cadman states that the flames have been 300 ft (100 m) high and burnt for $15^{1}/_{2}$ hours, [and] that the shock of the explosion was felt in Cedros (on the mainland).

Kugler, 1933

On the morning of August 4 the writers made their first landing on the island. The silhouette from the east was symmetric and flat-topped. A fairly well-defined 5-foot [1.5 m] cliff existed at the landing site. At the foot of this cliff was a narrow zone of sub-rounded and angular boulders of 12–18 inches [0.3–0.5 m] encrusted with shells, serpulids, and bryozoans showing that they must have been brought up from the floor of the sea as the island was formed. The boulders appeared to be composed predominantly of the type of sandstone cropping out along the adjacent coast.

A rough survey of the island showed it to be approximately circular in plan with a longer axis of approximately 600 feet [180 m] trending 100 west of north compared with 560 feet [170 m] for the shorter axis; the total area was estimated at 4.7 acres [1.9 ha]. The maximum height was estimated to be 20 feet [6 m] above sea-level at that time.

The form of the island was a low dome slightly asymmetric and tilted toward the west. The mud surface was irregular, hummocky, and scoriaceous with abundant pits up to several feet in diameter. The mud was dull brownish gray, being lighter colored where it had been dried by the sun, but otherwise appeared homogeneous.

Dry, odorless gas was issuing from a main vent southeast of center, the vent being about 3 feet [0.9 m] in diameter and 2 feet [0.6 m] deep. This main vent is clearly indicated by the lighter-colored small circular area on the aerial photographs. The gas blew strongly and intermittently with accompanying rumbling. Gas also issued from minor small vents surrounding the main one.

Higgins and Saunders, 1967

This mud volcano erupted again in May 2001 when 'Chatham Island' reappeared overnight (Seach, 2001). A similar mud volcano 'island', Malan Island, periodically appears off the Makran coast of Pakistan (Delisle, 2004).

These examples are from shallow water, but an example described from the Gulf of Mexico (Prior et al., 1989) lies in 2175 m of water. At the centre of this feature is a crater 400 m long, 280 m across, and 58 m deep. This lies within a central cone, surrounded by concentric faults, and from which there has been a debris flow; blocky debris is scattered across the seabed up to 430 m away. This debris includes individual blocks up to 35 m across. Prior et al. considered that '*the evidence suggests that the crater formed recently, probably within a few hundred years*'. They also noted that Soley (1910) had reported *an unusual sea-surface disturbance (possibly gas venting)*' in this vicinity in September 1910, but they were unable to prove a link. It must take enormous energy to form such a large crater, so the volume of gas released must have been considerable – particularly when it reached the sea surface!

II *Chikishlyar type:* mud volcanoes are characterised by continuous gentle activity. Gassy mud and water are emitted at a more or less constant rate from numerous vents, and shallow dome-like features or plate-shaped depressions are formed. Mud volcanoes of this type are common on the Kerch Peninsula, Ukraine (Dimitrov, 2002b).

III *Schugin type:* mud volcanoes are of an intermediate type and experience repeated phases of relatively weak activity. A variety of mud-volcano shapes may be formed, and these typically have composite craters. Dimitrov (2002b) considered this the most common type of mud volcano.

Dimitrov noted that there is no geographical distinction between these types of mud volcano. Some mud volcano areas on land contain all three types, and this probably applies to offshore areas as well. An alternative classification, suggested by Guliev (1992), based on the eruption kinetics of the onshore and offshore mud volcanoes in Azerbaijan, produced a similar range of mud volcano types.

Mud-volcano life cycles
The present day is no more than a snap-shot in geological time. Mud volcanoes that are active today will not be active forever, and there must be examples, like the ones described here, which are no longer active. Also, others must be forming, and will not become active until some time in the future.

COMPLEX MOUND, GULF OF MEXICO

The Complex Mound in Garden Banks Block 161, is about 3 km wide and rises 42 m above the surrounding seabed. It bears all the acoustic hallmarks of a mud diapir: (1) no coherent reflections, (2) varying strength of acoustic turbidity, and (3) distinct appearance of having risen up from below. A geotechnical borehole was completed and proved this interpretation to be partly wrong. There were no indications of gas in the sediments, and the sediment strength was not low, as is normally the case in extruded sediments. Roberts *et al.* (1999b) concluded that they had been extruded during mud-volcano activity during the late Pleistocene, but they have subsequently degassed, dewatered, and compacted. They attributed the acoustic turbidity to the absence of sediment layers in the feeder pipe.

RYUKU TRENCH, OFFSHORE JAPAN

In a study of mud volcanoes in the Ryuku Trench, Ujiié (2000) found that extruded sediments (identified by microfossils) became mixed with normal seabed sediments (hemipelagic mud) over a period of 300 to 400

years. Eruptions of the two mud volcanoes studied seem to have ended about 17 000 and about 34 000 years ago.

MILANO AND NAPOLI, MEDITERRANEAN RIDGE

Kopf *et al.* (1998) reported that drilling during ODP Leg 160 on two 'mud domes' (Milano and Napoli) demonstrated that mud-volcano activity began at least 1.5 My ago. They considered that activity was not continuous, nor was the 'style' of eruption consistent. They described an early explosive phase of eruptions that was probably accompanied by the emission of large volumes of fluid. During this phase a large clastic cone was built. This was followed by continued but intermittent extrusions which produced viscous mud-debris flows and turbidity currents. This occurred over long periods, and resulted in progressive subsidence of the mud volcano.

These intermittent eruptions are typical of the Lokbatan- and Schugin-type mud volcanoes described above. It may be that, as speculated for Bakhar and Dashgil, the feeder channel may be blocked pending another eruptive phase. The benthic chemosynthetic communities on the Napoli Mud Volcano indicate that, like Bakhar and Dashgil, it is still actively degassing (Corselli and Basso, 1996). However, Kopf *et al.* (1998) considered that Milano is 'apparently dormant'.

7.4 RELATED FEATURES

In the previous two sections we have looked in some detail at the features most commonly associated with seabed fluid flow. Now we turn our attention to other features, which, although less well known, are in some way related to pockmarks and mud volcanoes. The following advice, from an unpublished report on the identification of pockmarks and similar seafloor features, should be remembered when considering not only pockmark-like features, but also features similar to mud volcanoes.

> The three keys to identification of sea floor
> depressions and in particular pockmarks are
> experience, judgement and consistency.
> Experience, knowing what to look for in the survey
> records when an unfamiliar form is encountered.
> Judgement, using a set of criteria to categorize
> unknown targets seen in the data. Consistency,
> using the same criteria, reading the similar
> conclusions from similar targets.
>
> Ellis, 1998

7.4.1 Seabed doming

In Chapters 2 and 3 we noted several instances where the seabed is raised above the normal level to form a shallow, broad dome. These generally stand only a metre or so above the normal seabed, and may be a few hundred metres across. In most cases they have been reported as lying immediately above areas where evidence of gas (acoustic turbidity) rises to within a few metres of the seabed (Figures 2.25 and 3.16). Detailed inspection by ROV of the dome in Norwegian Block 25/7, North Sea (Section 2.3.3) revealed no indication of any evidence of seepage, nor has evidence been reported from elsewhere. It seems that these features form by the inflation of the seabed sediments as gas enters them from beneath, raising the pore fluid pressure.

7.4.2 Collapse depressions

Collapse depressions are similar in form to pockmarks, but they have a different origin. They were first described by Prior and Coleman (1982) from the shallow-water areas of the inter-distributary bays and immediate offshore regions of the Mississippi Delta. They are elongate, bowl-shaped depressions bounded by distinct scarps (< 3 m high). They range in diameter from 50 to 150 m, and typically have hummocky and irregular floors. Examples from the Alsek River area (Section 3.19.2) are a little larger (< 500 m in diameter) and deeper (<15 m). In both cases they occur in areas with rapidly deposited soft, organic-rich, sediments. Continuous microbial activity is presumed to explain the high levels of methane. Like other features (slumps, flows, bottle-neck depressions, etc.) associated with rapid sedimentation in a deltaic environment, they are thought to have formed by the collapse of the seabed as a result of liquefaction and pore-fluid expulsion, probably triggered by the passage of storm waves.

7.4.3 Freak sandwaves

In Section 3.5.2 we described freak sandwaves identified in the southern North Sea (Figures. 3.18 and 3.19). These are considerably steeper than the normal sandwaves in the surrounding area, and are apparently associated with gas seeps. Despite the coarseness of the seabed sediments, they are thought to have been produced by the removal of sediments. Invariably, gentle depressions (pockmarks) surround these freak waves. They are caused by the seeping gas preventing sand-grain settlement in what would otherwise be the favoured location.

7.4.4 Shallow mud diapirs and mud volcanoes

Unlike the true (deep-rooted) mud diapirs referred to above, shallow mud diapirs and associated mud volcanoes originate close to the seabed. They appear as relatively small topographic features a few metres or tens of metres in diameter and standing a few metres above the general level of the seabed. Examples have been described from the Mississippi Delta, the Adriatic Sea, the Arabian Gulf, and the Norwegian Sea (see Chapter 3). In previous literature they have been called 'mud diapirs', 'mud lumps', or 'mud spines'. However, remembering our definition of mud volcanoes (Section 7.3), if they extrude mud and/or fluid it would be more appropriate to call them 'shallow mud volcanoes'.

ADRIATIC SEA

Hovland and Curzi (1989) described features from the central Adriatic Sea, which are about 2 or 3 m tall and 20 to 40 m across (Figure 6.12). They rise from a seabed sediment sequence no more than about 10 m thick. The underlying sediments show no signs of disturbance, although acoustic turbidity is present at a depth of about 20 m. There are no gas chimneys, and there is no faulting to indicate sediment or fluid migration through the intervening layer of sediments. Detailed inspection of the seismic data suggests that the material from which they are made originates from the acoustically transparent (?muddy) sediments near the base of the topmost sediment sequence. They have pushed through younger, multilayered sediments; although they have not reached the surface, they are responsible for the surface topography. At least one (left side of Figure 6.12) is also associated with a gas seepage plume. This feature at least should be described as a shallow mud volcano.

Hovland and Curzi (1989) suggested that methane derived from organic-rich sediments at depth accumulates in the acoustically turbid sediments as small bubbles. These bubbles are able to rise through the overlying sediments, but become trapped beneath an impermeable clay layer at the base of the topmost sediment sequence where the bubbles accumulate, reducing the bulk density of the sediments and making them buoyant. Localised plastic deformation and diapirism produces the topographic features, and eventually the pore fluids break through the seabed to feed seeps. This model suggests that, like deep-rooted mud volcanoes and mud diapirs, the shallow features may be produced as a result of diapirism induced by density inversion. The acoustic turbidity and gas seeps indicate the importance of gas to this process.

MISSISSIPPI DELTA

The mud 'lumps' of the Mississippi Delta are thin, diapiric spines of mud that are squeezed upwards into and through the overlying sediments. They occur near each of the major distributaries of the delta. 'Normal' conditions result in the deposition of fine-grained sediments with high moisture contents and significant amounts of organic matter; methanogenesis is likely. In contrast, during flood conditions, the river rapidly deposits layers of coarse sediments. This sudden imposition of a load does not allow the now-buried fine-grained sediments to compact. Fluid (water and gas) is trapped in them, so a density inversion occurs, and fluid migration or mud diapirism is initiated. After major floods individual features may be pushed 6 to 9 m vertically, and they may break through the sea surface to form temporary islands (Prior and Coleman, 1982). Prior and Hooper (1999) presented a case study from a location in which it was proposed to build a complex of oil and gas structures (wells, platforms, and pipelines). There was concern that mud–diapiric activity would present problems during the 20 year planned life of these engineering structures.

Shallow mud diapirs like these are found in other deltas and areas with shallow gassy sediments.

ARABIAN GULF

Some of the mud diapirs described in Section 3.14.2 actually lie within pockmarks (Figure 3.35). These features seem to have shallow roots. The removal of the topmost sediment during pockmark formation relieved the stress on the underlying clays, which were squeezed up because of the load on the surrounding sediments.

VØRING PLATEAU, NORWEGIAN SEA

In the Vøring Plateau of the Norwegian Sea, off mid Norway mud diapirs and diapiric escarpments rise up to 70 m above the general seabed (at 1200 m water depth). The cores of some have been shown to consist of coccolith oozes deposited during the Palaeocene and Miocene. These features are concentrated over the Vema Dome, but are not found over the nearby Nyk High. It seems that extension and faulting over the dome allowed light hydrocarbons to migrate into the oozes, reducing their bulk density and making them buoyant and deformable (Hovland et al., 1998b). Vertical movement appears to have been very slow.

PECHORA SEA

Shallow mud diapirs in the Pechora Sea (Arctic Ocean) are associated with gas trapped beneath a clay layer (Bondarev et al., 2002; Figure 7.9). However, in this case the clay lies within a zone of permafrost. It seems probable that the permafrost at least contributes to the effectiveness of the gas seal. These ice-cored diapirs are like pingoes; perhaps gas and diapirism are responsible for pingo formation.

TERTIARY MUD DIAPIRISM IN THE NORTH SEA

Løseth et al. (2003) described extensive evidence of Tertiary mud mobilisation in the north-central North Sea. This includes near-circular diapirs typically 1 to 3 km in diameter, and massive features, the largest being 100 km long and 40 km wide. They do not have deep roots like the mud volcanoes of Azerbaijan; these are < 1000 m. However, there is evidence that some of the features, described as 'mounds', rose 200 to 300 m above what was then the seabed. Perhaps there were mud volcanoes in the North Sea during the Tertiary!

7.4.5 Red Sea diapirs

In the northern Red Sea there are numerous diapirs in water depths ranging between 700 and 1300 m. They are elongate features, some more than 30 km long and about 4 km wide. They are covered by 1 km or less overburden, and commonly ascend along rift-boundary faults or into the floor of the rift; both rifting and diapirism

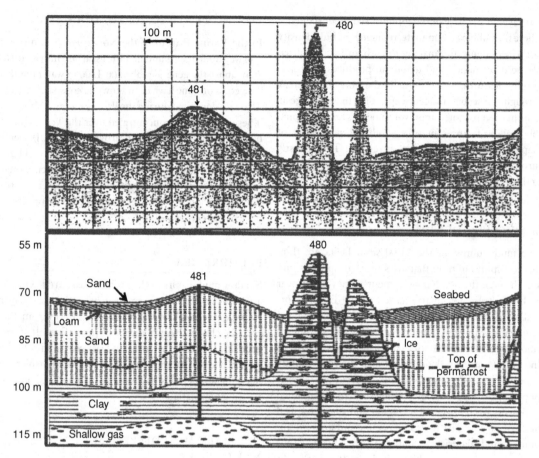

Figure 7.9 Shallow mud diapirs (pingoes?) in the Pechora Sea; seismic (sparker) profile (above) and interpretation (below) crossing shallow mud diapirs. The numbers 480 and 481 are site investigation boreholes; 480 blew out when it penetrated shallow gas at a depth of 49.5 m below the seabed. (Adapted from Bondarev *et al.*, 2002.)

are active at the present day (Mart and Ross, 1987). Mart and Ross assumed that the diapirs consist mainly of late Miocene evaporites (salts), some perhaps covered by an envelope of non-evaporitic sediments. One of the most pronounced diapiric features shown in one of their numerous examples (their Figure 11a) measures only about 500 m at the base and rises at least 250 m above the gently sloping sediment-covered seabed. Some diapirs are topped by negative 'collapse' structures, one example measuring about 2 km across and 75 m in depth.

Mart and Ross found it hard to explain how such diapirism could occur with less than 1000 m of sediments covering the salt beds. They suggested that a combination of an abnormally high temperature gradient and faulting facilitated the diapirism. The following possible mechanism, not considered by Mart and Ross,

may also be plausible. The northern Red Sea is a young ocean spreading centre where hot hydrothermal fluids are generated; deep-water basins in the Red Sea are occupied by hot brines which contain 1000 times more methane than normal seawater and have high metal contents (Section 3.13). These fluids are generated beneath the evaporites. Consequently, the diapirism may be initiated by heating and fluidisation. We suggest that the structures in the northern Red Sea are large mud diapirs associated with the seepage of light hydrocarbons.

7.4.6 Diatremes

Volcanic diatremes

'Maar' is a German word for a crater formed by a violent volcanic eruption not accompanied by the extrusion of magma. Diatremes were described by Lorenz (1975) as features formed when hot magma rising along a fissure contacts water. The resulting violent phreatomagmatic eruption gives rise to vent enlargement by slumping, ring-faulting, and, eventually, a crater at the surface. When the volcanic island of Surtsey was formed off Iceland, seawater entering the crater caused phreatomagmatic eruptions resulting in the deposition of wet and muddy ash and lapilli beds. Magma could not rise to form a normal lava lake until the crater wall was built to sea level and the supply of seawater was cut off (Lorenz, 1975; Thorarinsson, 1967).

On the basis of careful studies of extinct maars and diatremes at various levels of exposure, Lorenz was able to reconstruct their probable internal structure, and to understand the development process. He suggested that water entering a newly formed fissure forms a column heated at the base by the rising magma. Boiling produces bubbles that rise through the water, but increases in temperature at high pressure may result in water flashing to steam, which rises rapidly to the surface ejecting pyroclastic debris and steam. At the surface the cool three-phase system (steam, water, and solids) gives rise to wet or muddy air-fall and base-surge deposits. Following an eruption the pressure inside the vent is low, and pressure gradients between the vent and the surrounding rock cause the walls to spall and slump. The fissure is enlarged, but filled with wall-rock debris that will be expelled during the next eruptive cycle. The process is repeated, and a vent of considerable size may develop (Lorenz, 1975).

Sedimentary diatremes

Løseth et al. (2001) found a cluster of about 100 large (100 to 500 m diameter) pockmarks in a 65 km^2 area offshore Nigeria. Most of them were 15 to 25 m deep, the maximum being about 30 m. Using 3D-seismic data, they identified vertical pipes only a few metres in diameter (narrower than normal gas chimneys and the vertical blank zones described in Section 6.2.2) extending '*continuously from seabed and 1000 to 1300 m down to interpreted sandy horizons*' (Figure 7.10). These 'blowout pipes' appeared as circular features on time-slices. Like mud volcanoes, they contain clasts of sediments or rocks brought up from depth.

Løseth et al. compared the features they described to features seen in late Pleistocene sediments outcropping on Cape Vagia on the Greek island of Rhodes

(described by Hanken et al., 1999). They were found close to the beach over a dome-shaped structure in which gases are thought to have accumulated. Hanken et al. suggested that gas overpressure, increased as a result of tectonic uplift, was sufficient for the rocks (uncemented Upper Pliocene to Pleistocene limestones and clays) to be fractured. A heavily fractured zone about 20 cm in diameter (Figure 7.11) was found to be surrounded by about 4 m of less-fractured rock. This hydraulic fracturing enabled the gas to escape vertically through blowout pipes to escape violently to the seabed. After each gas escape, the pipes became blocked as clasts of the overlying rock fell into the newly created void and became embedded in a clay matrix. Subsequently, gas pressure release required the formation of new blowout pipes; a total of 28 were mapped.

These important examples demonstrate a significant, vigorous gas-escape process, and can best be described as a 'sedimentary diatreme'. Diatremes are most commonly considered to be vertical features piercing sedimentary rocks as a result of gassy, explosive eruptions of igneous volcanoes. In the present context the explosive eruption is of fluids with a sedimentary, not an igneous origin. Brown (1990) recognised that diatremes may be formed by gas emissions in features closely related to mud diapirs. Indeed, he suggested that '*diapirs (or parts of diapirs) may change into diatremes and vice versa*'. However, he also pointed out that, whilst diatremes indicate focussed fluid venting, they are not necessarily associated with mud diapirs: '*any type of conduit system (be it structural or stratigraphic) that rapidly expels fluid into a lower permeability sedimentary column can initiate diatremes, and this should be borne in mind before attributing all sedimentary intrusive phenomena to a purely diapiric origin*'.

The key feature about sedimentary diatremes is that they are formed by sediment fluidisation. Rising fluid (water and/or gas) entrains some sediment, carrying it upwards (Brown, 1990). In contrast, in a mud diapir the mud '*behaves as a single-phase viscous fluid*' (Kopf, 2002), whilst seeps produce no sediment, only fluid (although sediment may be lifted from the seabed sediments by the seeping fluid, for example during pockmark formation).

7.4.7 Sand intrusions and extrusions

Sandstone intrusions are comparable to igneous sills, dykes, etc. Some form close to the surface, and sand

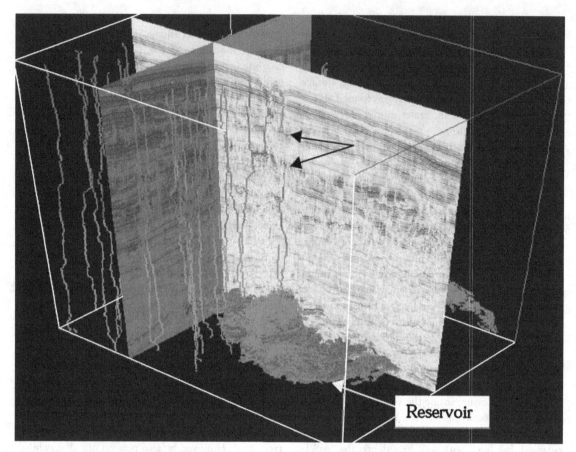

Figure 7.10* Three-dimensional visualisation of near-vertical 'blowout pipes' rising from an assumed reservoir towards the seabed; some actually reach the seabed, ending in pockmarks or craters – one of which is about 300 m across and 60 m deep. The reservoir is 2700 to 3000 ms below the seabed, so the pipes are 1000 to 1300 m long. Note that the pipes go in and out of vertical seismic sections (see arrows). (Image courtesy of Helge Løseth).

volcanoes with diameters of < 40 m have been reported (Figure 7.12). Although frequently treated as geological curiosities Jolly and Lonergan (2002) identified 68 references to them in the literature. It seems that they have been reported from most sedimentary environments, but they are most common in tectonically active environments with high sedimentation rates, in mud-dominated sedimentary systems, and where tectonic stresses result in high sediment pore fluid pressures. They form when high-pressure fluids entraining sand grains are injected along a fracture into sediments with a lower pore fluid pressure. Jolly and Lonergan envisaged a three-stage process. The first stage is characterised by the build-up of excess pore fluid pressure in an unconsolidated sandy sediment. In the second phase the seal retaining

the pressure fails, initiating the third phase in which the fluidised sand is injected. Once a pressure balance has been achieved, the intrusion 'freezes'.

Clearly these features are widespread, and in recent years their economic importance has been demonstrated. Extensive sand intrusion has been recognised in North Sea Tertiary sediments, for example in the Alba, Balder, Gryphon, and Forth-Harding petroleum fields (Jolly and Lonergan, 2002). Detailed descriptions have been published; Duranti et al. (2002) and Hillier and Cosgrove (2002) described examples from the Alba field where the mobilised sand was injected into the overlying sediments to form intrusions of three types: dykes emanating from the margins of the channel sands, sills injected along the bedding, and offshoots oblique to

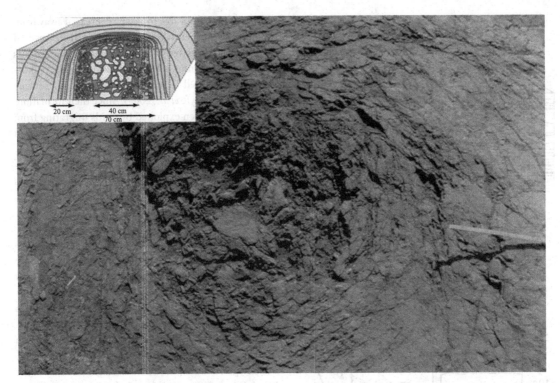

Figure 7.11* Photograph showing a horizontal cross section through a sedimentary blowout pipe of late-Pleistocene age on the Greek island of Rhodes; the inset sketch illustrates the feature in three dimensions. The inner central zone comprises relatively large (< 5 to 10 cm) angular mud clasts floating in a muddy matrix; the outer zone has smaller (< 5 cm) clasts. These zones are surrounded by about 20 cm of heavily fractured country rock (limestone) and a further 4 m of less fractured rock. (Images courtesy of Nils-Martin Hanken.)

the bedding. These range in size from a few centimetres to large dykes 400 m long by 30 m wide (Hillier and Cosgrove, 2002). Mazzini *et al.* (2003) described examples from the Gryphon field.

'Sand volcanoes', much smaller in size (up to a metre) have been reported from onshore Japan. However, these are genetically unrelated, and have been attributed to sediment liquefaction during earthquakes (see Section 7.5.5).

Fluid-escape structures
Lowe (1975) provided a comprehensive survey of 'water-escape structures' found in coarse-grained sediments. The example shown in Figure 7.13 depicts a flame-like, cylindrical pillar truncating laminated fine-grained sandstone. This is a vertical channel formed when localised fluid-escape velocities exceeded those required for minimum sediment fluidisation. Such pillars range from a few millimetres to sand intrusions many metres high (Lowe, 1975). Although these are often small, they resemble features seen on shallow seismic sections. It seems that Lowe's descriptions are as relevant to gas as well as water, so we prefer to call these 'fluid-escape features'.

7.4.8 Polygonal faults

Cartwright and colleagues (Cartwright, 1996; Cartwright and Lonergan, 1996; Lonergan *et al.*, 1998; Dewhurst *et al.*, 1999a; etc.) identified widespread polygonal fault systems on 3D seismic data in Lower Tertiary mudstones of the north-central North Sea. The system comprises a complex, closely-spaced network of almost randomly oriented extensional faults. Individual faults are typically 500–1000 m long with throws of 10–100 m; there is only 200–400 m between individual faults. The faults are restricted to Eocene, Oligocene, and Lower Miocene mudstone; neither

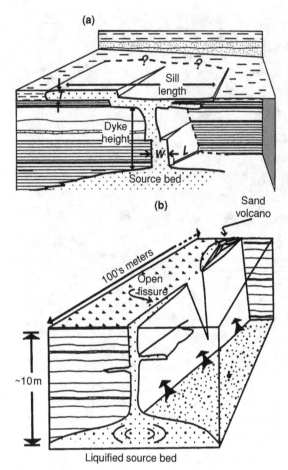

Figure 7.12 Schematic diagram illustrating the principal features of clastic sills and dykes: (a) intrusion in the subsurface; (b) intrusion associated with fissures and sand volcanoes. In (a) W is dyke width, L is dyke length (strike length of exposed outcrop), and T is sill thickness. Dykes associated with surface features and sand volcanoes that form as a result of earthquake shaking can have strike lengths up to several hundreds of metres; sand volcanoes typically have a small height to diameter ratio, and maximum diameters reported of 40 m. (Reproduced from Jolly and Lonergan, 2002 with permission from the Geological Society of London.)

the underlying nor the overlying sequences are affected (Cartwright, 1996).

It is generally thought that these vertical/subvertical faults formed during early burial as the sediments were compacted. They developed as dewatering pathways, enabling pore fluid escape and therefore consolidation of the muds. However, it seems that the pro-

portion of pore fluid able to escape decreased away from the faults, consequently the consolidation decreases, and thickness increases away from the faults. This accounts for the typical hummocky appearance of the top surface of the affected formations as seen on 3D seismic time-slices (see Fyfe *et al.*, 2003). Since their formation, the faults have been exploited by fluids migrating from lower levels. Polygonal fault systems are discussed in detail in various papers in van Rensbergen *et al.* (2003). Features described on the bed of Lake Superior (Section 3.28.1) may be relatively modern examples (Wattrus *et al.*, 2001).

7.4.9 Genetic relationships

The features described in the previous sections are clearly related because they all involve the movement of fluids within sediments, and in most cases through the seabed. It seems that this spectrum of features is produced by comparable processes, differences resulting from individual local conditions. An examination of the various features (mud diapirs, sand injections, and polygonal faults) described from the Tertiary of the North Sea demonstrates the relationships.

The features occur within the same area of the North Sea, and in sediments of the same stratigraphic age (the Lower Tertiary Hordaland Group sediments). Also, they all overlie the area in which the prolific Kimmeridge Clay Formation, the principal petroleum source rock in the northern North Sea, is well developed and mature. Gas chimneys rising through the Cretaceous provide evidence of vertical migration up to the Lower Tertiary sediments. Løseth *et al.* (2003) pointed out that the key characteristic of the Hordaland Group clays is their smectite content. Smectite is a clay mineral that holds water within its crystal lattice; it has a low density and low mechanical strength. Also, its low permeability inhibits pore fluid drainage, so smectite-rich clays, such as these, tend to be unconsolidated. Where compaction and dewatering was possible and polygonal faults were formed, fluids rising from lower levels were able to pass through the Hordaland Group sediments, using the fault systems as pathways. Where there were no such faults the introduction of fluids resulted in sediment mobilisation. The largest mud diapirs were formed '*in the area of the most extensive fluid flow*' (Løseth *et al.*, 2003). Where the gas chimneys are overlain by sands of Cretaceous or Palaeocene age, the fluids migrated

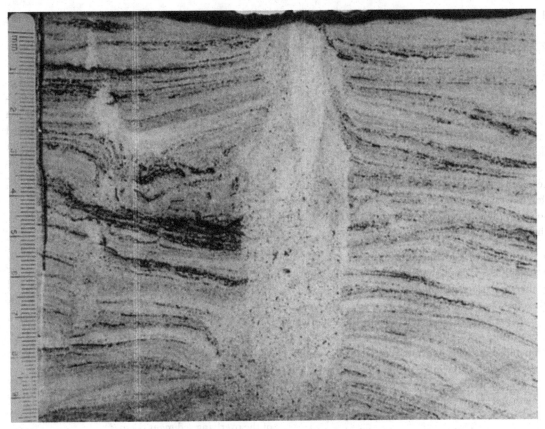

Figure 7.13 Photograph of a 'water' (fluid?) escape structure in layered sediments. (From Hovland and Judd, 1988; courtesy of D. R. Lowe.)

laterally, and the Hordaland Group sediments were not mobilised, except where the sands pinched out towards the edge of the basin. There are near-circular clay diapirs here (Løseth *et al.*, 2003). However, where the fluids entered intraformational sands within the lower Tertiary, these mobilised to form sandstone dykes, etc., although Løseth *et al.* also linked some circular mud diapirs with leakage from such sands.

We conclude from the above that the injection of fluids from the compacting, underlying Jurassic sediments, including the Kimmeridge Clay, provides the key to the mobilisation of these sediments over an extensive area of the North Sea. Where the fluids were able to pass through, none were formed. It is interesting to note that a large proportion of the area with polygonal faults is also associated with shallow gas and pockmarks!

We suggest that any thick, fluid-rich sediment sequences, particularly those with a large gas com-ponent, should be regarded as potentially important fluid sources, and that there is a high probability that there are fluid migration features in the overlying sediments.

7.5 MOVERS AND SHAKERS: INFLUENTIAL FACTORS

As can be seen from Sections 7.2 to 7.4, observations of seeps, pockmarks, mud volcanoes, mud diapirs, and related features show that emission products include solids, liquids, and gases. To understand the nature, distribution, and mode of formation of these features it is necessary to appreciate the mechanisms involved.

The differences between the seabed features described above are accounted for by four key factors:

1. the environment at the depth of origin of the mobile components (sediment and/or fluids);
2. the relative proportions of solids and fluids, particularly gas;
3. driving forces and triggers;
4. the availability of migration pathways.

There are relationships between these factors, and these influence the mechanism and speed of fluid/sediment migration from its depth of origin to the seabed. This will become clear as we look into the processes that lead to the formation of the seabed features described above. The essential ingredients are fluids and solids. All fluid types present beneath the seabed have the potential to migrate to the seabed:

- water expelled from muddy sediments;
- water released from compacting muds as clay mineral alteration (smectite dehydration) occurs;
- exotic waters (groundwaters: meteoric and connate) from aquifers, through fractures in the underlying basement;
- microbial methane generated in (relatively) shallow organic-rich sediments;
- petroleum fluids, including thermogenic gas (mainly methane) generated in (relatively) deep organic-rich sediments;
- methane exsolved from pore waters during ascent of fluids and muds;
- water and gases derived from hydrothermal activity;
- fluids derived from the serpentinisation of mantle-derived rocks (peridotite);
- mantle-derived fluids.

Although the nature and origin of the fluids is important, there is no significant difference between mud volcanoes formed by carbon dioxide or methane emissions, nor between pockmarks formed by water and those formed by gas. The solids involved are generally fine-grained, unlithified sediments, however uncemented sands are important in some cases, and mud-volcano breccia may include lithified rocks of considerable size. The features described range from those formed by the migration of sediment masses (diapirs), to those formed by the migration of fluids with no sediment (seeps). Fluids are important in all cases (fluids enable sediment masses

to become mobile), thus the spectrum of features is a function of the sediment:fluid ratio.

7.5.1 The deep environment

When considering migration mechanisms and velocities, the depth of origin of the mobile components is a fundamental parameter because of the changes imposed on sediments as they are buried to progressively greater depths, and the changes that affect fluids as they rise towards the surface. Before we discuss the circumstances that lead to fluid migration we must provide some geotechnical background.

The stress environment

In combination with the density of the overlying sediment, the depth determines the stress imposed on the sediment. In turn, this controls the driving forces, and consequently the relative proportions of the various mobile components. The stress environment beneath the seabed is defined as in Equation (7.1):

$$\sigma_v = \rho_{sed} \cdot g \cdot d_{ssb} \tag{7.1}$$

where: σ_v is the vertical stress (Pa); d_{ssb} is the depth subseabed (m); g is acceleration due to gravity ($= 9.807$ m s^{-1}); and ρ_{sed} is the saturated (bulk) density of the overlying sediment (kg m^{-3}).

This stress is imposed only on the solid components of the sediment transmitted by grain-to-grain contact. Normally σ_v is the dominant stress, but lateral (horizontal) stresses, σ_h, are significant as they confine the sediment, holding it together. In some situations, such as areas experiencing tectonic compression, the lateral stress may be important. In extreme cases $\sigma_h > \sigma_v$.

The pore-fluid pressure (P_{fluid})[1] is independent of the vertical stress, and is normally a function of the depth below sea level. It is normal to assume that the pore fluid is water, and that there is a continuous, connected water column extending from the sea surface, through the sediment pore spaces, to an indefinite depth; the height of this column defines the 'hydrostatic pressure'. As the atmosphere 'sits' on the sea surface, atmospheric pressure also contributes to hydrostatic pressure. Equation (7.2) defines the hydrostatic pressure

---

[1] Conventionally, the notation used for pore-fluid pressure is 'u'. In this text we use P_{fluid} and P_{hydro} as we need to distinguish between the pressures of individual fluids: P_{gas}, P_{water}, etc.

(P_{hydro}),

$$P_{hydro} = (\rho_{water} \cdot g \cdot d_{sl}) + P_{atmos} \qquad (7.2)$$

where d_{sl} is the depth below sea level (m); g is acceleration due to gravity ($= 9.807 \text{ m s}^{-1}$); ρ_{water} is the density of the overlying pore water *and* seawater (normally assumed to be about 1.024 g cm^{-3}); and P_{atmos} is the atmospheric pressure (at the sea surface this is normally about 101 kPa). In practice pore fluid pressure may vary from this 'normal' condition for various reasons; these are discussed below.

Intuitively, it may be difficult to envisage a continuous, connected water column extending from the sea surface deep into the sediments. However, subsurface tidal variations have been recorded by pressure sensors deployed down an ODP borehole (Carson *et al.*, 2000). Furnes *et al.* (1991) measured fluid pressure variations within the Gullfaks oil reservoir that corresponded to both tidal cycles and atmospheric-pressure variations. Both examples demonstrate hydraulic connectivity between the seawater and pore fluids.

The pore fluids work to keep the sediment grains apart, partially supporting them. Increases in the pore fluid pressure reduce the 'effective stress', σ'.

$$\sigma' = \sigma - P_{fluid} \qquad (7.3)$$

Equations (7.1), (7.2), and (7.3) define the stress environment beneath the seabed.

Sediment compaction
Simply, the shear strength of an unlithified sediment in Equation (7.4), can be expressed as

$$S_u = c + (\sigma - P_{fluid}) \tan \phi \qquad (7.4)$$

where S_u is the (undrained) shear strength; c is the cohesion (the electrochemical forces that act to hold clay particles together; coarse sediments do not possess this property); and ϕ is the angle of internal friction with respect to the effective stress. The pore-fluid pressure works to reduce grain-to-grain contact. It reduces the shear strength. However, as burial progresses and effective stress increases, so the pore water is squeezed out allowing the sediments to compact. Compaction is also achieved by various other processes, such as changes to the clay mineral species (fine sediments), improved packing efficiency of mineral grains (coarse sediments), etc. This means the strength increases with depth, until eventually cementation occurs (by the precipitation of

minerals within the pore spaces), converting the sediment into a lithified rock that has tensile strength (i.e. it resists being pulled apart, as well as being squashed).

The density gradient
The principal measure of sediment density is termed the 'bulk density' (ρ_{bulk} in g cm^{-3}), the density of the whole sediment – mineral grains plus the pore fluid;

$$\rho_{bulk} = M/V \qquad (7.5a)$$

that is

$$\rho_{bulk} = [M_{mineral} + M_{liquid} + M_{gas}]/V \qquad (7.5b)$$

where M is the mass and V is the volume.

Consequences of the compaction of sediments with burial are that porosity and pore-throat sizes are reduced, pore fluids are expelled, and the density of the sediments progressively increases with depth. Sediments compacted in this way are referred to as being 'normally consolidated'.

7.5.2 Driving forces

The two principal driving forces for mud-diapir formation and fluid migration are excess pore fluid pressure (overpressure) and buoyancy. They can work singly or together, and may affect the fluids alone, or the bulk sediment (i.e. both the solid and the fluid fraction), whether or not the fluid includes gas as well as water.

Overpressure
Under some circumstances, pore fluids are unable to drain from fine-grained sediments during burial. This may be because of one or, commonly, both of the following:

- the speed of sedimentation/burial – if the speed of burial outstrips the rate of compaction because the drainage of pore fluids is impeded by low permeability;
- tectonic compression – related to regions of plate convergence.

Also, fluid-bearing sand bodies may be isolated within impermeable clays, so the pore fluids are trapped. In these situations the pore-fluid pressure is unusually high, $P_{fluid} > P_{hydro}$ and the sediment is 'overpressured'.

$$P_o = P_{fluid} - P_{hydro} \qquad (7.6)$$

The generation of gas (methane) at a rate faster than it can be expelled would have the same effect.

There are several significant consequences of overpressure.

- During burial (as the stress is increased), the pore fluids support a greater than normal proportion of the load of the overlying sediments, minimising the contact between mineral grains. So, sediments with a high fluid content do not compact; they are referred to as being underconsolidated.
- Increases in shear strength are also inhibited, as are the processes leading to lithification, so increases in tensile strength are also inhibited.
- The bulk density remains low, especially if the fluid is gas.
- Where the trapped pore fluid is water, heat is retained.

So, overpressured sediments retain energy. This can be used to drive fluid and sediment flow.

Buoyancy of sediment masses

The buoyancy of a sediment body is a function of bulk-density contrast. Layers composed of low-density minerals such as halite (common salt – NaCl; $\rho_{bulk} = 1.154$ g cm^{-3}) are buoyant when overlain by 'normal' sediments: $\rho_{bulk} =$ about 3+ (clays) to about 2.7 g cm^{-3} (sandstone). Also, sediments that have retained their fluid content (i.e. which are underconsolidated), or in which buoyant fluids are accumulating, have a low bulk density compared to overlying 'normally consolidated' sediments. In both these cases bulk density *decreases* with depth; there is a density inversion, Equation (7.7),

$$BF_{parent} = (\rho_{parent} - \rho_{os}) \cdot g \cdot h_{parent} \qquad (7.7)$$

where BF_{parent} is the buoyancy of the sediment layer from which a diapir originates (the 'parent' formation); ρ_{parent} is the bulk density of the parent sediment; ρ_{os} is the bulk density of the overlying sediment; and h_{parent} is the height (thickness) of the parent sediment.

Kopf (2002) asserted that density inversion is a prerequisite for a buried body of sediment to ascend; but it makes no difference whether this inversion is primary (a function of the nature of the sediment) or secondary (resulting from changes caused by diagenesis, metamorphism, or changes to the fluid content or pressure). Typical mud-diapir parent sediments are fine grained, with a high water and/or gas content under stress from the overlying sediments and/or tectonic forces. Because gas has the lowest density, even when compressed, the presence of free gas, most likely in gas voids, is particularly influential. As gas-void volume increases, the bulk density of the sediment must decline. In near-seabed sediments beneath shallow waters where methane density is not much higher than it is at the surface (0.0007 g cm^{-3} at 0 °C and atmospheric pressure), the bulk density of typical sediments will be reduced by about 1% for every 1% by volume increase in gas content.

In practice there are two potential outcomes of density inversion. Either the underlying sediment will push through the capping sediment to flow as a coherent sediment mass, or the excess pore fluid will lead to fluidisation and escape as a fluid containing sediment particles (as described below). Before the underlying, buoyant sediment can push upwards, its buoyancy force must be sufficient to exceed the downward force of the overlying sediments, and any cohesion (of unlithified sediments) or tensile strength (of rocks).

Typically, migration is initiated by taking advantage of any existing features, if there are any: perhaps topographic highs (antiforms) in the interface with the overlying sediment, or faults. Mud diapirs are commonly associated with faults, and typically form at fault intersections. So, it seems that the availability of a conduit may provide the key to the initiation of mud diapirism. Actually, any vertical inhomogeneities in a sediment basin will be used for migration. Once started, the balance of forces may favour the continuation of upward movement.

GAS EXSOLUTION AND EXPANSION

A consequence of the rise of fluid, whether or not it is held within mobile sediments, is that it experiences a reduction in pressure. This may allow methane and other gases to come out of solution (reducing the bulk density). In addition, any free gas (gas in bubbles) will expand (further reducing the bulk density). Once initiated, gas migration may lead to a self-perpetuating situation (Figure 7.14).

7.5.3 Fluid migration

With the exception of hydrothermal fluids associated with volcanic activity, it is anticipated that fluid production will be at a steady rate, provided conditions

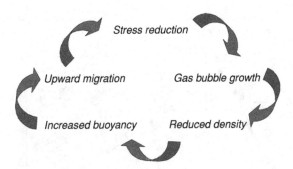

Figure 7.14 Cartoon indicating the feedback that perpetuates migration.

affecting the rate of organic decay, the generation of petroleum, or the head of water remain constant. Certainly, where seeping fluid originates from a deep, thermogenic source it is assumed that there will be a constant rate of fluid production lasting millions of years.

The movement of fluids through sedimentary rocks and unlithified sediments has concerned petroleum geologists for a long time. This is mainly because oil and hydrocarbon gases are not found in economic abundance within the source rocks where the hydrocarbons are generated, but in concentrations within porous and permeable reservoir rocks. The escape of petroleum from source rocks is called 'primary migration' (see Section 5.4.3). Subsequent flow, if unimpeded, eventually leads to escape to the surface and seabed fluid flow. However, 'cap' rocks and 'trap' structures, and the formation of reservoirs may impede this 'secondary migration'. These may occur at any depth from the source to the surface. Reservoirs do not trap all the migrating petroleum, and some overflow (or rather they underflow). Estimations of the petroleum produced in northern North Sea 'kitchens' suggest that generation far exceeds the volume of the available trap structures, many of which have been filled to the spill point (Thomas *et al.*, 1985). Also, some escapes through leaking cap rocks; as Tóth (1980) pointed out, rocks that appear impermeable in short-term laboratory or field tests '*may pass fluids easily during times significant on the geological time scale*'. This migration from reservoirs is called 'tertiary' migration. Both Tóth and Thomas *et al.* failed to mention what happens to the hydrocarbons that escape from the shallowest traps! They must eventually migrate towards the surface.

The frequency and duration of seabed fluid flow will vary with the continuity of supply. Where there

is a viable migration pathway direct from the source, or more likely, from a leaking or spilling reservoir, the supply of fluid will be stable. Where there are intermediate accumulations the supply is more likely to be intermittent. The overview provided by Thrasher *et al.* (1996) explained that migration is focussed along good migration pathways, and that these vary between short-distance fault-dominated pathways leading fluids directly from a reservoir, to long-distance lateral migration which serves to concentrate migrating fluids from basin-wide sources into a single pathway (Figure 7.15).

Flow mechanisms

The movement of gases through more or less impermeable sediments, such as clay, is a complex and poorly understood process. We know that it occurs, sometimes on a large scale, but in most cases we do not know exactly how. In this section we will look at some of the processes forming the basis for our current understanding of fluid movement through the subsurface. This discussion focusses on the flow of petroleum, and most particularly gases such as methane. However, we should not forget the other fluids that flow through the seabed; these are considered first.

GRAVITY

The flow of groundwater through the seabed is invariably driven by gravity. Water entering a permeable aquifer at the ground surface flows through the pore spaces of permeable sediments and rocks, to emerge as a spring (either on land or offshore) where the aquifer outcrops, provided the point of entry (the 'recharge zone') is higher than the outcrop, and the intervening distance is not too great. The rate of flow (Q) can be determined using Darcy's Law, expressed simply as in Equation (7.8),

$$Q = KAh/l \qquad (7.8)$$

where K is a coefficient that reflects the permeability of the aquifer; A is the cross-sectional area through which flow is possible; h is the height, relative to the spring, of the water table (the level at which pore spaces are filled with water) at the recharge zone; and l is the horizontal distance from recharge zone to spring.

Groundwater flow is discussed in detail in many texts. It is not necessary for us to say more, other than to

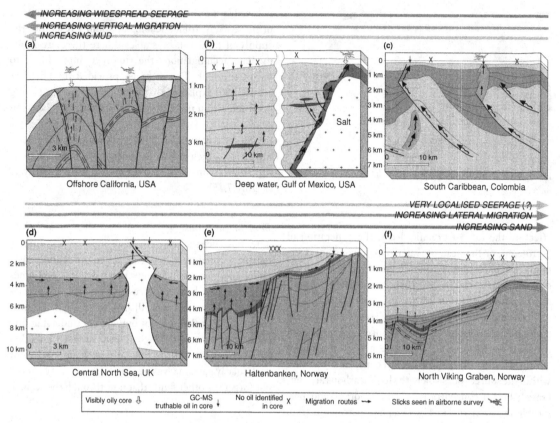

Figure 7.15 Spectrum of seepage styles dictated by geological controls illustrated by examples. (a) offshore California: prolific seepage with minimal lateral migration in an area of active tectonism (no vertical exaggeration); (b) deep water, Gulf of Mexico: rapid deposition of muddy sediment leading to near-vertical migration with focussing up salt walls and active faults leading to major point-source seeps (3× vertical exaggeration); (c) south Caribbean, Colombia: more localised seepage associated with mud diapirism (3.5 × vertical exaggeration); (d) central North Sea: laterally migrating fluids focussed upward over salt diapirs (no vertical exaggeration); (e) Haltenbanken, offshore Norway: lateral, updip migration within carrier beds (5.5 × vertical exaggeration); (f) north Viking Graben: no leakage to surface detected but significant lateral migration in a tectonically quiescent basin. (From Thrasher et al., 1996, AAPG Memoir, 66. Reprinted by permission of the AAPG whose permission is required for further use; AAPG©1996.)

point out that submarine groundwater discharge is only possible where there is a suitable recharge zone onshore, an aquifer, and a hydraulic gradient (h/l) sufficient to overcome the frictional resistance to flow through the aquifer. Generally these limitations restrict groundwater springs to coastal zones; the Florida Escarpment springs (Section 3.26) are exceptions.

CONVECTION

Large-scale convection cells feed the hydrothermal vents of ocean-spreading centres (as discussed in Section 5.2).

ADVECTION

An advective flux is simply the product of volumetric flow rate and the concentration of substance in solution. It is expected that the advective flux of gases such as methane through normal marine sediments (permeability 10^{-8} to 10^{-9} m^2) will be low because flow velocities – Equation (7.8) – are slow and solubility is low. In cracks or pipe-like features, or along highly permeable sand layers other equations, the 'cubic law', and Poiseuille's Law (Clennell et al., 2000) come into play, and flow rates become much higher, provided there is a source of excess pressure at depth. Indeed, if there is overpressure associated with the gas source at depth then a directed advective flux of water is also generated that can assist methane movement (Bjørkum *et al.*, 1998).

DIFFUSION

When porewater is not in motion, the steady-state diffusive flux of methane through water-saturated sediment can be computed from Fick's First Law. Thus, the diffusive flux of methane is the flow rate per unit area, and is proportional to the concentration gradient of the dissolved methane in the pore water. Diffusion is very inefficient at transporting gas over long distances; it '*acts to disperse rather than concentrate gas and is an exceedingly slow process*' (Hunt, 1997). Nevertheless, diffusion is omnipresent in systems that are out of chemical and thermal equilibrium (Clennell *et al.*, 2000).

DISPERSION

When the pore water is moving it is appropriate to consider dispersion. This includes the effects of mesoscale to macroscale fluctuations in the velocity field, and the random thermal molecular motions that constitute molecular diffusion. Together these effects combine into an effective dispersion coefficient (Xu and Ruppel, 1999). On a large scale of observation the effective dispersivity can be much greater than the normal diffusive flux of methane, particularly if flow makes use of fractures.

MULTIPHASE FLOW

The pressure required to force a two-phase fluid through a capillary sized fracture or pore is about double that required for a single-phase fluid, so migrating hydrocarbons may be trapped by capillary pressure while water migrating at the same pressure is not (Holbrook, 1999). Whereas pore water with dissolved methane might be free to migrate through the pores, gas bubbles might not. Nevertheless, the movement of dissolved methane via the porewater system is generally very inefficient; far more methane can be transported through the sediments by bubbles. Microbubble transport along fractures and bedding planes is considered by some to be a major mechanism of gas migration (Neglia, 1979; Saunders *et al.*, 1999). However, microbubbles are most important for short-distance travel in shallow sediments. In all other cases, gas must overcome capillary resistance if it is to move through the sediments (Clennell *et al.*, 2000).

BUOYANCY: THE MAIN DRIVING FORCE

Many mechanisms have been proposed for hydrocarbon migration, and many processes have been described that modify the composition of migrating hydrocarbons. Examination of subsurface and surface data indicates that all the proposed mechanisms and processes are active. However, many play minor roles only recognizable in special situations. The dominant migration mechansim is as a free phase, rising under the forces of buoyancy within carrier and reservoir rocks, and capillary imbibition in the transition from sources and seal into individual carrier rocks.

Matthews, 1996

Matthews and others (McAuliffe, 1980; K. M. Brown, 1990; Clayton and Hay, 1994; A. Brown, 2000; etc.) clearly state that, although diffusion and migration in solution in pore water may be effective, buoyancy is the dominant force driving gas bubbles (and other buoyant fluids) towards the surface. Clarke and Cleverly (1991) noted that oil seeps are actually driven by the buoyancy of the gas that comes with the oil as oil has insufficient buoyancy of its own.

Water, gas, and oil are immiscible, so they behave separately. Buoyancy force (BF) is a function of the density contrast between these fluids, Equation (7.9),

$$BF_b = \frac{4}{3}\pi \cdot r_b^3(\rho_{water} - \rho_{gas}) \cdot g \qquad (7.9)$$

where BF_b is the buoyancy of a gas bubble; r_b is the radius of a gas bubble; ρ_{water} is the density of water (generally about 1024 kg m^{-3}); ρ_{gas} is the density of gas (because of compressibility, this varies with temperature and pressure; Scholwater, 1979); and g is the acceleration due to gravity. Because there is a significant density contrast between gas and water, it is not difficult to see that buoyancy will drive gas bubbles upwards. Even at a depth of 3 km, methane has a density of only <170 kg m^{-3} (Clayton and Hay, 1994). Water saturated with dissolved methane (i.e. without gas bubbles) is also slightly less dense than normal pore water (Park *et al.*, 1990).

Fluid flow

Fluid may flow through either the normal pore spaces of 'carrier beds', or discontinuities such as open faults, fractures, or joints. To utilise these conduits the fluid pressure must either squeeze between the mineral

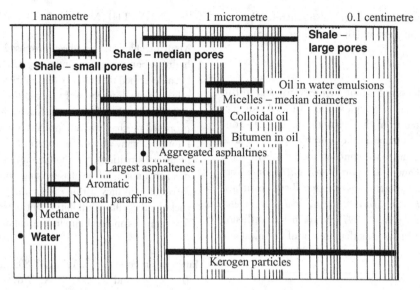

Figure 7.16 The sizes of the molecules of water, methane, and other hydrocarbons compared to the pore sizes of fine-grained sediments. (From Matthews, 1996, *AAPG Memoir*, **66**. Reprinted by permission of the AAPG whose permission is required for further use; AAPG © 1996.)

grains, or exert sufficient force to fail the sediment, forcing a migration pathway. The molecular size of fluids plays an important role in controlling how easily a fluid can migrate. Figure 7.16 shows that methane molecules, with a diameter of about 0.4 nm (0.4×10^{-9} m) are only slightly larger than water molecules, but much smaller than oil molecules. This means they fit easily through pore throats, the narrow passages between pore spaces, which prevent the migration of oil. In fact each fluid species has its own permeability, which is dependent on its viscosity, and compressible fluids (i.e. gases) behave differently from incompressible fluids (liquids). When more than one immiscible fluid (water, oil, and gas) is present each has a 'relative permeability' depending on the relative proportions of the fluids, and the nature of the wetting fluid (i.e. which one is in contact with the mineral grains). These matters are discussed in detail by numerous petroleum geology texts, so it is not necessary for us to discuss them here.

The mathematical concept underlying flux through a network of pores and throats is termed 'percolation theory' (Sahimi, 1994). A fluid moving through a rock or sediment must first overcome the capillary entry pressure and thereafter find the path of least resistance. Generally, capillary forces outweigh viscous forces since the flow is very slow, and so the favoured path is one that connects the largest throats, no matter how long and tortuous it may be. Although it might be thought that fluids flow through carrier beds en masse, occupying the full cross-sectional area of the bed, this is not necessarily so, and is probably rare. Instead, the migrating fluids pass through discrete pathways, or 'keyholes', that are a consequence of the heterogeneous nature of the rocks or sediments (Matthews, 1996). This is why gas seeps occur as numerous individual bubble streams, rather than affecting a whole area.

Impermeable barriers and trap structures may interrupt the progress of flowing fluids. The trapping ability of rocks and sediments depends on the continuity of the layers that have the highest capillary entry pressure (Ho and Webb, 1998). A thin layer of mud is just as effective a seal to two-phase flow as a thick layer, but thin muds do not stop diffusive methane transport. Gas can dissolve on one side of a capillary barrier, and without the entry pressure being exceeded, pass by diffusion through the layer of fine sediment, and reappear by evanescence on the other side (Miller, 1980). If the bubbles on the upper side can move away through buoyancy, then the process can continue *ad infinitum*. Furthermore, gas bubbles can exsolve in positions along the flow pathway, and eventually coalesce into stringers spanning across the capillary barrier, without the process of invasion percolation necessarily occurring (Li and Yortsos, 1995). Such bubble strings do not even need to merge completely for methane transport to be greatly enhanced. Theoretically, bubble clusters in a porous medium, like marine sediments, may form directly from a pore water that is supersaturated in, for example methane (Li and Yortsos, 1995). Such bubble clusters may be thought of as analogous to condensation nucleation in air when the air reaches dew-point temperature and becomes supersaturated with respect to water. Excess gas in supersaturated pore water can move freely by advection through throats that would block the same gas in bubble form. If more gas is in solution in a fine layer because of capillary supersaturation, there will be a gradient in its chemical potential that has the capacity to drive a diffusive flux towards a location with a normal level of saturation. On the other hand, if pore pressure drops due to mechanical disturbance, then gas can exsolve and become trapped in the new pores or fractures (Clennell et al., 2000).

SQUEEZING THROUGH THE KEYHOLES

Consider a water-saturated sediment invaded by bubbles of gas. The relationship between the pressure within gas bubbles (P_{gas}) and the pore water pressure (P_{water}) outside is related to surface tension (T) and the radius of curvature of the meniscus (R_c) at the bubble boundary. Because of the nature of the contact between the meniscus and the mineral grains, the gas pressure must be higher than the water pressure. The limits to gas pressure are given by Equation (7.10),

$$P_{water} + (2T/a) < P_{gas} < P_{water} + (2T/R_c)$$

(after Sills et al., 1991)   (7.10)

where a is the bubble radius. The radius R_c controls the upper gas-pressure limit; it is smallest in the 'throats' connecting the pore spaces, so these throats inhibit bubble movement. In clays the pore throats are very small, so the gas overpressure ($P_{gas} - P_{water}$) will be higher than in sands where pore throats are large. So, clay will hold gas at a higher pressure than sand.

Migration through pore spaces is termed 'capillary migration', and this will occur when the capillary pressure (P_{cap}) is exceeded; P_{cap} is defined from Equation (7.10).

$$P_{cap} = 2T/R_c \qquad (7.11)$$

Wheeler et al. (1990) considered that for clay (R_c) could be 0.5 µm, so $2T/R_c$ could exceed 300 kPa;

the capillary pressure that must be exceeded for gas to enter a water-saturated clay is 'simply too large' (Harrington and Horseman, 1999 – summarising Tissot and Pelet, 1971). Harrington and Horseman suggested that normal two-phase flow is impossible in clays and muds, consequently gas will not flow through the pore spaces of such sediments, so they are 'totally impermeable' unless there are 'pressure-induced cracks'. In contrast, the capillary pressure in sands is exceeded at a low relative gas pressure because pore throats are relatively large; gas is able to flow through the pore spaces. Of course, a migration route will take advantage of the widest available pore throats or keyholes. However, the very narrow pore throats of an overlying clay would prevent the bubbles from invading it, so, gas would accumulate in the sand – which is why gas reservoirs and shallow gas pockets occur at the high points of a sand/clay boundary. If the sand/clay boundary is even slightly inclined, then the gas will migrate laterally and updip; laboratory experiments have shown that this will occur at angles as low as 0.5° (Sim, 2001). This explains lateral migration through carrier beds, and why enhanced reflections (described in Section 6.2.2) often lead from areas of acoustic turbidity.

As fluid accumulates in a reservoir, the pressure of the buoyant fluid increases; P_{gas} can be estimated from the height of the gas column (see Figure 7.17).

$$P_{gas} = P_{hydro} + [(\rho_{water} - \rho_{gas})g\,h_{gas}] \quad (7.12a)$$

This can also be expressed as a gas overpressure (P_{go}).

$$P_{go} = (\rho_{water} - \rho_{gas})g\,h_{gas} \qquad (7.12b)$$

FORCED ENTRY

Intuitively, it is assumed that clays and other fine-grained sediments prevent the migration of fluids; the statements by Tissot and Pelet, and Harrington and Horseman quoted above confirm it. Typically, petroleum reservoirs are capped or sealed by such sediments. However, it is increasingly acknowledged that clay seals leak, at least over geological time periods. Gas chimneys (Section 6.2.2) provide clear evidence of vertical migration. These chimneys cut across the bedding, so migration is not along carrier beds, and they are generally not associated with faults; anyway, faults are usually not vertical, but gas chimneys are. Fractures, produced by gas pressure, explain this paradox (this is

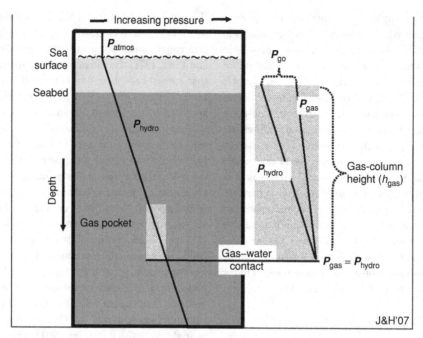

Figure 7.17* The pressure environment in a shallow gas pocket. Hydrostatic pressure (P_{hydro}) increases with depth from the sea surface – Equation (7.2); gas pressure (P_{gas} – expanded diagram on right) increases with height above the gas–water contact – Equation (7.12a); gas overpressure (P_{go}) depends on the density contrast between gas and water and the gas-column height – Equation (7.12b).

sometimes referred to as 'hydrofracturing' – but this is incorrect if gas, not water, is causing the fracturing).

In Section 6.1.1 we described gas voids formed in soft, fine-grained sediments as 'coin shaped', and we noted that they tend to be aligned with the long axis vertical. This indicates that they formed by lateral expansion, when P_{gas} has exceeded the horizontal effective stress (σ_h'), rather than the greater vertical effective stress (σ_v'), plus cohesion. The pressure at which this fluid-induced shear failure occurs, P_{shear}, can be expressed as in Equation (7.13).

$$P_{shear} = \sigma_h' + c \qquad (7.13)$$

Experimental work undertaken by the British Geological Survey (Harrington and Horseman, 1999; Horseman et al., 1999) has shown that gas passes through clays by forcing a pathway. Initial stages of their experiments resulted in the development of a gas overpressure. Then, at a certain critical point, break-through was achieved, the gas pressure spontaneously fell, and a flow of gas through the clay was measurable. This 'fracture event' occurred when the gas pressure was *fractionally larger than the effective stress of the clay*' (Harrington and Horse-

man, 1999). These experiments showed that fracturing of a clay may be induced by a high gas pressure, and that gas may pass through the clay within pressure-induced pathways, provided there is a suitable exit route.

At greater depths, where lithification has provided additional strength, the pressure at which a formation can be fractured hydraulically is known as the 'fracture pressure', P_{frac}; Equation (7.14),

$$P_{frac} = \sigma_h' + T \qquad (7.14)$$

where T is the tensile strength ($\sim$ zero in unlithified sediments). According to Clayton and Hay (1994), if the maximum effective stress is vertical, P_{gas} equivalent to 75–90% of the lithostatic pressure is sufficient to cause fracturing.

Tensile strength (resistance to being pulled apart) varies considerably between different rock types. Igneous rocks such as granites and basalts are the strongest ($T = \sim$15 MPa), and clays, shales, and mudstones the weakest ($T < 2$ MPa). However, these values can be considered no more than a guide because, for a single rock type, T varies considerably according to the degree of lithification (compaction, cementing,

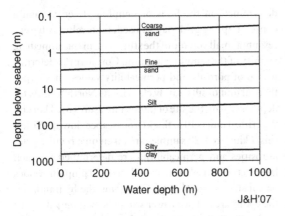

Figure 7.18* The transition from gas migration by capillary failure to shear failure in unlithified sediments (coarse sands, fine sands, silts, and silty clays) in water depths ranging from 1 to 1000 m. In each case capillary failure occurs beneath the line, shear failure above, and the transition occurs deeper beneath the seabed in shallow water. (Based on data from Sim, 2001.)

etc.). Hydraulic fracturing may occur during petroleum drilling if P_{frac} is exceeded by overpressured drilling mud, or by formation (pore) fluids during a kick (see Section 11.3.1). However, natural pore fluid pressures might also cause fracturing, leading to catastrophic fluid escapes.

Clayton and Hay (1994) suggested that the transition from capillary failure to fracture failure occurs at depths of between 500 and 1000 m in lithified rocks, but Sim (2001) showed that in unlithified sediments the transition (from capillary failure to shear failure) occurs much closer to the seabed, depending upon the grain size: > 500 m for silty clays, about 30 m for silts, but within the top few metres for sands (Figure 7.18).

Cooper et al. (1998) argued that gas migrates episodically when earthquakes dilate fractures. Nunn and Muelbroek (2002), on the other hand, thought that gas-filled fractures would have enough buoyancy to punch their way upward. Their results showed that methane-filled fractures a few metres long 'should propagate upward through geo-pressured sediments with velocities of hundreds of meters per year or higher'. As the gas rises into a lower-pressure environment the fractures increase in volume and they become able to entrain and transport considerable quantities of fluid (>1000 kg m^{-3} of oil or brine). Nunn and Muelbroek concluded that 'methane-filled fractures should propagate to the surface unless they are trapped beneath a layer that has high fracture toughness,

such as salt, or are absorbed when they intersect a permeable (> 10 mD) sand layer.'

At long last, there is a theoretical model that faces up to and explains the reality of observed gas migration and seepage!

FLOW ALONG FAULTS AND FRACTURES

Different people seem to have different understandings of the effects of faults, but most will accept that they dominate fluid-flow patterns. Some say that faults prevent migration, others that they facilitate it. In fact they are both true, in the right circumstances. Many petroleum reservoirs are closed by faults that trap the fluids. Even some shallow gas pockets are fault-bounded. Shear zones in mudrocks generally seal except at high pore fluid pressure (Dewhurst et al., 1999b). However, minerals (quartz, calcite, and many more) are commonly found lining faults seen in geological exposures. Because they were derived from migrating fluids, it is clear that at some time during the geological past these faults have acted as fluid-migration pathways.

Clayton and Hay (1994) argued that buried faults and fractures (> 1000 m below seabed) act as seals, not migration pathways, as the horizontal stress is normally sufficient to close them, unless they are kept open by sufficiently high pore fluid pressures, or the fault plane brings sand against sand. However, they envisaged transient activity associated with seismic activity. In shallower sediments things are different. The migration of fluids along faults must be controlled by the same factors as migration through the sediments themselves. Equation (7.11) probably applies, and the critical geometry is the width of individual discontinuities rather than pore-throat width. Once formed, fractures may also dominate fluid-flow patterns. Faults and fractures not only increase the effective permeability and breach barriers, but also sequester gas from the rock/sediment matrix because of the lower capillary pressure that exists there (Bethke et al. 1991). Fluid may flow freely when faults and fractures are held open by cements or asperities. On slopes, sediments are under tension in the upslope region, and compression at the base of the slope (Mandl, 1988; Jones, 1994; Bjørlykke and Hoeg, 1997). A consequence is that σ_h' is reduced in the upslope, making fracture failure easier. This may explain why seeps may be localised, and pockmarks are often found, at the headwall of slope failures (Yun et al., 1999). Doming caused by folding or diapir (salt or mud) movement has a similar effect, as is indicated by the frequency with which seeps are associated with salt diapirs.

Gas chimneys seem to indicate that gas is rising through a wide vertical column of otherwise uniform fine sediments. Does vertical migration through these chimneys start by following faults, or is it focussed by a more subtle geometry in the sediment layers?

CONTINUOUS OR INTERMITTENT FLOW?

A buoyant fluid may flow continuously from depth, but it may encounter migration barriers and traps that prevent the fluid escaping to the surface. Efficient traps may hold back the flow until a spill point is reached, or leaky seals may permit some fluids, most likely just gas, to pass. Such leakage may result in a relatively slow rate of upward migration. However, fluid overpressure continuously builds as the height of the reservoir or accumulation increases. Eventually a critical pressure (P_{cap} or P_{frac}) is reached, and the fluid is released. If a large reservoir has been held back, and the overpressure is significant, then the release may be catastrophic; but this is not always the case. As the reservoir is depleted, the overpressure decays, slowing the rate of flow. Eventually, the overpressure may be reduced until the barrier seals once more. This has been observed in experiments (Harrington and Horseman, 1999; Sim, 2001). Of course, mechanical failure may cause lasting 'damage' to the seal so that future accumulations are smaller, and a preferential migration pathway is established; Harrington and Horseman also observed this in experiments. We envisage that each seal failure releases a pulse of fluid to the shallower sediments, where it may be trapped again. Generally, the overpressure required to overcome barriers can be expected to decrease progressively towards the surface as weaker and weaker sediments are encountered. Consequently, overpressures will be less and each accumulation will tend to be smaller; shallow gas accumulations tend to hold less gas than deeper reservoirs. Also, the volume of gas released by each failure event will decrease towards the surface, but the frequency of release events will increase.

THE IMPORTANCE OF UNDERSTANDING THE PLUMBING

During routine site investigations when the porosity and permeability are determined, the values are most often determined on the basis of sampling. Even though a dense sampling grid may be used, each value only represents a small volume of the study area in the immediate vicinity of the sampling location. Consider the determination of porosity and permeability values for a hypothetical unit of silty, sandy clays in the North Sea. Water depth is 250 m, the site of interest measures 3×3 km and the critical unit is 10 m thick (from seabed down). Within this 9 km^2 area, 20 samples are taken on a regular grid; porosities and permeabilities are determined for each layer in the critical unit. Would these 20 pinpoint values be considered suitable for describing the hydraulics in the 9 km^2 area? The answer would be 'yes' only if the 10 m thick unit was completely homogeneous. If, within this large area, there was just one vertical keyhole or discontinuity through the unit, the 20 values would not be representative unless one of the samples, by chance, happened to be located at this critical zone. Experiments by Harrington and Horseman (1999) indicated that the gas permeability of a clay is dependent on: '*the number of pressure-induced pathways in the plane normal to flow, together with the width and aperture distribution of these features*' because water-saturated clays are impervious to gas. This also means that any (geotechnical or other) borehole that penetrates an impervious layer will have 'punctured' it; the 'wound' will probably remain for a very long time.

In some clay samples taken from depth in the North Sea the clay has been minutely fractured and flow structures exist. Indeed, fissures of different styles and scales, and with variable lateral distribution, occur in many overconsolidated clays. These samples demonstrate that structures that may convey fluids from depth to the surface do exist. We have also described vertical structuring on high-resolution sub-bottom profiler records and the scattered distribution of pockmarks and seep vents on the seabed. These all show that discrete migration pathways exist. Consequently we disregard porosity and permeability values achieved by pinpoint sampling; they are not valid or representative of large areas. We are of the opinion that in areas where such local vertical weakness zones exist vertical permeability is much greater than measurements indicate. If sampling is done at random, there is very little chance of hitting these vertical inhomogeneities. Even if one sample (in perhaps 20 or even 100) happened to hit such a zone, the result might well be treated as an anomaly and discarded.

THE INFLUENCE OF SUBSURFACE FEATURES

The distribution of many seabed features is affected, or controlled, by subsurface features, which determine the locations of migration pathways. These include: pockmarks associated with buried channels, canyons and faults, mud volcanoes associated with faults and fault intersections, and seeps concentrated above salt diapirs.

Lifting forces
UNHARNESSED ENERGY

Confined, overpressured fluids contain potential energy that is released when the seal retaining them is broken. The resulting fluid-flow rate (Q) is related to the pressure difference on either side of the seal, and can be estimated assuming Poiseuille flow, Equation (7.15),

$$Q = \frac{\pi \cdot r^4 \Delta p}{8 \eta l} \qquad (7.15)$$

where r and l are the radius and length of the migration pathway; Δp is the pressure differential; and η is the viscosity of the flowing fluid. Expansion also occurs with pressure reduction. Of the three possible fluid phases (water, oil, and gas), gas is by far the most compressible, so its expansion on release is the most energetic. Also, depending on the pressure (lithostatic and hydrostatic) and the presence or absence of oil, gas may come out of solution. This process is familiar to anyone who has opened a bottle of champagne (or any fizzy drink). When the bottle is opened there is an instant loss of pressure. Carbon dioxide comes out of solution and expands, but it can expand only in one direction – out of the bottle. The gas expands with sufficient energy to shoot the cork out, and then to carry some of the champagne out of the bottle with it. Kopf (2002), amongst others, called this process 'explosive decompression'; this is the mechanism responsible for the Lokbatan-type mud volcano eruptions described in Section 7.3.4.

These events are triggered by seal failure. After a period of activity during which overpressure was released and the source rocks achieved equilibrium; the feeder channel has 'sealed'. Subsequent gas generation or migration from below, and persistent pressure (whether from tectonic forces or continued sedimentation – or both) result in another build-up of overpressure. This can be released only by the removal of the blockage (the failure of the seal), which will herald a new eruptive phase. In some cases the removal of the blockage requires a great deal of energy; this is demonstrated by the angular, brecciated rock fragments found amongst the emission products of both onshore and offshore mud volcanoes.

Although explosive decompression is normally associated with mud volcanoes, it seems possible that sedimentary diatremes, which expel relatively little solid material, may be just as spectacular. The following example (based on a true, non-disclosed, blowout case) represents a less extreme, but no less interesting, scenario. During site investigation drilling at a water depth of about 20 m, the drill penetrated approximately 45 m of clays before entering a thin (~ 6 m thick) sand. At this point the well blew out. Gas escaped with so much force that the noise it made was described as a 'scream'. Gradually the rate of gas escape slowed, but gentle bubbling (accompanied by a *down*ward circulation of water) continued for several months. This event is explained by the rapid expansion of gas, and the exsolution of gas from the porewater. The borehole provided an expansion pathway. Once gas started to escape the pressure within the sand declined, leading to further gas expansion as a low pressure 'front' migrated away from the borehole. A self-perpetuating system was started, and gas escape continued until the excess pressure was dissipated.

USING THE ENERGY

The character of fluid flow, its velocity, and the effects it has on seabed sediments, are all fundamentally related to the availability of energy. The following, developed partly from ideas from Lowe (1975), define a spectrum of processes.

Seepage: the slow upward movement of pore fluids within existing voids, or rapid flow within lithified rocks, compacted and confined sediments, and coarse-grained sediments; it does not result in any movement of the sediment.

Hydroplastic deformation: characterised by grain-supported sediments with a significant yield stress. The yield stress may originate from either cohesive forces, as in partially compacted clays, muds, silts, and argillaceous sands, or frictional resistance, as in compacted clean sand and gravels. Presumably this state results from an increase in the pore-fluid pressure, and a consequent reduction in strength – see Equation (7.4).

Hydroplastic deformation of cohesive sediments is possible when the moisture content exceeds the Plastic Limit, but has not reached the Liquid Limit.

Liquefaction: a sudden breakdown of a metastable, loosely-packed grain framework, the grains becoming temporarily suspended in the pore fluid, settling rapidly through the fluid until a grain-supported structure is re-established. Liquefaction results in the sediment flowing like a liquid (for example, the muddy liquid flows extruded from active mud volcanoes). This happens when the effective stress (and therefore shear strength) is reduced to zero, a situation arising when the upward force of the flowing fluid equals the buoyant weight of the sediment particles (Brown, 1990).

Fluidisation: occurs when the drag exerted by moving pore fluids exceeds the buoyant weight of the grains; the particles are lifted, and the grain framework is destroyed. According to Nichols *et al.* (1994), the start of fluidisation occurs when the 'minimum fluidisation velocity' (U_{mf}) is reached.

$$U_{mf} = \frac{\varepsilon_{mf}^3 (\rho_s - \rho_f)g(\varphi d)^2}{5(1 - \varepsilon_{mf})\mu 36} \qquad (7.16)$$

Equation (7.16) was attributed by Nicolas *et al.* (1994) to Stanley-Wood *et al.* (1990).

Here, ε_{mf} is the intergranular porosity at minimum fluidisation; d and φ are the grain size and shape, respectively; ρ_s and ρ_f are the density of the grains and fluid, respectively; g is acceleration due to gravity; and μ is the fluid viscosity.

When the velocity (U) of the upward-flowing fluids equals U_{mf}, the flow balances the downward force of the individual sediment grains, allowing them to move freely under the influence of any small applied stress (Nichols *et al.*, 1994). When $U > U_{mf}$ the behaviour of the system depends on the density contrast between the solids and the fluids (i.e. whether the fluid is a gas or a liquid). In either case the distance between individual grains increases as U increases. Eventually, when the fluid velocity exceeds the settling velocity of the grains, the grains are lifted and carried away in the flow. Figure 7.19 shows that the onset of fluidisation is not just a function of the fluid flow, but also of the sediment type and its degree of consolidation. The most easily fluidised sediments are fine-grained sands and unconsolidated silts and clays, the sediment types in which pockmarks are generally formed (see Section 7.2.1).

Brown (1990) used the parameter λ to express the degree of overpressure,

$$\lambda = \frac{(P_{fluid} - P_{hydro})}{(P_d - P_{hydro})} \qquad (7.17a)$$

where P_d is the 'total mud pressure', i.e. the stress (σ) in the parent sediment. This can be re-stated in terms of the definitions we have used above – Equations (7.3) and (7.12b).

$$\lambda = P_{go}/\sigma \qquad (7.17b)$$

Brown stated that λ cannot fall below zero, diapiric activity (hydroplastic deformation) may occur when λ lies between 0 and 1, and liquefaction occurs when λ equals 1.

Explosive decompression: occurs in extreme cases where high fluid overpressure is released by seal failure. Lokbatan-type mud volcanoes are obvious examples, but sedimentary diatremes and giant pockmarks may also result from this process.

7.5.4 Modelling the processes

McKay (1983) reported a commonly seen but rarely noticed phenomenon, which also demonstrates, on model scale, how pockmarks form. He described 2–3 cm wide 'pockmarks' below the high-tide line on a sandy beach. They are formed when air escapes entrapment between a surface layer of swash-wet sand and the rising water table. First a dome is formed, but this eventually ruptures allowing the air, and some wet sand, to splutter out to produce a pockmark-like feature.

Although this is a natural example, it demonstrates that pockmark forming processes can be examined and understood more fully by modelling. Various experimental studies of buoyancy forces, reversed density gradients, and fluid migration have shed light on the migration of fluids beneath and through the seabed, and the formation of associated features.

Migration models

Experiments have consistently shown that when a low-density, low-viscosity fluid is overlain by a higher-density, higher-viscosity fluid the lower fluid rises in columns that are distributed in a pattern which describes a series of adjoining polygons, each rising column occurring at the nodes where three adjacent polygons meet (Figure 7.20). These nodes represent the points of

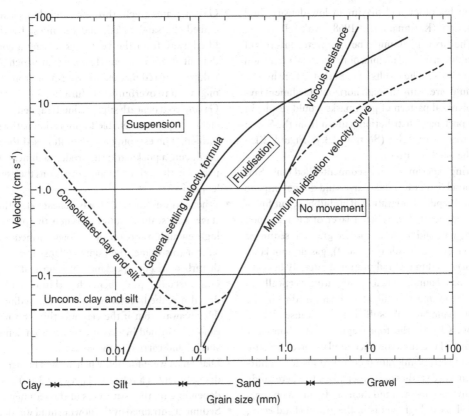

Figure 7.19 The relationship between sediment grain size and fluidisation velocity. The most easily fluidised sediments are sands with grain sizes between 0.1 and 0.5 mm. (From Hovland and Judd, 1988; redrawn from Lowe, 1975.)

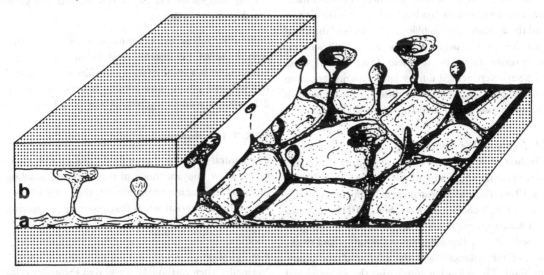

Figure 7.20 Experimental demonstration of plumes or diapirs of buoyant material rising through overlying material. Layer a is of low density and viscosity compared to layer b. Anketell *et al.* (1970) found that 'diapirs' of layer a material rise into layer b at each node of a megapolygonal pattern. (From Hovland and Judd, 1988; redrawn from Anketell *et al.*, 1970.)

maximum buoyancy and maximum low-density fluid accumulation (Krishnamurti, 1968; Anketell *et al.*, 1970). The experiments might be interpreted in terms of the distribution of mud diapirs, or the vertical migration of buoyant fluids. The results suggest that, where hydrocarbon fluids are rising through horizontal sediment layers, a polygonal pattern of pockmarks would result. In fact, the pockmark distribution pattern from the South Fladen area, North Sea (Section 2.3.1; Figure 2.15) approaches such a pattern.

During experiments in a sediment-filled tank, Sim (2001) found that gravel and coarse sand did not impede the vertical capillary migration of air bubbles, but finer-grained sediments did. When a layer of finer-grained sediment was laid over a coarser-grained sediment (sand over gravel, or silt over sand), gas accumulated in the coarse sediment, and migrated laterally beneath the sediment boundary before migrating vertically to the surface; even a very slight inclination (dip) of the sediment boundary is sufficient to cause lateral migration. Clearly, the topography of the boundary determines where reservoirs accumulate, overpressure builds, and vertical migration occurs. This was confirmed during field studies of an intertidal site where sandy gravel is overlain by sand and then sandy silt. Seep vents are located above high points in the gravel (Judd *et al.*, 2002b). In Sim's experiments, gas voids were formed in silts during the formation of vertical migration pathways, however, time delays between opening the air flow and the appearance of bubbles at the 'seabed' indicated that this did not occur until there was sufficient gas overpressure to fail the sediment. Once formed, these pathways were reused during later experiments, even when the air originated from different places beneath the underlying gravel 'reservoir' (Sim, 2001).

Modelling seabed features

The first laboratory experiments to produce pockmarks were performed by Wimpey Laboratories Ltd. (BP trading, 1974). These and subsequent experiments (Woolsey *et al.*, 1975; Nichols *et al.*, 1994; Sim, 2001) have shown that gas escape can result in the formation of a range of seabed features, depending upon the nature and thickness of the sediment, and the energy of the escaping fluid. The following summarise the experimental results.

- Gas escaped through gravel without disturbing the 'seabed'.

- Gas migrates freely through sand, but close to the seabed the sand failed, and the finest fraction was lifted away from the vent sites leaving a coarse lag deposit at the vent site. The point at which the sand failure occurred marked the transition from capillary migration to overburden failure.

- Gas injected beneath silt remained trapped until sufficient overpressure enabled a migration pathway to be formed. The escaping gas then fluidised the seabed, producing a pockmark; the eroded sediment was dispersed in the water. On some occasions the escape was 'explosive'.

- When gas was injected slowly beneath clay, the initial stage was 'seabed' doming. When the pressure was high enough a stream of bubbles formed from the side of this dome, the dome collapsed, and either a disturbed area or a pockmark was formed.

- Gas injected rapidly beneath a thin clay layer produced a seabed crater as the surface sediment was 'blown away', but if the clay was thicker a mud volcano was formed by clay mobilised from beneath the surface and carried to the surface.

- Diatremes were formed when air was blown through dry sediment at a velocity of between 3 and 50 cm s^{-1}, depending on the coarseness of the sediment.

- Sediment entrained by the flowing fluid was deposited at the 'seabed' as a sand or mud volcano, or, at high fluid-escape velocity it was emitted as a jet of sediment-laden fluid.

These relatively simple experiments demonstrated that seabed domes, pockmarks, mud volcanoes, and diatremes form as a result of fluid emission, and suggested that sediment plumes such as those described in Section 2.3.1 were formed by vigorous outbursts of fluid.

7.5.5 Triggering events

The natural build-up of pore fluid pressure will, eventually, allow buoyant fluids to pass barriers, enabling them to continue migrating towards the surface. However, various external forces may intervene, triggering an earlier release.

Earthquakes

Normally, when earthquake waves pass through a relatively loose, water-saturated sandy sediment, it tends to compact and decrease in volume, producing 'sand boils'. However, if the pore water cannot drain away, pore fluid pressure increases. If the pore fluid pressure exceeds the

Figure 7.21* An active sand boil caused by escaping groundwater and sediment liquefaction during and after the 1983 Nihonkai–Chubu earthquake, Japan. (From Hovland and Judd, 1988; photograph courtsey of Naofumi Matsumori.)

stress imposed by the overlying sediments, liquefaction occurs. In other words, the mass of particles previously supported by grain-to-grain contacts is transmitted to the pore fluid. Unsupported sand grains move together as the pore fluids escape and the soil structure fails and collapses. The sediment may deform and 'flow' under very small shear stress (Taylor, 1984), and may slide down even a gentle slope (see Section 11.2).

An incident that occurred on Concle Sands in Morecambe Bay on the coast of northwest England and reported in the local newspaper describes such an event. The landlord of the Concle Inn, Mr P. Singleton, and a local fisherman were walking across these sands, which are exposed at low tide, when an earthquake struck:

When we had got nearly halfway [between Fowla Island and Rampside], we saw at a distance from us a great mass of sand, water and stone, thrown up into the air higher than a man's head. When we got to the place there were two or three holes in the sand, large enough to bury a horse and cart, and in several places near them, the sand was so soft and puddly that they would have buried anyone if he had gone on to them.

Barrow Herald, February, 1865

Further along they found a crack in the ground, about thirty yards (~27 m) long, from which water was boiling up at about 500 gallons per minute (2270 l min^{-1}). There were more than 300 of them, and they extended more than half a mile (~0.8 km) along the sand (northwest), many of them being in a straight line, and only two or three yards (~1.75–2.75 m) from one another. Robert Muir Wood (1986, personal communication) argued that the craters described in the newspaper were more substantial than water-escape structures that would have immediately been overflowing with mucky water. He concluded that the holes (3–5 m in diameter, 1–1.5 m deep) were most probably formed by the sudden release of gas, triggered by the earthquake.

A report from the Tohoku District, Japan described similar features that formed during an M7.7 earthquake. The most severe damage was to lowlands composed of loose sandy soils of aeolian and fluvial origin of Holocene age. Later, the liquefaction sites could be identified by miniature sand volcanoes and large (8 m wide, 1.5 m deep) sand craters. Erupting groundwater and liquefaction can be seen in Figure 7.21. Resulting sand volcanoes are shown in Figure 7.22. The most severely damaged area was Shariki Village. In the 80 hectares of paddies, bordering onto sand dunes and alluvial lowland, there

Figure 7.22* Sand volcanoes scattered across a paddy field after the 1983 Nihonkai-Chubu earthquake, Japan. (From Hovland and Judd, 1988; photograph courtesy of Yukinori Fujita.)

were about 10 000 small sand volcanoes. They were also found on roads, house lots, and other cultivated areas:

> So much water spouted from the ground soon after the earthquake that there was flooding at Shiino Ushigata and 7 houses were inundated above floor level.
>
> Tohno and Shamoto, 1986

At some locations very large sand craters developed. These varied from 2 to 8 m in diameter and were about 1.5 m deep:

> A resident reported that the 7 m crater spouted water and sand to a maximum height of about 10 m and that this was continued through the next morning. Re-liquefaction produced by the maximum aftershock on June 21st also occurred.
>
> Tohno and Shamoto, 1986

The effect of the presence of gas bubbles within sediments subjected to earthquake loading is not fully understood. Is the earthquake load, which is generally cyclic, absorbed and dampened by the compression of the gas, or does the gas effectively reduce the strength of the sediment, making it more likely to fail? The following offshore examples suggest that gassy sediments are vulnerable.

MALIBU POINT, CALIFORNIA

On 9 February 1971 a M6.6 earthquake occurred at San Fernando some 50 km to the northeast of Malibu Point (Clifton *et al.*, 1971). A '*large volume*' of gas was observed bubbling from the sea surface about 500 m offshore. On the following day, when scubadivers inspected the seepage site, bubbles 0.5–2.0 cm across were still appearing at the sea surface. The gas was seen emanating from small craters (40 cm across and 10–15 cm deep) in a rippled, fine to very fine sand with an abundance of organic debris. A better-sorted, medium sand was present at a depth of 10 cm; it was found that bubble size was related to crater size and that the craters were sometimes several metres apart. They were arranged in a linear zone 12 m wide and 120 m long, aligned parallel to a geological structure; however, there was no evidence of faulting on the seabed.

KLAMATH DELTA, CALIFORNIA

An M7 earthquake occurred off the Klamath Delta, northern California, on 8 November 1980. This area had been covered by high-resolution geophysical surveys before the earthquake and was surveyed again afterwards. Comparative studies showed that a series of ridges, mounds, and pockmarks had formed on the seabed, as had a large number of gas vents; the incidence of shallow gas was significantly increased (Field *et al.*, 1982; Field and Jennings, 1987). Whereas pockmarks had been produced by single-point gas venting, we believe the ridges and mounds may represent mud diapirism associated with gas release. Field and Jennings (1987) concluded that a 20 km^2 area on the Klamath Delta had failed partly by liquefaction caused by sediment degassing triggered by the earthquake.

THE 1993 PATRAS EARTHQUAKE

The pockmarks off Patras and Aigion, northern Peleponnesos, Greece, represent some of the most spectacular and best-documented events in active pockmarks. Just by chance, a hydrographic recording station had been placed out in the Gulf of Patras, in connection with municipal construction. During July 1993 the water in Patras Gulf was hydrographically stable, with an upper layer of 23.5 °C and a bottom layer, 17 m thick, of about 16.6 °C.

The first sharp temperature increase was recorded at 22.40 on July 13th when the water temperature rose from 16.8 °C to 19.3 °C for a duration of 1 h 20 min. The second increase was recorded at 01.30 on July 14th, when the water temperature rose from 16.8 °C to 23 °C for 5 h 25 min. The third increase occurred at 10.00 on July 14th and the temperature rose from 17 °C to 22 °C for 5 h. During the same period the sea was calm, the wind was blowing from variable directions and was less than force 3.

Hasiotis *et al.*, 1996

Immediately *after* this last event, at 15.32 on 14 July 1993, there was an M5.4 earthquake, one of the largest to have occurred in the vicinity of Patras. For the next 70 hours the same temperature recorder, located 3 m above the bottom, recorded temperatures that only varied between 16.9 and 16.2 °C, before it was recovered (see Papatheodorou *et al.*, website, for more details of this earthquake event).

Patras Gulf covers an area of 800 km². Sediments at the transition zone between Pleistocene and Holocene are gas charged over about 70% of this area (Chronis *et al.*, 1991). Whereas gassy sediments had been mapped earlier (Chronis *et al.*, 1991; Papatheodorou *et al.*, 1993), side-scan sonar and sub-bottom profiler data recorded ten days after the earthquake showed pockmarks 25 to about 150 m in diameter and 0.5 to 15 m deep (Hasiotis *et al.*, 1996; Soter, 1998). Within the 1.7 km² area the pockmark density was 80 per km², but there were local clusters of < 150 per km² (Hasiotis *et al.*, 1996). Acoustic water-column targets indicated that most of the pockmarks were actively venting gas, but in some cases the extremely high reflectivity of the plumes suggested to Hasiotis *et al.* that either there was a very high density of gas bubbles and/or there was a sediment plume.

The researchers compared this event with an observation from another pockmark field, off Aigion, further to the east. Here, according to Papazachos and Papazachou (1989), Poucqueville, a French geographer had reported that during an earthquake in 1817 'the sea-water became hot enough to burn the hands of the fishermen and "bubbles together with smoke came out" of the sea'. Hasiotis *et al.* (1996) favoured the idea that the three warm-water pulses observed in Patras Gulf were caused by huge quantities of hot gas and water intermittently emitted through the pockmarks.

THE 1995 AIGION EARTHQUAKE

The city of Aigion is located some 30 km east of Patras, near the site of Helike, a city wiped out during an earthquake in 373 BC (Soter, 1999); this may be the 'lost city of Atlantis'. Several earthquakes have struck Aigion subsequently. During the earthquake of 23 August 1817, the Mageneitas River, west of the city, threw up very thick smoke before being engulfed by the sea. At the same time, Cape Aliki, at the eastern end of Aigion Bay, was covered by the sea, which suddenly became very hot (Soter, 1998). On the 15 June (03.15) 1995, a large earthquake hit the city of Aigion, killing 16 people and damaging 1000 buildings.

Soter (1998) performed a survey of unusual sitings, observations, etc., associated with this earthquake. He recorded the following reports from places in the area.

- [*Valimitika*]. Mme. Raymonde Bourge of Boulogne, France, a guest in the ill-fated Hotel Eliki, noticed and videotaped bubbling ('bouillonnements') at the edge of the sea at about 7 p.m. on the eve of the earthquake.
- [*Valimitika*]. Apostolis Sakelaropoulos, employed at the Hotel Eliki, reported that unusually large waves disturbed the Gulf for two days before the earthquake, in episodes lasting 2–3 hours and in the absence of winds or ships. Many large and unusual fish were also caught during this period.
- [*Diakopto*]. The owner of a restaurant in Diakopto, reported that on the day before the earthquake he caught more octopi than he was normally able to take in a year. Something was evidently forcing these animals out of their hiding places.
- [*Eratini*]. For a week before the earthquake, fishermen observed that the water off Eratini was warmer than normal and that many fish were being caught out of season. On the eve of the earthquake, the sea near the shore at Eratini was bubbling and many people saw it, according to Ioanna Katsiba.

Soter, 1998

Perhaps the best observation indicating gas emission immediately prior to the earthquake, was that from Nikolaiika Bay, 5 km east of Aigion:

> In the early morning hours of 15 June, Panayiotis Stephanopoulos, a graduate student in geophysics at the University of Patras, was analyzing data for the Helike Project in his room in the Hotel Poseidonia, on the beach near the Kerynites River. He became aware of a rushing sound like that of wind, which gradually intensified until it was 'louder than the sound of wind in a storm'. When he stepped outside to investigate, he saw that the trees were motionless and that the sound could not be due to wind. As soon as he returned, the earthquake struck. Stephanopoulos estimated that the sound began about five minutes before the earthquake.
>
> Soter, 1998

There are large pockmarks parallel to the coastline off Aliki Peninsula to the east of Aigion (Soter and Katsonopoulou, 1998; Hovland, 1998). They are about 80 m wide and 12 m deep. Enhanced subsurface reflections and water-column anomalies indicate the presence of gas in the sediments and water column. However, the pockmarks were not surveyed immediately prior to, or after the earthquake. It is therefore not possible to judge if there are detectable alterations on the seabed. Soter (1998) suggested that a line of smaller pockmarks coincides with the offshore portion of the major Aigion Fault, and that these provide a conduit for seismic outgassing events.

These examples demonstrate that pockmarks may be formed during earthquake activity, but Çifçi et al. (2003) showed that periodic earthquakes may result in a stacked sequence of buried and seabed pockmarks (Figure 3.30).

NORTH SEA EXAMPLES?

In the North Sea it might be expected that earthquakes are too infrequent and too weak to be significant. In fact, the present network of 13 seismic stations in and around the northern North Sea was completed only in 1985, and prior to 1977, when the British Geological Survey established new stations on the Shetland Islands and in the Beryl field, coverage was limited to a single insensitive seismic station at Bergen (Engell-Sørensen and Havskov, 1987). However, it is now known that there is seismic activity associated with the Viking and Central grabens, and the Øygarden Fault Zone (Musson et al., 1997). This suggests that earthquake events could act as triggers for gas escape even in the North Sea. Interestingly, in February 1995, there was a minor (M3.2) earthquake with an epicentre only a few kilometres from the South Fladen Pockmark Study Area (Dave Long, 1997, personal communication). Perhaps this event and others like it are responsible for triggering pockmark activity and the formation of new pockmarks.

EARTHQUAKES AND MUD VOLCANOES

Many people have reported relationships between earthquakes and mud volcanoes; perhaps this relationship is not surprising as mud volcanoes tend to be located in zones of tectonic compression (see Section 7.3.1). However, it may be a surprise to find that sometimes mud-volcano activity seems to be a result of an earthquake, but at other times the earthquakes follow the mud-volcano activity. Guliyev and Feizullayev (1997) explained that a statistical analysis of earthquakes and mud-volcano eruptions since 1810 had shown that earthquakes trigger eruptions. However, mud-volcano eruptions lead to local tremors. So there is a succession of events:

major earthquake → mud-volcano activity → minor earthquake.

The delicate relationship between pore fluid pressure and mud-volcano activity may make it possible for mud volcanoes to provide a warning of an imminent earthquake. Minor changes in stress conditions *before* an earthquake may affect pore fluid pressures, and these may be reflected in mud-volcano activity. For example, Martinelli and Judd (2004) reported that anomalously high radon (^{222}Rn) emissions have been recorded in association with earthquake activity in Italian mud volcanoes.

CAUSE *AND* EFFECT?

The reports cited here suggest that pockmark monitoring may provide evidence of precursor events, warning of imminent major earthquakes, and may help in mitigating material damage, and in saving human lives.

Soter was a strong advocate for pockmark monitoring, and suggested (in 1998) that the Helike Delta front (Greece) would be an obvious place to monitor bottom-water temperature and other outgassing phenomena. He argued that pockmarks are more suitable than equivalent onshore features, and that they could easily be monitored: '*submarine pockmarks may thus provide a natural laboratory for the study of seismic precursory phenomena*' (Soter, 1999).

Why should such gas-escape events act as a precursor to earthquake events? Soter's explanation is simple. Clearly, the gas cannot provide the energy for the earthquake. That comes from tectonic stresses in the rocks. But, an injection of gas, increasing the pore fluid pressure, will reduce the *effective* stress, and therefore the resistance of the rock to failure. The gas 'lubricates' the rock, triggering the earthquake (Gold and Soter, 1985; Soter, 1999). However, the link between seabed fluid flow and earthquakes is not necessarily restricted to precursor events. A consequence of at least some earthquakes is that movements result in a readjustment of the arrangement of the solid fabric of the rock, and also, probably, any surficial sediment affected. The consequence will be a reduction in the accommodation space available for the pore fluids. Porosity will be reduced and pore fluids will be expelled. Thus, pore fluid migration and seabed fluid flow may occur either in advance of earthquakes, or as a consequence of them (or both – or neither!), depending upon the particular characteristics of the affected area.

Waves and tides
Variations in the hydrostatic pressure caused by the passage of waves and tides affect the three phases of gassy sediments in different ways; the solids and liquids are effectively incompressible, but the gas is not. The poro–elastic behaviour of gassy sediments is controlled in response to changes in pressure according to two of the fundamental laws of physics. In accordance with Henry's Law gas solubility increases with pressure, and according to Boyle's Law gas volume decreases as pressure increases (Sills and Nageswaran, 1984). The expansion and contraction of bubbles may affect the ability of gas to pass through pore throats in three ways. First, contraction in response to a pressure increase will allow a bubble to pass through a pore throat previously too narrow; effectively enabling capillary flow – Equation (7.11) – in coarse-grained sediments. In fine-grained

sediments we envisage a different effect: reductions in hydrostatic pressure (P_{hydro}) will effectively increase the relative gas overpressure (P_{go}) encouraging shear failure – Equation (7.13). Third, the same pressure increase will drive gas into solution, creating a concentration gradient across the pore throat and therefore driving diffusive flow across the barrier. These effects convert narrow pore throats into one-way valves that open with each increase in pressure, and close when pressure declines. This tidal and wave pumping, which can drive gas flow through the sediments, is discussed in more detail by Sills *et al.* (1991), Fredlund and Rahardjo (1993), Wang *et al.* (1998), Holbrook (1999), and Hovland *et al.* (1999a).

STORM WAVES

Long-period waves induce a cyclic pressure field whatever the water depth, and these cyclic loads will be transmitted to layers deep beneath the seabed. Okusa (1985a) measured and computed wave-induced pore pressure in the silty, sandy sediments of Shimizu Harbour, central Japan under various marine conditions. He concluded that the cyclic hydrostatic pressure variations measured at the seabed were transmitted into the pore fluids, but the pore fluid pressure variations were '*dampened and delayed slightly in phase compared to the seafloor variations*'. He found the measured pore pressures were dampened more than expected by predictions made for water-saturated sediments, and ascribed this damping effect to air bubbles trapped in the sediments.

Okusa also performed a theoretical analysis of pore pressure variations induced by ocean waves in poroelastic sediment. The relevant parameters are the wavelength, the wave period, the permeability, elastic constants, and Skempton's B coefficient, the coefficient of pore-pressure build-up; B is a measure of the pore fluid's ability to stiffen a porous medium under compression (Skempton, 1954),

$$B = 1/(1 + n\beta/\alpha) \qquad (7.18)$$

where n is the porosity of the sediment; β is the compressibility of the pore fluid; and α is the compressibility of the porous skeleton.

For a completely water-saturated sediment, without gas, $B = 1$, but even a very small amount of gas results in a rapid decrease in B. Generally, the magnitude of the effective stress induced in the sediment by the wave pressure decreases as B is reduced, and low values (< 0.6) of B may induce liquefaction. Okusa (1985b) found that under certain conditions gassy sediments '*can be liquefied under the trough of the wave owing to the wave-induced negative effective stresses added on the stress state at rest*'. In a sandy cohesionless sediment with $B = 0.5$, liquefaction may extend to a depth of the order of the wave amplitude under certain wave conditions (Okusa, 1985b). Storm waves can attain amplitudes (trough–peak heights) of over 20 m. At a wave period of 15–18 s such waves would induce a significant high-frequency pressure cycle on soft seabed sediments in water depths as great as 200 m (S. Okusa, 1987, personal communication). However, the effects are dampened significantly where water depths exceed 150 m.

Okusa numerically modelled the seabed response to a 24 m high wave at 70 m water depth. From his calculations a wave with a period of 15 s and a length of 312 m would induce a pressure field medium extending at least 3 m into the sediments. This zone is about 150 m wide and travels across the seabed beneath the trough of the sea-surface wave; so any local zone of weakness on the seabed will experience a significant pressure reduction every time a trough passes overhead. With a period of, for example, 15 s the weak zone would experience the pressure cycle four times every minute. Such a 'hydraulic pumping' action, which may last for several hours, might cause any gas trapped within the top layer (3 m) to be pumped out of the sediments. The weak zone would thus act as a drain or vent.

In areas such as the North Sea the maximum flux of gas is expected during the first storm after a long spell of 'quiet' sea conditions, most probably the first of the winter depressions. Three things support this idea.

1. The BGS observations of suspended sediments and possible venting in the South Fladen area Witch Ground Basin were made soon after a storm (Section 2.3.1).
2. The occurrence of pockmarks or liquefaction pits on Yuigahama Beach in Japan after the crossing of typhoons (Section 3.18.5).
3. At least two of three observed anomalous water-column particulate matter distribution events reported by Brun-Cottan (2000) occurred after storms (discussed in Section 7.6.5).

Since all current seabed-survey systems, except autonomous underwater vehicles (AUVs), are depen-dent on surface support (i.e. a surface survey ship), surveys are weather dependent. If a strong gale develops, the ship either recovers all equipment and rides off the weather, or it heads for harbour. In an ideal situation, AUVs would be employed for seabed surveys during stormy weather. The AUVs could be used by oil companies and research institutions for seabed and installation inspection, and would be able to operate from sheltered harbours, travelling to site and back, irrespective of the sea state. Their use would significantly increase the chance of documenting gas-escape events.

TIDES

The idea that tides may trigger both earthquakes and volcanic eruptions has been discussed since the 1930s, when an earthquake swarm was observed on the Izu Peninsula, central Japan. For several days, it was found that the number of earthquakes was higher during low tide than during high tide (Kasahara, 2002). Century-long observations of mud volcanoes in Azerbaijan also show that there is a close relationship between tidal cycles and mud-volcanic eruptions (Aliyev *et al.*, 2002). Kasahara suggested that when all other critical factors for geo-disruption (earthquake, mud-volcanic or normal-volcanic eruption, and even submarine landslides) are fulfilled, the little force that finally tips the balance is likely to be the tidal force, even though it is minute.

Low-frequency pressure cycles are continually imposed on the seabed by tides. In the main pockmark areas of the northern North Sea they have a maximum height of only 2 m, which at 150 m water depth represents a negligible pressure variation. However, in shallow-water areas this effect may be significant, particularly where the tidal rise and fall is large. Certainly this is the case with the active pockmarks in the Arabian Gulf that were found to be '*turned off and on by the tide*' (Ellis and McGuinness, 1986). At their shallow (< 10 m) water site in Cape Lookout Bight, North Carolina, Martens and Klump (1980) found not only that bubbling increased as the tide fell, but also that bubbling was greater during the more extreme low spring tides.

Boles *et al.* (2001) were able to analyse natural seepage rates by monitoring flow rates from the Seep Tents in Santa Barbara Channel, California (Section 3.22.4). Their nine-month records clearly showed four principal tidal components with periods of 12.0, 12.5, 23.9, and 25.8 hours. Seepage rates increased at low tide and decreased at high tide: '*each meter increase of sea height results in a decrease of 10–15 m³ hr⁻¹ or 1.5–2.2% of the hourly flow rate*'. They also identified long-term cyclic fluctuations, and showed that minor fluctuations are superimposed on them by variations in hydrostatic pressure as the tide rises and falls.

There is some doubt about how the energy in ocean tides is dissipated (Egbert and Ray, 2000). One way of explaining a 'missing dissipation agent' is to consider parts of the seabed to be pliable or more compressible than previously believed. We now know that large portions of the seabed, at least in continental-shelf areas, are gas-charged and therefore compressible. These areas should react to the extra pressure load represented by the tidal peak wave, by contracting vertically; the seabed will be lowered by a few millimetres: '*tidally induced seafloor displacement is small, of the order of 1 mm, which is presently difficult to detect at tidal frequencies*' (Wang et al., 1998). Over large tracts of seabed such compliance might reduce the energy of tides.

It may be particularly interesting to know if the flux of gases and dissolved nutrient concentrations are both higher during spring tides when tidal falls are greatest. As we believe that nutrients are lifted into the water by escaping gas bubbles (see Section 8.3.4), this effect may help to explain some of the observed lunar cycles in marine fauna.

ATMOSPHERIC PRESSURE VARIATIONS

In Equation (7.2) we acknowledged that atmospheric pressure (P_{atmos}) is a contributor to hydrostatic pressure because the mass of the atmosphere sits on the sea surface. Atmospheric pressure varies with the weather (or rather, the weather varies with the atmospheric pressure). Mattson and Likens (1990) demonstrated this during work on methane bubble emissions from a shallow (11 m) lake. They had noted variations in flux rates and in their search for an explanation they found that bubbling increased during periods of low atmospheric pressure. Indeed, more than half the total methane flux occurred during two periods covering only 10 of the 64-day measuring period; both coincided with '*the two lowest air-pressure events of the summer*'. Although they recorded only a 1–3% variation in atmospheric pressure, this was enough to trigger gas escape. However,

it is not absolute pressure change that is important, but the rate at which the pressure drops. McQuaid and Mercer (1990) pointed this out. They explained that explosions in British coal mines had been linked to rapid falls in atmospheric pressure since the eighteenth century. When Michael Faraday and Charles Lyell investigated a coal-mine explosion in County Durham (northeast England) in 1844 they concluded that a fall in barometric pressure resulted in the emission of methane into the mine air stream. In 1879 a Royal Commission set up to investigate a spate of coal-mine explosions concluded that '*variations of atmospheric pressure exercise an undoubted effect on accumulations of gas in mines*' (McQuaid and Mercer, 1990).

Variations between low- and high-pressure systems (950 to 1040 m bar) are not great when compared to variations in water depth, and they are probably insignificant in deep water; however in shallow coastal waters, particularly in areas with little tidal rise/fall, rapid decreases in atmospheric pressure may be an important trigger.

OTHER WAVE TYPES

Besides very low-frequency tidal cycles (12 h cycle), low-frequency storm waves (15 s cycle), and tsunamis, there are other long-period or low-frequency waves that should be mentioned.

Infragravity waves are surface waves with a period in excess of 1 min (Kinsman, 1965). Their origin and wave heights are not well known, but their wavelengths are such that they induce pressure anomalies on the seabed in all water depths on the continental shelf (Suhayda, 1974). Estimated amplitudes of bottom-pressure anomalies associated with infrawaves may be as high as 10 kN m⁻¹ in water depths of 300 m (Watkins and Kraft, 1978).

Internal waves resulting from density stratification in the ocean may also induce low-frequency pressure variations in the sub-bottom layers. Near major river mouths, such as the Mississippi, internal waves may be well developed because the influx of fresh water causes a sharp density gradient. Internal waves, often manifested as glassy water on the sea surface have been detected on satellite images. They have also been noted to concentrate and transport surface debris and pollution (Watkins and Kraft, 1978; Shanks, 1987). Surfacing internal waves with a wavelength of 15 km were photographed in the Malacca Strait from space (the Apollo/Soyuz mission, 1976).

Several researchers have studied the effects of atmospheric pressure variations on the Earth. Van Dam and Wahr (1987), for example, found that the Earth's surface frequently responds with vertical displacements of 15–20 mm, which also affect the local gravity values by 3–6 µGal. They found that these perturbations were largest at high latitudes during winter months, when atmospheric mass contrasts were greatest. The largest pressure variations were associated with synoptic-scale storm systems. These long-wave pressure systems not only cause a change in sea-level height, but also an associated deformation in the Earth. Van Dam and Wahr (1987) found that vertical displacements *'can be greater than 10 mm at stations at the coast, and often 10 mm within several hundred kilometres of the coast'*. These long-wave pressure variations might be seen as important triggering mechanisms for a wide range of 'catastrophic' dynamic processes at the Earth's surface, including fluid release from the seabed.

Very little is known about low-frequency bottom-influencing pressure waves in deep water. Here there is evidently need for future high-resolution, long-period measurements. Since the pressure-cycle events we are looking for are probably transient and short lasting, it is necessary to perform the measurements over periods of more than a year, and with self-contained bottom recording stations.

Changes in eustatic sea level

Although changes in eustatic (global) sea level occur over much longer time periods, they may have significant implications for seabed fluid flow. When sea level falls, such as during the onset of glacial conditions, hydrostatic pressure – Equation (7.2) – falls, so effective stress – Equation (7.3), and therefore shear strength – Equation (7.4) – are increased, making shear failure – Equation (7.13) – less likely. However, if a buoyant fluid is overpressured before sea level falls, then the reduction in hydrostatic pressure will make capillary migration – Equation (7.11) – more likely. When sea level rises, the increase in hydrostatic pressure is likely to inhibit fluid flow (assuming that increases in shear strength induced by sea level fall are not reversible).

Sediment loading and unloading

Significant changes in lithostatic pressure can be caused by rapid movements of sediment: both rapid unloading

caused by slope failure ('landslides'), and loading where sediments that have moved downslope come to rest, or by rapid deposition (e.g. as a result of flood-water discharge from a river mouth). These negative and positive changes may both have significant implications, but for different reasons.

UNLOADING

In Section 7.5.3 we noted that buoyant fluid trapped beneath fine-grained sediment would need to be overpressured if it was to force entry through a seal – Equation (7.13). What if the vertical stress was suddenly reduced, for example when a slope fails? The horizontal stress (σ_h') is related to the vertical stress (σ_v'), so it might be expected that sediment unloading would trigger the fracturing of sediments by overpressured fluids. Certainly, this argument has been used to explain sudden releases of gas from dissociating gas hydrates (see Section 11.2.3).

LOADING

In Section 7.4.4 we described the 'mud lumps' of the Mississippi Delta, and explained that they formed after periods of rapid deposition when the river was in flood. This additional load produced a fluid overpressure in the underlying sediments that could not be drained because they were fine grained and had a low permeability. Without drainage and compaction, these sediments had a low density relative to the newly deposited sediments. The density inversion provided them with sufficient buoyancy force – Equation (7.9) – to pierce the overlying sediments to form shallow mud diapirs. The collapse depressions also described from the Mississippi Delta (Section 7.4.2) may have been triggered by rapid sediment loading, but in these cases the overpressured fluid managed to escape through the overlying sediments, leaving the 'parent' sediments to compact.

These are small-scale, rapidly formed features. Larger features on submarine fans (such as the Håkon Mosby Mud Volcano; Section 3.2.2) may have formed for the same reasons, but over longer periods of geological time. During periods of lowered sea level (i.e. during glacial periods), there is an increased rate of sedimentation and slope failure on the continental slopes. If the previously deposited sediments cannot drain easily, undercompaction, density inversion, and sediment buoyancy may result in the formation of mud diapirs and mud volcanoes.

Steam injection

Although it may be easy to appreciate that earthquakes can release gases trapped in submarine sediments it seems the reverse is also possible. According to scientists who have analysed data from a mysterious earthquake off Japan, the event could have been caused by a buried hydrothermal or volcanic eruption beneath the seabed. The M5.5 earthquake occurred near Tori Shima, Japan, on 13 June 1984 and produced a much larger tsunami than might be expected. It was 1.5 m high 160 km from the epicentre. It was estimated that a 20-second injection of magma blasted about 0.3 km^3 of a mixture of magma and water into the sediment beneath the seabed. At a temperature of about 1000 °C, the magma would have quickly superheated sediment pore water, increasing its volume 30-fold, and causing an explosion that lifted the seabed over a large area (Anon., 1986).

The intrusion of basic sills into wet sediments during the Palaeocene triggered the formation of hydrothermal-vent complexes in the Vøring and Møre basins, off Norway (Planke et al., 2002; Svensen, 2003). After detailed interpretation of seismic data, and field mapping of comparable features in the Karroo Desert of South Africa, Planke, Svensen et al. proposed the following model.

1. Intrusion of magma leads to heating, and locally boiling, of pore fluids in the intruded sediment.
2. Increased fluid pressure causes hydrofracturing, most likely starting at the tip of the intrusion.
3. Fluid decompression leads to an explosive hydrothermal eruption at the seabed, forming a hydrothermal vent complex. The explosive rise of fluids towards the surface causes brecciation and fluidisation of the sediments and commonly the formation of a seabed pockmark.
4. The fracture system created during the explosive phase is later reused for circulation of hydrothermal fluids during cooling of the magma. This stage is associated with sediment volcanism through pipes < 30 m wide that cut the brecciated sediments and infill seabed pockmarks;
5. Subsequently, the migration pathways may be reused by buoyant fluids.

Evidence of migration pathway reuse was provided by the presence of seep carbonates (MDAC; Planke et al.,

2002). Svensen et al. (2003) estimated that hydrocarbon fluid migration continued for about 50 million years!

Deep supercritical water?

In Section 5.2.3 we discussed supercritical water and its role in hydrothermal circulation systems. Conditions for the formation of supercritical water may occur also in deep sedimentary basins, such as the Southern Caspian Basin. Although we do not know what effect this water phase might have on deep sediments (how it affects porosity and permeability, for example) we think that its role should be considered, particularly with respect to deep-rooted mud volcanism.

Human intervention

Liquefaction of seabed sediments may also be caused indirectly when the stresses of wave loading, alternating in direction as a wave passes, are transmitted to the seabed by a fixed structure such as a gravity platform. The balance of the pressures between different phases of pore fluids may be critical; consequently any stresses imposed by the emplacement of structures on the seabed may be sufficient to trigger gas release.

Accidents that have occurred when shallow gas pockets have been pierced by drilling demonstrate that gas under pressure will utilise any available pathway to the seabed, even if it is man-made (see Section 11.3.1). An opposite human influence, the reduction of natural seep rates, results from the reduction in pore fluid pressure caused by the exploitation of shallow reservoirs (see Section 11.6.2).

7.5.6 Ice-related influences

Although pockmarks and other features associated with fluid flow are clearly not confined to areas affected by ice, either now or during recent glacial periods, there is a surprising number of ways in which permafrost, ice sheets, icebergs, and gas hydrates are implicated in their formation.

The impact of ice sheets on groundwater flow

Boulton et al. (1993, 1996) showed that the overburden stress imposed by European ice sheets during the Pleistocene induced pore fluid pressures far higher than 'modern' values. Consequently, flow regimes in underlying aquifers were completely different, and groundwater was expelled around the margins of the ice sheets. Various features, including 'diapirs, liquefaction structures and pipes, and forms such as glacial "dough-nuts" and pock marks', might have resulted (Boulton et al., 1993). They considered that these effects would occur beyond not only the ice sheets themselves, but also fringing areas of permafrost. As the ice sheets extended into the North Sea, they suggested that some fluid-flow features here might have been formed by this escaping groundwater. However, the pockmarks in the Witch Ground Basin and the Norwegian Trench (described in Chapter 2) are in post-glacial sediments, so they were not formed in this way.

Permafrost as a seal

Permafrost can affect seabed sediments, and it will reduce their permeability, possibly to zero. Bondarev et al. (2002) found seabed permafrost whilst drilling in the Pechora Sea (Russian Arctic). The drill passed through frozen sand (at 16 m below seabed) and frozen, ice-bound clay (at 20 m). Then, at 50 m, they entered an overpressured gas pocket and experienced a blowout. This was serious enough to generate a 'boiling kettle', and the flow continued for at least 10 days. Further evidence that gas can be trapped by permafrost came from a study undertaken by Kvenvolden et al. (1993) in Alaska. They collected cores from permafrost on land, and demonstrated that they contained significant quantities of microbial methane, which was released when the permafrost was melted.

Ice and gas hydrates are also capable of trapping gas, but these are discussed elsewhere (ice: Section 10.6.2; gas hydrates Section 6.3.3).

Iceberg scouring

On the Newfoundland Grand Banks, in the Barents and Norwegian seas, pockmarks have been found in far greater abundance inside iceberg ploughmarks than elsewhere on the seabed (see Harrington, 1985, for example). Thomas and Connell (1985) excavated an iceberg-ploughed furrow in Scotland and found numerous minor faults and fissures in the zone stressed during the passage of the iceberg (Figure 7.23). Perhaps even such localised increases in permeability are utilised by gas seeking a migration path, although it may be that this concentration of pockmarks is caused by the occurrence of relatively recent, soft sediments infilling the ploughmarks (Figure 7.24). However, iceberg grounding (discussed in Section 7.6.5) is an alternative explanation for these features.

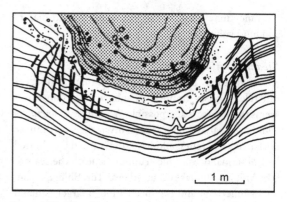

Figure 7.23 Microfissures, faults, and disturbed sediments below a relict iceberg furrow. (From Hovland and Judd, 1988; redrawn from Thomas and Connell, 1985.)

Melting ice

We have seen that gas can be trapped beneath and/or in subseabed permafrost, ice, and gas hydrates. But, what happens when the permafrost melts or the gas hydrates dissociate? Giant pockmarks in two areas, the North Sea and the Barents Sea, seem to provide evidence that the result might be spectacular gas-escape events.

BLOCK UK15/25 NORTH SEA

The giant pockmarks of Block UK15/25 have already been described (Section 2.3.7). Although these are clearly enormous features compared to the 'normal' pockmarks of this area, they used to be about 30% bigger according to Judd et al. (1994). They showed that the pockmarks were at their largest about 13 000 BP, since when they have been partially infilled by side-wall slumping, followed by the deposition of the Witch Member sediments (although gas-escape activity has kept the deepest areas open, revealing the underlying stiff Coal Pit Formation). The dating of pockmark formation is critical to understanding what happened. The boundary between the Witch Member and the underlying Fladen Member marks a period of rapid climatic amelioration as cold Arctic waters were flushed from the North Sea by warmer Atlantic waters. The result was the melting of subseabed permafrost. This was a period of relatively intense pockmark formation, as shown by the density of buried pockmarks (Long, 1992; also reviewed in Section 2.2.4). In a few places significant quantities of gas had been trapped beneath the permafrost. When the permafrost melted, this gas was released, probably in a

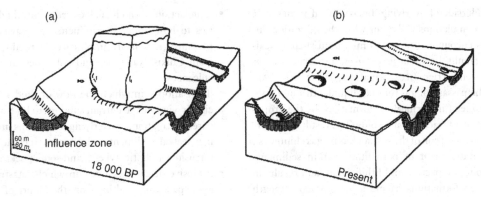

Figure 7.24 Pockmarks in iceberg furrows; schematic representation of Barents Sea features. (a) An iceberg ploughing through the seabed; note the zone of influence where the sediments are disturbed (as in Figure 7.23). (b) Present-day situation, with a thin layer of fine-grained sediments draped over the ice-scoured surface, and pockmarks formed mainly within the furrows. (From Hovland and Judd, 1988.)

single catastrophic event. The consequence was that the seabed sediment was blasted away to form these giant pockmarks (Judd *et al.*, 1994).

THE BARENTS SEA

A cluster of similar-sized giant pockmarks (described as 'craters') has been found in the central Barents Sea (see Section 3.2.1). Like those of Block UK15/25, they cut right through the surficial sediments, in this case revealing Triassic bedrock. Their origin is thought to be similar to that described above, but the presence of broken blocks of Triassic rocks in the bottom of these craters suggests that the gas escape was even more forceful. In this case it is thought that the gas release that caused the 'blowout' was derived from decomposing gas hydrates (Long *et al.*, 1998).

7.6 A UNIFIED EXPLANATION

In the Introduction to this chapter we said that there is a spectrum of seabed features associated with seabed fluid flow. In the previous section (7.5) we have tried to identify all the factors (processes, forces, and other influences) affecting the upward migration of buoyant fluids and buoyant sediment masses. In this section we outline the fundamental principles governing the rise of buoyant fluids and sediment masses. Then we provide conceptual explanations for the main types of seabed expression of seabed fluid flow: seeps, pockmarks, and mud volcanoes.

7.6.1 Fundamental principles

It seems to us that the processes responsible for forming seeps, pockmarks, mud volcanoes, and the various other features we have described, are guided by a relatively small number of common principles. Differences between features are attributed to variations in the influential factors; the dominance of one may result in the formation of a particular event or feature. A difference, perhaps only minor, in sediment type, fluid type, fluid concentration, or stress environment, or the influence of one or more external triggers may result in a completely different outcome. Some processes are slow, taking long periods of geological time from initiation to completion. Others are virtually instantaneous, catastrophic.

The fundamental principles are summarised below.

1. Buoyant fluids rise towards the surface provided there is a suitable migration pathway available. This may involve the following.
   - Flow along discontinuities (fractures and faults). However, significant fluid pressures may be required to keep discontinuities open, except relatively close to the seabed or where formations are under tension (for example in the vicinity of ocean spreading centres).
   - Flow through permeable formations, provided pressure gradients are sufficient to overcome the capillary pressure encountered in pore throats. However, multiphase flow may favour one fluid phase over others. This flow tends to be lateral and updip, and may be focussed along the

underside of overlying, finer-grained formations. Accumulations will occur when the migrating fluids encounter barriers to migration such as sealing faults, lateral facies changes, trap structures, or frozen barriers (ice, permafrost, or gas hydrates).

- The passage of gas (but not liquids) through fine-grained ('impermeable') formations by the propagation of fractures and voids. The result is the vertical migration of the gas, such as in 'gas chimneys'.

2. The entrapment of pore fluids within sediments inhibits compaction and lithification, and results in affected formations having a lower shear strength and bulk density than normally drained formations under similar stress conditions. Density inversions may result, and these may induce mobility. Affected rock types include:

- salt;
- serpentinite;
- fine-grained formations in areas of rapid sediment loading (by sedimentation or slope failure) or tectonic stress. Rapid sediment loading may affect formations at any depth below seabed.

Affected formations are generally characterised by a high fluid content. Because of its low density and its high compressibility, gas is by far more influential than oil or water.

3. Mobile formations may rise under the influence of buoyancy until:

- they reach a level of neutral buoyancy. This may be beneath, at, or above the seabed;
- retained fluids escape, leaving the formation able to compact, less able to deform, less buoyant, and less mobile;
- they reach a barrier, which prevents further migration.

4. Barriers to migration result in the build-up of fluid overpressures. Depending on the nature of the formation, the degree of overpressure, the volume of the accumulated fluid, and the compressibility of that fluid, the following may occur:

- the inflation of the formation to produce intrasedimentary or seabed doming;
- the breaching of the barrier and the escape of the fluid which then continues its migration without 'damaging' the barrier formation;
- the breaching of the barrier by fracturing;
- plastic deformation – which enables formations to flow *en masse*, for example in diapirism, including shallow mud diapirism;

- liquefaction – which enables sediment/fluid slurries to flow under the influence of gravity or a pressure gradient. Examples are the muddy flows from Chikishlyar-type mud volcanoes, and sand volcanoes;

- fluidisation – in which the flow of fluid is sufficiently vigorous to entrain solid particles from the parent formation, or overlying formations (including seabed sediments). Examples include sand intrusions ('clastic dykes'), and some pockmarks;

- explosive decompression – in which catastrophic gas expansion, resulting from the failure of a barrier or seal, leads to the explosive release of the gas and fractured rocks/sediments from the barrier or seal, or from overlying formations. Examples are diatremes (including sedimentary and hydrothermal diatremes), and Lokbatan-type mud volcanoes.

5. External triggers may induce fluid migration by:

- providing migration pathways, for example by the movement of faults, or the dilation of the rock/sediment matrix during cyclic earthquake loading;

- disturbing the pressure differential between trapped fluid phases, for example by the compression of gases during cyclic loading by earthquake, tidal or sea-surface gravity waves, or changes in water depth or atmospheric pressure;

- increasing the fluid content of the sediment, for example by steam injection, or the melting of ice, permafrost, or gas hydrates.

7.6.2 Explaining seeps

A distinguishing feature about seeps is that there is no significant solid component, only fluids, and the rock or sediment from which the seep flows remains undisturbed. We envisage that the following conceptual model explains a typical seep in a petroleum-bearing sedimentary basin such as the North Sea.

1. Petroleum fluids generated in thermally mature sediments in 'kitchen' areas, and microbial methane generated in shallower formations, continuously migrate into coarser-grained, more permeable formations.

2. Within the water-filled pores of these formations petroleum fluids, including methane, are buoyant. They rise towards the surface by capillary flow through migration pathways within 'carrier beds', gradually expelling and replacing excess water. Migration may be vertical, but when an overlying

finer-grained formation is encountered, migration becomes lateral and updip. Much of this is trapped in accumulations (reservoirs), which form in topographic highs in the top-carrier bed surface, or when other barriers (faults, facies variations, unconformities, etc.) are encountered.

3. Some hydrocarbons escape entrapment or spill from the reservoirs, and may migrate from reservoir to reservoir, filling each in turn, before continuing towards the surface. In shallower formations, where the horizontal stresses permit, migration may proceed along unsealed discontinuities (faults or fractures, i.e. conduits). Liquid components are restricted to this style of migration, and only reach the surface where conduits reach, or come close to the seabed. Even then it is generally the buoyancy of associated gas that enables oil to reach the seabed.

In the northern North Sea, the thick, clay-dominated Tertiary succession prevents the migration of oil. In contrast, the oil seeps of the Santa Barbara Channel, California occur where reservoirs in lithified rock are separated from the seabed by only a thin layer of surficial sediment.

4. Gas may pass through thin fine-grained sediment barriers by diffusion in pore water. It may escape from reservoirs capped by more substantial 'impermeable' fine-grained sealing formations by the propagation of fractures and gas voids, and/or by diffusion in pore water. In these ways gas is able to pass from porous layer to porous layer, using each as a temporary reservoir along its migration route. As each reservoir is filled the pore pressure increases until the gas is able to overcome the migration barrier; the pressure required to pass through successive impermeable layers will decrease as the gas rises to progressively younger and less compacted sediments.

Gas migration through fine-grained sediments is essentially vertical, for example through gas chimneys.

5. Like buoyant liquids, gas will take advantage of open discontinuities. So, sediments under tension, for example those overlying salt diapirs, provide migration opportunities tending to result in the focussing of gas seeps.

6. The duration of seeps depends on the continuity of fluid supply from deeper reservoirs, and the nature of the migration pathway.

Where migration is through coarse formations, capillary pressure is the limiting factor. Cycles of activity will start when overpressure exceeds the capillary pressure, but the overpressure is dissipated by flow. Flow will be rapid at first, but will slow progressively until the overpressure is reduced enough for the migration barrier to become effective again. The resulting reduction in pore pressure will increase the pressure gradient between this and deeper reservoirs, inducing the upward migration of a new pulse of fluid, and so initiating another cycle of escape.

When a pathway is forced through a barrier by deforming the sediment (by shear or fracture failure) the pathway will remain open unless the pore pressure declines to the point where the horizontal stress (confining pressure) is sufficient to close the barrier. If this happens, then seepage will be cyclic, if not then it will be continuous.

7. The durations of phases of activity and dormancy are difficult to establish without long-term monitoring of individual seeps. The 'return period' is dependent on the fluid supply and the characteristics of the 'plumbing'.

8. External forces may also influence seep activity. For example triggers (earthquakes, etc.) and variations in hydrostatic pressure may initiate seepage or affect flux rates.

9. Migrating fluids are utilised by microbial activity, particularly when they are close to the seabed (i.e. near and above the boundary between oxygen-free and oxygenated sediments; discussed in Chapter 8). The appearance of seeps at the seabed indicates that the rate of migration is greater than the rate of microbial utilisation. Consequently, seeps are restricted to locations where flow is focussed along well-established conduits (migration pathways). If no gas bubbles have been observed it may be that seepage takes the form of microscopic bubbles or pore water saturated with methane (as in deep-water sites), or that seabed utilisation accounts for the entire flux.

10. The presence of MDAC (discussed in detail in Section 9.2) indicates that seepage occurred over a considerable period of time. However, conduits (migration pathways) may become blocked by MDAC, causing the flow to be diverted to an alternative pathway; MDAC may indicate long-term active seeps, or former seeps.

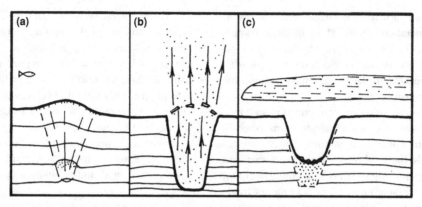

Figure 7.25 Conceptual model for pockmark formation. (a): Gas (or pore water) pressure builds in a shallow porous layer below an impermeable cohesive sealing layer. Excess pressure is relieved by deformation (doming) of the seabed. (b): Eventually, pore fluid pressure causes the seabed sediments to yield, allowing the fluid to erupt, fluidising the sediments. Gas, water, and sediment are ejected into the water column. (c): Fine-grained sediments are suspended in the water (as in Figure 2.18) and transported away by currents to be deposited elsewhere (depending on grain size and current speed). Coarse material falls back into or near the newly-formed pockmark. (From Hovland and Judd, 1988.)

7.6.3 The formation of pockmarks and related seabed features

Pockmarks and other seabed features associated with seeps are formed only when the rate of fluid emission is sufficient to affect the sediments, and when the seabed sediments are suitable.

Pockmarks

The following conceptual model is shown schematically in Figure 7.25.

1. Accumulations of gas in near-seabed sediments results in an increased pore pressure. This produces a dome-like swelling, seen first as the doming of subseabed reflections, then as seabed doming. This doming, which may cover a considerable area, represents a zone of tension where the seabed sediments are stretched to accommodate the new topography. Innumerable small fractures form both centrally and on the flanks of the dome (Withjack and Schemer, 1982).

2. Gas establishes a route to the seabed through these fractures. As hydraulic connection is established the pressure drop results in a violent burst of escaping gas. The gas expands as it rises from the shallow reservoir and the surrounding sediment fails by fluidisation and becomes entrained by the gas to form a gas – sediment plume rising into the water column. This process was termed 'gasturbation' by Josenhans *et al.* (1978), who envisaged that upwelling water would accompany the bubble escape, helping to disperse the sediment thrown into suspension. Any bottom currents would deflect the bubble stream and its attendant upwelling flow, causing turbulence and erosion of the pockmark wall. This would initiate the elongation of the pockmark. Once elongated, Josenhans *et al.* considered that the pockmark shape would so modify the current that turbulence at the downstream end, and possibly also the upstream end, would enable erosion and enhancement of the elongation to proceed even during periods when bubbles are not being emitted. This hypothesis is supported by the correlation between pockmark long axes and dominant tidal currents.

Unit pockmarks may be clustered around the site of the former dome, their number being greatest nearest the centre of the new pockmark. Where a small volume of gas is released a 'fully-grown' pockmark may not be produced, just a cluster of unit pockmarks, or a disturbed seabed.

3. The finest sediment particles (clays) entrained in the sediment may drift with the water current for some time before gradually settling back to the seabed, while coarser (silt and sand) particles settle within, or close to the pockmark. The scale of the seep-

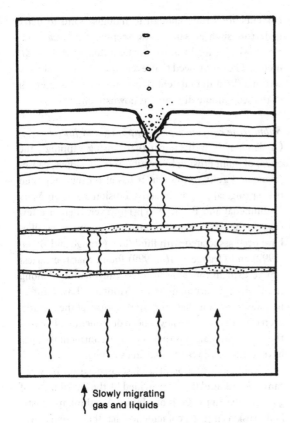

**Slowly migrating
gas and liquids**

Figure 7.26 If gas continues to migrate from below, it will accumulate in temporary reservoirs before migrating through established migration pathways to escape (periodically or continuously) through the pockmark floor. The overpressure required to escape from the reservoirs will determine the vigour of these escapes. (From Hovland and Judd, 1988.)

age, the particle-size distribution, and the strength of the water currents determine the precise distribution of sediment issuing from any particular pockmark; only on rare occasions do we see 'comet' marks, representing downcurrent deposition, near pockmarks.

4. The initial escape through to a virgin seabed may be quite violent, as a new migration path will have to be established. Subsequent events will not have to overcome the same resistance if the pathway is re-used; pockmarks are assumed to be the sites of repeated gas-escape events (Figure 7.26). After the first event, subsequent escapes of gas are likely to be of smaller scale and seabed doming will not occur. A probable consequence of pockmark formation is that

sidewall slopes are oversteepened, and liable to slump into the base of the pockmark.

5. Once a pockmark has been established it is envisaged that any gas within an area around it will be routed through it. In this way each pockmark will lie within, and will serve, a gas-drainage 'cell'. Normally the pockmark will be more or less centrally situated, but its exact location will depend on the topography of the shallowest trapping layer. Where the sediments are inclined, pockmarks occur at the upslope end of a cell. The sizes of these cells, and hence pockmark density, will depend upon the ease with which migration routes have been established, that is on the thickness, strength, and permeability of the near-seabed sediments; for example the lower the permeability the fewer the migration paths and hence the lower the density of pockmarks, unless the sub-seabed topography results in the concentration of gas in small, discrete pockets.

   Pockmark distribution will be defined by the distribution of migration paths and also by the regional net gas flux from below. Non-random patterns may be caused, for example, by faults within the shallow sediments.

6. The 'dormant' periods, during which shallow reservoirs are recharged, are characterised by infilling by the slumping of the sidewalls, and by general sedimentation. Most pockmarks are considered to be inactive or dormant features; those formed in areas of active sedimentation may be infilled rather than preserved, those in erosive environments will be removed, and those in shallow water where the seabed is affected by turbulence (e.g. associated with storms) will be destroyed.

7. Infilled pockmarks may shut down completely and become disused. However, the old conduit root system of the extinct pockmark may still work, and represent a viable conduit. This is indicated to be the case when we find pockmarks forming on the 'shoulders' of buried pockmarks, as for example in the Norwegian Trench.

An early concern of the oil industry was that pockmark activity would in some way pose a hazard to offshore installations. Where pockmarks occur close to existing or proposed structures (pipelines or platforms) an understanding of the 'return period' of phases of activity would therefore be of great benefit. Return periods inevitably vary from region to region, and even within individual areas, according to the continuity of fluid supply, the stability of the fluid supply pathway, and the frequency of triggering events. Whereas a continuous supply may result in a continuous gentle emanation of fluid, an infrequent supply may result if 'strong' migration barriers can only be breached when a large volume of buoyant fluid has accumulated. This may result in periodic catastrophic events. The nature (grain size and strength) of near-seabed sediments determines the overpressure required to overcome migration barriers. In turn, this affects the vigour with which fluids arrive at, and escape through the seabed. Periodic events may also be triggered by earthquake activity, as shown by Çifçi et al. (2003).

Pockmarks and diatremes

Pockmarks have been reported to mark the top of vertical gas migration pathways (gas chimneys and sedimentary diatremes). In these cases, the formation process is essentially as described here, apart from the delivery of gas from depth direct to the seabed. The seabed sediment has been fluidised and removed into the water column to form pockmarks in the way described above.

Pockmarks and water flow

Harrington (1985) suggested that pockmarks were formed by pore water expulsion from compacting sediments – another form of fluid flow. Whiticar (2002) argued that the pockmarks of Eckernförde Bay, Germany (described in Section 7.2.2) are 'submarine vents for freshwater and not structures from hydrocarbon outgassing'; pockmarks associated with groundwater flow also occur in Elaiona Bay, Gulf of Corinth (see Section 3.10.1). This mechanism is plausible for areas where aquifers outcrop at the seabed, but the hydraulic head must be sufficient to drive the flow so it must generally be restricted to near-shore areas; however, some notable exceptions (such as the Florida Escarpment seeps, described in Sections 3.26 and 8.1.5) are known.

Freak sandwaves

Strong fluid flows are required to fluidise sediments such as coarse sands (Figure 7.19), and large pore throats permit capillary flow. Consequently pockmarks are not normally found in such sediments. However, where water currents are strong enough to mobilise sand to make bedforms such as sandwaves, seeping fluids can create 'erosive' seabed features without fluidising the sediments. They only need to prevent the mobile sand from settling, the water current then transports the sand out of the seep plume dumping it downstream.

Submarine canyons, spring sapping, and faults

On land, it is well known that 'spring sapping', the effect of point erosion by groundwater in slopes, causes ravine and gully formation. It has also been suggested that spring sapping is an active erosion agent on Mars (Nummedal and Prior, 1981). However, there are few cases in the ocean, where submarine canyons and gullies have been associated with fluid flow. Orange and Breen (1992) and Orange et al. (1999) found that the easiest way to explain the large headless canyons found along the Florida Escarpment and in Monterey Bay, California, was to assume that there were springs at their heads, even though they have never been documented. Perhaps the submarine canyons that cut down continental slopes have formed as a result of spring sapping.

Another indicator of a close association between canyons and fluid flow is provided by the distribution of gas hydrates in the Gulf of Mexico. We think it is justified to ask: is it just a coincidence that there are massive gas hydrates nearly exposed on the seabed within the Green, Mississippi, and Desoto canyons of the Gulf of Mexico?

Many canyons cut across the continental slope of West Africa. Some are partially infilled with sediments. The meandering paths followed by turbidity currents sometimes change, and the abandoned channel sections get choked up with sediments. However, many of these infilled canyons are pockmarked, indeed on some seabed images it is only the strings of pockmarks that mark the courses of these abandoned canyons (Figure 3.23). It has been suggested (Cauquil et al., 2003) that the soft sediment infill provides an easy migration pathway for buoyant fluids, which may migrate vertically or laterally into the channel. Once in the soft sediments, the fluids escape to form pockmarks according to the mechanism described above. The presence of 'carbonate hardgrounds' and Lucinid bivalves (known to be associated with seeps – see Chapter 8) indicated to Cauquil et al. that methane was escaping through the pockmarks they studied. However, some strings of pockmarks were considered by Day (2002) to be too straight to mark preexisting channels. Do these mark fault-controlled gas migration, and are coalescing pockmarks an initial stage of canyon formation (a variation on spring sapping)? This apparent attraction of buoyant fluids towards canyon infill reminds us of the frequency with which infilled channels in the North Sea are associated with shallow gas.

Off southwest Taiwan, there are canyons and mud diapirs. Chow *et al.* (2001) discussed the possibility that the mud diapirs beneath these canyons formed by sediment offloading, and the reduction in stress caused by canyon erosion. However, we think it more likely that a more-or-less opposite model may apply, namely that the canyons formed because fluid escaping from the mud diapirs resulted in the formation of pockmarks. Eventually these pockmarks coalesced to form canyons aligned above the line of mud diapirs.

Another possibility that seems relevant to the above features is that both fluid migration (feeding seeps or springs) and the canyons are associated with faulting. This explanation has been presented to explain the focussing of seeps around the Danube Canyon (northwestern Black Sea), a feature aligned with a major deep-seated fault (see Section 3.11.3).

7.6.4 Mud volcanoes and diapirism

Deep-rooted mud diapir/volcano systems are complex. Others, most notably Brown (1990), Milkov (2000), Kholodov (2002b), and Kopf (2002), have discussed formation mechanisms in detail. Here we attempt no more than a summary:

1. In thick sedimentary basins, high pore fluid pressures are caused by the inability of pore fluids to drain from fine-grained sediments, despite the load of the overlying sediments and, in many cases, horizontal tectonic compression. The consequences are that the bulk density and shear strength of the sediments are lower than might be expected, and the sediments are capable of plastic deformation.
2. If the density inversion between the under-compacted sediment and the overlying formations provides sufficient buoyancy force, diapiric movement starts; this principle also applies to salt bodies and serpentinites. Generally, diapirs start to rise along faults.
3. Although density inversion may be enough to initiate migration of the sediment mass, the generation of gas within the sediments may provide additional buoyancy. Indeed, it seems difficult to exaggerate the importance of gas. As pressure declines with upward migration, expansion and exsolution provide a driving force, a decrease in bulk sediment density, and therefore increases in the buoyancy of the sediment mass.

Although there are deep sources of gas in some areas (e.g. the Southern Caspian Basin), the sediment and the fluid need not necessarily come from the same parent formation. Diapirism may be initiated by sediment buoyancy, and accelerated closer to the surface by the addition of fluids (such as shallow gas, or water derived from the dissociation of gas hydrates).

4. The pressure environment, and the relative proportions of sediment and fluid in a sediment mass, will determine how far a diapir rises. Diapirs may lose their momentum if the gas manages to escape; in extreme cases this leads to the formation of diatremes. Alternatively, the diapir itself may force its way right to the surface. There is a considerable variety of mud diapir/mud volcanic features lying somewhere between these extremes. The following were recognised by Milkov (2000).
   - Fluid (gas, oil, water, brine) flows, which entrain mud with a high fluid content and form flat cones or mud flows at the surface. For example, Atalante, offshore Barbados (Henry *et al.*, 1996); Håkon Mosby, Norwegian Sea (Vogt *et al.*, 1997); mud volcanoes in the Sorokin Trough, Black Sea (Woodside *et al.*, 1997).
   - Mud volcanoes fed by the migration of fluid and mud along faults and fractures, which stem from mud diapirs not reaching the surface. Examples are in the Gulf of Mexico (Prior *et al.*, 1989) and the Black Sea (Woodside *et al.*, 1997).
   - Mud volcanoes fed by the migration of fluid and mud along faults and fractures, which extend all the way from the source layer to the surface. There are examples in the deeper part of the Black Sea and offshore Barbados.
   - Mud diapirs that reach the surface, where a mud volcano is formed by fluid escaping along the body of the diapir. The Gelendzhik, Maidstone, and Moscow mud volcanoes of the

Mediterranean Ridge (Ivanov *et al.*, 1996) are examples.

- Mud diapirs that reach the surface but are not associated with mud volcanoes (i.e. there is no associated fluid escape). Milkov considered this type possible, but rare; there are examples offshore Barbados (Lance *et al.*, 1998).

To these we add: explosive mud volcanism of the Lokbatan type, and sedimentary diatremes such as those described from offshore Nigeria (Løseth *et al.*, 2001). These result from explosive decompression when a feeder-pipe blockage is cleared by gas overpressure.

5. The explanations above may account for the formation of deep-rooted mud volcanoes and diapirs. However, it is important to appreciate that:

- activity may be either continuous (Chikishlyar type) or intermittent (Lokbatan and Schugin types);
- some of these features endure for considerable periods of geological time;
- individual features may undergo more than one type of activity during their lifetime.

Shallow mud diapirs

The essential difference between deep-rooted and shallow mud diapirs is the depth of origin of the emission products. The driving force, sediment buoyancy due to density inversion, is essentially the same. Four scenarios explain how this situation may happen.

1. Gas (methane) rising from deeper sediments towards the surface becomes trapped in near-seabed sediments.
2. Microbial methane generated within organic-rich sediments near the seabed is unable to escape.
3. Rapid deposition of sediments, for example during river flooding, imposes a sudden load (vertical stress) on fine-grained sediments, inhibiting fluid escape and compaction.
4. Rapid sediment loading as a result of slope failure, inhibiting fluid escape and compaction.

The result, if any one (or more) of these processes occur, would be the plastic deformation of the fine-grained sediments to form a shallow mud diapir. (Fluidisation would result in the formation of pockmarks.)

7.6.5 Alternative explanations

Although we believe there is plenty of evidence to show that the features described in this chapter are formed by fluid flow, alternative explanations totally unrelated to fluid flow have been suggested.

Impulse sediment resuspension

Brun-Cottan *et al.* (2000) discussed observations on a vertical light transmissometer of clouds of suspended particulate matter hovering 10–50 m above the seabed at various locations (water depths down to 4800 m) in the north Mediterranean and New York Bight. They suggested that the clouds were of sediment resuspended from the seabed. In their opinion the resuspension process was triggered by '*a local input of an excess energy*' caused by '*local random fluctuations*' of the velocity of the bottom water that may have been caused by tidal currents, internal waves, bottom fishing gear, or even a '*variable intense wind field*'. Strangely, they did not consider action from below the seabed.

Two of the anomalous observations were made immediately after or during rough weather. We notice the similarity between these observations and those of McQuillin and Fannin (1979) (see Section 2.3.1). It seems to us that natural fluid seeps may have been the agent responsible for these sediment clouds.

Freshwater ice rafting

Paull *et al.* (1999, 2002) pointed out that there is no evidence of gas in many pockmark areas (Section 7.2.2). They also noted that there are many pockmark areas in high latitudes where sea bottom-water temperatures may periodically approach 0 °C, and suggested an alternative mechanism. They thought that freshwater seeping into seabed sediments might periodically freeze at the sediment–water interface to bind, and eventually float the sediment away from the seabed. They concluded that continued freshwater seepage at the same site and sustained incremental sediment removal at rates greater than the rate of deposition could excavate shallow pockmarks.

We certainly agree that some pockmarks (such as those in Eckernförde Bay, Germany) are associated with groundwater seepage. Also, we know that ice may form in seabed sediments (Section 7.5.6). It seems that this mechanism is possible, in the right circumstances. We therefore agree with Paull *et al.* that this is a possible '*additional mechanism*' for pockmark formation.

Iceberg grounding

We have explained that pockmarks may be associated with iceberg plough marks because fissuring provides migration pathways (see Section 7.5.6). However, there is another possible explanation for the association between pockmarks and iceberg plough marks. In shallow water, icebergs rising and falling with tides may be able to plough through the seabed sediments only at high tide. At low tide they may rest on the seabed, forming a footprint shaped depression, or if pushed by the wind or water currents, they may roll or swing round to form a near-circular depression. These features, correctly called 'iceberg pits' may be confused with pockmarks (Fader and King, 1981).

7.7 FOSSIL FEATURES

There is no reason to suppose that the processes we have described in this chapter are restricted to the present day. So, it is to be expected that 'fossil' features should be recognisable. A few have, mainly with the aid of high-quality 3D seismic data. For example, Cole *et al.* (2000) described large (500 m to 4 km wide) pockmarks of Palaeogene age in the UK sector of the North Sea. They are erosive, cutting 50–200 m into the sediments under the contemporary seabed. Cole *et al.* ruled out various other explanations for these features, and their location on top of gas chimneys persuaded them that they were indeed pockmark-like. Gas from '*deep thermogenic hydrocarbon kitchens*' seems to have accumulated in Palaeocene reservoirs before escaping to the surface. Bentonitic tuffs on the seabed formed a seal, which could only be breached after considerable gas overpressure had built up. This accounts for the large size of these features. Cole *et al.* concluded that '*these structures rank in scale alongside the largest known pockmarks and demon-*strate that natural fluid and hydrocarbon seeps have a long ancestry in these basins'. They also suggested that it is not a coincidence that these features are located beneath the pockmarked Witch Ground Basin where, they concluded, there is evidence for continued fluid migration.

Many other fossil examples of features associated with seabed fluid flow have been reported in the literature. Some, which are associated with chemosynthetic biological communities or mineralisation, are discussed in the following chapters (Sections 8.4.1, and 9.2.12).

7.8 RELATED FEATURES – LOOKING FURTHER AFIELD

Lunar surface features

Laboratory experiments were performed by Schumm (1970) to study the formation mechanism of some elongate lunar surface features. He concluded that some lunar craters, crater chains, crater clusters, and rilles are of endogenic origin, and that the features probably formed by fluidisation of the lunar regolith during emission of gas along fractures, *not* by impacting of external particles or bodies of rock (meteorites).

Seep features on Mars

After acquiring the splendid high-resolution photographs from the Mars Global Surveyor satellite, many new features have been documented on Mars. If there should be viable underground fluids (frozen water, hydrates, or carbon dioxide), then we expect also that some of the thousands of craters on Mars may have formed by fluid and sediment expulsion, much in the same way as pockmarks are formed on the seabed. However, probably only manned exploration of the Red Planet will be able to document such processes. In which case, the irony may be that they would be documented on Mars, before the submarine environment on the Blue Planet – our home . . .

8 · Seabed fluid flow and biology

Perhaps the most provocative consequence of the discovery of seafloor hydrothermal vents is the suggestion that these vents may have been the site where life originated.

Van Dover, 2000

The most unexpected and spectacular find of the detailed pockmark surveys described in Chapter 2 was the localised increase in biological activity. Since then many deep-water sites of seabed fluid flow have been shown to support rich ecosystems. We now take a more detailed look at the ecology of seeps. First we consider petroleum seeps (in both shallow and deep waters), groundwater flows, hot springs, and hydrothermal vents. It is the latter that have received the greatest attention from the marine research community, but the other habitats are also interesting and important. We also pay attention to carbonate reefs and mounds, and their fauna, as they seem to be unusual and particularly noteworthy. In the second section we consider specific aspects of the foodwebs associated with seabed fluid flow in these contexts. We then take a broader view of the ecology of seeps and their significance to marine biology. We conclude by reminding ourselves that seep-related communities are not new, but have been present for a considerable period of geological time. Were they the pioneers of life on the surface of our planet?

8.1 SEABED FLUID FLOW HABITATS

A close association between seabed fluid flow and 'anomalous' biological activity attracted attention in many of the places described in Chapters 2 and 3. Here we consider examples from each of the principal types of fluid flow demonstrating their impact on benthic ecology. Surprisingly, far more work has been done on communities associated with seabed fluid flow in deep water than on the more easily accessible sites on continental shelves.

8.1.1 Cold seeps on continental shelves

Ever since the first manned submersible dive on a pockmark in 1969 (offshore Nova Scotia, Canada, by the Bedford Institute of Oceanography (BIO); see King and MacLean, 1970) there have been reports of differences between the biology inside and outside pockmarks. The BIO dive reported reduced visibility because of myriads of krill and shrimp inside the pockmarks. We found the same thing during our investigation of pockmarks in the Norwegian Trench (see Section 2.3.5). However, it seems that this abundance is found in some pockmarks but not others.

North Sea pockmarks and seeps

Chapter 2 gives an account of the findings of our early work on North Sea seeps and pockmarks. Of particular relevance to this chapter were the following.

- Tommeliten: the 'bioherms', much richer in flora and fauna, and disarticulated bivalve shells, than the surrounding seabed (Section 2.3.2, Figures 2.23 and 2.24).
- Norwegian Block 25/7: a considerable diversity of sessile, planktonic and nektonic fauna, shells and shell debris, and bacterial mats within the pockmark, particularly living on the carbonate rocks (Section 2.3.3, Figures 2.28 to 2.32).
- The Holene: a dense concentration of the bivalve *Arctica islandica* (not in living position) in the bottom of a pockmark (Section 2.3.4, Figure 2.34);
- Norwegian Block 26/9: dramatic increases in the density of shrimp and other nekton inside the pockmarks compared to outside (Section 2.3.5, Figure 2.35).

- Gullfaks: bacterial mats, shells, sea anemones, shrimps and fish inside pockmarks, and fish congregating in the 'cavern' (Section 2.3.6, Figure 2.37).
- UK Block 15/25: a greater density and diversity of fauna, including fish, inside the pockmark than outside it (Section 2.3.7).

Others have done more detailed work subsequently. In particular, Paul Dando has worked with various colleagues on the ecology of the 'giant' pockmarks in UK Block 15/25. This involved bottom sampling during various cruises and visual inspections by the submarine Jago (Dando et al., 1991; Jones, 1993; 1996; these are summarised by Dando, 2001).

INFAUNA

Dando et al. (1991) reported that the macroinfauna of the pockmarks was similar to that of the surrounding area. It is dominated by four species of polychaetes (*Paramphinome jeffreysii*, *Levinsenia gracilis*, *Hetermastus filiformis*, and *Spiophanes kroyeri*), juvenile echinoderms, and the bivalve *Thyasira equalis*; together these make up 65% of the total number of individuals. However, Dando (2001) described the benthic infauna of the sulphidic sediments in the base of the pockmark as '*highly variable*' in terms of both the macrofauna and the nematode populations. Two samples from the deeper parts of the pockmark were completely barren of macroinfauna, and contained very few nematodes. In contrast two samples had a biomass exceeding those of the control samples. The bivalve *Thyasira sarsi* was found only in the pockmark, but not in the deeper parts; the nematode *Astonomena southwardorum* was found in the deeper parts, but was more prolific elsewhere in the pockmark. Both of these species are known to have microbial endosymbionts. One (dead) specimen of a third symbiont-hosting species, *Lucinoma borealis* was recovered attached to a carbonate rock. Despite the presence of these unusual organisms, Dando and co-workers (Dando et al., 1991; Dando, 2001) concluded that the infauna of the pockmark was generally '*impoverished*' relative to the 'normal' seabed outside.

EPIFAUNA

Dando (2001) said that there were '*conspicuous differences*' between the densities of fauna on the base and the sides of the pockmark and the surrounding seabed. In the base the major feature was the density of anthozoans (*Pennatula phosphorea*, *Virgularia mirabilis*, and *Cerianthus lloydii*); the sea anemones *Bolocera tuediae*, *Urticina feline*, and *Metridium senile*; the gastropod *Buccinium undatum*; the hermit crab *Pagarus* sp.; and the large echinoderm *Astropecten irregularis* were also relatively abundant. These epifaunal species were seen only occasionally away from the pockmark base. Dando also noted that bacterial mats, probably *Beggiatoa* sp., were common in the base of the pockmark. However, he concluded that the epifauna was abundant, not because of the seeps, but because of the presence of 'hardground', methane-derived authigenic carbonate (MDAC); hardground being rare in the centre of the North Sea.

FORAMINIFERA

The foraminiferal assemblage in a sample collected by Jago from a depth of 172 m in the Scanner pockmark was compared to a sample from a nearby 'control' site by Jones (1993). He concluded that the seep sample was statistically distinct from the control sample because it is characterised by a lower abundance (505 specimens compared with 594), lower diversity (21 species compared with 35), and higher dominance (23% compared with 21%) than the control sample. The relative abundances of taxa, both at order and species levels, also contrasted between the samples. Finally, the relative abundance of epifaunal/surficial, as opposed to infaunal, morphotypes was greater (66%) in the seep sample than in the control (48%). Jones concluded that it should be possible to detect seeps on the basis of the benthic foraminiferal communities.

FISH

The Jago videos showed that fish are much more common in the pockmark than they are outside. Only a few small demersal fish were seen outside the pockmarks, but inside there were shoals of large cod (*Gadus morhua*), ling (*Molva molva*), and various other species: Norway Haddock (*Sebastes vivaparus*), wolf-fish (*Anarhychus lupus*), and hagfish (*Myxine glutinosa*).

Perhaps the most amazing fish-related find in these pockmarks was the density of fish otoliths (ear bones), mainly of cod, in the sediments. Densities of up to 1550 per m^2 were reported by Dando et al. (1991) in the sediments at the base of the pockmark compared to less than 82 per m^2 in the sediments outside. Judd *et al.* (1994) wondered whether these could be a lag deposit, left inside the pockmark when finer sediments had been 'blown' out by escaping gas. However, they thought that these concentrations were too great to be accounted for by this mechanism. Their alternative suggestion was that the otoliths indicated that '*an abundance of fish*' had been attracted to the site. Subsequent sampling, reported by Dando (2001), found even greater concentrations: up to 42 188 otoliths per m^2. This lends support to the hypothesis, as does Dando's report of seeing a dying cod: '*an unusual occurrence but an illustration of how otolith accumulations may have arisen*' (Dando, 2001). Dando also reported that otoliths were so common that some polychaetes had made their tubes with them, rather than the usual sand grains!

Although these pockmarks are clearly attractive to fish, Dando (2001) considered that there was '*little evidence*' to support the hypothesis that the fish are in some way supported by the chemosynthesis associated with the seeps. He thought that the benthic infauna was not sufficient to support this fish population, and the absence of any methane-related carbon isotope signature argued against it. He therefore concluded that the '*fish are using the pockmarks for shelter, leaving the pockmarks to forage for food*'. This seems quite possible because fish are commonly attracted to artificial structures: pipelines, wrecks, etc.

CONCLUSION

Despite the abundance of fish and epifauna, and the presence of organisms associated with chemosynthesis, Dando rejected the idea that seeps support a local ecosystem.

> Any significant contribution to the food chain is likely to be confined to locations very close to the actual gas seep where the carbonate-cemented sandstone occurs (Hovland & Judd, 1988): more precise sampling will be needed to show this. This conclusion contrasts with the suggestion by Hovland & Judd (1988) and Hovland & Thomsen (1989) that the fauna around North Sea pockmarks and methane seeps is dependent on a hydrocarbon-based food chain.
>
> Dando *et al.*, 1991

We are not convinced. For example, this does not explain the large numbers of crustaceans (krill and shrimp) videoed swimming near the seabed in the Holene and the Norwegian Trench pockmarks. This question will be addressed again, once further evidence from other sites has been investigated.

Bubbling Reefs of the Kattegat

In Section 3.3.4 we mentioned the extensive area of gas seeps in the Kattegat, and the spectacular carbonate columns associated with them (Figure 3.13). Jensen *et al.* (1992) reported that clams, sponges, and endolithic green algae bore into these columns: '*forming an ecologically unique hardground community in an inner shelf sea which is otherwise dominated by soft bottom sediments*'.

More than a hundred macrofaunal species live on and around the columns, including species of polychaetes and bivalves rare elsewhere in Denmark, and prolific crabs (*Cancer pagarus*) and lobsters (*Homarus vulgaris*). Jensen *et al.* concluded that: '*The "bubbling reefs" with their rich fauna and flora are indeed oases of the seafloor*'. Yet, the fauna and flora they described seem to be there because of the hard substrate provided by the columns. When they examined the benthic fauna they found that nematode, polychaete, and oligochaete populations were low in abundance and biomass compared to control sites. However, they found extensive *Beggiatoa* mats, and dense films of rod-like, budding and filamentous methanotrophic bacteria on surfaces, including those of the macrobenthos, exposed to gas bubbles. Also, one nematode, *Leptonemella aphanothecae*, was found to harbour symbiotic sulphur-oxidising bacteria. Carbon-isotope studies identified a probable link between the bubbling gas ($\delta^{13}C$ −63 to −75‰), the carbonates ($\delta^{13}C$ −26 to −63‰), and the methanotrophic bacteria ($\delta^{13}C$ −43.4‰). However, the carbon of higher fauna ($\delta^{13}C$ −18 to −24‰) seems to be influenced little, if at all, by the gas seeps.

An active pockmark off Oregon

Juhl and Targon (1993) studied the epibenthic and pelagic microfauna in the water over a seeping pockmark in 132 m of water off the coast of Oregon. They found that:

• the mean abundance of microzooplankton was significantly higher in the near-bottom water (115 m) over the seep compared to the background, although

Dando (2001) thought that the difference may not be significant because these samples were taken 12 hours apart;

- $\delta^{13}C$ and $\delta^{15}N$ values of particulate organic matter in the water column and sediments were significantly lighter than backgound at the seep sites;
- $\delta^{13}C$ and $\delta^{15}N$ values of the body tissue of two species of shrimp (*Crangon communis* and *Pandalus jordani*), and two species of fish (rockfish, *Sebastes elongatus*, and slender sole, *Lyopsetta exilis*) were marginally lighter in samples caught near the seep than in samples from the background site.

These results seem to indicate that the seep '*provided a limited source of light C and N for the local food web*'. However, Juhl and Targon were not convinced that the influence of the seep was significant; although the relative increase in bacterioplankton over the pockmark was '*exactly what one would predict if the seep methane was promoting microbial activity*', it was too small considering the range in microbial populations elsewhere. They cited reports of small marine flagellates and free-living bacterioplankton that varied in abundance over several orders of magnitude, and noted that bacterial abundance over hydrothermal vents can be significantly greater than over this pockmark.

Northern California

Levin *et al.* (2000) identified 201 different organisms at three sites on the northern Californian Shelf. Fifty-five of them were found only near seeps, and 59 were not found at seeps. For example, one of the many annelids (*Capitella* sp.) and the crustaceans *Chirimedeia zotea* and *Synidotea angulata* were found only at seep sites on the shelf. Also, mysids (opossum shrimps) were concentrated in the bottom waters over seep vents. On the other hand, some species (for example the amphipods *Rhepoxynius abronius* and *Rhepoxynius dabious*) seem to be intolerant of sulphidic conditions and avoid seep sites. Of the remainder, some showed (statistically) a preference for seeps, and others seemed to show no preference. Multidimensional scaling of macrofaunal assemblage data suggested no significant difference between seep and non-seep sites on the shelf.

The Santa Barbara seeps

The vigorous Santa Barbara petroleum seeps, described in Section 3.22.4, have attracted considerable attention,

partly because of the concern expressed over marine pollution caused by the oil industry. Spies and Davis (1979) were the first to study the bacterial mats and other marine life associated with hydrocarbon fluid flow. They found a consistently greater density of infaunal benthic organisms at a shallow (16 m) natural oil and gas seep compared to a nearby area without seepage. At particularly active seeps *Beggiatoa* mats were found to be a few millimetres thick and 1 m or more in diameter. In a study of community structure Davis and Spies (1980) found that if patches of sediment become enriched by fluid flow they attract and support larger populations of deposit feeders, which in turn attract more predators (halibut, rays, etc.). Furthermore, the predators create localised disturbances and therefore induce still larger settlements of larval organisms.

8.1.2 Deep-water cold seeps

The first hydrocarbon-associated cold-seep community was discovered when Mahlon Kennicut *et al.* from Texas A & M University trawled across a seep site on the Louisiana continental slope. They expected to find diseased specimens from a naturally polluted site that would provide a natural analogue for sites polluted by man. Instead their first haul '*contained 800 kg of bivalve and gastropod shells, including both living and disarticulated bivalves ranging from 5 to 10 cm in length. The haul contained 30% living bivalves, 60% disarticulated bivalves and 10% living gastropods*' (Kennicut *et al.*, 1985). The second trawl contained dozens of brown, fibrous stalks that, at first, were thought to be reedy plants washed down from the Mississippi River. But, when they were broken, red blood spilled onto the deck; unexpectedly, they had found tubeworms (MacDonald, 1998). These were 2 m long and 1 cm diameter sibloginids (*Lamellibrachia* sp.) taxonomically different to those previously discovered at the hydrothermal vents of the Galapagos Rift and the Florida Escarpment groundwater seeps. These two trawls also contained significant numbers of fish, crab, and shrimps (Kennicut *et al.*, 1985).

In 1986, when Ian MacDonald and his team first made a submarine dive on an oil seep, they:

dropped right into the middle of a lush seafloor habitat, where we encountered large beds of mussels clustered around bubbling gas vents and extensive mats of brightly colored bacteria. Feeding on these exotic species was a diverse assemblage of fishes, crustaceans and other invertebrates that are commonly found in smaller numbers at shallower depths.

MacDonald, 1998

These early investigations of hydrocarbon-related deep-water cold seeps sparked the considerable interest that led to a large number of studies; some are mentioned in Chapter 3. In the following sections we have picked out just a few examples of communities in different contexts.

The Laurentian Fan

Cold-seep communties on the Laurentian Fan (Section 3.29.2), include vesicomyid clams, gastropods, tubeworms, galatheid crabs, and unidentified branched organisms. A white, powdery, filamentous material was found covering cobbles on the seabed; these are believed to be bacterial mats. The water depths are such that chemosynthetic processes must be sustaining these communities. Because the existence of communities was not anticipated, the submersible Alvin had not been prepared for the collection of the biological specimens, so details are not available. There is no conclusive evidence to indicate the nature of the primary energy source, apart from elevated hydrogen sulphide values (up to 180 µM) in a sediment core, but hydrocarbons and methane-rich sediments are known to occur further upslope. The absence of temperature anomalies and major sources of reduced sulphur compounds suggested to Mayer et al. (1988) that these communities are supported by methane.

Gulf of Mexico brine pools

Perhaps the most spectacular setting for a cold-seep community is at the brine-filled pockmark located at a depth of 640 m in the northern Gulf of Mexico (MacDonald et al., 1990; MacDonald, website). The pockmark is 15 m long, and is located at the summit of a 7 m high mound (a mud volcano?) overlying a buried salt stock. There is a continuous bed of methanotrophic mussels (Tamu childressi) in a 3 m wide strip defining the oxic–anoxic interface around the edge of the brine; larger individuals dominating the outer edge of the ring. Although the mussels can move over short distances, variations in the level of the brine surface are sometimes

too fast for them, so some die when the surface rises (Van Dover, 2000).

Hydrate Ridge

Most of the organisms (including siboglinid tubeworms) discovered at subduction-zone venting sites appear to be similar to those found in the hot vents of the spreading centres. However, Kulm et al. (1986) hypothesised that the bivalves and tubeworms (Lamellibrachia barhami) found in the cool vent areas of the Cascadia Subduction Zone have adapted to another type of energy metabolism, the capacity to utilise dissolved methane. This was based on the fact that a related genus of tubeworm is equipped with strands of bacteria capable of taking up methane (cf. Schmaljohann and Flügel, 1987), and that they found no evidence of hydrogen sulphide in the water or in the surficial sediment pore water near these seeps.

Subsequently, when the Ocean Drilling Project (ODP) drilled at Site 892 (1992), intense sulphide production in the upper sediments was discovered at Hydrate Ridge (Section 3.21.1). Because of a high flux of methane, diffusive seepage, and the presence of near-surface gas hydrates, the location is populated by large communities of clams (Calyptogena sp.) and by thick Beggiatoa bacterial mats (Boetius et al., 2000). Undisturbed cores with Beggiatoa mats collected in 1999 released gas bubbles on decompression during recovery to the sea surface. Sulphate-reduction rates in these sediments were extremely high: 140 mmol m^{-2} d^{-1} in the upper 15 cm, compared to levels below the detection limit (< 1 nmol cm^{-3} d^{-1}) at a nearby reference station (Boetius et al., 2000).

Subduction-related mud volcanoes

Along the accretionary wedge of the Barbados convergent margin, there are numerous mud volcanoes of up to 250 m relief. Seep-associated fauna also exist on at least one of them. The community includes large mussels (probably Bathymodiolus sp.), vesicomyid clams, and siboglinid tubeworms, but the most novel feature of these seeps is the large number of sponges and suspension-feeding gorgonian corals and bryozoans (Jollivet et al., 1990).

Mud volcanoes are also found in the Barbados Trench at about 5000 m, seaward of the accretionary wedge. These methane-rich sites are dominated by a species of bivalve (related to Calyptogena) and cladorhizid sponges (Olu et al., 1997). The sponges are extraordinary in that they host methanotrophic endosymbionts (Vacelet *et al.*, 1995, 1996) and occur as large 'bushes' where concentrations of methane are highest (see Section 8.2.2). Other components of the fauna include actinian anemones, which grow attached to the bivalve shells, galatheid squat lobsters, turrid gastropods, rare siboglinids, and '*large fields*' of maldanid and chaetopterid tube-dwelling polychaetes (Olu *et al.*, 1997).

Life in trenches
Cold-seep communities have been found in the Saganu Trough at a water depth of about 1300 m (Swinbanks, 1985), in the Nankai Trough at 3800 m, and in the Japan Trench at 5850 m. The Nankai Trough communities comprise a bed of the bivalve *Calyptogena* located at an assumed vent with a slightly elevated water temperature. Pore waters expelled along planes of thrust faulting in accretionary wedges attract tubeworm colonies and beds of bivalves. The 'oases' of the Japan Trench are characterised as circular spots 2–3 m in diameter densely populated by large bivalves *(Calyptogena)*. The total mesoscale biomass of each oasis, with up to 200 individuals per m², constitutes a huge quantity of organic matter (Horikoshi and Ishii, 1985). Henry *et al.* (1992) found that fluid-flow rates from clam colonies in the Nankai Trough are variable. The highest flow rates were measured where clam densities are highest (up to $100 \, m \, y^{-1}$), whereas bacterial mats occur where the fluids flow more slowly ($< 10 \, m \, y^{-1}$).

A similar setting on the actively subsiding Puerto Rico Trench was described as early as 1971 (Heezen and Hollister, 1971), although reported in less detail and at the time not properly understood. No vents were observed, but '*the biologists who dove were impressed with the abundance of life in the great depths (3100 m) of the trench, which they attributed to an ample supply of organic particles derived from the adjacent continental shelf*'. Here, too, slabs of carbonate rock were found. Today, these observations would probably classify the site as one of active venting.

The Håkon Mosby Mud Volcano
The Håkon Mosby Mud Volcano (Section 3.2.2) is an interesting biological anomaly, the first '*chemosynthetic oasis*' discovered in the Nordic Basin, more than 1000 km from the North Sea, the nearest known cold seeps (Vogt *et al.*, 1999a). Perhaps the most visually striking features are the extensive bacterial mats located inside the main

crater (Figure 8.1). They are 0.1–0.5 cm thick, and are dominated by large bacteria whose filaments are up to 100 µm long and 2 to 8 µm thick. They are morphologically similar to the non-sulphur filamentous bacteria of the genus *Leucothrix* and colourless filamentous sulphur bacteria of the genus *Thiothrix* (Pimenov *et al.*, 2000).

The fauna includes many other species, including tubeworms; two species, *Sclerolinum* sp. and *Oligobrachia* sp. form gigantic colonies spanning tens of metres. Pimenov *et al.* (2000) reported that *Sclerolinum*, which is the more abundant, lives in the oxidised sediments of the caldera, whilst bushes of *Oligobrachia* live in more reduced sediments. The fish population was also quite remarkable. Eelpout (*Zoarces vivaparus*) was present with a density of 1 per m², more than two orders of magnitude greater than normal. However, the most remarkable aspect of the varied Håkon Mosby fauna is that it thrives even though the ambient water temperature is consistently a chilling 0 °C!

8.1.3 The link between hydrocarbons and cold-seep communities

In order to determine whether it is true that chemosynthetic communities actually depend on petroleum seepage, MacDonald *et al.* (1996) described an ingeneous survey. They combined remote-sensing data from space with the knowledge of chemosynthetic locations in the Gulf of Mexico. Using Space Shuttle photographs, Landsat Thematic Mapper scenes, and European Radar Satellite data, they found 63 locations in the gulf with perennial oil flow to the sea surface. They verified the existence of 43 biological communities dependent on hydrocarbon seepage, yet they concluded that '*the coincidence of remotely sensed seeps with chemosynthetic communities is good, but far from universal*'. It seems that few chemosynthetic communites exist at the most vigorous seeps, and dense communities only form where seepage is protracted and the substrate has been 'prepared' by microbial activity (MacDonald *et al.*, 1996; MacDonald, website).

8.1.4 Shallow groundwater discharge sites

Because the ecological effects of SGD through the seabed are generally underestimated, we must investigate some examples, including places where groundwater, suspected to originate on 'dry land', emerges at the seabed, and examples of porewater expulsion.

Figure 8.1* Extensive bacterial mats at Håkon Mosby Mud Volcano. (Image courtesy of Rolf Birger Pedersen.)

Cape Henlopen, Delaware

Miller and Ullman (2001) were surprised to find benthic communities typical of estuarine conditions on the Atlantic Ocean shore-face of the sandflat at Cape Henlopen, Delaware, USA. Dense patches of the deep tube-dwelling polychaete worm, *Marenzelleria viridis*, an endemic species of the fresher, oligohaline conditions of estuaries, were found where sediment pore waters have a low salinity. *Marenzelleria viridis* feeds on diatoms in the sediment that flourish because of nutrients associated with discharging groundwater. In turn, predators such as bottom fish and shorebirds eat *Marenzelleria viridis*, so the groundwater discharge is linked to the higher food chain (Miller and Ullman, 2001).

Freshwater seeps in Western Australia

Johannes (1980) looked at the positive effect on marine life of freshwater flow through the seabed off Perth, Western Australia. He found that near-shore water is characterised by lower salinities than adjacent waters, and, in some cases, by clusters of small freshwater springs made visible by plumes of sand and by water that shimmers when waters of different salinities mix. He found that in his study area submarine seeps deliver several times more nitrate to the coastal waters than river runoff. Furthermore, Johannes considered that submarine groundwater discharge influences productivity, biomass, species composition, and zonation and is, in some areas, of greater ecological significance than surface runoff.

Lake Superior

Ring-formed depressions have been found in the deepest parts of Lake Superior (Section 3.28.1). In 1985 the manned submarine Johnson Sea Link II was used to dive into the deepest point of the lake (Clark, 1986) where long sand furrows 1 m deep and 4–6 m across were found. Pictures show a highly eroded and pitted lakebed surface where one would have expected thick, muddy sediments to occur. Perhaps the most remarkable discoveries were the lush 'fungal' beds growing on underwater reefs. These seem to be prime spots for spawning lake trout (Clark, 1986). Although Clark assumed '*resident sea life*' to have caused the 1 m high cliff features, we made another suggestion. The lakebed morphology resembles some of the features noted for North Sea pockmarks so we suggested that the deepest parts of Lake Superior represent sites of active groundwater and/or gas expulsion (Hovland and Judd, 1988). This would account both for the morphological lakebed features and for the high productivity of the area, assuming that lake life is stimulated by an extra input of nutrients from below.

8.1.5 Deep-water groundwater discharge sites

The first deep-water communites *not* associated with hydrothermal vents were discovered by Paull *et al.* off Florida in 1984. They are associated with submarine groundwater discharge (SGD), but are comparable to the chemosynthetic communities of cold seeps and hydrothermal seeps.

The Florida Escarpment seeps
Paull *et al.* (1984) described biological communities resembling hydrothermal-vent taxa where sulphide-rich, hypersaline water seeps from the foot of the Florida Escarpment at a water depth of 3266 m. Key members of these communities are mussels and tubeworms. The mussels are densely packed in patches covering about half the seabed. Small (~1 cm) trochid archeogastropods graze their valve surfaces, whilst stoloniferous soft corals encrust the siboglinid tubeworms, some of which occur in dense thickets of 100 or more per m². Other fauna include: large turrid neogastropods, small limpets, slender purple holothurians, small pink ophiurids, anemones, small transparent shrimp, zoarcid fish, galatheid crabs, and large half-buried clams (Paull *et al.*, 1984).

A slight increase in temperature was noticed in a mussel bed located on black sediment. Here the water temperature was 4.39 °C, while the sediment temperature was 4.80 °C. Nevertheless, no significant temperature anomalies or visible indications of seepage (shimmering water or bubbles) were noticed over the communities on any of the four dives with Alvin. Their absence led Paull *et al.* (1984) to conclude that they required only a source of reduced inorganic compounds and molecular oxygen; '*hydrothermal vents are but one vehicle to bring dissolved reduced inorganic compounds into contact with oxygenated seawater so that chemosynthesis can occur.*'

8.1.6 Coral reefs and seabed fluid flow

Deep-water corals build up mounds and colonise banks on the seabed, even in depths way beyond the photic zone. In some places the constructions are clearly associated with seeps, so they have been referred to as 'chemoherms' (e.g. Coleman and Ballard, 2001). However, most of the examples we describe in this section are not so clearly associated with seabed fluid flow, and have been given non-generic names. Undoubtedly, some of them would have been called reefs if they had been within 'scuba-range'. We call them 'coral reefs' unless coral species are not dominant, in which case we call them 'carbonate banks' or 'bioherms'. However, some features associated with microbial activity are hard to classify biologically, and are discussed in the next chapter (on mineral precipitation).

Norwegian deep-water corals
For over 200 years it has been known that dense coral communities can grow in the deep ocean (Gunnerus, 1768; Dons, 1944; Cairns and Stanley, 1981) but it is only over the last decade that they have become a 'hot' scientific topic (Freiwald, 1995). Norwegian fishermen working off mid Norway, for example, have known for generations that they can catch more fish in the areas with corals. Marine biologists have probably neglected them for so long because there has not been a viable explanation for their existence. In no way do they fit the generally accepted coral-reef domain. Tropical corals contain zooxanthellae, algal symbionts, which produce nutrients from sunlight, and which are highly beneficial to the host coral polyps. In contrast, the deep-water ('azooxanthellate' or 'ahermatypic') corals live in cold waters below the photic zone.

The faunal compositions of Norwegian and similar North Atlantic coral reefs vary, but generally consist of 2–300 different macrofaunal species, of which the frame-building stony ahermatypic corals *Lophelia pertusa* and *Madrepora oculata* are the most important. *Lophelia* reefs are known to occur in many parts of the Atlantic Ocean, and also in the Mediterranean Sea and the Indian, Pacific, and Southern oceans. Although it may be the most unexpected of locations, *Lophelia pertusa* colonies have also been documented at the Lucky Strike deep-sea hydrothermal-vent field on the Mid-Atlantic Ridge (Van Dover *et al.*, 1996; Van Dover, 2000).

During pipeline route surveys on the Norwegian Continental Shelf and in fjords, Statoil have come across several large *Lophelia* reefs, and many more have been reported by fishermen (Hovland and Mortensen, 1999; Fosså *et al.*, 2002). They occur in a variety of water depths, in fjords, on the continental shelf, and down to depths of nearly 500 m on the slope. They are not evenly distributed on the seabed, but occur in groups. An inventory of fauna associated with Norwegian reefs includes over 700 species (Mortensen *et al.*, 1998; Fosså and Mortensen, 1998; Hovland and Mortensen, 1999), most of which are well known from rocky near-shore areas.

HALTENPIPE REEF CLUSTER

So far, the best-documented cluster of reefs is the 'Haltenpipe Reef Cluster' (HRC), located about 75 km north of the town Kristiansund, on the western coast of Norway (Hovland *et al.*, 1998a). It lies about 1 km west of the Haltenpipe gas pipeline, constructed in 1996, and consists of 10 individual reefs, each up to 25 m high and 200 m in length. They are confined to a relatively small (500 m by 500 m) seabed area. The age structure of the HRC was investigated by sediment coring and radiocarbon dating. The oldest *Lophelia* remnants were found in sediment samples from about 2.4 m below the seabed adjacent to the living reefs. The oldest is 8620 ±70 years old; several others were grouped around 6000–6600 y BP and 2000–3000 y BP (Hovland and Mortensen, 1999).

The HRC grows on a gentle ridge on the seabed that delineates the transition from subcropping sedimentary rocks of Palaeocene age, in the west, to rocks of Cretaceous age in the east. Both dip at an angle of about 10° towards the west-northwest (Bugge *et al.*, 1984), and are covered by 0–15 m of glacial morainic material and post-glacial (Weichselian) silty clay. Geochemical surveys of sediments in pockmarks of the adjacent

Cretaceous basin and at the base of individual HRC reefs, have shown that the concentration of methane to butane in the sediment pore water increases towards the centre of pockmarks. They also increase towards the base of individual reefs of the HRC (Hovland *et al.*, 1998a).

Side-scan sonar, sub-bottom profiler, and multibeam echo sounder data acquired in 2000 show that there are numerous small pockmarks not only inside the Cretaceous depressions, but also on portions of the seabed where the Palaeocene subcrops and where the reefs grow. These pockmarks are between 3 m and 10 m in diameter and up to 0.5 m deep. 'Unit pockmarks' are most dense close to the base of the three westernmost coral reefs in the HRC, strongly suggesting a link between these reefs and seabed fluid flow.

SULA RIDGE

Some of the largest known Norwegian *Lophelia* reefs are on Sula Ridge, about 15 km north of the HRC. Here the individual *Lophelia* reefs are about 30 m high. As at HRC, dipping strata of Palaeocene age underlie them. There is clear seismic evidence of gas charging (Hovland *et al.*, 1998a), and methane concentrations in the water column nearby are of higher than normal background values (Michaelis, 2000).

POCKMARK-DWELLING CORAL REEFS AT KRISTIN

At the Kristin HTP (high temperature high pressure) hydrocarbon field, located about 100 km west of HRC, numerous *Lophelia* reefs up to 3 m high are found inside pockmarks that are up to 8 m deep (Figure 3.8). Because some of these reefs occur close to planned pipeline and cable routes, Statoil has started a local reef-monitoring programme. In contrast to what coral scientists might expect, there are only a few reefs on the topographically highest locations in the area; most of them are confined to the depressions. To us, these reefs suggest an intimate link to seabed seepage structures.

HYDRAULIC THEORY

After considerable debate, a 'hydraulic theory' has been developed to explain the existence of the paradoxical 'Norwegian coral reefs' (Hovland, 1990b; Hovland and Thomsen, 1997; Hovland *et al.*, 1998a; Hovland and

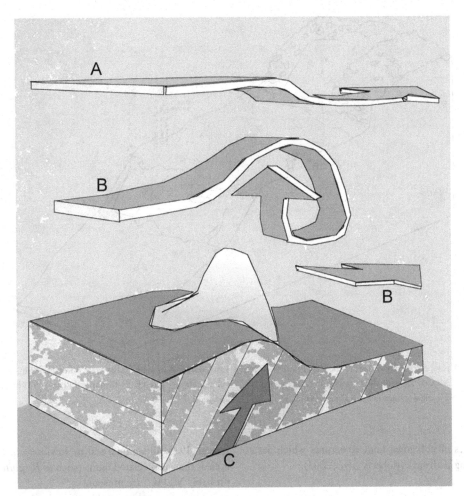

Figure 8.2* The 'hydraulic theory' for supporting deep-water corals (after Hovland and Mortensen, 1999). This cartoon shows a *Lophelia* reef standing on the seabed. Atlantic water flows over the reef (arrow A), and eddy turbulence (arrow B) is caused by the topography. The turbulence will tend to concentrate nutrients; however, the theory suggests that the decisive factor for the development of a 'permanent' deep-water reef is mineral-rich seabed fluid flow (shown by the arrow at C) which supports communities of chemosynthetic microbes in the seabed sediment and the overlying water. Larger organisms in the reef (copepods, foraminifera, benthic and pelagic plankton, etc.) feed on these primary producers, and are in turn preyed on by larger organisms. If the fluid flow were to stop, the theory suggests that the whole reef community would eventually die.

Mortensen, 1999; De Mol *et al.*, 2002; Hovland and Risk, 2003). According to this theory coral growth is promoted by fluid flow: the seepage of groundwater under the reefs in fjords, and of hydrocarbon-related fluids (including methane, carbon dioxide, and hydrogen sulphide) under the continental shelf reefs (Figure 8.2); it suggests that reefs are not necessarily dependent upon nutrients carried on currents in the water column. The following topographic, reflection seismic, geochemical, and chemical indicators support this theory:

- seaward-dipping permeable sedimentary strata;
- enhanced acoustic seismic reflectors;
- pockmark craters;
- locally elevated light-hydrocarbon porewater content;
- locally elevated methane seawater content.

Support for this theory also comes from the west of the British Isles where seabed mounds colonised by *Lophelia pertusa* are located over faults and the

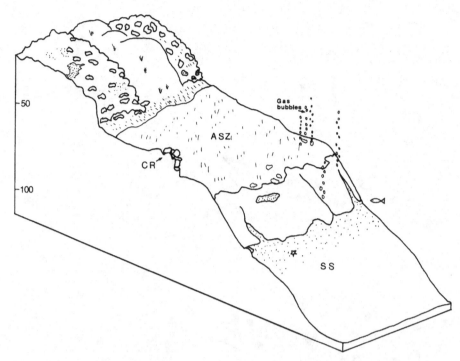

Figure 8.3 A perspective view of part of the East Flower Garden Bank, Gulf of Mexico, showing the presence of gas seeps. CR = coral reef; ASZ = algal sponge zone; SS = soft sand. The vertical scale is in metres. (Based on observations made from a submersible; adapted from Bright and Rezak, 1977 by Hovland and Judd, 1988.)

boundaries of polygonal fault structures which act as dewatering pathways (Roberts *et al.*, 2003).

Gulf of Mexico banks and reefs

In the northwest Gulf of Mexico, off Louisiana and Texas, there are a number of banks and reefs (Figure 8.3), many of which are important for both commercial and sport fishing. Extensive studies of the ecology of these banks and reefs and other features located on the Louisiana Slope, have been conducted by Texas A&M University and others, since the mid 1970s.

The East and West Flower Garden Banks, the most northerly tropical shallow-water coral reefs off the eastern coast of the Americas, support a rich fauna and flora. This was described in some detail by Bright and Rezak (1977), who noted that '*natural-gas seeps issue abundantly from Twenty-eight Fathom Bank and the East Flower Garden*'. They suggested that the seeps had no effect on the benthic populations, but from their description and other work (Powell and Bright, 1981; Powell *et al.*, 1983;

Jensen, 1986; all summarised by Levin *et al.*, 2000), it is clear that seep-related fauna (such as *Beggiatoa* – and probably MDAC) are present.

These and other neighbouring banks are located over shallow salt domes producing the topographic highs on which the reefs were founded. The salt-dome structures also produced traps for petroleum and in at least some cases (Bright, East Flower Garden, and Sackett) there are thermogenic gas seeps. There is a similar relationship between gas seepage and a reef on at least one of the banks off the southern Texas coast, the Baker Bank. This is one of a number of fishing banks in this area lying within the bounds of the Serendipity Gas Seep Area; it is therefore tempting to imply that these are all associated with gas seeps. It has been suggested that the reefs occupy these positions because of a topographic high caused by the presence of the salt structures. However, we consider it more likely that these ecosystems were established on these sites because fluids were focussed through the updomed and fractured sediments above the salt.

Tropical coral reefs and groundwater seeps?

Tropical coral reefs grow between latitudes of roughly 30° N and 30° S. However, not all coastlines in these waters have reefs. Why is this? Of course, in many places the answer is that currents and seawater turbidity prevent such developments. But, there remain thousands of kilometres of tropical coastlines where the seabed and water conditions seem suitable for coral-reef development. The answer may be hydrological or hydraulic offshore conditions.

We believe that there may be a close relationship between the flow of groundwater and the development of some tropical coral-reefs. Hovland (1999) commented on the idea that established reefs react to alterations in hydrological conditions. In response to a previous report that the soil-dwelling fungus *Aspergillus* was attacking and killing corals off the Florida Keys, it was suggested that the fungus was transported to the reefs in groundwater, probably along with other pollutants. Shinn (1993, unpublished report) reported that groundwater beneath the Florida Keys reefs contained nutrients, principally ammonia, at concentrations '*many times higher than that of normal seawater*', and suggested that tidal pumping drove groundwater seeps. Subsequently, Reich *et al.* (1998) conducted fluorescent dye studies which demonstrated a net flow of groundwater from Florida Bay to the Atlantic Ocean at a velocity of up to 2.5 m d^{-1}. So, the link between groundwater and this coral-reef ecosystem has been demonstrated. Hovland (1999) argued that both groundwater pollution and reductions in the hydraulic head caused by anthropogenic groundwater draw-down should be considered as a cause of the decline of coral reefs not only here but around the world.

Coral reefs and hydrocarbons

Carbonate reefs are often good hydrocarbon traps, mainly because of their high primary porosity. However, Link (1952) saw few problems in assuming that the trapped hydrocarbons were actually generated in the reefs '*because of the obvious concentration and accumulation of organisms in them*'.

Coates *et al.* (1986) found aliphatic hydrocarbons in surface sediments, water, and some organisms (seven species) on widely separated coral reefs on the Great Barrier Reef of Australia. They considered that they were of biogenic origin; consequently this information supports Link's argument. Coates *et al.* (1986) also presented evidence that may support our contention that some tropical reefs may be partially supported by fluid flow. An unresolved complex mixture, usually considered indicative of petroleum contamination, was found in greater than trace amounts only in *Holothuria* (sea cucumber) and *Acropora* (coral) from the Capricorn Group, and in some sediment samples from the Capricorn Group and Lizard Island area. Although Coates *et al.* argued that contamination by man is responsible for this 'pollution', we argue that natural seeps may provide an alternative explanation.

We do not dispute Link's contention that old reefs, buried to sufficient depth for thermal maturity to occur, could contain hydrocarbons generated from the organisms that had lived on the reef. However, we consider that existing (and previously existing) reefs and bioherms may be supported by light-hydrocarbon seeps (especially the gaseous components, methane to pentane); there are numerous coral reefs off the coast of Belize, many of which are closely associated with natural hydrocarbon seeps (Hovland and Mortensen, 1999). Although this may not be true in every case, we believe that this possibility should be considered. Indeed, we can envisage a situation in which successive generations of coral reef support the growth on new reefs, provided there are suitable seabed conditions (Figure 8.4).

Mud-volcano reefs

In the Sumba region of Indonesia there are shallow-water (tropical) corals growing on mud volcanoes (Barber *et al.*, 1986). When the mud volcanoes rise out of the sea, due to upheaval and accretionary processes, coral reefs occur attached to the sides of the emerging mud-volcano structures (Barber *et al.*, 1986). It is suspected that these corals rely on a mixed supply of nutrients, that is, both photosynthetically produced and perhaps chemosynthetically produced, from the fluids emitted by the mud volcano.

Volcanic corals

Pichler and Dix (1996) found a mixed trophic coral environment at a shallow hydrothermal vent, close by the volcanic Ambitle Island, Papua New Guinea. The strontium isotope ratio of a coral sample collected adjacent to a vent had a value (0.70746) abnormal for stony crabs, but intermediate between values of oceanic crust and modern seawater. Both focussed hot vents and diffusive venting occurs in the area. Vents 10–15 cm in diameter

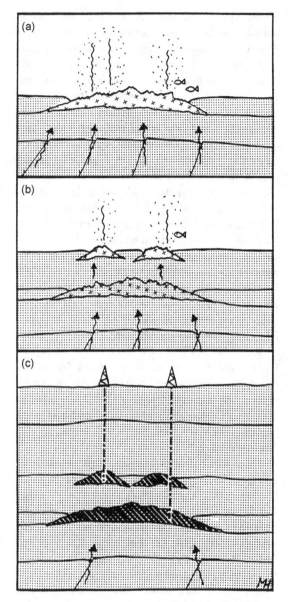

Figure 8.4 Cartoon illustrating the conceptual model for the formation of stacked reefs. (a) The original seep-related reef on the seabed. (b) A secondary reef forming from subsequent seabed fluid flow at a later seabed after the burial of the original reef. (c) Both reefs have been buried and their pore spaces filled with petroleum (oil and/or gas) which may be exploited by drilling. (Reproduced from Hovland, 1990b with permission from Blackwell Publishing.)

discharge a clear, two-phase fluid at such a rate (300–400 l min^{-1} – equivalent to a fire hose!) that there is a roaring sound underwater. Boiling occurs at the seabed where fluid temperatures are 94–98 °C, and the water is hot enough to cause shimmering several metres above the vents. But the gas condenses again 2–3 m above the seabed. In contrast, the streams of gas bubbles emerge through a sandy to pebbly seabed in the diffusive flow areas. Pichler and Dix assumed that the emitted gas was carbon dioxide.

Live corals were found in the apparently hostile environment adjacent to the vents. Dead corals are covered in precipitated minerals including euhedral aragonite crystals and microcrystalline crusts of iron oxyhydroxide, aragonite, and ferroan calcite.

8.1.7 Hydrothermal vents

According to Lutz (2000), Timothy Shank of Woods Hole Oceanographic Institute has calculated that, on average, a new species has been described every week and a half since biologists first visited the Galapagos Rift vents in 1979. Because of the rate of discovery, and the resulting stream of information on hydrothermally active areas, we cannot do justice to the important topic of the ecology of hydrothermal vents. Instead we would like to refer readers to three valuable books dedicated to this topic.

- *Handbook of Deep-Sea Hydrothermal Vent Fauna*; Desbruyères and Segonzac, 1997.
- *The Ecology of Deep-Sea Hydrothermal Vents*; Van Dover, 2000.
- *The Subseafloor Biosphere at Mid-Ocean Ridges*; Wilcock *et al.*, 2004.

Although we cannot cover this topic in depth, we feel that we must not ignore it.

The Galapagos Rift – where it all began

In February 1977, after noticing the seepage of warm, hydrogen sulphide-rich, water rising from depths of over 2600 m at the Galapagos Rift, west of Ecuador, the submersible Alvin was used to investigate the seabed. On this dive the investigators (geophysicists John Corliss and John Edmond) happened to make, perhaps, the biological discovery of the twentieth century – a unique, dense benthic community that, somehow, relied on active venting. In the words of Chevaldonné (1997): *'these communities challenged contemporary knowledge concerning deep-sea life at that time'*.

The following, based on reports by Corliss *et al.* (1979), Ballard and Grassle (1979), and others, summarises the, until then, inconceivable ecosystems revealed by these first dives on a hydrothermal-vent community.

- Extraordinary giant siboglinid tubeworms were found to colonise shimmering waters indicating the actual fluid flow (temperature range 8–13 °C; ambient seawater temperature nearly 2 °C). Some grow to more than 2 m in length, building sturdy cylindrical tubes which are closed at the base. They have no eyes, no mouth, no gut, no anus, but blood-filled red 'plumes'.
- Giant bivalves (vesicomyids) clustered in crevices of pillow lava where they may attain shell sizes of 30–40 cm. Due to abundant sulphate-reducing bacteria in the water and elevated temperatures, these filter feeders were found to grow 4 cm per year, which is about 500 times faster than normal deep-sea bivalves.
- Serpulids or bristle worms (polychaetes), which use their tentacles to capture microbes. They live at the lower limit of a steep temperature gradient (20–250 °C)
- Many types of crabs, including cancroid and other brachyurans, and squat lobster feeding on bacterial clumps and organic debris from the water.
- Brown bivalves (mytilids).
- Translucent sea anemones.
- Acorn worms (enteropneusts) resembling spaghetti lying on the volcanic rocks.
- A yellow 'dandelion–like' creature (a siphonophore).
- Also seen were reddish octocoral, leeches, barnacles, limpets, whelks, octopus (*Graneledon* sp.), brittle stars, sea cucumbers (*Benthodytes* sp.), skate, brotulid fish, and rattail (grenadier) fish.

Clearly, these reports preceded the realisation that some of the macrofauna described are supported by chemosynthetic symbionts. Since 1977 hydrothermal-vent communities have been discovered at ocean spreading centres at many locations in the Pacific, Atlantic, and Indian oceans (Map 33). Over the 17 years, from 1979 to 1996, a total of about 350 new animal species were found in the Pacific and Atlantic oceans; 93% of them are endemic to the deep-sea hydrothermal systems and

new to science (Tunnicliffe and Fowler, 1996). Molluscs, polychaetes and arthropods, dominate the species list, but some important groups (including echinoderms, sponges, brachiopods, and bryozoans) are absent or very rare at hydrothermal vents (Chevaldonné, 1997). There are important differences between vent communities in different geographical areas. Van Dover (2000) reported that the ecology of the Kairei vent field of the Indian Ocean, for example, has both similarities with and differences from vents in the Pacific and Atlantic oceans.

- One third of the 36 invertebrate taxa were previously only reported from the Pacific.
- Eleven genera are known from both Pacific and Atlantic vents.
- Only one species the dominant shrimp (*Rimicaris* aff. *exoculata*), belongs to a genus that is known exclusively from Atlantic vents.
- At least one species, the scaly-foot gastropod, may represent a new family, and many taxa may belong to new genera.
- Certain taxonomic groups (such as the alvinellid polyachaetes, siboglinid tubeworms, and small crustaceans – except copepods) that are common at Pacific vents are absent from Kairei.
- Amphipods, which are often abundant at Pacific and Atlantic vents, are also absent.

Shallow-water hydrothermal fluid flow

Hydrothermal fluid flow is not only confined to deep-water sites, but also occurs in shallow water, including the intertidal and photic zone, for example in Kagoshima Bay, Japan, where tubeworms and other typical vent fauna have been found (Section 3.18.5). Because nutrients obviously derived from such flow are difficult to distinguish from terrestrially sourced nutrients, we are faced with a severe problem when trying to assess the implications of shallow water seepage. Van Dover (2000) suggested that, generally, *'the fauna of shallow-water vents is a small-subset of the surrounding shallow-water species pool'*, and that intolerance and competition from shallow-water fauna has prevented deep-water vent fauna from colonising shallow-water vents. Yet, in some places there is a more complex interrelationship between 'normal' shallow-water and hydrothermal-vent fauna. For example, in Kraternaya Bight in the Kuril Islands photosynthesis and chemosynthesis run side-by-side so *'metabolic and energy cycles in the ecosystem appeared to be very complicated'* (Tarasov *et al.*, 1990).

Even where typical vent fauna are absent, there are several examples of enriched environments at shallow-water hydrothermal seeps. The most common biological indicators of shallow-water hydrothermal fluid flow are bacterial mats (Dando *et al.*, 1995a). An interesting and early recognised example is from the Palos Verdes Peninsula, California. Here, the vents support a white bacterial mat and emit warm (28 °C), fresh hydrogen sulphide-rich water into the intertidal and shallow (less than 20 m) subtidal zones. Accelerated growth rates in black abalone (*Haliotis cracherodii*) feeding on these mats were observed by Kleinschmidt and Tschauder (1985).

The largest field of shallow-water hydrothermal seeps studied to date is probably that of the Hellenic Volcanic Arc, in the Aegean Sea. The hydrothermal vents off the Island of Milos (Section 3.10.1) clearly have an impact on the local ecology, not least because of the introduction of phosphate-rich waters; Dando *et al.* (1995a) estimated an input of about 30 t per year. Apparently the microbial diversity of the vent sites was '*large*', and many new taxa were found. The epifaunal diversity was '*surprisingly high*', but no vent-specific species were found. For some organisms at least, the elevated temperatures seemed to be the attraction, rather than the introduction of exotic fluids (Dando *et al.*, 2000).

Seamounts and volcanoes

Rogers (1994) reported that vent-like communities had been reported from seamounts: Axial Seamount on the Juan de Fuca Ridge, Piip to the east of Kamchatka, Loihi off Hawaii, Red Volcano off the Pacific coast of Mexico, and Marsili in the Tyrrhenian Sea, north of Sicily. In some cases these communities are very similar to those of ocean spreading centres. For example, the community on Axial Seamount includes dense clumps of siboglinid tubeworms, limpets, and alvinellid polychaetes. Similarly, the vent-associated bivalve *Calyptogena* is present on the Piip Volcano. Other organisms here include sponges, soft corals (alcyonarians), stony corals (actinarians), polychaetes, crabs, amphipods, and sea cucumbers (holothurians) (Rogers, 1994). However, Rogers also identified marked differences between the fauna of off-axis seamounts and ocean spreading axes.

The thick grey mucus, probably secreted by the polychaetes, covering the tubeworms on Axial Seamount seems unusual. Bacterial mats, white and amber in colour, occur further away from these active vents, where there are also colonies of cobalt blue ciliates (Tunnicliffe *et al.*, 1985). Extensive bacterial mats also occur on Loihi, but these are red-brown in colour and have '*very high concentrations of iron, phosphorus and calcium*' (Rogers, 1994). Both here, and at Red Volcano, where bacteria seem to cover the mineral-rich chimneys that emit a warm clear fluid, the macrofaunal species normally associated with vents are absent. On Piip bacterial mats also seem unlike those of most seabed fluid-flow sites. They are very extensive, at one place covering 100% of an area of 100 m^2; they are mainly white, with patches of black, grey, and brown (Rogers, 1994). Gas bubbles through small holes in these mats.

It seems likely that this variety of bacterial mat colours can be attributed to the presence of various metals in the venting fluids, for example the iron oxide gives Red Volcano its distinctive colour, and at Piip there are chimneys made of aragonite and pyrite rich in mercury (Rogers, 1994).

8.2 FAUNA AND SEABED FLUID FLOW

Now that we have seen examples of seep-related biological communities, we must look at some of the organisms involved.

8.2.1 Microbes – where it all begins

Microbes are the foundation of the chemosynthetic communities found at vent and seep sites. They take part in all the key processes inside and outside organisms living there. Microbial activity occurs in the following situations where there is a transition from oxic to anoxic conditions (Jannasch, 1983; Aharon, 2000; Orcutt *et al.*, 2004):

- within the sediments of seeps;
- at elevated temperatures within vent systems;
- at subzero temperatures within seabed gas hydrates;
- within higher organisms that host microbes in a symbiotic relationship;
- in mats living on the seabed;
- in the water column above seeps.

The microbes at vents and seeps obtain their life-giving energy and cell-building carbon by processes that are fundamentally different from the 'normal' processes supporting life on Earth. To understand the significance of the role played by microbes it is necessary to appreciate how the 'normal' marine biological system works.

In Section 5.4.1 we explained how organic debris raining down to the seabed is broken down by microbial activity in the near-seabed sediments. Two of the main microbial groups involved, the sulphate-reducing bacteria (SRBs) and the methanogenic archaea, compete for the organic feast. It is perhaps surprising that these same groups co-operate to make use of methane when it is supplied by migration from microbial and/or thermogenic sources at greater depth. This co-operation takes place in the zone referred to as the sulphate–methane transition zone (SMTZ). This transition zone (Figure 5.5), where the concentration gradients of many compounds are sharp, was first described by Reeburgh and colleagues in the 1970s. The depth of the SMTZ varies with ecosystem and flux type. In some marine sediments it can be several decimetres thick, but, for example, it is only of sub-millimetre scale in certain stromatolitic hydrothermal environments (Whiticar, 1999).

The key to understanding the processes occurring in places with a high methane flux was the realisation that the seabed sediments at methane seeps are characteristically sulphide-rich, and colonised by sulphide-oxidising bacteria such as *Beggiatoa* sp. Due to the sharpness of the SMTZ, it was hypothesised that a process may be occurring which consumed both of these compounds and subsequently produced the sulphide observed. Iversen and Jørgensen (1985) speculated that methane was being oxidised at this interface by microorganisms using sulphate as the oxidant, a novel process called anaerobic methane oxidation.

$$CH_4 + SO_4^{2-} \rightarrow HCO_3^- + HS^- + H_2O \quad (8.1)$$

This process is investigated in detail in a following section.

An excess of methane results in the depletion of sulphate, the formation of excess bicarbonate and reduced sulphur, and a decrease in pH. In the presence of normal supplies of bicarbonate from seawater, the excess will be precipitated to form calcium carbonate, $CaCO_3$; carbon isotope values of these carbonates demonstrate that their carbon has been derived from methane rather than from seawater (see Section 9.2.5).

If there is an adequate supply of iron, then the excess reduced sulphur results in the formation of pyrite (iron sulphide; FeS_2). Sulphide not incorporated into pyrite will form hydrogen sulphide (H_2S), giving sediments at methane seeps the characteristic smell of rotten eggs. This sulphide is utilised by sulphide-oxidising bacteria (such as *Beggiatoa*). These processes are summarised in Figure 8.5.

Freshwater

In places like Eckernförde Bay, Germany, where freshwater is seeping from the seabed, the pore waters have a low sulphate content. The concentration of sulphate is normally significantly lower in freshwater than seawater, so methane generation rates are more likely to exceed consumption/oxidation rates, allowing methane, 'marsh gas', to escape into the water. However, unlike the water in marshes, the groundwaters seeping into Eckernförde Bay are aerobic. Methanogens are 'obligate anaerobes', they do not function in the presence of oxygen. Consequently these fresh groundwaters, despite their low sulphate content, are not suitable for methanogenesis. Indeed, their oxic nature makes them better suited to methanotrophs, so any methane present is likely to be utilised. These groundwater seepage sites contrast with other parts of Eckernförde Bay (including the sediments adjacent to some pockmarks) where, in the absence of groundwater but the presence of high concentrations of organic matter in the seabed sediments, methanogenesis does occur. Indeed, in much of the bay methane production results in supersaturation and bubble formation (Whiticar, 2002).

AOMs: a revolutionary consortium

Despite numerous attempts, and general agreement that anaerobic methane used sulphate as the terminal electron acceptor – see Equation (8.1), the processes at work in this zone were not fully understood, especially since the methanotrophs need oxygen, whereas the SRBs are strictly anoxic. This enigma persisted until 1999, when a group led by Antje Boetius found a symbiotic microbial consortium capable of anaerobic oxidation of methane; these anaerobic oxidisers of methane (AOMs) were first discovered beneath the *Beggiatoa* mats above hydrate-bearing sediments at Hydrate Ridge on the Cascadia Accretionary Wedge (Boetius *et al.*, 2000).

The nature of the AOMs was determined using a combination of stable-isotope analysis of microbial

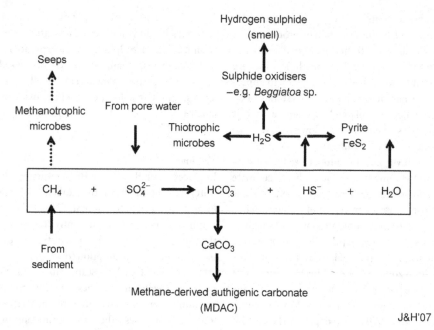

Figure 8.5* Anaerobic oxidation of methane: ingredients (methane and sulphate), products (hydrogencarbonate, hydrogen sulphide, and water), and by-products (MDAC and pyrite). Excess methane is available for utilisation by methanotrophs (free-living and symbiotic), and any still unused escapes through the seabed as seeps; H_2S (derived from HS^-) is available for utilisation for thiotrophic microbes (including *Beggiatoa*).

biomarkers, radioactive-tracer incubations, and fluorescence *in situ* hybridisation (FISH) by Boetius *et al.* (2000) and Orphan *et al.* (2001). Archaeal biomarkers identified include archaeol, hydroxyarchaeol, glycerol dialkyl glycerol tetraethers (GDGTs), and isoprenoid hydrocarbons. Bacterial biomarkers identified include hopanoids (such as diploptene) and fatty acids (Werne *et al.*, 2001). The average size of the microbial consortia found at Hydrate Ridge is only a few μm. They resemble a raspberry in shape in that they are globular with several sheaths of cells (Figure 8.6). In the centre of the globule there is a dense core consisting of about 100 archaea cells. The high depletion of ^{13}C (δ ^{13}C values of −96.2‰) in these cell aggregates proves that they are consuming methane (House *et al.*, 2001); in fact it is thought that the archaea are actually methanogenic species, but they are operating their normal process in reverse. Three groups of archaea have been identified in these consortia: ANME-1, -2, and -3 (ANME stands for ANaerobic MEthanotrophs). Surrounding the core, in several sheaths, there are layers of SRBs (Boetius *et al.*, 2001; Orphan *et al.*, 2001) which reduce sulphate to hydrogen sulphide. In methane-charged sediments

the number of individual organisms has been found to be extremely high; more than 10^{10} ml^{-1} (Boetius *et al.*, 2001). AOMs are described in more detail by Treude and Boetius, website.

The findings at Hydrate Ridge triggered the search for similar consortia elsewhere. Similar AOMs have now been discovered at methane seeps, above gas hydrates, and on mud volcanoes located as far apart as the Congo Basin, Guaymas Basin, Gulf of Mexico, the Mediterranean Ridge, the Norwegian slope, Eckernförde Bay, and at Tommeliten in the North Sea (Antje Boetius, 2002, 2003, personal communications).

The AOM consortia are relatively recently discovered. As with all important discoveries, there is some scepticism about them. Do the consortium members always work together, or are either the methanogens or the SRBs, or both, likely to operate independently? Orcutt *et al.* (2004) showed that, at their Gulf of Mexico study site, '*only a fraction*' of sulphate reduction could be attributed to AOMs; here abundant higher-hydrocarbon gases and crude oil also act as substrates for sulphate reduction (Joye *et al.*, 2004).

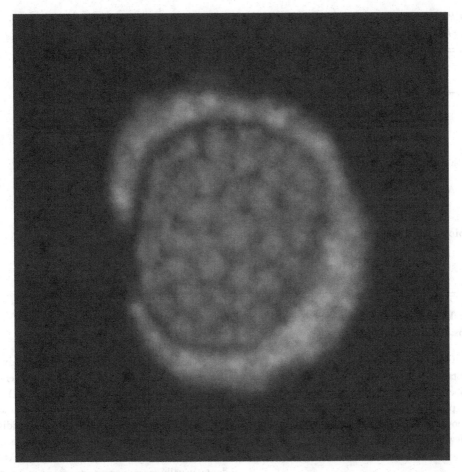

Figure 8.6* An AOM consortium (methanotrophic archaea surrounded by SRBs). [Image, produced by FISH analysis, courtsey of K. Knittel and A. Boetius, Max Planck Institute for Marine Microbiology, Bremen. For further details see the contribution by Treude and Boetius on the website.]

Live 'jellybabies' in the Black Sea

One of the weirdest discoveries from the deep sea was made by a German–Russian group working in the Black Sea in 2001. At a depth of 400 m, at the deep end of the Dnepr Canyon, west of the Crimea Peninsula, they found some ordinary looking chimneys from which gas was bubbling. When they took the Jago submarine in close for a better look, one of the chimneys fell over and bounced down the slope. It turned out to consist of a rubbery, fleshy, spongy substance which was pink on the inside (Antje Boetius, 2002, personal communication). These chimneys ('jellybabies') are up to 4 m tall, and are composed of microbial biomass supported by a meshwork of MDAC which acts as an internal skeleton. In places the structure contains hollow spheres which trap some gas, but, most of the gas going into the base of the chimneys is seen escaping from the mouth at the top (Reimer *et al.*, 2002; Boetius *et al.*, 2002; Treude and Boetius, website).

These jellybabies must represent an ancient type of community surviving here because, in these anoxic Black Sea waters there are none of the organisms that would normally graze on them.

Microfauna of hydrothermal vents

The microfauna of hot vents is likely to be more complicated than that of cold seeps because the temperatures may range from > 300 °C to about 0 °C. Different groups of microbes operate within different temperature ranges: at the extremes, superthermophiles live

Table 8.1 *Possible microbial metabolic processes at hydrothermal vents (adapted from Jannasch, 1995: Van Dover, 2000).*

	Metabolic process	Electron donor	Electron acceptor	Gibbs free energy*
Aerobic	Thiosulphate oxidation	$S_2O_3^{2-}$	O_2	−952
	Methane oxidation	CH_4	O_2	−810
	Sulphide oxidation	S^{2-}	O_2	−797
	Sulphur oxidation	S^0	O_2	−585
	Nitrification	NH_4^+	O_2	−275
	Hydrogen oxidation	H_2	O_2	−237
	Manganese oxidation	Mn^{2+}	O_2	−68.2
	Iron oxidation	Fe^{2+}	O_2	−44.3
Anaerobic	Denitrification	H_2	NO_3^{2-}	−239
	Sulphur reduction	H_2	S^0	−98.3
	Sulphate reduction	H_2	SO_4^{2-}	−38.1
	Methanogenesis	H_2	CO_2	−34.7

*$\Delta G°$ in kJ mol^{-1} electron donor, calculated for complete oxidation and normalised conditions: pH 7, 25 °C, 1 atm.

at > 115 °C and psychorophiles at <10 °C. Also, vent fluids contain various compounds that can be utilised by microbes (see Section 10.2.2), so a wide range of metabolic processes (see Table 8.1) is possible beneath, at, and above the seabed. Van Dover (2000) provides a good review of this topic.

The ubiquitous bacterial mat

Bacterial mats of sulphide-oxidising bacteria (thiotrophs) are perhaps the most obvious visual evidence of seepage, as they seem to occur wherever there are methane seeps. They have been reported from just about every cold-seep site around the world, from intertidal sites (e.g. Torry Bay, Scotland – Section 3.5; San Simón Bay, Ría de Vigo, Spain – Figure 8.7) to the deep oceans (such as the Aleutian Trench, Section 3.19.3). They are also found at other places with sulphide-rich sediments at the seabed, for example the Florida Escarpment groundwater seeps (see Section 3.26), around sulphide-rich hydrothermal vents, and in enclosed anoxic basins. They are described as filamentous, white to orange fluffy mats lying on the seabed at the interface between anoxic, sulphidic sediments and aerated seawater (Figure 8.1; also Figure 2.32). Figure 8.8 shows that they are not confined to fine-grained sediments. Generally the mats are less than 1 cm thick,

but some may be as thick as 10 cm. On muddy seabeds they may extend '*rootlike structures*' several centimetres into the sediment (Roberts and Carney, 1997).

The most commonly reported bacteria are from the genus *Beggiatoa*, but other sulphide oxidisers, for example *Thiothrix* and *Thioploca* spp. are also known. These microbes oxidise sulphide using a variety of oxidants, usually oxygen or nitrate stored in internal vacuoles (storing helps eliminate the problems of molecular diffusion in these 'huge' bacteria). The process is summarised by Equation (8.2);

$$HS^- + 2O_2 \rightarrow SO_4^{2-} + H^+ \qquad (8.2)$$

or, by Equation (8.3).

$$O_2 + 4H_2S + CO_2 \rightarrow SO_4^{2-} + H^+ \qquad (8.3)$$

Some of the most substantial mats of *Beggiatoa*, one of the largest known bacteria, have been found on sediments and amongst tubeworms at hydrocarbon fluid-flow locations in the Guaymas Basin. Here *Beggiatoa* filaments are up to 120 μm wide and 10 mm long, and the mats can be '*thick enough to ladle*' (Van Dover, 2000). Although *Beggiatoa* is characteristic of anoxic sediments, they are unlikely to occur in stagnant water

Figure 8.7* San Simón Bay, Ría de Vigo, northwest Spain looking towards the bridge over the Rande Strait. In the foreground white bacterial mats (possibly *Beggiatoa*) are exposed at low tide. (Photograph courtesy of Soledad García-Gil.)

Figure 8.8* Bacterial mats on coarse sediment off the coast of Svalbard. (Reproduced with permission from Knies *et al.*, 2004.)

as they need steep, opposed gradients of sulphide and oxygen (Møller *et al.*, 1985).

BACTERIAL MATS AND FLOCS

One of the remarkable aspects of bacterial mats is their ability to prevent fluids, including gases from escaping. They modify the local environment so that extreme chemical microenvironments occur beneath them. For example, the 594 m deep Crater Lake, Oregon, is located in a caldera of Mt Mazama in the Oregon Cascades, which last erupted 6850 years ago (Bacon and Lanphere, 1990). During nine dives with the one-person submersible 'Deep Rover', bacterial mats up to 3 m in diameter and 2–15 cm thick were observed. Measurements made beneath mats with a temperature probe found temperatures up to 6 °C higher than the typical bottom water, and the concentrations of the major ions in water sampled from within and immediately below the mats were approximately double those of any water sample collected within the caldera; manganese concen-

trations were about 1000 times higher than in deep lake water (Dymond *et al.*, 1989).

However, at numerous hydrothermal sites the venting fluids shimmer with heat and appear 'milky' because of the microbes they contain. According to Van Dover (2000), elevated microbial biomass concentrations were first observed by Winn *et al.* in 1986. Naganuma *et al.* (1989) reported that plumes over the North Fiji vents contained a microbial community with both heterotrophic (80%) and chemoautotrophic (20%) members, and these had much shorter generation times (hours to days) than most deep-sea microbes (about six–nine days). The microbial biomass can be four times that of the overlying surface waters, and suspended cell counts have been reported to be four orders of magnitude greater than background deep-sea microbial populations (Karl, 1995). However, during submarine volcanic eruptions huge quantities of white, fluffy, flocculent material ('floc') can be emitted.

Van Dover (2000) summarised the effects of an eruption of the East Pacific Rise in 1991. It seems that during the days and weeks after the eruption microbial floc was blown from seabed cracks to an altitude of about 50 m like a 'snow blower'. There was enough floc material 'to generate "white-out" conditions, obscuring visual navigation by the submersible pilots and requiring sonar guidance' (Van Dover, 2000). Over an area several kilometres long and 50 to 100 m wide there was a white floc blanket up to 5 cm thick (Haymon et al., 1993). This floc comprises inorganic sulphur filaments excreted by a vibroid bacterium and then colonised by swarms of microscopic cells (Van Dover, 2000).

8.2.2 Living together: symbiosis and seeps

The discovery of hydrothermal vents and seabed seeps presented a fundamental question to marine biologists. How can life exist in such profusion on the ocean floor, way below the photic zone? Many of the macrofaunal types living around vents and seeps belong to invertebrate groups (bivalves, for example) well known from shallower waters. However, organic matter ultimately derived from photosynthesis could not possibly support them in this deep-water environment. A similar question could be asked about the inhabitants of the sulphidic seabed sediments at cold seeps. These are anoxic and unfavourable for most of the organisms living in the seabed sediments, the benthic infauna. How have others adapted to tolerate these conditions?

The simple answer to both questions is: by symbiosis with microbes. The energy they live on was originally derived by microbial chemosynthesis and not photosynthesis. Sibuet and Olu (1998) listed 64 symbiont-hosting species reported from deep-water cold seeps, including siboglinids and bivalves (mytilids and vesicomyids), which also occur at hydrothermal vents.

Siboglinids and the identification of symbiosis
Some of the most memorable images of both hydrothermal-vent and cold-seep communities are those of the dense bushes of tubeworms stretching up into the water, with their vivid red plumes (MacDonald et al., website). The tubeworm, or siboglinid is one of the most fascinating organisms associated with seabed fluid flow; indeed it is amongst the strangest of our planet's 'wildlife'. The siboglinids are a diverse and geographically widespread phylum. They occur at many cold-seep sites, at depths ranging from the relatively shallow (300 m) waters of the Skagerrak to the deepest cold seeps (Peru Trench, Monterey Fan, Laurentian Fan, and Aleutian Trench), as well as at hydrothermal

vents ranging in depths from 85 m (at Kagoshima Bay, Japan) to the much deeper waters of the ocean spreading centres.

There have been various classifications of the tubeworms associated with cold seeps and hydrothermal vents. They had been assigned to separate phyla (the Pogonophora and the Vestimentifera), or to a single phylum (the Pogonophora) itself divided into three subphyla: Perviata (that live in reducing sediments), the Monilifera that live in decaying wood, shallow-water reducing sediments and near deep-water cold seeps, and the Obturata (or Vestimentifera) that live near hydrothermal vents and cold seeps. However, as a result of recent molecular studies (McHugh, 1997) they are currently recognised as being a single group of annelid worms, and have been assigned to a family named after the first-described genus.

Phylum: Annelida
Class: Polychaeta
Family: Siboglinidae

In this book we have adopted the term 'siboglinid tubeworms' to describe this family in preference to the 'old' (but still commonly used) names. Amongst others, this family includes the genera: *Lamellibrachia, Oligobrachia, Ridgeia, Riftia, Sclerolinum*, and *Siboglinum*.'

Some of the early work on marine chemosynthetic symbiosis was done on small (65 mm long) tubeworms with one tentacle, from Scandinavia. Southward et al. (1981) analysed several species (including *Siboglinum ekmani* and *Siboglinum fjordicum*) from fjords near Bergen (Norway) and *(Siboglinum poseidoni)* sampled from the Skagerrak. Flügel and Langhof (1983) worked on specimens dredged from a depth of 294 m in the southern Skagerrak. These two teams found these small worm-like animals living in 'normal' marine sediments in symbiosis with large numbers of endosymbiotic bacteria; these were found to be strikingly similar to free-living bacteria that utilise methane (Völker et al., 1977; Southward et al., 1981). Analyses of the stable carbon isotopes of soft tissues provided a hint of the energy source: $\delta^{13}C$ values ranged from -45.9 to -35.3%, at that time the lowest values reported for contemporary marine organic material (Southward et al., 1981). The fjord-dwelling species showed considerable depletion of ^{13}C, *Siboglinum ekmani* dredged from 700 m water depth had a $\delta^{13}C$ of -45.3%, while *Siboglinum fjordicum*, living at only 30 m, had a somewhat lesser depletion of -35.3%.

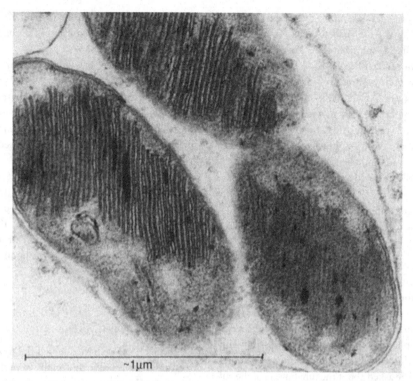

~1μm

Figure 8.9 Electron microphotograph of a bacteria-containing cell (bacteriocyte) inside a siboglinid (*Siboglinum* sp.) from the Skagerrak. The bacteria are seen to contain parallel rows of ribosomes. The bacteria are presumed to utilise methane, and to descend from free-living forms living in the sediment (courtesy of H. Flügel; from Hovland and Judd, 1988.)

After new sampling of *Siboglinum poseidoni* from the central Skagerrak, Schmaljohann *et al.* (1987, 1990) performed various culture experiments on the tubeworms and on their endosymbionts. They managed to keep the animals alive for up to three months in seawater under an atmosphere of air:methane (4:1), confirming that *Siboglinum poseidoni* takes up and oxidises methane. These investigators had demonstrated that there must be a flux of methane at the sites their samples came from.

Both internal and external bacteria contain intracellular membrane stacks (see Figure 8.9). It was therefore assumed that the animals become infected at the larval stage, after they have settled in the sediments. Furthermore, it is assumed that the host cells of *Siboglinum poseidoni* utilise excess organic compounds produced by the symbionts, thereby forming a true symbiosis. A similar bacterial core was found in the large siboglinid tubeworms from the Pacific vents (Cavanaugh *et al.*, 1981; Jones, 1981), and it is now known that bacterial endosymbionts are characteristic of the whole family, although some are thiotrophic and others methanotrophic.

The part of a tubeworm most deeply buried within the sediment is called the opisthosoma. The first descriptive account of a siboglinid was published in 1914, but even after extensive investigations of the animal, especially in the period 1950–63, it was not until 1963 that the previously unknown opisthosoma was found (Webb, 1964). In 1985, Jones reported that those of *Lamellibrachia barhami* and *Lamellibrachia luymesi*, species from hydrothermal vents, remained unknown. In most cases this part broke off during collection due to the fragility of the posterior part of the trunk. Nevertheless, the opisthosoma is probably a universal feature of the group. Even in authoritative books on the subject of vent ecology, there is only brief mention of this organ:

At the posterior end of the worm is the short opisthosome region, which secretes the basal lining of the tube and seems to serve as an anchor. Opisthosomal segments are small and separated internally by muscular septa.

Van Dover, 2000

Even though the opisthosoma contains two longitudinal blood vessels, pockets of secretory cells and multicellular glands, it is often referred to as a 'digging organ' or an 'anchoring organ'. However, to us it seems more likely that it takes up methane and/or hydrogen sulphide from the pore water of the sediments, satisfying the demand of endosymbionts living elsewhere in the organism. These symbionts also demand oxygen, indeed those of vent fauna demand oxygen at a rate greater by a factor of two than that of their hosts, according to Childress and Fisher (1992). Wittenberg (1985) described the distinctive red plume of the tubeworm *Riftia* as '*the most flamboyant oxygen-gathering device in nature*'. The red colour (Figure 8.10), which is not confined to this species, is provided by haemaglobin, which siboglinids have in great concentrations. From Van Dover's (2000) summary (based on previous work by, for example, Wittenberg *et al.*, 1981; Arp *et al.*, 1985; Childress and Fisher, 1992) it seems that siboglinid haemoglobin has a reduced affinity for oxygen at high temperatures, so oxygen gathered above the seabed by the plume and circulated in the blood will be released to the endosymbionts living in the trophosome, a warmer part of the anatomy.

Siboglinids are not the only vent fauna to have developed a specialised blood. Arp and Childress (1981) studied the blood of the brachyuran crab to establish how it can withstand the changing environment at the vent sites. It seems that, unlike that of the siboglinids, the crabs' blood has an affinity for oxygen over a wide temperature range, allowing a high level of metabolism in a wide range of temperature conditions, without the need for acclimatisation.

Bivalve symbiosis

Bivalves, like siboglinids, were part of the first hydrothermal-vent and cold-seep communities discovered. Although the mussels found during the Galapagos Rift discovery dive in 1977 were described as filter feeders, it was not long before it was realised that bivalves could also play host to symbiotic microbes. Childress *et al.* (1986) found bacteriocytes in the gills of Mytilid bivalves from the hydrocarbon-seep areas on the slope of the northwest Gulf of Mexico. They reported both very high rates of oxygen consumption and carbon dioxide production, found that carbon isotope ratios in the bivalves ($\delta^{13}C$ −51 to −57‰) correlated well with those of methane ($\delta^{13}C$ −45‰), and concluded that the bivalves were supported by methane consumption.

Figure 8.10* Tubeworms (*Lamellibrachia luymesi*) at 540 m water depth on the cold seep at Bush Hill, Gulf of Mexico. The distinctive red plumes are clearly visible on the colour version of this figure (see web). (Image courtesey of Ian MacDonald.)

In their review of deep-water cold-seep communities, Sibuet and Olu (1998) identified five families of bivalves known to inhabit cold seeps: the Vesicomyidae, Mytilidae, Solemyidae, Thyasiridae, and Lucinidae.

VESICOMYIDAE

The vesicomyid family includes more than 50 species (including the genus *Calyptogena*) of clams that are '*found nearly exclusively in sulphide-rich habitats such as cold seeps, hydrothermal vents and accumulations of organic debris (e.g. whale carcasses)*' (Barry and Kochevar, 1998). They live partly buried in soft sediment, extending a foot down into the sulphide-rich sediment but providing themselves with oxidants through a short siphon (Roberts and Carney, 1997). They host endosymbiotic sulphur-oxidising bacteria, but different species seem to inhabit locations with different sulphide concentrations. For example, Barry and Kochevar noted that some species at the Monterey Bay seeps favour sites with high sulphide levels or the central portion of low-concentration sites, whilst others prefer low concentration sites, or the margins of high-concentration seeps. Some species are known to be able to survive periods of reduced seepage flow rates, and even to live after seeps have become inactive (Sibuet and Olu, 1998). Barry *et al.* (1997) suggested that this ability may be explained by the fact that some members of the family live in non-seep habitats (e.g. whale carcasses) where sulphide levels are relatively low.

MYTILIDAE

Bathymodiolus, the single genus of this family that inhabits cold seeps, is a genus of mussels. Unlike the burrowing clams, these mussels require a hard substrate to live on. They have been reported from water depths from 400 to 3270 m in the Atlantic and western Pacific oceans (Sibuet and Olu, 1998), generally favouring sites with a high rate of seepage (Aharon, 1994). Childress *et al.* (1986) showed that members of this family are dependent upon methanotrophic bacterial symbionts, but Roberts and Carney (1997) noted that family members may use either methane or sulphide, depending upon the location, species, and symbionts; Van Dover *et al.* (2003) identified both methanotrophs and thiotrophs in *Bathymodiolus heckerae* collected from Blake Ridge. Some may be able to utilise methane from the water and sulphide from the sediments (Sibuet and Olu, 1998). However, they seem to occur at greater densities where methane levels are high, particularly where methane bubbles from the seabed, and around seeping brine pools. Barry *et al.* (1997) suggested that the absence of mytilids from the Pacific is related to the absence or rarity of evaporite deposits, and therefore seabed brines. They considered that brine mixes only slowly with normal seawater, so concentrations of mytilids around brine pools enjoy water with a relatively stable methane concentration.

SOLEMYIDAE

Two genera of the deep-burrowing Solemyidae, *Solemya* and *Acharax*, have been descibed from cold seeps in deep

and shallow water, and from hydrothermal vents (Sibuet and Olu, 1998). They have been found in the western Atlantic, the Gulf of Mexico, and on both sides of the Pacific Ocean. These are thought to be most common at sites with relatively low flux rates, and to be dependent upon sulphide-oxidising symbionts.

THYASIRIDAE

The thyasirids are well known from the sulphide-rich sediments of shallow-water seeps and other 'polluted' sites, but they have also been found in the deeper waters of the Atlantic, and the western and northern Pacific cold-seep sites (Sibuet and Olu, 1998; Kamenev *et al.*, 2001). They are suspension feeders with endosymbionts which utilise reduced sulphur. Like the lucinids, they burrow into the sediment, but the thyasirids make '*an extensive network*' of mucus-lined tunnels into which the sulphide-rich pore waters presumably seep. Thyasirids extend their burrows to about 20 cm below the seabed, making it unnecessary to reposition their inhalant siphons (as the lucinids do). Often a small mound of hydrated iron oxides marks the location of their exhalant flow.

Although they are normally described as hosting sulphide-oxidising bacteria, Kamenev *et al.* reported that *Conchocele bisecta*, which is widely distributed in the Sea of Okhotsk and the Japan Sea, hosted both sulphide-oxidising and methane-oxidising bacteria.

LUCINIDAE

This family is poorly represented compared to others. Sibuet and Olu (1998) listed only two genera (four species of *Lucinoma* and one of *Myrtea*) from sites in the western Pacific, the Gulf of Mexico, offshore West Africa, and the eastern Mediterranean Sea. The known geographical range was extended further when a new species, *Lucinoma kazani*, was described from the Anaximander Mountains in the eastern Mediterranean (Salas and Woodside, 2002). A shallow-water species, *Lucinoma borealis*, has been described from the North Sea (Dando, 2001).

Living habits and the dependence upon sulphide oxidisers are similar to the Solemyidae (Sibuet and Olu, 1998). Dando and Southward (1986) explained that they burrow into the sediment to obtain sulphide-rich pore waters, and extend a siphon upwards to the seabed to inhale oxygen-rich water. They reposition their inhalant siphon frequently in order partially to oxidise the overlying sediment. Their bacterial symbionts are located in their gills, which are enlarged as a consequence.

Carnivorous sponges

At the Atalante Mud Volcano, Barbados Accretionary Wedge, at a water depth of 4700–4900 m, there are dense bushes of carniverous sponges of a species of the genus *Cladorhiza*. Although it is thought that, like other members of this genus, they trap and eat small swimming prey such as small crustacea, Vacelet *et al.* (1996) found that sponge tissues contain at least two, and possibly three, separate types of bacteria. Carbon isotope values ($\delta^{13}C$ values of −48.4 to −48.8‰) show that these are methanotrophs, and it seems that they provide a significant proportion of the sponges' nutrition. The microbes seem to live outside the sponge cells, but they are absorbed and digested by certain sponge-cell types. It also seems that some sponge cells harbour bacterial embryos; so these sponges look after the developing bacteria, then eat them! Long *et al.* (1998) reported that sponges of this genus are present in the blowout craters of the Barents Sea described in Section 3.2.1. Are they indicators of active methane seepage?

Apparently it is usual for sponges to have symbiotic relationships with other organisms, indeed Vacelet *et al.* stated that *all* investigated sponge species have microsymbionts, and in some more than a third of the total tissue volume is actually composed of bacteria. However, this is one of only a small number of species found to have methanotrophs. Others have been reported from the Gulf of Mexico and Lake Baikal (Vacelet *et al.*, 1996; Sibuet and Olu, 1998).

Meiofaunal symbiosis

Bernhard *et al.* (2000) studied *Beggiatoa* mats in the anoxic, sulphidic waters of the Santa Barbara Basin, California; this is not a seep site but lies in the oxygen minimum zone of an isolated basin with poor water circulation. They found that the mats supported abundant communities of foraminifera, flagellates, ciliates, and nematodes, most of which had prokaryotic symbionts in both endo- and ectosymbiotic associations. They concluded that the Santa Barbara Basin is a *'symbiosis oasis'*. Similar associations had been reported previously (Buck and Barry, 1998) from Monterey Bay seeps. However, Buck and Barry reported that *'many species of euglenoids*

and ciliates possessed either endo- or epibiotic bacteria'. They also reported that nematodes and ciliates were more abundant (in terms of biomass) at the seeps compared to 'control' sites.

8.2.3 Non-symbiotic seep fauna

It is obvious that seep and vent communities in the deep sea rely on chemosynthesis as their primary source of energy. However, as well as the microbes and the higher organisms hosting some of them, there are other creatures that take advantage of the 'facilities' provided by seeps. Of course, the principal facility is a supply of food, but the presence of MDAC at both deep-water and continental-shelf seeps is attractive to species that require a 'hard substrate' to live on.

The non-symbiotic seep fauna belong to one or more of the following feeding groups.

- Detritivores: feeding on particles of decaying organic matter on and in the seabed.
- Filter feeders and suspension feeders: filtering particulate matter from the water. This may include material (including bacteria) lifted from the seabed by escaping fluids.
- Scavengers: consuming anything, including dead members of the community.
- Predators: eating other members of the community, such as bivalves.

These are not seep specialists, but opportunists that could live elsewhere on the seabed, but are taking advantage of the food supply either as permanent inhabitants ('colonists') or as casual visitors ('vagrants'). Unlike organisms that stay permanently attached to the seabed, free-swimming fauna, such as fish, may spend only a small part of their time at seeps. Alternatively, they may find seep-related features, such as pockmarks, attractive as they provide 'shelter'. In such cases it is impossible to assess the extent to which they rely on seeps for their food supply – this will be addressed in the next section.

Most of the meio- and macrofaunal species we observed in North Sea pockmarks, and that have been reported from other continental-shelf seep areas do not contain chemosynthesising microbial symbionts; Sibuet and Olu (1998) reported 147 non-symbiont-containing species identified from deep-ocean vents.

Macrofauna

It has been shown that higher organisms than nematodes may also feed on bacteria. Rivkin *et al.* (1986) noted that the larvae of an Antarctic asteroid consumed bacteria in preference to the expected algae. It seems that little is known about the feeding habits of some marine species. Perhaps a large number are bacteriovores under certain conditions.

Sibuet and Olu (1998) reported that deposit feeders (detritivores) and suspension (filter) feeders are the most common of the non-symbiont-hosting fauna at seeps. Some take advantage of the profusion of suspended particulate organic matter produced by the chemosynthesisers, whilst others graze on microbes attached to surfaces.

DETRITIVORES AND GRAZERS

According to Bernhard *et al.* (2000), *Beggiatoa* are generally toxic to mega- and macrofauna, although they reported that *Astryx permodesta*, a surface-dwelling gastropod found in their Californian study site was an exception. However, Trager and DeNiro (1990) reported that, at intertidal hydrothermal vents in southern California, limpets (*Lottia limatula*) grazed on bacterial mats, although they normally feed on algae. These limpets and various other macrofaunal species (including some crabs) which feed on thiotrophic and/or methanotrophic microbial matter and other organic detritus clearly spend considerable amounts of time by the vents as they themselves are covered in bacterial growths (Trager and DeNiro, 1990).

SUSPENSION (FILTER) FEEDERS

Several of the epifaunal species identified in North Sea pockmarks (Section 8.1.1) are suspension (or filter) feeders; many are attached to the hard substrate provided by MDAC. Of course, these species are not unique to seeps. However, if they are present near seeps they may well take advantage of seep-related food that comes their way. Indeed, it seems that filter feeders, including corals and sponges, are common around seeps and vents. Sibuet and Olu (1998) provided an example: sponges of the genera *Geodia* and *Stelletta* have been found to have '*numerous bacteria in their tissues*' suggesting that they rely on filtering free-living bacteria from the water. Similarly, Galkin (1997) reported that suspension feeders living around hydrothermal vents live off microbial growths from beneath the seabed that break loose and float to the surface in the venting fluids.

PREDATORS AND SCAVENGERS – AND FISH

There are specialist carnivores at hydrothermal vents, however at vents and seeps (in both deep and shallow water) it is probable that the majority of carnivores are vagrants, whilst a few are colonists. Deep-sea carnivores seen at seeps include gastropods, asteroids (starfish), octopus, shrimps, crabs, and fish (Sibuet and Olu, 1998). At seeps in the North Sea we found ophiuroids (brittle stars), echinoids (sea urchins), gastropods, crustaceans (shrimps, lobsters, and crabs), and fish.

CRUSTACEANS

Some of the most memorable video film of the hydrothermal vents shows dense accumulations of shrimps darting in and out of the super-heated fluids as they gush from the seabed. Smaller crustaceans (copepods and amphipods) also behave like this near vents, indeed Lutz (2000) suggested that clouds of amphipods swarming around vents at 9 °N on the East Pacific Rise '*may be the densest concentrations of invertebrate life on Earth*'.

Shrimp adaptions to hydrothermal vents include more than the trick of avoiding death by scalding; some have adapted to the darkness. The shrimp *Rimicaris exoculata*, found swarming over black smoker chimneys on the Mid-Atlantic Ridge, has developed novel photoreceptors in place of normal eyes. Van Dover *et al.* (1989) suggested that they can detect thermal radiation from hot vent fluids, enabling the shrimps to use them as '*a means of near-field navigation both to locate optimal micro-habitats and to avoid being cooked*'. Similar species such as *Alvinocaris* cf. *muricola*, *Alvinocaris stactophila*, and *Alvinocaris longirostris* have been reported from deep-sea cold seeps in both the Gulf of Mexico and the Barbados Accretionary Wedge. According to Sibuet and Olu (1998) they are not found beyond the seeps, so they must be specifically adapted for seeps. Presumably they are feeding on bacterial matter transported into the water by the venting fluids.

High hydrocarbon concentrations have also been found in shrimps in the Gulf of Mexico. When assessing the environmental damage from the Ixtoc blowout in the southern Gulf of Mexico, Boerman and Webster (1982) found significant levels of aromatic hydrocarbons in the sediments and chronic low-level petrogenic 'pollution' in shrimp populations both before and after the blowout incident. They concluded that the South Texas Continental Shelf region *'is not a pristine environment'*; at that time they were unaware of the natural fluid flow from seeps in the gulf. It was not possible for them to document any sediment contamination from Ixtoc oil. However, Ixtoc oil was found in sorbent pad samples, indicating that the oil existed in the 'system', associated with mobile sedimentary material: *'nepheloid layers, or surface flocculent layer'* (Boerman and Webster, 1982).

In view of this apparent affinity between shrimps and seeps, we are particularly intrigued by the dense populations of shrimps and krill in some North Sea pockmarks (see Section 8.1.1).

SHELL ACCUMULATIONS

The link between bivalve accumulations and pockmarks seems to be more than casual. As in the case of bacteria, we assume that the accumulations indicate some sort of nutrient concentration. Another suggestion is that the suspected micro-seeps occurring in pockmarks help in suspending organic matter (bacteria and other microorganisms) that may serve as food for the suspension-feeding fauna, including the bivalves.

Meiofauna

Populations of meiofauna such as nematode worms seem to be affected by seeps, or at least by sulphidic conditions. At the nearshore Kattegat seeps (Bangsbostrand, near Frederikshavn) Dando *et al.* (1994b, c) found *'major differences'* between depth of penetration, diversity, and abundance of the nematode populations at, close to, and away from the seeps. Biomass and abundance are actually lowest at the seep, but much higher 5 cm and 10 cm away from the seep; higher than at the 'control' site. The relative proportions of microbial feeders, predators, and omnivores did not vary between these locations; microbial feeders represented 79 to 82% of the populations. O'Hara *et al.* (1995) reported that gas bubbling from the seabed at this site entrained water which was, in turn, replaced by an influx of seawater. They found that this oxygenated and organic-rich water *'created conditions favourable to microbe-feeding nematodes, which in*

turn support a population of larger predatory nematodes'. A similar increase in population close to, but not at the seep was also reported by Straughan (1982) from work on shallow-water oil and gas seeps off California. However, this study indicated that the populations of bacteria, nematodes, and macrofauna all increased, but at varying distances from the seeps (Figure 8.11).

There have not been many studies of the meioinfauna (nematodes and other such lower organisms inhabiting the seabed sediments) at deep-ocean seeps, but where they have been studied no great difference has been seen between seep and non-seep communities. According to Dinet *et al.* 1988 (cited by Sibuet and Olu, 1998) the density of meiofauna at hydrothermal vents is reported to be one or two orders of magnitude lower than the 'normal' seabed. However, at some cold seeps, such as the Atalante Mud Volcano off Barbados, meiofaunal densities are much higher than 'normal'. In this example a population of *'unusual large nematodes'* occured in densities two orders of magnitude higher than in the abyssal plain away from the mud volcanoes. This may indicate that they feed on a locally enriched food supply: either free-living chemoautotrophic bacteria, or detrital organic matter from clusters of symbiotic fauna (Sibuet and Olu, 1998).

Panieri *et al.* (2000a) studied the effects of hydrocarbon seepage on benthic foraminifera in the Irish Rockall Trough. To test the relationship between benthic foraminifera and seeps they studied areas with high, moderate, and no hydrocarbon contents. They found fewer species in the hydrocarbon-rich environment where the dominant taxa (*Angulogerina, Bulimina,* and *Uvigerina*) were found to have anomalously negative carbon isotope ratios. This suggests that carbon from seeps could be used in construction of their tests (shells). Similar studies carried out in the Adriatic Sea actually proved that there is influence of methane on foraminiferal tests, which show a spread in isotopic signature from δ^{13}C 0.9‰ in a normal marine environment to -6.7‰ in the presence of methane seeps (Panieri *et al.*, 2000b). Sen Gupta and Aharon (1994) also found that foraminifera from seeps in the Gulf of Mexico were ^{13}C-depleted, and that they were adapted to living on *Beggiatoa* mats.

The 'ice worm'

One of the strangest seep-associated creatures is the so-called 'ice worm' (*Hesiocaeca methanicola*). This 2 cm

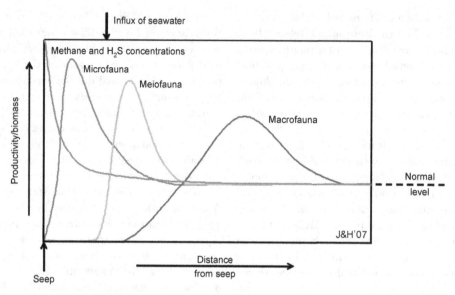

Figure 8.11* The effect of seeps on micro-, meio-, and macrofaunal abundances (developed from Straughan, 1982). According to O'Hara *et al.* (1995) the influx of oxygenated seawater, driven by seep outflow, supports the meiofaunal community.

long, pink polychaete worm lives inside small hollows in gas hydrate slabs at 700 m water depth in the Gulf of Mexico (Desbruyères and Toulmond, 1998; MacDonald, 1998; MacDonald *et al.*, website). It was discovered during an expedition to a gas hydrate mound at Bush Hill; it was noticed that one of the hydrate slabs had turned on its end due to buoyancy, exposing its underside. Numerous pink worms were seen, each living inside a small hollow in the hydrate, obviously living at the hydrate–sediment boundary. This probably demonstrates that hydrate is highly enriching for the general seabed environment, and that the polychaetes probably exist as bacteriovores. This unique observation also indicates that large blocks of gas hydrates will eventually float upwards through the water column and dissociate on the way, expelling nutrients, hydrocarbons, relatively fresh water, bacteria, and polychaetes in the process, thus producing a vertical columnar microenvironment in the water column, that can be utilised not only by bacteria, but also by plankton and fish.

8.3 SEEPS AND MARINE ECOLOGY

Deep-water hydrothermal, groundwater, and hydrocarbon seep communities were all reported for the first time within the last 30 years (by Lonsdale, 1977; Paull *et al.*, 1984; and Kennicut *et al.*, 1985, respectively). Seeps in shallow water were perhaps discovered earlier, but realisation of the processes involved has really taken place over a similar time-scale. The study of seep-related ecology has progressed very rapidly since then, but it is probably fair to say that the spectacular contrast between deep-sea seep communities and the 'normal' seabed has attracted more attention than the relatively subtle effects of their shallow-water equivalents.

The principal factors controlling the ecology of the seabed are: the depth, temperature, salinity, and movement of water, the availability of oxygen, and food availability. At sites of seabed fluid flow, the nature, rate, and duration of flow of the fluid are also important. As we have seen from the previous sections of this chapter, specific biological communities are related to seepages of the various types of seeping/venting fluid.

Sibuet and Olu (1998) summarised the communities of deep-sea environments as follows.

- The deep seabed: is '*generally characterised by low population densities, low biomass and high species diversity*'.

- Hydrothermal vents: are '*very high in biomass and relatively low in diversity with highly adapted, mainly novel megafaunal species. Hydrothermal vents are supported by high levels of in situ chemosynthesis*'.
- Cold seeps: are '*dominated by a small number of megafaunal invertebrate species, and in high densities and biomass produced via chemosynthetic bacterial activity*'.

In each of these cases microbes, both free-living and symbiotic, play a major role in supporting the communities. However, there is a significant contrast between the seeps and vents of the deep oceans and those of coastal and continental-shelf waters. Food availability is generally higher on the shallow seabed than on the deep seabed, so seep processes must do more to enable specialist organisms to out-compete 'normal' seabed processes in shallow water. Photosynthesis is progressively less competitive as water depth increases, so the contrast between 'normal' and seep-related seabed communities tends to increase with depth. This change from 'deep' to 'shallow' communities seems to occur at water depths of about 300 m.

Shallow-water communities were summarised by Tarasov *et al.* (1999):

> although there is venting or bubbling of gas, and bacterial mats develop, no specialized hydrothermal fauna (i.e. species known strictly from the hydrothermal vent environment and characterized by specific adaptions to this environment, e.g. symbiotrophic feeding) occur there. The background fauna may concentrate around sites where the production of chemosynthetic or methanotrophic bacteria is added to the photosynthetic primary production, but specific prokaryotic/eukaryotic associations and hydrothermal trophic pyramids (Gebruk *et al.*, 1997) are usually lacking and the level of endemism is usually low.
>
> Tarasov *et al.*, 1999

For cold seeps, the above comments about hydrothermal seeps were also considered applicable; however Tarasov believed that:

> single symbiont-containing species may occur in shallow-water vent/seep communities . . . Usually these examples belong to non-vent/seep fauna that

occurs also in other reducing habitats, like anoxic sediment, rotting organics.

> Tarasov *et al.*, 1999

An additional contrast between vents and seeps, noted by Roberts and Carney (1997), is that hydrothermal-vent sites, such as those at ocean spreading centres, are primarily rocky. Species requiring or tolerating a hard substrate are more likely to colonise these sites. Cold seeps, however, are initially likely to occur where there are soft sediments. Until MDAC develops, species requiring a hard substrate are therefore excluded.

8.3.1 Geographical distribution

Tunnicliffe (1991) looked into the spatial distribution of vent faunas. Because it is only the most active portion of the hydrothermal system that apparently builds up a complex vent fauna (complete with primary producers, scavengers, detritivores, grazers, and predators), they are patchily distributed. Venting tends to concentrate at the central topographic high of each spreading segment. The segments are separated by the suspected discontinuous pooling of magma chambers at separation distances of 30–100 km. Because transform faults cause even larger disjunctions, the vent fauna patchiness exists on scales of hundreds of kilometres and beyond (Tunnicliffe, 1991).

Van Dover *et al.* (2001) studied the relationship between the characteristics of endemic invertebrate vent fauna in the three main oceans: the Pacific, Indian, and Atlantic. They conclude that there are six biogeographic vent-fauna provinces in the world, despite the fact that the ecological settings and invertebrate–bacterial symbioses are similar to the Indian Ocean faunas in both western Pacific and Atlantic vents. Although the shrimp species that ecologically dominates the Indian Ocean vents closely resembles its mid-Atlantic counterpart (*Rimicaris exoculatis*), most organisms found at Indian Ocean vent fields have evolutionary affinities with western-Pacific vent faunas (Van Dover *et al.*, 2001). Besides this vent-fauna biogeographical type, the other five type locations are at the East Pacific Rise, Juan de Fuca Ridge, Explorer Ridge, Gorda Ridge, and at the Mid-Atlantic Ridge. We note that hydrothermal vents are not confined to ocean spreading centres (Section 4.3;

Map 33); vents in other contexts (submarine volca-noes, seamounts, etc.) also affect the geography of vent species.

Sibuet and Olu (1998) undertook a comparable study of the geographical distribution of deep-water cold-seep fauna. They found, for example, that the genus *Calyptogena* is very widespread. It has been reported from Barbados, the Gulf of Mexico, Florida, the Laurentian Fan in the Atlantic, as well as Ore-gon, California, Guaymas, Mexico (the Mid-American Trench), the Peru Trench, Chile, the Japan Trench, Nankai Trough, and Sagami Bay in the Pacific. The siboglinid *Lamellibrachia barhami* is also widespread, having been reported in both the Pacific and Atlantic oceans. However, the mytilids seem to be absent from the Pacific Ocean. Barry *et al.* (1996) compared and con-trasted cold-seep communities off the Pacific coast of USA with those of the Gulf of Mexico. In the Pacific they noted that seep fauna (vesicomyids, mytilids, sole-myids, lucinids, thysirids, and siboglinids) is dependent on thiotrophic endosymbionts, and they quote previ-ous studies (Fisher, 1990; Fiala-Médioni *et al.*, 1994) indicating that these groups are dependent only on sul-phur oxidation. No methane-dependent species have been reported from Pacific-coast sites such as Monterey Bay. In contrast, the Gulf of Mexico seep communi-ties include mytilid mussels. These can be thiotrophic, methanotrophic, or both, but have been reported in dense accumulations where methane levels are very high, and they are common where methane bubbles from the seabed.

On a smaller scale it seems there is evidence of species diversity within geographical areas. Sibuet and Olu found that individual symbiont-hosting species tend to be found only at single sites, or at sites close together. For example, their review found that each mytilid species is only known from a single seep area, whereas vesicomyids may be found at one or two geo-graphically close seeps.

Seep-related species are not necessarily confined to seeps. They will opportunistically utilise simi-lar 'polluted' environments whatever their cause, for example:

- the seep-related bivalve *Thyasira sarsi* has been found living on anoxic drill cuttings at an abandoned petroleum well in the North Sea;

- Dando *et al.* (1992) found tubeworms and a bivalve with symbionts living on the organic-rich cargo of a ship wrecked in the Atlantic Ocean;
- Sibuet and Olu (1998) reported finding six 'cold-seep' species inhabiting the rotting carcasses of dead whales.

Getting established

The distances between individual vents and seeps are such that the task of colonising new sites is remarkably difficult, yet essential if a species is to survive. The adults of most vent fauna are unable to migrate from one vent to another, so it is their larvae that are responsible for dis-persal. Two dispersal strategies are used. Some species have planktotrophic larvae able to feed during dispersal; others have lecithotrophic larvae that carry with them a food supply in the form of a yolk. For those using the second strategy, once the yolk is used up, starvation is inevitable. Whichever strategy is used, transport is provided by ocean currents (Hessler and Kaharl, 1995; Mullineaux and France, 1995).

Clearly, dispersal is hazardous for the larvae:

it often means travelling over long distances of hostile terrain. Many starve or are eaten before getting to where they need to be, and most never reach their destination, but drift away from an area suitable for habitation. As with so many marine animals, vent species are adapted to this great mortality. The plentiful supply of nutrients at vents results in high fecundity for some animals, and for a lineage to perpetuate itself, only two of the progeny need survive to reproductive age.

Hessler and Kaharl, 1995

8.3.2 Communities as indicators of seep activity and maturity

Measurements of fluid flow at the seabed are difficult (see Section 10.4.3), and tend to target selected seeps; these are probably chosen because they look 'good' rather than representative. Subjective inspection gener-ally suggests significant spatial variability in flow rates. Indeed, identification of individual seep vents is not a trivial task, and in our experience it is unlikely that the true number of individual vents at any given site can be accurately determined. Flux measurements gener-ally last for relatively short time periods (hours rather than days, weeks, or months), so temporal variations are also difficult to assess. However, Aharon (1994), Roberts and Carney (1997), and Olu *et al.* (1997) suggested that cold-seep communities could be used as indicators of the locations of seeps, and also the rate, spatial variability, and the longevity of the flow. The following comments about deep-water cold-seep communities are also applicable to vent communities.

LONGEVITY

Long-lived seeps enable a mature ecosystem to develop. However, there is, presumably, a gradual increase in species diversity once a new vent or seep starts. Roberts and Carney (1997) pointed out that nobody has witnessed the onset of hydrocarbon seepage; however they thought that colonisation of a new seep would be rapid, as it is at hydrothermal vents. Lutz *et al.* (1994) reported diving on a site in the east Pacific where there had been a new eruption. In April 1991 they found widespread bacterial mats, but not a well-developed community. In March 1992 a single species of tubeworm was established, and individuals had tubes < 30 cm long; the bacterial mats had 'retreated'. By December 1993 a second tubeworm species had colonised the site, with tubes > 1.5 m long; individuals were mature enough to reproduce (Roberts and Carney, 1997). Galkin (1997) found that tubeworms (siboglinids) were typical of the early stages of venting activity, and that during later stages bivalves and/or gastropods became dominant.

A mature ecosystem has a large species diversity, a range of feeding strategies, a complex trophic structure, a large biomass, and individuals of various sizes and ages – including dead animals. There will be a variety of symbiont-hosting species, and species without symbionts such as detritivores, filter feeders, predators, and scavengers (Olu *et al.*, 1996, Sibuet and Olu, 1998). To get to this stage, fluid flow must have been continuous for a considerable time period.

'Death assemblages', collections of dead seep-related organisms, may indicate that a seep has ceased flowing, although, as mentioned earlier, some vesicomyid bivalves seem able to survive periods of seep inactivity. However, in deep-ocean waters death assemblages may not survive very long – even after death. Below a certain depth, known as the 'carbonate compensation depth' (CCD), calcium carbonate is soluble

in seawater. The depth of the CCD varies according to the carbon dioxide content and biological productivity; generally it is about 4500 m, but it may be as shallow as 3500 m where productivity is low, or as deep as 6000 m (Thurman, 1993). Where cold-seep communities exist below the CCD, shell-forming organisms produce a film of soft tissue; this covers and protects the shell (and also attracts gastropods that feed on it!). Once the organism dies, the calcium carbonate dissolves, so at these depths the presence of shell material must indicate live organisms, and therefore also active seepage.

VIGOUR

Sibuet and Olu (1998) discussed the effects of fluid supply on the development of chemosynthetic communities, noting that fluid flow rates vary considerably from site to site, and indeed from place to place within an individual site. They considered that '*regular and diffusive expulsion*' is responsible for large communities, for example at Paita in the Peru Trench '*where clam beds which include patches of various sizes can reach 1000– 6000 m²*'; whereas communities with a limited extent (< 20 m², and often < 2 m²) indicate that fluid flow is restricted to discrete conduits. Ephemeral, poorly developed communities suggest discontinuous flow. Generally, the greater the flow the larger the community that can be supported, so biomass is an indicator of flux rate. However, rapid venting, in areas of fine-grained sediment, may be detrimental as suspended sediment blankets bacterial mats and clogs respiratory organisms (Roberts and Carney, 1997). Observations on mud volcanoes in the eastern Nile Fan suggested a two-fold (geographical and temporal) zonation. On recently active mud volcanoes, and in the immediate vicinity of active seeps the benthic fauna was found to be sparse, but on post-eruptive mud volcanoes with a diffusive flow there was more extensive microbial activity and macrobenthos (mainly siboglinid tubeworms, mytilid and lucinid bivalves, sea urchins and galatheid crabs), and planktonic fauna (including shrimps, medusae, and fish) in the water. Away from the mud volcanoes both benthic and planktonic fauna were sparse (Loncke *et al.*, 2004b; Dupré *et al.*, 2004).

A key factor, which at least partially explains this, is that AOMs obtain their methane supply in solution, not in free gas form. A diffusive flow provides more accessible gas than a vigorous stream of bubbles. Also, the rate of supply determines the rate of sulphate utilisation and sulphide production; high methane supply rates result in the SMTZ rising towards, and in places reaching, the seabed.

The importance of the plumbing

A further factor to be taken into consideration when studying seep and vent communities is the extremely rapid variations in conditions within a site. Depending upon the nature of the flowing fluid, one or more of the following may change considerably over very small (lateral or vertical) distances: temperature, salinity, and the concentrations of methane, sulphide, sulphate, and oxygen. Indeed, video surveys of Tommeliten have shown that carbonates, bioherms, bacterial mats, and bubble plumes rarely occur at the same place. This is because of the nature of the 'plumbing'. As discussed in Section 7.5.3, seabed fluid flow tends to be very focussed. This focussing can be considered at two scales. Firstly, seep sites seem to be quite distinct from surrounding 'normal' seabed over distances of metres to tens or even hundreds of metres. Seabed features such as mud volcanoes and pockmarks may make this distinction. Secondly, within each site there are normally very rapid variations over distances of less than a metre. A third scale, involving changes with time, is discussed in Section 10.4.3.

Perhaps an easily visualised example of these differences in scale is the contrast between the summit area of a mud volcano and the surrounding area on one scale, and on the other, the presence of individual salses and gryphons within the summit crater of the mud volcano. We have observed even smaller-scale changes at North Sea sites, but the following example comes from the Skagerrak. Dando et al. (1994a) collected seabed sediment samples with a 0.5 × 0.5 m box corer at a seep site in about 300 m of water. One sample was venting methane bubbles from about eight different vents on one side of the box, but there was no bubbling from the other side. Methane concentrations were about 0.7×10^3 to 3.4×10^3 µmol dm^{-3} on the gassy side, but less than 10 µmol dm^{-3} on the other side. Chemical gradients were very steep; a decrease in methane concentrations from 2.16×10^3 µmol dm^{-3} to 42 µmol dm^{-3} was measured over a distance of only 5 cm, and methane oxidation rates decreased from 740 to 8 µmol dm^{-3} d^{-1} over the same distance. There were comparable differences in the fauna, particularly the distribution of the seep-

related siboglinid *Siboglinum poseidoni* and bivalve *Thyasira sarsi*. The biomass in the gassy part of the core was three times higher than in the other half (122 g m^{-2} compared to 38.7 g m^{-2}); *Siboglinum poseidoni* accounted for 63% of this, *Thyasira sarsi* for the majority of the rest. In the non-seeping half, *Thyasira sarsi* was absent, and *Siboglinum poseidoni* sparse (Dando et al., 1994a).

With dramatic changes over such small distances, it is clear that seabed sampling strategies in seep areas must be more sophisticated than those normally employed in benthic ecology. Random sampling is just not good enough! Indeed, not only is precision positioning required, but samples should be taken with visual control. Even in the shallow waters of the North Sea, sampling should ideally be undertaken from an ROV or manned submarine; a TV-guided sampler is an alternative. Other alternatives leave a great deal to chance, and increase the risk of productive sites being dismissed as of no interest because sample locations missed the interesting bits by perhaps less than a metre.

8.3.3 Do shallow-water cold seeps support chemosynthetic communities?

There is nothing to stop the inhabitants of deep-water cold seeps utilising detrital matter that has fallen from higher in the water column. Similarly, even in shallow waters dominated by the 'normal' marine conditions (in which the benthos is supposed to be entirely dependent on material raining down from the photic zone) there is no reason to suppose that 'normal' organisms will not supplement their diets by taking advantage of any additional food source. If a seep happens to provide a supply of microbes, it is to be expected that these chemosynthetic organisms will supplement the diet of higher organisms. However, studies conducted so far have indicated that seeps on continental shelves are not inhabited by a diverse, specialist fauna in the way that deep-water seeps and vents are. Sahling et al. (2003) concluded their study of cold-seep communities at depths ranging from 160 to 1600 m in the Sea of Okhotsk by stating that: '*the number of chemoautotrophic species decreases dramatically with decreasing water depth.*' They found that at the shallowest site (160 m) '*the seeps lack chemoautotrophic macrofauna; their locations were indicated only by the patchy occurrence of bacterial mats*'.

There have been surprisingly few studies of the ecology of continental-shelf seeps; examples include offshore California (Spies and Davies, 1979; Davies and Spies, 1980; Straughan, 1982; Spies and Des Marais, 1983; Montagna *et al.*, 1989; Levin *et al.*, 2000), offshore Oregon (Juhl and Targon, 1993), in the North Sea (Dando *et al.*, 1991), and in the Kattegat (Jensen *et al.*, 1992; Dando *et al.*, 1994b, c). *Beggiatoa* mats seem to be found at every seep, but relatively few symbiont-hosting organisms (some nematodes, thyasirid bivalves, and a few siboglinids) have been identified. Dando *et al.* (1991), Juhl and Targon, and Levin *et al.* all concluded that any contribution made by chemosynthesis at shallow-water seeps is insignificant; too small to be detected. It is out-balanced by the normal, photosynthesis-based system. Levin *et al.* suggested that this may simply be a function of the greater availability of photosynthetically produced food in shallow water.

Shallow-water seeps do support some macroorganisms that are not typical of nearby non-seep areas – apart from the symbiont-hosting organisms mentioned above. Levin *et al.* (2000) identified the polychaete *Capitella* sp. as one benthic species that was regularly present in the sulphide-rich sediments of the three seep sites they studied on the northern Californian shelf. Dando (2001) concluded that deep pockmarks act as a refuge for fish (Figure 2.37, for example); cod (*Gadus morhua*), torsk (*Brosme brosme*), and ling (*Molva molva*) seem to be the most common fish in North Sea pockmarks (see Section 8.1.1). Also, Dando thought that seep sites are dominated by taxa attracted by a hard substrate, the MDAC, rather than by the seeps themselves (Dando *et al.*, 1991; Dando, 2001). Is this true?

Spies and Davis (1979) reported that the benthic communities of shallow-water (16 m) seeps in the Santa Barbara Channel, California were more prolific than those away from the seeps. However, this may be due to the presence of carbonates. In a subsequent study of seeps in this area, Spies and Des Marais (1983) found that the carbon of infaunal organisms was isotopically lighter at the seep than at a control station, proving that the food web is fuelled by isotopically light gases (methane, ethane, and propane). They argued that natural isotope data were consistent with a pathway of petroleum energy:

sulphate reducers $\rightarrow$ H_2S $\rightarrow$ *Beggiatoa* sp.

$\rightarrow$ nematodes and other infauna.

Carbon isotope ($\delta^{13}C$) studies of the fauna of seep areas have resulted in very different conclusions. Several authors have reported little evidence of methane-derived carbon:

- benthic infauna (with the exception of the bivalve *Thyasira sarsi* – see below) from a North Sea pockmark (Dando *et al.*, 1991): carbon isotope values between −16 and −20‰;
- fish from in/around the same North Sea pockmark (Dando *et al.*, 1991): carbon isotope values between −16.4 to −19.4‰;
- epifauna on carbonates in the Kattegat (Jensen *et al.*, 1992): −18 to −24‰;
- tissues of benthic fish and shrimps at a seep off Oregon (Juhl and Targon, 1993): −17.4 to −19.2‰;
- shelf macrofauna from seeps on the Eel shelf, northern California (Levin *et al.* 2000): −15.0 to −23‰.

These values are comparable to those of marine phytoplankton in productive coastal waters in temperate latitudes, which generally has a carbon isotope signature of −15 to −23‰ (Fry and Wainright, 1991, quoted by Levin *et al.*, 2000). Levin *et al.* (2000) considered their results to be *'consistent with a marine phytoplankton-based food chain'*. They went on to say that there was *'little evidence of lighter $\delta^{13}C$ signatures that would indicate chemosynthetically derived food sources in shelf sediments'*, although some polychaetes (*Mediomastus*, *Heteromastus*, *Capitella* sp., *Nephtys* sp., and *Sternaspis* sp.) exhibited lower $\delta^{13}C$ values (−19.2 to −23.0) than other infauna. It is argued that weak isotopic signatures suggest that seeps are *not* making an impact on the biological community. For example, Dando (2001) argued that stable-isotope studies of North Sea pockmark fauna *'gave no indication for carbon input through the food chain based on production by sulphur-oxidising or methane-oxidising bacteria'*. But, does this absence of a strong carbon-isotope signature necessarily mean that seeps are making no contribution to biological productivity at all?

With each step through the food chain the isotopic signature is diluted. For example, methane seeping from the pockmarks in UK Block 15/25, North Sea has an isotopic signature of −70‰ (Hovland and Irwin, 1989). This was depleted to −34.3 to −35.1‰ in tissues of *Thyasira sarsi* collected from this pockmark (data from Dando *et al.*, 1991). This species is known to be seep-related, but the chemoautotrophic endosymbiotic bacteria it hosts utilise reduced sulphur, rather than methane. Even the smaller step from gas to the tissues of macrofauna with methanotrophic endosymbionts results in a significant depletion. For example, a gas of -60 to $-65‰$ is utilised by sponges from the Barbados Accretionary Wedge whose tissues are -48.4 to $-48.8‰$ (Olu *et al.*, 1997).

Dando *et al.*, (1994b) argued that the ^{14}C content of living tissues would give an anomalous result if the tissue contained any carbon derived from methane. They tested filter-feeding bivalves (*Mytilus edulis*) and crabs (*Carcinus maenus*) living at the near-shore gas seeps near Frederikshavn on the Kattegat coast of Denmark. The crabs were '*concentrated around the seeps*'; some of them were observed '*darting in and out of the bubble stream*'. A ^{14}C value of 115.4% (where 1956 = 100%) for non-gut tissue from this crab is within the normal range. However, for *Mytilus edulis* a slight anomaly was detected: 115.1–115.6% modern compared to 116.9–117.6% modern for mussels living on a groyne 700 m from the seep site. This led Dando *et al.*, (1994b) to suggest '*a maximal contribution of 1–2% methane carbon to the tissue of the mussels.*'

These organisms obtain *some* carbon that originally came from methane, but do these carbon-isotope and ^{14}C data show that there is only a minor dependence on methane? Not necessarily. There is an alternative source of carbon available to them, hydrogencarbonate – also known as bicarbonate (HCO_3^-) – in seawater or pore water, so the carbon isotope signature may be significantly diluted, even if chemosynthetic energy is being utilised. Further, they use an enzyme, ribulose-bisphosphate carboxylase, which discriminates against ^{13}C (Dando and Spiro, 1993). It seems that carbon isotopes provide a very poor link between methane and seep-associated fauna. In most cases depletion to 1–10% of the original carbon isotope signature is to be expected, even for organisms living directly off methane (Antje Boetius, 2002, personal communication). Consequently, it would be premature to conclude that an absence of a depleted isotopic signature at the top of the food web indicates that there is no chemosynthetic contribution to productivity. Indeed, *any* deviation from the isotopic signature of phytoplankton (-15.0 to $-23‰$) suggests the utilisation of alternative sources of energy.

Regardless of the interpretation of carbon isotope data, the presence of organisms that utilise reduced sulphide provides irrefutable evidence that methane 'energy' is making its way into the food web. Let us consider the route by which chemosynthetic energy might progress through the web. Presumably it is derived initially by the microbial consortia that oxidise methane (the AOMs), and then by the sulphide oxidisers, *Beggiatoa*. Sediment feeders, grazers, and filter feeders (nematodes, some polychaetes, some crustaceans, etc.) that might consume these free-living microbes make the next step. The chemosynthetic energy will then progress through deposit feeders. In addition the endosymbiotic bacteria utilising reduced sulphur or methane are deriving chemosynthetic energy directly utilised by their meio- or macrofaunal hosts. Further steps are made by scavengers and predators, and finally to the highest levels of the web, that is fish (and fish-eating humans!), as is shown by Figure 8.12.

With progressive dilution at each step, it seems hardly surprising that there is little if any isotopic indication of the utilisation of methane in the macrofauna. It seems more logical to search lower down the foodweb for evidence that methane carbon is utilised. This is what Bauer *et al.* (1990) did at seep sites offshore California. They found that ^{14}C was more depleted in total organic carbon (TOC) and dissolved inorganic carbon (DIC) close to seeps, and that bacterial populations were larger close to seeps, indicating that methane-derived carbon was available for consumption. Carbon-14 was found to be slightly depleted (75 to 85%) in meiofauna, but more depleted (only 34.7% ±14.0 at one site) in deposit-feeding macrofauna (mainly polychaetes). Clearly the methane-derived carbon *was* being used.

Therefore we must conclude that, despite the apparently small number (relative to deep-water seeps) of macroorganisms depending directly on chemosynthesis, there is evidence that carbon from seeps does enter the foodweb. Nevertheless, it is obvious that shallow-water sites are not the chemosynthesis-based 'oases' found at sites below about 350 m. Carney offered an explanation.

Above the shelf-break, detritus input to the bottom is high, in situ photosynthesis becomes increasingly prevalent, and shelled epifauna are common. At this depth and shallower, predators are both abundant and amply supplied with alternate prey

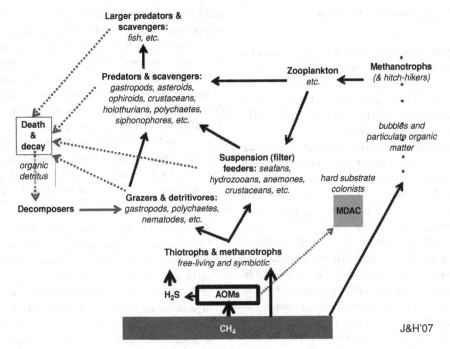

Figure 8.12 The chemosynthetic food web of cold seeps. Note that lower organisms (chemosynthesisers – free-living and symbiotic, bacteriophages, etc.) are colonists, dependent entirely on chemosynthetic energy derived from methane or hydrogen sulphide; higher organisms (scavengers, predators, etc.) may be vagrants taking advantage of this and other food supplies. Food webs associated with other sources of chemosynthetic energy (hydrothermal vents etc.) are comparable.

items. Quite simply, members of the VBV [vestimentiferan tube worm, bathymodiolid mussel, vesicomyid clam] association do not survive exploitation at these depths. Slightly deeper, production exceeds predatory loss, and deeper yet, predation is minimal.

Carney, 1994

Vestimentiferans (siboglinids) can survive in shallow water (e.g. in the volcanic caldera of Kagoshima Bay, Japan; see next section), but only under conditions hostile to most fauna, so Carney's explanation seems plausible, and probably applies quite generally. We, therefore, believe that chemosynthetic communities associated with either cold seeps or hot vents dominate benthic ecology provided they are not out-competed by faunal types better adapted to environments dominated by photosynthesis. This means that chemosynthetic communities dominate beneath about 350 m water depths, or at locations inhospitable to 'normal' communities, for reasons of temperature, salinity, or chemistry.

Shallow-water hydrothermal vents

The contrast between deep-water and shallow-water hydrothermal vents is similar to that between deep-water and shallow-water cold-seep communities. Tarasov et al. (1999) compared their findings, from Matupi Harbour, Papua New Guinea, with those of others who had worked on shallow-water hydrothermal vents: Kraternaya Bight, Kuril Islands (Tarasov, et al., 1986, 1990, 1993; Kamenev, 1991a,b), Whale Island, Bay of Plenty, New Zealand (Kamenev et al., 1993), Palos Verdes Peninsula, California (Kleinschmidt and Tschauder, 1985), and Milos Island, Aegean Sea (Dando et al., 2000). Their conclusion (cited early in Section 8.3) was that specialised hydrothermal communities have not developed, although specialist species are present, and 'normal' fauna may be attracted to vent sites. However, these conclusions do not seem to apply to all shallow-water hydrothermal vents; a 'typical' vent community was found at < 100 m in Kagoshima Bay, Japan (see Sections 3.18.5 and 8.1.7) where the conditions must be unfavourable to 'normal' fauna.

8.3.4 Do seeps contribute to the marine food web?

The occurrence of shallow gas in the sediments of the Irish Sea was documented in Section 3.5. Savidge *et al.* (1984) reported intense, localised increases in biological productivity in the Irish Sea. However, they did not find any dramatic change in other sea-surface characteristics to explain this increased productivity.

Essential in the understanding of marine ecological systems is knowledge of mineral and nutrient fluxes. A crucial question is 'which way the flux is directed – upwards or downwards?' If nutrients were somehow pumped into a shallow-shelf sea through the sea bottom, one might expect to see increased productivity in the water column above it. It is a well-established fact that the nutrient content of the interstitial water is many times higher than that of the water immediately overlying the sediments (Sholkowitz, 1973; Suess, 1977). Fisher *et al.* (1982) studied sediment–water column exchange of oxygen, ammonium, nitrate, and phosphate in three North Carolina estuaries. They showed that sediments supply, on annual average, 28–35% of the nitrogen and phosphorus required for the primary production of shallow marine systems. They assumed three mechanisms of recycling to be active.

1. Rapid, partly aerobic degradation in surficial sediments (0–10 cm) by macro- and microbenthos and return of nutrients to the water column in less than one year through the combined effects of bioturbation and diffusion.
2. Temporary burial, anaerobic decomposition by microorganisms and return to the water column over decades or centuries via diffusion and bulk flow of interstitial waters.
3. Deep burial (> 1 m) and recycling on time-scales equivalent to major geological periods of glaciation-modulated sea-level changes.

Smetacek *et al.* (1976) effectively demonstrated how nutrients were released from the sediments of Kiel Bight in the Baltic Sea by interstitial water flushing through the bottom of their 'plankton tower' (a bottle-shaped plastic cylinder open at both ends). By measuring the nutrient content outside and inside the plankton tower they found that nutrients were injected through the seabed at times when dense (high-salinity) North Sea water was replaced by low-salinity Baltic water. They found that this mechanism of nutrient release

could be of great ecological importance, both for the zooplankton and the benthos. This suggests that any mechanical flux of fluids upwards through the seabed would represent a nourishing environment for all types of microorganisms, detritus feeders, and suspension filter feeders. In other words, wherever sediment pore fluid is somehow pumped into the overlying water it will affect the marine environment in a very positive way in that nutrients are injected from below. The fact that marine biologists have been looking for elusive, significant upward-acting forces in the deep oceans may be inferred from the following passage about holothurians (sea cucumbers):

> Swimming holothurians and pelagic juveniles of both benthic and benthopelagic species could act as vehicles carrying organic matter from the seabed up into the water. These movements represent one of the very few routes for upward transport of organic matter from abyssal depths. Most of the vertical movement of organic matter in the oceans is downwards – as the wastes and remains of animals near the surface sink.
>
> Billett, 1986

We do not believe that it is necessary to invoke such a mechanism; seabed fluid flow can provide the transportation. Suess *et al.* (1999a), for example, reported that in an area of '*giant cold vent fields*' in the Derugin Basin, Sea of Okhotsk (500 m water depth), '*the lower water column is depleted in oxygen and enriched in nutrients*' (and enriched in methane and barium).

Because microseepages have been shown to enhance aquatic life, there are various corollaries. One is that of the variability of fluid flow with tidal effects. Fluid flow and microseeps are most probably grossly affected by tidal forces, as discussed previously (Section 7.5.5): at low tide the pressure on the seabed sediments is lower than at high tide; consequently, seeps are expected to be more vigorous during low tides than during high tides. This will also be the case for microseeps. Taking this a step further, it is logical to suppose that the highest fluid flow rates will occur during the lowest tides, namely low spring tides. Perhaps this mechanism could help explain the lunar cycles observed in biological productivity in marine (and lake) environments.

Methane oxidation in the water column

A considerable proportion of methane vented from the seabed is dissolved in the water column (see

Section 10.4.4). In all but the shallowest waters it is likely that the majority, if not all the methane bubbles dissolve (except during major venting episodes). In the absence of the hydroxyl ions that readily oxidise methane in the atmosphere, methane oxidation in the water column is microbially mediated; where there are methane seeps there will be methanotrophic microbes in the overlying water.

In Section 3.12.2 we commented on the abundance of microbial methane oxidisers in the sediments of Lake Baikal. Granina *et al.* (2001) studied methane oxidisers in the water column of the lake and found 100–1000 cells ml^{-1} water. A similar enhancement of bacterial productivity was found near the seabed at seep sites in the Gulf of Mexico (LaRock *et al.*, 1994), and heterotrophic organisms in near-seabed waters, where methane solution and oxidation is greatest, are isotopically lighter than their counterparts higher in the water (Samantha Joye, 2002, personal communication). Besides providing an environment for seabed chemosynthesis, the seeps are also responsible for providing extra organic matter to the water column (Namsaraev *et al.*, 2000), which may float downcurrent to increase the food resource over a wide area (LaRock *et al.*, 1994).

Hydrothermal vents and water-column biology

Some of the components of hydrothermal fluids are potential energy sources utilised by microbes in the water column (Table 8.1), consequently hydrothermal vents enhance biological activity in the water column as well as at the seabed. Deming and Baross (1993) reported bacteria concentrations of 340×10^4 ml^{-1} in a hydrothermal plume as it exits from a vent, but even 100 m above the vent concentrations are significantly higher (24×10^4 ml^{-1}) than adjacent to the vent (2 to 3×10^4 ml^{-1}). De Angelis *et al.* (1993a,b) suggested that, at methane-rich hydrothermal-vent sites, microbial methane oxidation may contribute as much as 1.5 times more carbon than the vertical flux sinking from the photic zone. Inevitably, such microbial activity will be beneficial to higher organisms, passing chemosynthetic energy into the food chain. An example is the large population of macrozooplankton found along the upper edge of the Juan de Fuca hydrothermal plume where methane oxidation rates are highest (de Angelis *et al.*, 1993a,b). Vents also provide nutrients such as phosphates and ammonia, which are beneficial to biological productivity.

If normal plume venting has such an effect, what about event plumes (Section 10.2.1)? They must result in significant 'blooms' affecting the productivity of the water.

Bubble deposition

We have already presented evidence of bacterial flocs being jetted into the water from hydrothermal vents (Section 8.2.1). Other, perhaps less spectacular, fluid-flow processes are also likely to lift particles from the sediment or the seabed. Leifer and Judd (2002) explained how bubbles might do this.

When a gas bubble rises from the seabed particles, which may include surfactants, minerals, nutrients, or microbes, adhere to its surface. These 'hitchhikers' are lifted up through the water column until either the bubble reaches the sea surface, or it dissolves. As is explained in Section 10.4.4, all the bubbles coming from the seabed at any given site tend to be of a similar size. Consequently, they all dissolve at more or less the same altitude above the seabed. This often occurs at the thermocline (if there is one), where there is a sudden change in temperature and salinity. When the bubbles dissolve, the hitchhikers are deposited – hence the term 'bubble deposition'. At the altitude of bubble solution there is a layer relatively rich in particles, nutrients, etc. Leifer and Judd (2002) suggested that the scattering layers above the Bush Hill site, Gulf of Mexico (Figure 8.13) are characterised by a relative abundance of organisms (zooplankton, etc.) taking advantage of the food supply delivered from the seabed by the bubbles. They also described an acoustic layer recorded above the Scanner pockmark in UK Block 15/25, North Sea on more than one occassion. During one survey concurrent visual observation with an ROV did not detect any visible bubbles in this 'acoustic' plume, so it was interpreted as representing a change in water density. On other occasions concentrations of the jellyfish *Cyanea capillata* had been recorded here at this altitude. Leifer's and Judd's modelling of seep bubble solution indicated that bubbles would dissolve at the same altitude. They surmised that at the acoustic layer identified where the bubbles dissolved, hitchhikers were deposited, and jellyfish were preying on smaller organisms that were themselves eating hitchhikers.

This idea needs further confirmation; however, it seems that the mechanism is viable. If confirmed, it provides a pathway by which seeps will make a real

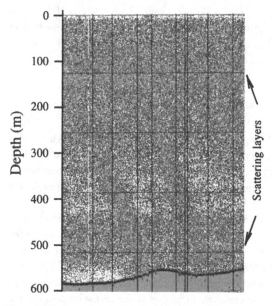

Figure 8.13* Seep plumes rising from the seabed at Bush Hill, Gulf of Mexico, to scattering layers. (Reprinted from Leifer and Judd, 2002 with permission from Blackwell Publishing.)

contribution to marine biological productivity, not just at the seabed.

Conclusion

In this section we have identified several ways in which nutrients of one kind or another may be lifted into the water column by seabed fluid flow. Lein *et al.* (1997) estimated the production of organic carbon (C_{org}) by chemosynthetic and photosynthetic primary producers in a 32 400 m^2 area over the Vienna Woods hydrothermal vents in the Manus Basin (Papua New Guinea). The C_{org} flux from chemosynthetic production into the bottom 350 m of water (2.7 kg d^{-1}) '*is of the same order as the C_{org} flux produced by photosynthesis in the surface layer*' (6.8 kg d^{-1}). This organic carbon is available for higher organisms to utilise. Assuming these results are not anomalous, it can be implied that hydrothermal-vent and cold-seep areas in the deep ocean, which together occupy large areas in various oceanographic and geological contexts (see Chapter 4), must make a significant contribution to marine biological productivity.

To return to the example we introduced at the start of this section, we suggest that the elusive, intensely productive patches in the Irish Sea might be located above seeps. In fact we go further. The presence of seabed fluid flow in an area may have a localised effect on the benthic ecology, but, because of water-column methane oxidation and the supply of hitchhikers, it may also have a more widespread effect on the ecology of the water column. In Section 11.6.5 we suggest that this has been a benefit to the fishing industry.

The simple photosynthesis-based food webs still accepted by some marine biologists are clearly over-simplifications; food webs associated with seeps and vents (Figure 8.12) must also be taken into account. The relative importance of the two primary sources of energy, photosynthesis and chemosynthesis, varies. Chemosynthesis is dominant in deep waters (below about 300–400 m). Photosynthesis dominates in shallower waters. However, shallow-shelf seas like the North Sea are such complex environments that it may be impossible to identify any one parameter as being responsible for a particular biological response. In one area one individual may be dominant, but elsewhere others may swamp its influence. Although photosynthesis is the dominant provider of energy, chemosynthesis may play a role, and it would be shortsighted to ignore its contribution to energy budgets without further research.

Once before we asked if we had been '*so blinded by the light of photosynthesis that the contribution of chemosynthesis has been overlooked?*' (Judd and Hovland, 1989). This question is still relevant.

8.3.5 Is fluid flow relevant to global biodiversity?

The science of biodiversity is the study of variety of life, including the determination of differences in biological variation. So far, there does not seem to have been much success, and there is a general call for a marked improvement in understanding the global distribution of biodiversity by ecologists and biogeographers.

Gaston (2000) suggested that understanding what causes the spatial heterogeneity of species richness would '*impinge on applied issues of major concern to humankind*': the spread of invasive species, the control of diseases, and the likely effects of global environmental change on the maintenance of biodiversity. On a global scale, the relationships between species richness and environmental energy have previously been found to be associated with latitudinal, elevational, and depth gradients (Rohde, 1992). But as Gaston (2000) rightly pointed out, there are many other influences, including longitude, topography, and aridity. Another key factor is the availability of energy:

> Greater energy availability is assumed to enable a greater biomass to be supported in an area. In turn, this enables more individual organisms to coexist, and thus more species at abundances that enable them to maintain viable populations. The result is an increasing species richness with energy availability. This assumes a basic equivalence between species in their energetic requirements at different levels of energy availability (Cousins, 1989).
>
> Gaston, 2000

We certainly agree that energy gradients are important. But, this must include the 'chemical energy' exploited during chemosynthesis associated with fluid flow, whether through the seabed or on terrestrial surfaces. If global biodiversity is only judged as a function of temperature (i.e. latitude, elevation, and depth) then the biodiversity described and discussed in this chapter will not be recognised; fluid flow is dependent on neither latitude nor depth in the ocean. Judging by the number of new species discovered in the marine environment over the last quarter century, and even by the number of new biotopes, we still have a lot to learn about life on this planet. Even though fluid-flow-related ecology is one of the most rapidly developing natural marine sciences, it still seems to be treated as a mere curiosity by many. However, this is not true of all marine scientists. Bernhard *et al.* (2000) pointed out that '*bacterial–eukaryotic symbiotic communities*' are present in a variety of marine environments; they identified silled basins, seeps, hydrothermal vents, and oxygen minimum zones. They argued that such communities '*have an underestimated and significant impact on oceanographic processes such as nutrient cycling*'. We do not disagree!

We judge the knowledge level of global marine biodiversity to be at an equivalent level to that of global terrestrial environments and biotopes over a century ago. Lutz (2000) remarked that we have spent the last 20 years cataloguing the rich life at vents; he hoped that the next 20 years will be spent understanding the history of this life, and speculated that the exploration of the origins of chemosynthetic organisms may '*point us towards the source of the earliest life on Earth*'.

Chicken and egg?

Comparisons between the fauna of hydrothermal vents and that of cold seeps raise an intriguing question: which evolved first? The following quote from Reysenbach and Shock (2002) suggests that hydrothermal-vent fauna was important for the development of life on our planet:

> Hydrothermal ecosystems are the most ancient continuously inhabited ecosystems on Earth. The geochemistry of hydrothermal systems directed the evolution of life on early Earth. In turn, as biological processes such as photosynthesis evolved, biological activity influenced geochemistry. This ecosystem evolution is recorded in hydrothermally altered rocks and potentially in the genomes of extant thermophiles. Numerous genome sequences of thermophiles are available that provide genetic information pertaining to their geochemical and ecological history and their metabolic potential. For example, *Archaeoglobus fulgidus* contains methanogen-specific genes, probably as a result of lateral gene transfer, which suggests that methanogens and sulfate reducers occupy similar ecological niches.
>
> Reysenbach and Shock, 2002

However, Sibuet and Olu thought that cold-seep communities came first:

> Molecular phylogeny can help to demonstrate the recent hypothesis, shown for mytilids, that the evolution of the cold seep, sulphide/methane dependent species was a critical antecedent to the hydrothermal-vent fauna. Did other components of the seep fauna play an ancestral source to hydrothermal-vent species?
>
> Sibuet and Olu, 1998

This 'chicken and egg' question is not trivial because it has been suggested that life on the surface of our planet may have begun at such places, as we will see below.

8.3.6 The 'deep biosphere' and the origins of life on Earth

The scientific revolution (paradigm shift) expected to result from the discovery of the AOMs is far reaching. Not only may they be responsible for a fundamental process at methane seeps, but it is even suggested that without these newly found 'bugs' *Homo sapiens* may not have existed. As Zimmer (2001) pointed out, volcanic activity and methanogenic microbes may once have generated atmospheric methane levels 1000 times higher than they are today. Although this had the benefit of preventing our planet from freezing, if methane production had continued unchecked Earth, like Venus, may have become too hot for life as we know it. '*We may have the evolution of methane-eating archaea to thank for saving us from that grim fate*'. (Zimmer, 2001).

Zimmer is not the only one to suggest that methane has been fundamental to the development of life on Earth. It was thought that life on Earth lived, primarily *on* Earth, but over recent years knowledge of the 'deep biosphere' has increased. The maximum depth from which microbial communites have been reported has steadily increased. It is thought that temperature and pressure are both limiting factors; Parkes (1999) reported that '*the current temperature maximum for bacterial life is 113 °[C]*'. Yet, despite this constraint, there is increasing evidence of a 'deep biosphere' comprising large bacterial populations at depths of up to a few kilometres in various subsurface environments: aquifers, granites and basalts, shales, marine sediments, etc.. The discovery that some North Sea oil reservoirs '*produce 3–16 kilograms of high-temperature bacteria along with the production fluid each day*' is, as Parkes said, '*startling*'. Research by Wellsbury *et al.* (2000) on samples from ODP drilling on Blake Ridge showed that bacteria are not only present in substantial numbers (1.8×10^6 cells ml^{-1} at a depth of up to 750 m below seabed), but reproducing in and beneath the GHSZ. Indeed, it seems that both bacterial activity and populations actually *increase* around and beneath the base of the GHSZ. This suggests that previous estimates of the subsurface microbial biomass may have been gross underestimates.

Clearly, as well as life on Earth, there is also a considerable microbial population *in* Earth. Does this mean that living organisms have adapted to progressively deeper and deeper environments? Gold (1999) argued that the reverse is true – that life evolved beneath the surface. He suggested that there is a 'deep, hot biosphere' fed by primordial methane from deep within the planet. He argued that microbial communities at seabed vents and seeps represent the pioneers who have ventured out of the deep, hot biosphere to establish a foothold in the 'borderlands'. Their successors pushed

further into the exterior, enabling life 'as we know it' to become established in the sea, and subsequently on land and in the air (Gold, 1999).

Although Parkes (1999) considered that the temperatures and pressures anticipated at the depths inhabited by Gold's deep, hot biosphere would be prohibitive, we have already seen that the biosphere is now known to extend much deeper beneath the surface than was once thought possible. Maybe Gold's ideas should not be dismissed prematurely. However, an alternative possibility has arisen thanks to the discovery of fluids originating from the serpentinisation of ultramafic rocks from the mantle (Section 5.2.4). Von Damm (2001) speculated that high concentrations of hydrogen and temperatures of 40–75 °C made such ultramafic sites ideal incubators for the evolution of life. The Lost City discovery demonstrated that hydrothermal-vent communities can exist without volcanic activity. Kelley *et al.* (2002) thought that these observations '*indicate that water-bearing planets, chondritic in composition, that have experienced tectonic processes are potential sites for Lost City type systems and life*'. Perhaps the quest for evidence of life on other planets should be targetted on vent sites on such planets!

8.4 A GLIMPSE INTO THE PAST

Although the previous section has hinted that seep- and vent-related microbial organisms played a fundamental role in the evolution of life on our planet, the preceding sections of this chapter have dealt exclusively with the present day. In this section we identify a few of the examples of evidence that seabed fluid flow has affected seabed ecology for long periods of geological time. Perhaps the oldest evidence is in the form of '*kinneyia*'. These are interpreted by some as impressions of gas bubbles in matground deposits, the fossil remains of bacterial mats. According to Pflüger (1999) many of these have been reported from rocks of Precambrian age, so we can deduce that life forms have been found in association with gas bubbles, and hence seabed fluid flow, since then.

8.4.1 Fossil cold-seep communities

Evidence of well-developed cold-seep communities has now been reported from several places, geological ages, and geological contexts. Campbell *et al.* (2002)

summarised data from seep-related carbonates ranging in age from the Middle Devonian to the present day (see Section 9.2.12); several of the sites identified by Campbell *et al.* also have fossil cold-seep communities. It seems that cold-seep communities have existed for at least the last 375 million years.

Cavagna *et al.* (1999) identified examples from convergent margins (including accretionary wedges, foredeep basins, and fore-arc basins) and passive margins. These range in age from the Jurassic to the Pliocene. They suggested that MDAC, identifiable by the carbon-isotope data and mineralogy, and biomarkers provide relatively easy demonstrations of an association with methane seeps. However, they also presented evidence from Miocene carbonates of Monferrato (northwest Italy) of fossil bacteria they identified as sulphur oxidisers – probably *Beggiatoa*. Peckman *et al.* (1999) reported tubeworms and lucinid bivalves from the same site. They also found lucinid bivalves in Lower Oxfordian (Jurassic) carbonates at Beauvoisin (southeast France). Peckman *et al.* said that the tubeworms '*closely resemble chemosynthetic pogonophora* [siboglinid] *tube worms from Recent cold seeps*'; they were embedded in isotopically light ($\delta^{13}C$ −30%) carbonate. The lucinids were found in '*very dense clusters*', similar to those described from modern deep-water cold-seep communities. Like *Beggiatoa*, lucinids and tubeworms are typical of seep environments.

Identifying seep communities becomes more difficult in older rocks that predate the macrofauna typical of modern seeps. For example, the fauna of Cretaceous seeps in the Canadian Arctic, described by Beauchamp and Savard (1992), comprise '*abundant bivalves and serpulid worm tubes associated with minor ammonites, gastropods, forminifers, and fish teeth*'. The fossil density is much higher than in surrounding rocks, but this fauna bears no direct resemblance to modern seep fauna, other than the dominance of bivalves and worm tubes. None of the bivalve genera have been reported from modern seep communities. Identification of these as seep-related communities is therefore dependent on other factors, principally the identification of MDAC. However, field relationships enabled Beauchamp and Savard to recognise the geological environments of fossil seep sites: one (Ellef Ringnes Island) was associated with faulting overlying a salt diapir; the other (Prince Patrick Island) was associated with listric faulting of a half-graben.

Geological mapping has revealed even greater details of Danian (early Palaeocene) cold seeps in the Panoche Hills, California. They have been traced over a distance of at least 15 km, in a stratigraphic zone up to 45 m thick; seep activity is thought to have persisted for about three million years. Here the fauna include several genera related to modern seep fauna (siboglinid tubeworms, lucinid, and solemyid bivalves), and there are isotopically light carbonates (Schwartz *et al.*, 2003). Excellent exposure means that the plumbing system beneath the seeps has been identified, extending for a vertical thickness of about 800 m (Weberling *et al.*, 2002). This may help to explain how modern seep systems work!

Perhaps the oldest macrofauna identified as having hosted microbial symbionts date from more than 440 million years ago. Fortey (2000), in his book about trilobites, described the members of the Olenidae family as '*the first known animals to live symbiotically with bacteria*'. These trilobites (e.g. *Cloacaspis*) had long bodies with a large number of thoracic segments: '*all the more space to put your bacteria*'! Fortey said that they were abundant in Ordovician black sulphurous shales of Svalbard. Was the environment in which they lived part of an anoxic sea, or did they colonise seeps?

9 · Seabed fluid flow and mineral precipitation

A major realization at this 'meeting of minds' was that many carbonate mounds, bioherms or reefs described in geological literature were in fact the product of chemosynthesis. Similarly, biologists became aware that chemosynthesis was deeply rooted in the geological past and may be traceable as far back as the origin of life itself.

Beauchamp and von Bitter, 1992

Mineral precipitation occurs at the seabed in two circumstances in association with seabed fluid flow: microbial utilisation of fluids, and in response to changes in physical and chemical conditions. This chapter starts with an investigation of the nature and origin of methane-derived authigenic carbonate (MDAC). This leads to a discussion of the influences of fluid flow on the formation of other types of marine carbonate including those associated with biological activity, such as stromatolites and deep-water coral reefs. Non-carbonate minerals are formed as a result of precipitation and fluid-flow processes. We discuss the metalliferous deposits formed by hydrothermal activity at ocean-spreading centres, and by cool-water submarine seepage. Finally there is a discussion of the possible modes of formation of ferromanganese nodules, including the hypothesis that they were formed as a result of seabed fluid flow.

9.1 INTRODUCTION

The balance between the liquid and solid states of the waters of seas and oceans, and the pore waters of sediments, is governed by the three variables: temperature, pressure, and solute concentration. A decrease in temperature or pressure, or an increase in concentration, generally leads to supersaturation and to the precipitation of excess solutes. In this chapter we examine a variety of situations in which mineral precipitation occurs in association with seabed fluid flow. We now know that methane-derived authigenic carbonate (MDAC) is a common feature of methane seeps at any water depth, yet, they are not sufficiently well known to be recognised by many carbonate sedimentologists. For example, Riding (2000), in his review of microbial carbonates, does not mention them at all! However, minerals precipitated from venting volcanic, hydrothermal, and geothermal fluids are better known because of their economic implications.

9.2 METHANE-DERIVED AUTHIGENIC CARBONATES

Niels Oluf Jørgensen was one of the first to undertake a detailed analysis of marine carbonate-cemented sediments. These rocks originated from the east (Kattegat) coast of Denmark, the Skagerrak, and from the North Sea, and had either been washed up onto beaches or provided by fishermen who had found them in their trawl nets (Jørgensen, 1976). Although the locations of origin were unknown to him, the samples analysed by Jørgensen had a characteristic carbon-isotope composition ($\delta^{13}C$ values of -30 to $-60‰$), which led him to conclude that they were formed by the oxidation of methane. These carbonates, which seem to have developed *in situ*, would now be described as 'MDAC'.

9.2.1 North Sea 'pockmark carbonates'

We first came across MDAC in 1983 during our study of North Sea pockmarks and seeps (Chapter 2; see also Hovland *et al.*, 1985, 1987). Our observations and analytical results are summarised as follows.

- The carbonates occur as slabs, crusts, and lumps in pockmarks and other seepage sites (e.g. Tommeliten and Gullfaks A – see Figures 2.28, 2.29, 2.32, and 2.43).

- They comprise the normal seabed sediment cemented by the precipitation of carbonate minerals: calcite, aragonite.
- Voids and cavities are present, and subvertical channels pass through them – some of these are lined with carbonate minerals such as botryoidal aragonite, others appear to have been corroded. These have the appearance of fluid migration channels.
- Small framboidal pyrite nodules are also present. For example beautiful minute pyrite framboids were found inside foraminifera tests and other cavities and 'channels'.
- There are indications of finely disseminated organic matter.
- Some samples contain evidence of brecciation.
- Carbon-isotope analyses of the carbonate cements revealed $\delta^{13}C$ values in the range -36.3 to -57.2‰.
- The occurrences were closely associated with evidence of seabed fluid flow (pockmarks, acoustic turbidity, elevated methane concentrations in sediment pore waters and seawater, and methane-rich seeps).

The carbon-isotope values are more closely correlated with those of methane gas collected from the same sites ($\delta^{13}C$ values in the ranges -26.7 to -47.7‰ at Tommeliten, -44.3 to -90.6‰ at Gullfaks, and -39.3 to -79.0‰ for seep and sediment gases at UK Block 15/25 – Section 2.3) than with normal marine carbonates (-7 to $+8$‰). This indicates that the carbon is derived from methane.

9.2.2 'Bubbling Reefs' in the Kattegat

Shallow gas, gas seepage, and carbonate-cemented sandstones are quite common along the northeast coast of Denmark, around the island of Læsø and further east in the Kattegat (Map 3); more than 40 individual sites have been identified (see Section 3.3.4). The carbonates are lithified sandstones similar, apart from the cementation, to the 'normal' sediments of the areas in which they occur. They generally take the form of individual slabs and (more commonly) thin, < 0.3 m, 'pavements', but there are also large blocky boulders and spectacular pillars and mushroom-shaped bodies which stand up to 2 m above the seabed (Figure 3.13); some pillars are almost 4 m tall and 1.5 m in diameter (Jensen

et al., 1992). The pillars are not solid, but have multiple, poorly cemented, vertical, pipe-like structures in them. It is probable that these features were originally formed beneath the seabed, and were subsequently exposed by the erosion of seabed sediments, probably during the post-glacial period of isostatic uplift experienced in this area (Jørgensen, 1992a,b).

Jørgensen (1992) explained that the carbonate cement is composed of the following.

- *High-magnesium calcite* occurring as an intergranular cement in the form of rhombohedral crystals. These vary in concentration '*from scattered individual rhombohedra to extensive crusts and almost complete infilling of the original pore space*'.
- *Dolomite*, typically similar in size and form to the calcite. However, this is found only on the pillars.
- *Aragonite* in the form of needles, clusters of needles, botryoidal clusters found mainly on the surfaces of skeletal matter; on some shells there are dense 'palisade' growths. Aragonite is found mainly in the lithified pavements, but sometimes on the pillars, but it is never found in association with dolomite.

Examples of these minerals are shown in Figure 9.1. Jørgensen (1992) reported that carbon-isotope values of the cements are generally between -40‰ and -50‰, although values of between -33‰ to -61‰ were measured. These values were taken to indicate that the carbon was derived partly from methane and partly from '*the inorganic bicarbonate pool*'.

9.2.3 Carbonate mineralogy

Methane-derived authigenic carbonate is composed of whatever seabed sediment is present at a seepage site lithified by carbonate cement. Typically this cement comprises one or more of the common carbonate minerals: high-magnesium calcite, aragonite, and dolomite. Other minerals, most commonly pyrite, may also be present. However, occurrences of these minerals as cements in modern seabed sediments are not confined to methane seeps.

Seawater contains abundant Ca^{2+} and Mg^{2+} ions. Warm (i.e. tropical/subtropical) surface seawater is usually supersaturated with respect to calcium carbonate ($CaCO_3$), consequently the precipitation of these minerals, and the formation of interstitial cements in seabed sediments, might be expected; however this is not

(a)

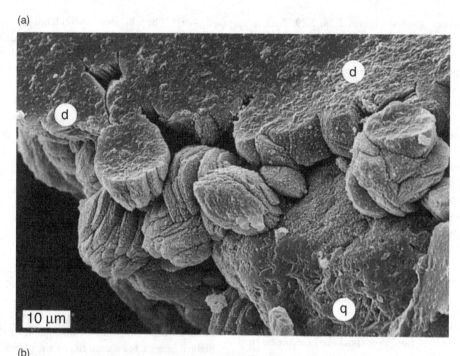

(b)

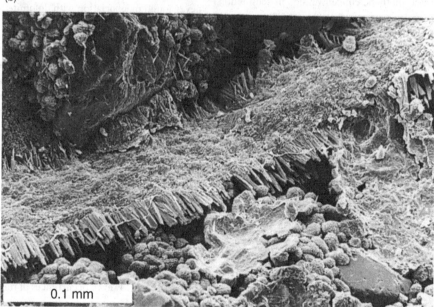

Figure 9.1 Scanning electron microscope images of fractured specimens of MDAC from the Kattegat: (a) quartz grains (q) and intergranular dolomite cement (d); (b) Dense palisade growth of aragonite cement on skeletal matter, and high-magnesuim calcite crusts enveloping quartz grains. (Reproduced from Jørgensen, 1992 with permission from Elsevier.)

true in cooler waters of the deep ocean, nor of shallow waters in mid to high latitudes, as these are undersaturated and CaCO$_3$ may suffer dissolution. However, carbonate cements are uncommon even in warmer climates and circumstances favouring the formation of these minerals are, surprisingly, not well understood (Morse and Mackenzie, 1990). Apart from CaCO$_3$ supersaturation, Tucker and Wright (1990) identified the following pre-requisites for the formation of carbonate cements on the modern seabed:

- the availability of a suitable substrate;
- a lack of mechanical abrasion;
- high rates of water exchange between the seawater and pore water;
- oxygenated pore water;
- time.

It seems that kinetic factors, some types of organic matter, and dissolved phosphate inhibit precipitation.

High-magnesium calcite and aragonite

The shells of many organisms and the majority of modern carbonate sediments are composed of calcite or aragonite, respectively the rhombohedral and orthorhombic polymorphs of CaCO$_3$. In cements, acicular (needle-like), botryoidal (grape-like) and micritic (microcrystalline) forms are most common. Precipitation may be summarised as shown in Equation (9.1).

$$Ca^{2+} + 2HCO_3^- \rightarrow CaCO_3 + H_2O + CO_2 \qquad (9.1)$$

Magnesium (Mg) may substitute for Ca in the crystal lattice; those with > 4 mole% MgCO$_3$ are termed 'magnesian calcite' or 'high-magnesium calcite'. Generally the Mg content decreases with decreasing temperature, and increasing water depth, so shallow-water marine calcites are generally high-magnesium calcites. Tucker and Wright (1990) stated: '*It is far from clear as to what controls whether aragonite or magnesian calcite is precipitated*'. However, the precipitation rate of aragonite increases with increasing temperature, so aragonite tends to dominate in warmer waters. Also, it seems that high sulphate and low phosphate concentrations, a high rate of the supply of CO$_3^{2-}$ ions, and well-oxygenated conditions favour the precipitation of aragonite over high-magnesium calcite; high-phosphate, low-sulphate conditions are more favourable

for high-magnesium calcite (Tucker and Wright, 1990, Jørgensen, 1992a,b; Burton, 1993; Peckmann *et al.*, 2001).

Dolomite

Dolomite, CaMg(CO$_3$)$_2$, is one of the most common sedimentary carbonate minerals, however its mode of formation remains 'controversial' (Morse and Mackenzie, 1990). Dolomite is most commonly found as a replacement for calcite or aragonite; that is, the original mineral is 'dolomitised' by magnesium-rich pore waters:

$$2CaCO_3 + Mg^{2+} \rightarrow CaMg(CO_3)_2 + Ca^{2+} \quad (9.2)$$

For kinetic reasons, dolomite rarely precipitates out of seawater, aragonite and high-magnesium calcite being precipitated in preference. When it does, it is from the carbonate anion (CO$_3^{2-}$) rather than the hydrogencarbonate anion (HCO$_3^-$).

$$Ca^{2+} + Mg^{2+} + 2\left(CO_3^{2-}\right) \rightarrow CaMg(CO_3)_2 \quad (9.3)$$

Conditions for dolomite formation usually result from seawater evaporation or dilution, increased temperatures and phosphate concentrations, lowered sulphate content, and anoxic conditions (Tucker and Wright, 1990, Jørgensen, 1992a,b). Microbial activity in methane-rich sediments lowers the sulphate concentration and, by the production of the carbonate and bicarbonate anions, raises the alkalinity, making the conditions more favourable for dolomite formation. According to Morse and Mackenzie (1990), dolomite formation has been reported from organic-rich sediments. Under these conditions calcite precipitation is favoured, but this reduces the Ca^{2+} concentration. The resulting change in the Mg^{2+}:Ca^{2+} ratio progressively improves conditions for dolomite formation.

9.2.4 Other modern authigenic carbonates

From the preceding section it is clear that authigenic carbonate cements, with mineral assemblages like those of MDAC, are not necessarily associated with methane seeps. Both Tucker and Bathurst (1990), and Morse and Mackenzie (1990) identified three principal situations in which carbonate cements may form:

- in voids in biogenic carbonates, especially reefs;
- on the exterior of carbonate particles (leading to the formation of hardened pellets, grapestones, crusts, hardground, and beachrock);
- as micritic cements associated with boring algae.

In each case the dominant cement minerals are high-magnesium calcite and aragonite, although it seems that aragonite cement is more likely to occur on preexisiting aragonite, and is more likely to form in high-energy environments (where endolithic algal filaments help to bind the sediment and provide sites for precipitation). It is thought that hardgrounds form close to the seabed in high-energy environments as large volumes of seawater, required to provide calcium and carbonate ions, are only available in such situations.

Reefs

Cementation is an important process in the formation and stability of reefs, being partly responsible for steep, wave-resistant reef profiles (Tucker and Wright, 1990). High-magnesium calcite is the dominant carbonate cement mineral, although aragonite also occurs in reefs. Morse and Mackenzie (1990) suggested that there is considerable variety in the openness of reef structures, and hence in the degree to which interstitial waters can flush through them. Poorly flushed reefs are more likely to have heterogeneous porewater compositions, and therefore the degree of cementation may vary within an individual reef, as well as from reef to reef. 'Some may be extensively cemented whereas others may exhibit scant cement development' (Morse and Mackenzie, 1990). According to Tucker and Wright (1990) cementation is concentrated where the flux of seawater is high, for example due to wave and tidal pumping, or upwelling currents. Cements are also proportionally more important where reef growth is slower.

We consider reefs and reef-like structures in Section 9.2.11.

Beachrock

'Beachrock', formed by the cementation of beach sediments near high-tide level, normally in tropical climates, occurs as a hard rock-like formation which may prove painful to the feet of unsuspecting bathers as they run into the sea. Tucker and Wright (1990) stated that formation 'mostly takes place below the beach surface', where the sediment grains are not being moved, but seawa-ter is constantly being flushed through. According to Lewis and McConchie (1994) rainwater or groundwa-ter may play a role in the process. However, Hanor (1978) demonstrated that, at a study site in the US Virgin Islands, aragonite precipitation occurred as a result of the degassing of carbon dioxide (CO_2). Tucker and Bathurst (1990) concluded that 'most beachrock cements are precipitated through evaporation and CO_2 degassing of seawater, although microbial effects may also be important, especially with regard to micritic cements'.

Tucker and Wright (1990) reported that beachrock can also be formed in temperate regions (they specifi-cally mentioned the southwest UK, but did not identify any sites). Apparently these are 'mostly cemented by low-Mg calcite precipitated from meteoric waters at the back of the beach'.

The lithified seabed of the Arabian Gulf

Large areas of seabed in the Arabian Gulf are covered in a rock-like 'hardground', known locally as 'faroush'. It occurs in intertidal and supratidal zones, lagoons, and in water depths to 30 m (Taylor and Illing, 1969; Shinn, 1969) over an area estimated by Shinn to cover about 70 000 km^2. The principal cementing minerals are aragonite (acicular needles and fibrous) and high-magnesium calcite. Shinn described polygonal frac-tures, buckling, and tepee structures. He also observed human artefacts embedded in the cement, and provided ^{14}C ages of up to 8000 BP demonstrating that these are indeed Holocene in age. Despite the close geographical relationship between these hardgrounds and petroleum reservoirs, carbon isotope data (δ^{13}C +2.2 to +4.5‰) demonstrate that these are 'normal marine carbonates'.

Brine-lake carbonates

Carbonates from the Nadir Brine Lake and a brine seep on the Napoli Dome (both eastern Mediterranean mud volcanoes) are seep-related, however they are excep-tional as they are composed 'entirely of low-Mg cal-cite' (Aloisi et al., 2000). The explanation for this lies in the composition of the brines. The Mg^{2+} concen-trations of the Nadir Brine Lake were found to be up to ten times less, and Ca^{2+} concentrations slightly higher than normal Mediterranean seawater, whilst brines seeping from Napoli Dome were depleted in sulphate by a factor of three with respect to nor-mal Mediterranean bottom water. In these conditions low-magnesium calcite precipitation is not inhibited, and dominates over the precipitation of aragonite and high-magnesium calcite (Aloisi *et al.*, 2000). These exceptional seep-related carbonates effectively confirm the explanation for the precipitation of high-magnesium calcite and aragonite under 'normal' seep conditions. Similarly, dolomite nodules concentrated in the same brine-seep areas demonstrate that sulphate-poor conditions are favourable for this mineral (Aloisi *et al.*, 2000).

9.2.5 Isotopic indications of origin

Carbon isotopes

Figure 5.7 shows that 'normal marine carbonates', which derive their carbon from seawater or pore water, generally have carbon isotope ratios within the range $\delta^{13}C -7$ to $+8$‰ (see Section 5.5). In contrast methane is significantly more depleted in ^{13}C, so carbonates that derived their carbon from methane (MDAC) have $\delta^{13}C$ values in the range -60 to -20‰. The considerable range of $\delta^{13}C$ values for MDAC in part reflects the range of $\delta^{13}C$ values in the 'parent' methane, but, according to Belenkaia (2000), it is also related to the '*intensity of gas flux and to the mineralogical diversity of carbonates*'. However, the formation process does not discriminate, but makes use of whatever carbon is available, provided the conditions are suitable for formation. This was shown by Hackworth and Aharon (2000) who investigated hydrocarbon-derived authigenic carbonates produced as a result of gas hydrate sublimation in the Gulf of Mexico. They demonstrated that the carbonates included carbon from seawater ($\delta^{13}C +0.6$‰) mixed with carbon from methane ($\delta^{13}C -45$‰) and higher hydrocarbon gases (ethane, propane, butane, and pentane: $\delta^{13}C -26$‰). Carbon-14 analyses showed that the carbonates contain 62–90% 'dead' carbon derived from hydrocarbons.

Oxygen isotopes

Oxygen has three stable isotopes, ^{16}O, ^{17}O, and ^{18}O. Although the vast majority (99.76%) of naturally occurring oxygen is ^{16}O, the ratio between ^{16}O and ^{18}O provides a sensitive indication of temperature. Oxygen-isotope ratios are reported by comparison to a standard.

$$\delta^{18}O‰ = \frac{[^{18}O/^{16}O_{sample} - ^{18}O/^{16}O_{standard}]}{[^{18}O/^{16}O_{standard}]} \times 1000 \tag{9.4}$$

The most commonly used standard is SMOW (Standard Mean Ocean Water).

Analyses of the $\delta^{18}O$ of MDAC indicate the temperature, and therefore the origin, of the water at the time of carbonate precipitation. Aloisi *et al.* (2002) equated $\delta^{18}O$ values of $+3.7$‰$_{SMOW}$ and $+1.5$‰$_{SMOW}$ for calcite and dolomite, respectively, to precipitation in seawater at 3 °C.

Generally, MDAC has $\delta^{18}O$ values indicating that they formed in pore water closely connected to seawater. However, variations have been reported, for example:

- Aloisi *et al.* (2000) reported $\delta^{18}O$ values of $+4.66$ to $+7.05$‰ from MDAC associated with brine seepage from mud volcanoes in the eastern Mediterranean;
- Belenkaia (2000) reported values as low as -9.45‰ from MDAC associated with gas hydrates;
- Naehr *et al.* (2000) reported values of -8.5 to $+6.8$‰, and suggested that these show the influence of meteoric waters and gas hydrates in MDAC from Monterey Bay, California.

9.2.6 MDAC formation mechanism

Until relatively recently it was accepted that seep-associated carbonates derived their carbon from methane because of the carbon-isotope ratios. It was accepted that oxidation of methane results in supersaturation of $CaCO_3$, whereby aragonite or calcite is produced by the chemical reaction approximated as in Equation (9.5),

$$CH_4 + 2O_2 \rightarrow CO_2 + 2H_2O \tag{9.5}$$

or by the action of sulphate-reducing bacteria in the anoxic subsurface zone.

$$SO_4^{2-} + CH_4 \rightarrow SO_3^{2-} + H_2S \tag{9.6}$$

This is in accordance with Jørgensen (1976), and the assertion of Reeburgh (1980) that hydrogencarbonate (bicarbonate) generation during this process resulted in pore waters being oversaturated with respect to $CaCO_3$, which then precipitated. However, the biochemical processes leading to its formation were not understood. The riddle was solved with the discovery of AOMs and the explanation of anaerobic oxidation of methane (Boetius *et al.*, 2000; see Section 8.2.1 and Equation (8.1)). Now, there seems to be compelling evidence that MDAC forms as a result of the combined effect of anaerobic oxidation of methane and sulphate reduction (i.e. the activities of AOMs) in anoxic sediments immediately below

(a)

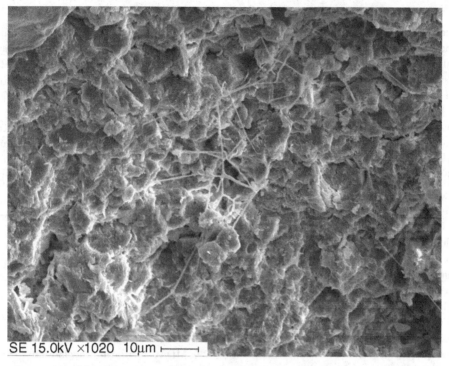

SE 15.0kV ×1020 10µm ┣━━━━━┫

(b)

DECV 5 25.0kV X30.0K 1.00µm

Figure 9.2 Scanning electron microscopy images of an MDAC chimney collected on the Iberico Mud Diapir in the Gulf of Cadiz. The authigenic carbonate cement is composed of high-magnesium calcite and dolomite micrite to microsparrite rhombs. The filamentous structures are interpreted as microbial filaments; some of those in (b) are being calcified and flattened against the dolomite crystals. (Images courtesy of Vitor Magalhães, University of Aveiro, Portugal.)

the seabed – summarised by Equation (8.1). The production of hydrogencarbonate (HCO_3^-) increases alkalinity and enables the precipitation of $CaCO_3$. There is evidence that the AOMs can be surrounded by calcite, indicating their role in the process (Boetius, 2002, personal communication). This shows how carbon from methane is included in carbonate minerals, and is confirmed by carbon isotope data.

Local conditions may favour the precipitation of either aragonite or high-magnesium calcite (see Section 9.2.4). It seems likely that the balance between various factors (temperature, methane supply, rates of sulphate reduction and anaerobic oxidation of methane, and the supply of oxygenated water) impose a delicate control on the mineral type(s) formed at a seep, and that temporal variations may result in changes at individual sites. For example, when sulphate concentrations are depleted by the activity of SRBs (sulphate-reducing bacteria), calcite will precipitate in preference to aragonite; variations in the magnesium content of individual calcite crystals were attributed, by Jørgensen (1992a,b), to variations of methane supply. Following a suggestion from Peckmann *et al.* (2001) we envisage that aragonite forms where the supply of seawater is better (perhaps closer to the seabed), so sulphate concentrations are higher and there is more oxygen. Luff *et al.* (2004) noted that burrowing organisms increase the flow of seawater through sediments, so 'bioirrigation' may also affect the aragonite/calcite balance.

Biomineralisation

Belenkaia's study of MDAC collected during TTR cruises in the Black Sea, the eastern Mediterranean, the Moroccan margin, and the Vøring Plateau enabled her to compare and contrast MDAC from a variety of settings. She reported that 'fossilised' bacteria are '*characteristic*' of MDAC, and that they are '*clearly visible*' with both optical and scanning microscopes; similar examples are shown in Figure 9.2. It seems that a number of species of bacteria play a prominent role in carbonate precipitation by attracting Ca^{2+} (and Mg^{2+}) during sporulation. Often, bacteria and newly formed minerals are enveloped in a mutual secretion of mucus, the remains of this biofilm are recognisable in some carbonates when they are etched with hydrochloric acid (HCl), even if bacteria are not visible (Belenkaia, 2000). Carbon isotope ratios are diagnostic of MDAC, and this

evidence of the association of biomarkers and bacteria held within MDAC conclusively demonstrates the role of microbes in their formation.

Carbonate minerals are probably not the only ones associated with biomineralisation. Framboidal aggregates of haematite (Fe_2O_3), rimmed by ankerite – $CaFe(CO_3)_2$, were described by Somoza (2001) from MDAC chimneys from the Gulf of Cadiz. He suggested that they were associated with the activity of AOMs. In previous sections we mentioned that phosphate concentration might influence which carbonate mineral precipitates. It seems that phosphate concentration is itself linked to the microbial processes. Phosphate concentrations of 500 ppm in aragonite cements, and < 3500 ppm in pure microcrystalline carbonates were measured in samples from the Black Sea by Peckmann *et al.* (2001). They found organic residues preserved in carbonates, indicated by autofluorescence. The intensity of the fluorescence was correlated with phosphate concentration, suggesting that phosphate is largely derived by the microbial breakdown of organic matter. So, another circle is squared; microbial activity affects phosphate content, which, in turn, affects the species of mineral that precipitates.

Biomarkers

The association between AOMs and MDAC is conclusively supported by the identification of fossilised microbes and biomarkers in MDAC. For example, in MDAC samples taken from the western Black Sea, Thiel *et al.* (2001) found biomarkers (isotopically depleted crocetane, PMI/PMIΔ, ether-bound biphytane, and cyclic C_{40}-isoprenoids) derived from benthic archaea that incorporate ^{13}C-depleted methane carbon in their biomass. These authors also referred to other publications that reported the presence of isotopically depleted archaeal biomarkers related to anaerobic methane oxidation in the Kattegat, Eel River Basin, Hydrate Ridge, and the Aleutian Subduction Zone; fossil biomarkers have also been identified (see Table 9.1).

9.2.7 Associated minerals

Local variations in the chemical composition of pore waters seem to have an impact on the formation of

Table 9.1 *Biomarker evidence of fossil seeps: isotopically-depleted biomarkers thought to be related to anaerobic methane oxidation reported from methane-derived authigenic carbonates. Figures are $\delta^{13}C$ values (‰PDB). Data derived from Thiel et al., 2001.*

Location	Age	Crocetane	PMI	Diethers	Source
Marmorito, Italy	Miocene	−116	−112	−108	Thiel *et al.*, 1999
Lincoln Creek, USA	Oligocene	−112	−120	Not reported	Thiel *et al.*, 2001
Beauvoisin, France	Jurassic	Absent	−76	Not reported	Peckmann *et al.*, 1999

various minerals in association with MDAC. Three examples are presented here, but there may well be more.

Pyrite (FeS₂)

The production by the AOMs of HS^- – Equation (8.1) – leading to the formation of hydrogen sulphide (H_2S), is responsible for anoxic, sulphurous sediments characteristic of seeps, as well as accounting for sulphide-supported microbes (e.g. *Beggiatoa*) and organisms hosting sulphide-oxidising endosymbionts described in Section 8.2.2. The production of H_2S also explains the frequent presence of pyrite in MDAC.

Dissolved hydrogen sulphide reacts with iron in oxygen-poor seabed sediments to precipitate iron monosulphides. These precede the formation of pyrite (FeS_2). The pyrite found in, or in association with MDAC is framboidal (raspberry-shaped) in form. Wilkin and Barnes (1996, 1997) undertook detailed studies of the formation of framboidal pyrite in both natural and laboratory environments. They noted that it is more or less always formed where sulphidic waters are saturated, or slightly undersaturated with respect to monosulphides, and saturated with respect to pyrite. Chemical gradients are normally steep in the immediate vicinity of microbial sulphate reduction, so reactions leading to pyrite formation from monosulphides can be expected to occur very close to microbial activity (i.e. where AOMs are operating). Furthermore, framboidal pyrite formation is *'spatially linked to redox interfaces that separate waters containing dissolved oxygen and sulfide, respectively'* (Wilkin and Barnes, 1997).

It seems that pyrite formation is achieved via *'progressive conversion of the thermodynamically unstable iron monosulfides'* $Fe_{1+x}S$ and Fe_3S_4 (Wilkin and Barnes, 1997). This evidently occurs very rapidly: *'iron monosulfides convert to pyrite nearly as soon as they form'* (Wilkin

and Barnes, 1997). They reported that this process could be summarised with the following half-reactions and coupled redox reaction.

$$Fe_3S_4 + 2H_2S \rightarrow 3FeS_2 + 4H^+ + 4e^- \quad (9.7a)$$
$$4H^+ + 4e^- + 2S^0 \leftrightarrow 2H_2S \quad (9.7b)$$
$$Fe_3S_4 + 2S^0 \leftrightarrow 3FeS_2 \quad (9.7c)$$

According to Wilkin and Barnes (1996), the sulphur-isotope ($\delta^{34}S$) composition of pyrite derived by this process is identical to that of the iron monosulphide precursor, although subsequent 'dilution' by later pyrite formation due to other surface processes may account for differences between the $\delta^{34}S$ of the pyrite and that of the original H_2S. Sulphur isotopes can therefore distinguish between pyrite formed as a result of seepage, and pyrite preexisting in the sediment. For example, Peckmann *et al.* (2001) concluded that pyrite in MDAC-cemented sediments in the Black Sea formed before seepage activity started because of the discrepancy between the $\delta^{34}S$ values of the H_2S (−38.7 to −4.9‰) and those of the framboidal pyrite (+16.8 to +19.7‰).

Although framboidal pyrite seems to be common in methane seep environments, Wilkin and Barnes (1997) pointed out that it is formed in other modern environments too. For example: anoxic environments where fine-grained sediments are deposited in marine and lacustrine environments, salt marshes, and in the water column of anoxic basins. Although they noted that pyrite framboids are also found in sedimentary rocks, for example, shales, carbonates, and coal, and in hydrothermal veins, we suggest that at least some of these ancient occurrences can be explained by the formation processes described above, that is as a result of microbial activity. Of course, the availability of iron is an essential prerequisite to pyrite formation.

Gypsum (CaSO₄. 2H₂O)

Gypsum (CaSO$_4$. 2H$_2$O)

Large, lenticular bodies of gypsum were found 1–3 m below the seabed in Recent carbonate-cemented sediments close to the shore of the Kattegat, Denmark (Jørgensen 1980). They were up to 75 cm long and 12–30 cm thick, lying subparallel to the sediment layers; '*like concretions rather than interstitial cement*' (Jørgensen, 1980). Sulphur isotope studies showed these gypsum bodies (δ^{34}S +27 to +33‰) are heavily enriched in ^{34}S compared to normal seawater (δ^{34}S about +20‰). The presence of MDAC demonstrates the flow of methane through these sediments. Jørgensen attributed these gypsum concretions to precipitation from sulphate-rich waters derived, like the methane, from Quaternary reservoirs. Jørgensen (1981) described authigenic potassium feldspar (adularia) enclosed within the gypsum.

Barytes (BaSO₄)

Barytes (BaSO$_4$)

Examples of barytes precipitation associated with cold seeps include the following.

- A cluster of reddish-brown-stained barytes chimneys found in an area of mud volcanoes and associated hydrocarbon seeps in a 400 m² area at 510–522 m water depth in the Garden Banks area of the Gulf of Mexico (Fu *et al.*, 1994).
- Torres *et al.* (1996) discovered light yellow to brown barytes in the form of concretions and chimneys (<15 cm high), and white crusts (a few millimetres thick) associated with communities of clams, serpulae, and venting fluids along a scarp failure on the Paita middle slope and on the walls of the Chiclayo Canyon on the Peru Slope. The white crusts were composed of '*very pure barium sulfate crystals*'.
- Barium-rich fluids vent at 1500–2000 m water depth over a distance of > 3 km along the San Clemente Fault in the San Clemente Basin, offshore California (Torres *et al.*, 2002). Some of the chimneys are 10 m tall (Lonsdale, 1979).
- Barium- and methane-rich fluids venting in an area of several square kilometres in the deeper (1500 m) parts of the Derugin Basin, Sea of Okhotsk, produced barytes 'build-ups' < 10 m high. Cold-seep communities of bivalves (*Calyptogena* sp., *Solemya* sp.) and gastropods were present near the seeps (Suess *et al.*, 1999a; Greinert *et al.*, 2002b). The barytes was found in the form of '*chimneys several meters high, travertine-*

like boulders, blocks, and mounds consisting of pure barite, barite sand and barite silt turbidites'.

- In the Guaymas Basin, Gulf of California, barium-rich fluids are somewhat different from those mentioned above because of the hydrothermal associations. Nevertheless, the venting fluids are 'cool' (< 100 °C). Lonsdale and Becker (1985) described '*forests of barite spires*'. Extensive '*carpets*' of *Beggiatoa* extend beyond the active vent sites, but suggest the presence of '*baked petroleum-saturated sediment with veins of barite*'.

The source of the barium in the venting fluids is a mystery. Fu *et al.* (1994) suggested a link with underlying salt deposits, but deep circulation of fluids seems more likely. Greinert *et al.* (2002a) concluded that barium enrichment occurred at a depth of 900 to 1800 m below seabed in the Derugin Basin. This seems to be compatible with the view of Torres *et al.* (1996, 2002) who proposed that non-detrital BaSO$_4$ is '*remobilized by sulfate depletion, coupled with a low-temperature hydrodynamic regime of fluid flow through the sediments and venting of these fluids at the seafloor*'. When these fluids encountered sulphate-bearing pore-fluids beneath the seabed, barytes precipitated in the sediments. Such deposits were found during ODP investigations at Site 684 (Torres *et al.*, 1996).

9.2.8 MDAC chimneys

We have seen that MDAC can occur in various shapes and sizes, however, some of the most intriguing features are seep-associated carbonate chimneys. They have been reported from the Kattegat (see Section 9.2.2), offshore Oregon (Ritger *et al.*, 1987; Kulm and Suess, 1990), Nankai Trough (Sakai *et al.*, 1992), Gulf of Mexico (Roberts and Aharon, 1994), off New Zealand (Orpin, 1997), Eel River Basin and Monterey Bay, California (Stakes *et al.*, 1999; Orange *et al.*, 2002), Black Sea (Peckmann *et al.*, 2001; Treude *et al.*, 2002), and Gulf of Cadiz (Díaz-del-Río *et al.*, 2003); see also Aeillo and Garrison (website) for a description of fossil examples. They occur as free-standing structures (for example, see Figure 3.13), but there are also many lying on the seabed – presumably these were standing, but fell over.

Some of the most spectacular chimney structures are found in the Gulf of Cadiz, where there are also mud volcanoes, pockmarks, and blowout structures.

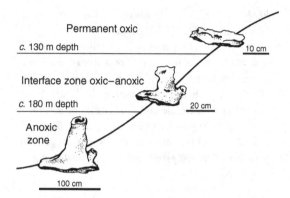

Figure 9.3 The variation in the shape and dimensions of MDAC on the shelf and slope of the northwestern Black Sea. This is determined by the chemistry of the bottom waters; MDAC in the oxic zone is confined to the sediments, but in the anoxic zone it is able to grow above the seabed. (Reproduced from Luth *et al.*, 1999 with permission from Blackwell Publishing.)

There is a variety of forms: spiral, cylindrical, conical and mounded, with numerous branches, protuberances, bifurcations, and mushroom-like structures. They are mostly composed of authigenic carbonates (ankerite, iron-bearing dolomite, and calcite) with abundance of iron oxides, forming agglomerates of pseudopyrite framboids with $\delta^{13}C$ values ranging from -46 to -20‰ (Somoza, 2001). Díaz-del-Río *et al.* interpreted this as evidence of a mixed deep-thermogenic and shallow-microbial carbon origin.

The effects of anoxia: examples from the Black Sea
Due to its peculiar oceanography the Black Sea offers a unique opportunity to study the influence of chemical environments on carbonate precipitation at methane seeps. It seems that the morphological type of carbonate varies across the oxic–anoxic boundary. Peckmann *et al.* (2001) described the types illustrated in Figure 9.3 as follows:

- 60 m water depth (oxic zone): flat, pancake-shaped crusts, 20–30 cm in diameter, confined to anoxic sediments and have no seabed relief; most have conspicuous, central, 1–2 cm wide channels through which gas has been observed to seep during ROV studies (Luth *et al.*, 1999);
- 110–130 m: carbonates thicker than in shallower water; they are porous, and cavity walls are lined with microbial films;

- 190 m (anoxic zone): irregular small chimneys, some of which stand on a flat <20 cm thick, 1 m diameter plate;
- 230 m (anoxic zone): chimneys, < 1 m tall and penetrated by active seeps, protrude from thick platforms.

These carbonates are clearly methane-related; $\delta^{13}C$ values as low as -41‰ were reported by Peckmann *et al.* The main carbonate mineral in the chimneys is aragonite, but high-magnesium calcite is also present.

Pink-brown microbial mats were found on carbonates at all depths, but they are thicker (up to several centimetres) in anoxic waters where they cover the outer surfaces, as well as the walls of channels and cavities. Peckmann *et al.* reported that samples from 190 m had been studied by Pimenov *et al.* (1997) who found the dominant microbes to be: '*very similar in ultrastructure to methanogenic archaea of the genus* Methanosaeta *(formerly referred to as* Methanothrix*)*'. They also demonstrated in the laboratory that these mats were able to perform methane oxidation and sulphate reduction under strictly anaerobic conditions. The fact that these mats are so extensive is probably explained by the absence of grazing organisms, which, like other seep-associated fauna, seem to be excluded from these inhospitable anoxic waters. The relationship between the carbonates and microbial activity is discussed in relation to Black Sea 'jellybabies' in Section 8.2.1 (see also Treude and Boetius, website).

Chimney formation above or below seabed?
The fact that MDAC chimneys are only found in the anoxic (deep) portion of the Black Sea may suggest that the anoxic environment allows the carbonate formation process to continue above the seabed. However, chimneys at other locations mentioned at the start of this section occur where seawater is not anoxic. Did they form in seawater and build up into the water during fluid-flow activity, or were they formed inside sediments around or as part of fluid-flow conduits, and subsequently became exposed by the erosion of seabed sediments? It is tempting to support the first mechanism. Apart from those in the anoxic Black Sea, chimneys seem to be dominated by dolomite or ankerite rather than aragonite or calcite, unlike crusts and other forms of MDAC. Perhaps particularly vigorous methane seepage favours the formation of dolomite, and enables it to grow beyond the anoxic sediments. If this were true, the chimneys

would be composed purely of freshly precipitated minerals. However, it seems that even the spectacular Gulf of Cadiz chimneys are formed by the cementation of preexisting sediment:

> Authigenic carbonates occur as very fine grained crystals (micrite) and as crystalline coatings over terrigenous grains. The terrigenous components, silt to sand sized, are not in contact with each other and are: quartz, feldspars, phyllosilicates (kaolinite, chlorite and muscovite) and minor oxides (Ti-oxides) and phosphates (apatite and monazite).
>
> Díaz-del-Río et al., 2003

Evidently, they are composed of preexisting sediments cemented by newly formed carbonates. In the previous discussion of MDAC formation processes it became clear that formation occurs beneath the seabed, and probably at or beneath the oxic–anoxic boundary. If this is true, how can there be MDAC chimneys standing above the seabed? The chimneys of the 'Bubbling Reefs' in the Kattegat are thought to have been exposed by the erosion of seabed sediments (see Section 9.2.2). Consequently we must favour the hypothesis that these, and other chimneys made out of cemented sediments, were formed beneath the seabed and have since been exposed. Loncke *et al.* (2004b) presented an intriguing explanation from the Nile Fan. It seems that where MDAC is forming whilst the seabed is being eroded, existing MDAC becomes exposed above the seabed. Concurrently, the sulphate–methane transition zone (SMTZ) progressively moves lower in the sediment, so new MDAC is formed at the *bottom* of existing MDAC. If erosion continues, this process might lead to the formation of columns in which the oldest carbonate is at the top, not the base.

9.2.9 Self-sealing seeps

Surveys of the prolific Tommeliten seeps in the North Sea, suggest that hydrocarbon seeps may seal themselves by a combination of carbonate precipitation and utilisation by organisms. When the seeps were resurveyed in 1998 and again in 2002, the bioherms identified during our 1983 ROV survey (Section 2.3.2) turned out to be located on top of large, flat slabs of MDAC. Other features mapped during these surveys clarified that three stages of seepage can be distinguished:

- new seeps, where gas comes directly from small vents in the sandy seabed;
- bacterial mats (probably *Beggiatoa* sp.) where gas accumulates in anoxic sediments below a thin layer of bacterial mat before occasionally venting through holes in the mats;
- MDAC 'bioherm' structures where no visible gas bubbling is evident. Sampling of carbonate nodules suggests that gas is migrating up to the lower part of structures within sediments, where utilisation results in the formation of carbonate cements.

These results suggest that the formation of bacterial mats may represent the first phase of natural sealing. The formation of bioherms, which host numerous sessile and filter-feeding organisms, represents the final phase in the natural seep-sealing process (Hovland, 2002).

Pore water is not normally anoxic in porous nearseabed sediments because of the circulation of oxygenated seawater. However, where light hydrocarbons (methane, ethane, propane, and butane) are present, perhaps partly dissolved in advecting pore water and partly in a free gas phase, the activity of AOMs – Equation (8.1) – results in anoxic conditions and hydrogen sulphide production. When this occurs inside the migration pathways we think microbes colonise oxylimnic parts of the plumbing. If this occurs it will affect the hydraulic conditions of sediments, impeding the flow of free gas, and (as pointed out by Luff *et al.*, 2004) separating sulphate-rich seawater from rising methane. Some gas will be retained inside conduits and free gas will probably start invading neighbouring sediments because of an increased relative pressure. When local gas pressure exceeds the holding capacity of the filaments of the bacterial mesh, gas will break out and vent through the mat. The initial seep stage is now over, and intermittent seepage occurs through the bacterial mat (Hovland, 2002).

The first sign of a bacterial mat on the surface will probably follow this stage. When all oxygen has been utilised, the oxylimnion probably migrates from around the conduits into the subseabed sediments, and gradually moves upwards to the sediment surface where it can only expand horizontally. When the seep throats are infested with bacteria, the rate of gas ebullition is expected to decrease, causing them to fill gradually with fine sediments. This will encourage the seep to change into one of diffusive flow, whereby most of the escaping gas molecules dissolve in pore water. This gas-charged (and anoxic) pore water will invade the sediments adjacent to the original conduit, so a larger area of seabed is affected by advecting anoxic pore water.

In the last phase carbonate precipitates, progressively blocking the migration pathway until no more visible ebullition occurs. The seepage is effectively sealed, and a bioherm structure develops. This process accounts for the formation of carbonate crusts found in various marine geological settings where seeps occur (Hovland, 2002).

9.2.10 MDAC: block formation

Methane-derived authigenic carbonate clearly occurs in many shapes and sizes: in chimneys, crusts, and pavements, but also sometimes in blocks or lumps (e.g. Figures 2.28 and 2.29). We know this because MDAC is normally observed lying on the seabed, although it is sometimes partially obscured by a veneer of sediment (such as in Figure 2.43). Yet, as discussed above, formation seems to occur beneath the seabed (except in anoxic waters, such as those of the Black Sea). So, how does it come to be lying on the seabed?

Peckmann et al. (2001) described evidence of physical deformation or cracking in MDAC from the western Black Sea. Cracks were found to be 'ubiquitous' in botryoidal aragonite, apparently starting in the centres and narrowing towards the edge of the botryoids. They are filled by microcrystalline aragonite. Although they considered desiccation as a possible cracking mechanism they favoured brecciation, suggesting 'a possible factor triggering brecciation is the gas seepage itself'. They thought that excess gas pressures created when conduits became blocked by carbonate might cause sufficient excess pressure to crack the mineral; are the 'self-sealing seeps' (described in the previous section) capable of this? Perhaps gas will divert around such blockages in relatively coarse sediment (as at Tommeliten), whereas in impervious sediments diversion is not possible, and excess gas pressure builds up. This may be a factor in pockmark formation (see Section 7.6.3). Is it realistic to suggest that seeps can be sealed so effectively that natural gas 'blowouts' occur, and that these are sufficiently powerful to lift large lumps of MDAC, and possibly even break them?

9.2.11 Carbonate mounds

A 'carbonate mud mound' was defined by Riding (2002) as a 'carbonate mud-dominated (micrite and fine silt) deposit with topographic relief and few or no stromatolites, thrombolites or in place skeletons'. He distinguished between 'low-relief' and 'high-relief carbonate mud mounds', the former having a relief of less than 5 m, the latter, more than 5 m. Deep-water coral reefs and carbonate banks, described in Section 8.1.6, are considered by some to be strictly biogenic, but we believe some at least to be related to mineralised fluid flow. As such they are not strictly biogenic, but dependent on the substratum and on fluids seeping through the sediments.

After the publication of a paper entitled 'Do carbonate reefs form due to fluid seepage?' (Hovland, 1990b), the author received two remarkably similar two-dimensional (2D)-seismic records showing some mounds on the seabed. One came from Peter Croker, a petroleum geologist from Ireland, and the other came only a few days later from Mike Martin, a consultant geophysicist from Western Australia. The result was another paper describing and discussing fault-associated carbonate mounds off western Ireland and northwest Australia (Hovland et al., 1994). This paper described 150 m high mounds and speculated that they formed because of hydrocarbon seepage.

In 1995, several academic and industry surveys were conducted in the Porcupine Basin and the Rockall area. These and subsequent surveys identified not only the 'Hovland Mounds', but also similar reef occurrences named after the vessels used to find them, the 'Logachev Mounds', the 'Belgica Mounds', and the 'Magellan Mounds' (Kenyon et al., 1998; Henriet et al., 1998; Croker and O'Loughlin, 1998; Hovland and Mortensen, 1999; De Mol et al., 2002; van Weering et al., 2003a; Henriet et al., 2003). Although Lophelia colonies are found to grow on the summits of these mounds (Henriet et al., 1998; De Mol et al., 2002), they are called carbonate mounds because their base and main body consist of carbonate sediments. The largest structures are the 'giant carbonate mounds' along the Porcupine and southwest Rockall Trough margins. According to van Weering et al. (2003a), these range in size and shape; some are 'steep pinnacles rising up to 350 m above the seabed with a diameter of 2–3 km', others occur in 'large, slightly less high, irregular shaped clusters with a diameter of up to 6 km. The total width of the clustered complex is about

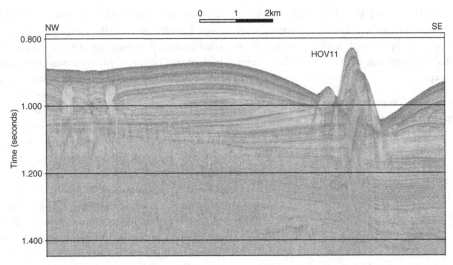

Figure 9.4 High resolution 2D-seismic profile across two buried Magellan Mounds (left) and a Hovland Mound (right) in the Porcupine Basin, offshore Ireland. This Hovland Mound is at least 120 m high and about 900 m across at its base, and it is located inside a seabed depression ('moat'). Also note the intrasedimentary structures beneath the mound; some are artefacts caused by lateral variations in water depth, sediment characteristics, and seismic velocities. (Courtesy of Ben De Mol, Renard Centre of Marine Geology, Ghent University; reproduced from De Mol *et al.*, 2004 with permission of the Oceanography Society.)

15 km. Those lying higher up the slope are '*lower in height, more isolated and are locally buried under a relatively thin sediment cover*'. These giant structures sit on normal marine and glacigenic sediments overlying the basalts of the Rockall Trough margins. Although there are faults and even reverse-polarity acoustic reflectors under some of these structures, no direct indicators of seepage have been documented.

The smaller Belgica, Hovland, and Magellan mounds occur on top of sediment drift and palaeo-slide sediments (Henriet *et al.*, 2003). The Hovland Mounds (e.g. Figures 3.20 and 9.4; see also Hovland *et al.*, 1994) are associated with large, pockmark-like depressions (called 'moats' by some researchers). Henriet *et al.* (2003) described a '*prolific cluster of relative small, buried*' Magellan Mounds 50 to 90 m high and 200 to 800 m wide (e.g. Figure 9.4). The Belgica Mounds commonly reach heights over 150 m and basal widths of 1000 m. Research on these mounds has been summarised in a special issue of *Marine Geology* (van Weering *et al.*, 2003b).

We speculate that associations with flowing fluids of different compositions could explain the presence of these mounds, and the differences between the four types of mound. In the Rockall Trough (but not the Porcupine Basin) carbon dioxide, hydrogen, and methane may have evolved during the intrusion of basalt sills; Reston *et al.* (2001) suggested that hydrogen and methane may have been released by mantle serpentinisation in association with an old transform fault in this region; petroleum fluids (especially methane) may have migrated from underlying source rocks. Basin modelling by Naeth *et al.* (2005) showed that fluids migrating from mature Jurassic source rocks would have been focussed above rotated Jurassic fault blocks where Cretaceous and Tertiary sediments pinch out. They found gas chimneys at such locations in the Connemara field (see Section 3.5.4), their models predicted significant fluid focussing associated with the Belgica Mounds, and they found evidence of Tertiary pinch-outs beneath the Hovland and Magellan Mounds.

Carbonate reefs and mounds in other areas
Although we have focussed on carbonate reefs and mounds of the northeast Atlantic Margin, they are known to occur elsewhere. In the Timor Sea, north of Australia, spectacular carbonate mounds and reefs are important for two reasons: because of their association with petroleum, and because they are important marine biological habitats; Ashmore Reef, for example, is a Marine Protected Area because of its rich

biodiversity and range of endemic species (Glenn, 2002). Integrated surveys, combining sea-surface seep detection, seabed mapping, and detailed seismics, have demonstrated a strong link between these banks and petroleum seeps:

> The key linking feature between a wide range of features, from small, seafloor build-ups, to isolated carbonate shoals such as Heywood Shoals, to large clusters of banks such the Karmt Shoals, is that they are all associated with modern seeps. Where seeps are absent, so (generally) is the evidence for significant contemporary carbonate bank formation.
>
> O'Brien *et al.*, 2002

Building carbonate mounds

There are strong similarities between these Australian examples and those of the Gulf of Mexico, which we discussed because of their fauna (Section 8.1.6). As previously suggested (Hovland and Judd, 1988), this 'association' between reefs and seeps is clearly not coincidental. Roberts *et al.* (1989) reported that reefs on the Louisiana Continental Slope were composed of ^{13}C-depleted carbonate. The Gulf of Mexico carbonate banks are generally located over salt diapirs, and are associated with seeps and bacterial mats. Australian examples are associated with seeps located above faults, along migration fairways, or where a regional sealing formation pinches out (O'Brien *et al.*, 2002).

O'Brien *et al.* (2002) suggested a model which seems generally applicable to seep-associated carbonate reefs and banks. They suggested a sequential process started during a period of relatively low sea level, when the water was shallow enough for photosynthesis to occur at the seabed. First, hydrocarbon seeps were colonised by a cold-seep benthic biological community. This eventually formed a positive topographic feature, colonised by various reef-building organisms whose growth rate enabled them to keep up with rising sea level.

9.2.12 Fossil seep carbonates

In recent years several fossil seeps have been identified. Early descriptions of fossil carbonate mounds came from Monferrato, northwest Italy (Clari *et al.*, 1988; Cavagna, *et al.*, 1999 – see also Section 8.4.1) where the Miocene-age Lucina and Marmorito limestones outcrop. Cavagna *et al.* (1999) explained that these had been identified as methane-derived carbonates on the basis of field evidence (patchy cementation, chemosymbiotic fossil communities, and '*a network of polyphase carbonate-filled veins not related to tectonics*'), carbon isotopes ('*as low as −50‰*'), and '*peculiar petrographic features*'. The latter suggested the presence of '*organic clumps or mats capable of trapping sediments and promoting carbonate precipitation: microcrystalline calcite peloids; dolomite crystals with irregular hollow cores; dolomite spheroids with dumbbell-shaped cores; laminated internal sediments lining cavities completely*'. They concluded that this evidence is all consistent with microbially-mediated sedimentary and diagenetic processes that '*can therefore be considered as an additional evidence of ancient methane seepage*'.

Following the suggestions of Cavagna *et al.* (1999), and others (Belenkaia, 2000; Thiel *et al.*, 2001), we consider the following features to be diagnostic of fossil MDAC.

1. Patchy distribution, which is clearly evidence of a phenomenon occurring only in some parts of a sedimentary basin.
2. Restricted lateral extent of carbonate cements, usually within siliciclastic sediments. Boundaries may be transitional or sharp: '*in a few clear cases the percentage of cement and the abundance of chemosynthetic taxa remains gradually decrease away from the paleoseep; more commonly, however, lateral relationships of limestone masses and blocks are sharp and more enigmatic*' (Cavagna *et al.*, 1999).
3. Evidence of fluid flow pathways, in the form of networks of fractures and cavities with '*complex and polyphase infillings by carbonate sediments and cements*' (Cavagna *et al.*, 1999).
4. Diagenetic products and fabrics anomalous in clastic sediments and contrasting with those shown by the surrounding sediments. Cavagna *et al.* identified the following: '*abundant intergranular cement, in "loosely" packed, uncompacted, sandstone or mudstones; cavity-filling cements showing different mineralogies (aragonite, calcite), morphologies (botryoidal, fibrous, sparry) and commonly displaying bandings due to a variable amount of inclusions*'.
5. Associated macro- and/or microfossils (discussed in Section 8.4.1).

6. Biomarkers (see Section 9.2.6 and Table 9.1).
7. Carbon isotope ratios in the range −60 to −20‰ (see Figure 5.7).

Fossil seeps are important, not only because their recognition helps to explain the formation of the related rocks, but also, remembering the 'catch phrase' of the eighteenth century Scottish geologist, James Hutton ('*the present is the key to the past*') we can take advantage of geological exposures to examine the whole system, including the subseabed plumbing system. Some fossil cold-seep communities, discussed in Section 8.4.1, have been identified because chemosymbiotic fauna, such as tubeworms and Lucinid bivalves, have been recognised. However, the identification of MDAC has been critical in many cases. Numerous fossil 'seep carbonates' are identified in Cavagna *et al.* (1999) and Campbell *et al.* (2002); here we have selected just a few examples.

Pobiti Kamani, Bulgaria

Eocene sand and silt lithified by a carbonate cement is exposed over an area of about 50 km^2 in northeast Bulgaria. The carbonates include lithified sand layers, concretions, fracture infills, and columns. At Pobiti Kamani there are some spectacular columns 3–8 m high and 0.50–3 m in diameter; many have hollow centres, so really they are chimneys (Figure 9.5). Many are lying on the ground, but it is not known when they fell over.

The carbonate cement is composed purely of low-magnesium calcite, according to Botz *et al.* (1993). They undertook carbon and oxygen isotope analyses to find out how the carbonates formed; $\delta^{13}C$ values of −1.2 to −29.2‰ and $\delta^{18}O$ of −7.7 to −0.9‰ suggested to them that there were several diagenetic phases. The isotopes of the carbonate from the outer edge of one column ($\delta^{13}C$ −29.2‰, $\delta^{18}O$ −0.9‰) suggest that methane was the source of this carbonate, and that it formed under marine conditions. However, values ($\delta^{13}C$ −14.1 to −1.2‰, $\delta^{18}O$ −7.7 to −1.6‰) from the inner parts of the columns, and other carbonate formations suggest that these were precipitated later during diagenesis; when the sediments were buried to a depth of about 1300 m and meteoric groundwaters percolated through them.

We suggest that initial chimney formation was associated with the seepage of methane or methane-rich water. Subsequently, fluid migration pathways, encased in MDAC were used by percolating groundwater. We

Figure 9.5* Fossil carbonate chimneys at Pobiti Kamani, Bulgaria. (Photo: Alan Judd.)

noticed dense accumulations of the giant foraminifera, *Numulites*, in some columns, but we are unsure whether these have any relevance to the origin of the columns.

Ahnet Basin, Algerian Sahara

Of all unlikely places, exposed fossil analogues to deep-water carbonate mounds and Norwegian deep-water coral reefs are found in the Algerian Sahara desert. After a team of German researchers visited the Ahnet Sedimentary Basin, they published a paper with the unlikely title: '*The World's most spectacular carbonate mounds (Middle Devonian, Algerian Sahara)*' (Wendt *et al.*, 1997). As can be seen from their photograph (Figure 9.6), the features are really spectacular. The haystack-shaped mounded structures consist of lithified clay with fossils of Devonian age (380 million years BP). These fossils include rugose corals, octocorals, sponges,

Figure 9.6* Fossilised carbonate mounds in the Algerian Sahara. About 400 million years ago these giant carbonate reefs formed at a water depth of about 400 m above faults near the edge of an extensive sedimentary basin. The Landrover in the foreground provides the scale [Reproduced from Wendt et al., 1997 with permission from SEPM (Society for Sedimentary Geology)]

mussels, and trilobites. They span a period of about 200 000 years. Based on the fact that there are no fossil traces of plants or algae, the structures must have been formed in the aphotic ocean zone, with an estimated water depth of about 400 m, which is of a similar depth to those found living off Norway and Ireland today (Wendt et al., 1997).

The fact that they form in lines, and because of the presence of framboidal pyrite, barytes, and apatite in particular zones of the structures (Neptunian dykes), has led to the conclusion that they formed at the periphery of the Ahnet sediment basin over deep-seated faults where mineralised pore water flowed through to the seabed; i.e. there is a clear relationship with fluid flow (Wendt et al., 1997). Similar mounds, although not as spectacular, have been found in the Mader Basin, Eastern Anti-Atlas, Morocco (Kaufmann, 1998).

A place called 'Cement'

In the state of Oklahoma, USA, there is a small town called 'Cement', so named because the ground consists of hard, grey, carbonate-cemented sandstone. Cement is also known because of the prolific Cement oil field: 'a giant multizone anticlinal accumulation of oil and gas in a series of clastic, evaporite and carbonate rocks of Early Pennsylvanian (Lower Carboniferous) to Early Permian age' (Donovan, 1974). The multistorey reservoirs lie along the crest of a northwest-trending, doubly plunging, slightly asymmetric anticline in the southeastern part of the Anadarko Basin. The anticlinal fold was caused by thrust faulting. Petroleum within the structure is able to migrate vertically from reservoir to reservoir through faults and a shallow unconformity. Near these vertical migration pathways gypsum overlying the productive zones is characterised by calcareous cement that is 'exceptionally deficient in ^{13}C and light carbon/heavy oxygen' (Donovan, 1974); δ^{13}C values ranged from −5 to −30‰. Donovan also noted: 'away from these avenues of leakage, the influence of hydrocarbons on the isotopic composition of carbonate cement decreases systematically'. Furthermore, he attributed colour changes in the sandstones to the 'reduction and dissolution of iron in the presence of hydrocarbons'. This seems to be a massive example of MDAC; the cemented zone covers an area approximately 20 km by 5 km, and in some areas it extends to a depth of at least 800 m.

Damon Mound: the link with salt

Sassen *et al.* (1994) estimated that there is 32.7×10^6 t of $CaCO_3$ over the Damon Mound salt diapir in Texas. Their detailed study demonstrated that these 'cap rocks', which have $\delta^{13}C$ values in the range -24 to $-32‰$, were derived by the microbial oxidation of migrating hydrocarbons. They estimated that 34.2×10^6 barrels of migrating crude oil must have been 'processed' by microbes over a 13 million year period during the Tertiary to produce these carbonates. However, there are also carbonates characterised by $\delta^{13}C$ values in the range -3 to $-13‰$ and a coral-reef fossil assemblage. This assemblage includes colonial corals and other photic-zone organisms, showing that these carbonates were formed in warm shallow water at a time when sea level was higher than it is today. Sassen *et al.* compared these carbonates to those on the seabed of the present Gulf of Mexico (see Section 8.1.6). Offshore, similar 'cap rocks' are large enough to affect seabed topography. Roberts *et al.* (1990) reported that there are dead ahermatypic corals and coralline algae over hydrocarbon-derived carbonate hardgrounds on the continental shelf in the northern Green Canyon area. These now lie in water too deep (140 m) for such organisms to grow, so they must date from glacial periods when sea level was lower. However, carbon dating (reported by Roberts and Aharon, 1994) indicated ages of between 1400 and 194 500 years, suggesting growth activity during more than one glacial/interglacial cycle.

These 'reefs' seem to be modern-day analogues of the carbonates of Damon Mound and similar onshore sites, so the presence of seeps and cold-seep communities here explains the formation of hydrocarbon-derived carbonates onshore. Carbonates in deeper waters on the slope appear to represent a more embryonic stage of cap-rock development. These features are examples of the common association between salt structures and carbonates (both fossil-free 'cap-rock' carbonates, and reef carbonates), anhydrites, and elemental sulphur deposits.

9.2.13 Summary of MDAC occurrences

Methane-derived authigenic carbonate has now been reported from many places around the world in conditions that differ quite considerably. For example, water depths range from zero (intertidal) on the Danish coast to > 4000 m in the Aleutian Trench. However, these carbonates are all associated with seeps and seep-related organisms. Methane-derived authigenic carbonate seems to occur in many shapes and sizes; as slabs, crusts, blocks, and chimneys. It seems that the form of the carbonate is related to the fluid-flow regime. Chimneys form where fluid is focussed into specific pathways. At the opposite extreme, diffusive flow leads to the formation of widespread crusts; Loncke *et al.* (2004b) described crusts extending over > 100 m^2 on the Nile Deep-Sea Fan, and Comrie *et al.* (2002) found extensive carbonates over a 30 km section of a cable route in the Veslefrikk area of the North Sea (Section 11.4.2). Although in many cases the carbonates were lying on the seabed, it seems that these have been exposed by the removal of sediment after formation beneath the sediment surface; much of the carbonate observed by Comrie *et al.* was completely covered by sediment, and there were no surface indications that it was present.

Luff *et al.* (2004) argued that areas with MDAC *'should not be considered as major contributors to benthic methane fluxes'* as MDAC formation clearly indicates that methane has been oxidised, and removed from the system. Although this may be true of areas of diffusive flux (i.e. areas with extensive MDAC crusts), it seems to us that, where flow is focussed, microbial communities are unable to utilise all the available methane. Methane-derived authigenic carbonate is also a useful indicator of the longevity of methane flux, thus large carbonate structures (e.g. carbonate mounds) demonstrate that methane has been available over extended time periods. The MDAC 'reservoir' also sequesters a significant amount of methane, effectively removing methane-derived carbon from the global carbon cycle for considerable periods of geological time.

Finally, it has been shown (Formolo *et al.*, 2004; Joye *et al.*, 2004) that the majority of the carbon in carbonates from oily seeps in the Gulf of Mexico comes from higher hydrocarbons, not methane. Despite this, we are sticking to the term MDAC; HCDAC (hydrocarbon-derived authigenic carbonate) may be more correct, but it is not universally applicable – and too difficult to say!

9.3 OTHER FLUID-FLOW-RELATED CARBONATES

It is not only methane-related fluid flow that causes precipitation of carbonates. Hydrothermal flow and the flow of low-salinity groundwater along beaches and nearshore can also cause cementation, either in association with microorganisms or by purely inorganic processes. We mentioned beachrock in Section 9.2.4; here we discuss some other carbonates we consider to be associated with fluid flow.

9.3.1 Microbialites and stromatolites

Microbialites and stromatolites are sedimentary structures formed as a result of microbial activity (Burne and Moore, 1987). Tucker and Wright (1990) quoted the following definition:

> A stromatolite is an organosedimentary structure produced by the sediment trapping, binding, and/or precipitation activity of microorganisms, primarily the cyanobacteria.
>
> attributed to Awramik and Margulis

In most situations growths of microbial mats are limited by grazing invertebrates (especially cerithid gastropods, according to Tucker and Wright). Where conditions are inhospitable for these grazers, for example in hypersaline bays, mineral particles trapped by microbial filaments accumulate to form stromatolites, according to the 'traditional' explanation found in sedimentary textbooks.

Stromatolites are important features since they are thought to have been responsible for the production of the atmosphere's free oxygen, and thus the evolution of animal life on Earth. They were widely distributed in Proterozoic seas and were still abundant in the Early Palaeozoic, but since then they have steadily declined (Playford *et al.*, 1976). Stromatolites are important biogenic carbonate structures, but are not traditionally regarded as a true fluid-flow feature. However, examples described in this section provide several lines of evidence supporting the concept that there may be a relationship, mainly with groundwater flow. The first example is somewhat different as the fluid (seawater) is flowing horizontally.

In current-swept channels between the Exuma Islands on the eastern Bahama Bank stromatolites (> 2 m high) occur as individuals and in long rows aligned perpendicular to the tidal current, by which nearly all of them are '*streamlined*'. Furthermore, '*many display higher growth rates on the side facing the incoming tide, which causes them to lean noticeably towards the*

clear water' (Dill *et al.*, 1986). Clearly they are benefiting from the flow of water, from which they must be taking nourishment. Also, microbial filaments trap oolitic sand grains, contributing to the growth of the structure.

Coastal salinas

Some of the most well-known present-day stromatolites are located in Shark Bay, Western Australia. This is a coastal salina, a lake characterised by its special hydrology. In winter and spring the water level in salinas is raised by rainwater input and a little marine-derived water. During summer and autumn the water level is lowered by evaporation to a level below that of the nearby ocean. The evaporitic carbonates forming in salinas are subdivided into three types of boundstone (boxwork, veneer, and algal boundstone) containing several interesting structures, including tepees, stromatolites, and mound springs. Warren (1982), who studied the coastal salinas of South Australia, found that these carbonates all form where marine-derived groundwater resurges and seeps from a surrounding dune aquifer into the margin of the salina.

Tepee structures derive their name from their characteristic pyramidal cross-sectional shape. They form in the 'veneer-boundstone' pavement as a response to groundwater-induced seasonal changes in the pore pressure of underlying boxwork sediments. Tepees are the overthrust margins of a series of large saucer-shaped structures, forming a network of openings in the pavement (Figure 9.7). They are sites of crustal leakage, groundwater seepage and, therefore, also sites of colonising halophytic vegetation (Warren, 1982). They are generally arranged in polygonal patterns, which, as we found in Section 7.5.4, are indicative of situations in which one fluid rises through another where three adjacent polygons meet. Warren found that stromatolites along the margins of Marion Lake, South Australia grow at the crest of tepee structures. Indeed, he found that all algal boundstones and all stromatolites in the South Australian coastal salinas occurred in zones of groundwater seepage: '*upwelling groundwater creates a micro-environment conducive to algal growth*'. Warren inferred that aragonite is precipitated by the evaporation of seeping groundwaters, and the upper surface of the algal tufa can only grow up to the maximum height of groundwater resurge (controlled by the surrounding water table; Figure 9.8). Unlike the Bahamian stromatolites, these seem to benefit from vertical fluid flow.

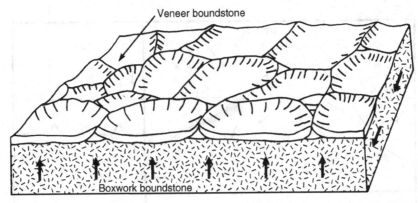

Figure 9.7 Tepee formation by groundwater seepage: the relationships between boxwork boundstone, tepees, and groundwater flow (arrows). [Redrawn by Hovland and Judd (1988) from Warren, 1982.]

As they became taller, by the growth of new layers of algae, seeping fluid must rise up within the columnar structure to the summit where new algae would use the nutrients. We suspect that they somehow act in the same way as a wick in an oil lamp or sap in a plant, drawing up fluid through an internal pore network by capillary force. This mechanism might explain their localised position on the seabed and why they continuously grow at the same spot, forming tall columns.

A study on the role of microbes in accretion, lamination, and early lithification of modern marine stromatolites does not add much about this, but documents periods of rapid sediment accretion on the stromatolites alternating with hiatal periods (Reid *et al.*, 2000). During periods of rapid sediment accretion, pioneer communities of gliding filamentous cyanobacteria dominate the stromatolite surfaces. During the hiatal periods, heterotrophic bacterial decomposition occurs, resulting in the formation of crusts of microcrystalline carbonate.

Lake Tanganyika

Four metres below the surface of Lake Tanganyika at Cape Banza hydrothermal water flows through several orifices at a rate of 2–$3\,\mathrm{l\,s^{-1}}$ and a temperature of $103\,^\circ\mathrm{C}$ (TANGANYDRO Group, 1992). The temperature and abundant bubbles indicate boiling. There are multiple-orifice aragonite chimneys, up to 70 cm high, and stromatolites grow along a cliff down to a depth of 30 m at the Luhanga hydrothermal field. Cohen *et al.* (1997) studied these modern stromatolites, but only in the context of palaeoclimate; they were not particularly concerned with their 'geobiology'. Even so, they noticed

that some of the lake water was '*ground-water input from small hydrothermal springs*'. We suggest that these may also be located where they can benefit from fluid flow.

Pavillion Lake, Canada

Pavillion Lake is only 800 m wide but 5.8 km long, and is located in the steep-walled limestone valley known as Marble Canyon, British Columbia. Because surface streams do not enter the remarkably clear lake, karst hydrology dominates. The microbialites in this lake, described by Laval *et al.* (2000), are up to 3 m high and occur along the lakesides in clusters aligned roughly perpendicular to the shoreline. They generally occur at three depths: shallow (~ 10 m), intermediate (~ 20 m), and deep (> 30 m). The shallow ones range in height from several centimetres to a few decimetres and comprise interconnected clusters of discrete round aggregates of calcite grains covered by photosynthetic microbial communities and their calcified remains. At intermediate depth, large microbialite domes (< 3 m high) consist of closely spaced aggregate clusters with a preferred orientation forming vertically ribbed structural components reminiscent of cones and leaves (Laval *et al.*, 2000); in deeper waters the structures are similar, but the individual 'cones' and 'leaves' are larger (20–35 cm in height). Laval *et al.* noted that the cones '*often have one or more internal conduit up to 5 mm in diameter*'. They concluded: '*Based on their appearance and the presence of internal conduits, it is probable that the distribution of the intermediate to deep, cone-topped microbialites correspond to regions of groundwater seepage into the lake*'. However, they also noted the presence

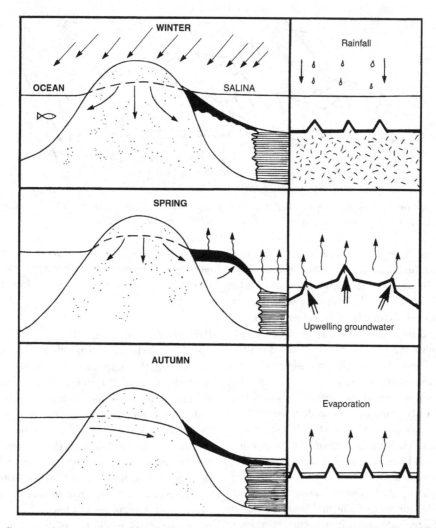

Figure 9.8 A diagrammatic representation of the role of groundwater seepage in the formation of tepee structures. During winter the boundstone is completely covered by water. In spring evaporation dries out the salina, the subsurface pore water pressure increases and water seeps up through the tepee cracks. During autumn the groundwater pressure decreases. [Redrawn and modified by Hovland and Judd (1988) from Warren, 1982.]

of calcified microbial fossils within internal conduits. It seems likely that calcification is a consequence of microbial activity below the surface bacterial mats. This seems analogous to the relationship between AOMs and MDAC.

Taupo Volcanic Zone, North Island, New Zealand
Another indication that stromatolites may rely on fluid flow, is the finding of so-called microstromatolites in hot volcanic pools of the Taupo Volcanic Zone. In 'Inferno Crater' and 'Champagne Pool', clusters of microstro-matolites, up to 10 mm high grow on small islands and on twigs (Jones *et al.*, 1997). The spring water of the pools has abundant carbon dioxide bubbles, and is of neutral chloride composition, but is enriched in silica, gold, silver, arsenic, and antimony.

Lake Van, Turkey
Kempe *et al.* (1991) described '*enormous (~ 40 m high) tower-like microbialites*' in Lake Van, in eastern Turkey. This is a remarkable lake, with a very high pH (9.7–9.8) and a salinity of 21.7‰. Mantle-derived gas enters

the lake together with other minerals and fluids by hydrothermal, mainly diffusive, seepage through the lake bottom. This fluid flow actually accounts for at least 0.04–0.06% of the total global helium flux (Kipfer *et al.*, 1994). Because the chemistry of the lake is similar to that of the Precambrian ocean, Kempe *et al.* speculated that the Lake Van microbialite structures are analogues of Pre-Cambrian stromatolites. If these structures can be proved to be fluid flow-related, then this, by analogy, more or less proves that stromatolites are seepage related!

Relationship to fluid flow?

The examples of stromatolites and microbialites described above suggest a strong link with fluid flow. The discovery by Greinert *et al.* (2002b) of stromatolitic fabric of authigenic-carbonate crusts at cold seeps in the Aleutian Accretionary Margin has further emphasised this link. Undoubtedly, future research will tell us the true importance of fluid flow for the formation of stromatolites, thrombolites, and microbialites, and thus also their importance for the oxygen in our atmosphere.

9.3.2 Ikaite

Inorganic minerals also occur in mounds and columns at lakebed and seabed fluid-flow locations. Ikaite ($CaCO_3.6H_2O$) is a tufa-like hydrated carbonate mineral formed by precipitation from waters rich in hydrogencarbonate (HCO_3^-) (Suess *et al.*, 1982; Schubert *et al.*, 1997). Its occurrence is limited by its instability at temperatures above 0 °C. It has been reported from the Antarctic (Suess *et al.*, 1982) and the Arctic (Laptev Sea, Schubert *et al.*, 1997; Greenland, Buchardt *et al.*, 1997, 2001), but also from Mono Lake, California. There are significant differences between these occurrences, but all seem to be associated with seabed (or lakebed) fluid flow.

Mono Lake, California

The 6 m high ikaite columns on the shore of Mono Lake are perhaps the most well-known examples of this unusual mineral. Mono Lake is an alkaline (pH 9.7) closed-basin lake, and the ikaite is thought to precipitate as a result of the mixing of lake waters and (hydrothermal) spring waters; both are rich in hydrogencarbonate. The ikaite forms during the winter months, but

decomposes to anhydrous $CaCO_3$ (gaylussite) in spring (Bischoff *et al.*, 1993).

Perhaps it is a coincidence, but there are also methane seeps in Mono Lake. Oremland and Miller (1993) identified more than 700 active seeps (61 to 98% methane) originating from an area occupying about a third of the lake's 150 km² area. Carbon isotope values (−72 to −55‰) indicate that this methane is of microbial origin. However, methane with a thermogenic isotopic signature (−55 to −45‰) was found to be present coming from thermal springs elsewhere in the lake.

Laptev Sea, offshore Siberia

Ikaite crystals up to 5.5 cm across were recovered from a depth of about 2.3 m below seabed during sediment coring in the Laptev Sea, offshore Siberia; this was on the upper continental slope, in a water depth of 240 m. Schubert *et al.* (1997) concluded that these crystals formed by precipitation from hydrogencarbonate associated with the microbial oxidation of methane. The methane may have been microbial, generated in the organic-rich silty clays; however, evidence of gas-migration features in the area suggested to Schubert *et al.* that it came from gas hydrates.

Ikaite tufa, Greenland

In Ikka Fjord, southwest Greenland, there is a spectacular area of about 0.75 km² where there are more than 500 ikaite columns (Figure 9.9) with heights of 1–20 m (Buchardt *et al.*, 1997, 2001). Carbonatite, an unusual carbonate-rich igneous rock, of the 1300 million year old Grønnedal-Ika igneous complex, occurs on either side of the 'garden' area of the fjord. Porewater sampling at the base of columns proved that groundwater (really a sodium hydrogencarbonate and sodium carbonate brine with high pH, high alkalinity, and high phosphate content), percolating through this carbonatite, enters the fjord as seeps.

9.3.3 Whitings

'Whitings' have puzzled marine scientists for many years. They are intermittent white patches occurring near the surface of some oceans, including the Florida Straits, the Bahamas, the Arabian Gulf, and the Dead Sea. They can be seen on photographs taken from space (Figure 9.10).

Figure 9.9 Ikaite columns in Ikka Column Garden, Ikka Fjord, Greenland. [Reproduced from Buchardt *et al.*, 2001 with permission from SEPM (Society for Sedimentary Geology).]

Whitings consist of clouds of suspended calcite and aragonite crystals. Whiting phenomena on the Great Bahama Bank were studied by Shinn *et al.* (1985), who found that dozens of long-lived whitings may exist at any one time, and that they continually 'rain' aragonitic sediment. Even though chemical changes in seawater could not be measured, their data indirectly suggested that precipitation of aragonite from seawater was responsible for the whitings. Morse and Mackenzie (1990) reviewed three main hypotheses that have been suggested to explain them:

- shoals of fish stir up the bottom sediments;
- $CaCO_3$ precipitation is triggered by algal blooms;
- $CaCO_3$ precipitation occurs on preexisiting nuclei concentrated by water circulation.

Loreau (1982) suggested that intense evaporation might be a trigger. In 1988 we proposed that seabed fluid flow might be responsible (Hovland and Judd, 1988). Direct precipitation of calcite or aragonite in seawater can only occur when the water is supersaturated with $CaCO_3$ and when triggered by rapid changes in temperature, pH values, or amount of dissolved CO_2. Since the temperature and the pH seem to have been stable, at least in the well-studied Bahamian whitings, we suggest that a sudden increase in CO_2 may act as a trigger. This could be achieved either by gas or carbonated groundwater flowing through the seabed; on the Great Bahama Bank and in the nearby Florida Strait there is also some evidence of gas and freshwater fluid flow (Manheim, 1967). This idea is supported by observations of whitings around microbialites associated with groundwater seeps in Pavillion Lake, Canada (see preceding section); Laval *et al.* (2000) reported that the whiting events occur during cyanobacteria population blooms, '*resulting in an accumulation of authigenic carbonate sediment around the microbialite fields*'.

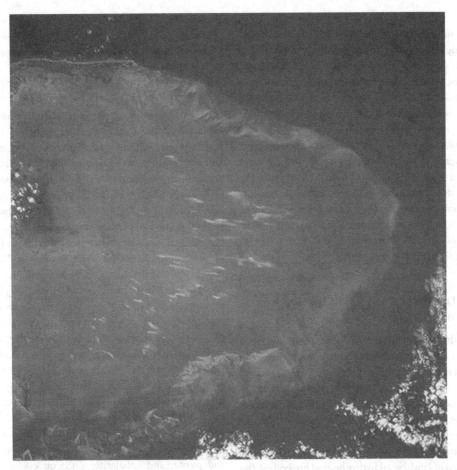

Figure 9.10* The bright, ellipitical white features in this Space Shuttle photograph are whitings in the Little Bahama Bank. Intricate patterns of sand ridges and channels that are formed by tidal currents ring the bank, marking the edge between the shallow bank waters (depths less than 10 m) and the much deeper offshore waters (dark). [NASA image STS088–719–29 taken on 6 December 1998; image courtesy of the Image Analysis Laboratory, NASA Johnson Space Center (http://eol.jsc.nasa.gov).]

In 1988 we also suggested the possibility that whitings may have been triggered by carbon dioxide or methane released from accumulations in shallow sediments. We noted that seeps have been documented in the Arabian Gulf (Section 3.14), and there is widespread acoustic turbidity in the Dead Sea (Friedman, 1965). Whitings off the Namibian coast (Figure 3.25) support this explanation. In this case the whitings are associated with gas (a mixture of methane, carbon dioxide and hydrogen sulphide) emissions from the sediments (discussed in Section 3.7.2), and the precipitation of elemental sulphur particles in the water column (Weeks *et al.*, 2004).

9.3.4 Carbonates and serpentinites

Enormous carbonate chimneys towering 60 m above the seabed at the Lost City site on the Mid-Atlantic Ridge (Section 3.8), and other less spectacular structures on seamounts in the southeastern Mariana Fore-arc (Section 3.18.2), formed by the precipitation of minerals from cool fluids derived from the serpentinisation of mantle peridotites (described in Section 5.2.4). The Lost City structures are composed of variable mixtures of calcite, aragonite (both $CaCO_3$), and brucite–$Mg(OH)_2$. The venting fluids are warm, 40–75 °C, and have a high pH 9.0–9.8, compared to 8.0 in normal seawater (Kelley *et al.*, 2001). They were found to be depleted in magnesium, enriched in calcium, and to contain high concentrations of dissolved methane and hydrogen. Cann and Morgan (2002) found it odd that calcium-rich fluids associated with peridotite, which is calcium-poor. They suggested that this is because serpentinite contains even less calcium, so the left-over calcium is removed in the water.

Chimneys found on the Mariana Fore-arc seamounts were composed of aragonite and calcite in varying proportions. This is quite remarkable because the site, at a depth of about 3150 m, lies below the carbonate compensation depth (CCD), below which carbonates are soluble in seawater. This accounts for the evidence of corrosion described by Fryer *et al.* (Fryer and Fryer, 1987; Fryer *et al.*, 1990).

9.4 HYDROTHERMAL SEEPS AND MINERALISATION

In the hydrothermal circulation systems we described in Section 5.2.3 seawater is cycled through the ocean crust, and emerges though hydrothermal vents, including black and white smokers. These, and similar structures, are of interest not only because they are a direct result of seabed fluid flow, but also because of their relevance to the mining of metallic-ore minerals, a subject we return to in Section 11.6.1. Despite centuries of exploitation of land-based hydrothermal ore bodies, the understanding of hydrothermal mineralisation is in its infancy:

> The distribution of sea-floor hydrothermal mineralization is an artefact of incomplete knowledge at this early stage of exploration. Less than one percent of the sea floor in the tectonic settings considered has been systematically investigated. Thus, the actual distribution of deposits along spreading centres is unknown. Recognition of such deposits is biased in favor of those that form prominent exposed topographic features, such as massive sulfide mounds at zones of focused high-temperature hydrothermal discharge. Relict stratiform deposits lacking appreciable relief, and all forms of deposits beneath the sea floor tend to remain unresolved by present exploration methods. Knowledge of size, shape, and especially, composition of sea-floor hydrothermal mineral

deposits is limited by inadequate measurement and sampling techniques. Information on these deposits is incomplete in two dimensions and almost unknown in the third dimension.

> Rona and Scott, 1993

Currently known occurrences are indicated on Map 33. Different types of mineralisation occur at hydrothermal vents, depending upon the chemical composition and temperature of the venting fluid, and whether the flow is diffuse or focussed (as discussed in a special issue of *Economic Geology* prefaced by Rona and Scott; 1993). The history of the fluid as it passes from recharge zone to vent, is critical. The maximum temperature the fluid has reached, and the occurrence (or not) of phase separation and supercritical water strongly influence the composition of the fluids (see Section 5.2.3).

Diffuse flow normally occurs at relatively low temperatures (< 10 to 50 °C) so fluids do not have a great capacity to carry dissolved minerals. However, amorphous iron and manganese hydroxides, authigenic clays, and silica may precipitate, and small unstable chimneys may form (Hannington *et al.*, 1995). The temperature of the venting fluid determines which minerals precipitate when flow is focussed through vents. High (> 300 °C)-temperature fluids contain metals such as iron, copper, and zinc, which precipitate as sulphides (chalcopyrite, $CuFeS_2$; the iron sulphides pyrrhotite and pyrite; and sphalerite, ZnS) on cooling to produce 'black smoke'. The minerals that precipitate from cooler (100–300 °C) fluids, silica, anhydrite ($CaSO_4$) and barytes ($BaSO_4$), are lighter in colour, hence they form 'white smokers'. If the fluid mixes with seawater before it reaches the seabed, precipitation occurs within the subsurface network of fissures, so the fluids are warm but almost translucent when they emerge (Chevaldonné, 1997).

When new vents appear the first mineral to be precipitated is anhydrite, precipitating from seawater at about 150 °C. Once an anhydrite chimney has formed, sulphides precipitate on the inside where the cooling effects of seawater are less pronounced. However, this heat insulation allows anhydrite on the outside of the chimney to cool down, and once below 150 °C it will start to dissolve. Substantial structures containing a mineral assemblage including metallic sulphides, oxides, silicates, barytes, and calcite may form. Some are enormous. For example at the TAG site (Mid-Atlantic Ridge) there is a complex structure about 30 m in diameter and 25 m high, and made primarily of anhydrite. These are small compared to the 'High-Rise' structures on the Endeavour section of the Juan de Fuca Ridge. Robigou *et al.* (1993) described one structure, 'Godzilla', standing about 45 m tall, rising directly from the seabed, and belching smoke from summit chimneys (Figure 1.3). Eventually vents become blocked or disused, and may collapse as a result of anhydrite solution. In many areas collapsed material has accumulated to form substantial mounds, through which new plumbing systems develop to maintain the flow. As the mounds grow they become cemented by silica which inhibits the escape of the fluids circulating within the mound, and adds to the concentration of crystallised minerals (Scott, 1997).

Often, black and white smokers are present in the same vent complex. In such cases it is probable that the cooling of the fluids to the temperatures typical of white smokers implies that higher-temperature minerals (sulphides, etc.) have already precipitated beneath the seabed.

Hekinian (1984) calculated that a 3 cm diameter vent discharging fluid at a rate of $10\,\mathrm{l\,s^{-1}}$ could result in the precipitation of 100 kg of metalliferous deposits in a single day. Hannington *et al.* (1995) reported that growth rates of '*at least 5–10 cm per day*' have been reported from several sites. Despite these rapid growth rates, the majority of dissolved sulphur and metals carried by venting fluids escape into the water column. The majority is 'lost', but some precipitate out and rain down close to the vents. Sulphide particles $> 50\,\mu\mathrm{m}$ are likely to be deposited within 1 km of the vent, but finer particles (perhaps 99% of the total mass) are dispersed over much greater distances. In time they oxidise or dissolve, releasing the metals back into the seawater (Hannington *et al.*, 1995).

Many individual hydrothermal systems do not remain active for long periods, but some seem to persist. The TAG field, for example, has been active for 40 to 50 thousand years (Hannington *et al.*, 1995). Evidence from ancient deposits provides a different perspective. For example, the North Pennine Orefield, in Northern England was the product of hydrothermal activity following the emplacement of the Weardale Granite. The granite crystallised about 410 million years ago, and hydrothermal activity continued for 45 million years. Clearly, over the expanse of geological time, seabed fluid flow of this type has played a significant role in determining the chemical composition of seawater (discussed in Section 10.2.3).

9.4.1 Sediment-filtered hydrothermal fluid flow

When hot hydrothermal fluids travel through sediments before entering the water column, they cool off, precipitate minerals, and have time to react with sediments and pore waters. These processes give rise to a number of different scenarios and interesting mineral deposits on the seabed. In Middle Valley of the northern Juan de Fuca Ridge, a relatively impermeable turbidite sediment blanket has capped a vigorous, very young oceanic spreading zone, which has resulted in the formation of a seabed mineral deposit similar in size and grade to ore deposits mined on land. Ocean Drilling Project Leg 169 drilled through the 'Bent Hill' massive sulphide deposit and penetrated the hydrothermal feeder zone through which metal-rich fluids reached the seabed. This was best represented in cores from 100 to 210 metres below seabed. Zierenberg *et al.* (1998) described three units within this interval.

1. The upper 45 m is intensely mineralised by subvertical veins of isocubanite–chalcopyrite and pyrrhotite, with vein density decreasing downcore. The host turbidites are altered to chlorite, quartz, and fine-grained rutile and titanite. Veins range from 1–8 mm, with the thickest veins showing crack–seal textures indicative of multiple dilation due to fluid overpressure followed by mineral precipitation.

2. The interval from 145 to 200 m below seabed is less intensely veined with numerous subhorizontal veins and disseminated, bedding-parallel sulphide increasing downcore. There are many subvertical veins branching off into subhorizontal sulphide impregnations of the more permeable horizons.

3. Core recovered between 200 and 210 m below seabed is also intensely mineralised, containing up to 50% by volume sulphide minerals; these are predominantly isocubanite containing coarse exsolution lamellae of iron-rich chalcopyrite. Pyrrhotite is much less abundant than in the overlying veins and other sulphide minerals are present in only trace amounts. A representative sample of high-grade mineralisation from this zone (the deep copper zone, or DCZ) contained 16 weight% copper. In contrast to the overlying intervals, sulphide veining is essentially absent. Sulphide

mineralisation occurs as impregnations and replacement of host sediments, and is strongly controlled by variations in the original sedimentary textures. Much of the mineralisation is developed in medium- to coarse-grained, locally cross-bedded, turbiditic sand, and preserves the original sedimentary structures.

The style and structure of the ore bodies currently studied on land can tell a great deal about the hydrology or hydraulics involved when hot fluids flow through porous, but nearly impermeable sediments. The transition from dominantly vertical crack–seal veins in the upper part of the feeder zone, to subhorizontal mineralisation controlled by sedimentary texture at the base of the feeder zone, indicates that cyclic overpressure capable of fracturing the rock only occurred near the seabed. Another possibility is that the fluids were supercritical in the veined feeder zone, and became subcritical at a shallower depth. Drilling on the east and west flanks of the deposit encountered a weakly mineralised, highly silicified mudstone horizon at the approximate depth of the top of the DCZ; this provided an important hydrological control on the high-temperature hydrothermal system that formed the massive sulphide. During periods when high-permeability pathways (represented by the veins) were sealed, fluid was forced to flow laterally into the more permeable sandy turbidite units. Conductive cooling of this ponded hydrothermal fluid facilitated silica deposition (Janecky and Seyfried, 1984), sealing the top of this interval, and the precipitation of isocubanite, which is the least soluble of the sulphide minerals occurring in this deposit (Zierenberg et al., 1998).

9.4.2 Anhydrite mounds

It came as a surprise to the investigators of a large midwater bubble plume located about 10 km east of Grimsey Island (off northern Iceland) to find hydrothermal vents and a massive anhydrite deposit on the seabed. The location, in 400 m of water, corresponds with the Grimsey Graben in the Tjornes Fracture Zone; the continuation of the Mid-Atlantic Ridge north of Iceland (Hannington et al., 2001). Measurements of the highest-temperature vent fluids showed a pH between 5.9 and 6.8. The sediments consist of interlayered marine mud and ash-rich layers. Gases in the water column are dominated by a mixture of carbon dioxide (up to 41%) and methane (up to 24%). Whereas these high concentrations of carbon dioxide indicate a magmatic source ($\delta^{13}C_{CO_2}$ values of –2.4 to –3.0), methane and higher hydrocarbons (up to butane) are suspected to be mainly derived from thermal decomposition of sedimentary organic material ($\delta^{13}C_{CH_4}$ values of –26.1 to –29.5‰; Hannington et al., 2001). High concentrations of dissolved methane (up to 560 000 nl l^{-1}) were found in the water close to the vents, and up to 1000 nl l^{-1} in the water column 10 m above them.

These anhydrite deposits cover an area more than 100 by 300 m. Assuming the mounds have a depth below seabed of 10 m and a bulk density of 3 t m^{-3}, this deposit contains 1 million t of anhydrite and up to 100 000 t of talc. Also, there may be significant buried deposits of sulphides (Hannington et al., 2001).

At Escanaba Trough, Southern Gorda Ridge, and at the area of active venting (AAV) in Middle Valley, Juan de Fuca Ridge there are deposits similar to those at Grimsey (Davis et al., 1992). At AAV the hydrothermal fluids diffuse through sediments but they are focussed where sediments are indurated and fractured. According to Hannington et al. (2001) the vent temperatures are between 184 and 287 °C, and the dominant minerals are anhydrite, barytes, magnesium silicates including talc, and minor sulphides (mainly pyrrhotite).

Anhydrite and barytes chimneys have also been found by Russian investigators employing the manned 'Mir' submarines on the Piip Submarine Volcano offshore Kamchakta (Torokhov and Taran, 1994; see Section 3.18.7). The Piip Volcano has two peaks, the North Peak at 380 m water depth and the South Peak at 450 m. At the Eastern Dome of the Northern Peak there are four sites where shimmering water at about 250 °C (i.e. near boiling point) exits through 1.5 m high, 10 to 15 cm diameter chimneys. Torokhov and Taran collected a sample of crust close to the base of the chimney. They described it as 'well crystallized gypsum looking like rice porridge'. Away from the vents the seabed was found to be 'covered with loose deposits with typical Fe-ochreous color', and they identified 'amorphous silica, pyrite and kaolinite' in a sample. At the South Peak the most intense hydrothermal activity is concentrated at the central part of the summit. Here also there are chimneys sitting on a 10 m high hydrothermal mound. One was found to be composed of aragonite and barytes, with a thin film of pyrite on the walls of an internal conduit.

Whereas hot ($> 300\,^{\circ}$C) hydrothermal flow through marine sediments tends to produce sulphide mounds, evidence from Iceland and a few other locations (Escanaba Trough, Guaymas, and Middle Valley) suggests that anhydrite–calcium sulphate, $CaSO_4$ – and talc – hydrous magnesium silicate, $Mg_3Si_4O_{10}(OH)_2$ – are produced when slightly lower-temperature (about 250 °C) hydrothermal fluids flow through sediments of volcanic origin.

9.4.3 Hydrothermal salt stocks

Like many other geologists, we find it very difficult to accept the conventional evaporation explanation for the formation of salt deposits. Also, it is very hard to understand how salt diapirs up to 18 km high formed. By applying new ODP results to other data, a group in Statoil has now come up with a brand new model for salt-stock development (Hovland et al., 2006). The new model, summarised below, was derived by combining knowledge of (a) petrophysical conditions beneath hydrothermally active deep-sea mineralising centres, (b) properties of supercritical fluids and, (c) external and internal observations of salt stocks.

Because halite (NaCl), anhydrite ($CaSO_4$), and gypsum ($CaSO_4.2H_2O$), are very soluble in seawater, they normally remain in solution during the cooling of hydrothermal fluids. However, laboratory experiments have shown that when normal seawater reaches its critical point on the P/T curve (at $P = 300$ bar – equivalent to about 2800 m water depth; $T = 405$ °C; see Figure 5.2), NaCl suddenly loses its solubility and is precipitated as aggregates of minute (10–100 µm) salt crystals. This effect was termed 'shock crystallisation' by Tester et al. (1993). Hovland et al. argued that this shows that salt stocks may be a special type of hydrothermal feature.

Investigations of the Brandon site, at 2834 m on the ultra-fast spreading southern East Pacific Rise, showed that supercritical seawater is present inside seabed vents (von Damm et al., 2003). The temperature of the venting fluids was measured from the submersible Alvin; it was 405 °C, extremely close to the critical point for seawater. It was also found that fluids from vents less than 2 m apart had compositions 'that differed by almost a factor of two'. Von Damm et al. remarked 'for the first time we have observed both the vapor and liquid phases venting simultaneously from a single structure'.

In some hydrothermal systems, supercritical conditions beneath the seabed may be favourable for the production of large amounts of salt. Because of the dependency on 'shock crystallisation', Hovland et al. predicted that favourable conditions occur within specific ranges of water depth, according to local thermal gradients. If the water is too deep, hydrothermal fluids may still be in a supercritical state as they pass into the water column (as in the example above), so the salts will precipitate there, together with other typical hydrothermal minerals.

This model is supported by evidence of seeps of gases, brines, and iron slurries coming from the apex of salt-stocks, together with numerous other observations of seeping fluids and overpressured compartments inside certain salt mines. On the basis of observations from salt-stock mines, Hovland et al. suggested that warm brines are transported vertically through specific conduit zones where dissolution and recrystallisation occurs. Within these 'anomaly' zones there may also be relatively rapid fluid flow through narrow channels or pipes. Because the pressurised fluids are constantly moving, they are kept open as the salt stock grows upwards. In steep-walled, high-temperature (400 °C) active sulphide chimneys at the Endeavour vent field (Juan de Fuca Ridge) quartz has been found to act as a sealing mineral for the interior precipitating sulphides. Hovland et al. suggested that anhydrite and gypsum are the equivalent sealing and stabilising minerals for halite precipitation within salt stocks.

Support for the hypothesis of hydrothermal salt formation came from chlorine isotope studies (Bach et al., 2002). It seems that $\delta^{37}Cl$ values of continental waters of various origins (oil field waters, freshwaters, and brines) cluster within 5‰ of 0‰. In contrast, fluids from the marine environment span a 15‰ range. Bach et al. quoted examples from hydrothermal vents (East Pacific Rise: +6.5 to +7.1‰; Logatchev: +4.6‰) and accretionary wedges (−7.7 to 2.0‰). They suggested 'that the heavy Cl isotope signature of the fluids is a result of seawater–rock interaction and/or mineral precipitation rather than phase separation of seawater. However, the specific mechanisms responsible for this enrichment are not yet understood'.

The new model for salt formation proposes that various dynamic physicochemical processes interact to produce intrasedimentary salt structures up to 18 km high, some of which pierce the seabed. It suggests that formation occurs in the following stages.

1. A focussed flow of ascending brine develops inside a major fault zone above a basinal heat source.
2. Salts ($NaCl$ and $CaSO_4$) start to precipitate and accumulate locally to form a salt stock on the seabed. When salt precipitates on or just below the seabed it is prone to dissolution; however, less-soluble minerals (for example $CaSO_4$, anhydrite) tend to form a protecting cap, minimising dissolution of $NaCl$.
3. A multiphase flow consisting of saturated brines, vapour, gases, and a slurry of $NaCl$ grains and other precipitated microscopic mineral grains, develops inside the stock.
4. Mineral grains are gradually deposited inside and above the salt stock as the ascending slurry cools adiabatically; because $NaCl$ maintains a high solubility in water, even at low temperatures, salt stocks may span a great temperature interval, and may become very high.
5. The internal pressure of developing salt stocks is always contained because of vapour and gas generation when water boils as it rises above the critical point inside the stock. This causes the stock continually to crack, but precipitation and recrystallisation of salts and minerals continually seal the cracks, so the stock is able to migrate upward.

The model predicts that the height of a developing salt stock will be determined by the difference between hydrostatic pressure in the seawater/sediments on the outside, and the pressure of the brine/vapour/gas/mineral slurry on the inside. The salt stock will therefore adjust its height (i.e. grow) according to subsidence of the surrounding sediments. The diameter of the stock will be partly determined by the difference between internal pressure and the confining pressure at the point where growth is occurring at any given time (normally at its apex). If the growth of the salt stock keeps pace with the sedimentation rate, it will always be surrounded by a constant crack-failure pressure, and will therefore tend to grow with a constant diameter.

This new model provides an explanation for some of the very difficult questions associated with the conventional evaporite/diapiric salt-stock theory; for example the following.

- Why is halite deficient in magnesium, when seawater contains abundant magnesium?

- What is the driving mechanism of 'salt diapirs' that penetrate the seabed?
- How were the thick 'evaporite' deposits formed when there was a limited supply of seawater?

However, numerous questions still remain unanswered. For example: what is the relationship between salt stocks and bedded evaporites? And, where do the Na and Cl originate from: seawater and basalts, via reactions with supercritical water?

9.5 OTHER MINERAL PRECIPITATES

9.5.1 Iron from submarine groundwater discharge

At the foot of the Florida Escarpment submarine groundwater discharge (SGD; see Section 3.26) comprises fluids at temperatures too low to carry in solution the metal assemblages found in hydrothermal vents. They are hydrogen sulphide-charged brines (we may ask, why are they brines?) from which minerals precipitate. The sediments contain mixtures of pyrite ($\sim 30\%$ by weight), barium–strontium sulphates ($\sim 4\%$), clays, and locally derived carbonates (shell hash), which are being cemented by iron sulphides. None of the minerals deposited at seep sites have been found to be of economic value. However, this type of SGD-related mineralisation could serve as a model to explain similar examples found in ancient strata.

9.5.2 Phosphates on seamounts, guyots, and atolls

In Section 4.3.4 we concluded that seamounts are important sites for various types of seabed fluid flow, so it is hardly surprising that they, and associated features (guyots and atolls), are favourable sites for mineral precipitation. Guyots are capped by carbonate (calcitic or dolomitic), fluorapatite, or phosphorite that may be as much as 1000 m thick (Heezen *et al.*, 1973; Cullen and Burnett, 1986); the phosphorites are particularly important to the economies of some island states. However, there is some debate about their mode of formation.

Cullen and Burnett (1986) argued that the dolomite–apatite association on one guyot, and the carbonate–fluorapatite and dolomite cap on two others in the north of the Fijian group of islands was formed in shallow water, lagoonal, or even subaerial environments prior to the subsidence and final inundation of the guyots. This matches one of the three hypotheses identified by Strizhov *et al.* (1985) for the origin of phosphates on seamounts.

1. Phosphates were formed by guano accumulation under subaerial conditions when the seamounts supposedly were uplifted above sea level as islands.
2. Phosphate was produced as a metasomatic replacement of carbonate muds when they entered phosphorous-rich bottom waters.
3. Phosphate material was produced by volcanogenetic-hydrothermal action.

They concluded that geochemical and isotope data supported a hydrothermal origin.

Aharon *et al.* (1987), studying the former atoll on Niue Island, South Pacific, found mineralogical and isotopic evidence that seawater was drawn through underlying volcano-derived rocks and deposits to the atoll. This would explain inclusions of the metals iron, copper, zinc, and manganese found in the dolomite. This seawater circulation was supposedly driven by thermal convection over a 'hot' volcanic pedestal. Our interpretation is that seawater could have been taken into the crust in the deep surrounding ocean. Then, like any other hydrothermal solution, it could have circulated, eventually emanating at the summit of the seamount or guyot underlying the atoll. We thus infer that there is a reef on this volcanic summit formed partly because of the nourishment carried by the exhaling fluids, and that this exhalation may explain the dolomitisation (Figure 9.11). A French research team who studied this process called it '*endo-upwelling*' (Rougerie and Wauthy, 1988).

The flow of fluids through seamounts may therefore be partially responsible for the formation of carbonates and phosphates that cap guyots. In addition, hydrothermal circulation occurs on active seamounts; consequently, hydrothermal seepage may result in the precipitation of a variety of metals on top of seamounts.

9.6 FERROMANGANESE NODULES

Ever since their discovery by the Challenger Expedition (Murray and Renard, 1891) ferromanganese nodules on the deep-ocean floor have fascinated Earth scientists, prospectors, politicians, and many other groups. Although they contain about 76 elements, the interest is principally because of their base-metal content, particularly the '*big four*': manganese, copper, nickel, and cobalt (Pryor, 1995). Iron and manganese are present in the highest concentrations (9–27% and 8–40%, respectively); nickel, copper and, to a lesser extent, cobalt may be present in concentrations of up to 1%. Nodules occur over vast areas of the Pacific, Atlantic, and Indian oceans, and although unevenly distributed, may cover more than 50% of the ocean floor. Their extent is such that they may represent a major (future) metal resource. Although the Clarion–Clipperton area south of Hawaii was long held as the most prolific nodule resource area, it is now thought that the Cook Islands EEZ (exclusive economic zone) contains the densest nodule population with 58 000 km^2 of seabed containing 20 kg to 60 kg m^{-2} (Pryor, 1995). According to Emery and Uchupi (1984) the areas favourable for manganese oxide deposition in the Atlantic Ocean are current-swept deep-water plateaus (especially Blake Plateau), sides of seamounts, flat abyssal plains, and lower continental rises underlain by red clays.

Spherical ferromanganese nodules range in size up to about 20 cm, but flatter ones may be 1 m or more in diameter. They are easily crushed and, when cut, the internal concentric structure indicates a colloidal mode of formation. The most common nucleus found at the centre of the nodule is volcanic material. However, fragments of older nodules, bones, and shark teeth have also been reported to serve as a nucleus (Seibold and Berger, 1982). One of the most difficult facts to explain in respect to their distribution is why they are concentrated at the water–sediment interface. Why are they not evenly distributed in the underlying sediments? This is a serious problem since the rate of deposition for nodules is between 1 and 8 mm per million years; that is about one thousandth the rate of deposition of most deep-ocean sediments (Emery and Uchupi, 1984). This requires a process of mechanical deburial in the sediments. Various modes of origin of ferromanganese nodules, crusts, and films have been proposed:

- deposition from seawater (hydrogenous);
- remobilisation of manganese within underlying sediment (diagenetic);
- deposition from vents of volcanic hot springs (hydrothermal);
- deposition from surface weathering of basalt (halmyrolytic).

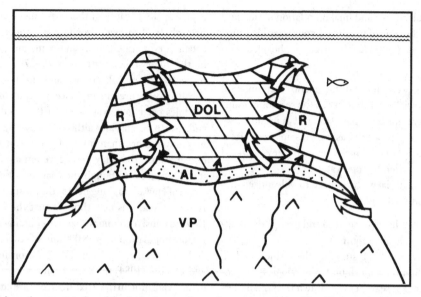

Figure 9.11 Schematic representation of the thermal convection model for dolomitisation of atolls. Assumed seawater circulation caused by upwelling around a 'hot' volcanic pedestal is shown by arrows. Our assumption that both reefs and dolomitisation may have a close relationship with slow volcanic exhalation is indicated by wavy arrows. DOL = dolomitised zone; R = reef; AL = ash layer; VP = volcanic pedestal. (From Hovland and Judd, 1988, who based the drawing on Aharon *et al.*, 1987).

Of these the first two are the most favoured. However, hydrothermally induced precipitation is the main process near and at hydrothermal vents. Kristiansson and Malmquist (1980) proposed another mechanism involving fluids seeping upwards through the sediments. They documented a general upward micro-flow of 'geogas', mainly composed of nitrogen, carbon dioxide, and minor parts of argon and helium, in a terrestrial environment. The presence of helium presumably implies a mantle origin. They suggested that this microseepage must be an important mechanism for transporting atoms and aggregates through the lithosphere. Kristiansson's and Malmquist's theory was mainly based on the observation that there are high concentrations of radon (^{222}Rn – with a half life of only 3.8 days) in seawater above manganese nodule fields (Sarmiento *et al.*, 1978). The only way to explain such an anomaly is that ^{222}Rn enters the water by some process much more efficient and rapid than diffusion. Widespread microseepage through the ocean floor is therefore one possible process. The hypothesis is, in fact, comparable to the formation model for MDAC. Kristiansson and Malmquist proposed that a carrier gas transports pore water with high concentrations of metal ions, which passes through '*cracks, micro cracks and fis-sures in the lithosphere*'. Growth of ferromanganese or polymetallic nodules takes place where this migrating fluid comes into contact with oxidising seawater; they called these '*momentary interaction zones*'. The complicated network of pores and fractures and internal laminar structures in large nodules was explained by seawater penetrating certain parts of the nodule, causing internal precipitation reactions with the ascending bubble stream.

Bubbles trapped within growing nodules may give sufficient buoyancy to enable them to remain on the seabed, even though their density, without bubbles, is higher than that of the supporting sediments. A simple calculation shows that the manganese nodules would have a high porosity, capable of considerable buoyancy if gas-filled: Menard (1964) reported a mean density of 2.1 g cm^{-3} and a mean porosity of 45%. A spherical nodule of 4 cm diameter has a volume of 12.6 cm^3 and mass of 26.5 g (in air). The density of the solid part of the nodule is 4.6 g cm^{-3}, between that of manganite (MnO = 4.4 g cm^{-3}) and ilmenite (FeTiO$_3$ = 4.75 g cm^{-3}). The nodule would weigh only 13.9 g in water if the pore spaces were water-filled (assuming a water density of 1 g cm^{-3}). However, if the pore spaces were filled with gas (with a negligible density), the weight in water would only be 8.9 g, and the total density would be $1.5 \, g \, cm^{-3}$, very close to the density of deep-sea sediments. Gas trapped in porous nodules may keep them 'floating' at the seabed. When a nodule eventually loses its contact with the feeder channel, it is sooner or later degassed, the bulk density rises and it sinks into the sediments, decomposes (dissolves) and vanishes (Malmquist and Kristiansson, 1981).

Uranium isotope ratios have also been studied in relation to deep-sea hydrothermal deposits. Reyss et al. (1987) found very high $^{234}U:^{238}U$ ratios in hydrothermal manganese crusts recovered from the Sangihe island arc system in the West Pacific. The ratios, which were twice those of normal seawater, indicate a low-temperature hydrothermal supply of uranium. They hypothesised that precipitation of the hydrothermal solution takes place inside, or at the top of, a thin sedimentary layer covering the basaltic crust before the solution mixes with seawater. This mechanism is more or less the same as that proposed by Malmquist and Kristiansson (1981).

Further evidence that polymetallic nodules and crusts are formed by mantle-derived material has been provided by hafnium (Hf) isotope analyses. White et al. (1986) studied the distribution of hafnium isotopes in marine sediments and ferromanganese nodules of the Indian and Pacific oceans. They expected that isotopic composition would show that hafnium in sediments and nodules was derived from the continental crust, the principal source of marine sediments. Surprisingly, however, the isotope ratios in both the marine sediments and the nodules appeared to reflect a large 'mantle' contribution. They concluded that ridge-crest hydrothermal activity and low-temperature basalt–seawater interactions are important sources for hafnium in sediments (seawater) and in ferromanganese nodules.

Usui et al. (1987) made a detailed study of the sediments and topography of a $20 \, km^2$ area of ferromanganese nodules in the equatorial zone of the Central Pacific Basin. The area is occupied by small abyssal hills with a local relief of up to 300 m. Although surrounding abyssal plains yield few ferromanganese nodules, they are abundant close to, and on, the hills (more than $10 \, kg \, m^{-2}$). The authors divided the nodules into three categories (rough, intermediate, and smooth) and correlated them with the shallow sediments; mainly radiolarian ooze and brown clay with volcanogenic ash. They

found few nodules in the flat-floored basins where the radiolarian ooze was > 100 m thick, but greater abundances where the ooze was thinner (< 70 m) around hills. This inverse relationship suggests that ferromanganese nodules are more common where sediments have a higher concentration of material derived volcanically from the mantle.

Usui et al. did not mention seepage or exhalation, but we consider that the evidence presented by other researchers (Sarmiento et al., 1978; Kristiansson and Malmquist, 1980; White et al., 1986) is of sufficient merit to ensure that these processes are considered as serious alternatives to the modes of origin that are more commonly accepted. Further evidence also indicates, to us at least, that ferromanganese crusts and nodules are closely related to low-temperature hydrothermal diffusive venting (or seepage). Muiños et al. (2002) described the petrology and geology of ferromanganese deposits acquired from 'Nameless Seamount' in the Gulf of Cadiz, sampled during the TTR-11 cruise. Samples collected using a TV-grab device at 1850 m water depth in the northwest corner of the seamount consisted of nodules and crusts with a variety of shapes and dimensions. One nodule had an inner core of calcite (5 mm diameter), then two layers of phosphatised limestone (P = 12%, 2 mm thickness), then three of iron-manganese oxides with laminated growth structures (Fe = 23%, Mn = 24%, Co = 1%, total of 16 mm). Because Muiños et al. had difficulty in explaining the growth of these layers using the commonly accepted modes of origin, we take this as further evidence of the nodule and crust growth environments being modified by microseepage.

9.7 FINAL THOUGHTS

From evidence presented in this chapter, it seems that several types of mineral precipitation are associated with seabed fluid flow in the last of the three important circulation zones: 'inflow', 'circulation', and 'outflow' zones (Figure 5.1). *Outflow* occurs when the fluid is emitted through the seabed. Key parameters are: temperature, dissolved minerals, pH, presence or absence of methane (and other hydrocarbons), and flow rate. These parameters are determined by the circulation history of the fluid (i.e. the previous two zones), but are critical in determining the type or types of minerals that form. Certain critical factors seem to be the most influential.

- The presence of methane (with or without other hydrocarbons) results in microbial activity and, ultimately, the formation of MDAC.
- Water that circulates and is exhaled without exceeding a temperature of about 130 °C retains its original magnesium content.
- Water that exceeds a temperature of 130 °C loses its magnesium to the 'country' rocks or sediments, resulting in the dolomitisation of $CaCO_3$, and the precipitation of $BaSO_4$.
- Water heated above the critical point results in the scavenging of various elements from the rock through which it passes. These elements are transported with the fluid until temperature, pressure, and/or pH changes result in their precipitation. As the steepest

temperature, pressure, and pH gradients occur at the seabed, it is here that most rapid mineral formation occurs.

Although the above seem to apply to most mineral formations related to fluid flow, there are exceptions. Perhaps most notable of these result from the flow of fluids that have not circulated at all, but have been derived from the mantle. Such 'primordial' fluids are still little understood and they have no inflow zones, just 'migration' pathways, and 'outflow' zones where mineral precipitation may happen.

All in all, the combinations of time, place, flux, mineral combinations, and fluids, are so enormous, that one starts wondering if each and every outflow zone in the ocean is unique. However, it is perhaps more correct to consider the spectrum of mineral precipitates as a reflection of a continuous range of natural processes.

10 • Impacts on the hydrosphere and atmosphere

Although no single, unequivocal proof exists for the hypothesis we present, a broad range of evidence suggests it. We expect reluctance by scientists to support such an idea at this stage without further testing, but if the hypothesis holds, it will require a major shift in thinking about what drives Quaternary climate change.

James Kennett, 2002

In an interview with Lifland about the reaction to his book: *Methane hydrates in Quaternary climate change*: *The Clathrate Gun Hypothesis* (Lifland, 2002)

Once through the seabed, fluids of any origin contribute to, and in some cases, significantly affect, the marine environment. The chemical, physical, and biological nature of seawater may be affected by seabed fluid flow. Even in the deep oceans, emissions from hydrothermal vents, cold seeps, and mud volcanoes are influential. However, a proportion of some components passes through the hydrosphere and enters the atmosphere. Methane is one of these components. Contributions to atmospheric methane by the oceans, and in particular the seabed, are poorly understood, but may be important not only to today's climate, but also to global climate change over geological time-scales.

10.1 INTRODUCTION

Despite recent advances in the understanding of hydrothermal vents, gas seeps, mud volcanoes, etc., the oceans are rarely, if ever, considered as a single system. It seems that many oceanographers are reluctant to acknowledge the potential significance of geological inputs; 'the ocean system' seems to stop close beneath the seabed – at about the redox boundary. To us this is nonsense.

Freshwater and petroleum seeps have been known for centuries, but hydrothermal vents are a relatively recent discovery. Many scientists may regard them as mere curiosities, but as the pace of discovery and research quickens their importance becomes increasingly obvious. In this chapter we consider how geological inputs, manifested as seabed fluid flow (hydrothermal vents, submarine groundwater discharge, and seeps), contribute to the ocean system, and whether they, in turn, influence interactions between the oceans and the atmosphere. We conclude by considering the place of seabed fluid flow in the global carbon cycle, and its role in global climate change.

10.2 HYDROTHERMAL VENTS AND PLUMES

Finding hydrothermal vents in the deep ocean is perhaps as easy as finding a needle in a haystack. Emitted fluids are rapidly diluted by normal ocean water, but elevated concentrations of various dissolved components (especially methane), light-attenuating particles, and elevated temperatures can be detected in the water above vents (Baker *et al.*, 1995; Kadko *et al.*, 1995). Baker *et al.* (1995) reported that dissolved manganese and methane are commonly used to identify hydrothermal plumes. They cited three reasons for this:

- compared to normal seawater, hydrothermal fluids are significantly enriched in both;
- both can be analysed precisely at sea;
- the contrasting residence times provide information about plume dynamics.

Plumes are useful exploration aids, showing the way to vent fields on the seabed. They spread out over wide areas, providing geochemists with '*a magnifying glass with which to prospect for new hydrothermal vent sites*' (Baker *et al.*, 1995). However, it is now appreciated that these plumes are significant in their own right because of the influence they have on the composition and heat of the oceans. They are also significant contributors to the biology of the oceans (discussed in Section 8.3.4).

Hydrothermal fluids pass into seawater in three ways: as a diffuse flow, as a focussed flow through vents (including black and white smokers), and in 'event plumes'. The first two types are regarded as 'steady-state' emanations as venting seems to continue for periods of years, decades or more, but 'event plumes' probably last no more than a few days. Baker *et al.* (1995) attempted to determine the frequency of venting activity at ocean spreading centres, both in time and space, by comparing spreading rates with plume incidence. They found a strong positive correlation, the largest number of plumes being found on the fast-moving southern part of the East Pacific Rise. However, they also found that *'vigorous hydrothermal activity can occur on segments of any spreading rate, so an instantaneous observation at a particular location may not be representative of its long-term hydrothermal history'*. As it is impossible to make observations over geological time periods, the best alternative is to observe many individual ridge segments spreading at about the same rate. Baker *et al.* studied ridge axes spreading at about 100 mm y^{-1}, and found that about one third of them are active at any one time, and that any area on an active part of a segment will be active for about one third of the time.

During the early and late stages of volcanic activity, diffuse flow dominates. Indeed, Hannington *et al.* (1995) thought that *'the heat and mass flux due to diffuse flow in some vent fields is substantial and may be an order of magnitude greater than that of focussed high-temperature flow'*. Scott (1997; citing Rona, 1984, and Morton and Sleep, 1985), suggested that diffuse flow at ocean spreading centres may account for as much as 90% of the oceanic convective heat flux, and significant amounts of the *'material flux'*. These conclusions must also apply to hydrothermal vents in other contexts.

10.2.1 Plumes

Hydrothermal plumes provide the most precise data about hydrothermal flows as, once located, flow rates can be measured, and both fluids and chimneys (discussed in Section 9.4) can be sampled and analysed. Typically, plumes from ocean spreading centres rise 100 to 500 m above the seabed (Kadko *et al.*, 1995), before attaining a level of neutral buoyancy. Then they spread

laterally. For example, Horibe *et al.* (1986) detected large plumes of methane-enriched water in the Mariana Valley back-arc spreading centre. The main hydrothermal methane plume rose 700–800 m above the seabed and extended several miles along the axial region of the Mariana Trough.

Event plumes

Event plumes, although short-lived, are far more spectacular than steady-state flows, rising as much as 1 km above the seabed before settling at a constant altitude and spreading laterally (Kadko *et al.*, 1995). The first event plume identified (in 1986) was described by Baker *et al.* (1987) as a 'megaplume'. It had a diameter of about 20 km, a thickness of about 600 m, and lay some 800 m above the Cleft Segment of the Juan de Fuca Ridge. Its volume was estimated as 10^8 m^3 (Baker *et al.*, 1995). The presence of anhydrite, a mineral that rapidly breaks down in normal seawater conditions, indicated that it vented from the seabed only a short time before it was found, and sure enough, it disappeared before the next survey, 60 days later. At first it seemed likely that a geologically rare event had been witnessed by good luck. But a second event, 'Megaplume II', was detected about one year later only 50 km from Megaplume I. Since then several others have been reported:

- the 9° 50′ N event on the East Pacific Rise, in 1991 (Haymon *et al.*, 1993; Lutz *et al.*, 1994);
- three individual events on the CoAxial segment of Juan de Fuca Ridge identified during a period of four months in 1993 (Baker *et al.*, 1995);
- the 42° 40′ N event on the Gorda Ridge, in early 1996 (Chevaldonné, 1997);
- the Loihi Seamount event, Hawaii, in July 1996 (Holden, 1996).

Event plumes seem to be linked to igneous activity. Hannington *et al.* (1995) and Van Dover (2000), summarising previous publications, reported that the CoAxial and Cleft Segment events were associated with dyke intrusions and magma extrusion; new lava flows associated with the CoAxial event covered an area 2.5 km long and 300 wide. The events were followed by widespread diffusive flow of low-temperature (≤ 60 °C) fluids from fractures in the newly and previously extrudsed lavas. In some places bacterial flocs, originating from underground warm-water chambers, flowed out of the seabed together with warm fluids, appearing in the water column like drifting snowflakes. These events have clearly injected a considerable microbial biomass, available as food for higher organisms, into the water column.

SHALLOW-WATER EVENTS

Events such as those described above are not confined to deep-water hydrothermal systems. Shallow-water hydrothermal 'seeps' have been reported from several places such as the Aegean Sea (Section 3.10.1), Matupi Harbour, New Britain (Section 3.17.3), and the Taupo Volcanic Zone, New Zealand (Section 3.17.4). In early November 2002 a vigorous degassing event was recorded close to Panarea Island (one of the Aeolian Islands of southern Italy). Widespread seabed carbon dioxide-rich gas and water emissions have been known from this area since Roman times. The 2002 event is thought to have started with a submarine explosion as a crater 10 by 20 m across and 7 m deep formed. This was followed by a period of vigorous bubbling which lifted sediment from the seabed forming a turbid cloud in the water. The composition of gases collected after the event are noticeably different from those collected about 20 years earlier, with increases in hydrogen and carbon monoxide concentrations, but a decrease in methane concentration. Chiodini *et al.* (2004) attributed this event to the injection of '*a large mass of magmatic gas into a pre-existing hydrothermal system*'. This event shows that quiescent volcanoes can be affected by periodic episodes of intense magmatic degassing.

The relative importance of event plumes and steady-state flux

Because event plumes are short-lived and restricted to relatively small areas, they are unlikely to be detected by large-scale oceanographic surveys such as those of the GEOSECS (Geochemical Ocean Section Study) programme and WOCE (the World Ocean Circulation Experiment). However, plumes gradually spread out over wide areas, so their chemical signature, most obviously the mantle-derived ^{3}He, provides an indication of the contribution they make to the oceans. Relative to those of steady-state plumes, the contributions of event plumes are injected high into the water column, and because there is little vertical mixing in the deep-ocean basins, the signatures of these two types of emission can be distinguished. Lupton (1995) noted that the ^{3}He maximum in the Pacific occurs at a depth of

about 2500 m, the average depth at which steady-state plumes spread laterally. He suggested that '*event plumes do not contribute a major fraction of the hydrothermal input to the deep Pacific*'. Nevertheless, Kadko *et al.* (1995) suggested that contributions from diffusive flow away from spreading-centre axes make an even greater contribution. It seems that event plumes really contribute only a drop in the ocean, compared to the steady-state flux.

10.2.2 Plume composition

Plumes rising from hydrothermal vents on ocean spreading centres may contain significant quantities of various elements and compounds. Relative to typical oceanic deep waters they may be enriched by up to a factor of 10^7 in manganese, iron, methane, hydrogen, helium–3 (^{3}He), and many other trace species (Baker *et al.*, 1995). Other trace species may include barium, calcium, cadmium, carbon dioxide, copper, lithium, lead, sulphur, silicon, zinc, and many others. These elements are leached from rocks through which fluids circulate in the ocean crust, but magnesium is removed (see Section 5.2.3), so vent fluids are depleted in magnesium relative to normal seawater.

Vent-fluid composition is affected by the composition of the rocks with which it comes into contact, whether they are typical 'mid-ocean-ridge basalts', the basalts found in intra-ocean-plate hot spots (such as the Hawaiian Islands, including Loihi Seamount), or the andesites, diorites, dacites, and rhyolites typical of island arcs. However, it is not just the rock type that affects composition because there is a remarkable difference in the composition of mid-ocean-ridge hydrothermal fluids. This might be accounted for by the depth of the reaction zone, particularly the depth relative to the critical point of seawater (see Section 5.2.3), which affects the concentration of elements such as silicon, iron, manganese, calcium, strontium, and hydrogen. The presence or absence of a sediment 'cap' also has a major influence (see Section 5.2.3). Subduction-related fluids may also show the influence of subducted sediments. This may result in the presence of thermogenic methane and higher hydrocarbon gases, however, abiogenic methane is a normal constituent of fluids from ocean spreading centres.

Hydrothermal gases from other geological settings are somewhat different from those of ocean spreading centres. For example, those of back-arc volcanic systems are influenced by the composition and water content of the subducted slab, and by the nature of the igneous activity. Dando *et al.* (2000) reported on hydrothermal vents in the Aegean back-arc basin. Here, because of the shallowness of the water, the emitted fluids are dominated by gas. The main component is carbon dioxide, but there is also significant hydrogen, methane, hydrogen sulphide, sulphur dioxide, and nitrogen. Dando *et al.* reported that gas fluxes from volcanic settings in the Mediterranean are large; they cited reports of 16.8×10^3 kg d^{-1} from a small cavern in the Souska area (Georgalas *et al.*, 1962) and 15×10^3 kg d^{-1} from an active area on Nea Kameni Island in the Santorini caldera (Chiodini *et al.*, 1995). Flux rates are clearly affected by earthquake events (see Section 3.10.1).

Not surprisingly, fluids resulting from the exothermic serpentinisation of peridotite (Section 5.2.4) are very different from those emerging after interaction with igneous intrusions. Charlou *et al.* (2002) reported that vent fluids from the Rainbow field on the Mid-Atlantic Ridge show extraordinary high hydrogen (16 mmol kg^{-1}), methane (2.5 mmol kg^{-1}), and carbon monoxide (5 µmol kg^{-1}) end-member concentrations. Approximations made by Dmitrievsky *et al.* (2001) suggest that 1.5×10^5 t methane and 3.0×10^5 t hydrogen are produced annually by serpentinisation along the Mid-Atlantic Ridge between the equator and the Rainbow vents (36° N).

10.2.3 Plumes and the composition of the oceans

Considering that ocean spreading centres, submarine volcanoes, and venting have been features of the history of our planet for billions of years, it seems inevitable to us that this form of seabed fluid flow has played a significant role in the development of the oceans' present composition.

Kadko *et al.* (1995) estimated that all the water in the world's oceans is cycled through hydrothermal systems every 1 to 10 million years; Fyfe (1994) estimated about 10 million years (see Section 5.2.3). However, rising plumes entrain water from the surrounding ocean, mixing with it, and exchanging heat, chemical properties, nutrients, and organisms. According to Kadko *et al.* (1995), it takes a much shorter period, about 100 000 years, for all ocean water to be affected by hydrothermal activity. Although this time-scale may be long compared to the 40 000 years taken to renew the oceans by the input of freshwater from rivers (from Broecker and Peng, 1982, quoted by Kadko *et al.*, 1995), the potency of hydrothermal plumes clearly suggests that they do make an impact.

These estimates sound very plausible, but let us remember that investigations of ocean hydrothermal systems have been going on for only a few decades, and for most of that time searches for hydrothermal plumes focussed only on ocean spreading centres. Baker *et al.* (1995) reported that plumes had been detected at > 24 locations at different spreading centres in the Pacific, Atlantic, and Indian oceans, yet only about 10% of the world's ridge systems had been surveyed by that time, and only a small part of it in '*enough detail to confidently locate all the major vent fields on a particular segment*'. Nevertheless, even this poor data set was enough to show that there is hydrothermal venting '*albeit at widely differing levels of activity, across the entire spectrum of ridge crest spreading rate, topographic expression, magmatic budget, and other geological and geophysical characteristics*' (Baker *et al.*, 1995).

The rate of exploration has been such that by 2001 Kelley *et al.* could report that the number of vents investigated on the ocean-spreading-centre system had risen from 24 to 40. They reported the discovery of the Lost City hydrothermal field (see Section 5.2.4), a previously unknown type of hydrothermal system, which they considered '*may be common along a significant portion of the ridge system*'. Since 1995 there have also been significant discoveries of hydrothermal plumes associated with island-arc volcanism, and back-arc and fore-arc basins. If the activity levels of the Kermadec and Mariana arcs (described in Section 4.3.2) are representative, then '*this setting may prove to be volumetrically significant to global hydrothermal fluxes of heat and chemicals to the oceans*' (de Ronde *et al.*, 2001). Seamounts also have a potentially significant impact, cycling a similar volume of fluid as the ocean-spreading-centre system (discussed in Section 4.3.1).

As hydrothermal systems associated with seamounts, island arcs, and serpentinisation, as well as ocean spreading centres, are now recognised (Map 33), we must suggest that Kadko *et al.* (1995) may have seriously overestimated the cycle time, and therefore underestimated the significance, of hydrothermal systems.

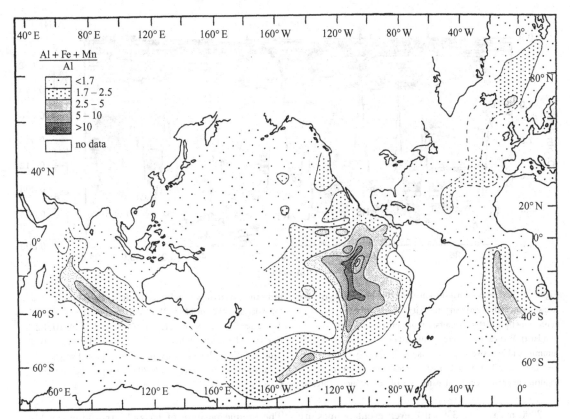

Figure 10.1 The fallout of metals from hydrothermal plumes arising from ocean spreading centres indicated by the (Fe + Mn + Al):Al ratio of surface sediments. Low values away from the spreading centres represent background pelagic sedimentation. (Reproduced with permission from Mills and Elderfield, 1995.)

Residence time and fate

The residence time of plume constituents differs considerably. Some survive only a short time, being precipitated out, dissolved, or utilised by microbial activity within, or close to the seabed. On the other hand, Lupton and Craig (1981) identified a huge, ^{3}He-rich, plume extending more than 2000 km away from the East Pacific Rise.

The residence time of methane is determined by oxidation rates, estimated by Welhan and Craig (1983) to be about 10 days in the open ocean. However, this time period will increase if the methane concentration of the 'background' water is higher. The oxidation of methane and other potential energy sources (hydrogen, hydrogen sulphide, Fe^{2+}, and Mn^{2+} ions) is microbially mediated. The Fe^{2+} ions and H_2S combine to form particles in the plume, and are oxidised relatively quickly. However, methane, hydrogen, and Mn^{2+} ions

survive longer and microbial activity related to them has been shown to increase as the plume drifts downcurrent (Lilley et al., 1995). Elements scavenged from the ocean water, along with fallout from plumes, are deposited downcurrent as metal-rich sediments. The effect of this fallout, seen on Figure 10.1, is regional and clearly related to the distribution and activity of ocean spreading centres. Although the significance was not appreciated at the time, metal-rich sediments in this fallout zone were identified from samples collected during the Challenger expedition more than 100 years ago (Murray and Renard, 1891).

The influence of altitude

Much of the literature about the influence of hydrothermal systems on the composition of the oceans has been concerned with systems on ocean spreading centres. Because the majority of these are in deep water

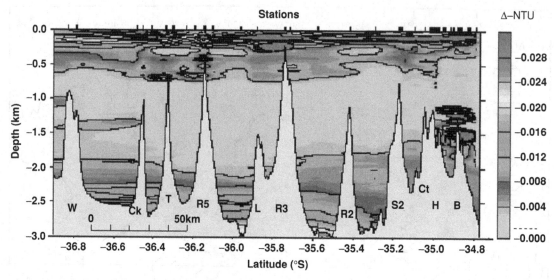

Figure 10.2* Longitudinal section along the southern Kermadec Arc showing light-scattering intensity (in Δ-NTU; nepholometryic turbidity units, a non-dimensional optical standard). Vertical casts were conducted over the volcano summits and in between the volcanoes. Deep tow-yo surveys were done over Tangaroa, Rumble V, Healy, and Brothers resulting in greater sample density over these volcanoes. Intense hydrothermal activity is indicated by the light-scattering intensity over the volcanoes. W = Whakatane; Ck = Clarke; T = Tangaroa; R5 = Rumble V; L = Lillie; R3 = Rumble III; R2 = Rumble II East; S2 = Silent II; Ct = Cotton; H = Healy; B = Brothers. [Reproduced (with minor modifications) with permission from de Ronde *et al.*, 2001 with permission from Elsevier.]

(> 2000 m; average 3300 m), even event plumes are unable to extend their influence above mid water. However, this is untrue of systems associated with seamounts and arcs, which inject their fluids direct into the middle or upper layers of the water column. This is clearly demonstrated by investigations of the Kermadec Arc by de Ronde *et al.* (2001), and that of the Mariana Arc by Merle *et al.* (2003), see Figure 10.2. Emissions of gases from shallow water vents, such as those off the Aegean Islands (Section 3.10.1) and the Matupi Harbour (3.17.3), are particularly significant as they are much more likely to pass through the water column and enter the atmosphere than those from deep-water sites. Major volcanic eruptions supply enormous quantities of gases more or less direct to the atmosphere (Section 10.6).

10.2.4 Heating the oceans

The vast volumes of water involved in hydrothermal systems make a significant contribution both to the cooling of the oceanic crust and to the warming of deep-ocean waters. New light on this aspect of deep-ocean circulation systems came from studies undertaken on Iceland.

The unique position of Iceland, straddling the Mid-Atlantic Ridge, makes it a natural onshore laboratory for studying a host of volcanic and hydrothermal processes. In 1973 the fishing town of Vestmannaeyjar hit the news headlines because of the volcanic eruption of Heimaey. The lava flow threatened the town, but an immense amount of cold water was pumped on to it to divert it, providing some valuable scientific evidence. At one place water was continuously sprayed onto the lava for 14 days at a rate of $100 \, l \, s^{-1}$ over an area of 7000 m² (Jonsson and Matthiasson, 1974). Subsequent drilling revealed that the lava had been cooled to 10 °C to a depth of 12 m, but within 1 m the temperature rose to 1050 °C. Bjørnsson *et al.* (1982) concluded that, assuming all the water had been evaporated, then heat was transferred at a rate of 40 kW m⁻², and the cold 'front' penetrated downwards at a rate of 0.9 m d⁻¹. The temperature increase of approximately 1000 °C within an interval of 1 m is striking, but in agreement with the theory of downward penetration of cold water into hot rock (Stefansson, 1983). However, such a locally high temperature gradient proves that solidified basalt has very low heat conductivity and therefore acts as a good insulator, probably due to its porous nature. Later excavations at this site confirmed that the water-cooled rock was intensely fractured and broken into cubes ($\sim$10–20 cm across), whereas lava not water-cooled consisted of larger blocks (Bjørnsson *et al.*, 1982).

If we translate the lesson from the Icelandic study to other volcanic centres in the ocean, we may conclude not only that circulating water will effectively cool newly formed igneous rock, but also that heat energy is transferred to hydrothermal fluids. Subsequently this is transferred to the ocean via hydrothermal vents as plumes of relatively hot ($< 350\ °$C) water, and on ridge flanks as a diffuse flow of warm water (Kadko *et al.*, 1995).

Event plumes inject enormous quantities of heat into the oceans. For example, over a ten-day eruptive phase, the 'January 1989 event' at Axial Seamount produced an estimated 200 GW, more than two orders of magnitude more than the typical heat flow from the caldera during quiescent periods (Cann and Strens, 1989). Estimates of the total heat output of even a small section of mid-ocean ridge are not easy. McDuff (1995) reported that five attempts made for the Endeavour Segment of the Juan de Fuca Ridge resulted in estimates of $\sim$100 to $\sim$10 000 MW. If the best-known area of hydrothermal activity produces such diverse estimates, then global estimates must be regarded as no more than 'first approximations', for the time being at least.

10.3 SUBMARINE GROUNDWATER DISCHARGE

In Section 5.3.1 we defined submarine groundwater discharge (SGD), and noted that it occurs on all the continental shelves of the world (Map 35). Although it is potentially very important, there have been relatively few studies of SGD (Taniguchi *et al.*, 2002; 2003; Burnett *et al.*, 2003). Whereas there are reasonably precise measures of the contributions to the oceans made by rivers, this cannot be said of SGD. Burnett *et al.* (2003) suggested not only that this was because the contributions of SGD are '*inherently very difficult to measure*', but also because hydrologists and oceanographers have adopted different approaches; '*hydrologists and oceanographers are literally approaching the same problem from different ends*' (Burnett *et al.*, 2003).

The scale of SGD varies considerably. Generally, discharge rates decline as water depth and distance from the coast increase (Taniguchi *et al.*, 2002). Apparently small springs can produce a surprising amount of freshwater. For example, the Ovacık springs, close to the Mediterranean coast of Turkey, occur at a water depth of only 1.5 m; they are visible because upwelling water produces 1.2 m wide cones at the sea surface. The three springs produce water at a rate of 0.75 m^3 s^{-1}, a total of 23.3×10^6 m^3 y^{-1} (Elhatip, 2003). In contrast, in the Gulf of Kastela in the Adriatic Sea, freshwater expulsion is so intense it poses a hazard to the navigation of small vessels. Zektzer *et al.* (1973) reported that springs often cause visible changes to the sea surface: '*the appearance of boils of water, the change of seawater colouring, the emergence of bubbles and the occurrence of a slick or smooth area over a submarine spring in a choppy sea*'. Some other springs are identified by similar sea-surface indications, and in some cases, for example in Cambridge Fjord, Baffin Island (see Section 3.29.3) spring water causes a polynya (ice-free area) to form.

These examples are all close to the coast, but in some places the distance from the coast, and the water depth at which SGD occurs is quite amazing. The Florida Escarpment seeps in $> 3\,200$ m of water are the deepest we have heard of; but one of the most startling reports was from a depth of 600 m on Blake Ridge, about 200 km off the east coast of the USA. In 1966, whilst surveying the manganese phosphate-paved seabed at a depth of about 510 m, the submersible Aluminaut suddenly lost buoyancy as it passed over a 50 m deep seabed depression. The water temperature here was about 10 °C compared to 12.5 °C outside the depression. Manheim (1967) concluded that this loss of buoyancy '*can only be explained by the outflow of water substantially less salty than seawater*'.

10.3.1 Detection and quantification

Although plumes from freshwater springs can be detected acoustically (Hay, 1984) other methods are more common, including direct measurement with seabed flow meters, tracer studies using natural components such as radon (^{222}Rn), artificial tracers such as sulphur hexafluoride and iodine, and modelling. Radium (Ra) isotope tagging, radon detection, and thermal imagery provide the most efficient and affordable reconnaissance method for SGD over large coastal areas. These methods rely on thermal infrared aerial surveys flown with light aircraft (Roseen *et al.*, 2001). In one example, nutrient loading from groundwater discharge was assessed by a thermal infrared aerial survey along a 50 mile shoreline over the Great Bay Estuary, New Hampshire (USA) one afternoon in 2000. On the basis of images available shortly afterwards, with no post-processing, suspected groundwater discharge zones were identified. These proved to show remarkable agreement with flow estimates predicted from piezometric mapping from nearly 300 onshore monitoring wells performed earlier (Roseen *et al.*, 2001). A similar agreement between predicted and measured SGD was not found, however, during a comparison between predictive hydrological modelling and measured seeps along a 100 m wide and 200 m offshore-extending zone at Turkey Point in Florida (Gulf of Mexico side). Here a SGD flux of 2300–3600 m^3 d^{-1} was measured with seepage meters, whereas the conceptual hydrological model provided an SGD flux estimate in the range of only 10–200 m^3 d^{-1} (Smith *et al.*, 2001).

Such discrepancies, and the use of a wide range of (not always compatible) measurement techniques, do not make reviews such as those of Taniguchi *et al.* (2002) and Burnett *et al.* (2003) easy. Global estimates of the input of water into the oceans by SGD are probably not reliable because of the small number of measurements; these are mainly concentrated in Europe, eastern USA, and Japan (Taniguchi *et al.*, 2002). They are mainly based on assumptions about the proportion of coastal rocks that are aquifers, and assume that SGD will approximate to 6–10% of surface water flow. It is difficult to judge whether or not this is realistic, but estimates range between hundreds and thousands of km^3 y^{-1}; Faure *et al.* (2002) noted that they vary between 400 and 4 000 km^3 y^{-1}. Burnett *et al.* (2003) suggested 2 400 km^3 y^{-1}. The true amount may be larger than this if the results published by Moore (1996) are accurate and widely applicable. He used ^{226}Ra enrichment studies to demonstrate SGD off the Atlantic coast of South Carolina, and concluded that groundwater flux into these waters must be about 40% of the river flux.

10.3.2 Water quality

The significance of SGD is not just because of the volume of water involved, but also its salinity, the chemical components it contains, and, in some cases, its temperature. The quality is very variable. In some places the water is good enough to drink. One example is a submarine spring in Chekka (Lebanon) where, < 2 km from the coast, a funnel-shaped dome was used to cover the outlet, and water was transported through plastic hoses (Kohout, 1966). A second example is from the Arabian Gulf; Williams (1946) and Fromant (1965) reported freshwater discharges from narrow crevices in the seabed in the shallow waters of the Bahrain Archipelago, and Chapman (1981) reported that pearl fishermen were able to stay at sea for long periods because they collected freshwater from submarine springs. However, water abstraction rates increased with the development of Bahrain, so that the flow has reversed, with seawater flowing in where freshwater used to flow out. Submarine groundwater discharge has also reversed in other heavily populated areas, southeast England and eastern USA being examples.

The concentrations of the following may be significantly different from the seawater into which it flows:

- CO$_2$;
- organic carbon;
- carbonates – a potentially significant source of Ca^{2+} ions;
- sulphate – low sulphate concentrations permit methanogenesis in seabed sediments, for example in Eckernförde Bay (Section 3.4.1);
- methane;
- trace metals;
- nutrients (nitrates, phosphates) – with potentially significant effects on biological productivity;
- anthropogenic pollutants (sewage, mining waste, agricultural products – including fertilisers).

Taniguchi *et al.* (2002) and Burnett *et al.* (2003) mentioned examples of enhanced faunal and floral productivity, including effects on coral reefs, and cold-water chemosynthetic communities associated with the deep-water Florida Escarpment seeps. In extreme cases increased eutrophication has had a damaging effect on local ecology. Because a substantial component of river flow actually occurs beneath the riverbed, SGD may be particularly important in estuaries. Here the emerging flow will have avoided the chemical mixing processes occurring above the riverbed in estuaries, so its composition may be significantly different.

Church (1996) recognised the potential of SGD '*to radically alter our understanding of oceanic chemical mass balance*', because of its influence on the chemistry and nutrient content of the oceans; '*it could rival hydrother-mal inputs in magnitude*'.

10.4 SEEPS

In some areas tar balls and oil slicks provide clear evidence that oil seeps do exist (provided anthropogenic sources can be eliminated!). However, finding seep plumes is no easier than finding hydrothermal-vent plumes, and the identification of shallow gas, pockmarks, or mud volcanoes does not necessarily mean that there will be active seeps. The ability of methane and other gases to escape through the seabed is dependent on the flux rate exceeding the microbial utilisation rate (Figure 8.5). Previously (Section 6.2.4) we explained that acoustic turbidity normally extends no closer than a metre or so from the seabed because methane oxidation rates in sulphate-rich pore waters of the topmost sediments exceed the rate of supply. Unlike hydrothermal plumes, there is not necessarily a significant diffusive flow even where there is gas close beneath the seabed. However, if seabed fluid flow is focussed, then utilisation rates may be exceeded, black sulphidic sediments extend to the seabed, and seepage occurs. Whether focussing is through seeps, pockmarks, or mud volcanoes makes no difference to the behaviour and influence of fluids once they are in the water column, so we will not distinguish between them here.

10.4.1 Identifying seeps

Most seeps of petroleum fluids, and seabed features associated with them, are located acoustically. However, acoustic characteristics are often subtle, and may be ambiguous. As with many other things, it is best to use as much evidence as possible. We have already discussed seabed features – pockmarks, methane-derived authigenic carbonate (MDAC), cold-seep chemosynthetic communities – associated with seeps. We now investigate evidence from the water column, but as seep detection is a tool for petroleum exploration we will discuss some aspects of this in Section 11.6.4.

Most seeps are dominated by gas (see Section 7.5.3), but oil seeps are more easily detected at the surface because they form slicks. Studies of oil droplets rising from great depths in the Gulf of Mexico (MacDonald *et al.*, 2002) suggest that oil is most likely to reach the sea surface when it forms as a coating around gas bubbles; oil droplets may be transported significant horizontal distances from their source in water currents because of their slow rate of ascent.

Acoustic plumes

Seep plumes (e.g. Figures 2.42, 3.10, 3.19, 3.27, 3.32, 3.34, 3.36) rising from the seabed can be detected by high-frequency acoustic systems such as echo sounders, side-scan sonar, and shallow sub-bottom profilers (the physics of gas bubble detection was discussed in Section 6.2.1). However, there is a fundamental problem. Gas may be present in the water column because of seeps, or because it is inside the swim bladders of fish. Acoustically it is very hard to distinguish between the two, indeed 'fish-finding sonars' are excellent seep detectors (Figure 3.15). Some records of water-column targets were shown to delegates at a conference who were asked what caused them. The geophysicists unanimously said 'fish', but the biologists all said 'certainly *not* fish'!

Some seep plumes and some fish shoals are easily distinguished. Depending upon the species, fish come together as shoals for three reasons: feeding, spawning, or as part of their normal behaviour. After examining examples of typical fish-shoal 'targets' Judd *et al.* (1997) concluded that the majority of shoals are diffuse and horizontally extended. A few species (such as sprats) tend to form vertically extended shoals, but these tend not to be geometrical or consistent in shape. In contrast, continuous seeps produce a column of bubbles that rises more or less vertically through the water, although water currents may cause some deflection. The conclusion that columnar targets represent bubble plumes has been supported by visual evidence from many places. Further confirmation was provided by Dimitrov (2002a) who presented a side-scan sonar image of gas plumes from divers' compressed air bottles. Similar experiments were undertaken by Hornafius *et al.* (1999) to calibrate a sonar system used for quantifying seep plumes.

Columnar targets seem to represent continuous seepage. In shallow water, these columns may extend to the sea surface, but more often the plume appears to fade away as the bubbles dissolve (or pass out of the acoustic beam). 'Detached' water-column targets may represent bubbles that only present a detectable signal when they resonate. As explained in Section 6.2.1, there is a relationship between bubble size and the frequency of the acoustic source; target strength varies significantly as bubble size changes during ascent. An alternative explanation for detached water-column targets is intermittent seepage. It seems reasonable to suggest that gas accumulations in soft fine-grained, and in coarse near-seabed sediments need acquire only a slight overpressure to escape, so some seeps may be discontinuous. Intermittent seepage would be very difficult to document on the seabed, but bubbling from individual seep vents has been observed to switch on and off over short time periods at an intertidal site (Judd et al., 2002b). In other areas isolated water-column targets have been recorded by repeat surveys. For example, over a period of several years, research cruises and student training exercises in the Ría de Vigo (northwest Spain) have consistently produced records with water-column targets in an area with extensive shallow gas (Soledad García-Gil, 1999, personal communication; see slide 13 of García-Gil, website). These are not columnar, but the consistency with which they are recorded suggests that the seabed is continuously emitting bursts of bubbles from different seep vents.

Although vigorous and continuous seeps may be easily detected using a range of acoustic systems operating at different frequencies, the detection of weaker and discontinuous seeps is more complicated. Inappropriate choices of acoustic sources and operator settings may result in seeps being missed. We have found that high-frequency systems operating within the 1–100 kHz range are best, particularly when the amplification is set high. Even within this preferred frequency range, results may vary considerably from one system to another, at least in part because results also depend on beam spreading – as well as bubble size. A 3.5 kHz source is better suited to bubbles 13–16 mm in diameter, but 4–14 mm diameter bubbles are more easily detected with a 12.5 kHz source (Heeschen et al., 2003).

The first evidence we saw of gas in the water column was on pinger and echo sounder sections across the Tommeliten seeps (see Section 2.3.2). In October 2002, profiles acquired here using the parametric sediment echosounder system Innomar SES-2000DS, developed by the underwater acoustics research group of the Rostock University, Germany, proved much more informative, especially when operated at 12 kHz. Unlike previous profiles, Figure 10.3 clearly shows the seep plume extending from the seabed (74 m water depth) at least to within 10 m of the sea surface (the shallowest depth visible using this hull-mounted system). Visual observations have confirmed that gas is bubbling from

the seabed at this location, and high methane concentrations (500 nM) were measured within the plume; however, concentrations returned to 'background' (5 nM) outside the seep area (Niemann et al., 2004). It seems from these profiles that gas is passing through the water to escape to the atmosphere.

Water sampling

Sampling water using various types of 'bottle' is common practice in scientific seep investigations, and provides sufficient quantities of water for detailed analyses to be carried out. Results from both shallow- and deep-water sites have made it possible to determine the composition of seeping fluids and the contrast between them and the 'normal' sea bottom water. Bottle casts have also shown the change in composition with altitude above the seabed, and have demonstrated the down-current drift of seep plumes; for example Figure 10.4 shows the methane-rich plume above Hydrate Ridge. This plume was described as being '*hundreds of metres high and several kilometres wide*' with methane concentrations of up to 74 000 nl l^{-1}, compared to < 20 nl l^{-1} in the 'background' (Suess et al., 1999b).

The shortcoming with this technique is the uncertainty of obtaining samples from the right place. Visually guided sampling by submarine, remotely operated vehicle (ROV), or TV-guided sampler provides more valuable data as the vent itself can be sampled. For seep detection, instruments that can continuously monitor methane concentrations are more valuable; Sniffer systems have proved useful in petroleum exploration (see Figure 3.41 and Section 11.6.4), but a new generation of light-weight, easily deployable continuous methane detectors (such as the METS Methane Sensor) is likely to revolutionise our understanding of the behaviour of seep plumes.

Listening for seeps

Pontoise and Hello (2002) described a novel approach to seep detection, although their discovery came about more or less by accident when they deployed 13 ocean-bottom seismometers (OBSs). The purpose of their research was to study the seismicity of the subduction zone offshore Ecuador. Over a continuous recording period of three weeks they recorded '*a series of unusual intense monochromatic signals*', which was different from anything else they had seen over their 20 years of experience. Their research into the origin of

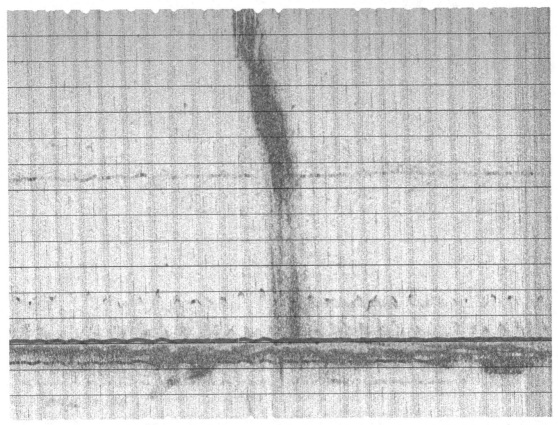

Figure 10.3* Parasound profile across the Tommeliten seeps, Norwegian North Sea. This image shows gas beneath the seabed, and a seepage plume extending from the seabed (74 m water depth) to the hull-mounted transducer – effectively to the sea surface. The mid-water targets represent equipment deployed from the ship. (Image recorded using an SES-2000DS parametric echo sounder operated at 12 kHz on Heincke cruise He180, October 2002; courtesy of Gert Wendt, University of Rostock.)

this 'noise' eliminated ship propeller noise and '*biological noise*' (from whales). They tentatively interpreted the noise as '*pressure-waves resulting from oscillating clouds of bubbles*'. They estimated that the bubble clouds were about 10 m across at depths of 5–15 m, and '*must contain more than 10^{10} bubbles of 1 mm radius to produce the observed frequency*'. Their study area was characterised by an extensive bottom-simulating reflector (BSR) indicating the presence of gas hydrates, so this explanation does seem plausible. If it is, then perhaps this research has found a new way of finding seeps!

Subsequently, diver-held video cameras filming the Shane Seeps in the Santa Barbara Channel, California recorded sound as well as pictures (Ira Leifer, 2001, personal communication). Yes, seeps make a distinct bubbling noise under water!

10.4.2 Eruptions and blowouts

In Section 10.2.1 we described 'event plumes' from hydrothermal systems. In Section 7.3.4 we described powerful (Lokbatan-type) mud-volcano eruptions, and noted that submarine examples of this type of eruption have been reported. Are there equivalent events, natural-gas blowouts, in seep systems? It has been argued that such events are myths, and that there is no evidence to support the hypothesis that they occur. Well, we certainly agree that the overwhelming majority of observations (visual and acoustic) indicate 'steady-state' seepage, whether it is gentle or vigorous. However, we do not concede that this rules out the possibility of more catastrophic events.

In Section 7.5.6 we described 'giant pockmarks' in the North Sea and the Barents Sea. These features seem

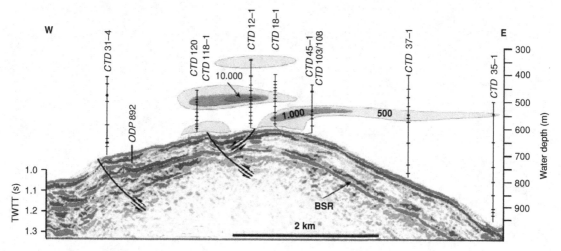

Figure 10.4* Shallow seismic section across the northern summit of Hydrate Ridge, offshore Oregon, showing methane plumes rising from the seabed. Methane concentrations in samples taken from CTD (conductivity salinity/temperature/ depth) casts are in nl l^{-1}; TWTT = two-way travel time. (Reproduced from Suess *et al.*, 1999b with permission from Elsevier.)

to have formed as a result of the release of gas accumulated under permafrost or gas hydrate in 'one-off' events when melting occurred during the climatic amelioration at the end of the last ice age. We think that other evidence supports the occurrence of occasional significant gas-release events, 'natural-gas blowouts', even under present-day conditions.

- We have heard several eye-witness accounts of large volumes of gas bubbles breaking the sea surface (and some from freshwater sites too), some from reputable scientists – yet there is no 'hard' and unambiguous evidence (video footage, etc.) to substantiate their claims and convince the sceptics. The following example is from the central Adriatic Sea:

  > in 1978, fishermen observed strange phenomena such as: 50 m high water column, vertical motion (wakes) in the water, anomalous big waves, red lights rising from the sea to 200–300 m in the sky before disappearing, anomalous radar signals, a diving black body, sunk fishing boat close to the coast.

  > Curzi, 1998

  Curzi suggested that at least some of these phenomena could be attributed to the escape of high-pressure gas from the seabed, possibly triggered by an episode of fault activity.

- The sediment clouds described by McQuillin and Fannin (see Figure 2.18 and Section 2.3.1), and others observed in the same area, surely indicate a significant gas-release event.
- During a survey of one of the giant pockmarks in Block UK15/25, North Sea (described in Section 2.3.7) an unusual water-column target was recorded (Figure 7.3). Gas bubble plumes like those recorded in this pockmark on other occasions are visible, along with what appears to be a massive cluster of bubbles.
- A network of seabed flux meters deployed at the Shane Seep, Santa Barbara Channel, recently quantified '*an intense gas ejection (500 l in 10 s). The flux virtually stopped both before and after this event, before resuming at a rate higher than the pre-event level. Meanwhile, 90 s after the ejection, the flux 5 m away decreased significantly, suggesting a shared plumbing system. Video of a separate event showed tar ejection as well. It seems that tar may have blocked the migration pathway, allowing pressure behind the blockage to increase until it was strong enough to "cough", clearing the blocked throat*' (Leifer et al., 2003; see also Leifer et al., 2006).

Although we can offer no further evidence to confirm the existence of natural-gas blowouts this is not entirely surprising. Unlike hydrothermal event plumes, the gas from a natural-gas blowout on the continental shelf would rapidly escape to the sea surface, leaving no detectable trace behind – apart from sediment clouds, which *have* been documented, and perhaps a new or enlarged pockmark. Also, there is a very small chance of being at the right place at the right time to record a rare event. To this we add that such events are most likely to be triggered by an earthquake or storm. Earthquake-triggered events *have* been documented (e.g. from Greece; see Section 7.5.5). Events triggered by storms (much more likely in the North Sea) may go unnoticed because normal survey operations have to be suspended in heavy weather.

Supportive evidence for natural blowouts comes from seabed craters caused by blowouts during offshore petroleum drilling. The formation of the Figge Maar (described in Section 11.3.1), for example, is convincing evidence that massive gas-release events can result in the formation of seabed craters.

10.4.3 Quantifying seeps

This book is devoted to explaining the significance of seabed fluid flow to the way our planet works, yet quantifying it is very difficult. So far, the number of reliable observations, measurements, and analyses is pathetically small. Sea-surface oil slicks are commonly associated with shipping and oil exploration/exploitation, yet significant quantities of crude oil enter the marine environment naturally: 200 000 to 2 000 000 (best estimate 60 000) $t\,y^{-1}$, > 45% of the total emission (Kvenvolden and Cooper, 2003; Committee on Oil in the Sea, 2003). Although this contribution of liquid hydrocarbon is important, gaseous hydrocarbons may be even more significant.

The composition of seep fluids

Methane unquestionably dominates the composition of cold-seep fluids. The proportion of methane, and the nature of the accompanying fluids varies considerably; methane-dominated seeps can be grouped into those:

- dominated by microbial methane: $\delta^{13}C$ values typically lie within the range −85 to −55‰;
- dominated by thermogenic methane: $\delta^{13}C$ values typically lie within the range −60 to −20‰, $C_1 : C_{2+}$ ratios 200–10;
- associated with gas hydrates – thermogenic-methane source;
- associated with gas hydrates – microbial-methane source.

Table 10.1 *Indicative composition of mud-volcano gases (after Judd, 2005a).*

Methane	Carbon dioxide	Nitrogen	Higher hydrocarbon gases
85.5%	9.5%	4.5%	0.5%

Based on data derived from the following published sources: Aliyev *et al.*, 2002; Etiope *et al.*, 2002; Dimitrov, 2003; Milkov *et al.*, 2003; Martinelli and Judd, 2004.

However, it seems that there are also 'mixed' seeps, for example those in which thermogenic and microbial gases generated at different depths are focussed towards the same migration pathway.

MUD-VOLCANO EMISSIONS

Mud volcanoes are a little different from seeps as they emit a combination of solids, liquids, and gases (see Section 7.3.3). This is known to be true offshore and onshore. Offshore the liquids may collect to form 'brine lakes' which, presumably, affect the salinity of the surrounding seawater. However, gas emissions are more likely to affect the composition of the water column.

The composition of gases from mud volcanoes of various regions (onshore and offshore) have been reported; Dimitrov (2003) and Milkov *et al.* (2003) provided global summaries. The majority of mud-volcano gases are dominated by methane, which may be accompanied by higher hydrocarbons (including crude oil) if the source is thermogenic. Carbon dioxide occurs in significant proportions in many cases, and hydrogen sulphide in a few; in south Alaska (onshore) there are a few mud volcanoes with nitrogen-dominated gases (Kopf, 2002). Table 10.1 provides an 'indicative' composition representing a global average (based on the data available). This is probably representative of submarine mud volcanoes.

Measuring flow rates

Apart from the logistics of deploying sensitive equipment on the seabed, there are some basic problems.

- Diffuse flow, focussed venting, and 'events' result in a considerable range of flux rates; a range of over eight orders of magnitude (< 0.1 mm to > 1000 m per year, according to Tryon *et al.*, 2001).
- Flux rates vary in space; the patchy distribution of seep-related benthic fauna shows that fluid flow is very localised. Indeed O'Hara *et al.* (1995) showed that, in porous sediments, bubble emissions set up a convection circulation of water, which is drawn into the sediment around the seep vent and entrained with rising bubbles; downward fluxes were also demonstrated around the deep-water seeps of Hydrate Ridge (Tryon *et al.*, 1999). Flux may be out of, or into the sediment, depending upon where you take your measurement. The unwary could get very confused if their equipment landed adjacent to a seep, rather than over it!
- Flux rates vary in time; the rate of flux from an individual vent may vary according to the rate of supply of fluid from below, pressure variations (e.g. tidal influences), and various triggering events (see Section 7.5.5). Flux-rate measurements reported in publications have varied in duration from a few minutes to a few days. Only the monitoring of the Seep Tents (Santa Barbara Channel, California – see Section 11.6.2) cover an extended period of time (nine months).
- The range of measuring strategies is such that data compatibility must be questioned.

Tryon *et al.* (2001) provided a useful review of the various methods used. These include simple mechanical devices (funnels that channel bubbles into graduated flasks), 'bubbleometers' (that count bubbles as they escape), and other meters for measuring the faster fluxes. More sophisticated techniques such as thermal and chemical profile modelling have been developed for slower flux rates. Deployment strategies have involved 'blind' placement on the seabed from surface ships, submarine and ROV installations, and TV-guided equipment. Although such devices have been deployed successfully at various locations, resulting data cannot be relied on to be representative of large areas, or over long time periods. There is an element of subjectivity in the placement of the device, and an element of chance; will the flux at the selected location remain active for the duration of the measuring period? Are data biased towards those that seem attractive because they are more vigorous than neighbouring seeps? Tryon *et al.* sug-

gested that the diffuse component of the flux from a seep area '*is probably as important as focussed flow in terms of the total mass balance of fluids because of the greater areas involved*'.

Despite the shortcomings in measurements, data from them are, for the time being at least, all there is to go on. However, a new generation of acoustic measuring techniques is being developed (Greinert and Nützel, 2004). By quantifying the volume of gas bubble plumes these techniques should provide an estimation of the flux from whole seep fields, rather than just from individual vents or areas of seabed.

FLUXES FROM MUD VOLCANOES

In Section 7.3.4 we explained that mud volcanoes can be categorised according to their style of activity. For the purpose of estimating fluxes, mud-volcano emissions can be usefully categorised as: continuous, 'normal' eruptions, and 'strong' eruptions. Estimates of the flux from each type of emission can be made for onshore mud volcanoes, and it is probably realistic to assume that comparable rates apply offshore. Milkov *et al.* (2003) collated estimates of gas emission rates from mud volcanoes around the world; the 36 mud volcanoes (a very small sample, but apparently all the data available) showed variations between 100 and 10^7 m^3 y^{-1}. Assuming a log-normal distribution, Milkov *et al.* suggested a mean flux of 3.6×10^6 m^3 for each quiescent (Chikishlyar-type) mud volcano.

At the other end of the scale are the strong, Lokbatan-type eruptions (Figure 7.8). A marine example photographed from Baku on 15 November 1958 produced a burning gas column estimated to be several kilometres high at first, later reducing to 120 m diameter and 500 m high (Figure 10.5). According to the Geological Institute of Azerbaijan, this flame was produced by a mud-volcano eruption on Makarov Bank, about 20 km southeast of Baku in a water depth of about 100 m. The volume of expelled gas was estimated to be 300×10^6 m^3, but such events are difficult to quantify. Dadashev (1963; cited by Milkov *et al.*, 2003) proposed 250×10^6 m^3 of gas per Lokbatan-type event; Dimitrov (2002b) suggested 340×10^6 m^3 per event, so the Makarov Bank event was about 'average'. Weaker, 'normal' eruptions probably produce about one tenth of this volume.

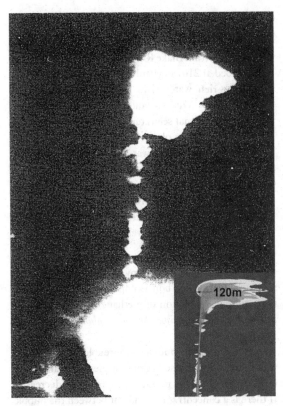

Figure 10.5* Flame from the eruption of a submarine mud volcano on Makarov Bank, photographed from Baku, Azerbaijan on 15 November 1958. Makarov Bank lies about 20 km offshore, in the Caspian Sea; the water is about 100 m deep. The height of the gas column was initially estimated to be several kilometres, but subsequently it stabilised at 500 m tall, and extended about 120 m laterally (as indicated by the inset interpretation). It was estimated that about 300×10^6 m³ of gas was expelled in this eruption. (Image courtesy of the Geological Institute of Azerbaijan.)

DIFFUSIVE FLOW FROM GASSY SEABED SEDIMENTS

The shallow, organic-rich waters of Cape Lookout Bight (described in Section 3.27.1) provide an extreme example of *in situ* methane production and diffusive flux through the seabed totalling about 9.6×10^4 m³ methane per year. During the winter months (November to May) the flux was found to average 49 μm m^{-2} h^{-1}, but in summer, when production rates are much higher, the flux increases to 163 μm m^{-2} h^{-1} because of the formation and escape of bubbles (Martens and Klump, 1980). As the '*highest anaerobic remineralization rates yet mea-*

sured *in a natural coastal environment*' were measured at the Cape Lookout site (Martens *et al.*, 1998), these flux rates can be assumed to represent a 'maximum' end member. It is more common for methane rising by diffusion to be oxidised within sulphate-rich pore waters immediately beneath the seabed (Whiticar, 2002), so the minimum end member is a zero flux.

FLUXES AT SEEPS

Exceptionally high bubble fluxes have been reported from offshore Georgia and California (the total gas emissions from each of these seep fields being approximately 6.4×10^7 m³ gas per m² per year; see Sections 3.11.6 and 3.22.4 respectively). These are spectacular emissions, with large bubbles rapidly and continuously flowing from closely spaced vents. Compared to these, the seeps we studied (and were so excited about) in the North Sea, are little more than pathetic dribbles, yet even these are significantly more vigorous than many seeps, and appear to be flowing continuously.

HYDRATE RIDGE GAS FLUX

Flux rate measurements at individual seep orifices can give a misleading impression of the total flux from an area, particularly as the selection of a vent to study tends to be subjective (the 'best' looking vent tends to be chosen) and flux rates vary over time. However, a different approach was taken by Klaucke *et al.* (2004). They mapped the whole of the Hydrate Ridge site according to various seabed 'habitats', then they estimated the flux rate from each. Their estimate, based on previously published methane flux data, suggested that of a total of 0.63 to 34 (best estimate 5.7) $\times 10^8$ mol y^{-1}, about 63% came from the area occupied by chemoherms, and a further 33% from active bubbling vents.

SHALLOW-GAS BLOWOUTS

In Section 10.4.2 we argued that natural seabed blowouts must occur. So far, it is not possible to quantify these events.

The global seabed flux

To evaluate the global flux of methane through the seabed it is necessary to take into account both the rate of emission from individual sources, and the distribution of those sources. Only a very small number of vents, seeps, and mud volcanoes have been studied for extended periods of time, but those that have show that flux rates vary considerably. When one seep in one place 'switches off' another probably 'switches on' elsewhere. It seems safe to assume that data available from short-term studies of individual sites represent a snapshot in time. Like other natural distributions, it seems that the distribution of seep rates is log-normal; there is a small number of places with a very large flux rate, and a large 'tail' of relatively minor flux rates (Hovland *et al.*, 1993). Also, we believe that, although some seeps are continuous, others are not. We envisage that some periodically active seeps emit short bursts of bubbles (which account for the detached water-column targets described in Section 10.4.1). We also think that occasional natural-gas blowouts occur. These may interrupt normal, continuous flow, but they might be responsible for the formation of new pockmarks (see Section 7.6.3).

Several authors have summarised available data (Cranston, 1994; Judd, 2000; Dimitrov, 2002a; Judd *et al.*, 2002a; Kopf, 2002; Milkov *et al.*, 2003; Etiope and Milkov, 2004; Judd, 2004), but to evaluate the global significance of the seabed flux it is also necessary to consider the distribution of seeps and mud volcanoes; we did this in Chapters 4 and 7. However, the principal motivation has been to assess the contribution to the atmosphere, rather than the hydrosphere, consequently we will return to this topic once we have considered the fate of the methane once it enters the water, and the distribution of methane in the 'normal' ocean.

10.4.4 The fate of the seabed flux

In deep water, although bubbling has been reported, the majority of methane escapes from the seabed dissolved in water. Unlike fluids from hydrothermal vents, seep fluids are not significantly warmer than normal seawater, yet these plumes are buoyant and rise before spreading laterally. For example, the methane-rich seep plume at Hydrate Ridge is '*hundreds of metres high and several kilometres wide*' (Figure 10.4), but it does not rise above 400 m below sea level (Suess *et al.*, 1999b). Although methane does not seem to be supplied directly to the atmosphere it clearly affects the composition of the water. Carbon-isotope ratios of dissolved inorganic carbon in the water column '*show a significant decrease*', and there is a small increase in dissolved inorganic

carbon dioxide, suggesting the effect of methane oxidation. Other examples include the following.

- The plume from Blake Ridge rises about 320 m above the seabed at 2167 m (Paull *et al.*, 1995).
- Methane-rich water rises about 800 m above the Håkon Mosby Mud Volcano, where its density equals that of the normal seawater. It then migrates downstream in a distinct plume with a vertical thickness of 50–100 m. This plume was detectable for a distance of at least 3–4 km (Damm and Budéus, 2003).
- A heterogeneous plume rising 200 m from a giant pockmark in the Congo–Angola Basin (Section 3.7.2). Charlou *et al.* (2004) showed that this plume was rich in particulate matter as well as methane, iron, and manganese.

Dissolved methane at any depth will be vulnerable to oxidation; carbon isotope studies have shown that microbial oxidation of methane is '*highly effective at removing methane from the water column*' (Grant and Whiticar, 2002).

Methane held in bubbles is more able to rise through water. Gas bubbles rise because of their buoyancy, and expand as hydrostatic pressure progressively decreases. If there is a concentration gradient between the bubble and the surrounding water (e.g. if the water has a lower partial pressure than the bubble) the content of the bubble (which we will assume to be methane) will migrate across the bubble boundary into the water. Atmospheric gases, oxygen and nitrogen, which are dissolved in the water but relatively depleted in the bubble, will invade the bubble. There is a competition between the outflow of methane (working to reduce the bubble size) and two processes working to expand the bubble: the inflow of atmospheric gases, and hydrostatic pressure release. If methane outflow wins, the bubbles dissolve completely; if expansion wins, they make it to the surface; surfacing bubbles transport at least some methane to the atmosphere, but they also contain other gases (mainly oxygen and nitrogen).

Ira Leifer, Jordan Clark, and their colleagues have undertaken detailed investigations of the survivability of methane bubbles (Leifer *et al.*, 2000; Leifer and Patro, 2002; Leifer and Clark, 2002; Clark *et al.*, 2003; etc.), and have identified the key parameters, apart from water depth, affecting bubble fate: initial bubble size, temperature, the concentrations of methane and other gases (oxygen and nitrogen) in the seawater. These parameters affect two key rates: the rate of ascent of the bubbles (V_b), and the rate of gas diffusion across the bubble boundary (F_g). The parameter V_b is primarily a function of buoyancy, but other factors affect it, as we discuss below; F_g, is defined as in Equation (10.1),

$$F_g = Ak_B \Delta C \qquad (10.1)$$

where A is the surface area of the bubble; k_B is the gas transfer rate, specific to each gas, and strongly dependent on gas diffusivity and bubble size; and ΔC is the concentration difference between the gas inside the bubble and the water outside it. So, for any gas, the rate of diffusion increases with bubble size, and with the concentration gradient. However, diffusion and rise rate are also affected if bubbles are 'dirty' (coated with oil, surfactants, etc.), and upwelling flows of water affect the time taken to reach the surface.

Bubble size

Observations have shown that bubbles of individual seeps tend to lie within a limited size range (generally < 5 cm diameter) when they leave the seabed. Bubble size is dependent on the nature of the seabed sediment, particularly the size of the pore throats. Very small bubbles (< 0.5 mm diameter) are spherical, but larger bubbles are spheroidal-, mushroom-, or cap-shaped; the size normally quoted is that of a sphere of the same volume. Although bubble buoyancy increases with volume, V_b is affected by shape; small bubbles rise vertically, larger ones oscillate (zigzag) and rise more slowly; eventually they deform and break up. The sizes at which these transitions occur depend largely on the nature of the bubble surface, whether it is 'clean' or 'dirty', but most seep bubbles rise at 20–30 cm s^{-1} (Leifer and Judd, 2002).

 Bubble size also has an important influence on gas inflow and outflow rates because it affects the ratio between surface area (A) and the volume (V).

$$A = 4\pi r^2 \qquad (10.2)$$

$$V = \frac{4\pi r^3}{3} \qquad (10.3)$$

So Equation (10.4) is obtained:

$$A = \frac{3V}{r} \qquad (10.4)$$

 As the radius increases, the volume increases ten times faster than the surface area; 'the quantity of gas a bubble can lose increases cubically with size' (MacDonald et al., 2002).

Dirty bubbles

Bubbles are efficient scavengers of microbes, particles, and surfactants. These adhere to bubble surfaces, providing a potentially important mechanism for transporting them into the water column (Leifer and Patro, 2002). However, they affect bubble behaviour (clean bubbles rise faster), and inhibit the migration of methane across the bubble boundary layer. Some of these 'hitchhikers' are lifted into the water column, but deposited when the bubbles dissolve (see Section 8.3.4). A film of oil may enable a bubble to survive until it reaches the sea surface, but it reduces buoyancy so it rises more slowly. Also, gas transfer is impeded, perhaps significantly, as this becomes a three-phase process in which gas must diffuse across the oily layer; oily bubbles are more likely than clean bubbles to transport methane to the sea surface (Leifer and MacDonald, 2003).

Upwelling flows

Once above the seabed, a rising bubble plume entrains water, which rises as an upwelling flow around the plume. This upwelling water is deflected outwards at the sea surface. At the Seep Tents site, offshore California, where the upwelling flow was estimated at 1–2 m s^{-1}, this radiating current 'was so intense that divers could neither swim into the area where bubbles were surfacing nor could they submerge' (Leifer et al., 2000). In contrast, divers approaching a shallow-water hydrothermal vent off the Aeolian Islands (Italy) were unexpectedly and unintentionally propelled to the surface (G. Chiodini, 2004, personal communication).

 These upwelling flows have two effects on bubble fate. Firstly, they increase the speed of ascent, as bubbles are rising through a mass of water that is itself rising. This decreases the time taken to reach the surface (by 50–75% in the case cited above). Secondly, as bubbles rise they increase the methane concentration of the column of water through which they are rising, so this water is enriched compared to the surrounding water, and the methane concentration gradient between the bubbles and the water is reduced. Together, these effects mean that upwelling flows significantly decrease the loss of methane from the bubbles, increasing the proportion that escapes to the atmosphere (Leifer and Judd, 2002; MacDonald et al., 2002; Leifer and MacDonald, 2003).

Armour plated bubbles

Acoustic plumes in the Sea of Okhotsk disappear at about 300–500 m water depth (see Section 3.18.6), and those mapped by Heeschen *et al.* (2003) at Hydrate Ridge consistently disappeared at a depth of about 480 m, regardless of the depth of the seabed (590–780 m) from which they came. It seems that the bubbles survive until they pass out of the gas hydrate stability zone (GHSZ).

Rehder *et al.* (2002) found that bubbles photographed at the seabed (750 m water depth) on Hydrate Ridge were *'almost spherical in shape'*, which is unusual for bubbles of 6–7 mm diameter. They attributed this to a skin of hydrate that gives the bubbles rigidity. This conclusion was confirmed by experiments undertaken in the deep (910 m) water off Monterey Bay, where Brewer *et al.* (1998) made hydrates simply by injecting gas into the seawater. Hydrate was seen to form a skin around gas bubbles, separating gas from seawater, so that gases would have to diffuse through the hydrate to enter or leave the bubble. Hydrate 'armour plates' gas bubbles, protecting them against solution for as long as they remain within the GHSZ. However, once above it the hydrate decomposes rapidly, leaving the bubble vulnerable to solution. The acoustic evidence is supported by carbon isotope data: $\delta^{13}C_{CH4}$ values of -60 ± 6‰ typical of the vent fluids were found within the GHSZ, but values typical of heavier upper-water-column sources, around -30‰, were found above it.

Floating hydrate

Gas hydrate, like ice, has a lower density than water, so any pieces of hydrate that detach from the seabed will rise towards the surface. Any sediment attached to the hydrate may reduce the buoyancy, although loosely held sediment may soon fall away. The rising hydrate will remain stable until it reaches the upper boundary of the GHSZ. Then it will start to decompose. However, the decomposition reaction is endothermic, and the poor transfer of heat from ocean water means that decomposition is slow. Also, a skin of normal water ice tends to form around the hydrate, providing some insulation. Buoyancy is increased by the formation and expansion of bubbles within the hydrate, causing acceleration. Larger pieces of hydrate are more efficient at trapping gas bubbles. This not only increases the proportion of hydrate surviving transit to the surface, but also increases the amount of gas that avoids solution. Indeed, Brewer *et al.* (2002) found that *'even small pieces of hydrate can survive transit through 800 m of water column and deliver methane to the atmosphere in less than one hour'*.

The largest direct venting of disrupted gas hydrates to the atmosphere so far documented is probably the remarkable catch in a trawl net reported from Barcley Canyon off northwest Canada, in November 2000 (Spence *et al.*, 2001). The fishing vessel 'Ocean Selector' (an appropriate name!) was trawling in 800 m of water, 50 km landward of ODP Leg 146, drill site 889, where gas hydrates and a BSR had been targetted in 1992. The trawl net was reported to snag on rugged topography, but came loose and floated to the surface with over 1000 kg of gas hydrates and some fish:

> The net fabric unexpectedly floated to the water surface, clearly indicating that the material in the net was buoyant. On breaking the surface, the material began to froth and hiss. It was said to be 'like Alka Seltzer.' The crew observed blocks of 'ice' floating around the net. The fisheries observer on the vessel identified the material as methane hydrate, similar to that seen on a public television program. Initially uncertain about safety, but wanting to recover the net, the crew brought the net back onboard and dumped the hydrate on the open-ended well at the stern of the boat. They noted a strong, pungent, petroleum-like smell. There was no rotten-egg smell, which is indicative of deadly hydrogen sulfide.
>
> Spence *et al.*, 2001

This ability of gas hydrate to float towards the surface may explain the bubbles seen breaking 700 m above the seabed gas hydrates in the Sea of Okhotsk (Cranston *et al.*, 1994). However, it is of even greater significance when hydrate-bearing sediments are involved in slope failures. The failure mode, the type of downslope movement (i.e. slide, slump, or turbidity current), and the distribution and concentration of hydrate within the sediment will all affect the efficiency with which hydrate breaks free from the sediment. Nevertheless, the question is not whether methane will be released to the atmosphere by seabed slope failure, but how much? It seems likely that significant quantities of methane may be released into the atmosphere by such events (Paull *et al.*, 2003 – see Section 11.2.1).

Bubble fate

In the preceding parts of this section we have outlined the various factors affecting the rate at which bubbles

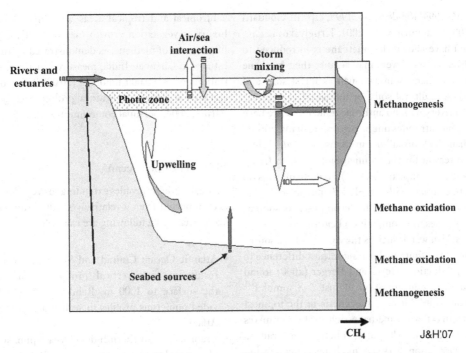

Figure 10.6* Methane cycling in the oceans. The column (right-hand side) indicates (schematically) variations in methane concentration with depth. Principal sources are influx from rivers and estuaries, and *in situ* methanogenesis (which occurs in microenvironments in particles falling from the photic zone). Principal sinks are microbial oxidation and flux to the atmosphere. The introduction of nutrients from coastal zones and upwelling systems enhances biological productivity in the photic zone. Methane (and nutrients) from seabed sources (seeps, mud volcanoes, and hydrothermal vents) complicate this system.

dissolve. Bubbles from a single seep area tend to lie within a restricted size range when they are emitted from the seabed. This means that they tend to dissolve at about the same altitude above the seabed, resulting in the formation of methane-rich layers within the water column (Leifer and Judd, 2002). Boundaries between water masses of contrasting temperature, salinity, and density (thermocline and pycnoclines) are influential, and methane concentration variations are often associated with them.

To understand the significance of methane injected into the water column by seeps it is necessary to understand the context; what is the concentration and distribution of methane in a seep-free ocean? This is the topic of the next section.

nature of the water into which they flow; the 'normal' ocean. This is the realm of oceanographers and biogeochemists who tend not to have any interest in the seabed, and who tend to study whole oceans or ocean regions, rather than specific areas or sites with a specific geological character. Instead of looking in detail at every aspect of the ocean we will consider only methane, as this is the most important of the fluids we have considered in previous chapters, and a major component of both hydrothermal and cold seeps.

The concentration of methane in the water column is determined by the balance between sources and sinks, as illustrated in Figure 10.6. In Sections 10.5.1 and 10.5.2 we briefly review the understanding of hydrospheric budgets, assuming no influence from the seabed sources discussed in the previous section.

10.5 METHANE IN THE 'NORMAL' OCEAN

Having considered the various types of seabed fluid flow we need to evaluate their significance by considering the

10.5.1 Rivers, estuaries, and lagoons

Studies by various authors (de Angelis and Lilley, 1987; Scranton and McShane, 1991; Jones and Amador, 1993;

Bange *et al.*, 1994; Rehder *et al.*, 1998; Upstill-Goddard *et al.*, 2000; Jayakumar *et al.*, 2001; Kruglyakova *et al.*, 2002, etc.) have shown that methane is introduced to the sea from rivers. Rivers may acquire their methane from natural inland sources (such as forest soils and wetlands), agricultural soils, or industrial contamination. Even rivers with no anthropogenic influence have methane concentrations one or two orders of magnitude higher than the 'normal' ocean (Sansone *et al.*, 1999). The main reason for this is microbiological. In freshwater there is no sulphate (SO_4^{2-}), so methanogens do not have to compete with the SRBs (sulphate-reducing bacteria) for organic matter to decompose. As soon as salinity increases, the competition begins.

In brackish water, such as the lagoons of the southern Baltic Sea, salinity makes a significant difference to methane production. Heyer and Berger (2000) found that a sulphate concentration of just 5–8 mmol l^{-1} is sufficient to inhibit methanogenesis in the topmost sediments. In estuaries, methane-rich freshwater mixes with methane-poor salt water. There is generally a negative correlation between methane concentration and salinity; de Angelis and Lilley (1987) reported a linear correlation in estuaries they studied in Oregon. They attributed this to '*simple two-point mixing between river water and seawater endmembers*'. In partially mixed estuaries (such as the River Tyne, England) methane concentrations (~ 650 nmol l^{-1}) significantly higher than those in the river water (< 17 nmol l^{-1}) occur at the '*turbidity maximum*' where river- and seawater mixing results in the trapping of sediments (Upstill-Goddard *et al.*, 2000). In addition to the river-borne methane, microbial methane generated in anoxic estuarine sediments may provide additional methane (Chanton *et al.*, 1989; Sansone *et al.*, 1999).

The fate of this methane is heavily dependent on the hydrodynamics of individual estuaries. Studies of North Sea estuaries led Upstill-Goddard *et al.* (2000) to suggest that 90% of river-borne methane is removed: about 6% by microbial oxidation, the remaining 94% escaping to the atmosphere. Despite this significant loss, estuaries generally inject methane into the sea. This accounts for the strong concentration gradients reported, for example, off the River Elbe, North Sea (Rehder *et al.*, 1998), the River Danube, northwest Black Sea (Amouroux *et al.*, 2002), and the Mandovi River on the west coast of India (Jayakumar *et al.*, 2001). Estuaries of rivers of subtropical and tropical areas with high rainfall and luxuriant vegetation growth may be particularly prolific sources of methane, as demonstrated by the Adyar estuary in Chennai, India; measured methane concentrations of $>190\,000$ nmol l^{-1} (10 000 000% of atmospheric saturation) indicate a prolific flux (Jonathan Barnes, 2004, personal communication).

10.5.2 The open ocean

Vertical methane profiles extending to great depths have been reported from a relatively small number of open ocean areas. The following are examples.

- Atlantic Ocean: Conrad and Seiler (1988) reported results of four vertical profiles extending from the surface to 4000 m; Rehder *et al.* (1999) provided numerous profiles to seabed across the North Atlantic.
- Arabian Sea/northern Indian Ocean: published profiles extend to 2000 m (Owens *et al.*, 1991) and 5000 m (Upstill-Goddard *et al.*, 1999).
- Offshore California: Tilbrook and Karl (1995) presented profiles to depths of 1500 m.

From these, and other studies it seems that three zones can generally be distinguished: a deep undersaturated zone, a shallow supersaturated zone, and a surface mixed layer. The actual depths differ from place to place, and probably vary over time.

Deep-ocean waters

The top of this zone is generally not clearly defined, but at depths below about 250 to 500 m methane concentrations fall to very low levels (< 50 nl l^{-1}), and are consistently undersaturated with respect to the atmosphere, whereas those above are supersaturated. This undersaturation is explained by the oxidation of methane derived from sources higher in the water; it '*has the signature of the remaining fraction of atmospheric methane partially oxidized in the water column by bacteria*' (Faber *et al.*, 1994). This signature is characterised by enrichment in ^{13}C ($\delta^{13}C > -20$‰) relative to atmospheric levels (see below). Rehder *et al.* (1999) estimated the residence time of methane entering the deep ocean to be about 50 years.

Shallow methane-saturated waters

This zone is typically saturated with methane, rich in phytoplankton (and therefore chlorophyll) and particulate organic matter, but relatively depleted in oxygen. The methane is generally assumed to have been generated *in situ* as there is no other obvious source, but an *in situ* source of methane has proved difficult to identify and hard to explain because microbially mediated methanogenesis requires anaerobic conditions (see Section 5.4.2). This is known as the 'oceanic methane paradox' (Tilbrook and Karl, 1995). Microenvironments in the guts of zooplankton and decaying particles may provide the right conditions. Sieburth (1987) pointed out that the interface between the anoxic interior and the oxic exterior of decaying particles would be a favourable habitat for methanotrophs. Conrad and Seiler (1988) suggested that there might be a syntrophic association *'in which the anaerobic methanogens provide the CH₄ to the aerobic methanotrophs that use it as a substrate, and thereby consume all the oxygen that would be toxic to their methanogenic partners'*. Surplus methane would be released into the water.

Support for *in situ* methane production comes from carbon isotope data; for example, $\delta^{13}C -46‰$ reported for methane in surface waters from the East Pacific Rise (24° S) is in equilibrium with the atmosphere (Faber *et al.*, 1994). The amount of methane derived as a result of photosynthesis in the oceans increases with temperature.

Surface waters

The reduction in methane concentrations towards the surface can be explained by a combination of methane oxidation and flux to the atmosphere, aided by mixing by wave action; Brewer *et al.* (2002) noted that the surface mixed layer, which may shallow in summer, might be driven to depths of 200–400 m by winter cooling and storms. Measurements of near-surface (< 5 m) methane concentrations are good indicators of the likely interchange between the oceans and the atmosphere. Several extensive studies have been undertaken, and most indicate that, even though methane concentrations are higher in underlying water, the ocean surface water is generally oversaturated relative to the atmosphere. The following are examples.

- Conrad and Seiler (1988) reported saturations of 110 (± 5)% relative to the atmosphere from the Atlantic Ocean (50° N to 35° S).

- Various studies of the northern Indian Ocean (Arabian Sea) have shown surface waters to be oversaturated, although the degree of oversaturation differs both geographically and over time, particularly as a result of the monsoon (Owens *et al.*, 1991; Patra *et al.*, 1998; Upstill-Goddard *et al.*, 1999; Jayakumar *et al.*, 2001).

- Rehder and Suess (2001) reported methane saturation on a transect in the southwest Pacific and the eastern South China Sea from Japan to Borneo was between 105 and 125% (higher concentrations and saturations were recorded in the coastal waters of Japan and Borneo).

A table presented by Bange *et al.* (1994) lists methane measurements from surface waters in more than 50 regions representing a wide range of latitudes, water depths, and methane-source types; in every case the surface waters are oversaturated with methane. However, work by Bates *et al.* (1996), in the Pacific Ocean, and by Rehder *et al.* (1999) in the North Atlantic provided evidence that oversaturation varies with water temperature; with the seasons, and with latitude. Ocean surface waters are oversaturated with methane where and when there is primary production by phytoplankton; non-tropical areas may be undersaturated during autumn and winter months.

Upwelling waters, typically nutrient-rich, are associated with above-normal productivity; typically they also have above-average methane concentrations (Cynar and Yayanos, 1992).

Continental shelves

We assume that competing *in situ* microbial production and oxidation, and air/sea interchange, occur on continental shelves, as they do further offshore, but evidence from continental shelves clearly shows the importance of coastal sources of methane, including rivers and estuaries (discussed in the previous subsection), and also methane produced in sediments in shallow water. For example, Bange *et al.* (1994) found that the Oder Bight (on the German Baltic coast) had higher methane saturations than regions further offshore because of both the plume from the River Oder, and also production in sediments.

Carbonate platforms and coral reefs

Sansone (1993) reported that pore waters in carbonate platforms and coral reefs may have elevated methane concentrations as a result of the decomposition of organic matter, at least locally, within them. He presented 'best-guess' estimated concentrations of 50 and 200 nM methane for the pore waters of reefs and carbonate platforms, respectively. Compared to 12 nM methane in tropical surface seawater, these represent a considerable enrichment that may have an important effect on the methane concentration of the hydrosphere, and the flux from seawater to the atmosphere.

The area covered worldwide by carbonate reefs was estimated by Milliman (1974) as 1.4×10^6 km^2; the area covered by coral reefs is believed to exceed 600 000 km^2, of which near-surface reefs cover an area of about 255 000 km^2 (CBD, 2001). Sansone (1993) presented 'best-guess' estimates of the contributions to the overlying seawater of 0.005 and 0.6 Tg methane per year (Tg $= 10^{12}$ g $= 10^6$ t), respectively, for carbonate platforms and coral reefs.

10.5.3 The influence of seabed methane sources

Previously (Sections 10.2 and 10.4) we considered three types of seabed fluid flow: diffusion, steady state, and event/blowout. Water depth also influences the significance of seabed fluid flow. Methane is more soluble in deep water than in shallow water, so bubbles dissolve more rapidly, and higher concentrations are possible (NB methane bubbles form only when the water is saturated in methane). However, the solution of bubbles supplies methane to water at any depth provided there is a concentration gradient, except where 'armour plating' (a hydrate-skin) or an oil-skin impedes solution (see Section 10.4.4). Background concentrations in deep water (the methane oxidising zone) are so low that any additional contribution, even from diffusive flow, is likely to have an impact on concentrations. For example, water emitted by the Håkon Mosby Mud Volcano (HMMV – see Section 3.2.2) has a methane concentration of about 340 nM, but this is rapidly diluted within the surrounding seawater to about 3 nM; even this is about twice the concentration found at 'reference' stations (Damm and Budéus, 2003). However, oxidation removes methane relatively quickly, probably within 10 to 50 days (Welhan and Craig, 1983; Rehder et al., 1999).

There are significant differences between biogeochemical studies of ocean waters (as described earlier in this section), and targetted studies of vent and seep plumes (described in Sections 10.2 and 10.4, respec-

tively). Whilst the former are trying to establish the 'background', the latter are specifically interested in the 'anomalies'. (Incidentally, exactly the same could be said about regional surveys of benthic ecology.) In regional studies, individual sample sites are selected more or less at random; in some cases published results make only passing mention of water depth, and rarely consider the possibility that seabed conditions and focussed seabed fluid flow might affect concentrations. By ignoring site-specific conditions it may be possible to take a sample and assume it is representative of a wide area, rather than recognising it as an anomaly. 'Profiles' based on discrete water samples may have a vertical spacing of between 100 and 1000 m between samples. Such sample intervals would completely miss the plume from HMMV (vertical thickness 50 to 100 m at 400 m depth; Damm and Budéus, 2003). Even continuous sampling of the sea surface waters provides only a coarse sampling grid; for example, Conrad and Seiler (1988) averaged their data over 1° of latitude, data collected by Bates et al. (1996) were '*binned into hourly intervals*', and Rehder et al. (1998) used a sample repetition rate of 30 minutes; these are functions of ship speed and analytical cycle time.

Acoustic images show that seep plumes are very narrow (see Figure 10.3, for example), and experience has taught us that it is very difficult to deploy a water-bottle rosette and successfully sample a plume, even in shallow water! Wernecke et al. (1994) made continuous measurements of methane concentrations in sea-bottom water at a North Sea gas seep, and found that although peak values were more than one hundred times higher than those away from the seep, anomalous values were restricted to within 40 m of the seep. Clearly, the chances of accurately representing variations in methane concentrations caused by seeps are very small. Despite these problems, regional studies by some authors have suggested that seabed sources may be supplying methane to the hydrosphere, for example:

- Rehder et al. (1999) suggested that high methane concentrations in two samples might be explained by previously undiscovered venting where the Mid-Atlantic Ridge is cut by the Gibbs Fracture Zone (52° N);
- prolific seep areas offshore California and Brunei were cited as explanations for high methane concentrations in surface waters by Cyanar and Yayanos (1992) and Rehder and Suess (2001), respectively.

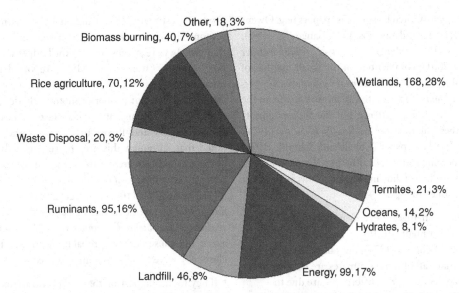

Figure 10.7* The IPCC atmospheric methane budget. Based on data presented in Table 4.2 of Ehhalt *et al.*, 2001; this table included data from seven publications, although not all publications provided estimates of the strength of all the sources. This diagram indicates the relative strength of each of the sources, based on the average of the provided strengths normalised to the total strength (598 Tg y^{-1}) accepted by IPCC. Figures indicate the estimated strengths expressed in Tg, then as a percentage of the total source.

Nevertheless, it would be easy to overlook, or completely miss individual vent or seep plumes, especially if those concerned are unaware that seabed sources exist. In contrast, studies of vents, seeps, and mud volcanoes have clearly demonstrated the contributions made by seabed fluid flow, as discussed in Section 10.4. Although we do not consider the flux of methane from the seabed in deep waters to be unimportant, we must emphasise the importance of the seabed flux on the continental shelves and in coastal waters. Because of losses of methane to the water during bubble rise, the shallower the water the greater the proportion of the seabed methane reaching the atmosphere.

10.6 EMISSIONS TO THE ATMOSPHERE

Gases from shallow-water submarine volcanic eruptions and hydrothermal vents enter the atmosphere (for example in the Aegean Sea – see Section 10.2.2). These are noteworthy, particularly because the dominant gas is carbon dioxide (as it is from some mud volcanoes). However, the most important emissions are of methane. Because methane is an important 'greenhouse gas', considerable international research efforts have

been devoted to measuring present-day and past atmospheric concentrations, and to identifying and quantifying sources; the 2001 report of Inter-Governmental Panel for Climate Change (IPCC) accepted that natural and anthropogenic emissions total 598 Tg y^{-1} (Ehhalt *et al.*, 2001). Their view is that '*the major source terms of atmospheric CH$_4$ have probably been identified*', however '*many of the source strengths are still uncertain*'; Figure 10.7 indicates the sources and their strengths.

10.6.1 Methane emissions from the oceans

Gas exchange across the sea surface occurs both when bubbles escape, and when the concentration of the gas in the sea-surface water exceeds (i.e. is >100% of) the concentration of that gas in the atmosphere. Mixing processes (particularly during storms) ensure that dissolved gases rising to within one or two hundred metres (or even more) of the sea surface are likely to escape to the atmosphere.

It is commonly stated that the oceans are a 'minor source' of atmospheric methane. This is certainly the impression given by the IPCC (Houghton *et al.*, 1996; Ehhalt *et al.*, 2001), and other reviews (e.g. Khalil, 2000; Wuebbles and Hayhoe, 2002). Cicerone and Oremland

(1988), frequently cited by specialist papers (e.g. Owens *et al.*, 1991; Cynar and Yayanos, 1993; Bange *et al.*, 1994; 1996; Bates *et al.*, 1996; Rehder *et al.*, 1998; Rehder and Suess, 2001), commented on the small number of data available; their Table 4 ('*annual methane release rates for identified sources*') suggested an annual flux from the oceans of 5 to 20 Tg methane, but the authors admitted that these figures '*are essentially Ehhalt's [1974] estimates*'. In a later paper (Ehhalt and Schmidt, 1978) where an oceanic source of 1.3 to 16.6 Tg methane per year was quoted, the following comment was made about the validity of estimates:

> Previous estimates by several authors differ by a factor of 5 (Ehhalt, 1974; Liss and Slater, 1974; Seiler and Schmidt, 1974), but fall in the same range. Since all of those estimates are derived from the same data base, the differences are due to discrepancies in the assumed values for diffusion coefficients, film thickness and solubility data, that were used in the calculations.
>
> Ehhalt and Schmidt, 1978

It seems that the widely held belief that the oceans are of only minor importance as a source of atmospheric methane is based on very thin evidence:

> Whereas numerous workers have attempted to evaluate the continental sources of methane, its oceanic source is still poorly documented. Only Ehhalt (1974), using pioneering measurements by Lamontagne *et al.* (1973), mentioned in its methane budget an oceanic source of 4.7 to 20.7 Tg a^{-1} (open ocean 4–6.7; shelf 0.7–14). This early evaluation was subsequently used by several authors without careful re-examination (Khalil and Rasmussen, 1983; Cicerone and Oremland, 1988). More recently, the model of Fung *et al.* (1991), once again used Ehhalt's values but without 'carrying out 3D simulations for scenarios of several methane sources and sinks that are extremely poorly known . . . These include oceans . . .'
>
> Lambert and Schmidt, 1993

Lambert and Schmidt then commented that there were still (in 1993) '*very few*' measurements of methane concentrations in near-surface seawater; they listed 93, of which only 63 were from '*open ocean*' sites. This hardly seems adequate as a basis for defining 67% of our planet as a minor source! Furthermore, some researchers have

produced estimates that individual regions might be responsible for the entire methane budget – although they clearly were suggesting that the budget was wrong. For example, Owens *et al.* (1991) reported that the northern Arabian Sea was not only a region with one of the highest rates of primary (photosynthetic) production, but also of unusually high surface-water methane concentrations and fluxes to the atmosphere. They noted that, although this area represents only 0.43% of the total surface area of the oceans, it '*can contribute between 1.3 and 133% of the total oceanic flux of methane*'. This conclusion led them to suggest: '*the current oceanic flux should be revised upwards and points to the need to evaluate other potential oceanic sources.*' We do not disagree, but these authors probably did not realise how close they were to other potential sources.

- They found highest methane concentrations at their three stations (9–11) closest to the Strait of Hormuz – where Uchupi *et al.* (1996) found extensive acoustic turbidity and pockmarking (see Section 3.14.3).
- Gas hydrates, mud volcanoes, and seeps have been shown to be associated with the Makran Accretionary Wedge (see Section 3.15.1) – which lies to the north of their stations 7 and 8. Delisle and Berner (2002) reported several plumes of methane-rich water extending > 20 km seaward from the continental slope (Figure 3.37).

Although we are not suggesting that these gas sources explain the data presented by Owens *et al.* (1991) or Upstill-Goddard *et al.* (1999) it does seem ironic that such sources are relatively close to their stations.

Oceanic methane measurements have also been reported by Cynar and Yayanos (1992), Bates *et al.* (1996), Rehder *et al.* (1998; 1999), and Rehder and Suess (2001). The general pattern emerging seems to support the contention that the oceans are a minor source, but there is increasing evidence that coastal waters are more important than the open ocean. Bange *et al.* (1994) compiled and reviewed available data to reassess the oceanic flux. They suggested a flux of between 11 and 18 Tg methane per year, with perhaps 75% coming from coastal waters.

Despite these advances, there are still shortcomings in the database. Most importantly, from our point of view, individual data points are generally widely separated, and many papers include data that have been 'averaged' or smoothed. Also, the majority of authors do not mention (are not aware of?) the possibility of seabed sources of methane. Does this provide a suitable basis for emission-strength estimations when, as we have shown in this chapter, there are many localised, and temporally variable sources of methane on the seabed? We think the paradigm that the oceans are unimportant as a source of atmospheric methane would be challenged if proper consideration was given to seabed sources.

10.6.2 Seabed sources of atmospheric methane

In 1955 P. N. Kropotkin pointed out a relationship between atmospheric methane and natural-gas seeps (Valyaev, 1998). However, atmospheric-methane budgets recognised by the IPCC do not identify seabed fluid flow as a source, although 'hydrates' were named in the 2001 report (Ehhalt et al., 2001). As we have shown, seabed sources of atmospheric methane include ancient and modern sediments, sedimentary rocks, mud volcanoes, hydrothermal vents, submarine volcanoes, and seamounts. Methane sources, including marine sources, can be divided into 'modern' and 'fossil' according to their ^{14}C content.

Modern methane

Most sources (wetlands, animals, agricultural, waste disposal, and biomass burning) produce 'modern' methane. Modern sources also include present-day microbial methanogenesis in the photic zone of the water column, and in brackish and marine organic-rich sediments. Published figures suggest that 'wetlands' are by far the largest modern source, accounting for >20% of the total emission, but estimates vary considerably (92 to 237 Tg y^{-1}; Cao et al., 1998, and Hein et al., 1997, respectively; both quoted by Ehhalt et al., 2001). Apparently this category includes only freshwater wetlands (Elaine Matthews, 2004, personal communication), so brackish and coastal marine sediment sources are not catered for at all!

Fossil methane

Sixteen to twenty-five per cent (between 96 and 150 Tg) of atmospheric methane is ^{14}C-free, 'fossil' methane (Lacroix, 1993). The IPCC and other authorities attribute the entire fossil methane budget to the fossil fuel (coal, gas, and oil) industries. Supporting evidence for this supposition is the increase in emissions since the Industrial Revolution, and an increase in ^{13}C-enrichment from a pre-industrial level of −49.7‰ to a current global average δ^{13}C value of −47.3 to −46.2‰ (Wuebbles and Hayhoe, 2002). But, studies of emissions and leakage by the fossil-fuel industries suggest that they are not responsible for the entire fossil methane source; estimates vary between 75 (Fung et al., 1991) and 110 Tg (Lelieveld et al., 1998). Lacroix (1993) suggested that other anthropogenic activities such as groundwater and peat extraction, and industrial processes such as petrochemical refining and geothermal electricity production should also be considered, but even then the shortfall may be as much as 75 Tg. This must be accounted for by natural geological emissions on land and offshore (Kvenvolden et al., 2001; Judd et al., 2002a; Etiope and Klusman, 2002; Etiope, 2004).

The majority of seabed methane is 'fossil' (^{14}C-free). Even methane from deep microbial sources from which there is an efficient migration pathway is likely to predate the Industrial Revolution; for example, Ivanov et al. (1993) found that methane from seeps in the Black Sea was between 3000 and 5500 years old, and Laier et al. (1992) reported ages of 2600 years for microbial gases seeping in Denmark and the Kattegat (see Section 3.3.4).

When considering the contributions to atmospheric methane from seabed sources it is necessary to take into account the distribution and strength of the source (seep, mud volcano, etc.), and the fate of methane once it is in the water, including contributions to the 'normal' oceanic methane budget – that is, all the things we have discussed so far in this chapter. Contributions from submarine hydrothermal vents and volcanoes are generally considered insignificant as even methane from event plumes dissolves in the water (see Section 10.2.3), yet shallow-water vents are potentially important, and should be included. We also consider that event plumes, natural blowouts, and Lokbatan-type mud-volcano eruptions deserve special attention as it seems probable that the proportion of methane escaping to the atmosphere will be significantly greater from these events than from 'steady-state' venting/seeping. Assessing the significance of these events is a challenge yet to be met. The coastal zone is particularly important, partly because of the variety of coastal environments in which methane is generated, but also because methane escaping from a shallow seabed is more likely to escape to the atmosphere.

Various attempts have been made to estimate the flux to the atmosphere from seabed sources. Of these estimates, we favour that of Kvenvolden et al. (2001). This suggested that seeps, gas hydrates, and mud volcanoes provide 30 Tg y^{-1} of methane to the seabed. Once losses to the water column were taken into account, a contribution to the atmosphere of 10 to 30 (best estimate, 20) Tg y^{-1} was suggested. Even though hydrothermal and (igneous) volcanic sources were not included, we feel that this is as reasonable an approximation of the seabed contribution to atmospheric methane as is possible with the data currently available; further data may well prove this to be a conservative estimate. We think that this 'seabed' contribution should be considered separate from the 'ocean' contribution of methane generated in the water column (see Section 10.5.2). We suggest that the oceans contribute at least 10 to 30 Tg from the seabed, plus the 11 to 18 Tg of Bange et al. (1994); a total of 21–48 Tg y^{-1}, or 4 to 9% of the global budget. Perhaps the oceans are not as insignificant as some authors would have us believe. But, if an increased seabed/ocean source is to be accepted, other sources must be reduced to compensate; the various slices of pie are not of fixed size, but the size of the pie is. These natural contributions, and those of geological sources on land (considered by Etiope and Klusman, 2002), clearly indicate that the hydrocarbon industries (oil, natural gas, and coal) are not responsible for the entire fossil methane budget.

Seasonal fluctuations

We have seen that, in the overwhelming majority of cases, flux measurements are short-term, and that estimates are really snapshots in time, which, hopefully, represent the natural temporal fluctuations. However, we have also shown that microbial methane production rates at shallow-water sites (Eckernförde Bay and Cape Lookout Bight, for example) vary seasonally, according to the bottom-water temperature. Such seasonal fluctuations are unexpected, except in shallow waters as, even on continental shelves, the temperature of the bottom water varies little, if at all. However, deep-water sites may be affected by the changing positions of gyres and consequent temperature fluctuation (see Section 6.3.5).

Seasonal and even shorter-term fluctuations in the supply of methane to the atmosphere may also result from water mixing and increases in wind speed; the

increase in the gas exchange rate (F) with increasing wind speed is clear from Equation (10.5),

$$F = 0.31v^2 \left(\frac{Sc}{660} \right)^{-1/2} \Delta C$$

(10.5; after Wanninkof, 1992)

where v is the wind velocity; Sc is the Schmidt number (specific to each gas); and ΔC is the concentration gradient across the air–sea interface. Whereas photic-zone methanogenesis increases in summer because of higher temperatures (see Section 10.5.2), winter storms result in more effective releases of methane to the atmosphere.

ICE AS A GAS CAP

The effectiveness of seabed permafrost as a seal was discussed in Section 7.5.6. Sea ice is also an effective seal, preventing the exchange of gases across the sea surface. Kvenvolden et al. (1993) found methane concentrations in water beneath sea ice in the Beaufort Sea were significantly supersaturated, and higher than measured in the region during ice-free summer months. A more detailed study of the effect of sea ice on sea-surface gas exchange was undertaken in the Sea of Okhotsk (Lammers et al., 1995b; Obzhirov et al., 2002). Here also it was found that, relative to summer levels (6 to 76 nmol l^{-1}), methane concentrations were high (95–385 nmol l^{-1}) beneath winter ice, and suggested that 74% of methane entering the atmosphere from their survey area was released during April, May, June, and July. This annual methane pulse must be a feature of all polar seas.

Sea ice is a barrier because it effectively reduces the wind effect – v of Equation (10.5) – to zero. It might be expected that ice would prevent gas bubbles escaping, and indeed this has been confirmed in Lake Baikal (see Section 3.12.2). However, unlike the freshwater of Lake Baikal, sea ice is partly permeable to gas (Gosink et al., 1976), and constant ice movements cause fracturing which may facilitate gas escape; so it seems that ice may not be completely effective as a barrier to gas bubbles, which may explain the '*episodic ebullition*' described from the Laptev Sea (Siberian Arctic) by Semiletov (1999). In fact surfacing gas bubbles may prevent ice from forming to produce an ice-free area, a polynya. The only examples of polynyas associated with seabed fluid flow we have come across are from lakes (Lake Baikal, Granin and Granina, 2002 – see Section 3.12.2;

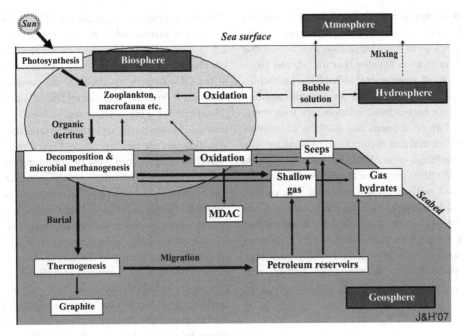

Figure 10.8* Schematic diagram indicating the role of marine methane in the carbon cycle.

and Siberian lakes, Semiletov, 1999), and the freshwater polynya in Cambridge Fjord, Baffin Island (described in Section 3.29.3). We see no reason to suppose that they could not form in shallow marine environments over active gas seeps.

The pre-industrial budget

Atmospheric methane concentrations have increased dramatically, by a factor of about 2.5, since the start of the Industrial Revolution (Chappellaz *et al.*, 2000; Ehhalt *et al.*, 2001). There is no reason to suppose that natural seabed emissions have varied significantly over this time period; consequently they represented a much more important part of the pre-industrial budget. Nisbet (2002) provided the following figures for atmospheric methane concentration to illustrate this point:

- present day: 1.75 ppm$v$ = ∼ 4000 Tg
- early Holocene: 0.65–0.8 ppm$v$ = ∼ 1500 Tg
- late glacial: 0.35 ppm$v$ = ∼800 Tg

Variations in seabed flux rates would have had a greater impact in the pre-industrial past than they would today.

10.7 GLOBAL CARBON CYCLE

Our consideration of natural seabed emissions of methane to the hydrosphere, biosphere, and atmosphere indicates to us that they are worthy of inclusion in the global carbon budget, yet standard texts do not recognise them. Indeed, it seems to be generally accepted that the only natural flux of carbon across the seabed interface is the burial of organic matter, which becomes incorporated in petroleum, gas hydrate, and limestone reservoirs. Commonly, the only recognised marine return pathway is extraction of petroleum by the oil and gas industry. Clearly this is not the case. Although there has been insufficient work to quantify all the fluxes and reservoirs, we consider that the following should be recognised as components (see Figure 10.8):

- the migration of methane to the seabed;
- sequestration of methane by the gas hydrate reservoir;
- methane utilisation at the seabed and the associated fluxes to the biosphere and the carbonate (MDAC) reservoir;
- the flux of methane into the water column;
- losses to the hydrosphere (and subsequent microbial oxidation and therefore flux to the biosphere);
- losses to the atmosphere.

In this chapter we have already discussed the fluxes to the hydrosphere and atmosphere. In Chapter 8 we discussed the biology of seabed fluid flow, and showed that the biosphere receives benefits beneath, at, and above the seabed. In all cases microbes hold the key, whilst higher organisms benefit either by the consumption of microbes, or by symbiotic relationships with them. As a general guide, it seems that 80–90% of methane production (natural and anthropogenic) is microbially oxidised (Reeburgh *et al.*, 1993; Reeburgh, 1996); de Angelis *et al.* (1993) suggested that 25% of the carbon from oxidised methane in hydrothermal vent plumes is converted to methanotrophic organic carbon. Considering the size of the methane flux, it seems that the benefit to the biosphere is sufficiently important to justify reconsideration of marine-biological energy balances, particularly in the vicinity of methane-rich hydrothermal and cold-seep plumes, and areas of prolific seepage.

10.8 LIMITING GLOBAL CLIMATE CHANGE

10.8.1 Quaternary ice ages

Milankovitch cycles, variations in the distance between the Sun and Earth, account for about 50% of global climate change (Raynaud *et al.*, 1993). If they were the only factor affecting climate change, transitions from glacials to interglacials would be smooth and balanced. In fact long, gradual slides into glacial periods contrast markedly with the return to interglacial conditions through rapid warming ('Dansgaard–Oeschger') events. The close relationship between global temperature, and atmospheric methane (and carbon dioxide) concentrations (Figure 10.9), as measured in ice cores from Greenland (over a 40 000 year period; Chappellaz *et al.*, 1993) and Antarctica (over 150 000 years; Jouzel *et al.*, 1993) suggests a link. It is unclear whether these greenhouse-gas concentrations vary in response to climate change, or whether they drive it; however, they do provide 'feedback', influencing the rate, and possibly the extent, of change.

The size of the reservoir of methane held in gas hydrates (Section 11.6.3) has been identified as a potential influence on climate change because hydrate stability is sensitive to variations in temperature and pressure (i.e. sea level). Sea level falls during global cooling,

affecting the hydrostatic pressure, and the stress environment in marine sediments (as explained in Section 7.5.1). The ~120 m fall in eustatic sea level during the last glacial period is thought to have resulted in the base of the GHSZ rising by about 20 m, causing the dissociation of hydrates, and the release of water and gas (Figure 10.10). This will have resulted in a reduction in strength. However, the lowering of hydrostatic pressure also increases the effective stress – see Equation (7.3). A likely consequence of these two opposing reactions is sediment failure, particularly on the continental slope (see Section 11.2). Such slope failures might be accompanied by the release of gas – both trapped beneath gas hydrate, and released by the decomposition of hydrate. In contrast, sea-level rise increases hydrate stability, and effectively raises the top of the GHSZ. However, a rise in water temperature associated with climatic amelioration causes hydrate dissociation. Vogt and Jung (2002), working on the Norwegian Margin, suggested that after the last glacial maximum this dissociation was delayed by several thousand years as the warm 'front' diffused down through the sediments to the base of the GHSZ; this process may have triggered the Storegga Slide about 7200 y BP, and other slides along the Norwegian margin.

It has been suggested (Nisbet, 1989; 1990; 2002; MacDonald, 1990; Kennett *et al.*, 1996, 2003) that sudden, massive releases of methane from decomposing hydrates during the early stages of warming were influential, either triggering global warming (Nisbet, 2002) or accelerating it (Kennett *et al.*, 2003). Kennett *et al.* referred to this as the 'Clathrate Gun Hypothesis'. Detailed studies of ice-core records have failed to identify any evidence to support the Clathrate Gun Hypothesis. This may be because records provide insufficient resolution, allowing a short-term spike to be missed, but the modellers assure us that major events would not be missed, and that there is no such evidence (Raynaud *et al.*, 1998; Brook *et al.*, 2000). The Clathrate Gun Hypothesis implicates a relatively small number of large events in global warming. It seems more realistic to expect a series of events spread over time as slope instability and hydrate dissociation were triggered separately in individual deposits according to local conditions; other gas-release mechanisms may also have played a part. Brook *et al.* (2000) argued '*we cannot exclude a slower and smaller release of clathrate methane as the source for the atmospheric concentration changes we observe*'.

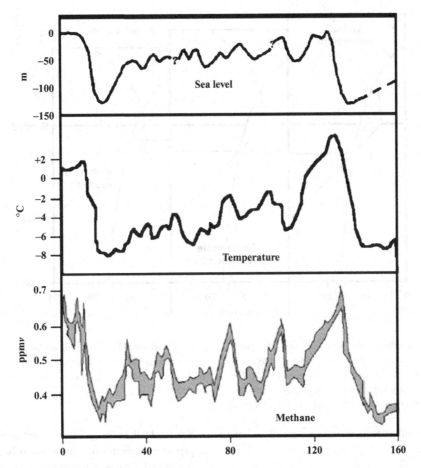

Figure 10.9 Correlation between eustatic sea level, temperature, and atmospheric-methane concentrations. Eustatic sea level (in metres relative to present sea level) taken as sea level of the Huon Peninsula, New Guinea, corrected for local tectonic effects (redrawn from Chappell and Shackleton, 1986). Temperature (in °C relative to present temperature) and atmospheric-methane concentrations (in ppm per volume, ppmv) are from the Vostok Ice Core, Antarctica. Horizontal scale in ky BP. (Redrawn from Houghton, 1997). (Reprinted from Judd *et al.*, 2002a with permission from Blackwell Publishing.)

Nisbet (2002) recognised that multiple small sources of methane might have the same effect as a single large source, and identified pockmarks in high latitudes as evidence that gas hydrates need not be the sole provider of methane. However, Judd *et al.* (2002a) explained that many forms of seabed fluid flow would have been affected by changes in surface conditions between glacial and interglacial periods, and that they would interact to provide positive *and* negative feedback to climate change. Variations in seabed emissions may be associated with processes other than climate change, such as seismic activity and changes in sedimentary regimes.

The geological thermostat
Conditions for generating the various fluids flowing through the seabed (hydrothermal fluids, thermogenic methane, and microbial methane from the deep biosphere) are generally unaffected by surface conditions. However, migration and escape through the seabed and the water column may be affected by:

- the distribution of gas hydrates and seabed permafrost;
- changes in water depth (and therefore hydrostatic pressure);
- the extent of sea ice and ice sheets;

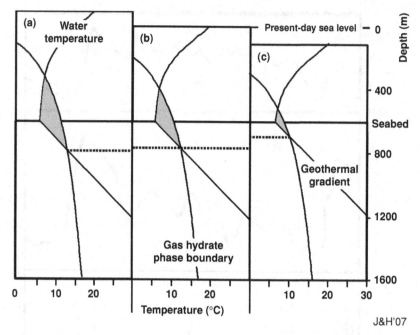

Figure 10.10* The effect of sea-level change on gas hydrate stability. The gas hydrate stability zone (shaded) lies between the intersections of the phase boundary (see Figure 6.18) and the temperature gradients: water temperature above the seabed, geothermal gradient. On diagrams (a) and (c) sea level is 100 m above (a) and 100 m below (c) 'present-day' sea level (b); other parameters are unchanged. Note how the base of the GHSZ (indicated by broken lines) rises towards the seabed when sea level falls.

- the location of coastlines and centres of sediment deposition.

Only deep-ocean processes (hydrothermal circulation associated with ocean spreading centres, arc volcanism, seamounts, etc.), and catastrophic gas–escape events (event plumes, natural blowouts, and Lokbatan-type mud-volcano eruptions) remain unaffected.

During global cooling ice sheets, sea ice, seabed permafrost, and polar gas hydrates advance, sea level falls, and coastlines migrate seaward across continental shelves. Each has implications for seabed fluid flow.

- Ice advance: Judd *et al.* (2002a) estimated that 10 000 000 km^2 of continental shelf in northern seas (Baltic, Barents, Bering, North, and Okhotsk seas, plus parts of the Arctic, Atlantic, and Pacific oceans) were covered by sea ice or seabed permafrost during the last glacial maximum. Much of this area is seep prone, and seeping methane will have been prevented from escaping to the atmosphere. During the glacial period this may have deprived the atmosphere of between 2 and 56 Tg of methane.

- Sea-level fall: the progressive reduction in hydrostatic pressure facilitates migration towards and through the seabed. Losses of methane from seeps, mud volcanoes, and hydrothermal vents to solution in seawater are reduced, significantly increasing the flux of methane from low-latitude continental shelves.
- Coastline advance: Seeps, vents, and mud volcanoes on emergent continental shelves inject methane directly to the atmosphere. However, the lowered baseline increases river erosion, sweeping sediments across the continental shelf and down the slope, reducing the volume of habitat for microbial methanogenesis.

During global warming, advances of ice and coasts, and sea-level changes are reversed. However, the pace of change will be affected by 'events' as well as progressive changes.

- Sea-level rise: the proportion of gas retained in the hydrosphere increases. Gas hydrates stabilise and replace methane lost during cooling. These changes are progressive, and result in a reduction in the methane flux.

- Coastline retreat: during marine transgressions river channels cut into the continental shelf are filled with sediments, and depocentres migrate landwards as continental shelves are flooded. Vegetated areas are flooded and buried in sediment. Once sea level stabilises, accommodation space for sediment deposition fills up; sediments accumulate in shallow coastal waters, and deltas advance. These conditions are favourable for the deposition of organic-rich sediments, and the generation of microbial methane. These changes are progressive, and result in an increase in the methane flux, particularly in shallow water.
- Ice retreat: as ice, seabed permafrost, and polar gas hydrates melt (progressively), methane is released. Various forms of 'release' are possible:

1. Gas trapped beneath the seabed escapes catastrophically once the confining permafrost or grounded sea ice melts sufficiently to be burst by the gas overpressure. The giant pockmarks of Block UK15/25 and the Barents Sea (described in Section 7.5.6) are examples of this type of release. Individual release events might involve large volumes of methane, but these events are unlikely to be synchronised, other than that they are most likely to occur during summer months. The impact of the release events depends on the number and size of gas reservoirs, and the rate at which they are 'uncorked'.

2. The concentration gradient between oversaturated seawater and the atmosphere drives gas exchange once sea ice has melted. Lammers *et al.* (1995b) described a similar release mechanism (see Section 10.6.2). Methane emissions will be concentrated in spring and summer periods when ice melts.

The processes described above (and in more detail by Judd *et al.*, 2002a) provide a mixture of positive and negative feedback. Some may counterbalance others, some are more influential than others, and the influence of individual sources may vary during the cooling–warming cycle. Changes in methane emissions during cooling seem to be progressive, whereas progressive change during warming is punctuated by 'events'. The most influential processes are thought to be those that

accumulate methane during one part of the cycle, and release it in the other. Deep-water gas hydrates release on cooling; seabed and surface ice, and polar hydrates release during warming. In both cases, the longer the accumulation period the greater the volume available for release.

It is significant that some of these processes are limited. During warming the rate of release, and therefore positive feedback, declines – once the majority of permafrost-covered continental-shelf ice, permafrost, and polar gas hydrate have melted. Perhaps these processes act like a thermostat, limiting climatic extremes. At this stage there is insufficient evidence to quantify the effectiveness of the 'geological thermostat'. However, the processes described here might effectively limit climatic extremes. Judd *et al.* (2002a) argued that gas hydrates alone are not responsible for the geological thermostat. Furthermore, other (non-geological) influences on atmospheric methane and other 'greenhouse gases' (especially carbon dioxide) must be recognised, and the whole must be considered within the context of orbital cycles; as glacial cycles match orbital cycles, their importance cannot be denied. It seems logical to suggest that orbital cycles are the primary drivers, whilst other factors, including the geological thermostat, determine the limits to climatic extremes and the rates at which climate changes from warm to cold and back again.

10.8.2 Earlier events

The potential influence of seabed fluid flow on climate change is not confined to the Quaternary. During geological history there have been many rapid changes in climate, some with associated mass extinctions. Might seabed fluid flow have played a role in initiating them? Dickens (1999, 2003), amongst others, argues that prominent negative $\delta^{13}C$ excursions at, for example, the Palaeocene/Eocene boundary (55.5 million years ago) can be explained by massive releases of methane from gas hydrates.

10.9 AFTERWORD

In this and previous chapters we have returned again and again to methane. Clearly it is of key importance to marine systems, both above and below the seabed. The complexity of the contributions made to the hydrosphere are caused by the following factors.

- Various methods of formation, at different depths below the seabed – probably also in microenvironments in the water column.
- Microbial utilisation beneath, at, and above the seabed.
- Flow through the seabed at different water depths, from intertidal to the deep ocean.
- Injection into the water at different rates, from slow diffusion to major, explosive releases.
- Mixing with water masses at different altitudes above the seabed, depending upon water depth and the rate of injection, as well as water circulation.
- Interchange with the atmosphere according to the relative concentrations above and below the sea surface.

Together these factors make the role of methane incredibly complex. Our attempts to understand this system are made on the basis of measurements made at a small number of locations over a few decades, with new discoveries continually demonstrating that we have yet to reveal all the secrets of all natural processes. Our feeble understanding might be no more successful than trying to understand the weather using data from a few dozen land-based weather stations and the odd weather balloon. Is it surprising that we have much to learn?

In the context of new discovery, we find it interesting that the atmospheric modellers always seem keen to dismiss geological processes, or is this just our imagination? Well, it seems that we are not alone:

> The methane hydrate reservoir is generally not considered an important component of late Quaternary climate change for several reasons: the reservoir is remote and poorly studied; little was known about methane hydrates until recently; and modern hydrates appear stable and show little evidence during the Holocene of prior instability earlier in the Quaternary. Finally, the upper continental margin, containing most of the reservoir, was considered a relatively stable oceanic environment, uninvolved with Quaternary climate behavior, especially on millenial time scales. For these and other reasons, methane hydrates have been considered 'a last resort' hypothesis for climate change [K. Kvenvolden, personal communication].
>
> Kennett *et al.*, 2003

This reluctance to accept new paradigms is not confined to the Clathrate Gun Hypothesis! We have encountered reluctance to accept the significance of seabed fluid flow in various other ways. Indeed, a look at most textbook entries on the global carbon cycle shows that few recognise the existence of seabed fluid flow. Yet, as we have tried to demonstrate, seabed fluid flow is influential in determining the chemical, physical, and biological character of the world's oceans, and affects the composition of the atmosphere.

11 · Implications for man

A look at the exploration history of the important oil areas of the world proves conclusively that oil and gas seeps gave the first clues to most oil-producing regions. Many great oil fields are the direct result of seepage drilling.

Link, 1952

Some marine geohazards, with the potential to affect offshore operations, are associated with seabed fluid flow. The petroleum industry, in particular, has learned from experience that these geohazards must be recognised and understood, and that safeguards must be in place if serious consequences are to be avoided. But seabed fluid flow is not always a negative thing. It provides resources such as methane and metals, and acts as a guide to others, particularly petroleum. However, the marine environment is sensitive and biological communities associated with vents and seeps are uncommon and fragile – they require protection.

11.1 INTRODUCTION

The implications of seabed fluid flow for offshore operations fall into two categories: those that are hazardous, and those that are beneficial. In this chapter we discuss the most important of these. Hazards associated with seabed fluid flow have been of concern since the beginning of offshore engineering for hydrocarbon development. Marine 'geohazards' include those associated with natural features and events (slope instability, gas escapes, and mud-volcano eruptions, etc.), and those (such as blowouts) that happen as a direct consequence of man's intervention with the natural seabed environment; Figure 11.1. Benefits include the direct value of seabed fluids (seep gases and gas hydrates) or their by-products (hydrothermal minerals) as resources, and the assistance provided to petroleum prospecting by seeps. There may

also be benefits to the fishing industry, and there is future promise for biotechnology.

Finally, we look at the other side of the coin; the impacts that human activities have, or may have, on seabed fluid flow and features associated with it. This includes activities that may trigger 'events', and those that may be harmful to delicate features associated with seabed fluid flow. It is good that international and national legislation is now affording some protection to some seep and vent sites.

11.2 SEABED SLOPE INSTABILITY

In 1929 trans-Atlantic submarine telephone cables were broken by slope failures when a turbidity current, triggered by an earthquake (magnitude 7.2), carried about 100 km^3 of sediment down the continental slope and onto the abyssal plain, reaching about 1000 km from the earthquake epicentre. Within 100 km of the epicentre cable breaks were instantaneous, but down the slope they broke one after another as the turbidity current swept across the Laurentian Fan and onto the Sohm Abyssal Plain. The timing of the breaks (the furthest break occurring 13 hours after the earthquake) enabled the speed of the current to be estimated; according to Piper *et al.* (1985) this was at least 65 km h^{-1} on the upper fan. This event provided dramatic evidence that seabed slopes may become unstable. As well as causing significant disruption to intercontinental communications, the associated tsunami killed 27 people in Newfoundland (Locat, 2001).

Seabed slopes may fail even on very slight gradients, as little as 0.5° on some deltas according to Prior and Hooper (1999). As the oil industry has extended its offshore interests into the deep waters of the continental rise, evidence of major slope failures has been discovered in many areas. However, slope instability may occur at any water depth. There have been several reports of gas associated with shallow-water slope failures, and there

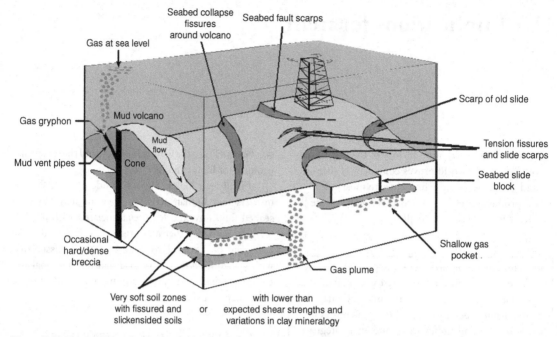

Figure 11.1* Cartoon illustrating the range of geohazards encountered in the Caspian Sea. (Image courtesy of Mike Sweeney, BP Exploration.)

is a strong correlation between regions of gas hydrates and major submarine slides (Laberg and Vorren, 1993). Is this coincidental, or does gas play a role in slope instability? We address this question by investigating some relevant examples.

11.2.1 Gas-related slope failures: case studies

Hampton *et al.* (1996) identified five environments in which seabed slope failures are common:

- fjords;
- active river deltas on the continental margin;
- open continental margin slopes;
- submarine canyons;
- oceanic volcanic islands and ridges.

Fjords: the 1996 Finneidfjord Slide
At about midnight on 20 June 1996 there was a shoreline slope-failure in Finneidfjord in northern Norway. The failure started 50–70 m from the shore. The following was based on eye-witness accounts:

Eye witnesses saw waves, bubbles and whirls moving away from the shore some time before

midnight. The slide developed retrogressively towards land. About 25 minutes after midnight a driver felt that his car and the road were shaking violently and stopped. The beach below the road was gone. Minutes later he witnessed 250 m of the road breaking in three parts and slumping into the sea. A car with one person also disappeared. Shortly afterwards, the nearest house started to move, then sank into the mud and disappeared into the sea. Three people inside did not manage to escape. All this happened within 5 minutes or less. Several minor mass movements occurred along the edges of the slide, but after 1 hour, no more slide activity was observed.

Longva *et al.*, 2003

Major clay slides like this are not uncommon in and around fjords in Norway, Alaska, and British Columbia; in many cases severe damage has been caused and lives have been lost (Hampton *et al.*, 1996; Longva *et al.*, 1998). They are caused by the sudden liquefaction of 'quick' clays, often as a result of an external trigger such as seismic activity, or an increase in pore fluid pressure. The Finneidfjord slide followed a period of heavy rain;

detonations from nearby tunnel construction works may have contributed (Longva *et al.*, 2003).

Beneath the fjord, a 'bright' layer on seismic profiles, representing relatively sandy sediments, was found at the presumed depth of the failure plane; this was underlain by acoustic turbidity (Best *et al.*, 2003). It seems that gas from underlying gassy sediments was accumulating in the sandy layer, resulting in excess pore fluid pressure that may have contributed to the failure.

Active river deltas: the Fraser Delta

Shallow gas is present over a large area of the Fraser Delta, British Columbia (see Section 3.20.2). There is evidence of slope failure in two areas of the delta front: off Sand Heads, and the 'Roberts Banks Failure Complex'. Both areas are characterised by sandy seabed sediments, and evidence of gas has been identified in both areas (Judd, 1995; Christian *et al.*, 1997). It seems that gas has made these sandy sediments susceptible to liquefaction failure (Atigh and Byrne, 2003; Grozic, 2003). Sand Heads is where the main distributary of the Fraser River crosses the delta flats, depositing much of its load of silty/sandy sediments. The delta front failed five times between 1970 and 1985; the 1985 event involved at least 1×10^6 m^3 of sediment (McKenna *et al.*, 1992). It is probable that slope failures in this area are largely a result of the rapid deposition, but sandy sediments such as these tend to be liable to liquefaction failure during cyclic loading; Atigh and Byrne (2003) and Grozic (2003) suggested that gas may have increased their susceptibility.

Further south on the delta foreslope the Roberts Bank Failure Complex is a distinctive area characterised by a hard (sandy) seabed, represented on the profiles by a strong (high-amplitude) reflection, and sandwaves. There is seismic evidence of slope failure (Hart and Olynyk, 1994). The sediments here contrast with those elsewhere on the slope because acoustic turbidity is not ubiquitous. However, small patches of acoustic turbidity and variable-amplitude reflections suggest there is some gas here. It could be said that gas is relatively scarce in the failure complex because the sediments are coarser than elsewhere, and therefore unlikely to be able to retain gas at the pressures that might be found in a finer-grained sediment. An alternative is that gas was present, but escaped when the slope failed.

Slope failures involving gassy sediments have been described from other deltas, including the Alsek Delta (Section 3.19.2), the Klamath Delta (Section 3.22.1), and the Mississippi Delta (Section 3.26). Some typical deltaic slope failure features were described in Section 7.4.2 (for further details see Prior and Coleman, 1982).

Open continental margins – upper slope: the Humboldt Slide

The Humboldt Slide, described by Gardner *et al.* (1999) as '*a large, complex slide zone*', is located on the upper slope of the Eel River Basin offshore northern California in water depths of 250 to 600 m. It affects an area of about 200 km^2, and a sediment volume of about 6 km^3. Although dated as late Pleistocene to early Holocene, it seems that the slide may still be active (Gardner *et al.*, 1999). Gas, gas hydrates, and related features are widespread in the Eel River Basin (see Section 3.22.1). Yun *et al.* (1999) questioned the role of gas in the slope failure. Was it cause or effect? Then, Gardner *et al.* (1999) suggested it may have at least contributed to the cause. This would explain, for example, why there is widespread gas in the underlying sediments but little near the surface. Gardner *et al.* thought that gas-induced increases in pore fluid pressure might '*contribute to the section's susceptibility to shearing and sliding*' by reducing the shear strength, S_u – see Equation (7.4). However, shallow gas and pockmarks are present over a large part of the slope and outer shelf of this region, but the Humboldt Slide is the only major failure. This suggests there must be a factor other than the presence of gas that made this particular location susceptible to failure. However, Lee *et al.* (1999) had a totally different interpretation. They actually questioned the concept that this is a slide, saying that '*many of its characteristics suggest a series of depositional bedforms*'.

Open continental margins: the Storegga Slide Complex

The Storegga Slide, on the Norwegian Margin (Map 3), is probably the world's largest sediment slide. It moved a volume of 3500 km^3 of sediment, and has a 290 km long headwall scar with slide material supposedly extending 750 km downslope to cover 90 000 km^2 of the Norwegian–Greenland Sea Abyssal Plain (Bryn *et al.*, 2003). Six failure events have occurred in this area during the last 500 000 years. Flood deposits found in lakes and bogs on the Norwegian coast and in the UK of the same age as the most recent (about 8200 BP) event suggest that at least this one caused a tsunami

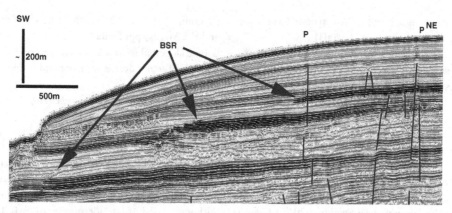

Figure 11.2* A multichannel 2D-seismic section across the northeast flank of the Storegga Slide. The letter 'P' indicates the locations of complex seabed pockmarks with carbonate ridges (described by Hovland *et al.*, 2005). They are associated with faults and 'pipes' (vertical gas migration pathways). The regional BSR, indicated by arrows, is probably apparent because of the enhancement of underlying reflections by gas trapped below the GHSZ (gas hydrate stability zone). (Reproduced from Hovland *et al.*, 2005 with permission from Elsevier.)

(Bondevik *et al.*, 1997). Because a huge gas field, the Ormen Lange field, lies under the slide scar a great deal of research effort has gone in to understanding this slide complex. The motivation is to ensure that the current seabed is stable, presenting no hazard to seabed installations.

The Storegga Slide lies in a depression between two depocentres for glacial sediments: the North Sea Fan and the Skjoldryggen areas (Bryn *et al.*, 2003). Although the detailed history of the slide is debatable, it is inferred that it was related to gas and gas hydrates; clearly identified BSRs (bottom-simulating reflectors) occur on the flanks of the current scar and project into the level of the slide-scar sole (Figure 11.2; Mienert *et al.*, 1998; Bouriak *et al.*, 2000), and gas-escape structures rise from the BSR to the present seabed upslope from the slide scar (see Section 6.3.3). The question 'what came first – the hydrates and/or the slide?' was discussed by Berndt *et al.* (2001). Although they failed to say much about the likelihood of free gas and gas hydrates providing the conditions necessary for failure in the first place, they found that the slide must have disturbed large volumes of buried gas hydrates causing it to dissociate and escape. They calculated that fluid must have escaped over a period of less than 250 years. Subsequent modelling by Sultan *et al.* (2003) demonstrated that various characteristics of the slide could not be explained if the influence of dissociating gas hydrate was excluded from the model. They concluded that the melting of gas hydrates may have initiated the failure,

and that '*the failure interface is initiated at the top of the hydrate layer and not at the level of the BSR*'.

Open continental margins – lower slope: the Cape Fear Slide

The first side-scan sonar images of the Cape Fear Slide, off the Carolinas, were published by Dillon *et al.* (1982) and Hutchinson *et al.* (1982). They show that, although this is the largest on the US Atlantic Margin, it is one of several similar slides in the area (Map 31), another being the Cape Lookout Slide (Popenoe *et al.*, 1993). The amphitheatre-shaped headwall scarp is located in the lower continental slope at a depth of about 2600 m. It is up to 120 m high and over 50 km long. A secondary complex of slumps and slide tracks extend 40 km upslope from the headwall scarp. Downslope, a broad trough, over 150 m deep and more than 40 km across, has been scoured into the seabed. The slide deposits in this trough extend for more than 250 km onto the Hatteras Abyssal Plain (Embley and Jacobi, 1986; Popenoe *et al.*, 1993).

The most intriguing fact about the Cape Fear and Cape Lookout slides is not that they start at great water depths, but that their headwall scarps are located close to salt diapirs. The Cape Fear headwall scarp encircles five diapirs; the largest is 8 km in diameter and its top protrudes above the seabed (Schmuck and Paull, 1993). These form part of the line of salt diapirs extending along the seaward side of the deep Carolina Trough; the Blake Ridge Diapir (Section 3.27.2) is also in this line.

An extensive BSR in this area is taken to indicate gas trapped below gas hydrates (Schmuck and Paull, 1993). Anomalous temperature and fluid-flow conditions associated with the diapirs cause the BSR to rise over them, indicating a thinning of the hydrate-stable layer (Paull *et al.*, 2000). The BSR also rises towards the slide scar's edges, and is less prominent near its centre; this may indicate gas escape. Gas venting that is thought to occur at the head of the slide (Schmuck *et al.*, 1992), and upslope of the slide scar (Dillon *et al.*, 1982) may be associated with the numerous normal faults seen on seismic profiles (Paull *et al.*, 2000). The combination of diapirs, gas, gas hydrates, and a major slide in this area seems to be more than a coincidence. Perhaps the slope failure and removal of sediment caused a pressure reduction and breakdown of the hydrates. Alternatively, in accordance with the hypothesis of salt-stock formation suggested by Hovland *et al.*, 2006 (outlined in Section 9.4.3), we suggest that warm, methane-charged fluids flowing out of the salt stocks might have been responsible for the dissociation of at least some of the gas hydrate. We think this weakened the sediments, triggering the slides.

11.2.2 Associated tsunamis

Tsunamis are thought to have been associated with seabed slope failure on at least two occasions. Numerical simulations and 3D (three-dimensional)-animations of large tsunami waves generated by the Storegga Slide clearly demonstrate the hazardous impact of large underwater slides. Another example, the Sissano tsunami, struck the north shore of Papua New Guinea in July 1998. Detailed seabed investigations reported by Tappin *et al.* (2001) identified a 5–10 km³ slump. On the seabed in the slump area there were fissures, brecciated angular blocks, vertical slopes, and talus deposits. Also, Tappin *et al.* reported '*active fluid expulsion that maintains a chemosynthetic vent fauna*'; evidence that seabed fluid flow was implicated in the failure event.

11.2.3 Why do submarine slopes fail?

The study of submarine slope failures is a major topic addressed in many specialist publications: Hampton *et al.* (1996), Mulder and Cochonat (1996), Locat (2001), Locat and Mienert (2003), to name but a few. These provide descriptions, classifications, analyses of the var-

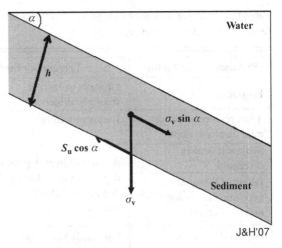

Figure 11.3 Forces acting on a submarine slope. α = slope; h = thickness of the sediment slice; σ_v = vertical stress; S_u = undrained shear strength of the sediment; $\sigma_v \sin \alpha$ = gravitational shear stress acting in the direction of potential movement; σ_v and S_u were defined in Equations (7.1) and (7.4) respectively. (Adapted from Hampton *et al.*, 1996.)

ious stages of movement, and guides to assessing risk. Our purpose here is only to discuss the roles of gas and gas hydrates.

Slope failure occurs when gravitational forces (vertical stress; $\sigma_v \sin \alpha$), which tend to pull a sediment mass downslope, exceed the resisting forces (shear strength; $S_u \cos \alpha$); see Figure 11.3. This happens in the weakest layer of sediment, allowing the slice of sediment above this 'failure plane' to move downslope. Table 11.1 lists factors that make a slope predisposed to failure, and others that trigger failure. Factors of interest here (shown in bold) are those that reduce sediment strength by generating excess pore fluid pressure – see Equation (7.6).

It is important to acknowledge that slope failures are not necessarily associated with gas or gas hydrates. McAdoo *et al.* (2000), who reviewed 83 deep-water slope failures offshore Oregon, California, Texas, and New Jersey, identified seismicity, active sedimentation and erosion, and salt tectonics as major factors. They did not mention gas hydrates, and mentioned gas only in connection with four slides, all near the pockmark field in California's Point Arena Basin. We are not trying to suggest that gas is a 'forgotten' factor (although this may be true in some cases). Rather, we wish to point out that where gas or gas hydrates are present they are likely to interact with other factors (rapid sedimentation,

Table 11.1 *Causes of submarine slope failure (including material from Hampton* et al., *1996; Prior and Hooper, 1999; Locat, 2001; and Leroueil* et al., *2003).*

Predisposition factors	Triggering factors	Causes of failure	
Erosion	Changes in current strength/direction	Slope over-steepening	Increasing stress
The presence of salt or mud diapirs at depth	Diapir movement		
Sediment density inversions	Rapid sediment deposition, e.g. deposition on deltas by river flood; deposition following sediment mass movement		
Potential for rapid deposition or upslope mass movement		Increased vertical stress	
Large tidal rise/fall	Unusually low tides	Reduction in pore fluid pressure leading to sediment failure	Reducing strength
Volcanically active zones	Volcanic activity	Cyclic loading ↓	
Seismically active zones	Earthquakes		
Susceptibility to storm waves	Storms		
The presence of aquifers	Heavy rain in aquifer catchment areas	Excess pore fluid pressure generation within porous layers capped by impervious layers	
The presence of thermogenic gas sources at depth	**Fluid migration**		
The presence of organic-rich sediments	**Gas generation**		
The presence of gas hydrates	**Gas hydrate decomposition**	Sediment liquefaction	

cyclic loading by waves, tides, earthquakes, etc.), and may facilitate slope failures which would not occur in their absence. Indeed, we have been struck by the frequency with which authors have suggested that more than one factor has influenced submarine slope failure. This seems to apply to failures at any water depth.

Seabed pore fluid pressure build up
In Chapter 7 we discussed pore fluid overpressures, and explained that overpressures may result from undercompaction – Section 7.5.2 – and the accumulation of buoyant fluids, particularly gas – Section 7.5.3;

Equation (7.12). An increase in pore fluid pressure results in a reduction in shear strength, and resistance to shear failure. Orange *et al.* (2003) suggested that seeps are evidence of the existence of fluid overpressure, and that overpressure provides an internal driving mechanism for failures characterised by a flat base and a steep, amphitheatre-shaped headscarp. They thought that persistent overpressure might lead to recurring failure within the original feature, leading to headward migration and linear failure morphology.

Hampton *et al.* (1996) noted that some major deltas are apparently unaffected by slope instability.

One example they mentioned, the Changjiang (Yangtze River), China, is the fourth largest contributor of suspended sediment in the world, delivering about 0.5×10^9 t of sediment annually. Yet, '*although significant deltaic deposits occur there, as well as pockmarks resulting from the expulsion of biogenic* [microbial] *gas, the seafloor is otherwise featureless*'. In this case we think the escape of gas has prevented the build-up of sufficient excess pore pressure, contributing to the stability of the delta.

It is interesting to compare the role of gas in marine slope failures to the role of water in onshore slope failures. Onshore, a typical failure plane occurs where water percolating downwards through a permeable formation encounters a fine-grained, impermeable layer. Offshore, failure occurs when gas migrating upwards is trapped beneath fine-grained, impermeable sediments. The roles of gas/air and water are reversed and the process is inverted.

The roles of gas hydrate

A relationship between gas hydrate and slope failure is suggested, if nothing else, by the frequency with which slope scarps coincide with the gas hydrate stability zone (GHSZ). McIver (1981) suggested they were not coincidences, and numerous studies have subsequently investigated the relationship. A particularly good example is the Atlantic Continental Slope of the USA where there are sufficient slide scars for an analysis to be meaningful (Paull *et al.*, 2000). There is a BSR right along the slope, and '*data clearly show that the slides are neither randomly distributed nor are they strongly associated with steep slopes*'. Paull *et al.* found that the majority of the slide headwall scarps occur at the updip limit of the GHSZ, about 500–700 m water depth. This led them to conclude: '*the observed distribution of slide scars is consistent with the distribution that is predicted to occur if gas hydrate decomposition has played a significant role in causing these sediment failures*'. Similar 'coincidences' have been reported from other continental slope areas. Paull *et al.* (2000) remarked that '*a great deal of circumstantial evidence strongly supports the concept that gas hydrate breakdown is often instrumental in triggering sediment mass movement on the sea floor*'.

The role of gas hydrate varies according to its stability. When stable it increases the mechanical strength and rigidity of the sediment, but dissociation releases both water and large volumes of gas. It is assumed that gas escapes through the seabed, but water may remain in the host sediment. The increased water content may seriously affect the sediment's strength, possibly reducing it to mush, and making slope failure more likely. Hydrate may dissociate from the seabed downwards as a result of seawater warming. However, when the dissociating hydrate is overlain by unaffected sediment, or when dissociation occurs at the base of the GHSZ (as explained in Section 10.8.1), it creates a weak layer; this is where failure might occur.

11.2.4 Predicting slope stability

Offshore geotechnical slope stability investigations have grown from the long tradition of onshore investigations. A 'factor of safety' (FoS), defined as the ratio between cyclic strength and induced dynamic stress in each element of the slope, is used to assess the stability of marine slopes. Chaney (1984) pointed out that, in the absence of a universally accepted FoS, it is necessary to employ '*applying judgement or an averaging process to the results*'. Of course, there may be uncertainties, indeed we sense a 'factor of uncertainty' when geotechnical engineers have to use 'judgement and averaging' in order to assess underwater slope stability – a situation not unlike uncertainties known within geophysical interpretation. This is, however, in no way a reassuring situation for the general public, who may think that we are dealing with exact science. Locat (2001) addressed this situation and came to conclusions we certainly agree with:

> Ultimately, the goal is to be able to carry out proper risk assessment analysis pertaining to submarine mass movement. This could be achieved by integrating the geotechnical characterization of mass movements into a risk assessment methodology, which can then be applied on a regional basis.
>
> Locat, 2001

Dugan and Flemings (2002) used such an assessment when considering the stability of the US Continental Slope offshore New Jersey. They suggested that the New Jersey continental slope was unstable because of high sedimentation rates approximately 0.5 million years ago, but that the modelled FoS is now approximately 1.5 (upper slope) to 3 (lower slope) – where FoS > 1 represents stability, and FoS ≤ 1 represents instability. The approach used by Lee *et al.* (1999), who studied the Eel Margin, offshore California, was to use a GIS (Geographical Information System) to map key parameters. Specifically, they determined the critical horizontal earthquake acceleration required to cause failure, k_c, and plotted the ratio of k_c to the peak seismic acceleration with a 10% probability of exceedance in 50 years. The resulting map effectively differentiates between areas that are, and are not susceptible to slope failure.

We are not qualified to comment on the reliability of these approaches to evaluating the risk of slope failure. However, we note that both Lee *et al.* and Dugan and Flemings made various assumptions about the subseabed conditions in their respective areas, and suggest that caution should be used when methods such as these are applied to areas in which there is fluid flow. As fluid flow tends to be focussed (Section 7.5.3), generalisations about fluid pressure conditions may not be valid, so locations of focussed fluid flow may be more susceptible to failure than predictions might suggest.

11.2.5 Impacts of slope failures on offshore operations

We introduced this section with a comment about the cutting of cables by a slope failure. Cables are not the only vulnerable installations. Any seabed structure, including pipelines, platforms, etc., would be affected by the failure of the seabed on which it was sitting. Site investigations must take into account not only the site of the installation being planned, but also the surrounding area. Seabed slope-failures, particularly those in deep water, can affect such enormous areas that an entire site may represent just a small fraction of a single slide, so it is important that the potential for the site to be affected by failures in the surrounding area should be considered. Prior and Hooper (1999) described an example from a deep-water (870 m) site in the Gulf of Mexico where gas hydrates were implicated in slope-failures. The site investigation for a Tension Leg Platform (TLP) identified evidence that slope-failure events had occurred close to the site since about 30 000 years ago; the most recent within the last 1000 years. Geophysical mapping showed that debris flows started from fluid-expulsion mounds and craters from which large volumes of overpressured gas had been expelled. The flows began on slopes of 10°–15° and extended up to 11 km downslope. However, more recent (<12 000 year old) flows had stopped at least 1800 m upslope from the installation

site, and were < 5 m thick. It was considered that the risk was acceptable, and the TLP was successfully installed.

11.3 DRILLING HAZARDS

Drilling for petroleum exploration and exploitation, site investigations, and scientific research are all hazardous undertakings. Some of the most potent hazards are associated with natural pore fluids: overpressured gas and water, and gas hydrates.

11.3.1 Blowouts

Overpressured gas is one of the main hazards in offshore drilling. As a pollutant, oil is more damaging to the environment than gas. However, as a hazard to drilling, personnel, and infrastructure, gas is more dangerous because of its mobility, flammability, negative effect on water buoyancy, and difficulties in control. According to Prince (1990) shallow gas is a responsible for '*approximately one third of all blowouts*', and has been responsible for the loss of both lives and drilling rigs.

Although geochemists and explorationists distinguish between microbial and thermogenic methane, when it comes to hazards, there should be no such distinction – both are flammable and difficult to control in large quantities. In any drilling operation, formation fluids (water, oil, or gas) will flow into the well bore when the formation fluid pressure exceeds the pressure in the hole. If the fluid entering the well bore is less dense than the drilling fluid, it will move upwards in response to buoyancy. Initially this is described as a 'kick'. The expansion of gas as it rises within the hole drives the expulsion of drilling fluid. This reduces the weight and pressure of the fluid column, encouraging more gas to enter the hole; a chain reaction results. If not controlled the result may be a blowout, a '*"wild", unrestrained flow of gas or gas-charged formation fluid at the surface*' (Graber, 2002). Gas blowouts put the drilling vessel or platform and the crew in peril in several ways: by releasing toxic gases, triggering fires, causing mechanical damage, and buoyancy loss. Situations may also occur where fluids from deep, overpressured zones flow up the borehole and encounter shallower, low-pressure zones. Under such conditions, the deep fluid (oil, gas, or water) may enter faults and fractures and permeable beds resulting in an uncontrollable leak – an 'underground blowout'. Such recharge and communication with overpressured zones can even make future shallow drilling in an area hazardous.

Shallow-gas blowouts

'Shallow gas' is defined differently by different people. Some regard 'shallow' as being within 1000 m (or some other arbitrary depth) of the seabed. A more pragmatic definition is that 'shallow' means above the first casing point, where a petroleum (exploration or production) well is lined with steel casing. Until this casing is set the well cannot be protected by a 'blowout preventer' (BOP), so the drilling operation is vulnerable to the influx of pore fluids (gas, oil, or water). The depth of the first casing differs according to various operational parameters, but is usually within the top few hundred metres of the well. Site investigation boreholes, which are not cased, may penetrate to 100 m or more, so they also are vulnerable to fluid overpressures in the topmost sediments (the case described in Section 7.5.3 being an example).

UNDERGROUND BLOWOUTS

Kicks can occur after the setting of casing whilst drilling deeper formations. In such cases it is normal for the well to be shut in using the BOP, and subsequently to 'kill' the kick, for example using overweighted drilling mud. However, this opens the possibility of an 'underground blowout'. Once it has been shut in, if the excess gas pressure fractures the formations at or beneath the casing shoe, gas will find a route to the seabed outside the casing. This is most likely to occur at the first (shallowest) casing point as this is probably in the least compacted and, therefore, weakest sediments. The consequence of exceeding the fracture pressure – defined in Equation (7.14) – can be that overpressured fluid forces a route to the seabed outside the casing, rendering the blowout 'uncontrolled and uncontrollable' (Grace, 1994). The route taken to the surface is not necessarily close to the casing, indeed sometimes gas has been recorded venting some distance from the rig. The seabed can be eroded to form pockmark-like craters, some of which have been large enough to consume jack-up rigs and platforms. Grace said that 'historically the most infamous and expensive blowouts in industry history were associated with fracturing to the surface from under the surface casing'.

COPING WITH PROBLEMS

There are two strategies for tackling shallow-gas kicks. The excess gas pressure may be passed through a 'diverter' at the seabed, to vent into the water column away from the rig. Alternatively, a diverter on the rig may handle the gas (Prince, 1990; Grace, 1994). By diverting the gas, the excess pressure is allowed to leak away. However, both of these strategies have their problems. In particular, the erosion of pipe work by sands produced with gas has been known to cause diverters to fail, increasing the risk of explosion or fire. When problems occur, the shallowness of the source of the problem means reaction times tend to be short, making control difficult. Adams and Kuhlman (1990) urged operators and drillers to place a high priority on ensuring equipment and procedures suitable for dealing with shallow-gas blowouts are in place. They noted that 'records show that failure to do so can result in loss of lives and major property damage'.

SHALLOW-GAS BLOWOUT EXAMPLES

During the development of the Gullfaks field it was known that pockmarks and shallow gas were present (see Section 2.3.6). Shallow gas was encountered in 57% of the exploration wells, and in one case resulted in an uncontrolled blowout. This gas is held in interlinked bodies (> 5 m thick) of unconsolidated silty sand at three levels between 130 and 230 m below seabed. Detailed investigations were undertaken after it was found that shallow gas extended beneath the site of the Gullfaks 'A' concrete gravity platform, and several wells were drilled solely to test or drain off this gas (Lukkien, 1985; Hovland, 1987). Pressure build-up tests undertaken in the topmost gas-bearing sand produced the following results.

- Formation fluid pressure was 3320–3370 kPa (i.e. 60–110 kPa overpressure) indicating a gas column of 6–10 m.
- The gas was 99% methane.
- The gas-bearing sands were 'highly permeable' (200 to 350 mD).
- During the test 400000 Sm^3 (m^3 at STP) was produced.
- A tentative estimate suggested that the volume of gas in place was 100×10^6 Sm^3.

The estimated porosity of sands in this zone was 30–40%, and the mean gas saturation was 50–60%. A total gas volume of 2.5×10^6 Sm^3 was produced over a 35-day period with a stable production rate of 64 000 Sm^3 d^{-1}, and a maximum of 115 000 Sm^3 d^{-1}. Continued production proved difficult, probably because the sand is likely to collapse, destroying the permeability close to the well.

It was thought that spontaneous kicks were unlikely during normal drilling because pore fluid pressure was only slightly above hydrostatic, but there was a danger of a loss of control either because of a loss of circulation (when drilling fluid invades the permeable formation), or during swabbing (when a pressure fluctuation is caused by equipment being moved up the well). Lukkien (1985) envisaged that, if a gas escape occurred, an underground blowout was likely.

The following examples further demonstrate the seriousness of the shallow-gas hazard.

1. During exploration drilling in the German Bight of the North Sea in 1963, the rig Mr Louie experienced a blowout resulting in the formation of a 400 m wide crater known now as the 'Figge Maar'. Apparently the crater was originally 31 m deeper than the surrounding seabed (34 m deep), but it has been progressively filling up with sediment so that its depth had reduced to 22 m by 1981, and 14 m by 1995 (Thatje *et al.*, 1999).

2. In 1969 a blowout in the Dos Cuadras field, offshore California, permitted high-pressure fluids from deeper regions of the reservoir to migrate to the shallower, low-pressure zone. Afterwards, several substantial oil and gas seeps occurred within 300 m of the drilling platform. They were of sufficient severity to cause repeated suspension of work and evacuation of the platform. By the time the well was under control, some 10 days after the blowout began, the seeps had gradually increased to an area of about 200 000 m^2. Observation of seep sites from a submersible showed that craters up to '*many feet*' in length and '*several feet*' deep were surrounded by angular rock debris which had apparently been blown from the seabed by the force of the flow (McCulloh, 1969).

3. In December 1972 the mat-supported jack-up rig 'J. Storm II' tilted. The rig was evacuated, and sank about 20 minutes later. A seabed survey conducted the following year revealed a flat-bottomed crater

nearly 500 m across and approximately 12 m deep. Abundant gas plumes in the water column were thought to be coming from both the broken drill stem and the floor of the crater (Worzel and Watkins, 1974).

4. A jack-up platform drilled through a fault zone extending from the seabed to a high-pressure gas pocket at depth. A vigorous stream of gas bubbles was seen escaping at the sea surface some 300 m from the rig, where the fault reached the seabed. These bubbles caused the sea surface to rise by between 12 and 22 m. A crater formed at this point and gasification caused the failure of a wedge of sediment. The rig was first set on fire by ignited gas that had escaped from the well casing; it foundered when the seabed sediments failed (Sieck, 1975).

5. In the South Pass Area, off the Mississippi Delta, a blowout occurred when shallow gas was encountered at a depth of about 210 m. The gas ignited, and the rig collapsed and sank. A seabed crater formed by the blowout was surveyed five days later when it measured about 600 m across and about 30 m deep. Gas was still escaping, the water column on a shallow seismic profile being '*literally covered*' with gas. The shallow sediments on the sides of the crater were apparently saturated with gas, and it was suggested that sediments between the drilling rig and the gas zone were blown into suspension when pressure increased beyond some critical point (Bryant and Roemer, 1983).

6. A blowout in the High Island area, Gulf of Mexico, resulted in the loss of an entire platform into a seabed crater when a well penetrated gas-charged sands at a depth of 1220 m. Subsequent surveys showed that the crater was 450 m wide and nearly 100 m deep; an estimated 4.4 million m^3 of sediment had been ejected to form the crater (Bryant and Roemer, 1983).

7. In 1985 the semi-submersible rig 'West Vanguard' was drilling a wildcat well at the Mikkel field on the Haltenbanken area off mid Norway. A blowout occurred when gas-charged channel sand was penetrated only 300 m below the seabed (Section 6.2.3). Attempts to divert the gas away from the rig floor and into the water failed because of sand in escaping fluids. The blowout caused one fatality, the evacuation of the rig and the following fire caused severe damage to the rig. An indication of the volume of gas released

Figure 11.4* A blowout in the Haltenbanken area of the Norwegian Sea in October 1985. Large volumes of gas (mainly methane) are seen escaping to the atmosphere from a shallow-gas reservoir; water depth 240 m. (Photographed by Leif Berge; from Hovland and Judd, 1988.)

is given in Figure 11.4. Gas bubbled continuously to the sea surface over a period of about two months.

8. In November 1990 Mobil experienced a blowout in the UK North Sea when a base-Quaternary shallow gas source was penetrated. Although the rate of gas release slowed after the first few days, gas was still escaping four years after the accident when it was predicted that, because of the size of the reservoir, it would continue for several years unless the vent collapsed and became blocked (Rehder et al., 1998).

9. Although most shallow-gas blowouts occur during exploration drilling, production platforms are not exempt. A report from the US Minerals Management Service (MMS, 2003) described an incident in which a sudden gas influx caught fire, even though a diverter system was used. It was thought that gas had been sucked out of a shallow sand as the drill string was being removed from the well. The blowout lasted only about ten minutes, but the platform had to be abandoned and damage was estimated to be two million dollars.

LESSONS LEARNED

There are obvious safety lessons to be learned from these case studies about the danger of fire, loss of rig buoyancy, and foundation failure, and it is clear that penetrating shallow gas reservoirs and drilling into migration pathways can induce blowouts. There are also several lessons relevant to natural seabed gas escapes to be learned from these man-made incidents.

- Gas escaping through the seabed is clearly capable of eroding the seabed to produce large craters (pockmarks) in a very short space of time.
- Craters may be infilled with sediment over very short periods of (geological) time.
- Gas escaping during catastrophic events passes through the water column to escape to the atmosphere – gas escaping after the West Vanguard blowout (Figure 11.4) passed through 240 m of water.
- After a catastrophic gas escape event, gas leakage may continue for a considerable period of time.

- Gas introduced into seabed sediments may cause sediment failure and mass movement.

Guarding against blowouts

In order to avoid blowouts, oil industry (and ODP) sites are carefully selected and surveyed before spudding in. (Although many oil companies have experienced blowouts, ODP has not – mainly because of a previous policy to stay away from hydrocarbon-bearing regions.) Regulations differ from country to country, but in many cases (including UK and Norway) pre-drilling 'hazard' surveys are mandatory. In the UK, for example, detailed guidelines drawn up by the United Kingdom Offshore Operators Association (UKOOA, 1997) clarify relevant regulations and describe *'good industry practice'* for the conduct of rig site surveys, including aspects related to the seismic identification of shallow gas. These guidelines provide details about suitable survey line density, equipment specifications, data-processing requirements, and gas indicators; we discussed the seismic indicators of gas in Section 6.2.2.

Townsend and Armstrong (1990) warned against interpreting *'any high amplitude seismic reflector as a potential "bright spot" and to infer the presence of shallow gas accumulations on this basis'*. They advised that unless predictions of gas were reliable the interpreters would lose their credibility with drillers. Other papers (e.g. Games, 1990; Walker, 1990) in the same book (Ardus and Green, 1990) emphasised the need to choose seismic data acquisition and processing parameters with care, as the need to identify shallow gas correctly may be frustrated if these are not appropriate. It is now common practice to use the top section of three-dimensional (3D) exploration seismic data sets (specially reprocessed) to provide initial indications of gas accumulations. Although it has been common to follow this with specialist high-resolution seismic surveys, particularly where it is thought that gas accumulations may be present, Sharp and Samuel (2004) concluded that this may not be necessary. It is clear that correct interpretation is still a matter of skill and experience.

ESTIMATING GAS PRESSURES AND VOLUMES

If there is evidence of a shallow gas accumulation on seismic data, it may be possible to estimate gas pressure and volume, providing a few important assumptions are made.

Pressure: In practice, gas pressure is most easily quoted as an overpressure (i.e. the pressure above hydrostatic), rather than an absolute value. Gas pressure can be assumed to equal hydrostatic pressure – defined in Equation (7.2) – at the bottom of the gas accumulation. The gas overpressure (P_{go}) at the top of the reservoir can then be calculated from the density contrast between gas and pore water and the height of the gas column (see Equation 7.12). Gas-column height may be estimated from seismic data if there is a 'flat spot' (see Section 6.2.2) to indicate the base of the gas column (the gas–water contact) and a 'bright spot' (see Section 6.2.2) indicating the top of the gas.

Volume: Gas reservoir volume can be estimated from gas column height and the areal extent of the reservoir if sediment porosity is known or can be estimated.

This approach may provide approximate values (as demonstrated by Salisbury, 1990), but uncertainties about values of the various parameters mean there are shortcomings (Salisbury, 1990). Another assumption made by this method is that the gas accumulation is not hydraulically connected to a deeper gas source, for example by a fault.

11.3.2 Hydrogen sulphide

In concentrations above 200 ppm, hydrogen sulphide is lethal. The gas is highly reactive and will render victims hopelessly suffocated within minutes of exposure. This colourless, flammable, and dense (heavier than air) gas often occurs in sediments at or near methane seep sites. In non-lethal concentrations just above 10 ppm, the gas smells like 'rotten eggs', but the ability to smell it is lost after only a few minutes of exposure when the concentration approaches the danger level of 100 ppm. Extreme care must therefore be taken when smelly cores are handled inside confined laboratories. During drilling and sampling on ODP Leg 146 (Cascadia Accretionary Prism, Hydrate Ridge), the hydrogen sulphide alarm went off and no cores were allowed inside before they had degassed outside. New safety equipment (air dilution fans, hose-fed air packs, and gas evacuation fans) and procedures have made handling hydrogen sulphide cores safer during ODP work (Graber, 2002).

11.3.3 Drilling and gas hydrates

So far, very few incidents due to *in situ* gas hydrates have been reported by the hydrocarbon industry. Indications of what may happen when gas hydrate-bearing sediments are disturbed have, however, been documented by scientific drilling. The ODP has drilled and cored through BSRs on at least five scientific legs and has sampled gas hydrates associated with BSRs and submarine mud volcanoes on several of its legs (Hovland *et al.*, 1999b). Gas hydrates in sediment could dissociate releasing gas, or the opposite may occur; free gas released by drilling may form gas hydrates elsewhere in sediments, next to the drill-hole.

Potential consequences of drilling gas hydrates

When drilling through natural hydrates the greatest concern is associated with the production of warm hydrocarbons, heating surrounding formations and causing hydrates to decompose. For example, '*soupy layers*' were repeatedly encountered when drilling through marine sediments above a prominent BSR during ODP Leg 146 (Westbrook *et al.*, 1994). These layers were firstly interpreted as having been severely disturbed by drilling operations, but later turned out to have been gas hydrate bearing. On depressurisation, dissociation led to the release of water, which affected the sediments and turned them into a totally structureless 'soup'. Potential consequences of this rapid reduction in sediment strength and pore fluid pressure increase (if evolved gas cannot escape) include damage to drilling equipment and seabed installations; examples of casing collapse due to excess local pressure caused by dissociating gas hydrates are known to have occurred. Consequently, any offshore operation in gas hydrate-prone areas must be sensitive to the potential for gas hydrate dissociation, and the possible consequences (Bouriak *et al.*, 2000; Hovland and Gudmestad, 2001). The assessment of a potential drilling site, deep-water construction site, or pipeline route must include an evaluation of the likelihood that gas hydrates may be present, or might form. It is necessary to use and interpret all available indications, including indirect (seismic, sonar, and topographic features), and direct (visual observation and seabed sampling) means (Max and Miles, 1999). An assessment of gas hydrate potential should also include theoretical considerations. Because the regional and local diffusive and focussed flux of light hydrocarbons through sediments is of such importance, it may also be important to quantify hydrocarbon flux from the seabed into the ocean, and because of the dynamic nature of hydrates, long-term monitoring of subcropping or outcropping hydrates could be useful at certain deep-water construction sites. McGee and Woolsey (1999) reported plans for performing such studies in the Gulf of Mexico.

Edmonds *et al.* (2001) considered that, when drilling through hydrate deposits, it is important to prevent dissociation. They recommended doing the following:

- reduce the temperature of the drilling mud;
- drill at a 'controlled' penetration rate;
- increase mud weight (but this may lead to lost circulation);
- increase mud circulation to ensure turbulent flow and high heat transfer and to remove any gas;
- use a chemical additive (e.g. lecithin) to stabilise the hydrate zone;
- run high-strength casing in the hydrate zone before drilling deeper.

In their opinion it is unwise to encourage '*controlled dissociation*', removing evolved gas. They considered this approach '*potentially more hazardous*'.

Gas hydrates inside cased wells

Barker and Gomez (1989) reported deep-water (350 m) drilling problems caused by gas hydrates off the US West coast. Gas entered the well, and the kill operation, which took seven days, was seriously hampered by hydrate ice forming on the BOP, choke line, kill line, and the riser. During another incident, in the Gulf of Mexico, in 950 m of water, the BOP failed to operate properly due to gas hydrates, causing a prolonged well control operation. If hydrates form in the drilling fluid they cause a change in mud properties, which can lead to barytes settling out.

Barker and Gomez (1989) summarised the adverse effects of hydrate formation during well control operations:

- choke and kill-line plugging, which prevents their use in well circulation;
- plug formation at or below the BOP, preventing well-pressure monitoring below the BOP;
- plug formation around the drill string in the riser, BOP, or casing, preventing drill-string movement;

- plug formation between the drill string and the BOP, preventing full BOP closure;
- plug formation in the ram cavity of a closed BOP, preventing the BOP from fully opening.

Gas hydrate formation in drilling mud

The likelihood of hydrate formation in water-based mud is higher than in oil-based mud, but water is always present in the mud system, and hydrate formation is possible with any mud formulation. Shut-in situations, and cooling, particularly in choke and kill-lines (which are usually not insulated) are risky. According to Edmonds et al., (2001), methods available for controlling hydrate in mud include:

- keeping the temperature above or the pressure below hydrate formation conditions;
- using chemicals to depress the hydrate formation point; these thermodynamic inhibitors include methanol, glycols, and salts;
- adding chemicals to reduce the rate of nucleation of hydrate crystals;
- adding chemicals to reduce the rate of growth of hydrate crystals which have nucleated;
- adding chemicals that modify the growth of hydrate crystals to prevent agglomeration, so that solid plugs do not form.

Shallow-water flows

Many petroleum wells have been 'lost' because of uncontrolled 'shallow-water flow' (SWF). Shallow-water-flow problems were first encountered in 1985 during drilling in the Gulf of Mexico. Since then they have cost several hundred million dollars, and have occurred in many deep-water petroleum provinces including: the Caspian Sea, Norwegian Sea, North Sea, offshore West Africa, Caribbean (Alberty, 1998; Ostermeier et al., 2000). This is mainly a deep (> 500 m) water condition found when drilling at least 400 m below the seabed, but sometimes problems have been encountered both in shallower water, and closer to the seabed. Although there have been exceptions, SWF most commonly occurs during open-hole drilling (i.e. before a marine riser and blowout preventer have been installed) into overpressured sands. Isolated sand bodies enclosed within finer sediments may be uncompacted as pore water has been unable to escape during burial (see Section 7.5.1). When such sands are penetrated, a sand slurry escapes up the

well. In extreme events there are long-lasting uncontrolled flows of overpressured water and sand which have caused well damage, casing damage, bent drill pipe, foundation failure, and complete loss of the hole. In some cases eruptions from overpressured sands have resulted in seabed cratering, mounds, and cracks (Schultz and Pickering, 2002).

As well as 'wet blowouts', geopressured sediments may also leak to the surface up the annulus between casing and formation if the casing is poorly cemented. Other causes, brought about by errors in drilling, are induced fractures (the equivalent of underground blowouts), and induced storage (when normally pressured sediments are charged and overpressured by excess drilling mud pressure, but discharge when circulation stops). Parallels with shallow-gas problems include the fact that SWF is most common when drilling into overpressured zones; these include isolated sand bodies: channel sands and amalgamated channel complexes, levee and lag deposits, debris flow deposits, and rotated slump blocks (Niemann et al., 1998; McConnell, 2000). Regional studies have shown certain formations to be particularly prone to SWF problems; Ostermeier et al. (2000) reported that the highest SWF risk in the Gulf of Mexico is on the continental slope associated with rapid late-Pleistocene sedimentation from the Mississippi River.

According to Alberty et al. (1999) the most common causes of overpressuring are differential compaction and compaction disequilibrium, but we suggest other processes may also be responsible. These might include tectonic pressure, hydraulic connection with underlying overpressured formations, etc., described in Chapter 7 as being responsible for gas overpressures and sand intrusions. There is no seismic signature of overpressured water, so identifying potential shallow-water-flow zones is difficult. If overpressured water is associated with gas, then seismic gas detection procedures are useful, but gas may not be present. Two main approaches to SWF prediction have emerged (see Ostermeier et al., 2000 for a review).

- Detailed 2D and 3D high-resolution seismics to identify and map sediment facies and feature types known to be associated with SWF (McConnell, 2000; Wood et al., 2000); shear wave analyses are also applicable (Schultz and Pickering, 2002).

- Site-specific investigations of the pore fluid pressure environment, including studies of geotechnical wells drilled before the petroleum well, and real-time measurements using MWD (measurement while drilling using gamma ray and multi-sensor resistivity tools to predict lithology), SWD (seismic while drilling in which either the source or receiver is deployed in the water column, and the other is downhole; Dutta and Nutt, 1998), and PWD (pressure while drilling). Ostermeier *et al.* considered that PWD was probably the most important technique as it provides almost instantaneous indications of downhole pressure variations.

The strategy for dealing with SWF problems used to be to move off site. Experience, particularly from the Gulf of Mexico, has shown that with careful planning and prognosis, it is possible to stay on site and cope with the problems. For example, the Garden Banks 785, No. 1 well was successful, even though conditions were very difficult and would normally have given shallow-water flow. Methods employed, using carefully weighted muds, nitrogen-foamed cement, and active use of the BOP, were described by Corthay (1998).

11.4 HAZARDS TO SEABED INSTALLATIONS

Some features associated with seabed fluid flow are obvious obstacles to seabed installations. Active mud volcanoes and hydrothermal vents should be avoided by both installations, such as platforms occupying a small area, and by pipelines and cables. However, there are more subtle hazards.

11.4.1 Pockmarks as seabed obstacles

Since we have been interested in pockmarks, the offshore industry has wanted to know the rate at which they form in order to assess their hazard potential. Unfortunately, as is clear from Section 7.6.3, there is no simple answer. The only guidance that can be given is as follows.

1. Catastrophic gas escape can form large, deep pockmarks in very short periods of time, as demonstrated by some of the shallow-gas blowouts mentioned in Section 11.3.1. Natural gas-escape events might lead to similar results.

2. Active pockmarks formed in the Arabian Gulf within a one-year period may have been triggered by the construction of a platform. Gas was not actually released by the construction but triggering was caused by the disruption of the pore pressure environment. Although Ellis and McGuinness (1986), who reported these pockmarks, did not divulge their sizes, their existence demonstrates that pockmarks can be formed within a span time that is short not only on a geological time-scale but also in human terms.

In both these cases human intervention triggered gas escape, but there may be natural triggers. Although it seems that pockmarks are potentially hazardous, the absence of shallow gas or a suitable groundwater source may suggest a very low risk. Perhaps the absence of incidents, even in heavily pockmark areas of the North Sea, suggests they pose little danger. Or have we just been lucky?

Even if pockmarks present no risk, they are obstacles that may affect installations, particularly pipelines. It is inadvisable for pipelines to span across pockmarks for several reasons: if a critical span length is exceeded the overstressed pipe may buckle; harmonic vibrations caused by water currents may result in the shedding of the concrete coating that provides negative buoyancy; and the chance of fishing gear or anchors snagging the pipeline is significantly increased. Modern pipe-laying techniques can position a pipe on the seabed with great accuracy (to within a metre of the predetermined optimum position) even in hundreds of metres of water. It is possible to avoid seabed obstructions such as pockmarks, provided a suitable route is available. However, there are alternative strategies, such as trenching the pipeline, and dumping rock in pockmarks.

11.4.2 Trenching through MDAC

The widespread MDAC (methane–derived authigenic carbonate) encountered during trenching operations in the Norwegian North Sea (see Section 3.5.3) seriously impeded trenching operations. Comrie *et al.* (2002) found that weakly cemented sediments, continuous over distances of several hundreds of metres, had no significant impact on trenching speed or depth of burial, but more competent blocks of MDAC did. Trenching speeds were reduced from 600 or 800 to 200 m h^{-1}, and burial depth was reduced. Where massive MDAC was encountered additional remedial trenching was necessary, and in some places even this did not achieve the required depth, so rock dumping was required.

11.4.3 Foundation problems

Gassy sediments ('soils' in engineering language) are potentially significant to the foundations of structures because of the effects gas has on their engineering properties and behaviour. This is largely a consequence of the way in which gas accumulates in sediment, for example the existence of gas voids in fine-grained, cohesive sediments (see Sections 6.1.1 and 7.5.1). Sills and Gonzalez (2001), summarising previous work by numerous authors, stated that '*the compressibility of a gassy soil is greater than that of the same soil without gas, and the undrained strength may be increased or decreased by the presence of gas, depending on the soil stiffness and the total stress*'. The effect on shear strength varies according to the consolidation history and hydrostatic pressure, weakening with increased water depth (Sills and Wheeler, 1992), as explained by Equation (7.4). Sills and Wheeler reported that decreases in undrained strength may be as much as 25% for a cohesive sediment at 50 m water depth with only 1 to 2% gas by volume, whereas sand with only 1% gas at 40 m water depth experienced a strength reduction of 60%. They also noted that compaction of silts and clays may double if 1 or 2% gas is present.

Effects on platforms

The implications of these effects of gas on the engineering behaviour of the foundations of structures are a matter for concern. However, engineering design seems to have coped.

GRAVITY PLATFORMS

Enormous gravity platforms, some of the largest structures to be moved over the face of our planet, have been successfully installed in Norway's Gullfaks and Troll fields on soft, silty sediments with extensive pockmarks and shallow gas. One of the techniques used to ensure the integrity of the foundations of Gullfaks 'A' platform was to extract shallow gas before installation (see Section 11.3.1), and to dewater the sediment beneath the skirts during and for some months after installation,

assisting with sediment compaction. Despite these precautions, in 1988 we expressed concern that gas might accumulate within the skirts of this type of platform (Hovland and Judd, 1988). Although sediment compaction will tend to prevent gas accumulation, we pointed out that production wells radiating down and away from the platforms may provide migration pathways for shallow gas, focussing it on the very place it was not wanted.

PILED STRUCTURES

Unlike gravity platforms, piled structures are supported by the skin friction and end-bearing capacity of piles, and anchored platforms (tension-legs, etc.) are held in place by pile anchors relying on skin friction alone. Like production wells, these piles, some of which penetrate as much as 100 m below the seabed, could provide vertical migration pathways for gas. A careful evaluation of this possibility is essential prior to installation as the presence of gas could reduce pile skin friction.

GAS DRAINAGE PILES

In order to reduce the risk of migrating shallow gas affecting seabed structures, Gudmestad and Hovland (1986) patented drainage piles that could be placed at suitable locations. These piles are hollow cylinders to be driven into the seabed and penetrate impermeable sediment horizons that trap shallow gas, thus providing a harmless vertical escape route. They could be installed around platform locations or shorter piles could be used along vulnerable sections of pipeline in order to divert migrating gas away from the installation. Such piles have been used for shallow-gas management at the site of a production unit offshore Malaysia.

11.4.4 Effects of gas hydrates

Deep-water foundations

We explained in Section 11.3.3 how the decomposition of gas hydrates may cause sediment to turn to 'soup', and that decreased sediment strength may result if gas from the decomposition of gas hydrates accumulates in sediments. Apart from the possibility of sediment instability, what might be the implications of these effects on deep-water foundations?

HEAVY STRUCTURES

Heavy structures placed on the seabed will exert additional pressure on the sediments, thus changing ambient conditions. Such structures are typically linked to hot hydrocarbons in casings, manifolds, or pipelines. Their stability would be threatened if gas release caused considerable soil movements or if gas pressure is allowed to build up under compartments of the structure when they protrude into the seabed. Such compartments should therefore be equipped with ventilation for gas pressure release.

SUCTION ANCHORS

Suction anchors are becoming increasingly popular in deep water for holding floating structures on station. As for the heavy structures discussed above, gas could accumulate inside the anchors threatening their safe operation. If the anchors exert a suction force, pressure reduction could cause local gas hydrate dissociation with possible gas build-up inside the anchor buckets and soil movements disturbing the friction forces along the walls of the anchor. In order to avoid hazardous situations, suction in the anchors should be kept to a minimum for locations where there is a danger of disturbing *in situ* gas hydrates. In such areas forces on the anchor should, whenever possible, be taken up horizontally rather than vertically, employing long mooring lines on the seabed. Because gas hydrate dissociation may lead to methane release and anoxic conditions in the sediments, it also leads to increases in sediment porewater sulphide content, which could represent an aggressive corrosive environment for steel structures. This could call for extra corrosion protection (Sahling *et al.*, 1999).

Pipelines and flowlines

The handling of gas hydrates inside pipelines, production units, etc., is a science, and indeed, a business in its own right, but they are fundamentally the same as those used to handle hydrates in drilling mud (Section 11.3.3). The risk of forming blocking gas hydrate plugs in flowlines and trunk pipelines increases with pressure, low temperature, and the amount of water mixed in with the hydrocarbons. Also any loss of flow (turbulence and mixing) will encourage hydrate formation. Electrical heating of pipes is one common method of controlling hydrates in subsea flowlines, adding kinetic inhibitors is another. Edmonds *et al.* (2001) suggested four basic methods of removing hydrate blockages:

- depressurisation to dissociate the hydrate;
- addition of chemical inhibitors such as methanol or glycols, which change the stability boundary and melt the hydrate;
- external (electrical) heating of pipes to dissociate the hydrate;
- mechanical (drilling).

Where depressurisation is used, the differential pressure across the plug should not be allowed to become too large or a 'projectile' may form in the pipework. In a pipeline this is best achieved by depressurising from both ends of the hydrate plug. Recent research suggests that heat tracing is a viable option for melting or preventing a hydrate blockage. Using coiled tubing to circulate hot water from the surface is another possibility. With both techniques, the lines would need to be insulated.

Flowlines transporting hot fluids are prone to upheaval buckling due to thermal expansion and failure of the stability design (gravel or concrete mattress cover). Heating could cause gas hydrates in the sediments to dissociate, rendering the seabed 'soupy', and causing a loss of pipeline stability and initiating buckling.

11.5 ERUPTIONS AND NATURAL BLOWOUTS

Submarine volcanic eruptions are known to be hazardous to shipping. Unfortunately, this was demonstrated very dramatically on 24 September 1953 when the research vessel No.5 KAIYO-MARU from the Hydrographic Department of the Japanese Maritime Safety Agency was sunk by an eruption whilst conducting a survey of Myojin-Sho Submarine Volcano on the Izu-Ogasawara (Bonin) Island Arc. All 31 people on board were killed (Morimoto, 1960). In this case it seems that the ship was holed by rocks blown through the water by the force of the eruption. To avoid a similar fate befalling shipping, there is a 1.5 km exclusion zone around a volcano called Kick 'em Jenny, 8 km north of Grenada, in the eastern Caribbean. The warning to shipping says that even when Kick 'em Jenny is quiet, there is a danger of a loss of buoyancy due to gas bubbles rising from the crater (SRU, 2003).

Another interesting discovery associated with a volcano was that of the wreck of a classical Greek ship. Divers found it in about 32 m of water off the tiny island of Dattilo, near Sicily. The wreck is reported to be *'actually lying on the soft bubbling mud of a living volcano'* (Yellowless, 1987). Was this ship also a victim of a volcanic eruption?

11.5.1 Gas-induced buoyancy loss

Vigorous gas release, whether from blow outs, gas hydrate decomposition, or natural venting could have consequences for the buoyancy and stability of surface vessels, and has been demonstrated by the loss of drilling rigs during blowouts. The idea that natural gas escapes from the seabed could be hazardous to shipping was first expressed by R. D. McIver in 1982. He suggested that large volumes of gas might *'rush to the surface'* if a gas hydrate seal was breached. He thought that a rapid and localised gas escape would have an effect *'identical with that of a blowout caused by marine drilling operations (i.e., there would be a patch of highly agitated frothy water of very low relative density)'*. He advised that *'any vessel accidentally encountering this patch would lose buoyancy and sink very quickly'*. McIver's comments were specifically concerned with the 'Bermuda Triangle'. Before we think about this we must examine the feasibility of sinking ships with gas. Surprisingly, this topic has not been studied by naval architects. However, Bruce Denardo of the Naval Postgraduate School, Monterey, California, investigated the reduction in buoyancy of spherical bodies in gassy water (Denardo *et al.*, 2001), and May and Monaghan (2003), of the School of Mathematics, Monash University, Australia, asked: *'Can a single bubble sink a ship?'*

According to Archimedes' Principle, a floating body floats when buoyancy force equals the weight of the displaced fluid. But, if the density of the water is reduced by gas bubbles, then buoyancy is reduced and the ship will sit lower in the water, displacing more water until equilibrium is regained or the ship sinks. However, there are other consequences, as explained by Bondarev *et al.* (2002). They reported that a shallow-gas blowout in the Pechora Sea (see Figure 7.9) caused the drillship's hydroacoustic positioning system to fail and stopped both the main engines, leaving the ship in a dangerous situation, tilting 5–7° to the stern. Fortunately, *'at the last strokes of the booster, the ship succeeded in leaving the dangerous zone'*.

Many drilling rigs have been lost because of a loss of buoyancy during blowouts. However, buoyancy loss is not necessarily only caused by drilling accidents or gas hydrate dissociation.

The mystery of the Witch's Hole

An echo sounder profile recorded during one of the first site investigations in the North Sea, for BP's Forties field, showed a pockmark with a small, near-vertical water-column target in it (Figure 2.2). In our first book we presented this as early evidence of gas escape from a pockmark and, like BP, we regarded this as support for the theory that pockmarks were formed by gas escape. Subsequently the British Geological Survey (BGS) identified an unusual pockmark in their South Fladen pockmark study area. This is probably the same feature. It is unusually large (about 120 m across and 2 or 3 m deep), but it is distinctive because of the large number of smaller pockmarks which surround it, giving it a pepper-pot appearance on side-scan sonar records (Figure 2.20). Also, there is acoustic turbidity indicating very shallow (about 10 m subseabed) gas beneath it (Figure 2.19). The BGS named this pockmark the Witch's Hole (see Section 2.3.1).

In 1987 Total Oil Marine plc agreed to run some survey lines across the Witch's Hole. We asked them to run side-scan sonar, and to keep the towfish close to the seabed to get a good look at the 'gas plume'. To their surprise, the towfish hit the water-column target, which proved to be not gas, but a shipwreck! A report of this wreck, based on the Total data, was published (Judd, 1990), together with some speculation about how it came to lie right in the middle of this unusual pockmark. Did it land there by chance, or was it sunk by gas escaping from the Witch's Hole? The riddle of the wreck in the Witch's Hole remained forgotten (almost!) until about 10 years later one of us (AGJ) was approached to make a television programme about methane (*The North Sea's Bermuda Triangle* – part of the *Savage Planet* series; Granada TV, 2002). Fugro UDI Ltd.'s ROV support ship, the Skandi Inspector, was used and TV cameras recorded footage of the wreck. This was identified by Robert Prescott and Mark Lawrence of the Scottish Institute for Maritime Studies (University of St Andrews) as an early twentieth-century steam trawler. Its exact identity has not been discovered.

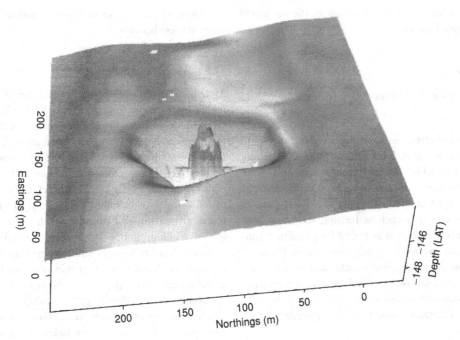

Figure 11.5* The Witch's Hole: multibeam echo sounder image of the Witch's Hole; the shipwreck is clearly seen standing upright in the centre of the pockmark. The pockmark is approximately 120 m across. LAT = lowest astronomical tide, i.e. depth (in m) below sea level. (Image courtesy of Richard Salisbury, Fugro Survey Ltd.)

The ROV survey, and subsequent multibeam echo sounder surveys, have shown that the wreck does lie right in the middle of the Witch's Hole (Figure 11.5). Apart from decay (rusting plates, etc.) visual inspection of the hull revealed no evidence of damage that might have caused the ship to sink; damage to the superstructure was probably caused later by fishing gear. However, no evidence of active gas seepage was seen either. The mystery remains unsolved. Is it possible that ships could be sunk by escaping gas? The probability of a ship landing by chance on the seabed right in the middle of a pockmark seems very small. In the South Fladen area only about 8% of the seabed is occupied by pockmarks, but there are no other pockmarks like the Witch's Hole in this 57 km^2 area, and < 7% of the South Fladen area is underlain by gassy sediments, so why is the wreck in this one? The probability is much less than 8% (0.08). However, the alternative is that the ship was sailing over the Witch's Hole precisely when it was leaking a large volume of gas. The probability of this is much smaller still! An alternative (and more probable) explanation is that the 'pockmark' was formed by the impact when the ship landed on the soft sediments.

We now know that the wreck in the Witch's Hole is not unique. Two other pockmarks, both similar in appearance to the Witch's Hole, and both with sonar targets looking like wrecks (yet to be confirmed) were identified in 2000 (Judd, 2001). Also, the Skagerrak seeps studied by Dando et al. (1994a) were reported to be close to a wreck. Are these coincidences, or is there really a 'Bermuda Triangle' in the North Sea?

The Bermuda Triangle

Tales of the loss of ships and aircraft in the so-called Bermuda Triangle are probably exaggerated. Some commentators consider that there is nothing unusual in this area or the number of losses here. However, McIver (1982) and others have pointed out that this area at least partially overlies the gas hydrate-bearing sediments of Blake Ridge (see Section 3.27.2). Consequently, if there was a sudden hydrate dissociation event, perhaps triggered by earthquake activity and an associated seabed slope failure, then massive amounts of methane might be released. McIver suggested that such events might affect not only ships in the area at the time, but also aircraft (concentrations of methane in the air would cause engine failure).

11.6 BENEFITS

The benefits of seabed fluid flow fall into three main groups.

1. **Direct benefits**: fluids and associated materials with resource potential. This category includes seeping freshwater, which has been used, at least on a minor scale, for centuries. Geothermal and hydrothermal fluids might be used to supply heat (as it does on land in countries such as Iceland, Italy, and New Zealand), and there is potential for exploiting minerals concentrated in hydrothermal fluids. However, the most significant direct benefits and potential benefits come from vast accumulations of ore minerals associated with hydrothermal venting, and the utilisation of methane in the form of seeping gas and gas hydrates.
2. **Benefits as an indicator**: because they disperse above the seabed, venting and seeping fluids indicate the presence of a source. Once the technological challenges of exploiting seabed hydrothermal minerals have been overcome, the detection of hydrothermal fluids may prove to be a valuable exploration tool. Whereas this is a tool for the future, natural oil and gas seeps have been used to guide petroleum exploration for centuries.
3. **Indirect benefits**: the principal indirect benefits of seabed fluid flow are biological. The enhancement of biological activity by natural vents and seeps is not restricted to hot and cold chemosynthetic communities. Biotechnologists and biochemists recognise that vent communities include microbes with capabilities not possessed by other organisms. Some of these are already being utilised, for example to process industrial sulphide wastes, producing biomass that is itself useable (Van Dover, 2000) and for production of pristine bioproteins (Hovland and Mortensen, 1999). The injection of microbes, substrates for microbes, and nutrients (including hitchhikers) into the water column have as yet unquantified benefits to biological productivity. Is it a coincidence that petroleum basins such as the North Sea and the Newfoundland Grand Banks are (or were, before overexploitation) prolific fishing grounds? We think not.

Here we can only provide a brief introduction to some of these benefits.

11.6.1 Metallic ore deposits

Metalliferous ore bodies have been exploited for millennia. For example, Romans mined the North Pennine Orefield, in Northern England. By the end of the twentieth century $> 4 \times 10^6$ t of lead concentrate, 2.4×10^6 t of barium minerals, and 2.1×10^6 t of fluorspar had been extracted, along with iron, copper, and zinc minerals (Dunham, 1990). Hannington *et al.* (1995) estimated global massive sulphide mineral production and reserves from 'fossil' hydrothermal deposits to be at least five billion tonnes. Volcanogenic-hosted massive sulphide deposits form in subduction-related island-arc settings (Kuroko type), at mid-ocean or back-arc spreading centres (Cyprus type), and in sediment-filled spreading centres (Besshi-type deposits); each has a characteristic mineral assemblage. Once it was understood that such ore bodies are linked to seabed fluid flow, the existence of ore bodies beneath the seabed was realised. For example, the ore body in the TAG field, which is of the Cyprus type, comprises nearly four million tonnes (Hannington *et al.*, 1995), and the metalliferous brines of the Atlantis II Deep in the Red Sea constitute a deposit of over 90 million tonnes.

Only a few seabed ore bodies have been found to date, but it is clear that they are representatives of a massive and widespread resource. Economic exploitation of these deposits has yet to take place, but they are significant to the present-day metalliferous mining industry because, now that their mode of formation is understood, 'fossil' occurrences can be searched for with greater efficiency. Equally, field studies on land provide evidence of the geological settings in which modern hydrothermal deposits are likely to be found. It seems that back-arc settings are particularly favourable for ancient economic ores (Scott, 1997).

Full coverage of this important topic is clearly beyond the scope of this book; we refer readers to specialist texts such as Scott (1997) and the Special Issue of *Economic Geology* prefaced by Rona and Scott (1993).

11.6.2 Exploiting gas seeps

In 1982 ARCO (Atlantic Richfield Company) installed two large steel pyramids to capture the seeping fluids from a site close to Platform Holly, in the Santa Barbara Channel, California (described in Section 3.22.4). The fluids are piped to a processing plant onshore (subsequently owned by Venoco). These 'seep tents', located in 67 m of water, cover a total seabed area of 1860 m^2 and weigh 318 t (Boles et al., 2001). Between 1982 and 1987 600 bbl of oil were produced, but since 1994 gas has been the only product. Average gas production rates have varied over time between 3×10^5 and 9×10^5 m^3 per month. Over one nine-month period in 1999/2000 hourly production rates were monitored, showing a significant correlation between flow rate and tides, as we discussed in Section 7.5.5.

Quigley et al. (1999) showed that in 1996 seeps above producing oil fields off Coal Oil Point, California (described in Section 3.22.4) covered less than half the area they had covered in 1973. In an earlier study of the same area, Fischer and Stevenson (1973) mapped the location of over 900 individual seeps within an 18 km^2 area off Coal Oil Point and compared the distribution with that in previous years. During the seven years between 1946–7 and 1953–4, the number of seeps had reduced to 30% of the original number, and by 1972 there were only about 8% of the original number. Together, these studies suggest a dramatic reduction from the natural (pre-1946) seep rates to the present day; no doubt, this has been caused by the progressive reduction in reservoir pressure during the production of petroleum over this period. Hornafius et al. (1999) noted that this has had a beneficial effect on the environment, reducing natural oil pollution in the sea, and hydrocarbons in the atmosphere: 'if the 50% reduction in natural seepage rate that occurred around Platform Holly also occurred because of future oil production from the oil field beneath the La Goleta seep, this would result in a reduction in nonmethane hydrocarbon emission rates equivalent to removing half the on-road vehicle traffic from Santa Barbara County'. Perhaps this is the best argument for developing the La Goleta oil field!

Such commercial seep exploitation is probably possible where there is only a thin cover of surficial sediment covering the reservoir rocks, as is the case here. We are not aware of any other exploitation of offshore seeps, but there are some onshore examples. A few farmers in the Tsilin valley, Taiwan use seep gas for cooking, and there has been some exploitation in northern Denmark (see Section 3.3.4).

11.6.3 Gas hydrates – fuel of the future?

The Big Prize

It has long been speculated that methane hydrates in ocean sediments represent the greatest reservoir of methane carbon on the planet, offering a clean and abundant energy source for the future (Trofimuk et al., 1973; Kvenvolden, 1983; Max, 2000c). Showstack (2000) reported that the total methane resource in hydrates 'exceeds the energy content of all other fossil fuel resources, such as coal, oil, and conventional gas'.

In fact, there is a great disparity between the various estimates, which range from 10^{15} to 10^{17} m^3 (approximately 7.25×10^4 to 7.25×10^6 Tg) of methane. Kvenvolden (1999) suggested the global amount of methane hydrate lies towards the lower or intermediate part of this range; his 'consensus value' was 21×10^{15} m^3. Hovland et al. (1997a) and Soloviev (2002) considered even this to be an exaggeration; Soloviev's estimate was 2×10^{14} m^3. Hovland et al. (1997a) explained that previous estimates had assumed the gas hydrate-bearing sediments contain massive hydrates, whereas sampling has shown that hydrates normally occur as small, finely dispersed crystals and aggregates within sediments; average concentrations may be as little as 3% by volume – more like the ore of a metallic mineral than a traditional hydrocarbon reservoir. Nevertheless, governments have been lured by the promise of vast resources that also have enormous strategic significance. Kleinberg and Brewer (2001) pointed out that the growing global demand for energy cannot be balanced by conservation measures, and that only fossil fuels are likely to meet this demand. In this context, the strategic importance of gas hydrates is difficult to overestimate, which is why 'Japan has long devoted a substantial effort to studying natural-gas hydrates, including the drilling and analysis of a number of test wells' (Kleinberg and Brewer, 2001). Strategic planning also explains why the majority of research programmes in countries like the USA, India, and Canada are government-funded, rather than being funded by industry. The USA's Methane Hydrate Research and Development Act of May 2000 established 'a new federal commitment to developing methane hydrates, which has been touted as a potential clean energy source that could make the U.S. less dependent on foreign sources of energy' (Showstack, 2000). It authorised research funding of 47.5 million US dollars over five years; Bil (2000) estimated the total international budget as > 100 million US dollars.

The distribution of gas hydrates
Hovland *et al.* (1997a) suggested that the most promising areas to look for gas hydrate resources are active deep-water mud volcanoes. Milkov and Sassen (2002) suggested that 'structural accumulations' (i.e. those associated with faulting, mud volcanoes, and other geological structures) such as those in northwestern Gulf of Mexico, and Hydrate Ridge, Cascadia Margin have the greatest commercial promise. They also thought some stratigraphic accumulations where the hydrate is widely disseminated in coarse sediments may be commercially viable; the massive methane hydrates held in thick deposits of sandy turbidites in the Nankai Trough off Japan are an example which is receiving close attention.

There have been many attempts to map the distribution of gas hydrates, all of which seem to be out of date by the time they are published as new sites continue to be discovered. Like others, Kvenvolden and Lorenson (2001) distinguished between locations where the presence of gas hydrates has been 'proved' by sampling, and those that are 'inferred', for example by the occurrence of a BSR; their inventory included 19 of the former and 77 of the latter. They said that most of the sampled hydrates were reported to be of microbial origin. Whatever the true distribution, it is clear from available data (some of which is reviewed in Chapter 3; see also Map 36) that gas hydrates are widely distributed around the world, from polar regions to equatorial regions. Soloviev (2001) estimated the distribution of gas hydrates by considering the extent of conditions suitable for the formation of gas: sedimentary basins, locations with high rates of Cenozoic sedimentation, subduction zones, and accretionary wedges. He considered that gas hydrate-prone areas must have a minimum sediment thickness of 2 km. He estimated that these covered a total of 35.7×10^6 km^2, about 10% of the area of the world's oceans (Figure 11.6), and he calculated that this area was distributed between the oceans as follows:

- Antarctic coastal regions: 19.7%;
- Arctic Ocean: 12.3%;
- Atlantic Ocean: 38.2%;
- Indian Ocean: 14.4%;
- Pacific Ocean: 15.4%.

The enormity of the gas hydrate reservoir is boosted by the fact that 1 m^3 of hydrate will, on dissociation, yield 164 m^3 of gas (assuming a 90% gas-filled lattice; Collett, 2002).

11.6.4 Technological challenge

Of course, gas hydrates will not be viable as a resource until the technology to exploit them has been developed. The techniques that have been investigated fall into two main types: those that decompose hydrates by pressure reduction, and those that favour heat injection (Sawyer *et al.*, 2000). Several years ago, Japanese researchers teamed up with Canadian and US scientists to explore commercial exploitation of gas hydrates by drilling through known, thick hydrate occurrences in Arctic Canada, the Mallik well (Dallimore *et al.*, 1999). In 2001 this project was expanded into a multinational campaign, including personnel from India and Germany. They have investigated both warm-water circulation and pressure release as production means of freeing up the subsurface hydrate-locked gas. Adam (2002) reported that the amount of methane produced was encouraging: '*enough to ignite a flare similar to those seen burning over oil rigs*'. However, he was not sure '*whether the yellow flame is symbolic or a genuine step forwards*'.

Releasing the gas from the hydrate is only one stage of exploitation. Many more technological challenges must be met before hydrate energy is produced.

Future perspectives
Despite strategic interest and obvious signs of progress, it will take several more years before the dream of hydrate energy is turned into reality. The results of a survey of industry specialists led Bil (2000) to conclude that onshore hydrates may be developed by 2015, but offshore hydrates will take much longer to exploit, 2060 was thought realistic. Nevertheless, the 'Big Prize', vast quantities of 'clean' energy, will surely drive research onwards. As at least 95% of hydrates are in continental-slope sediments '*the offshore represents the fundamental challenge and potentially the greatest reward*' (Bil, 2000).

Exploration for hydrocarbons
Seeps are effective tools for determining whether or not a sedimentary basin has petroleum potential. They show that the 'petroleum system' is working; that source rocks are present, and that they are mature. Link (1952), Hedberg (1981), and many others have commented on

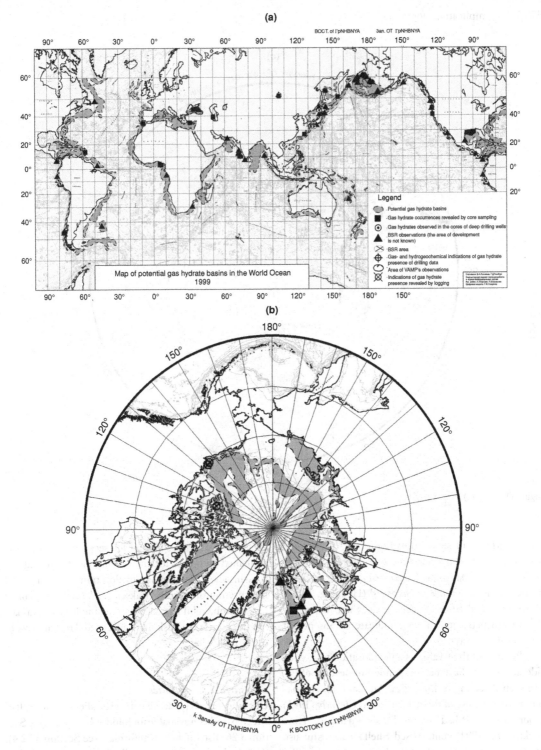

(a)

Map of potential gas hydrate basins in the World Ocean
1999

Legend

- Potential gas hydrate basins
- Gas hydrate occurrences revealed by core sampling
- Gas hydrates observed in the cores of deep drilling wells
- BSR observations (the area of development is not known)
- BSR area
- Gas- and hydrogeochemical indications of gas hydrate presence of drilling data
- Area of VAMP's observations
- Indications of gas hydrate presence revealed by logging

(b)

Figure 11.6* The global distribution of gas hydrate; based on the distribution, within the gas hydrate stability zone, of basins with thick sediments, areas of rapid Cenozoic sedimentation, and accretionary wedges associated with subduction zones. (a) Low latitudes; (b) North Pole; (c) South Pole. (Compiled by, and reproduced with permission of, Valery Soloviev, VNIIOkeangeologia, St Petersburg, Russia.)

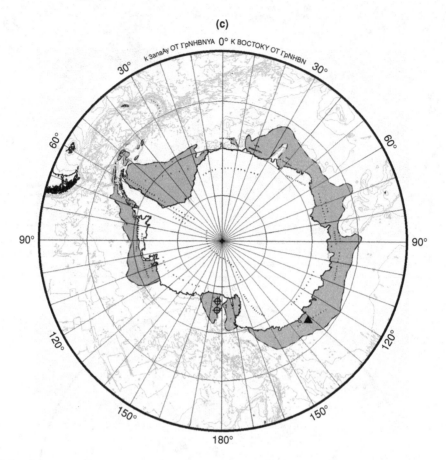

(c)

Figure 11.6* *(cont.)*

the importance of seeps to exploration: '*Historically, most of the world's major petroleum-bearing areas and many of its largest oil and gas fields were first called to attention because of visible oil and gas seepages*' (Hedberg, 1981). However, the absence of seepage does not mean that there is no petroleum; it may indicate an absence of migration pathways.

Because of their value as an exploration tool, considerable efforts have been expended to develop effective seep-detecting technologies. All major oil companies have made use of them. Clarke and Cleverly (1991) reported that BP had compiled a seep database. Kornacki *et al.* (1994) summarised Shell's successful use of seeps in evaluating the petroleum potential of the continental slope of the Gulf of Mexico. A benchmark paper by Thrasher *et al.* (1996) not only reviewed strate-

gies used by BP and Statoil for seep studies, but also explained the need for understanding the geology of a basin when designing an exploration seep study and interpreting the results. Isaksen *et al.* (2001) explained how ExxonMobil has used seep technology in evaluating the petroleum 'risk' in the Rockall Trough, west of the British Isles.

Sea-surface seep studies

Documentary evidence of natural oil slicks in the Gulf of Mexico has been available for hundreds of years (see Section 3.26; also offshore California – see Section 3.22.4), the first detailed map of slick distribution was published in 1910 (Soley, 1910; see also MacDonald, 1998). Nowadays, more sophisticated techniques are used to detect oil slicks (natural and man-made) from above the sea surface, taking advantage of characteristics summarised by Brown *et al.* (1995).

- Oil produces a surface sheen on water, reflecting light over a broad spectrum of wavelengths (certain thicknesses cause the familiar 'rainbow' colours).
- Oil absorbs solar radiation and re-emits some of the radiation as thermal energy; thick ($>150\,\mu m$) oil slicks appear 'hot' on infrared images, but thin ($<\sim 50\,\mu m$) slicks are not visible.
- Even thin ($<001\,\mu m$) oil layers are highly reflective in ultraviolet (UV) light.
- Oil fluoresces when excited by UV lasers.
- Because oil dampens capillary waves on the sea surface, the normally 'cluttered' and chaotic radar image of the sea surface changes when oil is present. Synthetic aperture radar (SAR) is particularly suited to oil detection. It can be used at night and through most clouds, although wave conditions can make data unusable, so a suitable weather window is required.
- Water and oil both emit microwave radiation, but at different intensities, oil's being about twice as strong as water's.

SEEP SPOTTING FROM SPACE

Oil seeps, including gas seeps with oily bubbles, produce sea-surface slicks visible from space (Figure 11.7). Examination of a single photograph from the space shuttle Atlantis enabled MacDonald *et al.* (1993) to identify at least 124 slicks within an area of 15 000 km² in the Gulf of Mexico. In a later paper MacDonald *et al.* (2002) used SAR (synthetic aperture radar) and Landsat data to identify seeps in the Gulf of Mexico, and to monitor their activity over time. They found that some are continuously active, and others are intermittent. De Beukelaer *et al.* (2003) provided further examples of slicks identified by SAR and linked to seabed features (recorded on side-scan sonar) by acoustic water-column plumes.

Satellite-based SAR surveys have been particularly successful in proving the presence of oil in frontier areas. 'Offshore Basin Screening', a remote-sensing technique developed by two UK companies, the NPA Group and TREICoL, provides confidence that more detailed exploration using surface-based techniques are likely

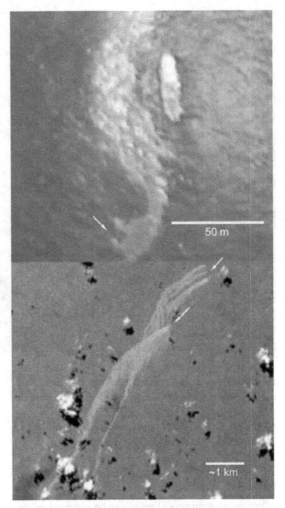

Figure 11.7* (Top) Oil drops reaching the sea surface in a discrete footprint (arrowed); sun-glint is enhanced in the area of floating oil. (Bottom) Photograph from the space shuttle shows sun-glint from slicks in four distinct places (arrowed). (Reproduced with permission from MacDonald *et al.*, 2002.)

to be worthwhile, particularly when combined with other regional studies such as gravity surveys (derived from satellite altimetry). At relatively low cost, even deep-water basins can be screened, as demonstrated by Williams and Lawrence (2002) who identified seeps from deep-water basins offshore Brazil and Angola. Of course, oil slicks are not always a result of natural seepage, but inspection of images taken during successive passes of a satellite can eliminate pollution from ships

Figure 11.8* Typical pancake-shaped oily bubbles surfacing in the southern Caspian Sea. (From Williams and Lawrence, 2002, *AAPG Studies in Geology* No. 48. Reprinted by permission of the AAPG whose permission is required for further use; AAPG©2002.)

and oil rigs etc., and thinner natural surface films of other natural substances (NPA Group, 2003).

ALF

More detailed surveys of seeps can be obtained by airborne studies. The most commonly used system in oil exploration is ALF, the airborne laser fluorosensor. Although data are not generally released to the public domain, ALF surveys have been extensively used worldwide, proving to be an effective tool for seep detection. They are able to detect much smaller targets than SAR.

GROUND TRUTHING

Confirmation that slicks spotted by satellite- or airborne-systems are caused by natural seeps is done by shipborne 'ground truthing'. When oil-covered bubbles break surface, the oil spreads out to form 'pancakes' (Figure 11.8). These coalesce to form continuous seeps that drift away from the point at which they surface because of wind, wave, tide, and current forces. Samples

of the seeping oil can be analysed by gas chromatography to provide details of composition.

Sniffer surveys

Sniffers essentially comprise a towfish containing a pump to lift water to the mother-ship, a stripper to extract gas from the water, and a gas chromatograph to analyse for methane and the higher hydrocarbon gases. These have proved extremely useful as seep-detection tools. The data density (illustrated in Figure 3.41) is far greater than that acquired from traditional water-bottle deployments (as described in Section 10.4.1).

Jones *et al.* (1999) presented a case study from the High Island area of the Gulf of Mexico. The Sniffer survey covered 385 km, including detailed grid surveys and regional lines. Analyses were performed at three-minute intervals on water pumped from about 9 m above the seabed, providing data points about 450 m apart. Anomalous concentrations of >500 nl l^{-1} methane, <5 nl l^{-1} ethane, and 0.5 to 1.0 nl l^{-1} propane contrasted with background values of ~ 100, <0.7, and <0.5 nl l^{-1}, respectively. Relative concentrations of these gases enabled the interpreters to distinguish between microbial and thermogenic gas sources, enabling them to identify significant anomalies, some of which have subsequently been converted to 'discoveries'.

Seabed geochemistry

Targetted seabed sampling and analysis of core headspace gases can provide valuable evidence of the likely nature of petroleum fluids in an individual prospect, providing indicators of the maturity of petroleum products present. Rather than detecting active seeps ('macroseeps') where gas bubbles into the seawater, seabed geochemical surveys detect '*microseeps*' (Horvitz, 1980) or '*passive seeps*' (Abrams, 1996).

Faber and Stahl (1984) described results of a geochemical survey including interstitial gases from 350 piston-core samples taken from an area in the northern North Sea. The survey area included parts of the Witch Ground Basin and the Norwegian Trench. The gas samples were analysed for a range of hydrocarbon gases, and carbon isotope determinations were also undertaken. By examining gas 'wetness', ratios of methane to ethane plus propane $[C_1/(C_2 + C_3)]$, and carbon isotope ratios, Faber and Stahl were able to demonstrate that, in some areas, sediment gases had formed by thermal hydrocarbon generation in deep source rocks, and had subsequently migrated to the surface.

Some doubt has been raised (for example by Whiticar *et al.*, 1985) about the validity of data from these surveys because a thermogenic isotope signature may be obtained from microbially produced light hydrocarbons which have been oxidised immediately beneath the seabed. Nevertheless, careful use of carbon isotope data in conjunction with hydrogen/deuterium ratios, and the ratio of methane to higher hydrocarbon gases (C_1/C_{2+}) has shown that valuable results can be obtained (as explained in Section 5.5). Brekke *et al.* (1997) compared results from analyses of adsorbed, interstitial, and headspace gases, and found the latter most useful; Bjorøy and Ferriday (2001) discussed different analytical and sampling techniques in offshore geochemical surface prospecting.

An integrated approach

Various authors have presented data from one or more of the techniques mentioned above. However, the techniques are most powerful when used together and in conjunction with other techniques such as seismics. Schumacher (2000) pointed out that seismic data, particularly 3D data, are '*unsurpassed for mapping trap and reservoir geometry*', but '*only surface geochemistry methods can consistently and reliably map hydrocarbon leakage associated with those traps*'. He concluded that, provided they are properly acquired and interpreted, '*the combination of surface geochemical data and sub-surface exploration data has the potential to reduce exploration risks and costs by improving success rates and shortening development time*'. The regional petroleum prospecting strategy employed by Geoscience Australia in the Timor Sea provides an excellent example of an integrated approach. Their programme has involved: SAR, ALF, Sniffer, side-scan sonar, multibeam echo sounder, seabed sediment sampling and seismics (Figure 11.9 is a good example). Apart from active seeps associated with live petroleum basins, they have identified reefs and carbonate banks that seem seep-related – and which are designated a marine park (O'Brien *et al.*, 2002; Struckmeyer, *et al.*, 2002 – see Section 3.17.2).

Although seabed fluid flow provides good guidance for prospecting, Kornacki *et al.* (1993) advised caution. Evidence from seeps and seabed geochemistry should not be relied on too heavily because it provides no information about the presence or absence of suitable reservoirs or trap structures, or seal integrity. This is true, of course.

Success stories

Geochemists, and others who promote the virtues of a particular exploration technique, are often accused of talking about how their post-discovery data would have identified the prospect, if only they had been used as part of the exploration effort. With this in mind, it is good to close this section with three case studies that proved the value of evidence of seabed fluid flow in advance.

COGNAC

Horvitz (1980) reported that an extensive programme of seabed coring was undertaken offshore Louisiana before the area was made available for drilling in 1974. Analyses of gas indicated a large halo-type anomaly over what is now known as the Cognac field. Seventy-five per cent

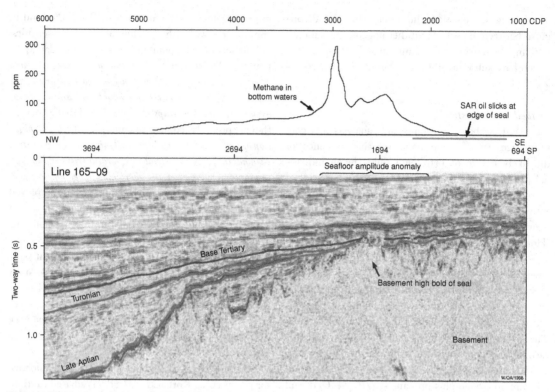

Figure 11.9* Seismic profile across the Yampi Shelf, offshore Australia, showing the relationship between the edge of an effective petroleum seal, an HRDZ (hydrocarbon-related diagenetic zone) of highly reflective seabed, water-column-gas methane anomalies, and satellite-detected sea-surface slicks. (Reproduced with permission from O'Brien *et al.*, 2002.)

of the wells drilled during the exploration of this field fell within the anomaly.

CANTARELL

The Cantarell field, a super giant oil field with 17 bbl reserves, was finally discovered in June 1977 thanks to a humble and persistent fisherman, Señor Cantarell. He noticed that there was always an oil seep at a certain spot every time he returned there. He figured that if it had come from a boat it could not keep reappearing. He took his story to the local office of Pemex (the Mexican oil company), but he was laughed out the door. He went to the regional office and the same thing happened. Finally, he took his story to the main Pemex building in Mexico City and they listened, shot some seismic to check, and the rest is history. About four years ago Pemex, drilled beneath the Cantarell structure to discover the Sihil structure, which alone is bigger than the largest US

offshore field (Paul Nadeau and Andrew Horbury, 2004, personal communications).

ESTRELLA

In 1995 a 3D seismic survey was undertaken in an area offshore Equatorial Guinea, and a '*suspected gas chimney*' was identified. Subsequently a seabed geo-chemical exploration programme was undertaken, and three piston-core samples were collected from the top of the chimney. Analyses proved the presence of ther-mogenic oil and gas in two of these cores. These results encouraged further action; a site survey was undertaken. Bathymetric mapping showed there was a '*large crater*' (400 m across and 17 m deep) above the gas chimney, and shallow seismic data showed the seabed sediments were gassy, and that there were seep plumes in the water. The prospect was drilled and '*the success of Estrella-1, which encountered about 200-ft* [60 m] *gross of hydrocarbon-bearing sand, confirmed this direct hydrocarbon indicator*' (Canales, 2002).

11.6.5 Benefits to fishing?

We have come across several anecdotes suggesting that fish productivity is enhanced in seep areas. Orange *et al.* (2002) described submarine dives on the off-shore part of the Eel River delta, northern California which documented active gas seeps. They reported that local fishermen provided '*anecdotal evidence for gas bubbling at the sea surface at this location as well as for increased fishing yields in this region*'. Conversations with North Sea fishermen revealed that many could identify MDAC (methane-derived authigenic carbonate), and some associated it with good fishing grounds; one location was even referred to as the 'gas bubble' (Louise Tizzard, 2004, personal communication).

Is it just a coincidence that the North Sea has two major petroleum provinces and a high biological productivity? Maybe so, but there are other coincidences; for example the Grand Banks of Newfoundland used to be (before serious overfishing) a major fishing ground, and is also a significant oil province (Levy and Lee, 1988). Similarly, fishing is important in numerous other shelf seas underlain by petroleum provinces. Another example of a 'coincidental' geographical link between seeps and biological productivity comes from the Rías Baixas, northwest Spain (see Section 3.6.1), an area noted for mussel farming. Although the success of mussel farms is normally attributed to upwelling along the Atlantic Continental Margin, it seems that mussels thrive best in the internal parts of the rías, where there is extensive shallow gas (Soledad García-Gil, 2002, personal communication). Is this a coincidence?

The link between petroleum accumulations and seeps is not so difficult. We discussed the jump from seeps to biological productivity in Section 8.3.4. Although further work is required to confirm (or deny) the link between seabed fluid flow and biological productivity, it seems that dismissing it may be premature. At the other extreme, evidence from offshore Namibia (see Section 3.7.2) suggests that massive releases of gas, including hydrogen sulphide, from the seabed are responsible for killing fish. Weeks *et al.* (2004) reported that the cape hake population was reduced to less than 30% of its normal level by an anoxic event in 1994, with serious economic implications for Namibia.

11.6.6 Seeps, vents, and biotechnology

The biotechnology industry is clearly aware that new 'resources' are to be found amongst the microbes of chemosynthetic communities of vents and seeps. Indeed it is seen as '*the most promising way*' to develop novel bioactive compounds '*for industrial, agricultural, environmental, pharmaceutical and medical uses*' (Querellou, 2003). Although discoveries of the last few decades present enormous potential, this remains largely unexplored. No doubt this will change rapidly.

About 75 km east of the Sula Ridge coral reefs, in mid Norway, Statoil has constructed a production plant for bioproteins. It is the first of its kind, and is located at the Tjeldbergodden processing plant for natural gas near Trondheim; gas is transported through the Haltenpipe trunk line from the Heidrun field, about 180 km offshore. The aim is to produce 10 000 t of bioproteins annually using methane (along with ammonia and oxygen) as a substrate for the bacterium *Methylococcus capsulatus*. The process partly mimics the natural process we suspect is occurring within the seabed beneath the large deep-water coral banks. The bacteria are grown, dried, and then mixed into animal (cattle, pig, and chicken) and salmon fodder. The proteins thus produced consist of 70% pure proteins, 5% water, 10% fat, the rest being fibres and minerals. They '*contain no toxins, pathogens, carcinogens, or other materials with unwanted side-effects*' (J. M. Huslid, 1998, personal communication). The plant started up in 1998 and reached its production goals by 2001.

11.7 IMPACTS OF HUMAN ACTIVITIES ON SEABED FLUID FLOW AND ASSOCIATED FEATURES

11.7.1 Potential triggers

Some human activities, mainly, but not exclusively those of the offshore petroleum industry, may affect seabed and subseabed conditions to the extent that fluid pressure regimes and migration are affected; the effect of petroleum production on seepage rates in the Santa Barbara Channel, California (Section 11.6.2) is an example. The following activities, most of them identified by Kvalstad *et al.* (2001), may affect stress and pore-pressure conditions, triggering, or potentially triggering the formation of new features, or 'events':

- drilling wells creating blowouts to the seabed, with the possible formation of craters (examples were described in Section 11.3.1);
- underground blowouts changing the pore pressure in shallow layers, with possible implications for slope stability;
- oil production leading to heat flow and temperature increase around wells and well clusters, possibly leading to the decomposition of gas hydrates (with implications for sediment strength, slope instability, etc.);
- depletion of reservoir pressure resulting in increased effective stress, and reservoir subsidence (a classic example of this being the Ekofisk field in the Norwegian North Sea), and stress changes in overlying sediments;
- installation activities (rock dumping, and the emplacement of structures, particularly gravity structures) increasing vertical stress, with implications for pore-pressure conditions within the sediments;
- mooring installations and anchoring forces imposing short- and long-term lateral forces.

A very real concern is that seabed installations on the continental slope will trigger a slope failure. Should such an event occur, natural examples described in Section 11.2.1 indicate the potential for catastrophe for structures located on an area that fails, in the pathway of the debris flow, or as a result of a tsunami (Kvalstad *et al.*, 2001). It would be prudent to take careful precautions. We consider that a good understanding of the fluid pressure and plumbing systems is essential before installations are put in place. This, of course, means identifying features associated with active migration and seabed fluid flow, and subseabed reservoirs, and taking into account existing natural triggers (as discussed in Section 7.5.5).

11.7.2 Environmental protection

Mapping of the Darwin Mounds, off the coast of northwest Scotland (see Section 3.5.4), revealed that deep-sea trawling is a threat to biological communities of seeps and vents. It seems that fishermen found the roundnose grenadier (*Coryphaenoides rupestris*) fish living around

the mounds, and had been exploiting them, causing damage to the reefs within two years of their discovery. Indeed, dwindling shallow-water fish stocks have driven trawlers into deeper and deeper waters, posing a severe threat to deep-water coral reefs and carbonate banks (Roberts *et al.*, 2003). This fishing practice is more-or-less equivalent to going logging in the Sahara, targetting all oasis-related date palms in the vast desert. If it is not restricted soon by regulation, we may be left only with remnants of deep-water reefs and banks, having to try to understand their ecology by reconstruction, rather than by true field observation and analysis. In recent years awareness of the need to protect the marine environment, including the deep seabed, has led to the introduction of various forms of protection for designated areas of the seabed, including some features associated with seabed fluid flow. Various organisations have jurisdiction within the marine environment. The following examples show how some important sites are (or will be) receiving protection.

International protection

Under provisions of the UN Convention on the Law of the Sea (UNCLOS), and other international law, every country has an obligation to protect the marine environment: to '*protect and preserve rare or fragile ecosystems as well as the habitat of depleted, threatened or endangered species, and other forms of marine life*' (UNCLOS, Article 194/5, 1982). In order to safeguard those parts of the marine environment beyond the limits of national jurisdiction (known as the 'Area'), UNCLOS established the International Seabed Authority (ISA), and determined that the Area and its resources were '*the common heritage of mankind*'; the ISA was made responsible for the Area, on behalf of mankind. At first, its principal function was concerned with the mining of polymetallic nodules, but according to the ISA web page (ISA, 2003), it now has responsibility for other marine activities including polymetallic sulphides (i.e. hydrothermal vents), cobalt-rich crusts, gas hydrates, and petroleum, and for the marine environment in general. The ISA may designate sensitive sites as no-mining areas. The UN World Summit on Sustainable Development (WSSD; Johannesburg, 2002) called for action to maintain the productivity and biodiversity of important and vulnerable marine areas both within and beyond national jurisdiction (UNDESA, 2005). It urged nations to adopt the ecosystem approach by 2010 and the establishment of representative networks of MPAs (Marine Protected Areas) by 2012. The General Assembly of the UN adopted Resolution a/57/L.48 (2002) endorsing the Plan of Implementation adopted at WSSD, and calling for *urgent and coordinated action to protect vulnerable benthic habitats* (UNCLOS, Article 194/5, 1982). To encourage this process, the WWF (World Wildlife Fund) have proposed some 'pilot case studies'. They proposed that the Logatchev hydrothermal-vent field, the largest (200 000 m²) vent field reported to date on the Mid-Atlantic Ridge should be designated as *a Showcase Example for a High Seas Network of Marine Protected Areas* (UNCLOS, Article 194/5, 1982).

European protection

Within Europe, the two main organisations with responsibilities for the marine environment are OSPAR and the European Commission. Each has played a different role.

OSPAR

The Convention for the Protection of the Marine Environment of the North-East Atlantic (the OSPAR Convention; a product of agreements made in Oslo and Paris) has been signed and ratified by: Belgium, Denmark, the European Commission, Finland, France, Germany, Iceland, Ireland, Luxembourg, the Netherlands, Norway, Portugal, Spain, Sweden, Switzerland, and the UK. It provides protection for a vast area, from the Greenland coast to the European coast, and from the Strait of Gibraltar to the North Pole. The work of OSPAR includes pollution prevention, environmental assessment, and the protection and conservation of ecosystems and biological diversity. The North-East Atlantic Programme of the WWF has contributed by reviewing selected habitats (Gubbay, 2002), and recommending individual sites for protection as MPAs. By 2003 they had identified 21 candidates, of which the following are associated with seabed fluid flow:

- Sula Ridge and Reef: deep-water corals off Norway;
- the Lucky Strike hydrothermal vents, within the Exclusive Economic Zone (EEZ) of the Portuguese Azores Islands;
- the Darwin Mounds;
- the Rainbow hydrothermal field: this lies in international waters, so the WWF are pushing for this

to become a 'High Seas Marine Protected Area' (HSMPA).

We are unsure if any of the other sites (which include seamounts and deep-water coral reefs) are associated with seabed fluid flow, but they may be (Hovland and Risk, 2003).

The urgency of the call for protection is not just a reaction to their obvious fragility, but also to the increase in deep-water fishing, and the discovery that fishing gear was causing serious damage to important sites. After a campaign by the WWF, the UK government finally secured official protection for the Darwin Mounds in August 2003. A news release (DEFRA, 2003) stated that *the use of bottom trawls or similar towed nets operating in contact with the bottom of the sea* is prohibited within the area bounded by the following coordinates: latitude 59°37' to 59°54' N, longitude 6°47' to 7°39' W.

THE EUROPEAN COMMISSION

In 1992 the European Community Council adopted 'Directive 92/43/EEC on the conservation of natural habitats and of wild fauna and flora' (EEC, 1992). It is generally known as the 'Habitats Directive'. The Habitats Directive requires member states to introduce legislation enforcing the directive, and to propose a list of national sites worthy of designation as 'Special Areas for Conservation' (SACs). These, together with sites designated for the protection of wild birds, will form the 'Natura 2000 Network'. Two habitats associated with seabed fluid flow have been recognised by the 'Interpretation Manual of European Union habitats (EC, 1999) Directive' (EUR 15/2).

1. Natura 2000 code 1170: Reefs

    Rocky substrates and biogenic concretions, which arise from the seafloor in the sublittoral zone, may extend into the littoral zone. These reefs generally support a zonation of benthic communities of algae and animal species including concretions, encrustations and corallogenic concretions.

2. Natura 2000 code 1180: Sub-marine structures made by leaking gases

Spectacular sub-marine complex structures consist of rocks, pavements and pillars up to four metres high. These formations are due to the aggregation of sandstone by carbonate cement resulting from microbial oxidation of gas emissions, mainly methane. The methane most probably originated from microbial decomposition of fossil plant materials. The formations are interspersed with gas vents that intermittently release gas. These formations shelter a highly diverse ecosystem with brightly coloured species.

NATIONAL PROTECTION

It is clear from preceding sections that, one way or another, individual nation states have an obligation to care for the marine environment. The UK government has responded with two initiatives: a response to the Habitats Directive (Johnstone *et al.*, 2002) and a series of 'Strategic Environmental Assessments' (SEAs) which will eventually cover the whole UK continental shelf and deeper waters to the west of the British Isles. Examples of features associated with seabed fluid flow are identified in both.

Natura 2000 Code 1180 clearly refers to the 'Bubbling Reefs' of the Kattegat (see Section 3.3.4), and several Danish seep sites with MDAC that have been designated as Sites of Conservation Interest (SCIs; HELCOM,[1] 1996). There are no known equivalent structures in UK waters, but Johnstone *et al.* recognised that MDAC is a *'variation of this habitat type'*, and named two pockmarks with *'carbonate structures formed by leaking gases'*: the Scanner pockmark in Block UK15/25 (see Section 2.3.7), and the Braemar pockmark in Block UK16/3. The second SEA (SEA2) report (DTI, 2001) included a section on pockmarks, and two technical reports that discussed them and their biology (Judd, 2001; Dando, 2001); SEA6 included a report on MDAC in the Irish Sea (Judd, 2005b).

It is reassuring that such attention is being paid to features associated with seabed fluid flow.

Implications for offshore operations
The protective environment described above places a great responsibility on those conducting offshore oper-

ations. Companies are no longer free to drive pipelines or cables across the seabed at will, nor to drill or place fixed structures exactly where they want without considering the environmental consequences, including the implications for benthic habitats. To be fair, many companies developed environmental policies, and conducted environmental impact assessments before legal pressures were imposed on them. At first these tended to be studies of specific sites, but in some areas consortia of operators undertook regional studies. An excellent example of this was the work of AFEN, the 'Atlantic Frontier Environmental Network', which undertook a regional environmental assessment of a large area to the northwest of the UK (AFEN, 2002). However, a significant difference has been made by legislation. For example in the EU and the USA, it is now the duty of the company to demonstrate that proposed work will not affect any specified habitats, even sites that have not been designated for protection. The following example illustrates this new philosophy:

Chemosynthetic communities are susceptible to physical impacts from structure placement (including templates or subsea completions), anchoring, and pipeline installation. NTL 98-11 prevents these physical impacts by requiring avoidance of potential chemosynthetic communities.

MMS, 2001

It is no longer acceptable to say *'we didn't know it was there'*; instead surveys must be undertaken to demonstrate the *absence* of sensitive sites before permission to proceed is granted. Statoil set a good example by re-routing the Haltenbanken pipeline to avoid coral reefs. This was recognised by the WWF who, in 2003, awarded a 'Gift to the Earth' award to Norway for coral-reef preservation measures implemented on the Norwegian Continental Shelf.

We welcome this acknowledgement that the last frontier of our precious planet requires tender loving care. We hope such measures will be adopted internationally, and that they will be enough to ensure the well-being of all sensitive sites – not only those associated with seabed fluid flow.

---

[1] HELCOM is the governing body of the Convention on the Protection of the Marine Environment of the Baltic Sea Area – more usually known as the Helsinki Convention.

References

AAPG (American Association of Petroleum Geologists), 2003. http://www.aapg.org/education/hedberg/london/

Abegg, F. and Anderson, A., 1997. The acoustic turbid layer in muddy sediments of Eckernförde Bay, western Baltic: methane concentration, saturation and bubble characteristics. *Marine Geology*, 137, 127–47.

Abegg, F., Anderson, A., Buzi, L., Lyons, A. P., and Orsi, T. H., 1994. Free methane concentrations and bubble characteristics in Eckernförde Bay, Germany. In Wever, T. F. (ed.), *Proceedings of the Gassy Mud Workshop, Kiel, 11–12 July, 1994.* Kiel, Forschungsanstalt der Bundeswehr für Wasserschall- und Geophysik (FWG) Report np 14.

Abrams, M. A., 1996. Distribution of subsurface hydrocarbon seepage in near-surface marine sediments. In Schumacher, D. and Abrams, M. A. (eds.), *Hydrocarbon Migration and its Near-Surface Expression.* American Association of Petroleum Geologists Memoir 66, 1–14.

Acosta, J., 1984. Occurrence of acoustic masking in sediments in two areas of the continental shelf of Spain: Ria de Muros (NW) and Gulf of Cadiz (SW). *Marine Geology*, 58, 427–34.

Acosta, J., Muñoz, A., Herranz, P., *et al.*, 2001. Pockmarks in the Ibiza Channel and western end of the Balearic Promontory (western Mediterranean) revealed by multibeam mapping. *Geo-Marine Letters*, 21, 123–30.

Adam, D., 2002. Fire from ice. *Nature*, 415, 913–14.

Adams, N. J. and Kuhlman, L. G., 1990. Case history analyses of shallow gas blowouts. IADC/SPE Drilling Conference, Paper 19919, 97–106.

Addy, S. K. and Worzel, J. L., 1979. Gas seeps and sub-surface structure off Panama City, Florida. *American Association of Petroleum Geologists (Bulletin)*, 63, 668–75.

AFEN, 2002. http://www.ukooa.co.uk/issues/Afen/ (accessed 16 January 2004).

Aharon, P., 1994. Geology and biology of modern and ancient submarine hydrocarbon seeps and vents – an introduction. *Geo-Marine Letters*, 14, 69–73.

Aharon, P., 2000. Microbial processes and products fueled by hydrocarbons at submarine seeps. In Riding, R. E. and Awramik, S. M. (eds.), *Microbial Sediments.* Berlin, Springer-Verlag, 270–81.

Aharon, P., Socki, R. A., and Chan, L., 1987. Dolomitization of atolls by sea water convection flow: test of a hypothesis at Niue, South Pacific. *Journal of Geology*, 95, 187–203.

Ahmed, S. S., 1972. Geology and petroleum prospects in eastern Red Sea. *American Association of Petroleum Geologists (Bulletin)*, 56, 707–19.

Alberty, M., 1998. Shallow waterflows – history, mechanisms, and intervention. Proceedings of the 1998 Shallow Water Flow Forum, Woodlands, TX, 24–25 June 1998 (presentation).

Alberty, M., Hafle, M. E., Mingle, J. C., and Byrd, T. M., 1999. Mechanisms of shallow waterflows and drilling practices for intervention. *SPE Drilling and Completion*, 14, 123–9.

Alexander, R. T. and Macdonald, K. C., 1996. Small off-axis volcanoes on the East Pacific Rise. *Earth & Planetary Science Letters*, 139, 387–94.

Aliyev, A., Guliyev, I. S., and Belov, I. S., 2002. *Catalogue of Recorded Eruptions of Mud Volcanoes of Azerbaijan (for Period of Years 1810 to 2001).* Baku, Azerbaijan, Nafta Press.

Allen, A. and Schlueter, R., 1970. Natural oil seepage at Coal Oil Point, Santa Barbara, California. *Science*, 170, 974–7.

Aloisi, G., Pierre, C., Rouchy, J.-M., Foucher, J.-P., and Faugères, J.-C., 2002. Isotopic evidence of methane-related diagenesis in the mud volcanic sediments of the Barbados accretionary prism. *Continental Shelf Research*, 22, 2355–72.

Aloisi, G., Pierre, C., Rouchy, J.-M., *et al.*, 2000. Methane and gas hydrate-related authigenic carbonate crusts in the mud volcanoes of the eastern Mediterranean Sea.

Abstracts of the Sixth International Conference on Gas in Marine Sediments. St Petersburg, VNIIOkeangeologia, 5–9, 6–8.

Amouroux, D., Roberts, G., Rapsomanikis, S., and Andreae, M. O., 2002. Biogenic gas (CH_4, N_2, DMS) emission to the atmosphere from near-shore and shelf waters of the north-western Black Sea. *Estuarine, Coastal and Shelf Science*, 54, 575–87.

Anderson, A. L., 1974. *Acoustics of Gas-Bearing Sediments*. Report Number ARL-TR-74-19, Applied Research Laboratories, Austin, TX, The University of Texas.

Anderson, A. L. and Hampton, L. D., 1980a. Acoustics of gas-bearing sediments I. Background. *Journal of the Acoustical Society of America*, 67, 1865–89.

Anderson, A. L. and Hampton, L. D., 1980b. Acoustics of gas-bearing sediments II. Measurements and models. *Journal of the Acoustical Society of America*, 67, 1890–1902.

Anderson, A. L., Abegg, F., Hawkins, J. A., Duncan, M. E., and Lyons, A. P., 1998. Bubble populations and acoustic interaction with the gassy floor of Eckernförde Bay. *Continental Shelf Research*, 18, 1807–38.

Andreassen, K., Hogstad, K., and Berteussen, K. A., 1990. Gas hydrate in the southern Barents Sea, indicated by a shallow seismic anomaly. *First Break*, 8, 235–45.

Andrews, I. J., Long, D., Richards, P. C., *et al.*, 1990. *United Kingdom Offshore Regional Report: the Geology of the Moray Firth*. London, HMSO, for the British Geological Survey.

Anketell, J. M., Cegla, J., and Dzulynski, S., 1970. On the deformational structures in systems with reversed density gradients. *Rocznik Polskiego Towarzystwa Geologicznego (Annals of the Geological Society of Poland)*, XL, 6–13.

Anonymous, 1986. Underwater volcanoes trigger earthquakes. *New Scientist*, 25 December 1986 / 1 January 1987, 24.

Anonymous, 1988. Geologic phenomona. *Geotimes*, January 1988, 17.

Apps, J. A. and Kamp, P. C. van de, 1993. Energy gases of abiogenic origin in the Earth's crust. In *The Future of Energy Gases*. United States Geological Survey, Professional Paper 1570, 81–132.

Ardus, D. A. and Green C. D. (eds.), 1990. *Safety in Offshore Drilling: the Role of Shallow Gas Surveys*. Dordrecht, Kluwer Academic Publishers.

Arp, A. J. and Childress, J. J., 1981. Blood function in the hydrothermal vent vestimentiferan tube worm. *Science*, 213, 324–44.

Arp, A. J., Childress, J. J., and Fisher, C. R., 1985. Blood gas transport in *Riftia pachyptila*. In *The Hydrothermal Vents of the Eastern Pacific: An Overview*. Bulletin of the Biological Society of Washington, No. 6, 289–300.

Atigh, E. and Byrne, P. M., 2003. Flow liquefaction failure of submarine slopes due to monotonic loadings – an effective stress approach. In Locat, J. and Mienert, J. (eds.), *Submarine Mass Movements and their Consequences*. Dordrecht, Kluwer Academic Publishers, 3–10.

Austvik, T., Li, X., and Gjertsen, L. H., 2000. Hydrate plug properties – formation and removal of plugs. In *Gas Hydrates: Challenges for the Future*. Annals of the New York Academy of Sciences, 912, 294–303.

Bach, W., Layne, G. L., and Von Damm, K. L., 2002. $\delta^{37}Cl$ of mid-ocean ridge vent fluids determined by a new SIMS method for stable chlorine isotope ratio measurements. *EOS – Transactions of the American Geophysical Union*, 83 (47), Fall Meeting Supplement, Abstract V61B–1367.

Bacon, C. and Lanphere, M. A., 1990. The geologic setting of Crater Lake, Oregon. In Drake, E. T., Larson, G. L., Dymond, J., and Collier, R. (eds.), *Crater Lake: An Ecosystem Study*. San Francisco, American Association for the Advancement of Science, pp. 19–27.

Bagirov, E., Nadirov, R., and Lerche, I., 1996. Flaming eruptions and ejections from mud volcanoes in Azerbaijan: statistical risk assessment from the historical records. *Energy Exploration and Exploitation*, 14, 535–83.

Baker, E. T., German, C. R., and Elderfield, H., 1995. Hydrothermal plumes over spreading-center axes: global distributions and geological inferences. In Humphris, S. E., Zierenberg, R. A., Mullineaux, L. S., and Thomsen, R. E. (eds.), *Seafloor Hydrothermal Systems: Physical, Chemical, Biological, and Geological Interactions*, American Geophysical Union, Geophysical Monograph 91, pp. 47–71.

Baker, E. T., Massoth, G. J., and Feely, R. A., 1987. Cataclysmic hydrothermal venting on the Juan de Fuca Ridge. *Nature*, 329, 149–51.

Ballard, R. D. and Grassle, J. F., 1979. Return to oases of the deep. *National Geographic Magazine*, 156, 680–705.

Bange, H. W., Bartell, U. H., Rapsomanikis, S., and Andreae, M. O., 1994. Methane in the Baltic and North seas and a reassessment of the marine emissions of methane. *Global Biogeochemical Cycles*, 8, 465–80.

Bange, H. W., Rapsomanikis, S., and Andreae, M. O., 1996. The Aegean Sea as a source of atmospheric nitrous oxide and methane. *Marine Chemistry*, **53**, 41–9.

Baraza, J. and Ercilla, G., 1996. Gas-charged sediments and large pockmark-like features on the Gulf of Cadiz slope (SW Spain). *Marine and Petroleum Geology*, **13**, 253–61.

Baraza, J., Ercilla, G., and Nelson, C. H., 1999. Potential geologic hazards on the eastern Gulf of Cadiz slope (SW Spain). *Marine Geology*, **155**, 191–215.

Barber, A. J., Tjokrosapoetro, S., and Charlton, T. R., 1986. Mud volcanoes, shale diapirs, wrench faults, and melanges in accretionary complexes, eastern Indonesia. *American Association of Petroleum Geologists (Bulletin)*, **70**, 1729–41.

Barker, W. and Ganez, R. K., 1989. Formation of hydrates during deepwater drilling operations. *Journal of Petroleum Technology*, **41**, 297–301.

Barriga, F. J. A. S., Binns, R. A., Miller, D. J., and the Shipboard Scientific Party, 2001. Hydrothermal corrosion: a major pre-ore forming process documented by ODP Leg 193 (PACMANUS, Manus Basin, Papua New Guinea). *EOS – Transactions of the American Geophysical Union*, **82**(47), Fall Meeting Supplement, abstract OS11A-MC.

Barry, J. M. and Kochevar, R. E., 1998. A tale of two clams: differing chemosynthetic life styles among vesicomyids in Monterey Bay cold seeps. *Cahiers de Biologie Marine*, **39**, 329–31.

Barry, J. P., Greene, H. G., Orange, D. L., *et al.*, 1996. Biologic and geologic characteristics of cold seeps in Monterey Bay, California. *Deep-Sea Research I*, **43**, 1739–62.

Barry, J. P., Kochevar, R. E., and Baxter, C. H., 1997. The influence of porewater chemistry and physiology on the distribution of vesicomyid clams at cold seeps in Monterey Bay: implications for patterns of chemosynthetic community organization. *Limnology and Oceanography*, **43**, 318–28.

Basov, E. I., Weering, T. C. E. van, Gaedike, C., *et al.*, 1996. Seismic facies and specific character of the bottom simulating reflector on the western margin of Paramushir Island, Sea of Okhotsk. *Geo-Marine Letters*, **16**, 297–304.

Bates, T. S., Kelly, K. C., Johnson, J. E., and Gammon, R. H., 1996. A reevaluation of the open ocean source of methane to the atmosphere. *Journal of Geophysical Research*, **101**, 6953–61.

Batiza, R., Fox, P. J., Vogt, P. R., Cande, S. C., and Grindlays, N. R., 1989. Abundant Pacific-type near-axis

seamounts in the vicinity of the Mid-Atlantic Ridge, 26 °S. *Journal of the Geological Society, London*, **97**, 209–20.

Bauer, C. and Fichler, C., 2002. Quaternary lithology and shallow gas from high resolution gravity and seismic data in the central North Sea. *Petroleum Geoscience*, **8**, 229–36.

Bauer, J. E., Spies, R. B., Vogel, J. S., Nelson, D. E., and Southon, J. R., 1990. Radiocarbon evidence of fossil carbon cycling in sediments of a nearshore hydrocarbon seep. *Nature*, **348**, 230–2.

Beauchamp, B. and Bitter, P. von, 1992. Chemo what? *Palaios*, **7**, 1.

Beauchamp, B. and Savard, M., 1992. Cretaceous chemosynthetic carbonate mounds in the Canadian Arctic. *Palaios*, **7**, 434–50.

Becker, K., the Leg 174B Scientific Party, and Davis, E. E., 1998. Leg 174B revisits Hole 395A: logging and long-term monitoring of off-axis hydrothermal processes in young ocean crust. *JOIDES Journal*, **24**(1), 1–3, 13.

Belenkaia, I., 2000. Gas-derived carbonates: reviews in morphology, mineralogy, chemistry and isotopes (data collected during the TTR programme cruises during 1995–1999). In *Abstracts of the Sixth International Conference on Gas in Marine Sediments*. St Petersburg, VNIIOkeangeologia, 9–10 (abstract).

Bellaiche, G., Loncke, L., Gaullier, V., *et al.*, 2001. The Nile Deep-Sea Fan: main results of the Fanil Cruise (October–November 2000). *European Union of Geosciences EUG XI Journal of Conference*, **6**(1), 521 (abstract).

Bellissent-Funel, M. C., 2001. Structure of supercritical water. *Journal of Molecular Liquids*, **90**, 313–22.

Ben-Avraham, Z., Smith, G., Reshef, M., and Jungslager, E., 2002. Gas hydrate and mud volcanoes on the southwest African continental margin off South Africa. *Geology*, **30**, 927–30.

Berkson, J. M. and Clay, C. S., 1973. Possible syneresis marine seismic study of late-Quaternary sedimentation origin of valleys on the floor of Lake Superior. *Nature*, **245**, 89–91.

Bernard, B. B. and Brooks, J. M., 2000. Gas hydrates on the Nigerian continental slope. In *Abstracts of the Sixth International Conference on Gas in Marine Sediments*, St Petersburg, VNIIOkeangeologia, 11.

Berndt, C., Mienert, J., Vanneste, M., Bünz, S., and Bryn, P., 2001. Submarine slope-failure offshore Norway triggers rapid gas hydrate decomposition. Proceedings of the

Fourth International Conference on Gas Hydrates, Yokohama, May 19–23.

Bernhard, J. M., Buck, K. R., Farmer, M. A., and Bowser, S. S., 2000. The Santa Barbara Basin is a symbiosis oasis. *Nature*, **403**, 77–80.

Best, A. I., Clayton, C. R. I., Longva, O., and Szuman, M., 2003. The role of free gas in the activation of submarine slides in Finneidfjord. In Locat, J. and Mienert, J. (eds.), *Submarine Mass Movements and their Consequences*. Dordrecht, Kluwer Academic Publishers, pp. 491–8.

Bethke, C. M., Reed, J. D., and Oltz, D. F., 1991. Long-range petroleum migration in the Illinois Basin. *American Association of Petroleum Geologists (Bulletin)*, **75**, 925–45.

Bett, B. J., 2001. UK Atlantic Margin Environmental Survey: introduction and overview of bathyal benthic ecology. *Continental Shelf Research*, **21**, 917–56.

Bett, B. J. and Cruise participants, 1999. *RRS Charles Darwin Cruise 112C, 19 May–24 June 1998. Atlantic Margin Environmental Survey: Seabed Survey of Deep-Water Areas (17th round Tranches) to the North and West of Scotland*. Southampton, Southampton Oceanography Centre Cruise Report No. 25.

Bialas, J. and Kukowski, N., 2001. R/V Sonne off Peru. *Sea Technology*, **42**(4), 29–32.

Biju-Duval, B., 2002. *Sedimentary Geology: Sedimentary Basins, Depositional Environments, Petroleum Formation*. Paris, Editions TECHNIP.

Biju-Duval, B., LeQuelle, P., Mascle, A., Renard, V., and Valery, P., 1982. Multibeam bathymetric survey and high resolution seismic investigations on the Barbados Ridge complex (eastern Caribbean): a key to the knowledge and interpretation of an accretionary wedge. *Tectonophysics*, **86**, 275–304.

Bil, K. J., 2000. Economic perspective of methane from hydrate. In Max, M. D. (ed.), *Natural Gas Hydrates in Oceanic and Permafrost Environments*, Dordrecht, Kluwer Academic Publishers, pp. 349–60.

Billett, D., 1986. The rise and fall of the sea cucumber. *New Scientist*, **109**(1500), 48–51.

Binns, R. A., Barriga, F. J. A. S., Miller, D. J., and the Shipboard Scientific Party, 2001. The third dimension of an active back-arc hydrothermal system: ODP Leg 193 at PACMANUS. *EOS – Transactions of the American Geophysical Union*, **82**, Fall Meeting Supplement, abstract F587.

Binns, R. A. Parr, J. M, Scott, S. D., Gemmell, J. B., and Herzig, P. M., 1995. PACMANUS: an active, siliceous volcanic-hosted hydrothermal field in the eastern Manus Basin, Papua New Guinea. Abstracts of the PACRIM conference, Auckland, New Zealand, 19–22 November 1995.

Bischoff, J. L., Stine, S., Rosenbauer, R. J., Fitzpatrick, J. A., and Stafford, T. W., 1993. Ikaite precipitation by mixing of shoreline springs and lake water, Mono Lake, California, USA. *Geochimica et cosmochimica acta*, **57**, 3855–65.

Bishop, R., 1978. Mechanism for the emplacement of piercement diapirs. *American Association of Petroleum Geologists (Bulletin)*, **62**, 1561–81.

Bitterli, P., 1958. Herrera subsurface structure of Penal field, Trinidad. *American Association of Petroleum Geologists (Bulletin)*, **42**, 145–58.

Bjørkum, P. A., Walderhaug, O., and Nadeau, P. H., 1998. Physical constraints on hydrocarbon leakage and trapping revisited. *Petroleum Geoscience*, **4**, 237–9.

Bjørlykke, K. and Hoeg, K., 1997. Effects of burial diagenesis on stresses, compaction and fluid flow in sedimentary basins. *Marine and Petroleum Geology*, **14**, 267–76.

Bjørnsson, H., Bjørnsson, S., and Sigurgeirsson, Th., 1982. Penetration of water into hot rock boundaries of magma in Grimsvotn. *Nature*, **295**, 580–1.

Bjorøy, M. and Ferriday, I. L., 2001. Surface geochemistry as an exploration tool: a comparison of results using different analytical techniques. Abstracts of the American Association of Petroleum Geologists Hedberg Conference, Vancouver, BC, Canada, 16–19 September 2001.

Blinova, B. N., Ivanov, M. K., and Bohrmann, G., 2003. Hydrocarbon gases in deposits from mud volcanoes in the Sorokin Trough, north-eastern Black Sea. *Geo-Marine Letters*, **23**, 250–7.

Blinova, V. and Stadnitskaya, A., 2001. Composition and origin of the hydrocarbon gases from the Gulf of Cadiz mud volcano area. In Akhamanov, G. and Suzyumov, A. (eds.), *Geological Processes on Deep-Water European Margins*. International Oceanographic Commission Workshop Report No. 175 on the International Conference and ninth post-cruise meeting of the Training Through Research Programme, Moscow-Mozhenka, Russia, 28 January–2 February 2001. Paris, UNESCO, pp. 45–6.

Boerman, S. and Webster, D. A., 1982. Control of heme content in *Vitreoscilla*-spp. by oxygen. *Journal of General and Applied Microbiology*, **28**, 35–43.

Boetius, A., Ravenschlag, K., Schubert, C. J., et al., 2000. A marine microbial consortium apparently mediating anaerobic oxidation of methane. *Nature*, **407**, 623–6.

Boetius, A., Elvert, M., Nauhaus, K., *et al.*, 2001. Anaerobic oxidation of methane mediated by a microbial consortium above marine gas hydrates. Conference Programmes with Abstracts, Earth System Processes conference, Geological Society of America and Geological Society of London, Edinburgh, 24–28 June, p. 48 (abstract).

Boetius, A. and the MUMM team, 2002. The enigmatic process of anaerobic oxidation of methane: first results of project MUMM. *Gas Hydrates in the Geosystem*. Status seminar, Geotechnologien science report No. 1, Research Centre for Marine Geosciences (GEOMAR), Kiel, pp. 105–6 (abstract).

Bohrmann, G., Greinert, J., Suess, E., and Torres, M., 1998. Authigenic carbonates from the Cascadia subduction zone and their relation to gas hydrate stability. *Geology*, **26**, 647–50.

Bohrmann, G., Heeschen, K., Jung, C., *et al.*, 2002. Widespread fluid expulsion along the seafloor of the Costa Rica convergent margin. *Terra Nova*, **14**, 69–79.

Bohrmann, G., Ivanov, M., Foucher, J.-P., *et al.*, 2003. Mud volcanoes and gas hydrates in the Black Sea: new data from Dvurechnskii and Odessa mud volcanoes. *Geo-Marine Letters*, **23**, 239–49.

Boles, J. R., Clark, J. F., Leifer, I., and Washburn, L., 2001. Temporal variation in natural methane seep rate due to tides, Coal Oil Point area, California. *Journal of Geophysical Research*, **106**, 27077–86.

Bondarev, V. N., Rokos, S. I., Kostin, D. A., Dlugach, A. G., and Polyakova, N. A., 2002. Underpermafrost accumulations of gas in the upper part of the sedimentary cover of the Pechora Sea. *Geologiya i Geofizika (Russian Geology and Geophysics)*, **43**, 587–98 (Russian edn.) / 545–56 (English edn.).

Bondevik, S., Svendsen, J. I., Johnsen, G., Mangerud, J., and Kaland, P. E., 1997. The Storegga tsunami along the Norwegian coast, its age and runup. *Boreas*, **26**, 29–53.

Both, R., Crook, K., Taylor, B., *et al.*, 1986. Hydrothermal chimneys and associated fauna in the Manus Back-arc Basin, Papua, New Guinea. *EOS – Transactions of the American Geophysical Union*, **67** (21), 489–90.

Botz, R. W., Georgiev, V., Stoffers, P., Khrischev, Kh., and Kostadinov, V., 1993. Stable isotope study of carbonate-cemented rocks from the Pobiti Kamani area, north-eastern Bulgaria. *Geologische Rundschau*, **82**, 663–6.

Botz, R., Stüben, D., Winckler, G., *et al.*, 1996. Hydrothermal gases offshore Milos Island, Greece. *Chemical Geology*, **130**, 161–73.

Boudreau, B. P., Algar, C., Johnson, B. D., *et al.*, 2005. Bubble growth and rise in soft sediment. *Geology*, **33**, 517–20.

Boulton, G. S., Caban, P. E., Gijssel, K. van, Leijnse, T., Punkari, M., Weert, F. H. A. van, 1996. The impact of glaciation on the groundwater regime of northwest Europe. *Global and Planetary Change*, **12**, 397–413.

Boulton, G. S., Chroston, P. N., and Jarvis, J., 1981. A marine seismic study of late-Quaternary sedimentation and inferred glacier fluctuations along western Inverness-shire, Scotland. *Boreas*, **10**, 39–51.

Boulton, G. S., Slot, T., Blessing, K., *et al.*, 1993. Deep circulation of groundwater in overpressured subglacial aquifers and its geological consequences. *Quaternary Science Reviews*, **12**, 739–45.

Bouma, A. H. and Rezak, R., 1969. Oil found on knolls on the Gulfs continental rise. *Ocean Industry*, **4**(5), 73–7.

Bouriak, S., Vanneste, M., and Saoutkine, A., 2000. Inferred gas hydrates, shallow gas accumulations and clay diapirs on the southern edge of the Vøring Plateau, offshore Norway. *Marine Geology*, **163**, 125–48.

Bourque, J., 2002. Shooting the moon. *Air & Space*, **17**, 54–61.

BP Trading Ltd, 1974. *Investigation into the formation of pockmarks*. Report on Pilot Studies, Lab. Ref. No. 5/10537, August 1974.

Brault, M., Simoneit, B. R. T., Marty, J. C., and Saliot, A., 1988. Hydrocarbons in waters and particulate material from hydrothermal environments at the East Pacific Rise, 13° N. *Organic Geochemistry*, **12**, 209–19.

Breen, N. A., Silver, E. A., and Hussong, D. M., 1986. Structural styles of an accretionary wedge south of the island of Sumba, Indonesia, revealed by Sea MARC II side scan sonar. *Geological Society of America (Bulletin)*, **97**, 1250–61.

Brekke, T., Lønne, Ø., and Ohm, S. E., 1997. Light hydrocarbon gases in shallow sediments in the northern North Sea. *Marine Geology*, **137**, 81–108.

Brewer, P. G., Orr, F. M., jr, Friederich, G., Kvenvolden, K. A., and Orange, D. L., 1998. Gas hydrate formation in the deep sea: *in situ* experiments with controlled release of methane, natual gas, and carbon dioxide. *Energy & Fuels*, **12**, 183–8.

Brewer, P. G., Paull, C., Peltzer, E. T., *et al.*, 2002. Measurement of the fate of gas hydrates during transit through the ocean water column. *Geophysical Research Letters*, **29**, 2081.

Bright, T. J. and Rezak, R., 1977. Reconnaissance of reefs and fishing banks of the Texas continental shelf. In Geyer, R. A. (ed.), *Submersibles and Their Use in Oceanography and Ocean Engineering*. Elsevier Oceanography Series, 17, Amsterdam, Elsevier, 113–50.

Broecker, W. S. and Peng, T. H., 1982. *Tracers in the Sea*. New York, Lamont-Doherty Geological Observatory.

Brook, E., Harder, S., Severinghaus, J., Steig, E., and Sucher, C., 2000. On the origin and timing of rapid changes in atmospheric methane during the last glacial period. *Global Biogeochemical Cycles*, **14**, 559–72.

Brooks, J. M., Anderson, A. L., Sassen, R., *et al.*, 1995. Hydrate occurrences in shallow subsurface cores from continental slope sediments. In Sloan E. D., jr, Happel, J., and Hantow, M. A. (eds.), *Natural Gas Hydrates*. Annals of the New York Academy of Science, New York, New York Academy of Sciences, pp. 381–91.

Brooks, J. M., Field, J. M., and Kennicut, M. C., II, 1991. Observations of gas hydrates offshore northern California. *Marine Geology*, **96**, 103–9.

Brooks, J. M., Kennicut, M. C., II, Fisher, C. R., *et al.*, 1987. Deep-sea hydrocarbon seep communities: evidence for energy and nutritional carbon sources. *Science*, **238**, 1138–42.

Brown, A., 2000. Evaluation of possible gas microseepage mechanisms. *American Association of Petroleum Geologists (Bulletin)*, **84**, 1775–89.

Brown, C. E., Fingas, M. F., Fruwirth, M., and Gamble, R. L., 1995. Oil spill remote sensing: a brief review of airborne and satellite sensors. Washington, DC, SPOT Image 1995 User Group, August.

Brown, K. M., 1990. The nature and hydrogeologic significance of mud diapirs and diatremes for accretionary systems. *Journal of Geophysical Research*, **95**, 8969–82.

Brown, K. M. and Westbrook, G. K., 1987. The tectonic fabric of the Barbados Ridge accretionary complex. *Marine and Petroleum Geology*, **4**, 71–81.

Brown, K. M. and Westbrook, G. K., 1988. Mud diapirism and subcretion in the Barbados Ridge accretionary complex. *Tectonics*, **7**, 613–40.

Brüchert, V., Currie, B., Peard, K. and Endler, R., 2004. Dynamics of methane and hydrogen sulphide in the water column and sediment of the Namibian shelf. Proceedings of the Goldschmidt Conference, Copenhagen, 5–11 June, abstract A340.

Brun-Cottan, J. C., Guillou, S., and Li, Z. H., 2000. Behaviour of a puff of resuspended sediments: a conceptual model. *Marine Geology*, **167**, 355–73.

Bryant, W. R. and Roemer, L. B., 1983. Structure of the continental shelf and slope of the northern Gulf of Mexico and its geohazards and engineering constraints. In Geyer, R. A. and Moore, J. R. (eds.), *CRC Handbook of Geophysical Exploration at Sea*. Florida, CRC Press, pp. 123–84.

Bryn, P., Solheim, A., Berg, K., *et al.*, 2003. The Storegga Slide Complex: repeated large scale sliding in response to climatic cyclicity. In Locat, J. and Mienert, J. (eds.), *Submarine Mass Movements and Their Consequences*. Dordrecht, Kluwer Academic Publishers, pp. 215–22.

Buchardt, B., Israelson, C., Seaman, P., and Stockmann, G., 2001. Ikaite tufa towers in Ikka Fjord, southwest Greenland: their formation by mixing of seawater and alkaline spring water. *Journal of Sedimentary Research*, **71**, 176–89.

Buchardt, B., Seaman, P., Stockmann, G., *et al.*, 1997. Submarine columns of ikaite tufa. *Nature*, **390**, 129–30.

Buck, K. R. and Barry, J. M., 1998. Monterey Bay cold seep infauna: quantitative comparison of bacterial mat meiofauna with non-seep control sites. *Cahiers de Biologie Marine*, **39**, 333–5.

Buffett, B. A. and Zatsepina, O. Y., 2000. Formation of gas hydrate from dissolved gas in natural porous media. *Marine Geology*, **164**, 69–77.

Bugge, T., Befring, S., Belderson, R. H., *et al.*, 1987. A giant three-stage submarine slide off Norway. *Geo-Marine Letters*, **7**, 191–8.

Bugge, T., Belderson, R. H., and Kenyon, N. H., 1988. The Storegga slide. *Philosophical Transactions of the Royal Society of London*, **A325**, 357–88.

Bugge, T., Knarud, R., and Mørk, A., 1984. Bedrock geology on the mid-Norwegian continental shelf. In Spencer, A. M. (ed.), *Petroleum Geology of the North European Margin*. London, Graham & Trotman Ltd. for the Norwegian Petroleum Society, pp. 271–83.

Burne, R. V. and Moore, L. S., 1987. Microbialites: organosedimentary deposits of benthic microbial communities. *Palaios*, **2**, 241–54.

Burnett, W. C., Bokuniewicz, H., Huettle, M., Moore, W. S., and Taniguchi, M., 2003. Groundwater and pore water inputs to the coastal zone. *Biogeochemistry*, **66**, 3–33.

Burruss, R. C., 1993. Stability and flux of methane in the deep crust – a review. In Howell, D. G. (ed.), *The Future of Energy Gases*. United States Geological Survey Professional Paper 1570, pp. 21–30.

Burton, A. B., 1993. Controls on marine carbonate cement mineralogy: review and reassessment. *Chemical Geology*, 105, 163–79.

Bussmann, I. and Suess, E., 1998. Groundwater seepage in Eckernförde Bay (western Baltic sea): effect on methane and salinity distribution of the water column. *Continental Shelf Research*, 18, 1795–806.

Butenko, J. and Barbot, J. P., 1979. Geological hazards related to offshore drilling and construction in the Orinoco River delta, Venezuela. Proceedings of the Offshore Technology Conference, Houston, TX, OTC Paper 3395.

Butenko, J., Milliman, J. D., and Ye, Y.-C., 1985. Geomorphology, shallow structure, and geological hazards in the East China Sea. *Continental Shelf Research*, 4, 121–41.

Butterfield, D. A., 2000. Deep ocean hydrothermal vents. In Sigurdsson, H. (ed.), *Encyclopedia of Volcanoes*. New York, Academic Press, pp. 857–75.

Cairns, S. D. and Stanley, G. D., jr, 1981. Ahermatypic coral banks: living and fossil counterparts. Proceedings of the Fourth International Coral Reef Symposium, volume 1, Manila, Philippines, 18–22 May, pp. 611–18.

Cameron, T. D. J., Laban, C., and Schuttenhelm, R. T. E., 1984. *Quaternary Geology, Flemish Bight Sheet 52° N –02° E*. 1:250 000 map series. London, HMSO for British Geological Survey and Rijks Geologische Dienst.

Cameron, T. D. J., Crosby, A., Balson, P. S., *et al.*, 1992. *The Geology of the Southern North Sea*. United Kingdom Offshore Regional Report. London, HMSO, for the British Geological Survey.

Campbell, K. A., Farmer, J. D., and Des Marais, D., 2002. Ancient hydrocarbon seeps from the Mesozoic convergent margin of California: carbonate geochemistry, fluids and palaeoenvironments. *Geofluids*, 2, 63–94.

Canales, J. A., 2002. How a venting feature offshore Equatorial Guinea led to a discovery. *World Oil*, 23(8), 93–4.

Cann, J. and Morgan, J., 2002. Secrets of the Lost City. *Geoscientist*, 12(11), 4–7.

Cann, J. R., and Strens, M. R., 1989. Modeling periodic megaplume emission by black smoker systems. *Journal of Geophysical Research*, 94, 27–37.

Cao, M., Gregson, K., and Marshall, S., 1998. Global methane emission from wetlands and its sensitivity to climate change. *Atmospheric Environment*, 32, 3293–9.

Carlson, P. R. and Marlow, M. S., 1984. Discovery of a gas plume in Navarin Basin. *Oil and Gas Journal*, 2 April, 157–8.

Carlson, P. R., Golan-Bac, M., Karl, H. A., and Kvenvolden, K. A., 1982. Geologic hazards in the Navarin Basin province, northern Bering Sea. Proceedings of the Offshore Technology Conference, Houston, TX, OTC Paper 4172.

Carlson, P. R., Golan-Bac, M., Karl, H. A., and Kvenvolden, K. A., 1985. Seismic and geochemical evidence for shallow gas in sediment on the Navarin continental margin, Bering Sea. *American Association of Petroleum Geologists (Bulletin)*, 69, 422–36.

Carney, R. S., 1994. Consideration of the oasis analogy for chemosynthetic communities at Gulf of Mexico hydrocarbon vents. *Geo-Marine Letters*, 14, 149–59.

Carson, B., Kastner, M., Bartlett, D., *et al.*, 2000. Active carbon flux on the Cascadia accretionary prism: preliminary results from long-term, *in-situ* measurements at ODP Site 892B. *EOS – Transactions of the American Geophysical Union*, 81(48), AGU Fall meeting, abstract OS51E-11.

Cartwright, J. A., 1996. Polygonal fault systems: a new type of fault structure revealed by 3D seismic data from the North Sea basin. In Weimer, P. and Davis, T. L. (eds.), *Applications of 3-D Seismic Data to Exploration*. American Association of Petroleum Geologists, Studies in Geology No. 42 and SEG Geophysical Developments Series No. 5, pp. 225–30.

Cartwright, J. A. and Dewhurst, D. N., 1998. Layer-bound compaction faults in fine grained sediments. *Geological Society of America (Bulletin)*, 110, 1242–57.

Cartwright, J. A. and Lonergan, L., 1996. Volumetric contraction during the compaction of mud rocks: a mechanism for the development of regional-scale polygonal systems. *Basin Research*, 8, 183–93.

Casas, D., Ercilla, G., and Baraza, J., 2003. Acoustic evidences for gas in the continental slope sediments of the Gulf of Cadiz (E Atlantic). *Continental Shelf Research*, 23, 300–10.

Caston, V. N. D., 1977. *Quaternary Deposits of the Central North Sea. 2: The Quaternary Deposits of the Forties Field, Northern North Sea*. Report no. 77/11. Edinburgh, Institute of Geological Sciences.

Cauquil, E., Stephane, L., George, R. A., and Shyu, J.-P., 2003. High resolution autonomous underwater vehicle (AUV) geophysical survey of a large, deep water pockmark, offshore Nigeria. Proceeding of the EAGE 65th Conference & Exhibition, Stavanger, 2–5 June, abstract P056.

Cavagna, S., Clari, P., and Martire, L., 1999. The role of bacteria in the formation of cold seep carbonates: geological evidence from Monferrato (Tertiary, NW Italy). *Sedimentary Geology*, **126**, 253–70.

Cavanaugh, C. M., Gardiner, S. L., Jones, M. L., Jannasch, H. W., and Waterbury, J. B., 1981. Prokaryotic cells in the hydrothermal vent tubeworm *Riftia pachyptila* Jones: possible chemoautotrophic symbionts. *Science*, **213**, 340–2.

CBD (Convention of Biological Diversity), 2001. *Global Diversity Outlook*. Montreal, Secretariat of the Convention on Biological Diversity, United Nations Environment Programme.

Chaney, R. C., 1984. Methods of predicting the deformation of the seabed due to cyclic loading. In Denness, B. (ed.), *Seabed mechanics*. London, Graham and Trotman, pp. 159–67.

Chanton, J. P., Martens, C. S., and Kelley, C. A., 1989. Gas transport from methane-saturated, tidal freshwater and wetland sediments. *Limnology and Oceanography*, **34**, 807–19.

Chapman, R. E., 1981. *Geology and Water*. The Hague, Martinus Nijhoff/Dr W. Junk Publishers.

Chappellaz, J., Blunier, T., Raynaud, D., *et al.*, 1993. Synchronous changes in atmospheric CH_4 and Greenland climate between 40 and 8 kyr BP. *Nature*, **366**, 443–5.

Chappellaz, J., Raynaud, D., Blunier, T., and Stauffer, B., 2000. The ice core record of atmospheric methane. In Khalil, M. A. K. (ed.), *Atmospheric Methane*, 2nd edn. Berlin, Springer-Verlag, pp. 9–24.

Charlou, J. L., Philippe, J.-B., Donval, J. P., *et al.*, 2000. High methane concentrations in plumes and brines associated with mud volcanoes of the eastern Mediterranean Sea (MEDINAUT diving cruise, Nov.–Dec. 1998). *Abstracts of the Sixth International Conference on Gas in Marine Sediments*, St Petersburg, VNIIOkeangeologia, 15.

Charlou, J. L., Donval, J. P., Fouquet, Y., Jean-Baptiste, P., and Holm, N., 2002. Geochemistry of high H_2 and CH_4 vent fluids issuing from ultramafic rocks at the Rainbow hydrothermal field (36°14′ N, MAR). *Chemical Geology*, **191**, 345–59.

Charlou, J. L., Donval, J. P., Fouquet, Y., *et al.*, 2004. Physical and chemical characterization of gas hydrates and associated methane plumes in the Congo–Angola Basin. *Chemical Geology*, **205**, 405–25.

Chester, R., 2000. *Marine Geochemistry*, 2nd edn. Oxford, Blackwell Science.

Chevaldonné, P., 1997. The fauna at deep-sea hydrothermal vents, an introduction. In Desbruyères, D. and Segonzac, M. (eds.), *Handbook of Deep-Sea Hydrothermal Vent Fauna*. Plouzane, IFREMER (Institut français de recherche pour l'exploitation de la mer), pp. 7–20.

Childress, J. J. and Fisher, C. R., 1992. The biology of hydrothermal vent animals: physiology, biochemistry, and autotrophic symbioses. *Oceanography and Marine Biology Annual Review*, **30**, 337–441.

Childress, J. J., Fisher, C. R., Brooks, J. M., *et al.*, 1986. A methanotrophic marine molluscan (bivalvia, Mytilidae) symbiosis: mussels fuelled by gas. *Science*, **233**, 1306–8.

Chiodini, G., Cioni, R., Marini, L., and Panichi, C., 1995. Origin of the fumarolic fluids of Vulcano Island, Italy and implications for volcanic surveillance. *Bulletin of Volcanology*, **57**, 99–110.

Chiodini, G., Caliro, S., Minopoli, C., *et al.*, 2004. Geochemistry of the submarine gaseous emissions of Panarea (Aeolian islands, southern Italy): evidence of a recent input of magmatic gases. First General Congress, European Geosciences Union, Nice, 25–30 April (poster). (See *Geophysical Research Abstracts*, **6**, 04251, 2004.)

Chow, J., Lee, J. S., Liu, C. S., Lee, B. D., and Watkins, J. S., 2001. A submarine canyon as the cause of a mud volcano – Liuchienuyu Island in Taiwan. *Marine Geology*, **176**, 55–63.

Christian, H. A., Barrie, J. V., Hunter, J. A., Luternauer, J., and Monahan, P. A., 1995. Deep hole geotechnical investigation, Westshore Terminal, Roberts Bank Superport, March 6 – 31. Draft Open File Report. Sidney, BC, Geological Survey of Canada.

Christian, H. A., Woeller, D. J., Robertson, P. K., and Courtney, R. C., 1997. Site investigations to evaluate flow liquefaction slides at Sand Heads, Fraser River delta. *Canadian Geotechnical Journal*, **34**, 384–97.

Christodoulou, D., Papatheodorou, G., Ferentinos, G., and Masson., M., 2003. Active seepage in two contrasting pockmark fields in the Patras and Corinth gulfs, Greece. *Continental Shelf Research*, **23**, 194–9.

Chronis, G., Piper, D., and Anagnostou, G., 1991. Late Quaternary evolution of the Gulf of Patras, Greece. Tectonism, deltaic sedimentation and sea level changes. *Marine Geology*, 97, 191–9.

Church, T. M., 1996. An underground route for the water cycle. *Nature*, 380, 579–80.

Cicerone, R. J. and Oremland, R. S., 1988. Biogeochemical aspects of atmospheric methane. *Global Biogeochemical Cycles*, 2, 299–327.

Çifçi, G., Dondurur, D., and Ergün, M., 2001. Gas-saturated sediments and their effects on the southern side of the eastern Black Sea. In Akhamanov, G. and Suzyumov, A. (eds.), *Geological Processes on Deep-Water European Margins*. International Oceanographic Commission Workshop Report No. 175 on the International Conference and ninth post-cruise meeting of the Training Through Research Programme, Moscow-Mozhenka, Russia, 28 January–2 February 2001. Paris, UNESCO, pp. 48–9.

Çifçi, G., Dondurur, D., and Ergün, M., 2003. Deep and shallow structures of large pockmarks in the Turkish shelf, eastern Black Sea. *Continental Shelf Research*, 23, 311–22.

Clari, P. A., Gagliardi, C., Governa, M. E., Ricci, B., and Zuppi, G. M., 1988. I calcari di Marmorito: una testimonianza di processi diagenetici in presenza di metano. *Bollettino del Museo Regionale di Scienze Naturali Torino*, 5, 197–216. (In Italian.)

Clark, J., 2002. Hydrocarbon seeps in the Santa Barbara Channel, California: an example of macro-seepage. Abstracts of the Seventh International Conference, Baku, Azerbaijan, 7th–12th October. *Gas in Marine Sediments, 2002*. Baku, Nafta Press, 28.

Clark, J., Leifer, I., Washburn, L., and Luyendyk, B. P., 2003. Compositional changes in natural gas bubble plumes: observations from the Coal Oil Point marine hydrocarbon seep field. *Continental Shelf Research*, 23, 187–93.

Clark, M. K., 1986. Continuation of a 'sedimental journey' planned in the Great Lakes. *Sea Technology*, May 1986, 23–6.

Clarke, R. H. and Cleverly, R. W., 1991. Petroleum seepage and postaccumulation migration. In England, W. A. and Fleet, A. J. (eds.), *Petroleum Migration*. Geological Society of London, Special Publication 59, 265–71.

Clayton, C., 1992. Source volumetrics of biogenic gas generation. In Vially, R. (ed.), *Bacterial Gas*. Paris, Editions Technip, pp. 191–204.

Clayton, C. J. and Dando, P. R., 1996. Comparison of seepage and seep leakage rates. In Schumacher, D. and Abrams, M. A. (eds.), *Hydrogen Migration and its Near-Surface Expression*. American Association of Petroleum Geologists Memoir 66, 169–71.

Clayton, C. J. and Hay, S. J., 1994. Gas migration mechanisms from accumulation to surface. *Bulletin of the Geological Society of Denmark*, 41, 12–23.

Clennell, M. B., 1992. The mélanges of Sabah, Malaysia. Ph.D. Thesis, University of London.

Clennell, M. B., Hovland, M., Booth, J. S., Henry, P., and Winters, W. J., 1999. Formation of natural gas hydrates in marine sediments. 1. Conceptual model of gas hydrate growth conditioned by host sediment properties. *Journal of Geophysical Research*, 104, 22985–3003.

Clennell, M. B., Judd, A., and Hovland, M., 2000. Movement and accumulation of methane in marine sediments: relation to gas hydrate systems. In Max, M. D. (ed.), *Natural Gas Hydrate in Oceanic and Permafrost Environments*. Dordrecht, Kluwer Academic Publishers, pp. 105–22.

Clifton, H. E., Greene, H. G., Moor, G. W., and Phillips, R. L., 1971. *Methane seep off Malibu Point Following San Fernando Earthquake*. United States Geological Survey Professional Paper 733, pp. 112–16.

Cloud, P. E., 1960. Gas as a sedimentary and diagenetic agent. *American Journal of Science*, 248A, 34–45.

Coates, M., Connell, D. W., Bodero, J., Miller, G. J., and Back, R., 1986. Aliphatic hydrocarbons in Great Barrier Reef organisms and environment. *Estuarine Coastal Shelf Science*, 23, 99–113.

Cohen, A. S., Talbot, M. R., Awramik, S. M., Dettman, D. L., and Abell, P., 1997. Lake level and paleoenvironmental history of Lake Tanganyika, Africa, as inferred from late Holocene and modern stromatolites. *Geological Society of America (Bulletin)*, 109, 444–60.

Colantoni, P., Gabbianelli, G., Ceffa, C., and Ceccolini, C., 1998. Bottom features and gas seepages in the Adriatic Sea. In Abstracts and Guidebook, Fifth International Conference on Gas in Marine Sediments, Bologna, 9–12 September, 28–31.

Cole, D., Stewart, S. A., and Cartwright, J. A., 2000. Giant irregular pockmark craters in the Palaeogene of the Outer Moray Firth Basin, UK North Sea. *Marine and Petroleum Geology*, 17, 563–77.

Coleman, D. F. and Ballard, R. D., 2001. A highly concentrated region of cold hydrocarbon seeps in the southeastern Mediterranean Sea. *Geo-Marine Letters*, **21**, 162–7.

Collett, T. S., 2002. Energy resource potential of natural gas hydrates. *American Association of Petroleum Geologists (Bulletin)*, **86**, 1971–92.

Collier, J. S. and White, R. S., 1990. Diapirism within Indus fan sediments: Murray Ridge, Gulf of Oman. *Geophysical Journal International*, **101**, 345–53.

Colman, S. M., Foster, D. S., and Harris, D. W., 1992. Depressions and other lake-floor morphologic features in deep water, southern Lake Michigan. *Journal of Great Lakes Research*, **18**, 267–79.

Commeau, R. F., Paull, C. K., Comineau, J. A., and Poppe, L. J., 1987. Chemistry and mineralogy of pyrite-enriched sediments at a passive margin sulfide brine seep: abyssal Gulf of Mexico. *Earth & Planetary Science Letters*, **82**, 62–74.

Committee on Oil in the Sea, 2003. *Oil in the Sea III: Inputs, Fates, and Effects*. Washington DC, The National Academies Press.

Comrie, R., Read, A., and Fletcher, T., 2002. Cemented hardgrounds on the Norwegian Continental Shelf and their impact on submarine cable installation. *Offshore Site Investigation and Geotechnics – Diversity and Sustainability*. London, Society for Underwater Technology.

Conrad, R. and Seiler, W., 1988. Methane and hydrogen in seawater (Atlantic Ocean). *Deep-Sea Research*, **35**, 1903–17.

Conti, A., Stefanon, A., and Zuppi, G., 2002. Gas seeps and rock formation in the northern Adriatic Sea. *Continental Shelf Research*, **22**, 2333–44.

Cooper, M. C., Selley, R. C., and Cartwright, J. A. 1998. Vertical gas migration mechanisms in the central North Sea, studied with ultra high resolution digital 2D and 3D seismic data. In Abstracts and Guidebook, Fifth International conference on Gas in Marine Sediments, Bologna, 9–12 September, pp. 163–5.

Corliss, J. B., Dymond, J. R., Gordon, L. I., *et al.*, 1979. Submarine thermal springs on the Galapagos Rift. *Science*, **203**, 1073–83.

Corselli, C. and Basso, D., 1996. First evidence of benthic communities based on chemosynthesis on the Napoli Mud Volcano (eastern Mediterranean). *Marine Geology*, **132**, 227–40.

Corthay J. E., II, 1998. Delineation of a massive seafloor hydrocarbon seep, overpressured aquifer sands, and shallow gas reservoirs, Louisiana continental slope. Proceedings of the Offshore Technology Conference, Houston, TX, OTC Paper 8594, 37–53.

Cousins, S. H., 1989. Species richness and the energy theory. *Nature*, **340**, 350–1.

Cragg, B. A., Parkes, R. J., Fry, J. C., *et al.*, 1996. Bacterial populations and processes in sediments containing gas hydrates (ODP Leg 146: Cascadia Margin). *Earth & Planetary Science Letters*, **139**, 497–507.

Craig, H., 1981. Hydrothermal plumes and tracer circulation along the East Pacific Rise; 20° N to 20° S. *EOS – Transactions of the American Geophysical Union*, **62**, 693.

Cranston, R. E., 1994. Marine sediments as a source of atmospheric methane. *Bulletin of the Geological Society of Denmark*, **41**, 101–9.

Cranston, R. E., Ginsburg, G. D., Soloviev, V. A., and Lorenson, T. D., 1994. Gas venting and hydrate deposits in the Okhotsk Sea. *Bulletin of the Geological Society of Denmark*, **41**, 80–5.

Croker, P. and García-Gil, S., 2002. A multibeam survey of the Codling Fault Zone, western Irish Sea. *Gas in Marine Sediments, Abstracts of the Seventh International Conference, Baku, Azerbaijan, 7th–12th October 2002*. Baku, Nafta Press, 29 (Poster).

Croker, P. and O'Loughlin, O., 1998. A catalogue of Irish offshore carbonate mud mounds. *Carbonate Mud Mounds and Cold Water Reefs*. Paris, IOC-UNESCO Workshop Report 143:11.

Crook, K. A. W., Lisitzin, A. P., and Borissova, I. A. (eds.), 1997. Results and prospects from the joint USSR Australia USA PNG Geological Study of the Manus Basin during the 21st Cruise of the R/V Akademik Mstislav Keldysh. *Marine Geology*, **142**, 1–209.

Cullen, D. J. and Burnett, W. C., 1986. Phosphorite associations on seamounts in the tropical southwest Pacific Ocean. *Marine Geology*, **71**, 215–236.

Cunha, M. R., Hilário, A. M., Teixeira, I. G., and all Shipboard Scientific Party aboard the TTR- 10 Cruise, 2001. The faunal community associated to mud volcanoes in the Gulf of Cadiz. In Akhamanov, G. and Suzyumov, A. (eds.), *Geological Processes on Deep-Water European Margins*. International Oceanographic Commission Workshop Report No. 175 on the International Conference and ninth post-cruise meeting of the Training Through Research Programme, Moscow-Mozhenka, Russia, 28 January–2 February. Paris, UNESCO pp. 61–2.

Curray, J. R., Moore, D. G., and the DSDP Scientific Party, 1982. *Initial Reports of the Deep Sea Drilling Project*,

Vol. 64 (I & II). Washington, DC, US Government Printing Office.

Curtis, C. D., 1983. Microorganisms and diagenesis of sediments. In Krumbein, W. W. (ed.), *Microbial Geochemistry*. Oxford, Blackwell, 263–86.

Curzi, P. V., 1998. Sedimentation, subsidence and tectonics affecting gas charged sediments in central Adriatic Sea. Abstracts and Guidebook, Fifth International Conference on Gas in Marine Sediments, Bologna, 9–12 September, 182–4.

Curzi, P. V. and Veggiani, A., 1985. I pockmarks nel mare Adriatico centrale. *Acta nat. 'Ateneo Parmense'*, **21**, 79–90. (In Italian)

Cynar, F. J. and Yayanos, A.A, 1992. The distribution of methane in the upper waters of the southern California Bight. *Journal of Geophysical Research*, **97**, 11269–85.

Cynar, F. J. and Yayanos, A. A, 1993. The oceanic distribution of methane and its flux to the atmosphere over southern Californian waters. In Oremland, R. S. (ed.), *Biogeochemistry of Global Change: Radiatively Active Trace Gases*. New York, Chapman & Hall, pp. 487–504.

D'Arrigo, J. S., 1986. *Stable Gas-in-Liquid Emulsions*. Amsterdam, Elsevier.

D'Heur, M. and Pekot, L. J., 1987. Tommeliten. In Spencer, A. M. (ed.), *Geology of the Norwegian Oil and Gas Fields*. London, Graham and Trotman for the Norwegian Petroleum Society, pp. 117–28.

Dadashev, F. G., 1963. *Hydrocarbon Gases of Mud Volcanoes in Azerbaijan*. Baku, Azerneshr, (In Russian).

Dählmann, A., Wallmann, K., Sahling, H., *et al.*, 2001. Hot vents in an ice-cold ocean: indications for phase separation at the southernmost area of hydrothermal activity, Bransfield Strait, Antarctica. *Earth & Planetary Science Letters*, **193**, 381–94.

Dallimore, S. R., Edwardson, K. A., Hunter, J. A., Clague, J. J., and Luternauer, J. L., 1995. *Composite Geotechnical Logs for Two Deep Boreholes in the Fraser River Delta, British Columbia*. Open File Report 3018 Sidney, BC, Geological Survey of Canada.

Dallimore, S. R., Uchida, T., and Collett, T. S., 1999. Summary. In Dallimore, S. R., Uchida, T., and Collett, T. S. (eds.), *Scientific Results from JAPEX/JNOC/GSC Mallik 2L-38 Gas Hydrate Research Well, Mackenzie Delta, Northwest Territories, Canada*. Geological Survey of Canada, Bulletin, 544, pp. 1–10.

Damm, E. and Budéus, G., 2003. Fate of vent-derived methane in seawater above Håkon Mosby Mud Volcano (Norwegian Sea). *Marine Chemistry*, **82**, 1–11.

Dando, P. R., 2001. *A Review of Pockmarks in the UK Part of the North Sea, with Particular Reference to their Biology*. Strategic Environmental Assessment – SEA2, Technical Report TR_001. London, Department of Trade and Industry.

Dando, P. R. and Southward, A. J., 1986. Chemoautotrophy in bivalve molluscs of the genus *Thyasira*. *Journal of the Marine Biological Association of the United Kingdom*, **66**, 915–29.

Dando, P. R. and Spiro, B., 1993. Varying nutrional dependence of the thyasiris bivalves *Thyasira sarsi* and *T. equalis* on chemoautotrophic symbiotic bacteria, demonstrated by isotope ratios of tissue carbon and shell carbonate. *Marine Ecology Progess Series*, **92**, 151–8.

Dando, P. R., Aliani, S., Arab, H., *et al.*, 2000. Hydrothermal studies in the Aegean Sea. *Physics and Chemistry of the Earth (B)*, **25**, 1–8.

Dando, P. R., Austen, M. C., Burke, R. A., jr, *et al.*, 1991. Ecology of a North Sea pockmark with an active methane seep. *Marine Ecology Progress Series*, **70**, 49–63.

Dando, P. R., Bussmann, I., Niven, S. J., *et al.*, 1994a. A methane seep area in the Skagerrak, the habitat of the pogonophore *Siboglinum poseidoni* and the bivalve mollusc *Thyasira sarsi*. *Marine Ecology Progress Series*, **107**, 157–67.

Dando, P. R., Hughes, J. A., Leahy, Y., Taylor, L. J., and Zivanovic, S., 1995a. Earthquakes increase hydrothermal venting and nutrient inputs into the Aegean. *Continental Shelf Research*, **15**, 655–62.

Dando, P. R., Hughes, J. A., Leahy, Y., *et al.*, 1995b. Gas venting from submarine hydrothermal areas around the island of Milos, Helenic Volcanic Arc. *Continental Shelf Research*, **15**, 913–29.

Dando, P. R., Jensen, P., O'Hara, S. C. M., *et al.*, 1994b. The effects of methane seepage at an intertidal/shallow subtidal site on the shore of the Kattegat, Vendsyssel, Denmark. *Bulletin of the Geological Society of Denmark*, **41**, 65–79.

Dando, P. R., O'Hara, S. C. M., Schuster, U., *et al.*, 1994c. Gas seepage from a carbonate-cemented sandstone reef on the Kattegat coast of Denmark, *Marine and Petroleum Geology*, **11**, 182–9.

Dando, P. R., Rees., E. I. S., Dando, M. A., Schlüter, M. and Sauter, E., 2001. Methane venting associated with submarine groundwater discharge in Eckernförde Bucht, Baltic Sea. *Abstracts of the Sixth International Conference on Gas in Marine Sediments*. St Petersburg, VNIIOkeangeologia, 21.

Dando, P. R., Southward, A. J., Southward, E. C., *et al.*, 1992. Shipwrecked tube worms. *Nature*, **356**, 667.

Dando, P. R., Stüben, D., and Varnavas, S. P., 1999. Hydrothermalism in the Mediterranean Sea. *Progress in Oceanography*, **44**, 333–67.

Darwin, C., 1839. *Journal of Researches into the Natural History and Geology of the Countries Visited During the Voyage of H. M. S. "Beagle" Round the World*. London, Henry Colburn.

Davies, R. J. and Stewart, S. A., 2004. 3-D architecture of mud diapirs and volcanoes: an example from the South Caspian Sea. Abstracts and Programme of the Conference on Seabed and Shallow Section Marine Geoscience: Shared Lessons and Technologies from Academia and Industry. Geological Society of London, London, 24–26 February, 25 (abstract).

Davis, E., 2000. Volcanic action at Axial Seamount. *Nature*, **403**, 379–80.

Davis, E. E. and Lister, C. R. B., 1977. Heat flow measured over the Juan de Fuca Ridge: evidence for widespread hydrothermal circulation in a highly heat transportive crust. *Journal of Geophysical Research*, **82**, 4845–60.

Davis, E. E., Becker, L., Pettigrew, T., Carson, B., and MacDonald, R., 1992. CORK: a hydrological seal and downhole observatory for deep-ocean boreholes. *Proceedings of the Ocean Drilling Program Initial Reports*, **139**, 43–53.

Davis, E. E., Goodfellow, W. D., Bornhold, B. D., *et al.*, 1987. Massive sulfides in a sedimented rift valley, northern Juan de Fuca Ridge. *Earth & Planetary Science Letters*, **82**, 49–61.

Davis, P. H. and Spies, R. B., 1980. Infaunal benthos of a natural petroleum seep. Study of community structure. *Marine Biology*, **59**, 31–41.

Davis, R. A., 1994. *The Evolving Coast*. New York, NY, Scientific American Library.

Day, K., 2002. Seabed canyons: slope instability problems, or just interesting features? Proceedings of the Offshore Technology Conference, Houston, TX, OTC paper 14101.

de Angelis, M. A. and Lilley, M. D., 1987. Methane in surface waters of Oregon estuaries and rivers. *Limnology and Oceanography*, **32**, 716–22.

de Angelis, M. A., Lilley, M. D., Olson, E. J., and Baross, J. A., 1993. Methane oxidation in deep-sea hydrothermal plumes of the Endeavour Segment of the Juan de Fuca Ridge. *Deep-Sea Research*, **40**, 1169–86.

de Batist, M., Klerkx, J., Vanneste, M., *et al.*, 2000. Tectonically induced gas hydrate destabilization and gas venting in Lake Baikal, Siberia. In *Abstracts of the Sixth International Conference* on Gas in Marine Sediments. St Petersburg, VNIIOkeangeologia, 22–3.

de Beukelaer, S. M., MacDonald, I. R., Guinasso, N. L., jr, and Murray, J. A., 2003. Distinct side-scan sonar, RADARSAT SAR, and acoustic profiler signatures of gas and oil seeps on the Gulf of Mexico slope. *Continental Shelf Research*, **23**, 177–86.

de Haas, H., Boer, W., and Weering, T. C. E. van, 1997. Recent sedimentation and organic carbon burial in a shelf sea: the North Sea. *Marine Geology*, **144**, 131–46.

De Mol, B., Rensbergen, P. van, Pillen, S., *et al.*, 2002. Large deep-water coral banks in the Porcupine Basin, southwest of Ireland. *Marine Geology*, **188**, 193–231.

De Mol, B., Huvenne, V., Bünz, S., *et al.*, 2004. EUROcean Deep Ocean Margins (EURODOM). *Oceanography*, **17**, 156–65.

de Ronde, C. E. J., Baker, E. T., Massoth, G. J., *et al.*, 2001. Intra-oceanic subduction-related hydrothermal venting, Kermadec Volcanic Arc, New Zealand. *Earth & Planetary Science Letters*, **193**, 359–69.

DEFRA (Department for Environment, Food and Rural Affairs, UK), 2003. UK gains emergency protection for the Darwin Mounds, News Release, 20 August. See http://www.defra.gov.uk/news/2003/030820a.htm

Degens, E. T. and Ross, D. A. (eds.), 1969. *Hot Brines and Recent Heavy Metal Deposits in the Red Sea*. New York, Springer-Verlag.

Delisle, G., 2004. The mud volcanoes of Pakistan. *Environmental Geology*, **46**, 1024–9.

Delisle, G. and Berner, U., 2002. Gas hydrates acting as cap rock to fluid discharge in the Makran accretionary prism? In Clift, P. D., Kroon, D., Gaedicke, C., and Craig, J. (eds.), *The Tectonic and Climatic Evolution of the Arabian Sea Region*. Geological Society of London, Special Publication 195, 137–46.

Delisle, G., Rad, U. van, Andruleit, H., *et al.*, 2002. Active mud volcanoes on- and offshore eastern Makran, Pakistan. Geologische Rundschau, **91**, 93–110.

Deming, J. W. and Baross, J. A., 1993. Deep-sea smokers: window to a subsurface biosphere? *Geochimica et cosmo chimica acta*, **57**, 3219–30.

Denardo, B., Pringle, L., and DeGrace, C., 2001. When do bubbles cause a floating body to sink? *American Journal of Physics*, **69**, 1064–72.

Des Marais, D. J., 1985. Carbon exchange between the mantle and the crust, and its affect upon the atmosphere today compared to Archaean time. *Geophysical Monograph*, **32**, 602–11.

Desbruyères, D. and Segonzac, M., 1997. *Handbook of Deep-Sea Hydrothermal Vent Fauna*. Brest, IFREMER (Institut français de recherche pour l'exploitation de la mer).

Desbruyères, D. and Toulmond, A., 1998. A new species of hesionid worm, *Hesiocaeca methanicola* sp. nov. (Polychaeta: Hesionidae), living in ice-like methane hydrates in the deep Gulf of Mexico. *Cahiers de Biologie Marine*, **39**, 93–8.

Detrick, R. S., Buhl, P., Vera, E., *et al.*, 1987. Multi-channel seismic imaging of a crustal magma chamber along the East Pacific Rise. *Nature*, **326**, 35–41.

Dewhurst, D. N., Cartwright, J. A., and Lonergan, L., 1999a. The development of polygonal fault systems by syneresis of colloidal sediments. *Marine & Petroleum Geology*, **16**, 793–810.

Dewhurst, D. N., Yang, Y., and Aplin, A. C., 1999b. Permeability and fluid flow in natural mudstones. In Aplin, A. C., Fleet, A. J., and Macquaker, J. H. S. (eds.), *Muds and Mudstones: Physical and Fluid Flow Properties*. Geological Society of London, Special Publication 158, 23–43.

Dia, A., Aquilina, L., Boulègue, J. Suess, E., and Torres, M., 1993. Origin of fluids and related barite deposits at vent sites along the Peru convergent margin. *Geology*, **21**, 1099–102.

Díaz-del-Río, V., Somoza, L., Martínez-Frias, J., *et al.*, 2001. Carbonate chimneys in the Gulf of Cadiz: initial report of their petrography and geochemistry. In Akhamanov, G. and Suzyumov, A. (eds.), *Geological Processes on Deep-Water European Margins*, International Oceanographic Commission Workshop Report No. 175 on the International Conference and ninth post-cruise meeting of the Training Through Research Programme, Moscow-Mozhenka, Russia, 28 January–2 February Paris, UNESCO, pp. 53–4.

Díaz-del-Río, V., Somoza, L., Martínez-Frias, J., 2003. Vast fields of hydrocarbon-derived carbonate chimneys related to the accretionary wedge/olistostrome of the Gulf of Cádiz. *Marine Geology*, **195**, 177–200.

Dickens, G. R., 1999. The blast in the past. *Nature*, **401**, 752–5.

Dickens, G. R., 2003. Rethinking the global carbon cycle with a large, dynamic and microbially mediated gas hydrate capacitor. *Earth & Planetary Science Letters*, **213**, 169–83.

Dill, R. F., Shinn, E. A., Jones, A. T., Kelly, K., and Steinen, R. P., 1986. Giant subtidal stromatolites forming in normal salinity waters. *Nature*, **324**, 55–8.

Dillon, W. P. and Max, M. D., 2000. The US Atlantic Continental Margin; the best-known hydrate locality. In Max, M. D. (ed.), *Natural Gas Hydrates in Oceanic and Permafrost Environments*. Dordrecht, Kluwer Academic Publishers, pp. 157–70.

Dillon, W. P. and Paull, C. K., 1983. Marine gas hydrates – II: geophysical evidence. In Cox, J. L. (ed.), *Natural Gas Hydrates: Properties, Occurrence and Recovery*. Boston, Butterworth, pp. 73–90.

Dillon, W. P., Danforth, W. W., Hutchinson, D. R., *et al.*, 1998. Evidence for faulting related to dissociation of gas hydrate and release of methane off the southeastern United States. In Henriet, J.-P., and Mienert, J. (eds.), *Gas Hydrates: Relevance to World Margin Stability and Climate Change*. Geological Society of London, Special Publication 137, 293–302.

Dillon, W. P., Lee, M. W., Fehlhaber, K., and Colemand, D. F., 1993. Gas hydrates on the Atlantic continental margin of the United States – controls on the concentration. In Howell, D. G. (ed.), *The Future of Energy Gases*. United States Geological Survey Professional Paper 1570, pp. 313–30.

Dillon, W. P., Popenoe, P., Grow, J. A., *et al.*, 1982. Growth faulting and salt diapirism: their relationship and control in the Carolina Trough, eastern North America. In Watkins, J. S. and Drake, C. L. (eds.), *Studies in Continental Margin Geology*. American Association of Petroleum Geologists Memoir 34, 21–46.

Dimitrov, L. I., 2002a. Contribution to atmospheric methane by natural seepages on the Bulgarian continental shelf. *Continental Shelf Research*, **22**, 2429–42.

Dimitrov, L. I., 2002b. Mud volcanoes – the most important pathway for degassing deeply buried sediments. *Earth-Science Reviews*, **59**, 49–76.

Dimitrov, L. I., 2003. Mud volcanoes – a sizeable source of atmospheric methane. *Geo-Marine Letters*, **23**, 155–61. doi: 10.1007/s00367-003-0140-3.

Dimitrov, L. I. and Dontcheva, V., 1994. Seabed pockmarks in the southern Bulgarian Black Sea zone. *Bulletin of the Geological Society of Denmark*, **41**, 24–33.

Dimitrov, L. and Woodside, J., 2003. Deep sea pockmark environments in the eastern Mediterranean. *Marine Geology*, **195**, 263–76.

Dinet, A. F., Grassle, J. F., and Tunnicliffe, V., 1988. Premières observations sur la méiofauna des sites hydrothermaux de la dorsale est-Pacifique (Guaymas,

21° N) et de l'Explorer Ridge. *Oceanologica acta*, No. SP.8, 7–14. (In French)

Dionne, J. C., 1973. Monroes: a type of so-called mud volcanoes in tidal flats. *Journal of Sedimentary Petrology*, 43, 848–56.

Dmitrievsky, A. N., Sagalevich, A. M., Balanyuk, I. E., Sorokhtin, O. G., and Matveenkov, V. V., 2001. *Gas Breath of the Oceans*, 31st International Geological Conference. Moscow, NIP SEA.

Donovan, T. J., 1974. Petroleum migration at Cement, Oklahoma: evidence and mechanism. *American Association of Petroleum Geologists (Bulletin)*, 58, 429–46.

Dons, C., 1944. Norges korallrev. *Norske Videnskabers Selskab, Trondheim, Forhandlinger*, 16A, 37–82. OR Kgl. Norsk vitenskapelig selskaps frorhandl., 16, 37–82. (In Norwegian.)

DTI (Department of Trade and Industry), 2001. *Strategic Environmental Assessment of the Mature Areas of the Offshore North Sea: SEA2*. Public domain Report to the DTI; see http://www.offshore-sea.org.uk/consultations/SEA_2/SEA/2_Assessment_Document.pdf

Dugan, B. and Flemings P. B., 2000. The New Jersey margin: compaction and fluid flow. *Journal of Geochemical Exploration*, 69–70, 477–81.

Dugan, B. and Flemings, P. B., 2002. Fluid flow and stability of the US Continental Slope offshore New Jersey from the Pleistocene to the present. *Geofluids*, 2, 137–46.

Duncan, A. R. and Pantin, H. M., 1969. Evidence of submarine geothermal activity in the Bay of Plenty. *New Zealand Journal of Marine and Freshwater Research*, 3, 602–6.

Dunham, K. C., 1990. *Geology of the Northern Pennine Orefiled. Volume 1: Tyne to Stainmore*, 2nd edn. Economic Memoir of the British Geological Survey, Sheets 19 and 25, and parts of 13, 24, 26, 31, 32 (England and Wales). London, HMSO.

Dupré, S., Loncke, L., Deville, E., *et al.*, 2004. Active seepage on top of the Nile Deep-sea fan mud pies. First General Congress, European Geosciences Union, Nice, April (poster). (See *Geophysical Research Abstracts*, 6, 05045.)

Duranti, D., Hurst, A., Bell, C., Groves, S., and Hanson, R., 2002. Injected and remobilized Eocene sandstones from the Alba field, UKCS: cores and wireline log characteristics. *Petroleum Geoscience*, 8, 99–108.

Dutta, N. C. and Nutt, L., 1998. The role of Seismic While Drilling measurements for SWF applications.

Proceeding of the 1998 Shallow Water Flow Forum, Woodlands, TX, 24–25 June (presentation).

Dymond, J., Collier, R. W., and Watwood, M. E., 1989. Bacterial mats from Crater Lake, Oregon and their relationship to possible deep-lake hydrothermal venting. *Nature*, 342, 673–5.

EC (European Commission), 1999. *Interpretation Manual of European Union Habitats, Version EUR 15/2*. Brussels, European Commission.

Edgerton, H. E., Seibold, E., Vollbrecht, K., and Werner, F., 1966. Morphologische Untersuchungen am Mittelgrund (Eckernförder Bucht, westliche Ostsee). *Meyniana*, 16, 37–50. (In German)

Edmonds, B., Moorwood, R. A. S., and Szczepanski, R., 2001. Controlling, remediation of fluid hydrates in deepwater drilling operations. *Ultradeep Engineering Supplement to offshore Magazine*, March, 7–10.

EEC (European Economic Council), 1992. Council Directive 92/43/EEC of 21 May 1992 on the conservation of natural habitats and of wild fauna and flora. *Official Journal*, L206 (22.7.1992), 7–50.

Egbert, G. D. and Ray, R. D., 2000. Significant dissipation of tidal energy in the deep ocean inferred from satellite altimeter data. *Nature*, 405, 775–8.

Egorov, A. V. and Trotsuik, V. Ya., 1990. The analysis of the subbottom gas hydrate formation from the microbiologically-produced methane. In Yaqutsey, V. P. (ed.), *Resources of Gas of Untraditional Sources, and the Problem of their Exploration*. VNIGNY, Leningrad, 201–11. (in Russian.)

Egorov, A. V., Crane, K., Rozhkov, A. N., and Vogt, P. R., 1999. Gas hydrates that outcrop on the sea floor: stability models. *Geo-Marine Letters*, 19, 68–75.

Egorov, V. N., Polikarpov, G. G., Gulin, S. B., *et al.*, 2003. Present-day views on the environmental forming and ecological role of the Black Sea methane gas seeps. *Marine Ecological Journal*, 2, 5–26. (In Russian.)

Ehhalt, D. H., 1974. The atmospheric cycle of methane. *Tellus*, 26, 58–70.

Ehhalt, D. H. and Schmidt, U., 1978. Sources and sinks of atmospheric methane. *Pure and Applied Geophysics*, 116, 452–64.

Ehhalt, D., Prather, M., Dentener, F., *et al.*, 2001. Atmospheric chemistry and greenhouse gases. In Houghton, J. T., Ding, Y., Griggs, D. J., *et al.* (eds.), *Climate Change 2001: the Scientific Basis. Contribution of Working Group I to the Third Assessment Report of the*

Intergovernmental Panel on Climate Change. Cambridge, Cambridge University Press, 241–87.

Eichelberger, J. C., Carrigan, C. R., Westrich, H. R., and Price, R. H., 1986. Non-explosive silicic volcanism. *Nature*, **323**, 598–602.

Eichhubl, P., Greene, H. G., Naehr, T., and Maher, N., 2000. Structural control of fluid flow: offshore fluid seepage in the Santa Barbara Basin, California. *Journal of Geochemical Exploration*, **69–70**, 545–9.

Eide, Ø. and Andersen, K. H., 1984. Foundation engineering for gravity structures in the northern North Sea. Report No. 154. Oslo, Norwegian Geotechnical Institute.

Elhatip, H., 2003. The use of hydrochemical techniques to estimate the discharge of Ovacık submarine springs on the Mediterranean coast of Turkey. *Environmental Geology*, **43**, 714–19.

Ellis, J. P. and McGuinness, W. T., 1986. Pockmarks of the northwestern Arabian Gulf. In *Proceedings of the Oceanology International Conference (Brighton, March 1986)*. Advances in Underwater Technology, Ocean Science and Offshore Engineering, vol. 6. London, Graham and Trotman, pp. 353–67.

Embley, R. W. and Jacobi, R. D., 1986. Mass wasting in the western North Atlantic. In Vogt, P. R. and Tucholke, B. E. (eds.), *The Geology of North America, the Western North Atlantic Region*. Boulder, CO, Geological Society of America, 479–90.

Emeis, K.-C., Brüchert, V., Currie, B., *et al.*, 2004. Shallow gas in shelf sediments of the Namibian coastal upwelling ecosystem. *Continental Shelf Research*, **24**, 627–42.

Emery, K. O., 1956. Sediments and water of the Persian Gulf. *American Association of Petroleum Geologists (Bulletin)*, **40**, 2354–83.

Emery, K. O., 1974. Pagoda structures in marine sediments. In Kaplan, I. R. (ed.), *Gases in Marine Sediments*. New York, Plenum Press, 309–17.

Emery, K. O. and Uchupi, E., 1984. *The Geology of the Atlantic Ocean*. New York, Springer-Verlag.

Engell-Sørensen, L. and Havskov, J., 1987. Recent North Sea seismicity studies. *Physics of Earth and Planetary Interiors*, **45**, 37–44.

Ergun, M. and Çifçi, G., 1999. Gas-saturated sediments in the eastern Black Sea and geohazard effects. Proceedings of the Offshore Technology Conference, Houston, TX, OTC Paper 10924.

Etiope, G., 2004. New directions: GEM – geologic emissions of methane, the missing source in the atmospheric methane budget. *Atmospheric Environment*, **38**, 3099–100.

Etiope, G. and Klusman, R. W., 2002. Geologic emissions of methane to the atmosphere. *Chemosphere*, **49**, 777–89.

Etiope, G. and Milkov, A. V., 2004. A new estimate of global methane flux from onshore and shallow submarine mud volcanoes to the atmosphere. *Environmental Geology*, **46**, 997–1002. doi: 10.1007/s00254-004-1085-1.

Etiope, G., Caracausi, A., Favara, R., Italiano, F., and Baciu, C., 2002. Methane emissions from the mud volcanoes of Sicily (Italy). *Geophysical Research Letters*, **29**, 14.

Evans, D., 1987. *Quaternary Geology, Tiree Sheet 56° N–08° W*. 1:250,000 map series. Edinburgh, British Geological Survey.

Faber, B. and Stahl, W., 1984. Geochemical surface exploration for hydrocarbons in the North Sea. *American Association of Petroleum Geologists (Bulletin)*, **68**, 363–86.

Faber, E., Gerling, P., Berner, U., and Sohns, E., 1994. Methane in ocean waters – concentration and carbon-isotope variability at East Pacific Rise and in the Arabian Sea. *Environmental Monitoring and Assessment*, **31**, 139–44.

Fader, G. B. J., 1985. Surficial and bedrock geology of the Grand Banks. *BIO Review '85*. Nova Scotia, Bedford Institute of Oceanography, 16–20.

Fader, G. B. J., 1991. Gas-related sedimentary features from the eastern Canadian continental shelf. *Continental Shelf Research*, **11**, 1123–53.

Fader, G. B. J. and King, L. H., 1981. *A Reconnaissance Study of the Surficial Geology of the Grand Banks Newfoundland*. Geological Survey of Canada, Paper 81-1A, 45–56.

Fader, G. B. J. and Miller, R. O., 1986. Regional geological constraints to resource development – Grand Banks of Newfoundland. Volume 1, Proceedings of the Third Canadian Conference on Marine Geotechnical Engineering, St John's, Newfoundland, June, 3–40.

Falconer, R. K. H., 1991. Shallow gas offshore Taranaki, New Zealand. Proceedings of the 1991 New Zealand Oil Exploration Conference, Christchurch, New Zealand, September, 426–33.

Fannin, N. G. T., 1979. The use of regional geological surveys in the North Sea and adjacent areas in the recognition of offshore hazards. In Ardus, D. A. (ed.), *Offshore Site Investigations*. London, Graham and Trotman, 5–21.

Farrow, G. E., 1978. Recent sediments and sedimentation in the Inner Hebrides. *Proceedings of the Royal Society of Edinburgh*, **83B**, 91–105.

Faugères, J. C., Gonthier, E., Bobier, C., and Griboulard, R., 1997. Tectonic control on sedimentary processes in the southern termination of the Barbados prism. *Marine Geology*, **140**, 117–40.

Faure, H., Walter, R. C., and Grant, D. R., 2002. The coastal oasis: ice age springs on emerged continental shelves. *Global and Planetary Change*, **3**, 47–56.

Faure, M., Lalevée, F., Gusokujima, Y., Iiyama, J.-T., and Cadet, J.-P., 1986. The pre-Cretaceous deep-seated tectonics of the Abukuma massif and its place in the structural framework in Japan. *Earth & Planetary Science Letters*, **77**, 384–98.

Fiala-Médioni, A., Pranal, V., and Colomines, J. C., 1994. Deep-sea symbiotic models chemosynthetic based: comparison of hydrothermal vents and cold seep bivalve molluscs. Proceedings of the Seventh Deep-Sea Biology Symposium, IMBC, Crete.

Field, M. E. and Jennings, A. E., 1987. Seafloor gas seeps triggered by a northern California earthquake. *Marine Geology*, **77**, 39–51.

Field, M. E., Gardner, J. V., Jennings, A. E., and Edwards, B. D., 1982. Earthquake-induced sediment failures on a 0.25° slope, Klamath River delta, California. *Geology*, **10**, 542–6.

Figueiredo, A. G., jr, Nittouer, C. A., and de Alencar Costa, E., 1996. Gas-charged sediments in the Amazon submarine delta. *Geo-Marine Letters*, **16**, 31–5.

Fischer, P. J. and Stevenson, A. J., 1973. Natural hydrocarbon seeps along the northern shelf of the Santa Barbara Basin, California. Proceedings of the Offshore Technology Conference, Houston, TX, OTC Paper 1738.

Fisher, C. R., 1990. Chemoautotrophic and methanotrophic symbioses in marine invertebrates. *CRC Critical Reviews in Aquatic Sciences*, **2**, 399–436.

Fisher, C. R., MacDonald, I. R., Joye, S., and Sassen, R., 1998. The lair of the ice worm: a clathrate dwelling polychaete from the Gulf of Mexico. *Proceedings of the 1998 Ocean Sciences Meeting*. San Diego, CA, American Geophysical Union.

Fisher, T. R., Carlson, P. R., and Barber, R. T., 1982. Sediment nutrient regeneration in three North Carolina estuaries. *Estuarine and Coastal Shelf Science*, **14**, 101–16.

Fleischer, P., Orsi, T. H., Richarson, M. D., and Anderson, A. L., 2001. Distribution of free gas in marine sediments: a global overview. *Geo-Marine Letters*, **21**, 103–22.

Flodén, T., and Söderberg, P., 1988. *Pockmarks and Related Seabed Structures in Some Areas of Precambrian Bedrock in Sweden*. Geological Survey of Finland, Special Paper No. 6, 163–9.

Flodén, T. and Söderberg, P., 1994. Shallow gas traps and gas migration models in crystalline bedrock areas offshore Sweden. *Baltica*, **8**, 50–6.

Flood, R. D., 1981. Pockmarks in the deep sea. *EOS – Transactions of the American Geophysical Union*, **62**, 304.

Flood, R. D. and Johnson, T. C., 1984. Side-scan targets in Lake Superior – evidence for bedforms and sediment transport. *Sedimentology*, **31**, 311–33.

Floodgate, G. D. and Judd, A. G., 1992. The origins of shallow gas. *Continental Shelf Research*, **12**, 1145–56.

Flügel, H. J. and Langhof, I., 1983. A new hermaphroditic pogonophore from the Skagerrak. *Sarsia*, **68**, 131–8.

Foote, R. Q., Martin, R. G., and Powers, R. B., 1983. Oil and gas potential of the Maritime Boundary region in the central Gulf of Mexico. *American Association of Petroleum Geologists (Bulletin)*, **67**, 1047–65.

Formolo, M. J., Lyins, T. W., Zhang, C., et al., 2004. Quantifying carbon sources in the formation of authigenic carbonates at gas hydrate sites in the Gulf of Mexico. *Chemical Geology*, **205**, 253–64.

Fornari, D. J., Batiza, R., and Allan, J. F., 1987. Irregularly shaped seamounts near the East Pacific Rise: implications for seamount origin and rise axis processes. In Keating B. H., Fryer, P., Batiza, R., and Boehlert, G. W. (eds.), *Seamounts, Islands, and Atolls*. American Geophysical Union, Geophysical Monograph 43, 13–21.

Fortey, R., 2000. *Trilobite! Eyewitness to Evolution*. London, Flamingo.

Fosså, J. H. and Mortensen, P. B., 1998. Artsmangfoldet på Lophelia-korallrev og metoder for kartlegging og overvåkning. *Fisken og Havet*, **17**, 1–95. (In Norwegian.)

Fosså, J. H., Mortensen, P. B., and Furevik, D. M., 2002. The deep-water coral *Lophelia pertusa* in Norwegian waters: distribution and fisheries impacts. *Hydrobiologia*, **471**, 1–12.

Fredlund, D. G. and Rahardjo, H., 1993. *Soil Mechanics for Unsaturated Soils*. New York, John Wiley Interscience.

Freiwald, A., 1995. Deep-water coral reef mounds on the Sula-Ridge, mid-Norway Shelf. Universität Bremen, Field Report, Cruise 24/95, R/V Victor Hensen.

Friedman, G. M., 1965. On the origin of aragonite in the Dead Sea. *Israel Journal of Earth-Science*, **14**, 79–85.

Fromant, A. C., 1965. The water supplies of Bahrain. *Journal of the Institute of Water Engineers*, **19**, 579–85.

Fry, B. and Wainwright, S. C., 1991. Diatom sources of ¹³C-rich carbon in marine food webs. *Marine Ecology Progress Series*, **76**, 149–57.

Fryer, P. and Fryer, G. J., 1987. Origins of nonvolcanic seamounts in a forearc environment. In Keating, B., Fryer, P., and Batiza, R. (eds.), *Seamounts, Islands, and Atolls*. American Geophysical Union, Geophysical Monograph 43, 61–9.

Fryer, P. and Mottl, M., 2000. Pollution Prevention and Safety Panel Information, Site MAF-4B, Leg 195. In *Scientific Prospectus of the Ocean Drilling Program (ODP) Leg 195*. College Station, TX, ODP, 1–12.

Fryer, P., Saboda, K. L., Johnson, L. E., *et al.*, 1990. Conical Seamount: SeaMARC II, *Alvin* submersible, and seismic-reflection studies. *Proceedings of the Ocean Drilling Programme, Initial Reports*, **125**, 69–80.

Fu, B., Aharon, P., Byerly, G. R., and Roberts, H. H., 1994. Barite chimneys on the Gulf of Mexico slope: initial report on their petrography and geochemistry. *Geo-Marine Letters*, **14**, 81–7.

Fujikura, K., Kojima, S., Tamaki, K., *et al.*, 1999. The deepest chemosynthesis-based community yet discovered from the hadal zone, 7326 m deep, in the Japan Trench. *Marine Ecology Progress Series*, **190**, 17–26.

Fung, I., John, J., Lerner, J., *et al.*, 1991. Three-dimensional model synthesis of the global methane cycle. *Journal of Geophysical Research*, **96**, 13033–65.

Furnes, G. K., Kvamme, O. B., and Nygaard, O., 1991. Tidal response on the reservoir pressure at the Gullfaks oil field. *Pure Applied Geophysics*, **135**, 421–46.

Fusi, N. and Kenyon, N. H., 1996. Distribution of mud diapirism and other geological structures from long-range sidescan sonar (GLORIA) data, in the eastern Mediterranean Sea. *Marine Geology*, **132**, 21–38.

Fyfe, J. A., Gregersen, U., Jordt, H., *et al.*, 2003. Oligocene to Holocene. In Evans, D. and Graham, C. (eds.), *The Millennium Atlas: Petroleum Geology of the Central and Northern North Sea*. Bath, Geological Society of London, 279–87.

Fyfe, W. S., 1992. Magma underplating of continental crust. *Journal of Volcanology and Geothermal Research*, **50**, 33–40.

Fyfe, W. S., 1994. The water inventory of the Earth: fluids and tectonics. In Parnell, J. (ed.), *Geofluids: Origin, Migration and Evolution of Fluids in Sedimentary Basins*. Geological Society of London, Special Publication 78, 1–7.

Gaedicke, C., Baranov, B. V., Obzhirov, A. I., *et al.*, 1997. Seismic stratigraphy, BSR distribution, and venting of methane-rich fluids west of Paramushir and Onekotan islands, northern Kurils. *Marine Geology*, **136**, 259–76.

Galindo-Zaldivar, J., Nieto, L. M., Robertson, A. H. F., and Woodside, J. M., 2001. Recent tectonics of Eratosthenes Seamount: an example of seamount deformation during incipient continental collision. *Geo-Marine Letters*, **20**, 233–42.

Galkin, S. V., 1997. Megafauna associated with hydrothermal vents in the Manus Back-arc Basin (Bismarck Sea). *Marine Geology*, **142**, 197–206.

Gallagher, J. W., Braaten, A. M., Hovland, M., and Kemp, A., 1989. Use of an interpretation station for the study of shallow gas sands on Haltenbanken. Proceedings of the Shallow Gas and Leaky Reservoirs Conference, Norwegian Petroleum Forening, Stavanger, April 10–11.

Games, K. P., 1990. Processing procedures for high resolution seismic data. In Ardus, D. A. and Green, C. D. (eds.), *Safety in Offshore Drilling: the Role of Shallow Gas Surveys*. Dordrecht, Kluwer Academic Publishers, 103–31.

Games, K. P., 2001. Evidence of shallow gas above the Connemara oil accumulation, Block 26/28, Porcupine Basin. In Shannon, P. M., Haughton, P. D. W., and Corcoran, D. V. (eds.), *The Petroleum Exploration of Ireland's Offshore Basins*. Geological Society of London, Special Publication 188, 361–73.

García-Gil, S., 2003. A natural laboratory for shallow gas: the Rías Baixas (NW Spain). *Geo-Marine Letters*, **23**, 215–29.

Gardner, J. M., 2001. Mud volcanoes revealed and sampled on the western Moroccan continental margin. *Geophysical Research Letters*, **28**, 339–42.

Gardner, J. V., Prior, D. B., and Field, M. E., 1999. Humboldt Slide – a large shear-dominated retrogressive slope failure. *Marine Geology*, **154**, 323–38.

Gaston, K. J., 2000. Global patterns of biodiversity. *Nature*, **405**, 220–7.

Gatcliff, R. W., Richards, P. C., Smith, K., *et al.*, 1994, *United Kingdom Offshore Regional Report: the Geology of the Central North Sea*. London, HMSO, for the British Geological Survey.

Gebruk, A. V., Galkin, S. V., Vereshchaka, A. L., Moskalev, L. I., and Southward, A. J., 1997. Ecology and biogeography of the hydrothermal vent fauna of the mid-Atlantic Ridge. *Advances in Marine Biology*, **32**, 94–146.

Geodekian, A. A., Trotsyuk, V. Ya., and Verkshovskaya, Z. I., 1976. Hydrocarbon gases in bottom sediments of the Sea of Okhotsk. *Doklady Akademii Nauk SSSR*, **226**, 228–30. (In Russian.)

Georgalas, G., Karageorghiou, E., and Papakis, N., 1962. Sur les fluctuations de la source thermominérale d'Ipati – Greece. *Memoires des 5me Réunion d'Athenes de Association International Hydrogéologiques*, 277–8. (In French.)

Gervitz, J. L., Carey, B. D., jr and Blanco, S. R., 1985. Regional geochemical analysis of the southern portion of the Norwegian Sector of the North Sea. In Thomas, B. M. (ed.), *Petroleum Geochemistry in Exploration of the Norwegian Shelf*. London, Graham & Trotman, 247–61.

Geyer, R. A., 1979. Naturally occurring hydrocarbon seeps in the Gulf of Mexico and the Caribbean Sea. Pamphlet, College Station, TX, Texas A & M University.

Ginsburg, G. D. and Soloviev, V. A., 1994. Mud volcano gas hydrates in the Caspian Sea. *Bulletin of the Geological Society of Denmark*, **41**, 95–100.

Ginsburg, G. D. and Soloviev, V. A., 1998. *Submarine Gas Hydrates*. St Petersburg, VNIIOkeangeologia.

Glasby, G. P., 1971. Direct observations of columnar scattering associated with geothermal gas bubbling in the Bay of Plenty, New Zealand. *New Zealand Journal of Marine and Freshwater Research*, **5**, 483–96.

Glenn, K., 2002. Coral reefs and hydrocarbon seeps. *Ausgeonews*, **68**, 4–7.

Glennie, K. W. and Underhill, J. R., 1998. Origin, development and evolution of structural styles. In Glennie, K. W. (ed.), *Petroleum Geology of the North Sea; Basic Concepts and Recent Advances*, 4th edn., Oxford, Blackwell, 42–84.

Gold, T., 1999. *The Deep Hot Biosphere*. New York, Copernicus Books.

Gold, T. and Soter, S., 1980. The deep-Earth-gas hypothesis. *Scientific American*, **242**, 154–61.

Gold, T. and Soter, S., 1985. Fluid ascent through the solid lithosphere and its relation to earthquakes. *Pure Applied Geophysics*, **122**, 492–530.

Gold, T., Gordon, B. E., Streett, W., Bilson, E., and Patnaik, P., 1986. Experimental study of the reaction of methane with petroleum hydrocarbons in geological conditions. *Geochimica et cosmochimica acta*, **50**, 2411–18.

Golmshtok, A. Ya., 2000. Influence of faulting and other factors on characteristics of the BSR, Lake Baikal, Siberia. In *Abstracts of the Sixth International Conference on Gas in Marine Sediments*. St Petersburg, VNIIOkeangeologia, 31–2.

Gontz, A. M., Belknap, D. F., and Kelley, J. T., 2001. Evidence for changes in the Belfast Bay pockmark field, Maine. Geological Society of America, Northeastern Section, 36th Annual Meeting, Burlington, VT, 12–14 March; see http://gsa.confex.com/gsa/2000NE/finalprogram/abstract_2141.htm

Gosink, T. A., Pearson, J. G., and Kelley, J. J., 1976. Gas movement through sea ice. *Nature*, **263**, 41–2.

Graber, K. K., 2002. Guidelines for site survey and safety. College Station, TX, Ocean Drilling Program ODP/TAMU Drilling Services Dept.

Grace, R. D. (with contributions by Cudd, B., Carden, R. S., and Shursen, J. L.), 1994. *Advanced Blowout and Well Control*. Houston, TX, Gulf Publishing Co.

Granin, N. G. and Granina, L. Z., 2002. Gas hydrates and gas venting in Lake Baikal. *Geologiya i Geofizika (Russian Geology and Geophysics)*, **43**, 629–37 (Russian edn.) / 589–97 (English edn.)

Granina, L. Z., Callender, E., Lomonosov, I. S., Mats, V. D., and Golobokova, L. P., 2001. Anomalies in the composition of Baikal pore waters. *Geologija i Geofyzika (Russian Geology and Geophysics)*, **42**, 360–70 (Russian edn.)/362–72 (English edn.)

Grant, A. C., Levy, E. M., Lee, K., and Moffat, J. D., 1986. Pisces IV research submersible finds oil on Baffin Shelf. *Current Research, Part A*, Geological Survey of Canada, Paper 86-1A, 65–9.

Grant, N. J. and Whiticar, M. J., 2002. Stable carbon isotope evidence for methane oxidation in plumes above Hydrate Ridge, Cascadia Oregon Margin. *Global Biogeochemical Cycles*, **16**, 1124.

Graue, K., 2000. Mud volcanoes in deepwater Nigeria. *Marine and Petroleum Geology*, **17**, 959–74.

Gravdal, A., 1999. Kvartære sedimentasjonsprosesser i Helland-Hansen området; Sidesøkende sonar (TOBI) og seismiske undersøkelser. M. Sc. Thesis, University of Bergen. (In Norwegian.)

Gravdal, A., Haflidason, H., and Evans, D., 2003. Seabed and subsurface features on the southern Vøring Plateau and northern Storegga Slide Escarpment. In Mienert, J. and

Weaver, P. (eds.), *European Margin Sediment Dynamics, Side-Scan Sonar and Seismic Images*. Berlin, Springer-Verlag, 111–17.

Green, C. D., Heijna, B., and Walker, P., 1985. An integrated approach to the investigation of new development areas. *Offshore Site Investigation*. Advances in Underwater Technology and Offshore Engineering Series, No. 3. London, Graham and Trotman, 99–120.

Greinert, J. and Nützel, B., 2004. Hydroacoustic experiments to establish a method for the flux determination of methane bubbles at cold seeps. *Geo-Marine Letters*, **24**, 75–85.

Greinert, J., Bollwerk, S. M., Derkachev, A., Bohrmann, G., and Suess, E., 2002a. Massive barite deposits and carbonate mineralization in the Derugin Basin, Sea of Okhotsk: precipitation processes at cold seep sites. *Earth & Planetary Science Letters*, **203**, 165–80.

Greinert, J., Bohrmann, G., and Elvert, M., 2002b. Stromatolitic fabric of authigenic carbonate crusts: results of anaerobic methane oxidation at cold seeps in 4850 m water depth. *Geologische Rundschau*, **91**, 698–711.

Grozic, J. L. H., 2003. Liquefaction potential of gassy marine sands. In Locat, J. and Mienert, J. (eds.), *Submarine Mass Movements and Their Consequences*. Dordrecht, Kluwer Academic Publishers, 37–45.

Gubbay, S., 2002. *The Offshore Directory: Review of a Selection of Habitats, Communities and Species of the North-East Atlantic*. Report for the World Wide Fund for Nature, October. Godalming, Surrey, WWF-UK.

Gudmestad, O. T. and Hovland, M., 1986. Procedure for draining off shallow gas from the seabed and an arrangement for execution of the procedure. United States Patent No. 4,569,618.

Guliev, I. S., 1992. *A Review of Mud Volcanism*. Baku, Geology Institute of Azerbaijan, Azerbaijan Academy of Sciences.

Guliev, I., 2002. South-Caspian depression – and intensive area of hydrocarbon fluid formation and migration. In *Gas in Marine Sediments, Abstracts of the Seventh International Conference, Baku, Azerbaijan, 7th–12th October, 2002*, 66–9.

Guliyev, I. S. and Feizullayev, A. A., 1997. *All About Mud Volcanoes*. Baku, Nafta Press.

Gunnerus, J. E., 1768. Om nogle Norske coraller, *KGL. Norske Videnskabers Selskabs Skrifter*, **4**, 38–73. (In Norwegian.)

Hackworth, M. and Aharon, P., 2000. Authigenic carbonate precipitation driven by episodic gas hydrate sublimation: evidence from the Gulf of Mexico. In *Abstracts of the Sixth International Conference on Gas in Marine Sediments*. St Petersburg, VNIIOkeangeologia, 46–7.

Hagen, R. A. and Vogt, P. R., 1999. Seasonal variability of shallow biogenic gas in Chesapeake Bay. *Marine Geology*, **158**, 75–88.

Hamilton, T. S. and Cameron, B. E. B., 1989. Hydrocarbon occurrence in the western margin of the Queen Charlotte Basin. *Bulletin of Canadian Petroleum Geology*, **37**, 443–66.

Hampton, L. D. and Anderson, A. L., 1974. Acoustics and gas in sediments. In Kaplan, I. R. (ed.), *Natural Gases in Marine Sediments*. Marine Science Series No. 3. New York, Plenum Press, 249–74.

Hampton, M. A. and Winters, W. J., 1981. Environmental geology of Shelikof Strait, OCS sale area 60, Alaska. Proceedings of the Offshore Technology Conference, Houston, TX, OTC Paper 4118.

Hampton, M. A., Lee, H. J., and Locat, J., 1996. Submarine landslides. *Reviews of Geophysics*, **34**, 33–59.

Hanken, N.-M., Rønholt, G., and Hovland, M., 1999. Dannelsen av "Blow-out pipes" basert på studier av Plio-Pleistocen sedimenter på Rhodos. Abstracts of the Norwegian Geological Union, Vintermøte Conference. Stavanger, Norway, January, 52. (In Norwegian.)

Hannington, M., Herzig, P., Stoffers, J., *et al.*, 2001. First observations of high-temperature submarine hydrothermal vents and massive anhydrite deposits off the north coast of Iceland. *Marine Geology*, **177**, 199–220.

Hannington, M. D., Jonasson, I. R., Herzig, P. M., and Petersen, S., 1995. Physical and chemical processes of seafloor mineralization at mid-ocean ridges. In Humphris, S. E., Zierenberg, R. A., Mullineaux, L. S., and Thomsen, R. E. (eds.), *Seafloor Hydrothermal Systems: Physical, Chemical, Biological, and Geological Interactions*. American Geophysical Union, Geophysical Monograph 91, 115–57.

Hanor, J. S., 1978. Precipitation of beach rock cements: mixing of marine and meteoric waters vs. CO_2-degassing. *Journal of Sedimentary Petrology*, **48**, 489–501.

Hansen, J. M., 1988. *Koraller i Kattegat. Kortlægning fase 1*. Internal report, Copenhagen, Miljøministeriet (Danish Ministry of the Environment). (In Danish.)

Harrington, J. F. and Horseman, S. T., 1999. Gas transport properties of clays and mudrocks. In Aplin, A. C., Fleet, A. J., and Macquaker, J. H. S. (eds.), *Muds and Mudstones: Physical and Fluid Flow Properties.* Geological Society of London, Special Publication 158, 107–24.

Harrington, P. K., 1985. Formation of pockmarks by pore-water escape. *Geo-Marine Letters*, **5**, 193–7.

Harris, R. N., Fisher, A. T., and Chapman, D. S., 2002. Fluid flow through seamounts: patterns of flow and implications for global fluid flux and heat loss. *EOS – Transactions of the American Geophysical Union*, Fall Meeting Supplement, Abstract T52F-05.

Hart, B. S. and Hamilton, T. S., 1993, High-resolution acoustic mapping of shallow gas in unconsolidated sediments beneath the Strait of Georgia, British Columbia. *Geo-Marine Letters*, **13**, 49–55.

Hart, B. S. and Olynyk, H. W. (Terra Surveys Ltd.), 1994, The Roberts Bank failure deposit, Fraser River Delta, British Columbia, Report to the Geological Survey of Canada. Sidney, BC, Terra Surveys Ltd.

Hartmann, M., Scholten, J. C., and Stoffers, P., 1998. Hydrographic structure of brine-filled deeps in the Red Sea: correction of Atlantis II Deep temperatures. *Marine Geology*, **144**, 331–2.

Hashimoto, J., Miura, T., Fujikura, K., and Ossaka, J., 1993. Discovery of vestimentiferan tube-worms in the euphotic zone. *Zoological Science*, **10**, 1063–7.

Hasiotis, T., Papatheodorou, G., Kastanos, N., and Ferentinos, G., 1996. A pockmark field in the Patras Gulf (Greece) and its activation during the 14/7/93 seismic event. *Marine Geology*, **130**, 333–44.

Hay, A. E., 1996. Remote acoustic image of the plume from a submarine spring in an arctic fjord. *Science*, **225**, 1154–6.

Haymon, R. M., Fornari, D. J., Damm, K. L. von, *et al.*, 1993. Volcanic eruption of the mid-ocean ridge along the East Pacific Rise crest at 9°45–52′ N: direct submersible observations of seafloor phenomena associated with an eruption event in April, 1991. *Earth & Planetary Science Letters*, **119**, 85–101.

Heaton, T. H. and Hartzell, S. H., 1987. Earthquake hazards in the Cascadia Subduction Zone. *Science*, **236**, 162–8.

Hedberg, H. D., 1974. Role of methane generation to undercompacted shales, shale diapirs and mud

volcanoes. *American Association of Petroleum Geologists (Bulletin)*, **58**, 661–73.

Hedberg, H. D., 1980. Methane generation and petroleum migration. In Roberts, W. H. III, and Cordell, R. J. (eds.), *Problems of Petroleum Migration.* American Association of Petroleum Geologists, Studies in Geology No. 10, 179–206.

Hedberg, H. D., 1996. Utilization of hydrocarbon seep information. In Schumacher, D. and Abrams, M. A. (eds.), *Hydrocarbon migration and its near-surface expression.* American Association of Petroleum Geologists Memoir 66, iii (Forward).

Heeschen, K. U., Tréhu, A. M., Collier, R. W., Suess, E., and Rehder, G., 2003. Distribution and height of methane bubble plumes on the Cascadia Margin characterized by acoustic imaging. *Geophysical Research Letters*, **30**, 1643.

Heezen, B. C. and Hollister, C. D., 1971. *The Face of the Deep.* New York, Oxford University Press.

Heezen, B. C., Matthews, J. L., Catalano, R., *et al.*, 1973. Western Pacific guyots. In Kaneps, A. G. (ed.), *Initial Reports of the Deep Sea Drilling Project.* Washington, DC, US Government Printing Office, vol. 20, part 3, 653–723.

Heggland, R., 1997. Detection of gas migration from a deep source by the use of exploration 3D seismic data. *Marine Geology*, **137**, 41–7.

Heggland, R., 1998. Gas seepage as an indicator of deeper prospective reservoirs. A study based on exploration 3D seismic data. *Marine and Petroleum Geology*, **15**, 1–9.

Heggland, R., Meldahl, P., de Groot, P., and Aminzadeh, F., 2000. Chimneys in the Gulf of Mexico. *The American Oil and Gas Reporter*, **43**, 78–83.

Heggland, R., Nygaard, E., and Gallagher, J. W., 1996. Techniques and experience using exploration 3D seismic data to map drilling hazards. Proceedings of the Offshore Technology Conference, Houston, TX, OTC Paper 7968.

Hein, R., Crutzen, P. J., and Heinmann, M., 1997. An inverse modeling approach to investigate the global atmospheric methane cycle. *Global Biogeochemical Cycles*, **11**, 43–76.

Hekinian, R., 1984. Undersea volcanoes. *Scientific American*, **251**(1), 46–55.

HELCOM, 1996. *Coastal and Marine Protected Areas in the Baltic Sea Region.* Baltic Sea Environment Proceedings, No. 63.

Hempel, P., Spiess, V., and Schreiber, R., 1994. Expulsion of shallow gas in the Skagerrak – evidence from sub-bottom profiling, seismic, hydroacoustical and geochemical data. *Estuarine, Coastal and Shelf Sciences*, **38**, 587–601.

Henriet, J.-P. and Mienert, J., 1998. *Gas Hydrates Relevance to World Margin Stability and Climatic Change.* Geological Society of London, Special Publication 137.

Henriet, J.-P., De Mol, B., Pillen, S., *et al.*, 1998. Gas hydrate crystals may help build reefs. *Nature*, **391**, 648.

Henriet, J.-P., De Mol, B., Vanneste, M., *et al.*, 2001. In Shannon, P. M., Haughton, P. D. W., and Corcoran, D. V. (eds.), *The Petroleum Exploration of Ireland's Offshore Basins.* Geological Society of London, Special Publication 188, 375–83.

Henriet, J.-P., Rooij, D. Van, Huvenne,V., De Mol, B., and Guidard, S., 2003. Mounds and sediment drift in the Porcupine Basin west of Ireland. In Mienert, J. and Weaver, P. (eds.), *European Margin Sediment Dynamics, Side-Scan Sonar and Seismic Images.* Berlin, Springer-Verlag, 217–20.

Henry, P., Le Pichon, X., Lallemant, S., *et al.*, 1990. Mud volcano field seaward of the Barbados accretionary complex: a deep-tow side-scan sonar survey. *Journal of Geophysical Research*, **95**, 8917–29.

Henry, P., Foucher, J.-P., Le Pichon, X., *et al.*, 1992. Interpretation of temperature measurements from the Kaiko–Nankai cruise: modeling fluid flow in clam colonies. *Earth & Planetary Science Letters*, **109**, 355–71.

Henry, P., Le Pichon, X., Lallemant, S., *et al.*, 1996. Fluid flow in and around a mud volcano field seaward of the Barbados accretionary wedge: results from Manon cruise. *Journal of Geophysical Research*, **101**, 20297–323.

Hess, H. H., 1946. Drowned ancient islands of the Pacific Basin. *American Journal of Science*, **244**, 722–91.

Hessler, R. R. and Kaharl, V. A., 1995. The deep-sea hydrothermal vent community: an overview. In Humphris, S. E., Zierenberg, R. A., Mullineaux, L. S., and Thomsen, R. E. (eds.), *Seafloor Hydrothermal Systems: Physical, Chemical, Biological, and Geological Interactions.* American Geophysical Union, Geophysical Monograph 91, 72–84.

Heyer, J. and Berger, U., 2000. Methane emission from the coastal area in the southern Baltic Sea. *Estuarine, Coastal and Shelf Science*, **51**, 13–30.

Higgins, G. E. and Saunders, J. B., 1967. Report on 1964 Chatham Mud Island, Erin Bay, Trinidad, West Indies.

American Association of Petroleum Geologists (Bulletin), **51**, 55–64.

Hill, J. M., Halka, J. P., Conkwright, R., Koczot, K., and Coleman, S., 1992. Distribution and effects of shallow gas on bulk estuarine sediment properties. *Continental Shelf Research*, **12**, 1219–29.

Hillier, R. D. and Cosgrove, J. W., 2002. Core and seismic observations of overpressure-related deformation within Eocene sediments of the Outer Moray Firth. *Petroleum Geoscience*, **8**, 141–9.

Hinchcliffe, J. C., 1978. Death stalks the secret coast. *Triton*, **23**, 56–7.

Ho, C. K. and Webb, S. W., 1998. Capillary barrier performance in heterogeneous porous media. *Water Resources Research*, **34**, 603–9.

Holbrook, P., 1999. A simple closed form force balanced solution for pore pressure, overburden and the principal effective stresses in the Earth. *Marine and Petroleum Geology*, **16**, 303–19.

Holbrook, W. S., Hoskins, H., Wood, W. T., *et al.*, 1996. Methane hydrate and free gas on the Blake Ridge from vertical seismic profiling. *Science*, **273**, 1840–3.

Holbrook, W. S., Lizarralde, D., Pecher, I. A., *et al.*, 2002. Escape of methane gas through sediment waves in a large methane hydrate province. *Geology*, **30**, 467–70.

Holden, C., 1996. The next Hawaiian island? *Science*, **273**, 1177.

Holmes, R., 1977. *Quaternary Deposits of the Central North Sea, 5: the Quaternary Geology of the UK Sector of the North Sea Between 56° and 58° N.* Institute of Geological Sciences, Report No. 77/14.

Holtedahl, H., 1993. *Marine Geology of the Norwegian Continental Margin.* Trondheim, Norges Geologiske Undersøkelse, Special Publication No. 6.

Horibe, Y., Kim, K., and Craig, H., 1983. Off-ridge submarine hydrothermal vents: back-arc spreading centres and hotspot seamounts. *EOS – Transactions of the American Geophysical Union*, **64**, 724.

Horibe, Y., Kim, K., and Craig, H., 1986. Hydrothermal methane plumes in the Mariana Back-arc spreading centre. *Nature*, **324**, 131–3.

Horikoshi, M. and Ishii, T., 1985. Cold seep communities at 3800 and 5850 m off Japan. *Deep-Sea Newsletter*, 16–17.

Hornafius, J. S., Quigley, D., and Luyendyk, B. P., 1999. The world's most spectacular marine hydrocarbon seeps (Coal Oil Point, Santa Barbara Channel, California): quantification of emissions. *Journal of Geophysical Research*, **104**, 20703–11.

Horseman, S. T., Harrington, J. F., and Sellin, P., 1999. Gas migration in clay barriers. *Engineering Geology*, **54**, 139–49.

Horvitz, L., 1980. Near-surface evidence of hydrocarbon movement from depth. In Roberts, W. H., III, and Cordell, R. J. (eds.), *Problems of Petroleum Migration*. American Association of Petroleum Geologists, Studies in Geology, 10, 241–69.

Hotita, J. and Berndt, M. E., 1999. Abiogenic methane formation and isotopic fractionation under hydrothermal conditions. *Science*, **285**, 1055–7.

Houghton, J. T., Meira Filho, L. G., Callander, B. A., *et al.*, 1996. *Climate Change 1995: the Science of Climate Change*, Cambridge, Cambridge University Press (For the Inter-governmental Panel on Climate Change).

House, C. H., Orphan, V. J., McKeegan, K. D., and Hinrichs, K.-U., 2001. Methane-consuming microbial consortia identified and studied using a novel combination of fluorescent *in-situ* hybridization and ion microprobe δ^{13}C analysis. Conference Programmes with Abstracts, Earth System Processes Conference, June, Geological Society of America and Geological Society of London, Edinburgh, 24–28 June, (abstract) 65–6.

Hovland, M., 1981. Characteristics of pockmarks in the Norwegian Trench. *Marine Geology*, **39**, 103–17.

Hovland, M., 1983. Elongated depressions associated with pockmarks in the western slope of the Norwegian Trench. *Marine Geology*, **51**, 35–46.

Hovland, M., 1987. Shallow gas drainage. The Norwegian Petroleum Directorate, Seminar on Shallow Gas, Stavanger, 12 September.

Hovland, M., 1988. Organisms: the only cause of scattering layers? *EOS – Transactions of the American Geophysical Union*, **69**, 760.

Hovland, M., 1990a. Suspected gas-associated clay diapirism on the seabed off mid Norway. *Marine and Petroleum Geology*, **7**, 267–76.

Hovland, M., 1990b. Do carbonate reefs form due to fluid seepage? *Terra Nova*, **2**, 8–18.

Hovland, M., 1991. Large pockmarks, gas-charged sediments and possible clay diapirs in the Skagerrak. *Marine and Petroleum Geology*, **8**, 311–16.

Hovland, M., 1998. Seabed pockmarks on the Helike Delta front. In Katsonopoulou, D., Schildardi, D., and Soter, S. (eds.), *Helike II, Ancient Helike and Aigialeia*. Aigon, Dora Kastopoulou.

Hovland, M., 1999. Coral culprits. *New Scientist*, 27 February, 54–5.

Hovland, M., 2002. On the self-sealing nature of marine seeps. *Continental Shelf Research*, **22**, 2387–94.

Hovland, M., 2005. Gas Hydrates. In Selley, R. C., Cocks, L. R. M., and Plimer, I. R. (eds.), *Encyclopedia of Geology*. Amsterdam, Elsevier, vol. 4, 261–8.

Hovland, M. and Curzi, P., 1989. Gas seepage and assumed mud diapirism in the Italian central Adriatic Sea. *Marine and Petroleum Geology*, **6**, 161–9.

Hovland, M. and Gudmestad, O. T., 2001. Potential influence of gas hydrates on seabed installations. In Paull, C. K. and Dillon, W. P. (eds.), *Natural Gas Hydrates*. American Geophysical Union, Geophysical Monograph 124, 300–9.

Hovland, M. and Irwin, H., 1989. Hydrocarbon leakage, biodegradation and the occurrence of shallow gas and carbonate cement. Proceedings of the Shallow Gas and Leaky Reservoirs Conference, Norwegian Petroleum Society, Stavanger, Norway, 10–11 April.

Hovland, M. and Judd, A. G., 1988. *Seabed Pockmarks and Seepages: Impact on Geology, Biology and the Marine Environment*. London, Graham and Trotman Ltd.

Hovland, M. and Mortensen, P. B., 1999. *Norske korallrev og prosesser i havbunnen (Norwegian coral reefs and seabed processes)*. Bergen, John Grieg Forlag. (In Norwegian with an English summary.)

Hovland, M. and Risk, M., 2003. Do Norwegian deep-water coral reefs rely on seeping fluids? *Marine Geology*, **198**, 83–96.

Hovland, M. and Sommerville, J. H., 1985. Characteristics of two natural gas seepages in the North Sea. *Marine and Petroleum Geology*, **20**, 319–26.

Hovland, M. and Thomsen, E., 1989. Hydrocarbon–based communities in the North Sea? *Sarsia*, **74**, 29–42.

Hovland, M. and Thomsen, E., 1997. Cold-water corals – are they hydrocarbon seep related? *Marine Geology*, **137**, 159–64.

Hovland, M., Judd, A. G., and King, L. H., 1984. Characteristic features of pockmarks on the North Sea floor and Scotian Shelf. *Sedimentology*, **31**, 471–80.

Hovland, M., Talbot, M., Olaussen, S., and Aasberg, L., 1985. Recently-formed methane-derived carbonate from the North Sea floor. In Thomas, B. M. (ed.), *Petroleum Geochemistry in Exploration of the Norwegian Shelf*. London, Graham and Trotman, for the Norwegian Petroleum Society, 263–6.

Hovland, M., Talbot, M., Qvale, H., Olaussen, S., and Aasberg, L., 1987. Methane-related carbonate cements in pockmarks of the North Sea. *Journal of Sedimentary Petrology*, **57**, 881–92.

Hovland, M., Judd, A. G., and Burke, R. A., jr, 1993. The global flux of methane from shallow submarine sediments. *Chemosphere*, **26**, 559–78.

Hovland, M., Croker, P., and Martin, M., 1994. Fault-associated seabed mounds (carbonate knolls?) off western Ireland and north-west Australia. *Marine and Petroleum Geology*, **11**, 232–46.

Hovland, M., Lysne, D., and Whiticar, M. J., 1995. Gas hydrate and sediment gas composition, ODP Hole 892A, offshore Oregon, USA. In Carson, B., Westbrook, G. K., and Musgrave, R. J. (eds.), *Proceedings Ocean Drilling Program, Scientific Results, Leg 146*. College Station, TX, ODP, 151–61.

Hovland, M., Gallagher, J. W., Clennell, M. B., and Lekvam, K., 1997a. Gas hydrate and free gas volumes in marine sediments: example from the Niger Delta front. *Marine and Petroleum Geology*, **14**, 245–55.

Hovland, M., Hill, A., and Stokes, D., 1997b. The structure and geomorphology of the Dashgil Mud Volcano, Azerbaijan. *Geomorphology*, **21**, 1–15.

Hovland, M., Mortensen, P. B., Brattegard, T., Strass, P., and Rokoengen, K., 1998a. Ahermatypic coral banks off mid-Norway: evidence for a link with seepage of light hydrocarbons. *Palaios*, **13**, 189–200.

Hovland, M., Nygaard, E., and Thorbjørnsen, S., 1998b. Piercement shale diapirism in the deep-water Vema Dome area, Vøring Basin, offshore Norway. *Marine and Petroleum Geology*, **15**, 191–201.

Hovland, M., Løseth, H., Bjørkum, P. A., Wensaas, L., and Arntsen, B., 1999a. Seismic detection of shallow high pressure zones. *Offshore*, December, 94–6.

Hovland, M., Francis, T. J. G., Claypool, G. E., and Ball, M. M., 1999b. Strategy for scientific drilling of marine gas hydrates. *Joides Journal*, **25**, 20–4.

Hovland, M., Svensen, H., Forsberg, C. F., *et al.*, 2005. Complex pockmarks with carbonate-ridges off mid-Norway: products of sediment degassing. *Marine Geology*, **218**, 191–206.

Hovland, M., Kuznetsova, T., Rueslåtten, H., *et al.*, 2006. Sub-surface precipitation of salts in supercritical seawater. *Basin Research*, DOI: 10.1111/j1365–2117. 2006. 00290.

Howe, J. A., Shimmield, T., Austin, W. E. N., and Longva, O., 2002. Post-glacial depositional environments in a mid-high latitude glacially overdeepened sea loch, inner Loch Etive, western Scotland. *Marine Geology*, **185**, 417–33.

Howell, D. G. (ed.), 1993. *The Future of Energy Gases*. United States Geological Survey Professional Paper 1570.

Huang, B. J., Xiao, X. M., and Dong, W. L., 2002. Multiphase natural gas migration and accumulation and its relationship to diapir structures in the DF1-1 gas field, South China Sea. *Marine and Petroleum Geology*, **19**, 861–72.

Hunt, J. M., 1997. *Petroleum Geochemistry and Geology*. New York, W. H. Freeman.

Hurdle, B. G. (ed.), 1986. *The Nordic Seas*. New York, Springer.

Hutchinson, D. R., Grow, J. A., Klitgord, K. D., and Swift, B. A., 1982. Deep structure and evolution of the Carolina Trough. In Watkins, J. S. and Drake, C. L. (eds.), *Studies in Continental Margin Geology*. American Association of Petroleum Geologists Memoir 34, 129–52.

Huvenne, V. A. I., Blondel, Ph., and Henriet, J.-P., 2002. Textural analyses of sidescan sonar imagery from two mound provinces in the Porcupine Seabight. *Marine Geology*, **189**, 323–41.

Hyndman, R. and Davis, E., 1992. A mechanism for the formation of methane hydrate and seafloor bottom simulating reflectors by vertical fluid expulsion. *Journal of Geophysical Research*, **97**, 7025–41.

Imray, J. F., 1868. *Sailing Directions for the West Coast of North America, Part 1*. London, James Imray.

ISA (International Seabed Authority), 2003. http://www.isa.org.jm/ (accessed 16th January 2004.)

Isacks, B. and Barazangi, M., 1977. Geometry of Benioff zones: lateral segmentation and downwards bending of the subducted lithosphere. In Talwani, M. and Pitman, W. C., III (eds.), *Island Arcs, Deep Sea Trenches and Back Arc Basins*. Washington, DC, American Geophysical Union, 99–114.

Isaksen, G. H., Wall, G. R., Thomsen, M. A., *et al.*, 2001. Application of petroleum seep technology in mitigating the risk of source-rock adequacy and yield-timing in a frontier basin: the Rockall Trough, UK. Conference Proceedings with Abstracts, Earth System Processes Conference, Geological Society of America and Geological Society of London, Edinburgh, 24–28 June, IOS (abstract).

Ishibashi, J.-I. and Urabe, T., 1995. Hydrothermal activity related to arc-backarc magmatism in the western Pacific.

In Taylor, B. (ed.), *Backarc Basins: Tectonics and Magmatism*. New York, Plenum Press, 451–95.

Ivanov, M. V., Polikarpov, G. G., Lein, A.Yu., *et al.*, 1991. Biogeochemistry of the carbon cycle in the region of methane seeps of the Black Sea. *Doklady Akademii Nauk SSSR*, **320**, 1235–40. (In Russian.)

Ivanov, M. V., Lien, A.Yu., and Galchenko, V. F., 1993. The oceanic global methane cycle. In Oremland, R. S. (ed.), *Biogeochemistry of Global Change: Radiatively Active Trace Gases*. New York, Chapman & Hall, 505–20.

Ivanov, M. V., Limonov, A. F., and Weering, T. C. E. Van, 1996. Comparative characteristics of the Black Sea and Mediterranean Ridge mud volcanoes. *Marine Geology*, **132**, 253–71.

Iversen, N. and Jørgensen, B. B., 1985. Anaerobic methane oxidation rates at the sulfate–methane transition in marine sediments from Kattegat and Skagerrak (Denmark). *Limnology and Oceanography*, **30**, 944–55.

Jackson, D. I., Jackson, A. A., Evans, D., *et al.*, 1995. *United Kingdom Offshore Regional Report: The Geology of the Irish Sea*. London, HMSO, for the British Geological Survey.

Jackson, D. R., Williams, K. L., Wever, T. F., Friedrichs, C. T., and Wright, L. D., 1998. Sonar evidence for methane ebullition in Eckernförde Bay. *Continental Shelf Research*, **18**, 1893–916.

Jakubov, A. A., Ali-Zade, A. A., and Zeinalov, M. M., 1971. *Mud Volcanoes of the Azerbaijan SSR: Atlas*. Baku, Academy of Sciences of Azerbaijan. (In Russian.)

Jakubov, A. A., Dadashev, F. G., and Mekhtiev, A. K., 1983. Zakonomernosti razmeshchenia gryazevykh vulkanov na dne Kaspiyskogo morya (Regular features of gas hydrate distribution on the *Caspian Sea* floor). In *Geologo-geomorfologischeskie issedovania Kaspiyskogo morya* (*Geological and Geomorphological Studies in the Caspian Sea*). Moscow, Nauka Press, 70–2. (In Russian.)

Jamtveit, B., Svensen, H., Podladchikov, Y., and Planke, S., 2004. Hydrothermal vent complexes associated with sill intrusions in sedimentary basins. In Breitkreutz, C. and Petford, N. (eds.), *Physical Geology of High-Level Magmatic Systems*. Geological Society of London, Special Publication 234, 233–41.

Janecky, D. J. and Seyfried, W. E., jr, 1984. Formation of massive sulfide deposits on oceanic ridge crests: incremental reaction models for mixing between hydrothermal solutions and seawater. *Geochimica et cosmochimica acta*, **48**, 2723–38.

Jannasch, H. W., 1983. Microbial processes at deep sea hydrothermal vents. In Rona, P. A., Bostrøm, K., Laubier, L., and Smith, K. L., jr (eds.), *Hydrothermal Processes at Seafloor Spreading Centers*. New York, Plenum Press, 677–709.

Jannasch, H. W., 1995. Microbial interactions with hydrothermal fluids. In Humphris, S. E., Zierenberg, R. A., Mullineaux, L. S., and Thomsen, R. E. (eds.), *Seafloor Hydrothermal Systems: Physical, Chemical, Biological, and Geological Interactions*. American Geophysical Union, Geophysical Monograph **91**, 273–96.

Jansen, J. H. F., 1976. Late Pleistocene and Holocene history of the northern North Sea, based on acoustic reflection records. *Netherlands Journal of Sea Research*, **10**, 1–43.

Jayakumar, D. A., Naqvi, S. W. A., Narvekar, P. V., and George, M. D., 2001. Methane in coastal and offshore waters of the Arabian Sea. *Marine Chemistry*, **74**, 1–13.

Jenden, P. D., Hilton, D. R., Kaplan, I. R., and Craig, H., 1993. In Howells, D. G. (ed.), *The Future of Energy Gases*. United States Geological Survey Professional Paper 1570, 31–56.

Jensen, P., 1986. Nematode fauna in the sulphide-rich brine seep and adjacent bottoms of the East Flower Garden, NW Gulf of Mexico. *Marine Biology*, **92**, 489–503.

Jensen, P., Aagaard, I., Burke, R. A., jr, *et al.*, 1992. 'Bubbling reefs' in the Kattegat: submarine landscapes of carbonate-cemented rocks support a diverse ecosystem at methane seeps. *Marine Ecology Progress Series*, **83**, 103–12.

Jerosch, K., Allais, A.-G., Schlüter, M., and Foucher, J.-P., 2004. Geo-referenced video-mosaicking as a means to GIS-supported sea floor community mapping at Håkon Mosby mud volcano. First General Congress, European Geosciences Union, Nice, 25–30 April (poster). (See *Geophysical Research Abstracts*, **6**, 00669.)

Johannes, R. E., 1980. The ecological significance of the submarine discharge of groundwater. *Marine Ecology Progress Series*, **3**, 365–73.

Johnson, H., Richards, P. C., Long, D., and Graham, C. C., 1993. *United Kingdom Offshore Regional Report: The Geology of the Northern North Sea*. London, HMSO, for the British Geological Survey.

Johnson, T. C., 1980. Late-glacial and post glacial sedimentation in Lake Superior based on seismic reflection profiles. *Quaternary Research*, **13**, 380–91.

Johnstone, C. M., Turnbull, C. G., and Tasker, M. L., 2002. Natura 2000 in UK offshore waters: advice to support the implementation of the EC Habitats and Birds Directives in UK offshore waters. Peterborough, Joint Nature Conservation Council Report 325.

Jollivet, D., Blanc, G., Faugeres, J.-C., Griboulard, R., and Desbruyères, D., 1990. Composition and spatial organization of a cold seep community on the South Barbados Accretionary Prism: tectonic, geochemical and sedimentary context. *Progress in Oceanography*, 24, 25–45.

Jolly, R. J. H. and Lonergan, L., 2002. Mechanisms and controls on the formation of sand intrusions. *Journal of the Geological Society, London*, 159, 605–17.

Jones, B., Renaut, R. W., and Rosen, M. R., 1997. Vertical zonation of biota in microstromatolites associated with hot springs, North Island, New Zealand. *Palaios*, 12, 220–36.

Jones, G. B., Floodgate, G. D., and Bennell, J. D., 1986. Chemical and microbiological aspects of acoustically turbid sediments: preliminary investigations. *Marine Geotechnolology*, 6, 315–32.

Jones, M. E., 1994. Mechanical principles of sediment deformation. In Maltman, A. J. (ed.), *Geological Deformation of Sediments*. London, Chapman and Hall, 36–71.

Jones, M. L., 1981. *Riftia pachyptila*, a new genus, new species: the vestimentiferan worm from the Galapagos Rift geothermal vents (Pogonophora). *Proceedings of the Biological Society of Washington*, 93, 1295–313.

Jones, M. L., 1985. On the Vestimentifera, new phylum: six new species, and other taxa, from hydrothermal vents and elsewhere. In Jones, M. L. (ed.), *Hydrothermal Vents of the Eastern Pacific: an Overview*. Bulletin of the Biological Society of Washington, No. 6, 117–58.

Jones, R. D. and Amador, J. A., 1993. Methane and carbon monoxide production, oxidation, and turnover times in the Caribbean Sea as influenced by the Orinoco River. *Journal of Geophysical Research*, 98, 2353–9.

Jones, R. W., 1993. Preliminary observations on benthonic formanifera associated with biogenic gas seep in the North Sea. In Jenkins, D. G. and Graham, D. (eds.), *Applied Micropaleontology*. Dordrecht, Kluwer Academic Publishers, 69–91.

Jones, R. W., 1996. *Micropalaeontology in Petroleum Exploration*. Oxford, Oxford University Press.

Jones, V. T., III, Matthews, M. D., and Richers, D. M., 1999. Light hydrocarbons for petroleum and gas prospecting.

In Hale, M. and Govett, G. J. S. (eds.), *Geochemical Remote Sensing of the Subsurface*. Handbook of Exploration Geochemistry, 7, 133–211.

Jonsson, V. K. and Matthiasson, M., 1974. Cooling the Heimaey lava with water – report on the operation. *Journal of the Engineering Association of Iceland*, 59, 70–83. (In Icelandic.)

Josenhans, H. W. and Zevenhuizen, J., 1983. Pockmarks on the Labrador Shelf triggered or caused by iceberg scouring. Proceedings of the Conference on Geotechnical Practice in Offshore Engineering, Austin, TX, 27–29 April.

Josenhans, H. W., King, L. H., and Fader, G. B. J., 1978. A side scan sonar mosaic of pockmarks on the Scotian Shelf. *Canadian Journal of Earth Sciences*, 15, 831–40.

Jouzel, J., Barkov, N. I., Barnola, J. M., *et al.*, 1993. Extending the Vostok ice-record of palaeo-climate to the penultimate glacial period. *Nature*, 364, 407–11.

Joye, S. B., Boetius, A., Orcutt, B. N., *et al.*, 2004. The anaerobic oxidation of methane and sulfate reduction in sediments from Gulf of Mexico cold seeps. *Chemical Geology*, 205, 219–38.

Judd, A. G., 1982. Computerised marine geophysical and geotechnical mapping techniques and their application to a study of pockmarks. Ph.D. thesis, University of Newcastle upon Tyne.

Judd, A. G., 1990. Shallow gas and gas seepages: a dynamic process? In Ardus, D. A. and Green, C. D. (eds.), *Safety in Offshore Drilling: the Role of Shallow Gas Surveys*. Dordrecht, Kluwer Academic Publishers, 27–50.

Judd, A. G., 1995. Gas and Gas Mobility in the Offshore Sediments of the Fraser Delta, British Columbia. Canada, Sidney, BC, Pacific Geoscience Centre, Geological Survey of Canada, Open File Report.

Judd, A. G., 2000. Geological sources of methane. In Khalil, M. A. K. (ed.), *Atmospheric Methane: Its Role in the Global Environment*. Berlin, Springer-Verlag, 280–303.

Judd, A. G., 2001. Pockmarks in the UK sector of the North Sea. Technical report TR_002, Strategic Environmental Assessment of parts of the North Sea (SEA2), Department of Trade and Industry, September.

Judd, A. G., 2004. Natural seabed gas seeps as sources of atmospheric methane. *Environmental Geology*, 46, 988–96.

Judd, A. G., 2005a. Gas emissions from mud volcanoes. Significance to global climate change. In Martinelli, G. and Panahi, B. (eds.), *Mud Volcanoes, Geodynamics and Seismicity*. NATO Science Series, IV. Earth and

Environmental Sciences, 51, Dordrecht, Springer, 147–57.

Judd, A. G., 2005b. The distribution and extent of methane-derived authigenic carbonate. Technical report, Strategic Environmental Assessment of the Irish Sea (SEA6), Department of Trade and Industry, March.

Judd, A. G. and Hovland, M., 1989. The role of chemosynthesis in supporting fish stocks in the North Sea. *Journal of Fish Biology*, **35** (Supplement A), 329–30.

Judd, A. G. and Hovland, M., 1992. The evidence of shallow gas in marine sediments. *Continental Shelf Research*, **12**, 1081–96.

Judd, A. G., Long, D., and Sankey, M., 1994. Pockmark formation and activity, UK block 15/25, North Sea. *Bulletin of the Geological Society of Denmark*, **41**, 34–49.

Judd, A. G., Davies, G., Wilson, J., *et al.*, 1997. Contributions to atmospheric methane by natural seepages on the UK continental shelf. *Marine Geology*, **140**, 427–55.

Judd, A. G., Hovland, M., Dimitrov, L. I., García-Gil, S., and Jukes, V., 2002a. The geological methane budget at continental margins and its influence on climate change. *Geofluids*, **2**, 109–26.

Judd, A. G., Sim, R., Kingston, P., and McNally, J., 2002b. Gas seepage on an intertidal site: Torry Bay, Firth of Forth, Scotland. *Continental Shelf Research*, **22**, 2317–31.

Juhl, A. R. and Targon, G. L., 1993. Biology of an active methane seep on the Oregon continental shelf. *Marine Ecology Progress Series*, **102**, 287–94.

Jørgensen, B. B., Bang, M., and Blackburn, T. H., 1990. Anaerobic mineralization in marine sediments from the Baltic Sea – North Sea transition. *Marine Ecology Progress Series*, **59**, 39–54.

Jørgensen, N. O., 1976. Recent high magnesian calcite aragonite cementation of beach and submarine sediments from Denmark. *Journal of Sedimentary Petrology*, **46**, 940–51.

Jørgensen, N. O., 1979. Magnesium incorporation in Recent marine calcite cement from Denmark. *Journal of Sedimentary Petrology*, **49**, 945–50.

Jørgensen, N. O., 1980. Gypsum formation in Recent submarine sediments from Kattegat, Denmark. *Chemical Geology*, **28**, 349–53.

Jørgensen, N. O., 1981. Authigenic K-feldspar in Recent submarine gypsum concretions from Denmark. *Marine Geology*, **39**, M21–M25.

Jørgensen, N. O., 1989. Holocene methane-derived dolomite-cemented sandstone pillars from the Kattegat, Denmark. *Marine Geology*, **88**, 71–81.

Jørgensen, N. O., 1992a. Methane-derived carbonate cementation of marine sediments from the Kattegat, Denmark: geochemical and geological evidence. *Marine Geology*, **103**, 1–13.

Jørgensen, N. O., 1992b. Methane-derived carbonate cementation of marine sediments from the Kattegat, Denmark. *Continental Shelf Research*, **12**, 1209–18.

Kadko, D., Baross, J., and Alt, J., 1995. The magnitude and global implications of hydrothermal flux. In Humphris, S. E., Zierenberg, R. A., Mullineaux, L. S., and Thomsen, R. E. (eds.), *Seafloor Hydrothermal Systems: Physical, Chemical, Biological, and Geological Interactions*. American Geophysical Union, Geophysical Monograph 91, 446–66.

Kahn, L. M., Silver, E. A., Orange, D., Kochevar, R., and McAdoo, B., 1996. Surficial evidence of fluid expulsion from the Costa Rica accretionary prism. *Geophysical Research Letters*, **23**, 887–90.

Kalinko, M., 1964. Mud volcanoes, reasons of their origin, development and fading. *VNIGRI*, **40**, 30–54. (In Russian.)

Kamenev, G. M., 1991a. Macrobenthos of sublittoral Kraternaya Bight. 1. Qualitative composition and species distribution. In Zhirmusky, A. V. and Tarasov, V. G. (eds.), *Shallow-Water Hydrothermal Vents and the Ecosystem of Kraternaya Bight (Ushishir Volcano, Kurile Islands). Book II Biota*. Vladivostok, Far East Branch, USSR Academy of Sciences, 48–91. (In Russian.)

Kamenev, G. M., 1991b. Macrobenthos of sublittoral Kraternaya Bight. 2. Quantitative distribution of species and bottom communities. In Zhirmusky, A. V. and Tarasov, V. G. (eds.), *Shallow-Water Hydrothermal Vents and the Ecosystem of Kraternaya Bight (Ushishir Volcano, Kurile Islands). Book II Biota*. Vladivostok, Far East Branch, USSR Academy of Sciences, 92–137. (In Russian.)

Kamenev, G. M., Fadeev, V. I., Selein, N. I., Tarasov, V. G., and Malakhov, V. V., 1993. Composition and distribution of macro- and meiobenthos around hydrothermal vents in the Bay of Plenty (New Zealand). *New Zealand Journal of Marine and Freshwater Research*, **27**, 407–18.

Kamenev, G. M., Nadtochy, V. A., and Kuznetsov, A. P., 2001. *Conchocele bisecta* (Conrad, 1849) (Bivalvia: Thyasiridae) from cold-water methane-rich areas of the Sea of Okhotsk. *The Veliger*, **44**, 84–94.

Karisiddaiah, S. M. and Veerayya, M., 1994. Methane-bearing shallow gas-charged sediments in the eastern Arabian Sea: a probable source for greenhouse gas. *Continental Shelf Research*, **14**, 1361–70.

Karisiddaiah, S. M. and Veerayya, M., 2002. Occurrence of pockmarks and gas seepages along the central western continental margin of India. *Current Science*, **82**, 52–7.

Karisiddaiah, S. M., Veerayya, M., Vora, K. H., and Wagle, B. G., 1993. Gas-charged sediments on the inner continental shelf off western India. *Marine Geology*, **110**, 143–52.

Karl, D. M., 1995. Ecology of free-living, hydrothermal vent microbial communities. In Karl, D. M. (ed.), *The Microbiology of Deep-Sea Hydrothermal Vents*. New York, CRC Press, 35–124.

Karlsson, W., 1986. The Snorre, Statfjord and Gullfaks oil fields and the habitat of hydrocarbons on the Tampen Spur, offshore Norway. In Spencer, A. M. (ed.), *Habitat of Hydrocarbons on the Norwegian Continental Shelf*. London, Graham and Trotman for the Norwegian Petroleum Society, 181–97.

Kasahara, J., 2002. Tides, earthquakes, and volcanoes. *Science*, **297**, 348–9.

Kasten, S., Hensen, C., Zabel, M., *et al.*, 2001. Gas hydrates in surface sediments of the northern Congo Fan – geochemical and microbiological characterization of the top of the gas hydrate stability zone. *European Union of Geosciences EUG XI Journal of Conference*, **6** (1), 151.

Kastner, M., Kvenvolden, K. A., and Lorenson, T. D., 1998. Chemistry, isotopic composition, and origin of a methane–hydrogen sulphide hydrate at the Cascadia subduction zone. *Earth & Planetary Science Letters*, **156**, 173–83.

Kastner, M., Kvenvolden, K. A., Whiticar, M. J., Camerlenghi, A., and Lorenson, T. D., 1995. Relation between pore fluid chemistry and gas hydrates associated with bottom-simulating reflectors at the Cascadia Margin, Sites 889 and 892. *ODP, Scientific Results*, **146**, 175–87.

Kaufmann, B., 1998. Diagenesis of middle Devonian carbonate mounds Mader basin (eastern Anti-Atlas, Morocco). *Journal of Sedimentary Research*, **67**, 945–56.

Kelley, D. S., Karson, J. A., Blackman, D. K., *et al.*, 2001. An off-axis hydrothermal vent field near the Mid-Atlantic Ridge at 30° N. *Nature*, **412**, 145–9.

Kelley, D. S., Baross, J. A., Frueh-Green, G. L., Schrenk, M. O., and Karson, J. A., 2002. The ultramafic-hosted Lost City Hydrothermal Field: clues in the search for life elsewhere in the Solar System? *EOS – Transactions of the American Geophysical Union*, **83**, abstract B62A-03.

Kelley, J. T., Dickson, S. M., Belknap, D. F., Barnhardt, W. A., and Henderson, M., 1994. Giant sea-bed pockmarks: evidence for gas escape from Belfast Bay, Maine. *Geology*, **22**, 59–62.

Kempe, S., Kazmierczak, J., Landmann, G., *et al.*, 1991. Largest known microbialites discovered in Lake Van, Turkey. *Nature*, **349**, 605–8.

Kennett, J., Cannariato, K. G., Hendy, I. L., and Behl, R. J., 2003. *Methane Hydrates in Quaternary Climate Change: the Clathrate Gun Hypothesis*. Washington, DC, American Geophysical Union.

Kennett, J., Hendy, I., and Behl, R., 1996. Late Quaternary foraminiferal carbon isotope record of Santa Barbara Basin: implications for rapid climate change. Abstracts of the Annual Meeting of the American Geophysical Union, San Francisco 15–19 December, F292 (abstract).

Kennicut, M. C., Brooks, J. M., Bidigare, R. R., *et al.*, 1985. Vent type taxa in a hydrocarbon seep region on the Louisiana Slope. *Nature*, **317**, 351–3.

Kennicut, M. C., II, Brooks, J. M., Bidigare, R. R., *et al.*, 1989. An upper slope 'cold' seep community: northern California. *Limnology and Oceanography*, **34**, 635–40.

Kenyon, N. H., Belderson, R. H., and Stride, A. H., 1982. Detailed tectonic trends on the central part of the Hellenic Outer Ridge and in the Hellenic Trench System. In Legget, J. (ed.), *Trench-Forearc Geology*. Geological Society of London, Special Publication 10, 335–43.

Kenyon, N. *et al.*, 1997. RRS Charles Darwin cruise 104 leg 2, 21 Mar– 19 Apr 1997. Geological processes in the Strait of Hormuz, Arabian Gulf: a contribution to the Scheherezade Programme. Southampton, Southampton Oceanography Centre Cruise Report No. 11.

Kenyon, N. H., Ivanov, M. K., and Akhmetzhanov, A. M., 1998. *Cold Water Carbonate Mounds and Sediment Transport on the Northeast Atlantic Margin*. Intergovernmental Oceanographic Commission, Technical series, No. 52.

Kerr, P. F., Drew, I. M., and Richardson, D. S., 1970. Mud volcano clay, Trinidad, West Indies. *American Association of Petroleum Geologists (Bulletin)*, **54**, 2101–10.

Khalil, M. A. K., 2000. Atmospheric methane: an introduction. In Khalil, M. A. K. (ed.), *Atmospheric Methane: its Role in the Global Environment*. Berlin, Springer-Verlag, 1–8.

Khalil, M. A. K. and Rasmussen, R. A., 1983. Sources, sinks and seasonal cycles of atmospheric methane. *Journal of Geophysical Research*, **88**, 5131–44.

Kholodov, V. N., 2002a. Mud volcanoes, their distribution regularities and genesis: communication I. Mud volcanic provinces and morphology of mud volcanoes. *Lithology and Mineral Resources*, **37**, 197–209. (Translated from *Litologiya i Polezne Iskopaemye*, **37**, 227–41.)

Kholodov, V. N., 2002b. Mud volcanoes, their distribution regularities and genesis: communication II. Geological–geochemical peculiarities and formation model. *Lithology and Mineral Resources*, **37**, 293–309. (Translated from *Litologiya i Polezne Iskopaemye*, **37**, 339–58.)

Kim, D., Park, S., Lee, G., and Seo, Y., 2002. Compressional wave velocity and electrical resistivity of gassy sediment in the southeastern shelf of Korea. In *Gas In Marine Sediments, Seventh International Conference, Baku, Azerbaijan, 7th–12th October 2002*. Baku, Nafta Press, 90–1.

Kim, K., Craig, H., and Horibe, Y., 1983. Methane in 'realtime' tracer for submarine hydrothermal systems. *EOS – Transactions of the American Geophysical Union*, **64**, 724 (abstract).

King, L. H. and MacLean, B., 1970. Pockmarks on the Scotian Shelf. *Geological Society of America (Bulletin)*, **81**, 3141–8.

Kinsman, B., 1965. *Wind Waves*. Englewood Cliffs, NJ Prentice-Hall.

Kipfer, R., Aeschebach-Hertig, W., Baur, H., *et al.*, 1994. Injection of mantle type helium into Lake Van (Turkey): the clue for quantifying deep water renewal. *Earth & Planetary Science Letters*, **125**, 357–70.

Kipphut, G. W. and Martens, C. S., 1982. Biogeochemical cycling in an organic-rich coastal marine basin – 3. Dissolved gas transport in methane-saturated sediments. *Geochimica et cosmochimica acta*, **46**, 2049–60.

Klaucke, I., Bohrmann, G., and Weinrebe, W., 2004. Estimation of the regional methane efflux on Hydrate Ridge, Oregon. First General Assembly of the European Geosciences Union, Nice, 25–30 April (poster). (See *Geophysical Research Abstracts*, **6**, 03489.)

Kleinberg, R. L. and Brewer, P. G., 2001. Probing gas hydrate deposits. *American Scientist*, **89**, 244–51.

Kleinschmidt, M. and Tschauder, R., 1985. Shallow crater hydrothermal vent systems off the Palos Verdes Peninsula, Los Angeles County, California. In Jones,

M. L. (ed.), *Hydrothermal Vents of the Eastern Pacific: an Overview*. Bulletin of the Biological Society of Washington, No. 6, 485–8.

Knies, J., Damm, E., Gutt, J., Mann, U., and Pinturier, L., 2004. Near-surface hydrocarbon anomalies in shelf sediments off Spitsbergen: evidences for past seepages. *Geochemistry, Geophysics, Geosystems*, **5** (6); Q06003; DOI:10.1029/2003GC000687.

Kohout, F. A., 1966. Submarine springs: a neglected phenomenon of coastal hydrology. *Hydrology*, **26**, 391–413.

Kopf, A. J., 2002, Significance of mud volcanism. *Reviews of Geophysics*, **40**, 2-1–2-52, 1005.

Kopf, A., Robertson, A. H. F., Clennell, M. B., and Flecker, R., 1998. Mechanisms of mud extrusion on the Mediterranean Ridge accretionary prism. *Geo-Marine Letters*, **18**, 97–114.

Kornacki, A. S., Kendrick, J. W., and Berry, J. L., 1994. Impact of oil and gas vents and slicks on petroleum exploration in the deepwater Gulf of Mexico. *Geo-Marine Letters*, **14**, 160–9.

Koski, R. V., Clague, D. A., Rosenbauer, R. A., *et al.*, 2002. Hydrothermal tar mounds in Escabana Trough, Southern Gorda Ridge. *EOS – Transactions of the American Geophysical Union*, **83**, Fall Meeting Supplement, Abstract V72A-1293.

Krastel, S., Spiess, V., Ivanov, M., *et al.*, 2003. Acoustic investigations of mud volcanoes in the Sorokin Trough, Black Sea. *Geo-Marine Letters*, **23**, 230–8.

Krishnamurti, R., 1968. Finite amplitude convection with changing mean temperature. *Journal of Fluid Mechanics*, **33**, 445–63.

Kristiansson, K. and Malmquist, L., 1980. A new model mechanism for the transportation of radon through the ground. Proceedings of the Society of Exploration Geophysicists 15th International Meeting, Houston, TX, 16–20 November.

Kruglyakova, R., Gubano, Y., Kruglyakov, V., and Prokoptsev, G., 2002. Assessment of technogenic and natural hydrocarbon supply into the Black Sea and seabed sediments. *Continental Shelf Research*, **22**, 2395–408.

Kugler, H. G., 1933. Contribution to the knowledge of sedimentary volcanism in Trinidad. *Journal of the Institute of Petroleum Technology, Trinidad*, **19**, 743–60.

Kuhn, T. S., 1970. *The Structure of Scientific Revolutions*. Chicago, University of Chicago Press.

Kukowski, N. and Pecher, I., 1999. Thermo-hydraulics of the Peruvian accretionary complex at 12° S. *Geodynamics*, **27**, 373–402

Kukowski, N. and Pecher, I. A., 2000. Gas hydrates in nature: results from geophysical and geochemical studies (editorial). *Marine Geology*, **164**, 1.

Kukowski, N., Schillhorn, T., Huhn, K., *et al.*, 2001. Morphotectonics and mechanics of the central Makran accretionary wedge off Pakistan. *Marine Geology*, **173**, 1–19.

Kulm, L. D. and Suess, E., 1990. Relationship between carbonate deposits and fluid venting: Oregon accretionary prism. *Journal of Geophysical Research*, **95**, 8899–915.

Kulm, L. D., Suess, E., Moore, J. C., *et al.*, 1986. Oregon Subduction Zone: venting, fauna, and carbonates. *Science*, **231**, 561–6.

Kutas, R. I., Rusakov, O. M., and Kobolev, V. P., 2002. Gas seeps in northwestern Black Sea: geological and geophysical studies. *Geologiya i Geofizika (Russian Geology and Geophysics)*, **43**, 698–705 (Russian edn.) / 664–70 (English edn.).

Kvalstad, T. J., Nadim, F., and Harbitz, C. B., 2001. Deepwater geohazards: geotechnical concerns and solutions. Proceedings of the Offshore Technology Conference, Houston, TX, OTC Paper 12958.

Kvenvolden, K. A., 1983. Marine gas hydrates – I: geochemical evidence. In Cox, J. L. (ed.), *Natural Gas Hydrates: Properties, Occurrence and Recovery*. Boston, Butterworth, 63–72.

Kvenvolden, K. A., 1999. Potential effects of gas hydrate on human welfare. *Proceedings of the National Academy of Sciences, USA*, **96**, 3420–6.

Kvenvolden, K. A., 2000. Natural gas hydrate: introduction and history of discovery. In Max, M. D. (ed.), *Natural Gas Hydrate in Oceanic and Permafrost Environments*. Dordrecht, Kluwer Academic Publishers, 9–16.

Kvenvolden, K. A., 2002. Methane hydrate in the global organic carbon cycle. *Terra Nova*, **14**, 302–6.

Kvenvolden, K. A. and Cooper, C. K., 2003. Natural seepage of crude oil into the marine environment. *Geo-Marine Letters*, **23**, 140–6.

Kvenvolden, K. A. and Lorenson, T. D., 2001. In Paull, C. K. and Dillon, W. P. (eds.), *Natural Gas Hydrates*. American Geophysical Union, Geophysical Monograph 124, 5–7.

Kvenvolden, K. A. and Redden, G. D., 1980. Hydrocarbon gas in sediment from the shelf, slope and basin of the Bering Sea. *Geochimica et cosmochimica acta*, **44**, 1145–50.

Kvenvolden, K. A., Nelson, C. H., Thor, D. R., *et al.*, 1979. Biogenic and thermogenic gas in gas-charged sediments of Norton Sound, Alaska, Proceedings of the Offshore Technology Conference, Houston, TX, OTC Paper 3412.

Kvenvolden, K. A., Lilley, M. D., Lorenson, T. D., Barnes, P. W., and McLaughlin, E., 1993. The Beaufort Sea continental shelf as a seasonal source of atmospheric methane. *Geophysical Research Letters*, **20**, 2459–62.

Kvenvolden, K. A., Lorenson, T. D., and Reeburgh, W. S., 2001. Attention turns to naturally occurring methane seepage. *EOS – Transactions of the American Geophysical Union*, **82**, 457.

Laberg, J. S. and Vorren, T. O., 1993. A late Pleistocene submarine slide on the Bear Island Trough mouth fan. *Geo-Marine Letters*, **13**, 227–34.

Lackschewitz, K. S., Kummetz, M., Ackermand, D., *et al.*, 2001. Hydrothermal alteration in the PACMANUS hydrothermal field: implications from secondary mineral assemblages and mineral chemistry, ODP Leg 193. *EOS – Transactions of the American Geophysical Union*, **82**, Fall Meeting Supplement, Abstract F589.

Lacroix, A. V., 1993. Unaccounted-for sources of fossil and isotopically-enriched methane and their contribution to the emissions inventory: a review and synthesis. *Chemosphere*, **26**, 507–57.

Laier, T., Jørgensen, N. O., Buchardt, B., Cederberg, T., and Kuijpers, A., 1992. Accumulation and seepages of biogenic gas in northern Denmark. *Continental Shelf Research*, **12**, 1173–86.

Laier, T., Kuijpers, A., Dennegård, B., and Heier-Nielsen, S., 1996. Origin of shallow gas in Skagerrak and Kattegat – evidence from stable isotopic analyses and radiocarbon dating. *NGU (Norges Geologiske Undersøkelse) Bulletin*, **430**, 129–36.

Lambert, G. and Schmidt, S., 1993. Reevaluation of the oceanic flux of methane: uncertainties and long term variations. *Chemosphere*, **26**, 579–89.

Lammers, S., Suess, E., and Hovland, M., 1995a. A large methane plume east of Bear Island (Barents Sea): implications for the marine methane cycle. *Geologische Rundschau*, **84**, 59–66.

Lammers, S., Suess, E., Mansurov, M. N., and Anikiev, V. V., 1995b. Variations in atmospheric methane supply from the Sea of Okhotsk induced by seasonal ice cover. *Global Biogeochemical Cycles*, **9**, 351–8.

Lamontagne, R. A., Swinnerton, J. W., Linnenbon, V. J., and Smith, W. D., 1973. Methane concentrations in various marine environments. *Journal of Geophysical Research*, 78, 5317–24.

Lance, S., Henry, P., Le Pichon, X., *et al.*, 1998. Submersible study of mud volcanoes seaward of the Barbados accretionary wedge: sedimentology, structure and rheology. *Marine Geology*, 145, 255–92.

Lancelot, Y. and Ewing, J. I., 1973. In Hollister, C. D., Ewing, J. I., *et al.* (eds.), *Initial Reports on Deep Sea Drilling Project*. Washington, DC, US Government Printing Office, vol. XI, 791–9.

Landes, K. K., 1973. Mother Nature as oil polluter. *American Association of Petroleum Geologists (Bulletin)*, 57, 637–41.

LaRock, P. A., Hyun, J.-H., and Bennison, B. W., 1994. Bacterioplankton growth and production at the Louisiana hydrocarbon seeps. *Geo-Marine Letters*, 14, 104–9.

Larsen, M. C., Nelson, C. H., and Thor, D. R., 1980. Sedimentary processes and potential hazards on the sea floor of northern Bering Sea. *USGS Open File Report* 80–979.

Laval, B., Cady, S. L., Pollack, J. C., *et al.*, 2000. Modern freshwater microbialite analogues for ancient dendritic reef structures. *Nature*, 407, 626–9.

Le Pichon, X., Foucher, J.-P., Boulègue, J., *et al.*, 1990. Mud volcano field seaward of the Barbados accretionary complex: a submersible survey. *Journal of Geophysical Research*, 95, 8931–43.

Le Pichon, X., Kobayashi, K., and Kaiko–Nankai Scientific Crew, 1992. Fluid venting activity within the eastern Nankai Trough accretionary wedge: a summary of the 1989 Kaiko–Nankai results. *Earth & Planetary Science Letters*, 109, 303–18.

Lee, H., Locat, J., Dartnell, P., Israel, K., and Wong, F., 1999. Regional variability of slope stability: application to the Eel Margin, California. *Marine Geology*, 154, 305–21.

Lee, M. W. and Dillon, W. P., 2001. Amplitude blanking related to pore-filling of gas hydrate in sediments. *Marine Geophysical Research*, 22, 101–9.

Lee, S. H. and Chough, S. K., 2003. Distribution and origin of shallow gas in deep-sea sediments of the Ulleung Basin, East Sea (Sea of Japan). *Geo-Marine Letters*, 22, 204–9.

Lehner, P., 1969. Salt tectonics and Pleistocene stratigraphy on continental slope of northern Gulf of Mexico. *American Association of Petroleum Geologists (Bulletin)*, 53, 2431–79.

Leifer, I. and Clark, J. F., 2002. Modelling trace gases in hydrocarbon seep bubbles: application to marine hydrocarbon seeps in the Santa Barbara Channel. *Geologiya i Geofizika (Russian Geology and Geophysics)*, 43, 613–21 (Russian edn.) / 572–9 (English edn.).

Leifer, I. and Judd, A. G., 2002. Oceanic methane layers: the hydrocarbon seep bubble deposition hypothesis. *Terra Nova*, 14, 417–24.

Leifer, I. and MacDonald, I. R., 2003. Dynamics of the gas flux from shallow gas hydrate deposits: interaction between oily hydrate bubbles and the oceanic environment. *Earth & Planetary Science Letters*, 210, 410–24.

Leifer, I. and Patro, R. K., 2002. The bubble mechanism for methane transport from the shallow sea bed to the surface: a review and sensitivity study. *Continental Shelf Research*, 22, 2409–28.

Leifer, I., Clarke, J. F., and Chen, R. F., 2000. Modifications of the local environment by natural hydrocarbon seeps. *Geophysical Research Letters*, 27, 3711–14.

Leifer, I., Clark, J. F., Luyendyk, B., and Valentine, D., 2003. Identifying future directions for subsurface hydrocarbon migration research. *EOS – Transactions of the American Geophysical Union*, 84, 364.

Leifer, I., Luyendyk, B. P., Boles, J., and Clark, J. F., 2006. Natural marine seepage blowout: contribution to atmospheric methane. *Global Biogeochemical Cycles*, 20, GB3008.10.1029/2005GB002668.

Lein, A. Yu., Pimenov, N. V., and Galchenko, V. F., 1997. Bacterial chemosynthesis and methanotrophy in the Manus and Lau basins ecosystems. *Marine Geology*, 142, 47–56.

Lelieveld, J., Crutzen, P., and Dentener, F. J., 1998. Changing concentration, lifetime and climate forcing of atmospheric methane. *Tellus*, 50B, 128–50.

León, R., Somoza, L., Ivanov, M., *et al.*, 2001. Seabed morphology and gas venting in the Gulf of Cadiz mud volcano area: imagery of multibeam data and ultra-high resolution seismic data. In Akhamanov, G. and Suzyumov, A. (eds), *Geological Processes on Deep-Water European Margins*. International Oceanographic Commission Workshop Report No. 175 on the International Conference and ninth post-cruise meeting of the Training Through Research Programme, Moscow-Mozhenka, Russia, 28 January–2 February 2001, Paris, UNESCO, 43–4.

Leroueil, S., Locat, J., Levesque, C., and Lee., C. F., 2003. Towards an approach for the assessment of risk associated with submarine mass movements. In Locat, J.

and Mienert, J. (eds.), *Submarine Mass Movements and Their Consequences*. Dordrecht, Kluwer Academic Publishers, 59–68.

Levin, L. A., James, D. W., Martin, C. M., *et al.*, 2000. Do methane seeps support distinct macrofaunal assemblages? Observations on community structure and nutrition from the northern California slope and shelf. *Marine Ecology Progress Series*, **208**, 21–39.

Levy, E. M. and Lee, K., 1988. Potential contribution of natural hydrocarbon seepage to benthic productivity and the fisheries of Atlantic Canada. *Canadian Journal of Fisheries and Aquatic Sciences*, **45**, 349–52.

Lewis, D. W. and McConchie, D., 1994. *Practical Sedimentology*, 2nd edn. New York, Chapman & Hall.

Lewis, K. B. and Marshall, B. C., 1996. Seep faunas and other indicators of methane-rich dewatering on New Zealand convergent margins. *New Zealand Journal of Geology and Geophysics*, **39**, 181–200.

Li, X. and Yortsos, Y. C., 1995. Theory of multiple bubble growth in porous media by solute diffusion. *Chemical Engineering Science*, **50**, 1247–71.

Lifland, J., 2002. Methane hydrates in Quaternary climate change: the clathrate gun hypothesis. *EOS – Transactions of the American Geophysical Union*, **83**, 513, 516.

Lilley, M. D., Feeley, R. A., and Trefry, J. H., 1995. Chemical and biochemical transformations in hydrothermal plumes. In Humphris, S. E., Zierenberg, R. A., Mullineaux, L. S., and Thomsen, R. E. (eds.), *Seafloor Hydrothermal Systems: Physical, Chemical, Biological, and Geological Interactions*. American Geophysical Union, Geophysical Monograph 91, 369–91.

Limonov, A. F., Woodside, J. M., Cita, M. B., and Ivanov, M. K., 1996. The Mediterranean Ridge and related mud diapirism: a background. *Marine Geology*, **132**, 7–19.

Limonov, A. F., Weering, T. C. E. van, Kenyon, N. H., Ivanov, M. K., and Meisner, L. B., 1997. Seabed morphology and gas venting in the Black Sea mud volcano area: observations with the MAK-1 deep-tow sidescan sonar and bottom profiler. *Marine Geology*, **137**, 121–36.

Limonov, A. F., Ivanov, M. K., and Foucher, J.-P., 1998. Deep-towed side-scan sonar surveys of the United Nations Rise, eastern Mediterranean. *Geo-Marine Letters*, **18**, 115–26.

Link, W. K., 1952. Significance of oil and gas seeps in world oil exploration. *American Association of Petroleum Geologists (Bulletin)*, **36**, 1505–40.

Lisitzin, A. P., Lukashsin, V. N., Gordeev, V. V., *et al.*, 1997. Hydrological and geochemical anomalies associated with hydrothermal activity in SW Pacific marginal and back-arc basins. *Marine Geology*, **142**, 7–45.

Liss, P. S. and Slater, P. G., 1974. Flux of gases across the air–sea interface. *Nature*, **247**, 181–4.

Locat, J., 2001. Instabilities along ocean margins: a geomorphological and geotechnical perspective. *Marine and Petroleum Geology*, **18**, 503–12.

Locat, J. and Mienert, J., 2003. *Submarine Mass Movements and Their Consequences*. Dordrecht, Kluwer Academic Publishers.

Loncke, L., Huguen, C., Mascle, J., *et al.*, 2001. Pock-marks, mud volcanoes and gas chimneys: evidences from the Nile deep-sea fan. *European Union of Geosciences EUG XI Journal of Conference*, **6** (1), 517 (abstract).

Loncke, L., Mascle, J., and Fanil Scientific Parties, 2004a. Mud volcanoes, gas chimneys, pockmarks and mounds in the Nile deep-sea fan (eastern Mediterranean): geophysical evidences. *Marine and Petroleum Geology*, **21**, 669–89.

Loncke, L., Bayon, G., Duperron, S., and Mascle, J., 2004b. Pockmarks from the Nile Deep-Sea Fan: *in situ* observations from NAUTINIL Expedition. First General Assembly of the European Geosciences Union, Nice, 25–30 April (poster). (See *Geophysical Research Abstracts*, **6**, 05559.)

Lonergan, L., Cartwright, J. A., and Jolly, R., 1998. The geometry of polygonal fault systems in Tertiary mudrocks of the North Sea. *Journal of Structural Geology*, **20**, 529–48.

Long, D., 1986. *Seabed Sediments, Fladen Sheet 58° N– 00°*. British Geological Survey, 1:250,000 map series.

Long, D., 1992. Devensian Late-glacial gas escape in the central North Sea. *Continental Shelf Research*, **12**, 1097–110.

Long, D., Lammers, S., and Linke, P., 1998. Possible hydrate mounds within large sea-floor craters in the Barents Sea. In Henriet, J.-P. and Mienert, J. (eds.), *Gas Hydrates: Relevance to World Margin Stability and Climate Change*. Geological Society of London, Special Publication 137, 223–37.

Longva, O. and Thorsnes, T. (eds.), 1997. *Skagerrak in the Past and at the Present. An Integrated Study of Geology, Chemistry, Hydrography and Microfossil Ecology*. Norges Geologiske Undersøkelse, Special Publication 8.

Longva, O., Thorsnes, T., Mauring, E., and Blikra, L. H., 1998. High-resolution seismic and bathymetric data from clay slides in fjords in northern Norway.

Proceedings of the International Workshop on Sedimentary Processes and Palaeoenvironment in Fjords, University of Tromsø, Norway, 22–24 April (abstract). See www.ibg.uit.no/geologi/spinof/ws-long.html

Longva, O., Janbu, N., Blikra, L. H., and Bøe, R., 2003. The 1996 Finneidfjord slide: seafloor failure and slide dynamics. In Locat, J. and Mienert, J. (eds.), *Submarine Mass Movements and their Consequences*. Dordrecht, Kluwer Academic Publishers, 531–8.

Lonsdale, P., 1977. Clustering of suspension-feeding macrobenthos near abyssal hydrothermal vents at oceanic spreading centres. *Deep-Sea Research*, 24, 857–63.

Lonsdale, P., 1979. A deep hydrothermal site on a strike-slip fault. *Nature*, 281, 531–5.

Lonsdale, P., 1986. A multibeam reconnaissance of the Tonga Trench axis and its intersection with the Louisville guyot chain. *Marine Geophysical Research*, 8, 295–327.

Lonsdale, P. and Becker, K., 1985. Hydrothermal plumes, hot springs, and conductive heat flow in the Southern Trough of Guaymas Basin. *Earth & Planetary Science Letters*, 73, 211–25.

Lonsdale, P. and Hawkins, J., 1985. Silicic volcanism at an off-axis geothermal field in the Mariana Trough Back-arc Basin. *Geological Society of America (Bulletin)*, 96, 940–51.

Loreau, J. P., 1982. Sediments aragonitiques et leur genése. *Mémoires du Museum National d'Histoire Naturelle, Série C, Sciences de la terre*, vol. *XLVII*. (In French.)

Lorenson, T. D., Kvenvolden, K. A., Hostettler, F. D., *et al.*, 2002. Hydrocarbon geochemistry of cold seeps in the Monterey Bay National Marine Sanctuary. *Marine Geology*, 181, 285–304.

Lorenz, V., 1975. Formation of phreatomagmatic maardiatreme volcanoes and its relevances to kimberlite diatremes. *Physics and Chemistry of the Earth*, 9, 17–27.

Løseth, H., Wensaas, L., Arnsten, B., *et al.*, 2001. 1000 m long gas blow-out pipes. Extended Abstracts, 63rd EAGE Conference and Exhibition, Amsterdam, 11–15 June, 524–7 (abstract).

Løseth, H., Wensaas, L., Arnsten, B., and Hovland, M., 2003. Gas and fluid injection triggering shallow mud mobilization in the Hordaland Group, North Sea. In Rensbergen, P. van, Hillis, R. R., Maltmann, A. J., and Morley, C. K. (eds.), *Subsurface Sediment Mobilization*. Geological Society of London, Special Publication 216, 139–57.

Lowe, D. R., 1975. Water escape structures in coarsegrained sediments. *Sedimentology*, 22, 157–204.

Lowell, R. P., Rona, P. A., and Herzen, R. P. von, 1995. Seafloor hydrothermal systems. *Journal of Geophysical Research*, 100, 327–52.

Lucas, A. L., 1974. A high resolution marine seismic survey. *Geophysical Prospecting*, 22, 667–82.

Lüdman, T. and Wong, H. K., 2003. Characteristics of gas hydrate occurrences associated with mud diapirism and gas escape structures in the northwestern Sea of Okhotsk. *Marine Geology*, 210, 269–86.

Luff, R., Wallmann, K., and Aloisi, G., 2004. Numerical modeling of carbonate crust formation at cold vent sites: significance for fluid and methane budgets and chemosynthetic biological communities. *Earth & Planetary Science Letters*, 221, 337–53.

Lukkien, H. B., 1985. Shallow gas: coping with its hazards offshore. *World Oil*, July, 59–63.

Lupton, J. E., 1995. Hydrothermal plumes: near and far field. In Humphris, S. E., Zierenberg, R. A., Mullineaux, L. S., and Thomsen, R. E. (eds.), *Seafloor Hydrothermal Systems: Physical, Chemical, Biological, and Geological Interactions*. American Geophysical Union, Geophysical Monograph 91, 317–46.

Lupton, J. E., Klinkhammer, G. P., Normark, W. R., *et al.*, 1980. Helium-3 and manganese at the 210° N East Pacific Rise hydrothermal site. *Earth & Planetary Science Letters*, 50, 115–27.

Lupton, J. E. and Craig, H., 1981. A major ^{3}He source on the East Pacific Rise. *Science*, 214, 13–18.

Luternauer, J. L., Barrie, J. V., Christian, H. A., *et al.*, 1994. Fraser River delta: geology, geohazards and human impact. In Monger, J. W. H. (ed.), *Geology and Geological Hazards of the Vancouver Region, Southwestern British Columbia*. Geological Survey of Canada, Bulletin 481, 197–220.

Luth, C., Luth, U., Gebruk, A. V., and Thiel, H., 1999. Methane gas seeps along the oxic/anoxic gradient in the Black Sea: manifestations, biogenic sediment compounds, and preliminary results on benthic ecology. *Marine Ecology*, 20, 221–49.

Lutz, R. A., 2000. Deep sea vents. *National Geographic Magazine*, 198 (4) 116–27.

Lutz, R. A., Shank, T. M., Fornari, D. J., *et al.*, 1994. Rapid growth at deep-sea vents. *Nature*, 371, 663–4.

MacDonald, G. J., 1983. The many origins of natural gas. *Journal of Petroleum Geology*, 5, 341–62.

MacDonald, G. J., 1990. Role of clathrates in past and future climate change. *Climate Change*, **16**, 247–82.

MacDonald, I. R., 1998. Natural oil spills. *Scientific American*, November, 30–35.

MacDonald, I. R. and Leifer, I., 2002. Constraining rates of carbon flux from natural seeps on the northern Gulf of Mexico slope. In *Gas in Marine Sediments, Seventh International Conference, Baku, Azerbaijan, 7th–12th October 2002*. Baku, Nafta Press, 119.

MacDonald, I. R., Reilly, J. F., Guinasso, N. L., jr, *et al.*, 1990. Chemosynthetic mussels at a brine-filled pockmark in the northern Gulf of Mexico. *Science*, **248**, 1096–9.

MacDonald, I. R., Guinasso, N. R., jr, Ackleson, S. G., *et al.*, 1993. Natural oil slicks in the Gulf of Mexico visible from space. *Journal of Geophysical Research*, **98**, 16351–64.

MacDonald, I. R., Guinasso, N. L., jr, Sassen, R., *et al.*, 1994. Gas hydrate that breaches the sea floor on continental slope of the Gulf of Mexico. *Geology*, **22**, 699–702.

MacDonald, I. R., Reilly, J. G., jr, Best, S. E., *et al.*, 1996. Remote sensing inventory of active oil seeps and chemosynthetic communities in the northern Gulf of Mexico. In Schumacher, D. and Abrams, M.A. (eds.), *Hydrocarbon Migration and its Near-Surface Expression*. American Association of Petroleum Geologists, Memoir 66, 27–37.

MacDonald, I. R., Leifer, I., Sassen, R., *et al.*, 2002. Transfer of hydrocarbons from natural seeps to the water column and atmosphere. *Geofluids*, **2**, 95–107.

MacDonald, I. R., Bohrmann, G., Escobar, E., *et al.*, 2004. Asphalt volcanism and chemosynthetic life in the Campeche Knolls, Gulf of Mexico. *Science*, **304**, 999–1002.

MacKay, M. E., Jarrard, R. D., Westbrook, G. K., *et al.*, 1995. Origin of bottom-simulating reflectors: geophysical evidence from the Cascadia accretionary prism. *Geology*, **22**, 459–62.

MacLean, B., Falconer, R. K. H., and Levy, B. M., 1981. Geological, geophysical and chemical evidence for natural seepage of petroleum off the northeast coast of Baffin Island. *Canadian Petroleum Geology*, **29**, 75–95.

Mahfoud, R. F. and Beck, J. N., 1995. Why the Middle East fields may produce oil for ever. *Offshore*, April, 58–64, 106.

Mahmood, A., Ehlers, C. J., Butenko, J., and Randall, A. G., 1980. Seafloor sediments in Orinoco delta, Venezuela.

Proceedings of the Offshore Technology Conference, Houston, TX, OTC Paper 3774.

Maisey, G. H., Rokoengen, K., and Raaen, K., 1980. Pock-marks formed by seep of petrogenic gas in the southern part of the Norwegian Trench, IKU, Trondheim Report No. P-258/1/80.

Maldonado, A., Somoza, L., and Pallarés, L., 1999. The Betic orogen and the Iberian–African boundary in the Gulf of Cadiz: geological evolution (central North Atlantic). *Marine Geology*, **155**, 9–43.

Malmquist, L. and Kristiansson, K., 1981. Microflow of geogas – a possible formation mechanism for deep-sea nodules. *Marine Geology*, **40**, M18.

Mandl, G., 1988. *Mechanics of Tectonic Faulting*. Amsterdam, Elsevier.

Manheim, F. T., 1967. Evidence for submarine discharge of water on the Atlantic continental slope of the southern United States, and suggestions for further research. *Transactions of the New York Academy of Sciences (Series II)*, **29**, 839–53.

Manley, P. L. and Flood, R. D., 1988. Cyclic sediment deposition within Amazon Deep-Sea Fan. *American Association of Petroleum Geologists (Bulletin)*, **72**, 912–25.

Manley, P. L. and Flood, R. D., 1989. Anomalous sound velocities in near-surface, organic-rich, gassy sediments in the central Argentinian Basin. *Deep-Sea Research I*, **36**, 611–23.

Mart, Y. and Ross, D. A., 1987. Post-Miocene rifting and diapirism in the northern Red Sea. *Marine Geology*, **74**, 173–90.

Martens, C. S. and Klump, J. V., 1980. Biogeochemical cycling in an organic-rich coastal marine basin – I. Methane sediment–water exchange processes. *Geochimica et cosmochimica acta*, **44**, 471–90.

Martens, C. S. and Klump, J. V., 1984. Biogeochemical cycling in an organic-rich coastal marine basin – 4. An organic carbon budget for sediments dominated by sulfate reduction and methanogenesis. *Geochimica et cosmochimica acta*, **48**, 1987–2004.

Martens, C. S., Albert, D. B., and Alperin, M. J., 1998. Biogeochemical processes controlling methane in gassy coastal sediments – Part I. A model coupling organic matter flux to gas production, oxidation and transportation. *Continental Shelf Research*, **18**, 1741–70.

Martin, J. B., Kastner, M., Henry, P., Le Pichon, X., and Lallemant, S., 1996. Chemical and isotopic evidence for sources of fluids in a mud volcano field seaward of the

Barbados accretionary wedge. *Journal of Geophysical Research*, **101**, 20325–45.

Martinelli, G. and Judd, A. G., 2004. Mud volcanoes of Italy. *Geological Journal*, **39**, 49–61.

Marty, B., Jambon, A., and Sano, Y., 1989. Helium isotopes and CO_2 in volcanic gases of Japan. *Chemical Geology*, **76**, 25–40.

Marty, D. G., 1992. Ecology and metabolism of methanogens. In Vially, R. (ed.), *Bacterial Gas*. Paris, Editions Technip, 13–26.

Mascle, J., Huguen, C., Benkhelil, J., *et al.*, 1999. Images may show start of European–African plate collision. *EOS – Transactions of the American Geophysical Union*, **80**, 421, 425, 428.

Masson, D. G., Bett, B. J., Billett, D. S. M., *et al.*, 2003. The origin of deep-water, coral-topped mounds in the northern Rockall Trough, northeast Atlantic. *Marine Geology*, **194**, 159–80.

Massoth, G. J., Butterfield, D. A., Lupton, J. E., *et al.*, 1989. Submarine venting of phase-separated hydrothermal fluids at Axial Volcano, Juan de Fuca Ridge. *Nature*, **340**, 702–5.

Matsumoto, R., Watanabe, Y., Satoh, M., *et al.*, 1996. Distribution and occurrence of marine gas hydrates. Preliminary results of ODP Leg 164: Blake Ridge drilling. *Journal Geological Society of Japan*, **102**, 932–44. (In Japanese with English abstract.)

Matthews, M. D., 1996. Migration – a view from the top. In Schumacher, D. and Abrams, M. A. (eds.), *Hydrocarbon Migration and its Near-Surface Expression*. American Association of Petroleum Geologists, Memoir 66, 139–55.

Mattson, M. D. and Likens, G. E., 1990. Air pressure and methane fluxes. *Nature*, **347**, 718–19.

Matveeva, T. V., Kaulio, V. V., Mazurenko, L. L., *et al.*, 2000. Geological and geochemical characteristics of near-bottom gas hydrate occurrence in the southern basin of the Lake Baikal, eastern Siberia. In *Sixth International Conference on Gas in Marine Sediments*. St Petersburg, VNIIOkeangeologia, 91–3.

Max, M. D., 2000a. Hydrate as a future energy resource for Japan. In Max, M. D. (ed.), *Natural Gas Hydrates in Oceanic and Permafrost Environments*. Dordrecht, Kluwer Academic Publishers, 225–38.

Max, M. D., 2000b. *Natural Gas Hydrates in Ocean and Permafrost Environments*. Dordrecht, Kluwer Academic Publishers.

Max, M. D., 2000c. Hydrate resource, methane fuel, and a gas-based economy? In Max, M. D. (ed.), *Natural Gas*

Hydrates in Oceanic and Permafrost Environments, Dordrecht, Kluwer Academic Publishers, 361–70.

Max, M. D. and Miles, P. R., 1999. Marine survey for gas hydrates. Proceedings of the Offshore Technology Conference, Houston, TX, OTC Paper 10769.

May, D. A. and Monaghan, J. J., 2003. Can a single bubble sink a ship? *American Journal of Physics*, **71**, 842–9.

Mayer, L., 1981. Erosional troughs in deep sea carbonates and their relationship to basement structure. *Marine Geology*, **39**, 59–80.

Mayer, L., 2004. New tools for fusion and visualization of ocean mapping data – the incubator of insights. Abstracts and Programme of the Seabed and Shallow Section Marine Geoscience: Shared Lessons and Technologies from Academia and Industry Conference, Geological Society of London, 24–26 February, 6.

Mayer, L. A., Shor, A. N., Hughes Clark, J., and Piper, D. J. W., 1988. Dense biological communities at 3850 m on the Laurentian Fan and their relationship to the deposits of the 1929 Grand Banks earthquake. *Deep-Sea Research*, **35**, 1235–46.

Mazurenko, L., Soloviev, V., Ivanov, M., Pinheiro, L., and Gardner, J., 2001. Geochemical features of gas hydrate-forming fluids of the Gulf of Cadiz. In Akhamanov, G. and Suzyumov, A. (eds.), *Geological Processes on Deep-Water European Margins*. International Oceanographic Commission Workshop Report No. 175 on the International Conference and ninth post-cruise meeting of the Training Through Research Programme, Moscow-Mozhenka, Russia, 28 January–2 February. Paris, UNESCO, 50.

Mazurenko, L. L. and Soloviev, V. A., 2003. Worldwide distribution of deep-water fluid venting and potential occurrences of gas hydrate accumulation. *Continental Shelf Research*, **23**, 162–76.

Mazzini, A., Duranti, D., Jonk, R., *et al.*, 2003. Palaeo-carbonate seep structures above an oil reservoir, Gryphon field, Tertiary, North Sea. *Continental Shelf Research*, **23**, 323–39.

Mazzotti, L., Segantini, S., Tramontana, M., and Wezel, F.-C., 1987. Classification and distribution of pockmarks in the Jabuka Trough (central Adriatic). *Bollettino de Oceanologia Teorica ed Applicata*, **5**, 237–50.

McAdoo, B. G., Orange, D. L., Silver, E. A., *et al.*, 1996. Seafloor structural observations, Costa Rica Accretionary Prism. *Geophysical Research Letters*, **23**, 883–6.

McAdoo, B. G., Pratson, L. F., and Orange, D. L., 2000. Submarine landslide geomorphology, US Continental Slope. *Marine Geology*, **169**, 103–36.

McAuliffe, C. D., 1980. Oil and gas migration: chemical and physical constraints. In Roberts, W. H., III and Cordell, R. J. (eds.), *Problems of Petroleum Migration*. American Association of Petroleum Geologists, Studies in Geology No. 10, 89–107.

McClennen, C. E., 1989. Microtopography and surficial sediment patterns in the central Gulf of Maine: a 3.5 kHz survey and interpretation. *Marine Geology*, **89**, 69–85.

McClusky, S., Balassanian, S., Barka, A., *et al.*, 2000. Global Positioning System constraints on plate kinematics and dynamics in the eastern Mediterranean and Caucasus. *Journal of Geophysical Research*, **105**, 5695–719.

McConnell, D. R., 2000. Optimizing deepwater well locations to reduce the risk of shallow-water-flow using high-resolution 2D and 3D seismic data. Proceedings of the Offshore Technology Conference, Houston, TX, OTC Paper 11973.

McCulloch, D. S., 1989. Geologic map of the south-central California continental margin, Map N-4A. In Greene, H.G. and Kennedy, M. P. (eds.), *Geologic Map Series of the California Continental Margin*. Map Scale 1: 250 000. California Division of Mines and Geology.

McCulloh, T. H., 1969. *Geologic Characteristics of the Dos Cuadras Offshore Oil Field*. United States Geological Survey, Professional Paper 679-c.

McDuff, R. E., 1995. Physical dynamics of deep-sea hydrothermal plumes. In Humphris, S. E., Zierenberg, R. A., Mullineaux, L. S., and Thomsen, R. E. (eds.), *Seafloor Hydrothermal Systems: Physical, Chemical, Biological, and Geological Interactions*. American Geophysical Union, Geophysical Monograph 91, 357–68.

McGee, T. M. and Woolsey, J. R., 1999. An installation in the northern Gulf of Mexico for monitoring interactions between the water column and sea-floor sediments containing gas hydrates. Proceedings of the Offshore Technology Conference, Houston, TX, OTC Paper 10771.

McHugh, D., 1997. Molecular evidence that echiurans and pogonophorans are derived Annelids, *Proceedings of the National Academy of Sciences, USA*, **94**, 8006–9.

McIver, R. D., 1981. Gas hydrate. In Meyer, R. F. and Olson, J. C. (eds.), *Long-Term Energy Resources*. Boston, MA, Pitman Publishing, 713–26.

McIver, R. D., 1982. Role of naturally occurring gas hydrates in sediment transport. *American Association of Petroleum Geologists (Bulletin)* **66**, 789–92.

McKay, A. G., 1983. Acoustic observations in seabed materials. Ph.D. thesis, University of Durham.

McKenna, G. T., Luternauer, J. L., and Kostaschuk, R. A., 1992. Large-scale mass-wasting events on the Fraser River delta front near Sand Heads, British Columbia. *Canadian Geotechnical Journal*, **29**, 151–6.

McQuaid, J. and Mercer, A., 1990. Air pressure and methane fluxes. *Nature*, **351**, 528.

McQuillin, R. and Fannin, N. G. T., 1979. Explaining the North Sea's lunar floor. *New Scientist*, **83**(1163), 90–2.

McQuillin, R., Fannin, N. G. T., and Judd, A. G., 1979. *IGS Pockmark Investigations 1974–1978*. Institute of Geological Sciences, Marine Geophysics Unit, Report No. 98.

MEDINAUT/MEDINETH Shipboard Scientific Parties (Aloisi, G., Asjes, S., Bakker, K., *et al.*), 2000. Linking Mediterranean brine pools and mud volcanism. *EOS – Transactions of the American Geophysical Union*, **81**, 625, 631–3.

Meldahl, P., Heggland, R., Bril, A. H., and de Groot, P. F. M., 1999. The chimney cube, an example of semi-automated detection of seismic objects by directive attributes and neural networks. Part I: method. Expanded Abstracts, Society of Exploration Geophysicists 69th Annual Meeting, Houston, TX, 31 October – 5 November, 931–4.

Menard, H. W., 1964. *Marine Geology of the Pacific*. New York, NY, McGraw-Hill.

Merewether, R., Olsson, M. S., and Lonsdale, P., 1985. Acoustically detected hydrocarbon plumes rising from 2-km depths in Guaymas Basin, Gulf of California. *Journal of Geophysical Research*, **90**, 3975–85.

Merle, S., Embley, R., Baker, E., and Chadwick, W., 2003. Submarine Ring of Fire 2003 – Mariana Arc. R/V T. G. Thompson Cruise TN-153. Newport, OR, Cruise Report, NOAA.

Michaelis, W., 2000. Boreale Schwämme als marine Naturstoffquelle (BOSMAN). Hamburg, The University of Hamburg, Technical cruise report Pos 254. (In German.)

Michaelis, W., Jenisch, A., and Richnow, H. H., 1990. Hydrothermal petroleum generation in Red Sea sediments from the Kebrit and Shaban deeps. *Applied Geochemistry*, **5**, 103–14.

Mienert, J., Posewang, J., and Baumann, M., 1998. Gas hydrates along the northeastern Atlantic Margin: possible hydrate-bound margin instabilities and possible release of methane. In Henriet, J.-P. and Mienert, J. (eds.), *Gas Hydrates: Relevance to World Margin Stability and Climate Change*. Geological Society of London, Special Publication 137, 275–91.

Miles, J. A., 1994. *Illustrated Glossary of Petroleum Geochemistry*. Oxford, Clarendon Press.

Milkov, A. V., 2000. Worldwide distribution of submarine mud volcanoes and associated gas hydrates. *Marine Geology*, **167**, 29–42.

Milkov, A. V. and Sassen, R., 2001. Estimate of gas hydrate resource, northwestern Gulf of Mexico continental slope. *Marine Geology*, **179**, 71–83.

Milkov, A. V. and Sassen, R., 2002. Economic geology of offshore gas hydrate accumulations and provinces. *Marine and Petroleum Geology*, **19**, 1–11.

Milkov, A. V., Sassen, R., Apanasovich, T. V., and Dadashev, F. G., 2003. Global gas flux from mud volcanoes: a significant source of fossil methane in the atmosphere and ocean. *Geophysical Research Letters*, **30**, DOI: 10,1029/2002GL016358.

Miller, D. C. and Ullman, W. J., 2001. Ecological consequences of estuarine groundwater discharge at Cape Henlopen, Delaware Bay, USA. Proceedings of the Geological Society of America Annual Meeting, Boston, MA, November 5–8 (abstract).

Miller, R. D., 1980. Freezing phenomena in soils. In Hillel, D. (ed.), *Introduction to Soil Physics*. San Diego, Academic Press, 254–99.

Milliman, J. D., 1974. *Marine Carbonates*. New York, NY, Springer-Verlag.

Milliman, J. D., Qin, Y. S., and Butenko, J., 1985. Geohazards in the Yellow Sea and East China Sea. Proceedings of the Offshore Technology Conference, Houston, TX, OTC Paper 4965.

Mills, R. A. and Elderfield, H., 1995. Hydrothermal activity and the geochemistry of metalliferous sediments. In Humphris, S. E., Zierenberg, R. A., Mullineaux, L. S., and Thomsen, R. E. (eds.), *Seafloor Hydrothermal Systems: Physical, Chemical, Biological, and Geological Interactions*. American Geophysical Union, Geophysical Monograph 91, 392–407.

Minshull, T. and White, R., 1989. Sediment compaction and fluid migration in the Makran Accretionary Prism. *Journal of Geophysical Research*, **94**, 7387–402.

Missiaen, T., Murphy, S., Loncke, L., and Henriet, J.-P., 2002. Very high resolution seismic mapping of shallow gas in the Belgian coastal zone. *Continental Shelf Research*, **22**, 2291–302.

MMS (US Department of the Interior, Minerals Management Service, Gulf of Mexico OCS Region), 2001. *Gulf of Mexico Deepwater Operations and Activities: Environmental Assessment* (May 2000). New Orleans, US Department of the Interior, Minerals Management Service, Gulf of Mexico OCS Region.

MMS (US Department of the Interior, Minerals Management Service, Gulf of Mexico OCS Region), 2003. *Shallow Gas Blowout and Rig Evacuation*. New Orleans, US Department of the Interior Minerals Management Service, Gulf of Mexico OCS Region, Safety Alert No. 214 (7 April).

Møller, M. M., Nielsen, L. P., and Jørgensen, B. B., 1985. Oxygen responses and mat formation by *Beggiatoa* spp. *Applied Environmental Microbiology*, **50**, 373–82.

Molnia, B. F., 1979. Origin of gas pockmarks and craters. *Geological Society of America, Abstracts with Programs*, **11**, 481–2.

Molnia, B. F. and Rappeport, M. L., 1984. Mosaic of the Alsek Sediment Instability Area. *USGS Open File Report*, 84–397.

Molnia, B. F., Carlson, P. R., and Kvenvolden, K. A., 1978. Gas-charged sediment areas in the northern Gulf of Alaska. *Geological Society of America, Abstracts with Programs*, **10**, 458–9.

Montagna, P. A., Bauer, J. E., Hardin, D., and Spies, R. B., 1989. Vertical distribution of microbial and meiofaunal populations in sediments of a natural coastal hydrocarbon seep. *Journal of Marine Research*, **47**, 657–80.

Moore, J. C., Mascle, A., Taylor, E., *et al.*, 1987. Expulsion of fluids from depth along a subduction-zone decollement horizon. *Nature*, **326**, 785–8.

Moore, W. S., 1996. Large groundwater inputs to coastal waters revealed by ^{226}Ra enrichments. *Nature*, **380**, 612–14.

Morimoto, R., 1960. Submarine eruption of the Myojin Reef. *Bulletin of Volcanology*, **23**, 151–60.

Morrissey, M. M. and Mastin, L. G., 2000. Phreatomagmatic fragmentation. In Sigurdsson, H. (ed.), *Encyclopedia of Volcanoes*. London Academic Press, 431–45.

Morse, J. W. and Mackenzie, F. T., 1990. *Geochemistry of Sedimentary Carbonates*. Developments in Sedimentology, 48. Amsterdam, Elsevier.

Mortensen, P. B., Hovland, M., Fosså, J. H., and Furevik, D. M., 1998. Size and abundance of *Lophelia* banks in mid-Norwegian waters. Bergen, Intergovernmental Oceanographic Commission Workshop Report 143.

Morton, J. L. and Sleep, N. H., 1985. A mid-ocean ridge thermal model: constraints on the volume of axial hydrothermal heat flux. *Journal of Geophysical Research*, **90**, 11345–53.

Mottl, M. J., Wheat, C. G., Fryer, P., Gharib, J., and Martin, J. B., 2004. Chemistry of springs across the Mariana forearc shows progressive devolatilization of the subducting plate. *Geochimica et cosmochimica acta*, **68**, 4915–33.

Muiños, S., Gaspar, L., Monteiro, J. H., *et al.*, 2002. Ferromanganese deposits from the Nameless Seamount. In Cunha, M., Pinheiro, L., and Suzyumov, A. (eds.), *Geosphere/Biosphere/Hydrosphere Coupling Processes, Fluid Escape Structures and Tectonics at Continental Margins and Ocean Ridges*. International Conference Abstracts, Intergovernmental Oceanographic Commission. Paris, UNESCO, Workshop Report No. 183, 27–30.

Mukhopadhyay, R., Iyer, S. D., and Ghosh, A. K., 2002. The Indian Ocean Nodule Field: petrotectonic evolution and ferromanganese deposits. *Earth-Science Reviews*, **60**, 67–130.

Mukhtarov, A. Sh., Kadirov, F. A., Guliyev, I. S., Feyzullayev, A., and Lerche, I., 2003. Temperature evolution in the Lokbatan Mud Volcano crater (Azerbaijan) after the eruption of 25 October 2001. *Energy Exploration & Exploitation*, **21**, 187–207.

Mulder, T. and Cochonat, P., 1996. Classification of offshore mass movements. *Journal of Sedimentary Research*, **66**, 43–57.

Mullineaux, L. S. and France, S., 1995. Dispersal of deep-sea hydrothermal vent fauna: mechanisms and implications for species distributions. In Humphris, S. E., Zierenberg, R. A., Mullineaux, L. S., and Thomsen, R. E. (eds.), *Seafloor Hydrothermal Systems: Physical, Chemical, Biological, and Geological Interactions*. American Geophysical Union, Geophysical Monograph 91, 408–24.

Muradov, Ch. and Javadova, R., 2002. Estimation of resources of gas in gas hydrate accumulations of southern Caspian Sea. In *Gas in Marine Sediments, Seventh International Conference, Baku, Azerbaijan, 7th–12th October 2002*. Baku, Nafta Press, 149–50.

Murray, J. and Renard, A. F., 1891. Report on deep-sea deposits. Challenger Expedition Reports, 3. London, HMSO.

Murthy, K. S. R. and Rao, T. C. S., 1990. Acoustic wipeouts over the continental margins off Krishna, Godavari and Mahanadi River basins, east coast of India. *Journal of the Geological Society of India*, **35**, 559–68.

Musson, R. M. W., Pappin, J., Lubkowski, Z., Booth, E., and Long, D. 1997. *UK Continental Shelf Seismic Hazard*. Sudbury, HSE Books, Offshore Technology Report OTH 93./416.

Naehr, T. H., Rodriguez, N. M., Bohrman, G., Paull, C. K., and Botz, R., 2000. Methane-derived authigenic carbonates associated with gas hydrate decomposition and fluid venting above the Blake Ridge Diapir. *Proceedings of the Ocean Drilling Program, Scientific Results*, **164**, 285–300.

Naeth, J., di Primio, R., Horsfield, B., *et al.*, 2005. Hydrocarbon seepage and carbonate mound formation: a basin modelling study from the Porcupine Basin (offshore Ireland). *Journal of Petroleum Geology*, **28**, 147–66.

Naganuma, T., Otsuki, A., and Seki, H., 1989. Abundance and growth rate of bacterioplankton community in hydrothermal vent plumes of the North Fiji Basin. *Deep-Sea Research*, **36**, 1379–90.

Namsaraev, B. B., Zemskaya, T. I., Dagurova, O. P., *et al.*, 2000. Biological communities in the sediments: regions of hydrothermal venting (northern Baikal) and near-surface occurrence of gas hydrates (southern Baikal). *Abstracts of the Sixth International Conference on Gas in Marine Sediments*. St Petersburg, VNIIOkeangeologia, 102–3.

Namsaraev, B. B., Zemskaya, T. I., Dagurova, O. P., *et al.*, 2002. Bacterial communities of the bottom sediments near a hydrothermal source in Frolikha Bay, northern Baikal. *Geologiya i Geofizika (Russian Geology and Geophysics)*, **43**, 644–7 (Russian edn.) / 604–8 (English edn.).

Nardin, T. R. and Henyey, T. L., 1978. Pliocene–Pleistocene diastrophism of Santa Monica and San Pedro shelves, California continental borderland. *American Association of Petroleum Geologists (Bulletin)*, **62**, 247–72.

Neglia, S., 1979. Migration of fluids in sedimentary basins. *American Association of Petroleum Geologists (Bulletin)*, **63**, 573–97.

Nelson, C. H. and Johnson, K. R., 1987. Whales and walruses as tillers of the seafloor. *Scientific American*, 256(2), 74–81.

Nelson, C. H., Thor, D. R., Sandstrom, M. W., and Kvenvolden, K A., 1979. Modern biogenic gas-generated craters (sea-floor 'pockmarks') on the Bering Shelf, Alaska. *Geological Society of America (Bulletin)*, 90, 1144–52.

Nelson, C. S. and Healy, T. R., 1984. Pockmark-like structures on the Poverty Bay sea bed – possible evidence for submarine mud volcanism. *New Zealand Geology and Geophysics*, 27, 225–30.

Newton, R. S., Cunningham, R. C., and Schubert, C. E., 1980. Mud volcanoes and pockmarks: seafloor engineering hazards or geological curiosities? Proceedings of the Offshore Technology Conference, Houston, TX, OTC Paper 3729.

Nichols, R. J., Sparks, R. S. J., and Wilson, C. J. N., 1994. Experimental studies of the fluidization of layered sediments and the formation of fluid escape structures. *Sedimentology*, 41, 233–53.

Niemann, H., Orcutt, B., Suck, I., *et al.*, 2004. Methane seeps in the North Sea: Tommeliten revisited. First General Congress, European Geosciences Union, Nice, 25–30 April (poster). (See *Geophysical Research Abstracts*, 6, 06365.)

Niemann, J., Grace, G., Ropp, C., and Burch, D., 1998. Shallow water flow evaluation: current technology and capabilities. Proceedings of the 1998 Shallow Water Flow Forum, Woodlands, TX, 24–25 June (presentation).

Nisbet, E. G., 1989. Some northern sources of atmospheric methane: production, history and future implications. *Canadian Journal of Earth Science*, 26, 1603–11.

Nisbet, E. G., 1990. The end of the ice-age. *Canadian Journal of Earth Science*, 27, 148–57.

Nisbet, E. G., 2002. Have sudden large releases of methane from geological reservoirs occurred since the last glacial maximum, and could such releases occur again? *Philosophical Transactions of the Royal Society of London, A*, 360, 581–607.

Nittrouer, C. A., 1999. STRATOFORM: overview of its design and synthesis of its results. *Marine Geology*, 154, 3–12.

NPA Group, 2003. *Offshore Basin Screening: Key features of OBS Methodology*. NPA-TREICoL Join Venture documentation. Edenbridge, Kent, NPA Group.

Nummedal, D. and Prior, D. B., 1981. Generation of Martian chaos and channels by debris flows. *Icarus*, 45, 77–86.

Nunn, J. A. and Muelbroek, P., 2002. Kilometer-scale upward migration of hydrocarbons in geopressured sediments by buoyancy-driven propagation of methane-filled fractures. *American Association of Petroleum Geologists (Bulletin)*, 86, 907–18.

O'Brien, G. W., Lisk, M., Duddy, I. R., *et al.*, 1999. Plate convergence, foreland development and fault reactivation: primary controls on brine migration, thermal histories and trap breach in the Timor Sea, Australia. *Marine and Petroleum Geology*, 16, 533–560.

O'Brien, G. W., Glenn, K., Lawrence, G., *et al.*, 2002. Influence of hydrocarbon migration and seepage on benthic communities in the Timor Sea, Australia. *APPEA Journal*, 2002, 1–14.

O'Connell, S., 1985. *Anatomy of Modern Submarine Depositional and Distributary Systems*. Unpublished thesis, Texas A&M University.

O'Hara, S. C. M., Dando, P. R., Schuster, U., *et al.*, 1995. Gas seep induced interstitial water circulation: observations and environmental implications. *Continental Shelf Research*, 15, 931–48.

Obzhirov, A., Suess, E., Salyuk, A., *et al.*, 2000. Methane flares of the Okhotsk Sea. In *Abstracts of the Sixth International Conference on Gas in Marine Sediments*. St Petersburg, VNIIOkeangeologia, 104.

Obzhirov, A. I., Vereshchagina, O. F., Sosnin, V. A., *et al.*, 2002. Methane monitoring in waters of eastern shelf and slope of Sakhalin. *Geologiya i Geofizika (Russian Geology and Geophysics)*, 43, 605–12 (Russian edn.) / 564–71 (English edn.).

Obzhirov, A., Shakirov, R., Salyuk, A., *et al.*, 2004. Relations between methane venting, geological structure and seismo-tectonics in the Okhotsk Sea. *Geo-Marine Letters*, 24, 135–9.

Okusa, S., 1985a. Measurements of wave-induced pore pressure in submarine sediments under various marine conditions. *Marine Geotechnology*, 6, 119–44.

Okusa, S., 1985b. Wave-induced stresses in unsaturated submarine sediments. *Géotechnique*, 35, 517–32.

Olu, K., Duperret, A., Sibuet, M., Foucher, J.-P., and Fiala-Médiono, A., 1996. Structure and distribution of cold seep communities along the Peruvian active margin: relationship to geological and fluid patterns. *Marine Ecology Progress Series*, 132, 109–25.

Olu, K., Lance, S., Sibuet, M., *et al.*, 1997. Cold seep communities as indicators of fluid expulsion patterns through mud volcanoes seaward of the Barbados Accretionary Prism. *Deep-Sea Research*, **44**, 811–41.

Orange, D. L. and Breen, N. A., 1992. The effects of fluid escape on accretionary wedges, 2. Seepage force, slope failure, headless submarine canyons, and vents. *Journal of Geophysical Research*, **97**, 9277–95.

Orange, D. L., Greene, H. G., Reed, D., *et al.*, 1999. Widespread fluid expulsion on a translational continental margin: mud volcanoes, fault zones, headless canyons, and organic-rich substrate in Monterey Bay, California. *Geological Society of America (Bulletin)*, **111**, 992–1009.

Orange, D. L., Yun, J., Maher, N., Barry, J., and Greene, G., 2002. Tracking California seafloor seeps with bathymetry, backscatter and ROVs. *Continental Shelf Research*, **22**, 2273–90.

Orange, D. L., Hovland, M., Greene, G. H., *et al.*, 2003. The implications of hydrocarbon seepage, gas migration and fluid overpressures to frontier exploration and geohazards. American Association of Petroleum Geologists Annual Meeting, Salt Lake City, 11–14 May (abstract). See www.searchanddiscovery.com/documents/abstracts/annual2003/short/80432.PDF

Orcutt, B. N., Boetius, A., Lugo, S. J., *et al.*, 2004. Life at the edge of methane ice: microbial cycling of carbon and sulfur in Gulf of Mexico hydrates. *Chemical Geology*, **205**, 239–51.

Oremland, R. S. and Miller, L. G., 1993. Biogeochemistry of natural gases in three alkaline, permanently stratified (meromictic) lakes. In Howell, D. G. (ed.), *The Future of Energy Gases*, United States Geological Survey Professional Paper 1570, 439–52.

Orphan, V. J., House, C. H., Hinrichs, K.-U., McKeegan, K. D., and DeLong, E. F., 2001. Methane-consuming archaea revealed by directly coupled isotopic and phylogenetic analysis. *Science*, **293**, 484–7.

Orpin, A. R., 1997. Dolomite chimneys as possible evidence of coastal fluid expulsion, upper most Otago Continental Slope, southern New Zealand. *Marine Geology*, **138**, 51–67.

Ostermeier, R. M., Pelletier, J. H., Winker, C. D., *et al.*, 2000. Dealing with shallow-water flow in the deepwater Gulf of Mexico. Proceedings of the Offshore Technology Conference, Houston, TX, OTC Paper 11972.

Owens, N. J. P., Law, C. S., Mantoura, R. F. C., Burkill, P. H., and Llewellyn, C. A., 1991. Methane flux to the atmosphere from the Arabian Sea. *Nature*, **354**, 293–6.

Oxburgh, E. R., O'Nions, R. K., and Hill, R. I., 1986. Helium isotopes in sedimentary basins. *Nature*, **324**, 632–5.

Panieri, G., Ricchiuto, T., Martinenghi, C., and D'Onofrio, S., 2000a. Effects of hydrocarbon seepage on benthic foraminifera assemblage in Irish Rockall Trough. *Abstracts of the Sixth International Conference on Gas in Marine Sediments*, St Petersburg, VNIIOkeangeologia, Sept. 5–9, 107.

Panieri, G., Ricchiuto, T., D'Onofrio, S., *et al.*, 2000b. Benthic foraminifera associated with methane seeps in Adriatic Sea. *Abstracts of the Sixth International Conference on Gas in Marine Sediments*, St Petersburg, VNIIOkeangeologia, 106.

Papatheodorou, G., Hasiotis, T., and Ferentinos, G., 1993. Gas-charged sediments in the Aegean and Ionian seas, Greece. *Marine Geology*, **112**, 171–84.

Papatheodorou, G., Lavrentaki, M., Mourelatos, P., Voutsinas, K., and Xenos, K., 2001. Pockmarks on the seabed of the Aetoliko Lagoon, Greece. *Alieftika Nea (Fishing News)*, No. 238, April, 73–87. (In Greek.)

Papazachos, B. and Papazachou, K., 1989. *Earthquakes in Greece*. Thessaloniki, Ziti Publications. (In Greek.)

Park, A., Dewers, T., and Ortoleva, P., 1990. Cellular and oscillatory self-induced methane migration. *Earth Science Reviews*, **29**, 249–65.

Parkes, R. J., 1999. Oiling the wheels of controversy. A review of *The Deep Hot Biosphere*, by Thomas Gold. *Nature*, **401**, 644.

Patra, P. K., Lal, S., Venkataramani, S., Gauns, M., and Sarma, V. V. S. S., 1998. Seasonal variability in distribution and fluxes of methane in the Arabian Sea. *Journal of Geophysical Research*, **103**, 1167–76.

Paull, C. K., 1997. Drilling for gas hydrates: offshore drilling program Leg 164. Proceedings of the Offshore Technology Conference, Houston, TX, OTC Paper 8294.

Paull, C. K. and Dillon, W. P., 1981. Appearance and distribution of the gas hydrate reflection in the Blake Ridge region, offshore South Carolina. In *US Geological Survey Miscellaneous Field Investigations Map MF-1252*. Woods Hole, MA, USGS.

Paull, C. K. and Dillon, W. P., 2001. *Natural Gas Hydrates*. American Geophysical Union, Geophysical Monograph 124.

Paull, C. K., Hecker, B., Commeau, R., *et al.*, 1984. Biological communities at the Florida escarpment resemble hydrothermal vent taxa. *Science*, **226**, 965–7.

Paull, C. K., Ussler, W., III, Borowski, W. S., and Speiss, F. N., 1995. Methane-rich plumes on the Carolina Continental Rise: associations with gas hydrates. *Geology*, 23, 89–92.

Paull, C. K., Borowski, W. S., Rodriguez, N.M., and the ODP Leg 164 Shipboard Scientific Party, 1998. Marine gas hydrate inventory: preliminary results of ODP Leg 164 and implications for gas venting and slumping associated with the Blake Ridge gas hydrate field. In Henriet, J.-P. and Mienert, J. (eds.), *Gas Hydrates: Relevance to World Margin Stability and Climate Change.* Geological Society of London, Special Publication 137, 153–60.

Paull, C., Ussler, W., III, and Borowski, W. S., 1999. Freshwater ice rafting: an additional mechanism for the formation of some high-latitude submarine pockmarks. *Geo-Marine Letters*, 19, 164–8.

Paull, C. K., Ussler, W., III, and Dillon, W. P., 2000. Potential role of gas hydrate decomposition in generating submarine slope failures. In Max, M. D. (ed.), *Natural Gas Hydrate in Oceanic and Permafrost Environments.* Dordrecht, Kluwer Academic Publishers, 149–56.

Paull, C., Ussler, W., III, Maher, N., et al., 2002. Pockmarks off Big Sur, California. *Marine Geology*, 181, 323–35.

Paull, C. K., Brewer, P. G., Ussler, W., III, et al., 2003. An experiment demonstrating that methane slumping is a mechanism to transfer methane from seafloor gas-hydrate deposits into the upper ocean and atmosphere. *Geo-Marine Letters*, 22, 189–203.

Pecher, I. A., Ranero, C. R., Huene, R. von, Minshull, T. A., and Singh, S. C., 1998. The nature and distribution of bottom simulating reflectors at the Costa Rican convergent margin. *Geophysical Journal International*, 133, 219–29.

Pecher, I., Henrys, S., and Zhu, H., 2003. Methane focussing on the Hikurangi Margin, New Zealand – inferences from the reflection strength of BSRs. European Geophysical Society–American Geophysical Union–European Union of Geosciences Joint Assembly, Nice, 6–11 April, Abstract 8018.

Pecher, I., Kukowski, N., Chiswell, S., et al., 2004. Seafloor erosion from repeated dissociation and formation of gas hydrates? Evidence from the Hikurangi Margin, New Zealand. *Geophysical Research Abstracts*, 6, 01698.

Peckman, J., Thiel, V., Michaelis, W., et al., 1999. Cold seep deposits of Beauvoisin (Oxfordian; southeastern France)

and Marmorito (Miocene; northern Italy): microbially induced authigenic carbonates. *International Journal of Earth Sciences*, 88, 60–75.

Peckmann, J., Reimer, A., Luth, U., et al., 2001. Methane-derived carbonates and authigenic pyrite from the northwestern Black Sea. *Marine Geology*, 177, 129–50.

Perfit, M. R. and Davidson, J. P., 2000. Plate tectonics and volcanism. In Sigurdsson, H. (ed.), *Encyclopedia of Volcanoes.* New York, Academic Press, 89–131.

Petford, N. and McCaffery, K. J. W., 2003. *Hydrocarbons in Crystalline Rocks.* Geological Society of London, Special Publication 214.

Pflüger, F., 1999. Matground structures and redox facies. *Palaios*, 14, 25–39.

Pichler, T. and Dix, G. R., 1996. Hydrothermal venting within a coral reef ecosystem, Ambitle Island, Papua New Guinea. *Geology*, 24, 435–8.

Pieri, M. and Mattavelli, L., 1986. Geologic framework of Italian petroleum resources. *American Association of Petroleum Geologists (Bulletin)*, 70, 103–30.

Pimenov, N. V., Rusanov, I. I., Poglazova, M. N., et al., 1997. Bacterial mats on coral-shaped carbonate structures in methane seep areas of the Black Sea. *Mikrobiologiya*, 66, 354–60. (In Russian.)

Pimenov, N., Savvichev, A., Lein, A., and Ivanov, M., 2000. Microbiology of North Atlantic cold seeps. *Abstracts of the Sixth International Conference on Gas in Marine Sediments*, St Petersburg VNIIOkeangeologia, 111.

Pinheiro, L., Ivanov, M., Sautkin, A., et al., 2003. Mud volcanism in the Gulf of Cadiz: results from the TTR-10 cruise. *Marine Geology*, 195, 131–51.

Piper, D. J. W., Shor, A. N., Farre, J. A., O'Connell, S., and Jacobi, R., 1985. Sediment slides and turbidity currents on the Laurentian Fan: sidescan sonar investigations near the epicenter of the 1929 Grand Banks earthquake. *Geology*, 13, 538–41.

Planke, S., Svensen, H., Jamtveit, B., and Podladtchikov, Y., 2002. Seeps and sediment volcanism in volcanic basins, part II: marine seismic and borehole analysis in the Vøring and Møre basins, Norway. In Gas in Marine Sediments, *Seventh International Conference, Baku, Azerbaijan, 7th-12th October 2002.* Baku, Nafta Press, 159–60.

Planke, S., Svensen, H., Hovland, M., Banks, D. A., and Jamtveit, B., 2003. Mud and fluid migration in active mud volcanoes in Azerbaijan. *Geo-Marine Letters*, 23, 258–68.

Platt, J., 1977. Significance of pockmarks for engineers. *Offshore Engineer*, August 1977, 45.

Playford, P. E., Cockbain, A. E., and Low, O. H., 1976. *Geology of the Perth Basin, Western Australia*. Geological Survey of Western Australia, Bulletin 124.

Pontoise, B. and Hello, Y., 2002. Monochromatic infra-sound waves recorded offshore Ecuador: possible evidence of methane release. *Terra Nova*, **14**, 425–35.

Pontoppidan, E., 1752. *Det første forsøg paa Norges Naturlige Historie*. (The first attempt on Norway's natural history.) Copenhagen. (In Norwegian.)

Popenoe, P., Schmuck, E. A., and Dillon, W. P., 1993. The Cape Fear landslide: slope failure associated with salt diapirism and gas hydrate decomposition. In Schwab, W. C., Lee, H. J., and Twichell, D. C. (eds.), *Submarine Landslides: Selected Studies in the US Exclusive Economic Zone*. United States Geological Survey, Bulletin, 2002, 40–53.

Popescu, I., Lericolais, G., Panin, N., *et al.*, 2004. The Danube submarine canyon (Black Sea: morphology and sedimentary processes). *Marine Geology*, **206**, 249–65.

Porfir'ev, V. B., 1974. Inorganic origin of petroleum. *American Association of Petroleum Geologists (Bulletin)*, **58**, 3–33.

Potter, J. and Konnerup-Madison, J., 2003. Hydrocarbon occurrences in igneous rocks. In Petford, N. and McCaffery, K. J. W., 2003. *Hydrocarbons in Crystalline Rocks*. Geological Society of London, Special Publication 214, 151–73.

Powell, E. N. and Bright, T. J., 1981. A thiobios does exist – Gnathostomulid domination of the canyon community at the East Flower Garden Brine Seep. *Internationale Revue der Gesamten Hydrobiologie*, **66**, 675–83.

Powell, E. N., Bright, T. J., Woods, A., and Gittings, S., 1983. Meiofauna and the thiobios in the East Flower Garden Brine Seep. *Marine Biology*, **73**, 269–83.

Premchitt, J., Rad, N. S., To, P., Shaw, R., and James, J. W. C., 1992. A study of gas in marine sediments in Hong Kong. *Continental Shelf Research*, **12**, 1251–64.

Prince, P. K., 1990. Current drilling practice and the occurrence of shallow gas. In Ardus, D. A. and Green, C. D. (eds.), *Safety in Offshore Drilling: the Role of Shallow Gas Surveys*. Dordrecht, Kluwer Academic Publishers, 3–25.

Prior, D. B. and Coleman, J. M., 1982. Active slides and flows in underconsolidated marine sediments on the slopes of Mississippi Delta. In Saxov, S. and Nieuwenhuis, J. K. (eds.), *Marine Slides and Other Mass Movements*. New York, Plenum Press, 21–49.

Prior, D. B. and Hooper, J. R., 1999. Sea floor engineering geomorphology: recent achievements and future directions. *Geomorphology*, **31**, 411–39.

Prior, D. B., Doyle, E. H., and Kaluza, M. J., 1989. Evidence for sediment eruption on deep sea floor, Gulf of Mexico. *Science*, **243**, 517–19.

Pryor, T. A., 1995. New described super-nodule resource. *Sea Technology*, September, 15–18.

Querellou, J., 2003. Biotechnology of marine extremophiles. Book of Abstracts, International Conference on the sustainable development of the Mediterranean and Black Sea environment, Thessaloniki, Greece, 28 May – 1 June (extended abstract).

Quigley, D. C., Hornafius, J. S., Luyendyk, B. P., *et al.*, 1999. Decrease in natural marine hydrocarbon seepage near Coal Oil Point, California, associated with offshore oil production. *Geology*, **27**, 1047–50.

Rad, U. von, Rösch, H., Berner, U., *et al.*, 1996. Authigenic carbonates derived from oxidized methane vented from the Makran accretionary prism off Pakistan. *Marine Geology*, **136**, 55–77.

Rad, U. von, Berner, U., Delisle, G., *et al.*, 2000. Gas and fluid venting at the Makran accretionary wedge off Pakistan. *Geo-Marine Letters*, **20**, 10–19.

Ranero, C. R. and Huene, R. von, 2000. Subduction erosion along the Middle America Convergent Margin. *Nature*, **404**, 748–52.

Rao, Y. H., Subrahmanyam, C., Rastogi, A., and Deka, B., 2001. Anomalous seismic reflections related to gas/gas hydrate occurrences along the western continental margin of India. *Geo-Marine Letters*, **21**, 1–8.

Rathburn, A. E., Levin, L. A., Held, Z., and Lohmann, K. C., 2000. Benthic foraminifera associated with cold methane seeps on the northern California margin: ecology and stable isotopic composition. *Marine Micropaleontology*, **38**, 247–66.

Raynaud, D., Chappellaz, J., and Blünier, T., 1998. Ice-core record of atmospheric methane changes: relevance to climatic changes and possible gas hydrate sources. In Henriet, J.-P. and Mienert, J. (eds.), *Gas hydrates: Relevance to World Margin Stability and Climate Change*. Geological Society of London, Special Publication 137, 293–302.

Raynaud, D., Jouzel, J., Barnola, J. M., *et al.*, 1993. The ice record of greenhouse gases. *Science*, **259**, 926–34.

Reeburgh, W. S., 1969. Observations of gases in Chesapeake Bay sediments. *Limnology and Oceanography*, **14**, 368–75.

Reeburgh, W. S., 1980. Anaerobic methane oxidation: rate depth distribution in Skan Bay sediments. *Earth & Planetary Science Letters*, **47**, 345–52.

Reeburgh, W. S., 1996. 'Soft spots' in the global methane budget. In Lidstrom, M. F. and Tabita, F. R. (eds.), *Microbial Growth on C₁ Compounds*. Dordrecht, Kluwer Academic Publishing, 334–42.

Reeburgh, W. S., Whalen, S. C., and Alperin, M. J., 1993. The role of methylotrophy in the global methane budget. In Murrell, J. C. and Kelly, D. P. (eds.), *Microbial Growth on C₁ Compounds*. Andover, Intercept, 1–14.

Rehder, G. and Suess, E., 2001. Methane and pCO₂ in the Kuroshio and the South China Sea during maximum summer surface temperatures. *Marine Chemistry*, **75**, 89–108.

Rehder, G., Brewer, P. W., Pletzer, E. T., and Friederich, G., 2002. Enhanced lifetime of methane bubble stream within the deep ocean. *Geophysical Research Letters*, **29**, DOI: 10.1029/2001GL013966.

Rehder, G., Keir, R. S., Suess, E., and Pohlmann, T., 1998. The multiple sources and patterns of methane in North Sea waters. *Aquatic Geochemistry*, **4**, 403–27.

Rehder, G., Keir, R. S., Suess, E., and Rhein, M., 1999. Methane in the northern Atlantic controlled by oxidation and atmospheric history. *Geophysical Research Letters*, **26**, 587–90.

Reich, C., Shinn, E. A., Hickey, T. D., and Tihansky, A. B., 1998. Shallow groundwater transport in highly permeable limestones in the Florida Keys: a tracer experiment. National Ground Water Association National Convention, Association of Ground Water Scientists and Engineers, Programme, 179–80 (abstract).

Reid, R. P., Visscher, P. T., Decho, A. W., *et al.*, 2000. The role of microbes in accretion, lamination and early lithification of modern marine stromatolites. *Nature*, **406**, 989–92.

Reimer, A., Peckmann, J., and Reitner, J., 2002. Methane-derived carbonate mineralisation in the northwestern Black Sea. *Gas Hydrates in the Geosystem, Status seminar*. Geotechnologien Science Report No. 1. Kiel, GEOMAR, 105–6 (abstract).

Reston, T. J., Pennell, J., Stubenrauch, A., Walker, I., and Perez-Gussinye, M., 2001. Detachment faulting, mantle serpentinization, and serpentinite-mud volcanism beneath the Porcupine Basin, southwest of Ireland. *Geology*, **29**, 587–90.

Reysenbach, A.-L. and Shock, E., 2002. Merging genomes with geochemistry in hydrothermal ecosystems. *Science*, **296**, 1077–82.

Reyss, J. L., Lemaitre, N., Bonte, P., and Franck, D., 1987. Anomalous ^{234}U/^{238}U ratios in deep-sea hydrothermal deposits. *Nature*, **325**, 798–800.

Rice, D. D., 1992. Controls, habitat, and resource potential of ancient bacterial gas. In Vially, R. (ed.), *Bacterial Gas*. Paris, Editions Technip, 91–118.

Rice, D. D., 1993. Biogenic gas: controls, habitat, and resource potential. In Howells, D. G. (ed.), *The Future of Energy Gases*. United States Geological Survey Professional Paper 1570, 583–606.

Rice, D. D. and Claypool, G. E., 1981. Generation, accumulation, and resource potential of biogenic gas. *American Association of Petroleum Geologists*, **65**, 5–25.

Richardson, M. D. and Davis, A. M. (eds.), 1998. Modeling gassy sediment structure and behavior. *Continental Shelf Research*, **18**, Nos. 14–15, 1669–964.

Richmond, W. C. and Burdick, D. J., 1981. Geologic hazards and constraints of offshore northern and central California. Proceedings of the Offshore Technology Conference, Houston, TX, OTC Paper 4117.

Riding, R., 2000. Microbial carbonates: the geological record of calcified bacterial–algal mats and biofilms. *Sedimentology*, **47**, Supplement 1, 179–214.

Riding, R., 2002. Structure and composition of organic reefs and carbonate mud mounds: concepts and categories. *Earth-Science Reviews*, **58**, 163–231.

Riedel, M., Spence, G. D., Chapman, N. R., and Hyndman, R. D., 2002. Seismic investigations of a vent field associated with gas hydrates, offshore Vancouver Island. *Journal of Geophysical Research*, **107**, 2200.

Rise, L., Sættem, J., Fanavoll, S., *et al.*, 1999. Seabed pockmarks related to fluid migration from Mesozoic bedrock strata in the Skagerrak offshore Norway. *Marine and Petroleum Geology*, **16**, 619–31.

Ritger, S., Carson, B., and Suess, E., 1987. Methane-derived authigenic carbonates formed by subduction-induced pore-water expulsion along the Oregon/Washington margin. *Geological Society of America, (Bulletin)*, **98**, 147–56.

Rivkin, R. B., Bosch, I., Pearse, J. S., and Lassard, E. J., 1986. Bacteriovory: a novel feeding mode for Asteroid larvae. *Science*, **233**, 1311–14.

Roberts, H. H. and Aharon, P., 1994. Hydrocarbon-derived carbonate build-ups of the northern Gulf of Mexico continental slope: a review of submersible investigations. *Geo-Marine Letters*, 14, 135–48.

Roberts, H. H. and Carney, R. S., 1997. Evidence of episodic fluid, gas, and sediment venting on the northern Gulf of Mexico continental slope. *Economic Geology*, 92, 863–79.

Roberts, H. H., Sassen, R., Carney, R., and Aharon, P., 1989. ^{13}C-depleted authigenic carbonate buildups from hydrocarbon seeps, Louisiana Continental Slope. *Transactions of the Gulf Coast Association of Geological Societies*, XXXIX, 523–30.

Roberts, H. H., Aharon, P., Carney, R., Larkin, J., and Sassen, R., 1990. Seafloor responses to hydrocarbon seeps, Louisiana continental slope. *Geo-Marine Letters*, 10, 232–43.

Roberts, H. H., Wiseman, W. J., jr, Hooper, J., and Humphrey, G. D., 1999a. Surficial gas hydrates of the Louisiana continental slope – initial results of direct observations and *in situ* data collection. Proceedings of the Offshore Technology Conference, Houston, TX, OTC Paper 10770.

Roberts, H. H., Menzies, D., and Humphrey, G. D., 1999b. Acoustic wipe-out zones – a paradox for interpreting seafloor geologic/geotechnical characteristics (an example from Garden Banks 161). Proceedings of the Offshore Technology Conference, Houston, TX, OTC Paper 10921.

Roberts, J. M., Long, D., Wilson, J. B., Mortensen, P. B., and Gage, J. D., 2003. The cold-water coral *Lophelia pertusa* (Scleractinia) and enigmatic seabed mounds along the north-east Atlantic margin: are they related? *Marine Pollution Bulletin*, 46, 7–20.

Robertson, A. H. F., Emeis, K.-C., Richter, C., *et al.*, 1998. Collision-related break-up of a carbonate platform (Eratosthenes Seamount) and mud volcanism on the Mediterranean Ridge: preliminary synthesis and implications of tectonic results of ODP Leg 160 in the eastern Mediterranean Sea. In Cramp, A., MacLead, C. J., Lee, S. V., and Jones, E. J. W. (eds.), *Geological Evolution of Ocean Basins: Results from the Ocean Drilling Program*. Geological Society of London, Special Publication 131, 243–71.

Robigou, V., Delaney, J. R., and Stakes, D. S., 1993. Large massive sulphide deposits in a newly discovered active hydrothermal system, the High-Rise Field, Endeavour Segment, Juan de Fuca Ridge. *Geophysical Research Letters*, 20, 1887–90.

Rogers, A. D., 1994. The biology of seamounts. *Advances in Marine Biology*, 30, 305–50.

Rohde, K., 1992. Latitudinal gradients in species diversity: the search for the primary cause. *Oikos*, 65, 514–27.

Rona, P. A., 1984. Hydrothermal mineralization at ocean ridges. *Canadian Mineralogist*, 26, 431–65.

Rona, P. A. and Scott, S. D., 1993. Preface to a special issue on sea-floor hydrothermal mineralization: new perspectives. *Economic Geology*, 88, 1935–76.

Roseen, R., Brannaka, L., and Ballestero, T. P., 2001. Assessing estuarine groundwater nutrient loading by thermal imagery and field techniques verified by piezometric mapping: a methodology evaluation. Geological Society of America Annual Meeting, Boston, MA, November 5–8. (http://gsa.confrenc.com/gsa2001AM/finalprogram/)

Rougerie, F. and Wauthy, B., 1988. The endo-upwelling concept: a new paradigm for solving an old paradox. *Proceedings of the 6th International Coral Reef Symposium, Australia*, 3, 1–6.

Sadler, H. E. and Serson, H. V., 1981. Freshwater anchor ice along an arctic beach. *Arctic*, 34, 62–3.

Saffer, D. M. and Bekins, B. A., 1998. Episodic fluid flow in the Nankai accretionary complex: timescale, geochemistry, flow rates, and fluid budget. *Journal of Geophysical Research*, 103, 30351–70.

Sahimi, M., 1994. *Applications of Percolation Theory*. London, Taylor & Francis.

Sahling, H., Galkin, S. V., Salyuk, A., *et al.*, 2003. Depth-related structure and ecological significance of cold-seep communities – a case study from the Sea of Okhotsk. *Deep-Sea Research*, 550, 1391–409.

Sahling, H., Rickert, D., and Suess, E., 1999. Faunal community structure along a sulphide gradient: interrelationship between porewater chemistry and organisms associated with gas hydrates, Oregon subduction zone. *EOS – Transactions of the American Geophysical Union*, 80, Abstract F510.

Sain, K., Minshull, T. A., Singh, S. C., and Hobbs, R. W., 2000. Evidence for a thick free gas layer beneath the bottom simulating reflector in the Makran accretionary prism. *Marine Geology*, 164, 3–12.

Sakai, H., Gamo, T., Ogawa, Y., and Boulegue, J., 1992. Stable isotope ratios and origins of carbonates associated with cold seepage at the eastern Nankai Trough. *Earth & Planetary Science Letters*, **109**, 391–404.

Salas, C. and Woodside, J., 2002. *Lucinoma kazani* n. sp. (Mollusca: Bivalvia): evidence of a living benthic community associated with a cold seep in the eastern Mediterranean Sea. *Deep-Sea Research I*, **49**, 991–1005.

Salisbury, R. S. K., 1990. Shallow gas reservoirs and migration paths over a central North Sea diapir. In Ardus, D. A. and Green, C. D. (eds.), *Safety in Offshore Drilling: the Role of Shallow Gas Surveys*. Dordrecht, Kluwer Academic Publishers, 167–80.

Salisbury, R. S. K., Denley, M. R., and Douglas, G., 1996. The value of integrating existing 3D seismic into shallow gas studies. Proceedings of the Offshore Technology Conference, Houston, TX, OTC Paper 7990.

Sandstrom, M. W., Meredith, D., and Kaplan, I. R., 1983. Hydrocarbon geochemistry in surface sediments of Alaskan outer continental shelf: Part 2. Distribution of hydrocarbon gases. *American Association of Petroleum Geologists (Bulletin)*, **67**, 2047–52.

Sansone, F. J., 1993. Global carbon dioxide and methane fluxes from shallow-water marine carbonate frameworks. In Oremland, R. S. (ed.), *Biogeochemistry of Global Change: Radiatively Active Trace Gases*. New York, Chapman and Hall, 521–9.

Sansone, F. J., Holmes, M. E., and Popp, B. N., 1999. Methane stable isotopic ratios and concentrations as indicators of methane dynamics in estuaries. *Global Biogeochemical Cycles*, **13**, 463–74.

Sarmiento, J. L., Broecker, W. S., and Biscaye, P. B., 1978. Excess bottom radon 222 distribution in deep ocean passages. *Journal of Geophysical Research*, **83**, 5068–86.

Sassen, R. and MacDonald, I., 1997. Hydrocarbons of experimental and natural gas hydrates, Gulf of Mexico continental slope. *Organic Geochemistry*, **26**, 289–93

Sassen, R., Brooks, J. M., MacDonald, I. R., Kennicut, M. C., II, and Guinasso, N. L., jr, 1993a. How oil seeps, discoveries relate in deepwater Gulf of Mexico. *Oil and Gas Journal*, 19 April, 64–9.

Sassen, R., Brooks, J. M., MacDonald, I. R., *et al.*, 1993b. Association of oil seeps and chemosynthetic communities with oil discoveries, upper continental slope, Gulf of Mexico. *Transactions of the Gulf Coast Association of Geological Societies*, **43**, 349–55.

Sassen, R., Cole, G. A., Drozd, R., and Roberts, H. H., 1994. Oligocene to Holocene hydrocarbon migration and

salt-dome carbonates, northern Gulf of Mexico. *Marine and Petroleum Geology*, **11**, 55–65.

Saunders, D. F., Burston, K. R., and Thompson, C. K., 1999. Model for hydrocarbon microseepage and related near-surface alterations. *American Association of Petroleum Geologists (Bulletin)*, **83**, 170–85.

Sautkin, A., Talukder, A. R., Comas, M. C., Soto, J. I., and Alekseev, A., 2003. Mud volcanoes in the Alboran Sea: evidence from micropalaeontological and geophysical data. *Marine Geology*, **195**, 237–61.

Savidge, G., Forster, P., and Voltolina, D., 1984. Intense localised productivity in the Irish Sea. *Estuarine, Coastal and Marine Sciences*, **18**, 157–64.

Sawyer, W. K., Boyer, C. M., II, Frantz, J. H., jr, and Yost, A. B., II, 2000. Comparative assessment of natural gas hydrate production models. SPE/CERI Gas Technology Symposium, Calgary, Alberta, Canada, April 3–5 SPE Paper No. 62513.

Scanlon, K. M. and Knebel, H. J., 1985. Pockmarks on the floor of Penobscot Bay, Maine. *Geological Society of America (Bulletin)*, **17**, 62 (abstract).

Scanlon, K. and Knebel, H. J., 1989. Pockmarks in the floor of Penobscot Bay, Maine. *Geo-Marine Letters*, **9**, 53–8.

Schmaljohann, R. and Flügel, H. J., 1987. Methane-oxidizing bacteria in Pogonophora. *Sarsia*, **72**, 91–8.

Schmaljohann, R., Faber, E., Whiticar, M. J., and Dando, P. R., 1990. Co-existence of methane- and sulphur-based endosymbioses between bacteria and invertebrates at a site in the Skagerrak. *Marine Ecology Progress Series*, **61**, 119–24.

Schmidt, M., Botz, R., Faber, E., *et al.*, 2003. High-resolution methane profiles across anoxic brine–seawater boundaries in the Atlantis-II, Discovery, and Kebrit deeps (Red Sea). *Chemical Geology*, **200**, 359–75.

Schmuck, E. A. and Paull, C. K., 1993. Evidence for gas accumulation associated with diapirism and gas hydrates at the head of the Cape Fear Slide. *Geo-Marine Letters*, **13**, 145–52.

Schmuck, E. A., Poponoe, P., and Paull, C. K., 1992. Gas venting at the head of the Cape Fear Slide? *Geological Association of America, Southeastern Section, Abstracts with Programs*, **24**, 2 (abstract).

Schoell, M., 1980. The hydrogen and carbon isotope composition of methane from natural gases of various origins. *Geochimica et cosmochimica acta*, **44**, 649–61.

Scholwater, T. T., 1979. Mechanics of secondary hydrocarbon migration and entrapment. *American Association of Petroleum Geologists (Bulletin)*, **63**, 723–60.

Schroot, B. M. and Schüttenhelm, R. T. E., 2003.
Expressions of shallow gas in the Netherlands North
Sea. *Netherlands Journal of Geoscience / Geologie en
Mijnbouw*, **82**, 91–105.

Schubert, C. J., Nürnberg, D., Scheele, N., Pauer, F., and
Kriews, M., 1997. ^{13}C isotope depletion in ikaite
crystals: evidence for methane release from the Siberian
shelves? *Geo-Marine Letters*, **17**, 169–74.

Schüler, F., 1952. Untersuchungen über die Machtigkeit von
Schlikschichten mit Hilfe des Echographen. *Deutsche
Hydrographische Zeitschrift*, **5**, 220–31. (In German.)

Schultz, G. and Pickering, S., 2002. Geohazard prediction
from seismic applications helps prevent drilling losses.
Offshore, October, 100.

Schulz, H.-M., Emeis, K.-C., and Volkman, N., 1997.
Organic carbon provenance and maturity in the mud
breccia from the Napoli mud volcano: indicators of
origin and depth of burial. *Earth & Planetary Science
Letters*, **147**, 141–51.

Schumacher, D., 2000. Surface geochemical exploration for
oil and gas: new life for old technology. *The Leading
Edge*, **19**, 258–61.

Schumm, S. A., 1970. Experimental studies on the formation
of lunar surface features by fluidization. *Geological
Society of America (Bulletin)*, **81**, 2539–52.

Schüttenhelm, R. T. E., Kuijpers, A., and Duin, E. J. Th.,
1985. The geology of some Atlantic abyssal plains and
the engineering implications. In *Advances in Underwater
Technology and Offshore Engineering*. London, Graham
& Trotman, **3**, 29–43.

Schwartz, H., Sample, J., Weberling, K. D., Minisini, D.,
and Moore, J. C., 2003. An ancient linked fluid
migration system: cold seep deposits and sandstone
intrusions in the Panoche Hills, California, USA.
Geo-Marine Letters, **23**, 340–50.

Scoffin, T. P., 1988. The environments of production
and deposition of calcareous sediments on the shelf
west of Scotland. *Sedimentary Geology*, **60**,
107–24.

Scott, S. D., 1997. Submarine hydrothermal systems and
deposits. In Barnes, H. K. (ed.), *Geochemistry of
Hydrothermal Ore Deposits*, 3rd edn. New York, John
Wiley, 797–876.

Scott, S. D., 2001. Deep ocean mining. *Geoscience Canada*,
28, 87–96.

Scranton, M. I. and McShane, K., 1991. Methane fluxes in
the southern North Sea: the role of European rivers.
Continental Shelf Research, **11**, 37–52.

Seach, J., 2001. Mud volcano offshore Trinidad.
http://www.volcanolive.com (accessed June
2005).

Seibold, B. and Berger, W. H., 1982. *The Sea Floor: an
Introduction to Marine Geology*. Berlin, Springer-Verlag.

Seiler, W. and Schmidt, U., 1974. Dissolved nonconservative
gases in seawater. In Goldberg, E. (ed.), *The Sea*. New
York, Wiley, vol. V, 219–43.

Seliverstov, N. I., Torokhov, P. I., Egorov, Yu. O., *et al.*,
1994. Active seeps and carbonates from the Kamchatsky
Gulf (east Kamchatka). *Bulletin of the Geological Society
of Denmark*, **41**, 50–4.

Selley, R. C., 1998. *Elements of Petroleum Geology*, 2nd edn.
London, Academic Press.

Semiletov, I. P., 1999. Aquatic sources of CO_2 and CH_4 in
the ploar regions. *Journal of the Atmospheric Sciences*, **56**,
286–306.

Sen Gupta, B. K. and Aharon, P., 1994. Benthic foraminifera
of bathyal hydrocarbon vents of the Gulf of Mexico:
initial report on communities and stable isotopes.
Geo-Marine Letters, **14**, 88–96.

Shakirov, R., Obzhirov, A., Suess, E., Salyuk, A., and
Biebow, N., 2004. Mud volcanoes and gas vents
in the Okhotsk Sea area. *Geo-Marine Letters*, **24**,
140–9.

Shanks, A. L., 1987. The onshore transport of an oil spill by
internal waves. *Science*, **235**, 1198–200.

Sharp, A. and Samuel, A., 2004. An example study using
conventional 3D seismic data to delineate shallow gas
drilling hazards from the West Nile Delta Deep Marine
Concession, offshore Nile Delta, Egypt. *Petroleum
Geoscience*, **10**, 121–9.

Sherwood Lollar, B., Westgate, T. D., Ward, J. A., Slater,
G. F., and Lacrampe-Couloume, G., 2002. Abiogenic
formation of alkanes in the Earth's crust as a minor
source for global hydrocarbon reservoirs. *Nature*, **416**,
522–4.

Shinn, B. U., Steinen, R. P., Lidz, B. H., and Halley, R. B.,
1985. Bahamian whitings – no fish story. *American
Association of Petroleum Geologists (Bulletin)*, **69**, 307
(abstract).

Shinn, E. A., 1969. Submarine lithification of Holocene
carbonate sediments in the Persian Gulf. *Sedimentology*,
12, 109–44.

Shirayama, Y. and Ohta, S., 1990. Meiofauna in a cold-seep
community off Hatsushima, central Japan. *Journal
of the Oceanographical Society of Japan*, **46**,
118–24.

Sholkowitz, E., 1973. Interstitial water chemistry of the Santa Barbara Basin sediments. *Geochimica et cosmochimica acta*, **37**, 2043–73.

Showstack, R., 2000. Harnessing methane. *EOS – Transactions of the American Geophysical Union*, **81**(20), 222.

Sibuet, M. and Olu, K., 1998. Biogeography, biodiversity and fluid dependence of deep-sea cold-seep communities at active and passive margins. *Deep-Sea Research II*, **45**, 517–67.

Sibuet, M. and Olu-Le Roy, K., 2002. Peculiar benthic ecosystems of continental margins and recent discoveries: major ecological patterns of methane seeps and coral communities from submersible observations. In Cunha, M., Pinheiro, L., and Suzyumov, A. (eds.), *Geosphere/Biosphere/Hydrosphere coupling processes, Fluid Escape Structures, and Tectonics at Continental Ridges*. Paris, UNESCO, IOC Workshop Report No. 183.

Sibuet, M., Juniper, S. K., and Pautot, G., 1988. Cold-seep benthic communities in the Japan subduction zones: geological control of community development. *Journal of Marine Research*, **46**, 333–48.

Siddiquie, H. N., Rao, D. G., and Vora, K. H., 1985. An appraisal of the seabed conditions on the northwestern continental shelf of India. Proceedings of the Offshore Technology Conference, Houston, TX, OTC Paper 4966.

Sieburth, J. M., 1987. Contrary habits for redox-specific processes: methanogenesis in oxic waters and oxidation in anoxic waters. In Sleigh, M. A. (ed.), *Microbes in the Sea*. Chichester, Ellis Horwood, 11–38.

Sieck, H. C., 1973. Gas charged sediment cones pose a possible hazard to offshore drilling. *Oil and Gas Journal*, July 16, 148–63.

Sieck, H. C., 1975. High resolution geophysical studies for resource development and environmental protection. Proceedings of the Offshore Technology Conference, Houston, TX, OTC Paper 2179.

Sills, G. C. and Gonzalez, R., 2001. Consolidation of naturally gassy soft soil. *Géotechnique*, **51**, 629–39.

Sills, G. C. and Nageswaran, S., 1984. Compressibility of gassy soil. Proceedings, Oceanology International Conference, Brighton, London, Society for Underwater Technology, Paper 012.6.

Sills, G. C. and Wheeler, S. J., 1992. The significance of gas for offshore operations. *Continental Shelf Research*, **12**, 1239–50.

Sills, G. C., Wheeler, S. J., Thomas, S. D., and Gardner, T. N., 1991. Behaviour of offshore soils containing gas bubbles. *Géotechnique*, **41**, 227–41.

Silver, E. A., 1996. Introduction to the special section on fluid flow in the Costa Rica accretionary prism. *Geophysical Research Letters*, **23**, 881.

Sim, R. H., 2001. The migration of gas through shallow marine sediments. Unpublished thesis, University of Sunderland.

Simoneit, B. R., 1993. Aqueous high-temperature and high-pressure organic geochemistry of hydrothermal vent systems. *Geochimica et cosmochimica acta*, **57**, 3231–43.

Simoneit, B. R., Lonsdale, P. F., Edmond, J. M., and Shanks, W. C., III, 1990. Deep-water hydrocarbon seeps in Guaymas Basin, Gulf of California. *Applied Geochemistry*, **5**, 41–9.

Sinton, J., 1997. The Manus spreading center near 3°22'S and the Worm Garden hydrothermal site: results of *Mir2* submersible dive 15. *Marine Geology*, **142**, 207–9.

Skempton, A. W., 1954. The pore-pressure coefficients A and B. *Géotechnique*, **4**, 143–7.

Sloan, E. D., 1998. *Clathrate Hydrates of Natural Gas*, 2nd edn. New York, NY, Marcel Dekker.

Smetacek, V., Bodungen, von B., Bröckel, von K., and Zeitzschel, B., 1976. The plankton tower. II. Release of nutrients from sediments due to changes in the density of bottom water. *Marine Biology*, **34**, 373–8.

Smith, L., Zawadzki, W., and Findlater, L., 2001. Modeling of submarine groundwater discharge at Turkey Point, Florida. Geological Society of America, Annual Meeting, Boston, November 5–8. http://gsa.confex.com/gsa/2001AM/finalprogram/

Söderberg, P., 1993. Marine geological investigations in the Åland Sea and the Stockholm Archipelago, Sweden – seismic stratigraphy, tectonics and occurrences of gas related structures. Doctoral thesis, University of Stockholm.

Söderberg, P., 1997. Nature, origin and occurrences of hydrocarbons in a crystalline bedrock environment, Stockholm Archipelago, Sweden. Course Notes, Methane in Marine Sediments, Advanced Study Course, Longhirst, 7–25 July. Brussels, European Commission, MAST III Programme.

Söderberg, P. and Flodén, T., 1991. Pockmark developments along a deep crustal structure in the northern Stockholm Archipelago, Baltic Sea. *Beiträge zur Meereskunde*, **62**, 79–102.

Söderberg, P. and Flodén, T., 1992. Gas seepages, gas eruptions and degassing structures in the seafloor along the Strömma tectonic lineament in the crystalline Stockholm Archipelago, east Sweden. *Continental Shelf Research*, **12**, 1157–71.

Sokolov, V. A., Buniat-Zade, Z. A., Goedekian, A. A., and Dadashev, F. G., 1969. The origin of gases of mud volcanoes and the regularities of their powerful eruptions. In Schenck, P. A. and Havemar, I. (eds.), *Advances in Organic Geochemistry 1968*. Oxford, Pergamon, 473–83.

Soley, J., 1910. The oilfields of the Gulf of Mexico. *Scientific American*, Supplement No. 1788, 9 April.

Solheim, A. and Elverhøi, A., 1985. A pockmark field in the central Barents Sea; gas from a petrogenic source? *Polar Research*, **3**, 11–19.

Solheim, A. and Elverhøi, A., 1993. Gas related sea-floor craters in the Barents Sea. *Geo-Marine Letters*, **13**, 235–43.

Soloviev, V. A., 2001. Gas-hydrate-prone areas of the ocean and gas hydrate accumulations. *Journal of Conference Abstracts*, **6**(1), 158 (abstract/poster).

Soloviev, V. A., 2002. Global estimation of gas content in submarine gas hydrate accumulations. *Geologiya i Geofizika (Russian Geology and Geophysics)*, **43**, 648–61 (Russian edn.) / 609–24 (English edn.).

Soloviev, V. A. and Ginsburg, G. D., 1994. Formation of submarine gas hydrates. *Bulletin of the Geological Society of Denmark*, **41**, 86–94.

Soloviev, V. A. and Ginsburg, G. D., 1997. Water segregation in the course of gas hydrate formation and accumulation in submarine gas-seepage fields. *Marine Geology*, **137**, 59–68.

Somoza, L., 2001. Hydrocarbon seeps, gas hydrate and carbonate chimneys in the Gulf of Cadiz: An example of interaction between tectonic and oceanographic controlling factors. Proceedings, Natural hydrocarbon seeps, global tectonics and greenhouse gas emissions, European Science Foundation, Exploratory Workshop, 27–28 August, 18 (abstract).

Somoza, L., Díaz-del-Río, V., León, R., *et al.*, 2003. Seabed morphology and hydrocarbon seepage in the Gulf of Cadiz mud volcano area: acoustic imagery, multibeam and ultra-high resolution seismic data. *Marine Geology*, **195**, 152–76.

Soter, S., 1998. The Aigion earthquake of 1995: macroscopic anomalies. In Katsonopoulou, D., Schildardi, D., and Soter, S. (eds.), *Helike II, Ancient Helike and Aigialeia*. Aigion, Dora Katsonopoulou, 495–517.

Soter, S., 1999. Macroscopic seismic precursors and submarine pockmarks in the Corinth–Patras Rift, Greece. *Tectonophysics*, **308**, 275–90.

Soter, S. and Katsonopoulou, D., 1998. The search for ancient Helike, 1988–1995: geological, sonar and bore hole studies. In Katsonopoulou, D., Schildardi, D., and Soter, S. (eds.), *Helike II, Ancient Helike and Aigialeia*. Aigion, Dora Katsonopoulou, 67–116.

Southward, A. J., Southward, E. C., Dando, P. R., *et al.*, 1981. Bacterial symbionts and low $^{13}C/^{12}C$ ratios in tissues of Pogonophora indicate unusual nutrition and metabolism. *Nature*, **293**, 616–20.

Spence, G. D., Chapman, N. R., Hyndman, R. D., and Cleary, C., 2001. Fishing trawler nets massive 'catch' of methane hydrates. *EOS – Transactions of the American Geophysical Union*, **82**(50), 621, 627.

Spies, R. B. and Davis, P. H., 1979. The infaunal benthos of a natural oil seep in the Santa Barbara Channel. *Marine Biology*, **50**, 227–37.

Spies, R. B. and Des Marais, D. J., 1983. Natural isotope study of trophic enrichment of marine benthic communities by petroleum seepage. *Marine Biology*, **73**, 67–71.

Spiess, V., Krastel, S., Wagner, M., and Ivanov, M., 2004. Seismoacoustic imaging of gas hydrate and fluid migration structures in the Sorokin Trough, Black Sea. METROL workshop on methane fluxes in ocean margin sediments, Utrecht, June.

Spivack, A. J., Kastner, M., and Ransom, B., 2002. Elemental and isotopic chloride geochemistry and fluid flow in the Nankai Trough. *Geophysical Research Letters*, **29**, 1661–4.

SRU (Seismic Research Unit, University of the West Indies), 2003. Kick 'em Jenny Submarine Volcano. http://www.uwiseismic.com/SRU_Site01/KeJ/kejhome.html (accessed 19 November 2003).

Stadnitskaia, A., Ivanov, M., and Gardner, J., 2000. Hydrocarbon gas composition and distribution in surface sediments of mud volcanic province, Gulf of Cadiz, NE Atlantic. *Sixth International Conference on Gas in Marine Sediments, Abstracts Book*. St Petersburg, VNIIOkeangeologia, 126–9 (abstract).

Stakes, D., Orange, D., Paduan, J. B., Salamy, K. A., and Maher, N., 1999. Cold-seeps and authigenic carbonate formation in Monterey Bay, California. *Marine Geology*, **159**, 93–109.

Stanley, D. J. and Warne, A. G., 1994. Worldwide initiation of Holocene marine deltas by deceleration of sea level rise. *Science*, **265**, 228–31.

Stanley-Wood, N. G., Obata, E., Takahasi, J., and Ando, K., 1990. Liquid fluidisation curves. *Powder Technology*, **60**, 61–70.

Stefanon, A., 1981. Pockmarks in the Adriatic Sea? In Ricci Lucchi, F. (ed.), Excursion Guide Book, International Association of Sedimentologists 2nd European Regional Meeting, Bologna. Bologna, Technoprint, 189–92 (abstract).

Stefansson, V., 1983. Physical environment of hydrothermal systems in Iceland and on submerged oceanic ridges. In Rona, P. A., Bostrom, K., Laubier, L., and Smith, K. L., jr (eds.), *Hydrothermal Processes at Seafloor Spreading Centers*. New York, Plenum Press, 321–60.

Stewart, S. A., 1999. Seismic interpretation of circular geological structures. *Petroleum Geoscience*, **5**, 273–85.

Stoker, M. S., 1981. *Pockmark Morphology: a Preliminary Description. Evidence for Slumping and Doming*. Institute of Geological Sciences, Marine Geophysics Unit, Report 81/10.

Stoker, M. S., Hitchen, K., and Graham, C. C., 1993. *The Geology of the Hebrides and West Shetland Shelves, and Adjacent Deep Water Areas*. United Kingdom Offshore Regional Report. London, HMSO, for the British Geological Survey.

Stoll, R. D. and Bautista, E. O., 1998. Using the Biot theory to establish a baseline acoustic model for seafloor sediments. *Continental Shelf Research*, **18**, 1839–58.

Straughan, D., 1982. Observations on the effects of natural oil seeps in the Coal Oil Point area. *Philosophical Transactions of the Royal Society of London*, B, **297**, 269–83.

Strizhov, V. P., Ustinov, V. I., Nikolayev, S. D., and Isayeva, A. B., 1985. Conditions of formation of vein phosphates on Pacific Ocean seamounts as indicated by isotope data. *Oceanology*, **25**, 90–4.

Strong, S. W. S., 1933. The Sponge Bay uplift, and the Hangaroa mud blowout. *New Zealand Journal of Science and Technology*, **15**, 76–8.

Struckmeyer, H. I. M., Williams, A. K., Cowley, R., et al., 2002. Evaluation of hydrocarbon seepage in the Great Australian Bight. *APPEA Journal*, **2002**, 371–85.

Suess, E., 1977. Nutrients near the depositional interface. In McCave, J. (ed.), *Benthic Boundary Layer*. New York, Plenum Press, 57–9.

Suess, E. and Massoth, G. J., 1984. Evidence for venting of pore waters from subducted sediments of the Oregon continental margin. *EOS – Transactions of the American Geophysical Union*, **65**, 1089 (abstract).

Suess, E., Balzer, W., Hesse, K.-F., et al., 1982. Calcium carbonate hexahydrate from organic-rich sediments of the Antarctic Shelf: precursors of glendonites. *Science*, **216**, 1128–31.

Suess, E., Bohrmann, G., Huene, R. von, et al., 1998. Fluid venting in the eastern Aleutian Subduction Zone. *Journal of Geophysical Research*, **103**, 2597–614.

Suess, E., Bohrmann, G., Greinert, J., et al., 1999a. Giant cold vents and barite mineralization in the Derugin Basin. Fluid seeps on active continental margins. Convention Program, American Association of Petroleum Geologists, Pacific Section Convention, Monterey, CA, 28 April–1 May, 44 (abstract).

Suess, E., Torres, M. E., Bohrmann, G., et al., 1999b. Gas hydrate destabilization: enhanced dewatering, benthic material turnover and large methane plumes at the Cascadia Convergent Margin. *Earth & Planetary Science Letters*, **170**, 1–15.

Sugisaki, R. and Mimura, K., 1994. Mantle hydrocarbons: abiotic or biotic? *Geochimica et cosmochimica acta*, **58**, 2537–42.

Suhayda, J. N., 1974. Determining nearshore infragravity wave spectra. *Proceedings, International Symposium on Ocean Wave Measurement and Analysis. New Orleans, LA, September 9–11, 1974*. New York, American Society of Civil Engineers, 54–63.

Sultan, N., Cochonat, P., Foucher, J.-P., et al., 2003. Effect of gas hydrate dissociation on seafloor slope stability. In Locat, J. and Mienert, J. (eds.), *Submarine Mass Movements and Their Consequences*. Dordrecht, Kluwer Academic Publishers, 103–11.

Sumida, P. Y. G., Yoshinaga, M. Y., Madureira, L. A. S.-P., and Hovland, M., 2004. Seabed pockmarks associated with deepwater corals off SE Brazilian continental slope, Santos Basin. *Marine Geology*, **207**, 159–67.

Sun, S.-C. and Liu, C.-S., 1993. Mud diapirs and submarine channel deposits in offshore Kaohsiung-Hengchun, southwest Taiwan. *Petroleum Geology of Taiwan*, **28**, 1–14.

Svensen, H., Planke, S., Jamtveit, B., and Pedersen, T., 2003. Seep carbonate formation controlled by hydrothermal vent complexes: a case study from the Vøring Basin, the Norwegian Sea. *Geo-Marine Letters*, **23**, 351–8.

Sverdrup, H. U., Johnson, M. W., and Fleming, R. H., 1942. *The Oceans, their Physics, Chemistry and General Biology.* New York, NY, Prentice-Hall.

Swinbanks, D., 1985. Japan finds clams and trouble. *Nature,* 315, 624.

Syvitski, J. P. M., Burrell, D. C., and Skei, J. M., 1987. *Fjords; Processes and Products.* New York, Springer-Verlag.

TANGANYDRO Group, 1992. Sublacustrine hydrothermal seeps in northern lake Tanganyika, east African Rift: 1991 Tanganydro expedition. *BCREDP, Elf Aquitaine Production,* 16, 55–81.

Taniguchi, M., Burnett, W. C., Cable, J. E., and Turner, J. V., 2002. Investigation of submarine groundwater discharge. *Hydrological Processes,* 16, 2115–29.

Taniguchi, M., Wang, K., and Gamo, T., 2003. *Land and Marine Hydrogeology.* Amsterdam, Elsevier.

Tappin, D. R., Watts, P., McMurtry, G. M., Lafoy, Y., and Matsumotu, T., 2001. The Sissano, Papua New Guinea tsunami of July 1998 – offshore evidence on the source mechanism. *Marine Geology,* 175, 1–23.

Tarasov, V. G., Propp, M. V., Propp, L. N., Blinov, S. V., and Kamenev, G. M., 1986. Shallow-water hydrothermal vents and a unique ecosystem of Kraternaya Caldera (Kurile Islands). *Soviet Journal of Marine Biology,* 2, 72–4. (In Russian.)

Tarasov, V. G., Propp, M. V., Propp, L. N., *et al.,* 1990. Shallow-water gasohydrothermal vents of Ushishir Volcano and the ecosystem of Kraternaya Bight (the Kurile Islands). *Marine Ecology,* 11, 1–23.

Tarasov, V. G., Sorokin, Y. I., Propp, M. V., *et al.,* 1993. Structural and functional characteristics of marine ecosystems in zones of shallow-water gasohydrothermal activity in western Pacific Ocean. *Izvestiya RAN, Seriya Biologicheskaya,* 6, 914–26. (In Russian.)

Tarasov, V. G., Gebruk, A. V., Shulkin, V. M., *et al.,* 1999. Effect of shallow-water hydrothermal venting on the biota of Matupi Harbour (Rabaul Caldera, New Britain Island, Papua New Guinea). *Continental Shelf Research,* 19, 79–116.

Taylor, D. I., 1992. Nearshore shallow gas around the UK coast. *Continental Shelf Research,* 12, 1135–44.

Taylor, J. C. M. and Illing, L. V., 1969. Holocene intertidal calcium carbonate cementation, Qatar, Persian Gulf. *Sedimentology,* 12, 69–107.

Taylor, M. H., Dillon, W. P., and Pecher, I. A., 2000. Trapping and migration of methane associated with the gas hydrate stability zone at the Blake Ridge diapir: new insights from seismic data. *Marine Geology,* 164, 79–89.

Taylor, R. K., 1984. Liquefaction of seabed sediments: Tri-axial test simulations. In Denness, B. (ed.), *Seabed Mechanics.* London, Graham and Trotman, 131–8.

Tester, J., Holgate, H. R., Armellini, F. J., *et al.,* 1993. Supercritical water oxidation technology. In Tedler, D. W. and Pohland, F. G. (eds.), *Emerging Technologies in Hazardous Waste Management III.* Washington, DC, American Chemical Society, 35–76.

Thatje, S., Gerdes, D., and Rachor, E., 1999. A seafloor crater in the German Bight and its effects on the benthos. *Helgoland Marine Research,* 53, 36–44.

Thiel, V., Peckmann, J., Seifert, R., *et al.,* 1999. Highly isotopically depleted isoprenoids: molecular markers for ancient methane venting. *Geochimica et cosmochimica acta,* 63, 3959–66.

Thiel, V., Peckmann, J., Richnow, H.-H., *et al.,* 2001. Molecular signals for anaerobic methane oxidation in Black Sea seep carbonates and a microbial mat. *Marine Chemistry,* 73, 97–112.

Thießen, O., Schmidt, M., Botz, R., Schmitt, M., and Stoffers, P., 2004. Methane venting into the water column above Pitcairn and the Society–Austral seamounts, South Pacific. In Hekinian, R., Stoffers, P., and Chemineé, J.-L. (eds.), *Oceanic Hotspots: Intraplate Submarine Magmatism and Tectonism.* Berlin, Springer-Verlag, 407–29.

Thomas, B. M., Møller-Pedersen, P., Whiticar, M. F., and Shaw, N. D., 1985. Organic facies and hydrocarbon distributions in the Norwegian North Sea. In Thomas, B. M. *et al.* (eds.), *Petroleum Geochemistry in the Exploration of the Norwegian Shelf.* London, Graham and Trotman, 3–26, for Norwegian Petroleum Society.

Thomas, G. S. P. and Connell, R. J., 1985. Iceberg drop, dump, and grounding structures from Pleistocene Glacio-lacustrine sediments, Scotland. *Journal of Sedimentary Petrology,* 55, 243–9.

Thor, D. R. and Nelson, C. H., 1979. A summary of interacting, surficial geologic processes and potential geologic hazards in the Norton Basin, northern Bering Sea. Proceedings of the Offshore Technology Conference, Houston, TX, OTC Paper 3400.

Thor, D. R. and Nelson, C. H., 1980. Ice gouging on the subarctic Bering Shelf. In United States Geological Survey Open File Report 80–979.

Thorarinsson, S., 1967. Some problems of volcanism in Iceland. *Geologische Rundschau,* 57, 1–20.

Thrasher, J., Fleet, A. J., Hay, S., Hovland, M., and Düppenbecker, S., 1996. Understanding geology as the key to using seepage in exploration: the spectrum of seepage styles. In Schumacher, D. and Abrams, M. A. (eds.), *Hydrocarbon Migration and its Near-Surface Expression*. American Association of Petroleum Geologists, Memoir 66, 223–41.

Thurman, H. V., 1993. *Essentials of Oceanography*, 4th edn. New York, Macmillan.

Tilbrook, B. D. and Karl, D. M., 1995. Methane sources, distributions and sinks from California coastal waters to the oligotrophic North Pacific gyre. *Marine Chemistry*, 49, 51–64.

Tinivella, U., Lodolo, E., Camerlenghi, A., and Boehm, G., 1998. Seismic tomography study of a bottom simulating reflector off the south Shetland Islands (Antarctica). In Henriet, J-P. and Mienert, J. (eds.), *Gas Hydrates: Relevance to World Margin Stability and Climate Change*. Geological Society of London, Special Publication 137, 141–51.

Tissot, B. and Pelet, R., 1971. Nouvelles données sur les méchanismes de genèse et de migration de pétrole: simulation mathématique et application à la prospection. In *Proceedings of the 8th World Petroleum Congress*. London, Applied Science Publishers, 35–46. (In French.)

Tissot, B. P. and Welte, D. H., 1984. *Petroleum Formation and Occurrence*, 2nd edn. Berlin, Springer-Verlag.

Tkeshelashvili, G. I., Egorov, V. N., Mestrivishvili, Sh. A., et al., 1997. Methane emissions from the Black Sea bottom in the mouth zone of the Supsa River at the coast of Georgia. *Geochemistry International*, 35, 284–8.

Tohno, I. and Shamoto, Y., 1986. Liquefaction damage to the ground during the 1983 Nihonkai-Chubu (Japan Sea) earthquake in Aomori prefecture, Tohoku, Japan. *Natural Disaster Science*, 8, 85–116.

Torokhov, P. V. and Taran, Y. A., 1994. Hydrothermal fields of the Piip Submarine Volcano, Komandorsky Back-arc Basin: chemistry and origin of vent mineralization and bubbling gas. *Bulletin of the Geological Society of Denmark*, 41, 55–64.

Torres, M. E., Bohrmann, G., and Suess, E., 1996. Authigenic barites and fluxes of barium associated with fluid seeps in the Peru subduction zone. *Earth & Planetary Science Letters*, 144, 469–81.

Torres, M. E., Linke, P., Trehu, A., Brown, K. M., and Heeschen, K., 1999. Gas hydrate dynamics at Hydrate Ridge, Cascadia. *EOS – Transactions of the American Geophysical Union*, 80, Abstract F527.

Torres, M. E., McManus, J., and Huh, C.-A., 2002. Fluid seepage along the San Clemente Fault scarp: basin-wide impact on barium cycling. *Earth & Planetary Science Letters*, 203, 181–94.

Tóth, J., 1980. Cross-formational gravity-flow of groundwater: a mechanism of the transport and accumulation of petroleum (the generalized hydraulic theory of petroleum migration). In Roberts, W. H., III and Cordell, R. J. (eds.), *Problems of Petroleum Migration*. American Association of Petroleum Geologists, Studies in Geology No. 10, 121–67.

Townsend, A. R. and Armstrong, T. L., 1990. Shallow gas detection using AVO processing of high resolution seismic data. In Ardus, D. A. and Green, C. D. (eds.), *Safety in Offshore Drilling: the Role of Shallow Gas Surveys*. Dordrecht, Kluwer Academic Publishers, 133–65.

Trager, G. C. and de Niro, M. J., 1990. Chemoautotrophic sulfur bacteria as a food source for mollusks at intertidal hydrothermal vents: evidence from stable isotopes. *The Veliger*, 33, 359–62.

Traynor, J. J. and Sladen, C., 1997. Seepage in Vietnam – onshore and offshore examples. *Marine and Petroleum Geology*, 14, 345–62.

Treude, T., Nauhaus, K., Knittel, K., et al., 2002. A carbonate landscape in the anoxic Black Sea formed by massive mats of methane oxidising Archeae. In *Gas in Marine Sediments*, Seventh International Conference, Baku, Azerbaijan, 7th–12th October 2002. Baku, Nafta Press, 1185–6.

Trofimuk, A. A., Cherskiy, N. V., and Tsarev, V. P., 1973. Osobennosti nakopleniya prirodnykh gazov v zonakh gidratoobrazovaniya Mirovogo okeana. (Accumulation features of natural gases in zones of hydrate development of the ocean.) *Doklady Akademii Nauk SSSR*, 212, 931–4. (In Russian.)

Tryon, M., Brown, K., Dorman, L., and Sauter, A., 2001. A new benthic aqueous flux meter for very low to moderate discharge rates. *Deep-Sea Research I*, 48, 2121–46.

Tryon, M. D., Brown, K. M., Torres, M. E., et al., 1999. Measurements of transience and downward fluid flow near episodic methane gas vents, Hydrate Ridge, Cascadia. *Geology*, 27, 1075–8.

Tucker, M. E. and Bathurst, R. G. C., 1990. Marine diagenesis: modern and ancient. In Tucker, M. E. and Bathurst, R. G. C. (eds.), *Carbonate Diagenesis*. *International Association of Sedimentologists*, Reprint series, 1, 1–9.

Tucker, M. E. and Wright, V. P., 1990. *Carbonate Sedimentology*. Oxford, Blackwell Scientific.

Tunnicliffe, V., 1991. The biology of hydrothermal vents: ecology and evolution. *Oceanography and Marine Biology Annual Review 1991*, **29**, 319–407.

Tunnicliffe, V. and Fowler, C. M. R., 1996. Influence of sea-floor spreading on the global hydrothermal vent fauna. *Nature*, **379**, 531–3.

Tunnicliffe, V., Juniper, S. K., and de Burgh, M. E., 1985. The hydrothermal vent community on Axis Seamount, Juan de Fuca Ridge. In Jones, M. L. (ed.), *Hydrothermal Vents of the Eastern Pacific: an Overview*. Biological Society of Washington, Bulletin No. 6, 453–64.

Uchida, T., 2002. Detection and evaluation of subsurface natural gas hydrates in the Nankai Trough, offshore Japan: a future energy resource? Proceedings, 2002 Denver Annual Meeting, Geological Society of America. Denver, CO, October, Paper No. 156-2 (abstract).

Uchupi, E., Swift, S. A., and Ross, D. A., 1996. Gas venting and late Quaternary sedimentation in the Persian (Arabian) Gulf. *Marine Geology*, **129**, 237–69.

Uenzelmann-Neben, G., Spiess, V., and Bleil, U., 1997. A seismic reconnaissance survey of the northern Congo Fan. *Marine Geology*, **140**, 283–306.

Ujiié, Y., 2000. Mud diapirs observed in two piston cores from the landward slope of the northern Ryukyu Trench, northwestern Pacific Ocean. *Marine Geology*, **163**, 149–67.

UKOOA (United Kingdom Offshore Operators Association), 1997. *Guidelines for the Conduct of Mobile Drilling Rig Site Surveys*. United Kingdom Offshore Operators Association, Report OPS13.

UNDESA (United Nations Department of Economic and Social Affairs), 2005. *Plan of Implementation of the World Summit on Sustainable Development*. New York, Division for Sustainable Development, United Nations Department of Economic and Social Affairs.

Upstill-Goddard, R., Barnes, J., and Owens, N. J. P., 1999. Nitrous oxide and methane during the 1994 SW monsoon in the Arabian Sea/northwestern Indian Ocean. *Journal of Geophysical Research*, **104**, 30067–84.

Upstill-Goddard, R., Barnes, J., Frost, T., Punshon, S., and Owens, N. J. P., 2000. Methane in the southern North Sea: low-salinity inputs, estuarine removal, and atmospheric flux. *Global Biogeochemical Cycles*, **14**, 1205–17.

USGS (United States Geological Survey), 1999. *Natural Oil and Gas Seeps in California*. http://seeps.wr.usgs.gov/ (accessed 23rd January 2004).

Usui, A., Nishimura, A., Tanahashi, M., and Terashima, S., 1987. Local variability of manganese nodule facies on small abyssal hills of the Central Pacific Basin. *Marine Geology*, **74**, 237–75.

Vacelet, J. N., Boury-Esnault, N., Fiala-Médioni, A., and Fisher, C. R., 1995. A methanotrophic carnivorous sponge. *Nature*, **377**, 296.

Vacelet, J. N., Fiala-Médioni, A., Fisher, C. R., and Boury-Esnault, N., 1996. Symbiosis between methane-oxidising bacteria and a deep-sea carnivorous cladorhizid sponge. *Marine Ecology Progress Series*, **145**, 77–85.

Valyaev, B., 1998. Earth hydrocarbon degassing and oil/gas/condensate field genesis. *Gas Industry*, 6–10.

Van Dam, T. M. and Wahr, J. M., 1987. Displacements of the Earth's surface due to atmospheric loading: effects on gravity and baseline measurements. *Journal of Geophysical Research*, **92**, 1281–6.

Van Dover, C. L., 2000. *The Ecology of Deep-Sea Hydrothermal Vents*. Princeton, NJ, Princeton University Press.

Van Dover, C. L., Szuts, E. Z., Chamberlain, S. C., and Cann, J. R., 1989. A novel eye in 'eyeless' shrimp from hydrothermal vents of the mid-Atlantic Ridge. *Nature*, **337**, 458–60.

Van Dover, C. L., Desbruyères, D., Segonzac, M., *et al.*, 1996. Biology at the Lucky Strike hydrothermal field. *Deep-Sea Research*, **43**, 1509–29.

Van Dover, C. L., Humphris, S. E., Fornaria, D., *et al.*, 2001. Biogeography and ecological setting of Indian Ocean hydrothermal vents. *Science*, **294**, 818–23.

Van Dover, C. L., Aharon, P., Bernhard, J. M., *et al.*, 2003. Blake Ridge methane seeps: characterization of a soft-sediment, chemosynthetically based ecosystem. *Deep-Sea Research*, **50**, 281–300.

van Rensbergen, P., and Morley, C. K., 2001. Fluid expulsion from overpressured shale, an alternative for shale diapirism. Examples from offshore Brunei. Conference Proceedings, Subsurface Sediment Mobilization Conference, University of Gent, September 11–13, 73 (abstract).

van Rensbergen, P., de Batist, M., Klerkx, J., *et al.*, 2000. Side-scan sonar evidence of cold seeps in Lake Baikal, Siberia. In *Abstracts of the Sixth International Conference* on *Gas in Marine Sediments.* St Petersburg, VNIIOkeangeologia, 138–9.

van Rensbergen, P., Hills, R. R., Maltman, A. J., and Morley, C. K. (eds.), 2003. *Surface Sediment Mobilization.* Geological Society of London, Special Publication 216.

van Weering, Tj., Jansen, J. H. F., and Eisma, D., 1973. Acoustic reflection profiles of the Norwegian Channel between Oslo and Bergen. *Netherlands Journal of Sea Research,* 6, 214–63.

van Weering, Tj. C. E., de Haas, H., Akhmetzanov, A. M., and Kenyon, N. H., 2003a. Giant carbonate mounds along the Porcupine and SW Rockall Trough margins. In Mienert, J. and Weaver, P. (eds.), *European Margin Sediment Dynamics, Side-Scan Sonar and Seismic Images.* Berlin, Springer-Verlag, 211–16.

van Weering, Tj. C. E., Dullo, C., and Henriet, J.-P., 2003. Geosphere–biosphere coupling: cold seep related carbonate and mound formation and ecology. *Marine Geology,* 198, 1–2.

Vanneste, M., De Batist, M., Golmshtok, A., Kremlev, A., and Versteeg, W., 2001. Multi-frequency seismic sudy of gas hydrate-bearing sediments in Lake Baikal, Siberia. *Marine Geology,* 172, 1–21.

Vasshus, S., 1998. A system for automatic detection of pockmarks in digital terrain models. M.Sc. Thesis, Kingston University.

Vassilev, A. and Dimitrov, L. I., 2000. Spatial and quantity evaluation of the Black Sea gas hydrates. Abstracts of the Sixth International Conference on *Gas in Marine Sediments.* St Petersburg, September 5–9, 140.

Venkatesan, M. I., Kaplan, I. R., and Ruth, E., 1983. Hydrocarbon geochemistry in surface sediments of Alaskan outer continental shelf: Part 1. C_{15+} hydrocarbons. *American Association of Petroleum Geologists (Bulletin),* 67, 831–40.

Vernon, J. W. and Slater, R. A., 1963. Submarine tar mounds, Santa Barbara County, California. *American Association of Petroleum Geologists (Bulletin),* 47, 1624–7.

Vilks, G. and Rashid, M. A., 1975. Foraminifera and organic geochemistry of two sedimentary cores from a pockmarked basin of the Scotian Shelf. In *Report of Activities, Part C.* Geological Survey of Canada, Paper 75–1C, 5–8.

Vogt, P. R. and Jung, W.-Y., 2002. Holocene mass wasting on upper non-Polar continental slopes – due to post-Glacial ocean warming and hydrate dissociation. *Geophysical Research Letters,* 29, 551–4.

Vogt, P. R., Cherkashev, G., Ginsburg, G., *et al.,* 1997. Håkon Mosby mud volcano provides unusual example of venting. *EOS – Transactions of the American Geophysical Union,* 78, 556–7.

Vogt, P. R., Gardner, J., and Crane, K., 1999a. The Norwegian–Barents–Svalbard (NBS) Continental Margin: introducing a natural laboratory of mass wasting, hydrates, and ascent of sediment, pore water, and methane. *Geo-Marine Letters,* 19, 2–21.

Vogt, P. R., Crane, K., Sundvor, E., *et al.,* 1999b. Ground-truthing 11- to 12-kHz side-scan sonar imagery in the Norwegian–Greenland Sea. Part II: probable diapirs on the Bear Island Fan slide valley margins and the Vøring Plateau. *Geo-Marine Letters,* 19, 111–30.

Von Damm, K. L., 2001. Lost City found. *Nature,* 412, 127–8.

Von Damm, K. L., Brockington, M., Bray, A. M., *et al.,* 2003. Extraordinary phase separation and segregation in vent fluids from the southern East Pacific Rise. *Earth & Planetary Science Letters,* 206, 365–78.

Völker, H., Schweisfurth, R., and Hirsch, P., 1977. Morphology and ultrastructure of *Crenothrix polyspora* Cohn. *Journal of Bacteriology,* 131, 306–13.

Walker, P., 1990. UKOOA recommended procedures for mobile drilling rig site surveys (geophysical and hydrographic) – shallow gas aspects. In Ardus, D. A. and Green, C. D. (eds.), *Safety in Offshore Drilling: the Role of Shallow Gas Surveys.* Dordrecht, Kluwer Academic Publishers, 257–89.

Wallmann, K., Linke, P., Suess, E., *et al.,* 1997. Quantifying fluid flow, solute mixing, and biogeochemical turnover at cold vents of the eastern Aleutian subduction zone. *Geochimica et cosmochimica acta,* 61, 5209–19.

Wang, K., Davis, E. E., and Kamp, G. van der, 1998. Theory for the effects of free gas in subsea formations in tidal pore pressure variations and seafloor displacements. *Journal of Geophysical Research,* 103, 12339–53.

Wanninkof, R., 1992. Relationship between windspeed and gas exchange over the ocean. *Journal of Geophysical Research,* 97, 7373–82.

Warren, J. K., 1982. The hydrological significance of Holocene tepees, stromatolites, and boxwork limestones in coastal salinas in South Australia. *Journal of Sedimentary Petrology,* 52, 1171–201.

Watkins, D. J. and Kraft, L. M., jr, 1978. Stability of continental shelf and slope of Louisiana and Texas: geotechnical aspects. In Bouma, A. H. (ed.),

Framework, Facies and Oil-Trapping Characteristics of the Upper Continental Margin. American Association of Petroleum Geologists, Studies in Geology No. 7, 267–86.

Watkins, J. S. and Worzel, J. L., 1978. Serendipity gas seep area, south Texas offshore. *American Association of Petroleum Geologists (Bulletin)*, 62, 1067–74.

Wattrus, N. J., Rausch, D. E., Cartwright, J. A., and Bolton, A., 2001. Lake Superior's rings: an expression of soft-sediment deformation? Conference Proceedings, Subsurface Sediment Mobilisation Conference, University of Gent, September 11–13, 2 (abstract).

Webb, M., 1964. Additional notes on *Sclerolinum brattstromi* (Pogonophora) and the establishment of a new family, Sclerolinidae. *Sarsia*, 15, 33–6.

Weberling, K., Moore, C., Sample, J., Schwartz, H., and Vrolijk, P., 2002. Clastic intrusions and cold seeps in the Cretaceous–Paleocene Great Valley forearc basin, Panoche Hills, CA: structural context of a linked fluid system. In *Gas in Marine Sediments, Seventh International Conference, Baku, Azerbaijan, 7th–12th, October 2002*. Baku, Nafta Press, 192–4 (abstract).

Weeks, S. J., Currie, B., Bakun, A., and Peard, K. R., 2004. Hydrogen sulphide eruptions in the Atlantic Ocean off southern Africa: implications of a new view based on SeaWiFS satellite imagery. *Deep-Sea Research I*, 51, 153–72.

Weiland, C. M., Barth, G. A., Chadwick, B., Embley, R. W., and Getsiv, J., 2000. Virtual exploration of the New Millennium Observatory (NeMO) at Axial Volcano, Juan de Fuca Ridge. *EOS – Transactions of the American Geophysical Union*, 81, Supplement, 1265–6.

Welhan, J. A. and Craig, H., 1983. Methane, hydrogen, and helium in hydrothermal fluids at 21° N, East Pacific Rise. In Rona, P. A., Bostrom, K., and Smith, K. L. (eds.), *Hydrothermal Processes at Seafloor Spreading Centres*. New York, NY, Plenum, 225–78.

Welhan, J. A. and Lupton, J. E., 1987. Light hydrocarbon gases in Guaymas Basin hydrothermal fluids: thermogenic versus abiogenic origin. *American Association of Petroleum Geologists (Bulletin)*, 71, 215–23.

Wells, R. E., Engebretson, D. C., Snavely, P. D., jr, and Coe, R. S., 1984. Cenozoic plate motions and the volcanotectonic evolution of western Oregon and Washington. *Tectonics*, 3, 275–94.

Wellsbury, P., Goodman, K., Cragg, B. A., and Parkes, R. J., 2000. The geomicrobiology of deep marine sediments from Blake Ridge containing methane hydrate (Sites

994, 995, and 997). In Paull, C. K., Matsumoto, R., Wallace, P. J., and Dillon, W. P. (eds.), *Proceedings of the Ocean Drilling Program, Scientific Results*, 164, 379–91.

Wendt, J., Belka, Z., Kaufmann, B., Kostrewa, R., and Hayer, J., 1997. The world's most spectacular carbonate mounds (Middle Devonian, Algerian Sahara). *Journal of Sedimentary Research*, 67, 424–36.

Werne, J. P., Pancost, R. D., Hopmans, E. C., and Damsté, J. S. S., 2001. Spatial variability in the anaerobic methane-oxidizing microbial community in Mediterranean mud volcanoes: evidence from lipid biomarkers and carbon isotopic compositions. Procdings of the European Union of Geologists 11th Conference, Strasbourg, 8–12 April. (See *Journal of Conference Abstracts*, 6(1), 161.)

Wernecke, G., Flöser, G., Korn, S., Weitkamp, C., and Michaelis, W., 1994. First measurements of the methane concentration with a new *in-situ* device. *Bulletin of the Geological Society of Denmark*, 41, 5–11.

Werner, F., 1978. Depressions in mud sediments (Eckernförde Bay, Baltic Sea), related to sub-bottom and currents. *Meyniana*, 30, 99–104.

Westbrook, G. K., Carson, B., Musgrave, R. J., and Suess, E., 1994. *Proceedings of the Ocean Drilling Program, Initial Reports*, 146.

Wever, T. F. and Fiedler, H. M., 1995. Variability of acoustic turbidity in Eckernförde Bay (southwest Baltic Sea) related to the annual temperature cycle. *Marine Geology*, 125, 21–7.

Wever, Th. F., Abegg, F., Fiedler, H. M., Fechner, G., and Stender, I. H., 1998. Shallow gas in the muddy sediments of Eckernförde Bay, Germany. *Continental Shelf Research*, 18, 1715–40.

Wheeler, S. J., 1988. A conceptual model for soils containing large gas bubbles. *Géotechnique*, 38, 389–97.

Wheeler, S. J., Sham, W. K., and Thomas, S. D., 1990. Gas pressure in unsaturated offshore soils. *Canadian Geotechnical Journal*, 27, 79–89.

White, K., 2002. Leg 204 Ocean Drilling Program explores large gas hydrate field offshore Oregon. http: www.oceandrilling.org/Newsroom/Releases/ 204_hydrates.html (accessed December 2002).

White, W. M., Patchett, J., and Ben Othman, D., 1986. Hf isotope ratios of marine sediments and Mn nodules: evidence for a mantle source of Hf in seawater. *Earth & Planetary Science Letters*, 79, 46–54.

Whitfield, J., 1999. Mercurial vents. *Nature*, 401, 755.

Whiticar, M. J., 1978. Relationships of interstitial gases and fluids during early diagenesis in some marine sediments. Doctoral thesis, University of Kiel.

Whiticar, M. J., 1999. Carbon and hydrogen isotope systematics of bacterial formation and oxidation of methane. *Chemical Geology*, **161**, 291–314.

Whiticar, M. J., 2002. Diagenetic relationships of methanogenesis, nutrients, acoustic turbidity, pockmarks and freshwater seepages in Eckernförde Bay. *Marine Geology*, **182**, 29–53.

Whiticar, M. J. and Werner, F., 1981. Pockmarks: submarine vents of natural gas or freshwater seeps? *Geo-Marine Letters*, **1**, 193–9.

Whiticar, M. J., Suess, B., and Wehner, H., 1985. Thermogenic hydrocarbons in surface sediments of the Bransfield Strait, Antarctic Peninsula. *Nature*, **314**, 87–90.

Whiticar, M. J., Faber, E., and Schoell, M., 1986. Biogenic methane formation in marine and freshwater environments: CO1/2 reduction vs acetate fermentation; isotope evidence. *Geochimica et cosmochimica acta*, **50**, 693–709.

Wiedicke, M., Neben, S., and Spiess, V., 2001. Mud volcanoes at the front of the Makran accretionary complex, Pakistan. *Marine Geology*, **172**, 57–73.

Wiese, K. and Kvenvolden, K. A., 1993. Introduction to microbial and thermal methane. In Howell, D. G. (ed.), *The Future of Energy Gases*, United States Geological Survey Professional Paper 1570, 13–20.

Wilcock, W. S. D., Delong, E. F., Kelley, D. S., Barross, J. A., and Carry, S. C. (eds.), 2004. *The Subseafloor Biosphere at Mid-Ocean Ridges*, American Geophysical Union, Geophysical Monograph 144.

Wilken, R. T. and Barnes, H. L., 1996. Pyrite formation by reactions of iron monosulphides with dissolved inorganic and organic sulphur species. *Geochimica et cosmochimica acta*, **60**, 4167–79.

Wilken, R. T. and Barnes, H. L., 1997. Formation processes of framboidal pyrite. *Geochimica et cosmochimica acta*, **61**, 323–39.

Wilkens, R. H. and Richardson, M. D., 1998. The influence of gas bubbles on sediment acoustic properties: *in situ*, laboratory, and theoretical results from Eckernförde Bay, Baltic Sea. *Continental Shelf Research*, **18**, 1859–92.

Williams, A. and Lawrence, G., 2002. The role of satellite seep detection in exploring the South Atlantic's ultradeep water. In Schumacher, D. and LeSchack, L. A. (eds.), *Surface Exploration Case Histories:*

Applications of Geochemistry, Magnetics, and Remote Sensing. American Association of Petroleum Geologists, Studies in Geology No. 48 and SEG Geophysical References Series No. 11, 327–44.

Williams, M. O., 1946. Bahrain: port of pearls and petroleum. *National Geographic*, **89**, 194–210.

Williams, P. R., Pingram, C. J., and Dow, C. B., 1984. Mélange production and the importance of shale diapirism in accretionary terranes. *Nature*, **309**, 145–6.

Wilson, R. D., Monaghan, P. H., Osanik, A., Price, L. C., and Rogers, M. A., 1974. Natural marine oil seepage. *Science*, **184**, 857–65.

Winn, C. D., Karl, D. M., and Massoth, G. J., 1986. Microorganisms in deep-sea hydrothermal plumes. *Nature*, **320**, 744–6.

Withjack, M. O. and Schemer, C., 1982. Fault patterns associated with domes - an experimental and analytical study. *American Association of Petroleum Geologists (Bulletin)*, **66**, 302–16.

Wittenberg, J. B., 1985. Oxygen supply to intracellular bacterial symbionts. In Jones, M. L. (ed.), *Hydrothermal Vents of the Eastern Pacific: An Overview*. Bulletin of the Biological Society of Washington, No. 6, 301–10.

Wittenberg, J. B., Morris, R. J., Gibson, Q. H., and Jones, M. L., 1981. Hemoglobin kinetics of the Galapagos Rift vent worm, *Riftia pachyptila* Jones (Pogonophora: Vestimentifera). *Science*, **213**, 344–6.

Wood, G. A., Orren, R. J., and Conway, A. M., 2000. 3D high resolution seismic; case study deepwater west Africa. Proceedings of the Offshore Technology Conference, Houston, TX, OTC Paper 12068.

Wood, W. T., Gettrust, J. F., Chapman, N. R., Spence, G. D., and Hyndman, R. D., 2002. Decreased stability of methane hydrates in marine sediments owing to phase-boundary roughness. *Nature*, **420**, 656–60.

Woodside, J. and Ivanov, M., 2002. Is there a shallow BSR in the eastern Mediterranean? In *Gas in Marine Sediments, Seventh International Conference, Baku, Azerbaijan, 7th–12th October 2002*. Baku, Nafta Press, 189–91.

Woodside, J. M. and Shipboard Scientists of the MEDINAUT / MEDINETH Projects, 2001. Nautile observations of eastern Mediterranean mud volcanoes and gas seeps – results from the MEDINAUT and MEDINETH projects. In Akhamanov, G. and Suzyumov, A. (eds.), *Geological processes on deep-water European margins*. Paris, UNESCO, International Oceanographic Commission Workshop Report No. 175 on the International Conference and ninth post-cruise meeting of the Training Through Research Programme, Moscow-Mozhenka, Russia, 28 January–2 February, 39–40.

Woodside, J. M., Ivanov, M. K., and Limonov, A. F. (eds.), 1997. *Neotectonics and Fluid Flow through Seafloor Sediments in the Eastern Mediterranean and Black Seas.* Intergovernmental Oceanographic Commission Technical Series, **48** (parts I and II).

Woodside, J. M., Ivanov, M. K., Limonov, A. F. and Shipboard Scientists of the Anaxiprobe Expeditions, 1998. Shallow gas and gas hydrates in the Anaximander Mountains regions, eastern Mediterranean Sea. In Henriet, J.-P. and Mienert, J. (eds.), *Gas Hydrates: Relevance to World Margin Stability and Climate Change.* Geological Society of London, Special Publication 137, 177–93.

Woolsey, T. S., McCallum, M. E., and Schumm, S. A., 1975. Modelling of diatreme emplacement by fluidization. *Physics and Chemistry of the Earth*, **9**, 29–42.

Worzel, J. L. and Watkins, J. S., 1974. Location of a lost drilling platform. Proceedings of the Offshore Technology Conference, Houston, TX, OTC Paper 2016.

Wuebbles, D. J. and Hayhoe, K., 2002. Atmospheric methane and global change. *Earth-Science Reviews*, **57**, 177–210.

Xu, W. and Ruppel, C., 1999. Predicting the occurrence, distribution, and evolution of methane gas hydrate in porous marine sediments. *Journal of Geophysical Research*, **104**, 5081–95.

Yassir, N. A., 1989. Mud volcanoes and the behaviour of overpressured clays and silts. Ph.D. thesis, University of London.

Yellowless, J., 1987. Ancient wreck in the belly of a live volcano. *Diver*, July, 16–17.

Yuan, F., Bennell, J. D., and Davis, A. M., 1992. Acoustic and physical characteristics of gassy sediments in the western Irish Sea. *Continental Shelf Research*, **12**, 1121–34.

Yun, J. W., Orange, D. L., and Field, M. E., 1999. Subsurface gas offshore of northern California and its link to submarine geomorphology. *Marine Geology*, **154**, 357–68.

Zektzer, I. S., Ivanov, V. A., and Meskheteli, A. V., 1973. The problem of groundwater discharge to the seas. *Journal of Hydrology*, **20**, 1–36.

Zhao, W.-L., Davis, D. M., Dahlen, F. A., and Suppe, J., 1986. Origin of convex accretionary wedges: evidence from Barbados. *Journal of Geophysical Research*, **91**, 10246–58.

Zhou, Y., Goldfinger, C., Johnson, J. E., *et al.*, 1999. Distribution and morphology of venting-related carbonates near Hydrate Ridge, Oregon Margin, based on sidescan sonar and multibeam imagery. *EOS – Transactions of the American Geophysical Union*, **80**, Abstract F510.

Zierenberg, R. A., Fouquet, Y., Miller, D. J., *et al.*, 1998. The deep structure of a sea-floor hydrothermal deposit. *Nature*, **392**, 485–8.

Zimmer, C., 2001. 'Inconvincible' bugs eat methane on the ocean floor. *Science*, **293**, 418–19.

Zonenshayn, L. P., Murdmaa, I. O., Baranov, B. V., *et al.*, 1988. An underwater gas source in the Sea of Okhotsk west of Paramushir Island. *Oceanology*, **27**, 598–602.

Index

Pages containing dedicated sections and definitions are in **bold**.

abalone (*Haliotis* sp.; archeogastropod) 262

abiogenic (inorganic) 4
  methane 87, 150, 151, 157–60, 161, 162, 190, 303, 325
  petroleum 157–62

abyssal
  crater 140–1
  hill 321
  plain 70, 76, 79, 80, 84, 92, 120, 125, 126, 131, 134, 136, 141, 173, 275, 319, 321, 355, 357, 358

accretionary wedge (prism) 4, 138–9, 181, 197, 289, 317, 376, 377
  Aleutian 311
  Banda Arc 94
  Barbados 121–2, 182, 252–3, 273, 274, 282
  Cascadia 109, 111, 181, 184–6, 252, 263, 366–7
  Gulf of Cadiz 68
  Hikurangi 97
  Makran 91, 92, 182, 187, 346
  Mediterranean Ridge 75, 76
  offshore Japan 101, 253
  Peru 119, 139
  South Shetland 118

Acharax sp. (Solemyid bivalve) 91, 272

acid/acidity 76, 96, 100, 144, 148, 149, 152, 297
  amino 152
  pH 96, 100, 148

acorn worms (enteropneusts) 261

attenuation (acoustic) 166–7

acoustic blanking 17, 106, 123, 124, 167, **172–8**

acoustic impedance (Z) 95, **172**, 176, 183

acoustic plume 24–5, 27, 43, 44, 54, 61, 65, 67, 77, 89, 90, 97, 102, 103, 104, 105, 118, 132, 142, 193, 194, 195, 231, 285, **331–2**, 333, 340, 372, 379

acoustic scattering 177

acoustic turbidity (seismic indicator of gas) 8, 15, 16, 17, 21, 22, 24, 25,

30, 33, 35, 37, 50, 51, 53, 54, 56, 57, 58, 60, 62, 66, 67, 88, 92, 93, 99, 104, 106, 108, 109, 111, 120, 123, 124, 127, 128, 129, 131, 153, 165, **167–9**, 174, 178, 193, 196, 205, 206, 207, 221, 291, 313, 331, 346, 357, 372

acoustic velocity (seismic velocity; v_p) 9, 27, 99, 109, 166, 167, 169, 170, 172, 175, 177, 183, 185, 186, 187

acoustic void 51, 74, 94, **172**, 175

acoustic wave length (λ) **166**, 167 (*see* Equation 6.1)

acoustic wipeout (*see* acoustic blanking) 93, 123, **172**

Acropora sp. (coral) 259

actinarian (sea anemone; order Actinaria) 27, 36, 39, 100, 249, 261, 262
  Bolocera tuediae 249
  Metridium senile 249
  Urticina feline 249

adiabatic decompression 149

Adriatic Sea **72–4**, 172, 174, 206–7, 275, 329, 334

Adyar Estuary (India) 342

Aegean
  back-arc basin 326
  Islands 328
  micro-plate 74, 75
  Sea 75, 262, 283, 325, 345

Aeolian Islands (Italy) 325, 339
  Panarea Island 325

aerobic 124, 153, 263, 266, 284, 343

Aetoliko Lagoon (Greece) 75

AFEN (Atlantic Frontier Environmental Network) 386

Africa **69–72**, 136
  African Plate 67, 68, 74
  Algeria 305–6
  Angola 338, 379
  South Africa 70, 140, 145, 237
  West Africa 69–72, 136, 172–4, 244, 272, 368

agriculture 150, 330, 342, 345, 347, 383

ahermatypic coral 255–6, 307

Ahnet Basin (Algeria) 305–6

Aigion (Greece) 75, 230–2
  earthquake 231–2
  fault 232

airgun (seismic source) 43, 105, 167, 184

Alabama (USA) 123

Alaska (USA) **103–6**, 139, 238, 335, 356

Alba field (UK North Sea) 210

Alboran Sea 69

Alboran mud diapir belt 69

alcohol 152

alcyonarian (colonial anthozoan) 262

Aleutian margin (trench/subduction zone) **106–7**, 266, 267, 269, 297, 307, 311

ALF (airborne laser fluorosensor) 94, **380**, 381

algae 58, 155, 250, 255, 258, 274, 294, 306, 307, 308, 309, 385
  Halimeda sp. 94
  Vaucheria dichterma 58

algal bloom 312

algal boundstone 308, 309

algal reef 73, 94

Algeria 305–6

Almazán mud volcano (Gulf of Cadiz) 68

Alsek River (Alaska) 107, 206

Alsek Delta/River Instability Area (Alaska) **107**, 206, 357

Alvin (submarine) 5, 98, 99, 118, 123, 127, 131, 252, 255, 260, 317

alvinellid polychaetes 110, 261, 262

Alvinocaris sp. (shrimp) 274

Amazon river/fan (offshore Brazil) **120**

Ambitle Island (Papua New Guinea) 259

amino acid 152

ammonia (NH_3) 139, 149, 153, 259, 285, 383

ammonite (extinct subclass of cephalopods) 289

ampeliscid amphipods 105

Ampharetidae (polychaetes)

amphipods (crustaceans; order Amphipoda) 87, 100, 105, 192, 261, 262, 274

Chirimedeia zotea 251
Rhepoxynius sp. 251
amplitude blanking (*see also* acoustic
    blanking) 172, 176
Anadarko Basin (Oklahoma) 306
anaerobe 153, 263–4, 266, 284
anaerobic oxidation of methane 91, 263–4,
    295, 297, 298, 300, 337
anaerobic oxidisers of methane (AOM)
    263–5, 295
anaerobic methanotrophs (ANME) 264
Anarhychus lupus (wolf fish) 249
Anastasya Mud Volcano (Gulf of Cadiz)
    68
Anatolian Microplate 74
Anaximander Mountains (Mediterranean
    Sea) 79, 272
Andean 119
andesite/andesitic 96, 102, 148, 325
anemones (sea; order Actinaria) 27, 36,
    39, 46, 93, 96, 249, 253, 255, 261,
    283
    Bolocera tuediae 249
    Metridium senile 249
    Urticina feline 249
Angola 338, 379
anhydrite ($CaSO_4$; mineral) 96, 98, 103,
    118, 147, 148, 307, 314–15,
    316–17, 318, 324
    chimney 103, 314–15
    mound **316–17**
ankerite ($Ca(Fe,Mg,Mn)(CO_3)_2$; mineral)
    68, 297, 300
ANME (anaerobic methanotrophs) 264
annelid (phylum Annelida) 251 (*see also*
    polychaetes)
anoxia/anoxic 52, 75, 81, 83, 124, 136,
    148, 153, 154, 155, 252, 262, 265,
    266, 269, 273, 277, 278, 289, 293,
    295, 298, 300, 301, 302, 342, 343,
    371, 383
Antarctic 118, 148, 274, 311, 350, 351,
    376
Antarctic Plate 118
Anthozoan (class within the phylum
    Cnidaria) 249 (*see also* sea
    anemone, coral and sea pen)
anthracite 156
anthropogenic 55, 259, 330, 331, 342, 345,
    347, 350
anticline/anticlinal 70, 94, 105, 109, 112,
    115, 117, 118, 122, 306
Anvil Point (seeps; UK) 59
AOM (anaerobic oxidisers of methane)
    263–5, 295
apatite($Ca_5(PO_4)_3(F,Cl,OH)$; mineral)
    301, 306, 319

aquifer 57, 72, 88, 100, 101, 113, 120, 129,
    134, 145, 150, 174, 193, 214, 217,
    218, 237, 244, 288, 308, 330, 360
Aquifex (bacterium) 93
Arabian Gulf **88–91**, 158, 193, 194, 197,
    206, 207, 234, 294, 311, 313, 330,
    369
Arabian Plate 91
Arabian Sea 92, 342, 343, 346
aragonite/aragonitic ($CaCO_3$: mineral)
    44, 79, 99, 103, 110, 118, 260, 262,
    291–5, 297, 300, 302, 304, 308,
    309, 312, 313, 314, 316
archaea (single-celled microbes; one of
    the three domains of living things:
    archaea, bacteria, and eukaryotes)
    4, 96, 153, 263–5, 288, 297, 300
    Methanothrix sp. (now *Methanosaeta*
        sp.) 300
    Thermococcus 96
archeogastropods 255
    Haliotis cracherodii 262
    trochid 255
Arctic **45–9**, 132, 238, 289, 311, 348,
    376
    Canada 289, 376
    Ocean 207, 352, 376
Arctica islandica (arcticid bivalve) 35, 248
Area (The international waters under the
    jurisdiction of the International
    Seabed Authority) **384**
area of active venting (AAV; Juan de Fuca
    Ridge) 316
Argentina 120
Argentina silus (fish) 32
Argentine Basin (South Atlantic Ocean)
    120, 136
argon (Ar) 103, 158, 320
armour plating (of gas hydrate) 344
aromatic hydrocarbons 131, 155, 159, 220,
    275
arsenic (As) 98, 310
artesian 88, 113, 134
arthropod (phylum Arthropoda –
    includes crustaceans, etc.) 261
Ascension Fault (offshore California) 113,
    140
ASHES (Axial Seamount Hydrothermal
    Emissions Study) 110
Ashkelon (Israel) 79
Ashmore Reef (Timor Sea) 303
asphalt/tar 1, 2, 115, 116, 117, 120, 122,
    123, 124, 125, 155, 331, 334
    mounds/mud volcanoes 116, 122,
        125
Astonomena southwardorum (nematode)
    249

Astropecten irregularis (echinoderm) 249
Astryx permodesta (gastropod) 274
Atalante Mud Volcano (offshore
    Barbados) 121–2, 245, 273, 275
Atlantic Ocean 47, 67, 93, 118, 141, 146,
    149, 255, 256, 259, 261, 272, 277,
    278, 319, 326, 342, 343, 352, 355,
    376, 385
    water 37, 68, 238, 257
Atlantic Margin
    European **62–6**, 136, 303, 383
    US **126–7**, 253, 330, 358, 361
Atlantic Plate 121
Atlantis 213
Atlantis II Deep (Red Sea) 87, 374
Atlantis Massif (Atlantic Ocean; *see also*
    Lost City) 72
atmosphere 1, 6, 58, 67, 85, 102, 116, 134,
    203, 214, 236, 285, 308, 311,
    323–8, 332, 338, 339, 341–4,
    345–9, 350, 352, 354, 365, 375
atmospheric methane 6, 58, 85, 103, 127,
    161, 187, 285, 288, 323, **345–9**,
    350–3
atmospheric pressure (P_{atmos}) 72, 214–15,
    216, 235, 236, 240
atoll **141**, 318–20
attenuation (acoustic) 166–7
Auger field (US Gulf of Mexico) 124
Australasia **94–8**
Australia 94, 254, 259, 302, 303, 304, 308,
    372, 381–2
Austalian Plate 94, 97
authigenic (formed *in situ*)
    carbonate (which may/may not be
        MDAC) 79, 112, 113, 115, 116,
        120, 127, 183, 311, 312
    clay 314
    methane-derived authigenic carbonate
        (MDAC) 6, 15, 27, 32, 50, 62, 118,
        145, 161, 175, 195, 200, 249,
        290–307, 331, 369–70, 383
AUV (autonomous underwater vehicle) 2,
    69, 71, 234
Aveiro Mud Volcano (Gulf of Cadiz) 68
awaruite (Ni_3Fe; mineral) 150
Axial Seamount (Juan de Fuca Ridge)
    110–11, 142, 262, 329
Azerbaijan 1, 85–6, 122, 189, 196,
    198–200, 201, 202, 203, 205, 207,
    234, 337
Azeri field (Azerbaijani Caspian Sea) 86
azooxanthellate coral (*see also* ahermatypic
    coral) 255
Azores Islands 385
Åsgard field (offshore Norway) 50,
    51

Bach Ho oil field (offshore Vietnam) 157
back-arc
    basin 98, 101, 102, **140**, 148, 326
    extension/spreading 7, 95, 97, 145,
        324, 374
    volcanism 75, 76, 326
backscatter (acoustic) 64, 65, 66, 110, 113,
    132, 200
bacteria (single-celled microbes; one of
        the three domains of living things:
        archaea, bacteria, and eukaryotes)
    4, 56, 76, 78, 87, 92, 93, 96, 102,
    158, 250, 251, 252, 261, 262, 264,
    266–74, 275, 277, 282, 285, 288,
    289, 297, 301, 309, 324, 383
    Aquifex 93
    Beggiatoa sp. 35, 91, 96, 102, 113, 125,
        132, 193, 200, 249, 250, 251, 252,
        258, 263, 264, 266, 267, 273, 274,
        275, 281, 282, 289, 298, 299, 301
    Desulfovibrio sp. 96
    fossil 288, 289, 297
    Leucothrix sp. 253
    methanotrophic (methane-oxidising;
        see also individual genera) 56, 87,
        102, 250, 252, 261, 269, 270, 272,
        281, 282, 342
    Methylococcus capsulatus 383
    Spirochaeta sp. 96
    sulphate-reducing bacteria (SRBs) 69,
        153, 261, 263–5, 295, 297, 342
    symbiotic 269, 270, 271–4, 277, 282,
        287, 289
    Thioploca sp. 87, 91, 92, 266
    Thiothrix sp. 253, 266
    Thiotroph (*see also individual genera*)
        102, 114, 250, 253, 263, 266, 272,
        281, 282
    Thiovulum sp. 96
bacterial mat 27, 33, 35, 39, 40, 49, 58, 62,
    83, 84, 91, 95, 98, 102, 103, 107,
    109, 110, 112, 113, 114, 119, 125,
    131, 132, 193, 200, 248, 249, 250,
    251, 252, 253, 254, 262, 266–9,
    274, 277, 279, 280, 285, 301, 304,
    310
bacterioplankton 251
bacterivores 274, 276
Baffin Island/Shelf (E Canada) **131–2**,
    329, 349
Bahamas 308, 311, 312, 313
Bahrain 2, 88, 330
Baja California 105, 142
Baker Bank (offshore Texas) 258
Bakhar Mud Volcano (Azerbaijan) 200,
    201, 205
Balder field (Norwegian North Sea) 211

Balearic Islands (Spain) 69
Balkanides 81
Baltic Sea **56–9**, 284, 342, 343, 352,
    386
Balmoral field (UK North Sea) 9
Banda Arc 94
Bantry Bay (Ireland) 60
Baraza Mud Volcano (Gulf of Cadiz) 67,
    68
Barbados 203, 245, 246, 275, 278
    accretionary wedge **121–2**, 182, 252,
        273, 274, 282
    trench 252
Barcley Canyon (offshore British
    Columbia) 340
Barents Sea 13, **45–7**, 48, 49, 50, 52, 192,
    238, 239, 273, 334, 352, 353
barium (Ba) 76, 107, 119, 148, 284, 299,
    318, 325, 374
barnacle (crustaceans; infraclass
    Cirripedia) 261
Barra (Outer Hebrides, UK) 62
baryte/barium sulphate (BaSO₄; mineral)
    98, 103, 107, 119, 299, 306, 314,
    316, 322, 367
basalt/basaltic (type of basic rock) 98,
    136, 137, 138, 140, 142, 145, 146,
    147, 148, 149, 157, 158, 222, 288,
    303, 318, 319, 321, 325, 328
basement 7, 58, 59, 119, 121, 132, 146,
    157, 160, 178, 214, 382
Bathymodiolus sp. (mytilid bivalve) 97, 99,
    252, 272
Bay of Plenty (New Zealand) 283
beachrock 73, **294**, 308
Bear Island Fan (Barents Sea/northeast
    Atlantic Ocean) 47, 50
Beaufort Sea 103, 348
Beauvoisin (France) 289, 298
Becken Effekt 56, 165
Beggiatoa sp. (sulphur-oxidising
        bacterium) 35, 91, 96, 102, 113,
        125, 132, 193, 200, 249, 250, 251,
        252, 258, 263, 264, 266, 267, 273,
        274, 275, 281, 282, 289, 298, 299,
        301
Belfast Bay (Maine) 128, 129, 191, 195
Belgica Mounds (offshore Ireland) 64,
    302, 303
Belgium 385
Belize 259
Bent Hill (Juan de Fuca Ridge) 315
benthic fauna 3, 27, 28, 29, 36, 102, 249,
    250, 279, 336
Benthodytes sp. (holothurian) 261
Bering Sea **103–5**, 142, 192, 352
Bermuda Triangle 372, **373–4**

Berri oilfield (offshore Saudi Arabia) 88,
    89, 90
Beryl field (UK North Sea) 232
Besshi-type mineral deposits 374
bicarbonate (HCO₃⁻; *see*
    hydrogencarbonate)
Big Sur (California) **114–15**, 193, 194
bioactive compounds 383
biodiversity
    species 31, 44, 99, 248, 249, 262, 275,
        276–8, 279, 304, 384, 385
    global **286–7**
biofilm 297
biogenic 4, 151, 160, 162, 259, 294, 302,
    308, 385 (*see also* microbial
    methane and thermogenic
    methane)
    methane 4, 151, 160, 162
biogeochemical/biogeochemistry 92, 107,
    126, 186, 341, 344
bioherm 27, 29, 248, 255, 259, 280, 290,
    301, 302
bioirrigation 107, 297
biological oxygen demand (BOD) 152
biomarker 154, 264, 289, **297**, 298
    hopanoid 264
biomass 100, 102, 120, 249, 250, 253, 254,
    265, 268, 273, 276, 277, 279, 280,
    287, 288, 297, 325, 347, 374
biopolymers 152
bioproteins 374, 383
biotechnology 355, 383
Boyle's Law 233
bitumen (*see also* asphalt/tar) 155, 158,
    220
bivalve (class Bivalvia of Phylum
        Mollusca) 27, 31, 35, 36, 53, 69,
        70, 78, 97, 100, 102, 103, 107, 109,
        112, 118, 120, 122, 244, 248, 249,
        250, 251, 252, 253, 261, 262, 269,
        271–3, 275, 278, 279, 280, 281,
        282, 289, 299, 305
    Arcticid
        Arctica islandica 35, 248
    clam (general) 79, 99, 101, 105, 107,
        112–13, 114, 119, 120, 122, 125,
        250, 252, 253, 255, **271–3**, 279,
        283, 299
    lucinid (clams) 70, 125, 244, 271,
        272–3, 278, 279, 289, 305
        Lucinoma sp. 249, 272
        Myrtea sp. 272
    mytilid (mussels) 119, 261, 269, 271,
        272, 278, 279, 287
        Bathymodiolus sp. 97, 99, 252, 272
        Mytilus edulis 282
        Tamu childressi 252

solemyid (clams) 100, 113, 119, 271, **272**, 278, 289
 Acharax sp. 91, 272
 Solemya sp. 272, 299
Symbiosis **271**
thyasirid (clams) 69, 97, 102, 271, **272**, 281
 Conchocele bisecta 102, 272
 Maorithyas sp. 97
 Thyasira sp. 249, 278, 280, 281
vesicomyid (clams) 100, 107, 112, 113, 114, 119, 120, 122, 125, 252, 261, 269, **271–2**, 278, 279, 283
 Calyptogena sp. 70, 91, 97, 99, 100, 103, 112, 113, 118, 120, 122, 252, 253, 262, 271, 278, 299
Black Sea 2, **81–5**, 182, 183, 203, 245, 265, 297, 299, 300, 302, 342, 347
black smoker 96, 103, 274, 314, 315, 324
Blake Ridge (offshore South Carolina) **126–37**, 136, 181, 182, 183, 272, 288, 329, 338, 373
 Depression **127**
 Diapir **127**, 358
Blake Outer Ridge (offshore South Carolina) 184, 185
blank zone (*see also* acoustic blanking) 172, 174, 177, 209
Block NO1/9 – Tommeliten (Norwegian North Sea) 3, 20, **25–9**, 59, 60, 191, 248, 264, 280, 290, 291, 301, 302, 332, 333
Block NO24/9 – The Holene (Norwegian North Sea) 14, 20, **33–5**, 36, 248, 250
Block NO25/7 (Norwegian North Sea) 14, 15, 20, **30–3**, 34, 35, 206, 248
Block NO26/8 (Norwegian North Sea) 20, **36**, 37
Block NO26/9 (Norwegian North Sea) 20, **36**, 38, 248
Block UK15/25 (UK North Sea) 13, 20, 40, **41–4**, 59, 189, 191, 195, **238–9**, 249, 281, 285, 291, 334, 353, 386
 Challenger pockmark 40, 43, 44
 Scanner pockmark 20, 40, 41, 42, 43, 44, 189, 190, 195, 249, 285, 386
 Scotia pockmark 40, 43, 44
Block UK16/3 – Braemar pockmark (UK North Sea) 386
blowout (during drilling) 73, 108, 123, 178, 184, 225, 238, 275, 355, **362–6**, 368, 369, 372, 384
blowout (natural) 194, 302, **333–5**, 337–8, 344, 347, 352, **371–2**

blowout crater 46, 273, 299
blowout pipe 174, 209, 210, 211
blowout preventer (BOP) 363, 367, 368, 369
BOD (biological oxygen demand) 152
boiling 103, 145, 146, 147, 148, 149, 209, 237, 260, 309, 316
boiling mud 84
'boiling water' 72, 86, 229, 238
Bolocera tuediae (sea anemone) 249
Bolshoi Mud Volcano (Lake Baikal) 87
Bonaccia gas field (Italian Adriatic Sea) 73
Bonaparte Basin (Timor Sea) 94
Bonin (Izu-Ogasawara) island arc 371
Bonin Trench 100
Bonjardim Mud Volcano (Gulf of Cadiz) 68, 69
boomer (seismic profiling system) 11, 17, 19, 22, 23, 24, 40, 43, 51, 54, 90, 104, 168, 169
BOP (blow-out preventer) 363, 367, 368, 369
Borneo 93, 343
botryoids/botryoidal 110, 291, 293, 302, 304
bottleneck depressions 106
bottom-simulating reflector (BSR) 51, 70, 86, 87, 91, 92, 94, 101, 102, 105, 109, 111, 114, 118, 119, 120, 126, 127, 174, 177, 182, **183–6**, 333, 334, 340, 358, 359, 361, 367, 376, 377
Boundary Bay (British Columbia) 108
Bounty Seamount (South Pacific Ocean) 142
boxwork 308, 309
boundstone 308, 309, 310
Brachyurans (infraorder of crustaceans; includes the true crabs) 261, 271
Braemar pockmark (UK16/3, North Sea) 386
Brandon vents (Rapa Nui hydrothermal field, East Pacific Rise) 317
Bransfield Strait, Antarctica 118, 148
Brazil 120, 379
breccia/brecciation 77, 84, 86, 96, 97, 98, 99, 109, 140, 145, 180, 196, 197, 200, 202–4, 214, 225, 237, 291, 302, 359
Brent Delta (Jurassic, North Sea) 135
bresiliid shrimp 96
Bridgwater Bay (UK) 60
bright spot (seismic indicator of gas) 15, 16, 21, 27, 105, 111, 112, 124, 169, **170–2**, 173, 174, 176, 177, 186, 193, 366

brine 78, 79, 87, 124, 146, 147, 149, 182, 197, 202, 208, 223, 245, 252, 272, 294, 311, 318, 374
brine lake/pool 68, 78, 79, 80, 81, 87, 198, 252, 272, 294, 335
 Nadir Brine Lake (Mediterranean Ridge) 78, 294
brine seepage 68, 252, 294–5, 317
bristle worms (*see* polychaete)
Britannia field (UK North Sea) 9
British Columbia (Canada) **107–9**, 309–10, 312, 340, 356, 357
British Isles **59–66**, 136, 257, 378, 386
brittle stars (ophiuroids) 261, 274
Brosme brosme (torsk; fish) 32, 33, 39, 281
brotulid fish (family Ophidiidae) 261
Browse Basin (Timor Sea) 94
brucite – Mg(OH)$_2$; mineral 149, 313
Brunei 93, 344
BSR (bottom-simulating reflector) 51, 70, 86, 87, 91, 92, 94, 101, 102, 105, 109, 111, 114, 118, 119, 120, 126, 127, 174, 177, 182, **183–6**, 333, 334, 340, 358, 359, 361, 367, 376, 377
bubble 3, 6, 15, 21, 27, 31, 40, 52, 58, 60, 70, 76, 79, 83, 85, 86, 89, 98, 102, 107, 110, 126, 128, 129, 131, 132, 142, 144, 147, 153, 166, 194, 198, 200, 209, 216–17, 219, 231, 235, 241, 242, 252, 255, 260, 262, 263, 272, 278, 280, 283, 285, 288, 309, 310, 329, 331–4, 336–41, 344, 345, 348, 349, 356, 364, 365, 371, 372, 279, 380, 381, 383
 at sea surface 58, 72, 76, 82, 98, 100, 102, 112, 118, 131, 142, 230, 285, 329, 334, 339, 348, 364, 365, 383
 buoyancy of 219, 338, 339, 340 (*see* Equation 7.9)
 dirty **339**
 in sediments 58, 66, 109, 110, 120, **163–5**, 166–9, 176, 178, 182, 206–7, 220–1, 228, 230, 233, 337 (*see also* gas void)
 oily 339, 379, 380
bubble boundary 221, 338, 339
bubble deposition **285–6**
bubble noise 332–3
bubble plume/stream 1, 42, 43, 51, 66, 102, **104, 109**, 123, 125, **194, 195**, 220, 231, 242, 280, 282, 316, 331–2, 333, 334, 336, 339, **364**
bubble size 43, 85, 102, 117, 167, 168, 181, **198, 200**, 221, 230, 332–3, 338, 339, 340, 341
bubblemarks 58, 89

bubbleometer 125, 335–6
Bubbling Reefs (Danish Kattegat) 55, 56, **250, 291, 301,** 386
Buccinium undatum (gastropod) 249
Bulgaria **81,** 305
Bulimina sp. (forminifera) 275
bulk density (ρ_{bulk}) 147, 196, 207, 214, **215,** 216, **240,** 245, 316, 321 (*see also* Equation 7.5)
Bullwinkle oilfield (US Gulf of Mexico) 124
buoyancy 132, 142, 181, 190, 215, 216, 219, 226–8, 320, 324, 369
  bubble/gas 153, 164, 217, **219,** 220, 223, 241, **320,** 338, 339 (*see also* Equation 7.9)
  hydrate 276, 340
  sediment 196, **216,** 236–7, 240, 245, 246 (*see also* Equation 7.7)
  rig/ship 329, 362, 365, 371, **372–4**
buried pockmarks 17, 24, 73–4, 106, 238, 243
Burrard Inlet (British Columbia) 108
Bush Hill (US Gulf of Mexico) 125, 188, 271, 276, 286
Bushveld igneous complex (South Africa) 145
butane (C_4H_{10}; hydrocarbon gas) 27, 40, 44, 84, 102, 103, 125, 157, 158, 160, 161, 180, 256, 295, 301, 316
butyrate ($CH_3CH_2CH_2$-COOH; a carboxylic acid) 152
Buzdag Mud Volcano (Azerbaijani Caspian Sea) 86

Cabos Blanco Slide (offshore Costa Rica) 119
Cabrillo Canyon (Monterey Canyon) 114
cadmium (Cd) 148, 325
Cadiz, Gulf of (offshore Morocco/Portugal/Spain) **67–9,** 181, 296, 297, 299, 301, 321
calcite ($CaCO_3$; mineral) 44, 99, 103, 120, 223, 260, 291, 293, 295, 297, 300, 304, 309, 312–14, 321
  high-magnesium calcite 103, 109, 291–2, **293,** 294–7, 300, 305
caldera 80, 96, 100, 110, 111, 198, 253, 268, 283, 326, 329
California 1, 98, **111–17,** 234–5, 241, 251, 273, 275, 278, 281, 333, 336, 375
  Baja California 105, 142
  Big Sur **114–15,** 194
  Gulf of California **117–18,** 137, 299
  Malibu Point **117,** 230
  Mono Lake 311

Monterey Bay/Canyon **112–14,** 140, 244, 269, 272, 273, 278, 295, 299, 340, 372
Northern California **111–12,** 251, 281
  Eel shelf and slope **111–12,** 139, 281, 299, 357, 362–3, 383
  Klamath Delta 111, **112,** 230, 375
  Obsidian Dome 98
  Palos Verdes 262, 283
  Panoche Hills 289
  Santa Barbara Basin/Channel 2, **115–17,** 194, 234–5, 241, 251, 273, 281, 333, 336, 375
California grey whales 105
Callianassa subterranean (burrowing shrimp) 192
Calyptogena Canyon (offshore Makran Coast, Pakistan) 91
Calyptogena sp. (vesicomyid bivalve) 70, 91, 97, 99, 100, 103, 112, 113, 118, 120, 122, 252, 253, 262, 271, 278, 299
  C. kilmeri 113
  C. pacifica 112, 113
  C. soyoae 99
Cambridge Fjord (Baffin Island, Canada) 132, 329, 349
Camp Cup (Monterey Canyon) 113
Campeche, Gulf of (southern Gulf of Mexico) 123
Campeche Knolls (Mexican Gulf of Mexico) 125
Canada 45, 375
  Arctic 289, 376
  Baffin Island **131–2,** 329, 349
  British Columbia **107–9,** 309–10, 312, 340, 356, 357
  Labrador 129–30
  Newfoundland 131, 132, 238, 355, 374, 383
  Nova Scotia 7–8, 130, 248
Cancer pagarus (edible crab) 250
Cantarell oilfield (Mexican Gulf of Mexico) 399
Canyon 70, 97, 119, 136, 225, 244–5, 356
  Barcley (offshore British Columbia) 340
  Cabrillo (offshore California) 114
  Calyptogena (offshore Pakistan) 91
  Chiclayo (offshore Peru) 299
  Danube (northwest Black Sea) 245
  Desoto Canyons (US Gulf of Mexico) 244
  Dnepr (Black Sea) 82, 265
  Green (US Gulf of Mexico) 244, 307
  Kaoping (offshore Taiwan) 94
  Marble (British Columbia) 309–10

Mississippi (US Gulf of Mexico) 244
Monterey (offshore California) 112–14, 140
Navarinsky (Bering Sea) 105
Supsa (offshore Supsa River, Georgia; eastern Black Sea) 85
Zhemchung (Bering Sea) 105
Cape Aliki (Greece) 231
Cape Banza (Lake Tanganyika) 309
Cape Comorin (India) 92
Cape Fear Diapir (North Carolina) 182
Cape Fear Slide (offshore North Carolina) 126, **127, 358–9**
cape hake (*Merluccius capensis*; fish) 383
Cape Hatteras Slide (offshore North Carolina) 126
Cape Henlopen (Delaware) **253–4**
Cape Lookout Bight (North Carolina) **126,** 136, 234, 337, 348
Cape Lookout Slide (offshore North Carolina) 358–9
Cape Mendocino (California) 111
Cape Vagia (Rhodes, Greece) 209
capillary 219, 220, 309, 379
  (entry) pressure (P_{cap}) 219, 220, **221,** 223, 239, 241 (*see also* Equation 7.11)
  failure 223
  flow 233, 240, 244
  migration 221, 228, 236
  resistance 219
Capitella sp. (polychaete) 251, 281
Captain Soley 123
carbohydrate 152
carbon (C) 76, 84, 87, 101, 114, 122, 128, 153, 154, 155, 156, 158, 161, 162, 186, 250, 263, 275, 281, 282, 285, 291, 295, 297, 300, 307, 349–50, 375
  dating (^{14}C) 101, 115, 256, 294, 307
  organic 59, 84, 92, 96, 104, 118, 120, 136, 154, 156, 160, 161, 186, 282, 286, 330, 350
  inert (graphite) 150, 156, 160, 349
  inorganic 160, 161, 282, 338
  TOC (total organic carbon) 59, 154, 282
carbon dioxide (CO_2) 76, 84, 93, 96, 100, 104, 109, 125, 144, 148, 150, 153, 158, 160, 161, 180, 186, 202, 214, 225, 247, 257, 260, 271, 2294, 303, 310, 313, 316, 320, 325, 326, 330, 335, 338, 345, 350, 353
carbon isotopes
  ^{12}C 162
  ^{13}C 162, 295, 297, 306
  ^{14}C 101, 115, 162, 282, 295, 307

ratio $^{12}C:^{13}C$ ($^{13}\delta$) 27, 40, 44, 57, 62,
    73, 84, 88, 91, 94, 101, 102, 103,
    104, 107, 109, 114, 115, 122, 125,
    149, 155, 160, 161, **162**, 250, 263,
    264, 269, 271, 273, 275, 278, 281,
    282, 290, 291, 294, 295, 297, 304,
    305, 306, 311, 316, 335, 338, 340,
    343, 381
carbon monoxide (CO) 160, 325, 326
carbonate 27, 31, 44, 62, 64, 68, 70, 72,
    78, 79, 88, 94, 97, 101, 107, 109,
    110, 112, 113, 114, 115, 118, 120,
    122, 132, 141, 177, 183, 244, 248,
    249, 253, 255, 289, 290–314, 318,
    319, 343–4, 358, 381
  faroush 88, 90, 294
  fossil **304–7**
  hydrocarbon-derived authigenic
    carbonate (HCDAC) 307
  methane-derived (MDAC) 6, 15, 27,
    32, 50, 62, 114, 118, 145, 161, 175,
    195, 200, 249, **290–307**, 331, 349,
    369–70, 383
  normal marine 88, 114, 124, 161, 290
carbonate cement 36, 64, 73, 88, 94, 250,
    290, 291, 293, 296, 301, 304, 305,
    306, 386
carbonate chimney/column 6, 68–9, 83,
    97, 99, 101, 103, 112, 114, 149,
    250, 262, 265, 296, 297, **299–301**,
    302, 305, 307, 309, 313, 314, 382
carbonate minerals – *see* aragonite, calcite,
    dolomite and ikaite
  ankerite – $Ca(Fe,Mg,Mn)(CO_3)_2$ 68,
    297, 300
  aragonite – $CaCO_3$ 44, 79, 99, 103, 110,
    118, 260, 262, 291–2, **293**, 294–5,
    297, 300, 302, 304, 308, 309, 312,
    313, 314, 316
  calcite – $CaCO_3$ 44, 99, 103, 120, 223,
    260, 291, 293, 295, 297, 300, 304,
    309, 312–14, 321
  dolomite – $CaMg(CO_3)_2$ 68, 69, 291–2,
    293, 295–6, 300, 304, 318–19
  high-magnesium calcite 103, 109,
    291–2, **293**, 294–7, 300, 305
  ikaite – $CaCO_3.6H_2O$ **311**
  microbial 290
carbonate mound/reef 60, 61, 64, 68, 79,
    94–5, 175, 248, 259, 290, **294**,
    302–4, 305, 306, 307, 344, 381
  Ashmore Reef (Timor Sea) 303
  Belgica Mounds (Porcupine Bight,
    offshore Ireland) 64, 302, 303
  Heywood Shoals (Timor Sea) 94, 304
  Hovland Mounds (Porcupine Bight,
    offshore Ireland) 65, 303

Karmt Shoals (Timor Sea) 94, 304
  Logachev Mounds (Porcupine Bight,
    offshore Ireland) 302
  Magellan Mounds (Porcupine Bight,
    offshore Ireland) 64, 302–3
  Sahul Shoals (Timor Sea) 94
carbonate compensation depth (CCD) 99,
    279, 314
carbonatite (carbonate igneous rock) 311
Carcinus maenus (green crab) 282
Cardiff Bay (UK) 60
Caribbean Sea **120–2**, 140, 218, 368, 372
Caribbean Plate 118, 121
carnivore 274
carnivorous sponge **273**
  Cladorhiza sp. 46, 122, 273
Carolinas (North and South; USA) 126,
    234, 284, 330, 358
Carpinteria (California) 115, 116
Cascadia Margin **109–11**, 181, 376
  accretionary wedge/subduction zone
    109, 111, 181, 184–6, 252, 263,
    366–7
  Hydrate Ridge **109–10**, 180, 182–5,
    252, 263–4, 297, 332, 334, 336,
    337, 338, 340, 366, 376
casing (steel tube lining a well) 16, 163,
    363–4, 367–8, 371
Caspian Sea **85–6**, 140, 159, 198, 201,
    203, 337, 356, 368, 380
  Southern Caspian Basin 85–6, 197,
    237, 245
cavern 39–40, 102, 249, 326
CCD (carbonate compensation depth) 99,
    279, 314
Cedros (Trinidad) 204
Cellulose 152, 155
Cement (Oklahoma) **306–7**
Cenozoic
  Pleistocene 10, 59, 93, 94, 104, 123,
    129, 135, 151, 192, 205, 209, 211,
    231, 237, 357, 368
  Recent (or Holocene) 7, 45, 56, 57, 59,
    72, 76, 79, 81, 88, 93, 99, 104, 106,
    111, 115, 128, 129, 135, 161, 193,
    229, 231, 289, 294, 299, 349, 354,
    357
  Tertiary 7, 10, 16, 22, 43, 44, 50, 59, 60,
    66, 93, 94, 104, 105, 107, 115, 123,
    129, 131, 136, 159, 207, 210, 211,
    212, 213, 241, 303, 307, 382
  Quaternary (Pleistocene and Holocene)
    19, 31, 45, 51, 52, 59, 60, 66, 80,
    81, 94, 97, 104, 105, 107, 114, 121,
    159, 299, 323, 350, 353, 354, 365
Central Adriatic Depression 74
Central America 118–19

Central Black Sea Mud Volcanoes 83–4
Central Graben (North Sea) 232
Cerianthus lloydii (anthozoan: tube
    anenome) 249
chalcopyrite ($CuFeS_2$; mineral) 149, 314,
    315
Challenger pockmark (UK15/25, North
    Sea) 40, 43, 44
Challi field (Australian Timor Sea) 95
Chamorro Seamount (southwest Pacific)
    99
Champagne Pool (Taupo Volcanic Zone)
    311
Chandragup Mud Volcano (Pakistan) 92,
    199
channel sands 210, 364, 368
Chatham Island (Trinidad) **122**, 204
Chek Lap Kok airport (Hong Kong) 93
Chekka (Lebanon) 330
chemoherm 255, 337
chemosynthesis 2, 4, 96, 125, 193, 250,
    255, 261, 269, 273–4, 277, 280–3,
    285, 286, 287, 290
Chennai (India) 342
Cheirimedeia zotea (amphipod) 251
Chesapeake Bay (eastern USA) **127–8**,
    178
Chiclayo Canyon (offshore Peru) 299
Chikishlyar-type mud volcanoes **204–5**,
    240, 246, 336
Chile 118, 183, 278
chimney 1, 3
  acoustic 25, 30, 87, 176, 178
  anhydrite 314–15, 316
  baryte 299, 316
  carbonate 6, 68–9, 83, 97, 99, 101, 103,
    112, 114, 149, 250, 262, 265, 296,
    297, **299–301**, 302, 305, 307, 309,
    313, 314, 382
  dolomite 68
  gas 4, 27, 62, 80, 95, 111, 168, **170**, 171,
    178, 182, 206, 209, 212, 221, 224,
    240, 241, 244, 303
  hydrothermal 3, 6, 72, 95–6, 97, 103,
    118, 142, 262, 274, 299, 309, 314,
    315, 316, 317, 324
China (Peoples Republic of) 93, 99, 361
Chirag Mud Volcano (Azerbaijani
    Caspian Sea) 201
Chirag field (Azerbaijani Caspian Sea)
    86
Chirikov Basin (Bering Sea) **103–4**, 105
Chironomidae (insects) 87
chirp (seismic profiling system) 53, 99,
    169
chiton (class Polyplacophora of phylum
    Mollusca) 113

chloride 84, 121, 185, 310
   depletion 185
chlorine (Cl) 144, 202, 317
   chlorine isotope (^{37}Cl) 317
chlorite (clay mineral) 97, 147, 301, 315
chocolate mousse 131
Chonelasma sp. (sponge) 46
chromatography 21, 27, 105, 380
Cibeles Mud Volcano (Gulf of Cadiz) 68
ciliates (protists) 223, 262
Cladorhiza sp. (sponge) 46, 122, 273
clam (see under *bivalve*)
Clam Flat (Monterey Canyon) 113–14
Clamfield (Monterey Canyon) 113
Clarion-Clipperton area (Pacific Ocean)
   319
Clathrate Gun Hypothesis 323, 350
clathrate (see also *hydrate*) 179, 350
Cleft Segment (Juan de Fuca Ridge) 324
climate 178, 293, 294, 323
   climate change 6, 323, 345, **350–3**, 354
   palaeoclimate 309
Cloacaspis (trilobite) 289
Co-Axial Segment (Juan de Fuca Ridge)
   324
coal/coal-bearing 7, 8, 59, 81, 108, 136,
   155, **156**, 159, 161, 162, 235, 298,
   347, 348, 375
Coal Oil Point (California) 115–16, 375
Coal Pit Formation 11, 43, 238
cobalt (Co) 319
cobalt-rich crusts 384
Cocos Plate 118, 140, 142
cod (*Gadus morhua*) 46, 249, 251, 281
Codling Fault (Irish Sea) 60, 61
Cognac field (US Gulf of Mexico) 381–2
cohesion (*c*) **215**, 216, 222, 225
cohesionless (sediments/soils without
   cohesion) 30, 88, 104, 234
cohesive (sediments/soils with cohesion)
   36, 39, 67, 88, 126, 226, 242, 370
cold seep communities 4, 68, 79, 81, 91,
   99, 100, 109, 112, 113, 114, 118,
   118, 120, 122, 124, 125, 139, 145,
   195, 200, 251, 252–3, 255, 265,
   267, 269, 271, 272, 274, 275, 277,
   278–80, 283, 287, 290, 304, 305,
   307, 331
   fossil **288–9**
cold-water coral – *see* deep-water coral
collapse depression 68, 106, 191, **206**,
   236
collapse structure 66–7, 84, 208
Colombia 118, 199, 218
colonists 46, 273, 274, 283
columbellid gastropods (family
   Columbellidae) 113

columnar disturbance 15, 19, 21, 30, 37,
   39, 51, 170, **172**
compression
   of gas 216, 219, 220, 225, 230, 233, 240
   of sediment (soil) 107, 140, 216, 223,
      233, 235, 370
   tectonic 67, 68, 94, 97, 111, 113, 122,
      140, 150, 181, 197, 214, 215, 232,
      245
Complex Mound (US Gulf of Mexico)
   205
Conchocele sp. (thyasirid bivalve) 102, 272
Concle Sands (UK) 229
concrete gravity platform 363
condensate 25, 118, **155**
Congo Basin (offshore West Africa) 70,
   264, 338
Congo Fan (offshore West Africa) 70
Conical Seamount (Pacific Ocean) 98,
   149, 150, 198
connate water **150**, 202, 214
Connemara field (offshore Ireland) 62,
   303
conservation 375, 385–6
Consub (submarine) 9
contour current 136
contourite 126, 136
Cook Inlet (Alaska) 1103
Cook Islands 319
Coondapur (India) 92
Cooper field (US Gulf of Mexico) 124
copepods (crustaceans: subclass
   Copepoda) 122, 257, 261, 274
copper (Cu) 87, 97, 148, 149, 314, 315,
   319, 325, 374
   DCZ (deep copper zone) 315
coral 72, 83, 88, 119, **255–60**, 274 (*see also*
   soft coral, stony coral)
   Acropora sp. 259
   ahermatypic 255–6, 307
   cold-water (*see* deep-water)
   debris 64
   deep-water (cold-water) 49, 53, 64, 66,
      69, 120, 255–8, 290, 302, 305, 383,
      384, 385
   octocoral 261, 305
   reef 141, **255–60**, 305, 307, 320, 343–4,
      383, 386
   reef protection 386
   rugose 305
   tropical 141, 255, 258, **259–60**
Corinth, Gulf of (Greece) 75, 76, 244
Cornea field (Australian Timor Sea) 95
corrosion/corrosive 99, 148, 314, 371
Coryphaenoides rupestris (roundnose
   grenadier; fish) 384
Costa Rica **118–19**, 142, 181

crab (brachyurans) 27, 36, 72, 76, 96, 110,
   250, 251, 259, 261, 262, 271, 274,
   279, 282
   Cancer pagarus (edible crab) 250
   Carcinus maenus (green crab) 282
   Galatheid 99, 100, 113, 119, 252, 255
   Pagarus sp. (hermit crab) 249, 250
Crangon communis (shrimp) 251
Crater Lake (Oregon) 268
Cretaceous (Mesozoic) 7, 50, 51, 81, 107,
   113, 126, 132, 140, 159, 212, 256,
   289, 303
Crimea Peninsula (Ukraine) 81, 83, 84,
   265
Croatia 72
crustacean (subphylum Crustacea; *see also*
   barnacle, crab, isopod, krill,
   lobster, ostracod, shrimp, squat
   lobster) 30, 31, 36, 47, 105, 250,
   251, 252, 261, **274–5**, 282, 283
CT scan (X-ray computer-aided
   tomography) 57, 164, 166, 169
CTD (conductivity-temperature-depth)
   106, 334
current (contour) 136
current (water) 8, 11, 19, 27, 52, 60, 61,
   65, 67, 81, 91, 104, 105, 110, 126,
   135, 136, 242, 243, 244, 246, 247,
   257, 259, 278, 285, 294, 308, 313,
   319, 327, 331, 332, 339, 360, 369,
   380
current (turbidity) 131, 136, 205, 340,
   355
Cyanea capillata (jellyfish) 285
cyanobacteria (photosynthesising
   bacteria) 308, 309, 312
cycloalkanes (C_nH_{2n}; also known as
   naphthenes) 156, 158
cyclops (crustacean: copepod) 87
Cyclops Mud Volcano (offshore
   Barbados) 121

Damon Mound (Texas) 307
Damon Salt Diapir (Texas) 307
Dansgaard–Oeschger events 351
Danube, river 81, 342
Danube Canyon (northwest Black Sea)
   245
Danube pro-delta (northwest Black Sea)
   81
Darwin Mounds (offshore UK) **64–6**,
   384, 385
Dashgil Mud Volcano (Azerbaijan) 198,
   205
dating 54
   carbon (^{14}C) 101, 115, 256, 294, 307
   uranium: thorium (U:Th) 101

Dattilo Island (near Panarea Island, Italy)
DCZ (deep copper zone) 315
Dead Sea 1, 2, 311, 313
decarboxylation (fermentation) 153
decomposition (of gas hydrate) 182, 340, 350, 360, 361, 370, 372, 384
decomposition (of organic matter) 56, 58, 92, 104, 106, 128, 149, **152**, 284, 309, 316, 344, 349, 386
deep biosphere 1, 154, **287–8**, 351
deep copper zone (DCZ) 315
Deep-Earth Gas Hypothesis **159–60**
Deep Hot Biosphere 160, 288
Deep Portuguese Margin field 68, 69
Deep Rover (submarine) 268
Deep Sea Drilling Programme (DSDP) 91, 126, 146, 180, 183
    Leg 11 183
    Site 570 180
deep-sea fan (*see* fan)
deep-water basins 86, 135, 208, 379
deep-water construction 367
deep-water coral 49, 53, 64, 66, 69, 120, 255–8, 290, 302, 305, 383, 384, 385
    Lophelia pertusa 51, 53, 64, 65, 69, 255–8, 302
    Madrepora oculata 64, 69, 256
    Norwegian **255–8**, 305, 383, 385, 386
degassing 67, 78, 101, 106, 112, 120, 138, 144, 159, 205, 230, 294, 325
Delaware (USA) 254–5
Dellwood Knolls (Juan de Fuca Ridge) 138
Delta 59, 66, **135**, 136, 197, 206, 353, 355, 356, 360, 361
    Alsek (Alaska) **107**, 206, 357
    Amazon (submarine delta; Brazil) **120**
    Brent (Jurassic; North Sea) 135
    Cambridge Fjord (Baffin Island, Canada) 132
    Fraser (British Columbia) **107–9**, 357
    Helike (Greece) 233
    Huang Ho (People's Republic of China) 99
    Klamath (California) 111, **112**, 230, 374
    Mekong (Vietnam) 93
    Mississippi (USA) 106, 123, 124, 135, 198, 206, 207, 236, 357, 364
    Niger (Nigeria) **69–70**, 135, 174
    Nile (Egypt) **79–81**, 135
    Orinoco (Venezuela) 118
    Selenga (Lake Baikal) 86
    Ur-Frisia (Pleistocene; North Sea) 135
    Yangtse (People's Republic of China) 99

Denmark **52–5**, 250, 282, 290, 291, 299, 308, 347, 375, 385, 386
density (ρ) 147, 172, 196, 198, 212, 214–17, 219, 222, 226–8, 236, 245, 320–1
    bulk (ρ_{bulk}) 147, 178, 196, 207, 214, **215**, 216, 240, 243, 245, 316, 321
    (*see also* Equation 7.5)
    gas (ρ_{gas}) 216, 219, 240, 320, 366
    inversion 207, 216, 236, 240, 245, 246, 360
    methane 163, 216, 219
    pockmark 8, 10–11, 18, 19, 24, 33, 43, 88, 97, 104, 195, 231, 238
    water (ρ_{water}) 215, 219, 240, 285, 320, 338, 340, 341, 366, 372
Derugin Basin (Okhotsk Sea) 284, 299
DESMOS (hydrothermal site, east Manus Basin) 96
Desoto Canyon (US Gulf of Mexico) 244
Desulfovibrio sp. (bacterium) 96
detrivore 152, 277
deuterium (^{2}H) 84, 162, 381 (*see also* hydrogen)
Devonian (Palaeozoic) 8, 59, 81, 289, 305
dewatering 67, 101, 106, 113, 129, 180, 182, 212, 258
dew-point 221
Diakopto (Greece) 231
diapir
    diapirism/sediment buoyancy **216**, **226**, 227, 236–7, 238, 240, **245–6**, 360
    mud (including clay and shale) 4, 51, 52, 54, 55, 67–9, 76–7, 84, 91–2, 93, 94, 97, 101, 112, 119, 120, 121, 122, 172, 183, **195–8**, 200–1, 202, 205, 207–8, 209, 212, 213, 214, 215, 216, 218, 223, 228, 230, 245, 296, 360
    salt 3, 25, 27, 28, 57, 59, 60, 123, 126, 127, 182, 197, 208, 218, 223–4, 225, 241, 289, 304, 317–18, 358–9
    serpentinite 144
    shallow 73–4, 172, 174, 206–7, 208, 236–7, **246**
diatomaceous ooze 72
diatreme 4, 190, 196, **208–9**, 228, 240, 244–6
    sedimentary **209**, 225–6, 240, 263
    volcanic **208–9**, 240
DIC (dissolved inorganic carbon) 282, 338
diffusion (of gas) 126, 181, 182, **219**, 220, 241, 266, 284, 320, 337, 339, 344, 346, 354
dimethylsulphide (DMS; (CH$_3$)$_2$S) 153

Discovery Deep (Red Sea) 87
disturbed seabed 13, 50, 242
divers/diving 55, 58, 59, 75, 81, 90, 116, 117, 164, 230, 279, 331, 333, 339, 372
Dnepr Canyon (Black Sea) 82, 265
dolomite (CaMg(CO$_3$)$_2$; mineral) 68, 69, 291, 292, **293**, 295, 296, 300, 304, 318, 319
dome (*see* seabed doming and intra-sedimentary doming)
Dos Cuadras field (offshore California) 116, 364
Downing Basin (offshore Newfoundland) 131
Drake Passage (Antarctica) 118
Draugen field (offshore Norway) 50
drilling
    DSDP/ODP/IODP 95, 99, 101, 109, 118, 126, 127, 145, 148, 149, 151, 183–4, 186, 205, 288, 316, 266, 367
    hazards 6, 107, 237–8, **362–9**, 372 (*see also* geohazards)
    hydrate **367**, 371, 375–8
    petroleum exploration/production 1, 8, 37, 73, 75, 107, 115, 156, 178, 223, 260, 335, 355, **362–8**, 372, 381, 384
    site investigation 9, 91, 108, 172, 225
    other 141, 159, 163, 328
drilling fluid (mud) 59, 156, 223, 362–4, 367, 368, 371
dropstone, glacial 176, 192
DSDP (Deep Sea Drilling Program) 91, 126, 146, 180, 184
    Leg 11 184
    Site 570 180
dunite (type of ultramafic rock) 98
Dutch North Sea 60, 192
Dvurechenskii Mud Volcano (Sorokin Trough, Black Sea) 84
dyke
    igneous 138, 140, 145, 306, 325
    clastic/sedimentary 140, 201, 209–11, 212, 213, 240

earthquake 58, 59, 69, 73–4, 75–6, 104, 106, 108, 110, 111, 112, 117, 131, 189, 194, 211, 212, 223, **228–33**, 234, 237, 240, 241, 244, 326, 335, 355, 360, 362, 373
East Cameron (licence block, US Gulf of Mexico) 124
East Cape (New Zealand) 97
East Caribbean island arc 140

East China Sea 99
East Flower Garden Bank (US Gulf of Mexico) 258
East Manus Basin (Papua New Guinea) 95
East Pacific Rise (east Pacific Ocean) 1, 117, 141, 142, 158–9, 268, 274, 277, 317, 324, 327, 343
Eastern Caribbean Arc 140
Eastern Mediterranean Sea 74–81, 272, 295
Echinoderm (sea urchin; class Echinoidea) 72, 249, 261, 274, 279
Astropecten irregularis 261
echo sounder 7, 8, 9, 10, 15, 23, 25, 27, 30, 43, 61, 65, 75, 76, 77, 81, 82, 83, 101, 102, 104, 110, 118, 132, 167, 331, 332, 172, 373
fish-finding echo sounder 61
multibeam echo sounder (MBES) 2, 10, 11, 12, 14, 16, 19, 22, 40, 41, 50, 52, 61, 64, 68, 71, 80, 82, 83, 373, 381
parametric echo sounder 44, 84, 332, 333
Eckernförde Bay (Germany) 56–8, 153, 164, 165, 166, 169, 178, 193, 194, 244, 246, 263, 264, 330, 348
ectosymbiosys 273
Ecuador 140, 260, 332
Edmond hydrothermal field (Indian Ocean) 93
Eel Margin/Shelf (offshore California) 111–12, 361
Eel River/Delta (California) 111, 112, 383
Eel River Basin (offshore California) 111, 139, 297, 299, 357
eelpout (Zoarces vivaparus) 253
Eemian (Quaternary interglacial interval; Sangamon in North America) 54
El Pilar Mud Volcano (offshore Barbados) 122
Elaiona Bay (Greece) 76, 244
Elbe, river (Germany) 342
Ellef Ringnes Island (Canada) 289
Elm Mud Volcano (Caspian Sea) 86
Ekofisk field (Norwegian North Sea) 25, 384
Emerald Basin (Scotian Shelf) 7–8
Emma field (Italian Adriatic Sea) 73
Endeavour Ridge (Cascadia Margin) 146
Endeavour Segment (Juan de Fuca Ridge) 315, 317, 329
endosymbionts
zooxanthellae 255
endosymbiosis 193, 249, 269–72

methane 122, 253, 282
sulphur 114, 271–2, 278, 282, 298
energy
acoustic 166, 167, 169, 175–8
Gibbs free energy 266
kinetic 144, 189, 200, 204, 216, 225–6, 228, 233, 235, 246, 294
metabolic 107, 125, 153, 252, 262, 263, 266, 269, 273, 281, 282, 283, 285, 286–7, 327, 350
resource – hydrate 6, 101, 126, 375–6
resource – petroleum 158, 160, 345, 375
solar 4, 151, 379
thermal 110, 328–9, 379
engineering 190, 207, 355, 370
England (UK) 59, 229, 235, 315, 330, 342, 374
enhanced reflection (seismic indicator of gas) 15, 17, 21, 53, 124, 169, 172, 174, 221
Ensenada San Simón (Spain) 66, 266, 267
enteropneusts (acorn worms; hemichordates – class Enteropneusta) 261
environment/environmental 107, 202, 275, 286, 287, 298, 318, 320, 362, 375, 383–6
assessment/protection 62, 384–6
freshwater and lacustrine 128, 155, 284, 298
geological 86, 97, 120, 145, 148, 155–6, 197, 206, 210, 288, 289
marine 1, 2, 6, 115, 123, 134, 150, 151, 155, 179, 183, 247, 260, 262, 263, 268, 269, 271, 275, 276, 277, 278, 283, 284, 285, 286, 288, 289, 298, 309, 317, 323, 335, 337, 341, 343, 347, 349, 354, 355, 371, 384–6
pressure/stress 4, 179, 214–15, 222, 223, 239, 245, 350, 369
seabed 4, 10, 108, 124, 125, 134, 152, 153–4, 190, 210, 214, 243, 276, 288, 294, 298, 300, 355, 357
EPSP (Environmental Protection and Safety Panel) 183–4, 185
Eocene (Tertiary) 65, 94, 136, 140, 145, 211, 305, 353
epifauna 249, 250, 262, 274, 281, 282
EPSP (Environmental Protection and Safety Panel) 183–4, 185
equation
5.1 149
5.2 150
5.3 150
5.4 150
5.5 150

5.6 151
5.7 153
5.8 153
5.9 156
5.10 156
5.11 160
5.12 162
5.13 162
6.1 166
6.2 172
7.1 214, 215, 350
7.2 214, 215, 222, 235, 236, 366
7.3 215, 226, 236, 350
7.4 215, 225, 236, 357, 359, 370
7.5 215
7.6 216, 359
7.7 216
7.8 217, 218
7.9 219, 236
7.10 221
7.11 221, 223, 233, 236
7.12 221, 222, 226, 360, 366
7.13 222, 233, 236
7.14 222, 363
7.15 225
7.16 226
7.17 226
7.18 233
8.1 263, 295, 297, 298, 301
8.2 266
8.3 266
9.1 293
9.2 293
9.3 293
9.4 295
9.5 295
9.6 295
9.7 298
10.1 339
10.2 339
10.3 339
10.4 339
10.5 348
Equatorial Guinea 382
Eratosthenes Seamount (Mediterranean Sea) 79, 80
Eratini (Greece) 231
Erin Bay (Trinidad) 122
erosion 8, 27, 40, 73, 100, 104, 105, 120, 127, 132, 141, 188, 191, 192, 193, 195, 242, 244, 245, 291, 300, 301, 352, 359, 360, 363
eruption 6
gas 58, 73, 98, 145, 209, 333–5, 368
mud volcano 122, 196, 198, 200, 202–5, 225, 232, 234, 333, 336, 337, 347, 352, 355

volcano 96, 110, 142, 145, 148, 158, 208–9, 234, 237, 268–9, 279, 328, 333, 345, 371–2
Escanaba Trough (Gorda Ridge) 138, 158, 316, 317
Escarpia sp. (siboglinid tubeworm) 70
Estrella oilfield (offshore Equatorial Guinea) 382–3
estuary 60, 135, **341–2**
  Adyar (India) 342
  Danube (Romania) 342
  Elbe 342
  Great Bay (New Hampshire) 330
  Mersey (UK) 60
  Plym (UK) 60
  Tamar (UK) 60
  Tees (UK) 60
  Thames (UK) 59, 60, 135
  Tyne (UK) 342
ethane (C_2H_6; hydrocarbon gas) 27, 40, 44, 68, 84, 85, 96, 102, 103, 105, 109, 116, 120, 123, 125, 131, 132, 158, 160, 161, 180, 202, 281, 295, 301, 380, 381
ethanol (C_2H_6O) 152
Eugene Island (licence block, US Gulf of Mexico) 124
euglenoid (protists) 273
euphausiid (see *krill*)
Eurasian Plate 67, 74, 91, 94, 100
eustatic sea level 134, 236, 350, 351
evaporation 85, 293, 294, 308, 310, 312, 317, 329
evaporate 87, 202, 208, 272, 306, 308, 318
event (mega) plume 285, **324–5**, 328, 329, 333, 334, 347, 352
exploration 40, 69, 95, 111, 115, 122, 123, 132, 154, 247, 287, 323, 326
  drilling 8, 37, 73, 75, 87, 178, 200, 362–6
  mineral 314
  petroleum (oil and gas) 6, 16, 41, 62, 70, 93, 107, 123, 134, 157, 160, 178, 331, 332, 335, 355, 362–3, 374, **376**, **378–83**
Explorer Plate (east Pacific Ocean) 109
Explorer Ridge (Juan de Fuca Plate) 278
EXPLOS (TV sled) 106
explosive decompression 225, **226**, 240, 246
Exuma Island (eastern Bahama Bank) 308

Faroe–Shetland Basin 140
Fan (continental rise sediment accumulations) 136, 197, 236
  Amazon (offshore Brazil) 120
  Bear Island (offshore Norway) 47, 50

Congo (offshore Congo) 70
Danube (NW Black Sea) 81
Indus (offshore Pakistan) 91–2
Laurentian (offshore E Canada) **131**, 252, 269, 278, 355
Mississippi (US Gulf of Mexico) 124
Monterey (offshore California) 269
Niger (offshore Nigeria) **69–70**
Nile (offshore Egypt) **79–81**, 301, 307
North Sea (offshore Norway) 358
Orange River (offshore Namibia/South Africa) 70
Faro Mud Volcano (Gulf of Cadiz) 68
faroush (carbonate) 88, 90, 294
fatty acid 152, 264
fault 49, 52, 64, 65, 69, 76, 77, 79, 80, 82, 84, 87, 91–2, 94–5, 97, 100, 104, 109, 115, 117, 119, 120, 123, 135, 145, 158, 159, 170, 173, 175, 178, 181, 182, 190, 197, 200, 206, 212, 216, 217, 218, 219, 221, 223–5, 230, 238, 239, 240, 241, 243, 244, 245, 257, 302–3, 304, 306, 356, 362, 364, 376
  active 69, 73–4, 97, 101, 103, 112, 119, 218, 334
  compressional 97
  concentric/ring 204, 209
  extensional (inc. graben-related) 80, 98, 107, 138, 207, 211, 289
  detachment 119
  growth 80
  intersecting 76, 80, 197, 216, 225
  mud diapir-volcano-associated 78–9, 84, 93, 124, 127, 204, 207, 216, 225, 245
  normal 79, 119, 129, 359
  polygonal 129, **211–12**, 213, 258
  salt-related 26–7, 60, 78, 80, 123, 124, 125, 218, 289, 318
  strike-slip 61, 67
  transform 95, 113, 140, 141, 277, 303
  subduction-related/thrust 112, 119, 121, 139, 253, 306
Aigion (Greece) 232
Ascension (offshore California) 113, 140
Codling (Irish Sea) 60, 61
Little Salmon (offshore California) 112
Monterey Bay (offshore California) 113, 140
Øygarden (offshore Norway) 232
Posolskiy (Lake Baikal) 87
San Andreas (California) 113, 140
San Clemente (offshore California) 299

San Gregorio (offshore California) 113, 140
Tanakura (Japan) 99
fauna 27, 31, 34, 35, 36, 44, 46, 49, 65, 72, 87, 95, 97, 100, 107, 112, 113, 114, 118, 119, 122, 125, 131, 140, 235, 248–89, 300, 304, 305, 349, 359, 385
  benthic 3, 27, 28, 29, 36, 102, 249, 250, 279, 336
  cold (inc. hydrate) seep-related 3, 27, 28, 29, 31, 33, 34, 35, 36, 44, 46, 87, 91, 97, 99, 100, 102, 107, 112, 113, 118, 119, 121, 122, 124, 125, 131, 193, **248–53**, 262–83, 300, 336, 349
  epi- 249, 250, 262, 274, 281
  fossil 121, **288–9**, 305
  hydrothermal vent-related 2, 3, 72, 87, 93, 95, 96, 99, 100, 103, 110, 140, 142, 248, 251, 255, 256, 259, **260–2**, 263, **265–6**, 268, 269–71, 274, 275, 276, 277, 279, 283, 285, 286, 287, 299, 324, 327, 359
  infauna 124, 249, 250, 251, 269, 275, 281
  macro- 69, 72, 87, 91, 96, 99, 100, 112, 122, 125, 193, 249, 250, 251, 256, 261, 262, 269, 273, **274–5**, 276, 280, 281, 282, 289, 349
  meio- 96, 99, 122, 273, **275**, 276
  micro- 249, 250, **262–9**
  nekton 33, 36, 248, 273
  planktonic 27, 33, 87, 105, 248, 257, 276, 279, 284
  SGD-related **253–5**, 259, 263, 266, 276, 330
  sessile 27, 33, 248, 301
fayalite (Fe_2SiO_4; olivine mineral) 149–50
feldspar (group of minerals) 299, 301
Feni Drift (northwest Atlantic) 136
Fennoscandian Shield 58
fermentation 153, 161
ferromanganese nodules 290, **319–21**
Fick's First Law 219
Figge Maar (German North Sea) 335, 364
Fiji 268
Finland 385
Finneidfjord (Norway) 356–7
  Slide **357–8**
Firth of Forth (UK) 59, 60
Fischer–Tropsch reaction 150, 157, 158
FISH (fluorescence in situ hybridisation) 264, 265

fish 22, 23, 27, 36, 39, 46, 72, 75, 78, 89,
    102, 131, 166, 231, 249–50, 251,
    252, 253, 254, 261, 273, 274, 276,
    279, 281, 282, 283, 289, 312, 340,
    383, 384
  Argentine (*Argentina silus*) 32
  brotulid fish (family Ophidiidae) 261
  cape hake (*Merluccius capensis*) 383
  cod (*Gadus morhua*) 46, 249, 251, 281
  eelpout (*Zoarces vivaparus*) 253
  hagfish (*Myxine glutinosa*) 249
  ling (*Molva molva*) 39, 249, 280
  Norway haddock (*Sebastes vivaparus*)
    32, 249
  rattail (grenadier; *Macroudae* sp.) 261
  ray (superorder Batoidea) 78, 251
  rockfish (*Sebastes elongates*) 251
  roundnose grenadier (*Coryphaenoides
    rupestris*) 384
  skate (family Rajidae) 261
  slender sole (*Lyopsetta exilis*) 251
  torsk (*Brosme brosme*) 32, 33, 39, 281
  wolf fish (*Anarhychus lupus*) 249
fish otoliths (ear bone) 249–50
fishermen/fishing 54, 66, 72, 73, 74, 75,
    81, 88, 101, 112, 129, 229, 231,
    246, 255, 256, 258, 286, 290, 328,
    330, 334, 340, 355, 369, 373, 374,
    382, **383**, 384, 385
fish-finding echo sounder/sonar 61, 102,
    331
fjord 49–50, 60, **135**, 256, 257, 269, 356
  Cambridge Fjord (Baffin Island,
    Canada) 132, 329, 349
  Finneidfjord (Norway) 356–7
  Ikka Fjord (Greenland) 311, 312
  Loch Eribol (UK) 60
  Loch Etive (UK) 60
  Lyngenfjord (Norway) 49–50, 51, 52
flagellate 251, 273
Flags Formation (Quaternary, North Sea)
    10
Flammability 362, 366
Flandrian (Quaternary) 193
flare (intense acoustic (gas) plume) 82, 84,
    101, 102, 103
  Zonenshayn Flares 102
flat spot (seismic indicator of gas) 170,
    172, 173, 176, 366
Flemish Bight (North Sea) 60
floc/flocculent (bacterial matter) 96,
    268–9, 275, 285, 324
flora (plants) 28, 75, 135, 151, 152, 156,
    248, 250, 251, 258, 306, 309, 330,
    385, 386
Flores (Indonesia) 94, 175
Florida (USA) 123, 124, 255, 278, 330

Florida Bay (Florida) 259
Florida Escarpment (groundwater) seeps
    (US Gulf of Mexico) **125–6**, 218,
    244, 251, 255, 266, 318, 329, 330
Florida Keys (USA) 259
Florida Platform (offshore USA) 126
Florida Strait (USA) 311, 312
flowline 182, 371
fluidisation 145, 208, 209, 211, 216, **226**,
    227, 237, 240, 242, 246, 247 (*see
    also* Equation 7.16)
fluorapatite ((Ca₅(PO₄)₃F); mineral) 318
fluvial 10, 66, 107, 229
foraminifera 69, 94, 249, 257, 273, 275,
    291, 305
  Bulimina sp. 275
  Numulites sp. 305
fore-arc 98, 101, 313, 314
fore-arc basin 97, 105, 111, **139–40**, 148,
    289, 326
formate (or methanoate; HCOO⁻) 152,
    153
forsterite (Mg₂SiO₄; olivine mineral)
    149–50
Fort Walton Beach (Florida) 123
Forth-Harding field (UK North Sea) 210
Forties field (UK North Sea) 8, 9, 15, 22,
    372
fossil 305, 306, 386
  bacteria/bacterial mat 288, 289, 297
  biomarker 297
  carbonate 299–300, **304–7**
  fauna 121, **288–9**, 305
  methane 347–8
  microfossil 205, 304, 310
  pockmark (*see also* buried pockmark)
    247
  seep/seep community **288–9**, 298, 304,
    305
  water 150
foundation (of offshore structures) 365,
    368, **370**
fractionation (of methane) 41, 123, 162
fracture pressure (P_f) 222, 363 (*see also*
    Equation 7.14)
framboidal pyrite 291, 298, 300, 306 (*see
    also* pyrite)
framboidal haematite 69, 297
Franklin Seamount (offshore Papua New
    Guinea) 95–6
Fraser Delta (British Columbia) **107–9**,
    357
Fraser River (British Columbia) 107, 108,
    357
freak sandwave **60–2**, 64, 65, 206, 244
Fredrikshavn (Denmark) 52
freshwater 128, 134, 150, 154, 263, 342

lake/lacustrine 86–7, 155, 189, 348
  hydrate-associated 187
  ice-rafting **246–7**
  polynya 349
  river 326
  seep/spring 2, 5, 76, 88, 129, 132, 145,
    193, 244, 254, 263, 312, 323,
    329–30, 334, 374 (*see also*
    submarine groundwater
    discharge)
  wetland 347
Frolikha Bay (Lake Baikal) 87
fumarole 145, 158

gabbro/gabbroic (type of ultra-basic
    rock) 149, 157
Gades Mud Volcano (Gulf of Cadiz) 68
Gadus morhua (cod; fish) 46, 249, 251, 281
Galapagos Rift (east Pacific Ocean) 4, 117,
    251, **260–1**, 271
galatheid crabs (family Galatheidae) 99,
    100, 113, 119, 252, 255
Galicia (Spain) **66–7**
gamma ray 369
Garden Banks (Gulf of Mexico) 205, 258,
    299, 369
gas (*see also* individual gases CO₂, ethane,
    H₂S, methane, etc.)
  blanking – *see* acoustic turbidity
  brightening 27, 120, 169 (*see also* bright
    spot and enhanced reflection)
  bubbles 3, 6, 15, 21, 27, 31, 40, 52, 58,
    60, 70, 76, 79, 83, 85, 86, 89, 98,
    102, 107, 110, 126, 128, 129, 131,
    132, 142, 144, 147, 153, 166, 194,
    198, 200, 209, 216–17, 219, 231,
    235, 241, 242, 252, 255, 260, 262,
    263, 272, 278, 280, 283, 285, 288,
    309, 310, 329, 331–4, 336–41, 344,
    345, 348, 349, 356, 364, 365, 371,
    372, 379, 380, 381, 383
  chimney 4, 27, 62, 80, 95, 111, 168,
    170, 171, 178, 182, 206, 209, 212,
    221, 224, 240, 241, 244, 303
  column 91, 172, 221, 222, 363, 366
  compressibility 216, 219, 220, 225, 230,
    233, 240
  density (ρ_{gas}) 216, 219, 240, 320, 366
  drainage pile 370
  exsolution 165, 214, 216, 220, 221, 225,
    245
  fields (listed *under* Petroleum fields)
  front 17, 37, 62, 108, 128, 153, **168**, 178
  hydrate 1, 2, 4, 6, 45, 46, 47, 49, 51, 52,
    67, 68, 69, 70, 76–7, 81, 82, 84, 86,
    87, 91, 92, 94, 97, 101, 102, 103,
    105, 109, 111, 114, 118, 119, 120,

121, 122, 124, 125, 126–7, 139, 147, 163, 174, 175, 176, 177, **178–88**, 197, 200, 201, 236, 237, 238, 239, 240, 244, 245, 247, 252, 262, 263, 264, 276, 295, 311, 323, 333, 334, 335, 344, 345, 346, 347, 348, 349, 350, 351, 352, 353, 354, 355, 356, 357, 358, 359, 360, **361**, 362, **367–8**, **370–1**, 372, 373, 374, **375–6**, 377–8, 384 (*for further details see under* hydrate)

overpressure (of gas; P_{go}) 67, 209, 218, **221**, 222, 233, 238, 246, 247, 332, 353, 362, 366, 368 (*see also* Equation 7.12b and overpressure)

pressure (P_{gas}) 97, 203, 209, 214, **221**, 222, 301, 302, 363, 366, 371 (*see also* Equation 7.12a)

seep 1–6, 9–10, 25, 27–8, 40, 43–4, 45, 46, 51, 52, 53–6, 58, 59, 60, 61, 62, 63, 64, 65, 66, 67, 68, 70, 72, 73, 74, 76, 77, 79, 80, 81, 82, 83, 84, 85, 86, 87, 91, 92, 93, 94, 97, 98, 99, 102, 103, 104, 105, 106, 107, 108, 109, 110, 111, 112, 115–17,118, 123, 134, 135, 126, 127, 131,, 163, 168, 174, 182, 189, 190, 193, 194, 197, 200, 206, 207, 208, 219, 220, 228, 241, 250, 251, 258, 275, 282, 291, 301, 302, 317, 323, 331–41, 344, 347, 348, 349, 355, 364, 373, 374–5, 383

void 21, 23, 120, 127–8, **163–5**, 166, 169, 216, 222, 228, 241, 370

gassy/gas charged sediment (shallow gas) 17, 27, 28, 38, 50, 52, 53, 57, 58, 60, 66, 67, 70, 72, 73, 77, 79, 81, 82, 84, 91, 92, 93, 108, 112, 120, 124, 127, 128, 131, 165, 167, 169, 170, 176, 183, 204, 207, 230, 231, 233, 234, 235, 256, 302, 337, 357, 362, 364, 370, 373, 382

gastropod (class Gastropoda of phylum Mollusca) 27, 78, 96, 99, 112, 113, 119, 249, 251, 252, 253, 261, 274, 279, 283, 289, 299, 308

   Archeogastropods 255
    Haliotis cracherodii 262
   trochid 255
   Astryx permodesta 274
   Buccinium undatum 249
   columbellid (family Columbellidae) 113
   limpet (family Acmaeidae) 110, 113, 119, 255, 261, 262, 274
   Neogastropod 255
   Neptunea sp. 112

turrid (family Turridae) 253, 255
whelk (neogastropod) 261

gas–water interface 172, 173, 176, 222

Gelendzhik Mud Volcano (Mediterranean Ridge) 76, 202, 245

geochemical/geochemistry 9, 18, 50, 68, 151, 154, 157, 193, 256, 257, 287, 319, 323, 325, 362, 377, **381**, 382

Geodia sp. (sponge) 46, 274

geohazard 6, 86, 178, 355, 356
   active faults 69, 73–4, 97, 101, 103, 112, 119, 218, 334
   gas-induced buoyancy loss 329, 362, 365, 371, **372–4**
   hydrate as a drilling hazard 183–4, 185, **367–8**
   hydrate-induced slope instability 361
   hydrogen sulphide (H_2S) **366** (*see also entries under* hydrogen sulphide)
   increased sediment (soil) compressibility 370 (*see also* sediment compressibility)
   reduced sediment (soil) strength 215, 216, 225–6, 230, 240, 357, **370** (*see also entries under* shear strength, and Equations 7.4 and 7.13)
   seismic (earthquake) activity 228–33 (*see also entries under* earthquake)
   shallow gas 73, 108, 123, 178, 184, 225, 238, 275, 355, **362–6**, 368, 369, 372, 384
   shallow water flow (SWF) **368–9**
   slope failure/instability 355–62 (*see also entries under* slope failure/instability)

geopolymer 154

Georgia (Republic of) 81, **85**, 337

Georgia Strait (British Columbia) 107

geothermal 98, **145**, 158, 160, 161, 190, 290, 347, 374

geothermal gradient 93, 118, 145, 146, 154, 156, 159, **180**, 181, 352

Germany 56, 153, 165, 166, 193, 194, 244, 246, 263, 332, 376, 385

Gharniarigh-Tapeh Mud Volcano (Iran) 199

GHSZ (gas hydrate stability zone) 68, 82, 119, 128, **180**, 181, 182, 183, 184, 186, 188, 288, 340, 350, 352, 358, 361, 377

Gibraltar Strait (Mediterranean Sea) 67, 69, 385

Ginsburg Mud Volcano (Gulf of Cadiz) 67, 68

glacial (ice age) 59, 115, 120, 134, 135, 136, 192, 236, 237, 256, 307, 334, 340, 350, 351, 352 (*see also* Pleistocene and Weichselian)

dropstone 176, 192

last glacial maximum (LGM) 135, 250, 352

sediment 7, 10, 56, 57, 58, 131, 192, 358

global
   carbon cycle 6, 87, 307, 323, **349–50**, 354
   climate change 6, 323, 345, **350–3**, 354
   warming 68, 188, 350, 352, 353

GLORIA (Geological Long-Range Inclined ASDIC; a long-range side-scan sonar) 76, 121, 141

glycerol ($C_3H_8O_3$; a sugar alcohol) 152

glycerol dialkyl glycerol tetraethers (GDGTs) 264

glycol ((CH_2)$_n$(OH)$_2$; a dihidric acid) 368, 371

Godavari River (India) 93

Godzilla hydrothermal chimney (Juan de Fuca Ridge) 5, 6, 315

gold (Au) 97, 310

Goleta (California) 115

Gorda Plate (east Pacific Ocean) 109, 111

Gorda Ridge (east Pacific Ocean) 138, 277, 316, 324

graben (down-faulted block) 78, 80, 97
   Central (North Sea) 232
   Grimsey (offshore Iceland) 316
   Rhine (Germany) 159
   Viking (North Sea) 36, 218, 232
   Whakatane (New Zealand) 98

Grand Banks, Newfoundland 129, 131, 238, 374, 383

Grand Banks Earthquake (offshore Newfoundland) 131

Grand Isle (licence block, US Gulf of Mexico) 124

Graneledon sp. (octopus) 261

granite/granitic (type of acid rock) 8, 66, 113, 136, 157, 222, 288, 315

graphite (C; mineral) 150, 156, 160, 349

gravity (g) 214, 215, 217–18, 219, 226, 236, 240
   core 9, 14, 40, 66, 69, 86, 105, 108
   infragravity waves 235
   platforms 237, 363, 370, 384
   survey 132, 178, 379

Great Bahama Bank 312

Great Barrier Reef (Australia) 259

Great Bay Estuary (New Hampshire) 330

Great Lakes (Canada/USA) **129** (*see also* Lake Michigan and Lake Superior)

Greater Ekofisk (Norwegian North Sea) 25

Greece **75–6**, 230, 233, 335

Green Canyon area (Gulf of Mexico) 244, 307

greenhouse gas 202, 345, 350, 353

Greenland 311, 312, 350, 385

Greenland Sea 357

Grenada (Eastern Caribbean) 140, 371

grenadier (or rattail; deepwater fish; family Macrouridae) 261

Grimsey Island (Iceland) 316

graben 316

Grønnedal-Ika igneous complex (Greenland) 311

groundwater 4, 57, 150, 159, 190, 214, 229, 237–8, 294, 305, 308, 322, 347, 369

(*see also* submarine groundwater discharge)

gryphon (mud volcano feature) 79, **198**, 199, 200, 201, 202, 280, 356

Gryphon field (UK North Sea) 210, 211

Gualdalquivir Diapiric Ridge field (Gulf of Cadiz) 68

Guaymas Basin (Gulf of California) **118–19**, 158, 264, 266, 278, 299, 317

Gulf of

Alaska (northeast Pacific Ocean) 103, **105–6**

Arabian (Persian) **88–91**, 158, 193, 194, 197, 206, 207, 234, 294, 311, 313, 330, 369

Cadiz (offshore Morocco/Portugal/Spain) **67–9**, 181, 296, 297, 299, 301, 321

California (Mexico) **117–18**, 137, 299

Campeche (Mexico) 123

Corinth (Greece) 75, 76, 244

Kastela (Croatia) 72, 329

Kavala (Greece) 75

Kutch (India) 92

Maine (USA) **128–9**, 190, 191, 193

Mexico 1, 115, **122–6**, 157, 159, 176, 182, 183, 188, 202, 204, 205, 218, 244, 245, 252, 253, 258, 264, 271, 272, 273, 274–5, 276, 278, 285, 286, 295, 299, 304, 307, 330, 362, 364, 367, 368, 369, 376, 378, 379, 380

Patras (Greece) 75, 231–2

St Lawrence (east Canada) 130

Suez (north Red Sea) 1

Gullfaks (Norwegian North Sea) 9, 20, **36–41**, 43, 102, 171, 191, 215, 249, 291, 363, 370

Guneshli field (Azerbaijani Caspian Sea) 86

guyot **141**, 142, 318–19

gypsum ($CaSO_4.2H_2O$; mineral) 88, 103, 299, 306, 316, 317

Habitats Directive (European Commission) 6, 385, 386

Haemoglobin 271

haematite (Fe_2O_3; mineral) 69, 297

hafnium (Hf) 321

hagfish (*Myxine glutinosa*) 249

Hainan (People's Republic of China) 93

Halibut Channel (offshore Newfoundland) 130

Halimeda sp. (reef-forming alga) 94

Haliotis cracherodii (abalone; archeogastropod) 262

halite (rock salt – NaCl; mineral) 216, 317, 318

halophytic vegetation 308

Haltenpipe gas pipeline (offshore Norway) 16, 256, 383

Haltenpipe Reef Cluster (HRC; offshore Norway) 53, **256**

Hancock Seamount (Pacific Ocean) 142

Hardground 62, 76, 88, 90, 244, 249, 250, 290, 307

Hatsushima cold seep site (Japan) 99

Hatteras Abyssal Plain (Atlantic Ocean) 358

Hawaiian Islands (USA) 141, 142, 148, 262, 319, 325

Hawaiian-Emperor chain (Pacific Ocean) 141

Håkon Mosby Mud Volcano (HMMV; offshore Norway) **47**, 49, 50, 181, 197, 200, 236, 245, 253, 254, 338, 344

heat flow 49, 87, 120, 121, 127, 158, 159, 181, 182, 329, 384

heavy metal 87

Hebrides (islands, UK) 60, 62

Hecate Strait (British Columbia) 107

Heidrun field (offshore Norway) 53, 383

Helike Delta (Greece) 233

helium (He) 59, 76, 103, 157, 158, 159, 160, 311, 320

^{3}He 148, 158, 325, 327

^{3}He:heat ratio 110

mantle-derived (juvenile) 157, 158, 159, 325

Hellenic Trench system (Mediterranean Sea) 76, 79, 262

Hellenic Volcanic Arc (Mediterranean Sea) 76

Henry's Law 233

Hesiocaeca methanicola (the 'ice worm'; polychaete) 275

Heteromastus filiformis (polychaete) 281

Heywood Shoals (Australian Timor Sea) 94, 304

HF2 hydrothermal site (offshore Papua New Guinea) 96

HF3 hydrothermal site (offshore Papua New Guinea) 96

High Island area (US Gulf of Mexico) 364, 380

High Seas Marine Protected Area (HSMPA) 386

High Seas Network of Marine Protected Areas 386

highly reflective seabed 27, 382

high-magnesium calcite (mineral) 103, 109, 291–2, **293**, 294–7, 300, 305

Hikurangi Margin (offshore New Zealand) 97

Hingol Island (offshore Pakistan) 91

hitchhiker 285, 286, 339, 374

Holene (NO24/9; Norwegian North Sea) 14, 20, **33–5**, 36, 248, 250

Holocene (or Recent; Quaternary) 7, 45, 56, 57, 59, 72, 76, 79, 81, 88, 93, 99, 104, 106, 111, 115, 128, 129, 135, 161, 193, 229, 231, 289, 294, 299, 349, 354, 357

holothuria (sea cucumber) 100, 255, 259, 262, 283, 284

Benthodytes sp. 261

Homarus vulgaris (common lobster) 250

Hong Kong **93**

Hook Ridge (Bransfield Strait, Antarctica) 118, 148

hopanoids (biomarker) 264

Hopen Island (Barents Sea) 45

Hormuz Strait (Arabian Gulf) **91–2**, 346

hot spring 111, 145, 158, 248, 319

Hovland Mounds (offshore Ireland) 65, 303

HRDZ (hydrocarbon-related diagenetic zone) 94, 382

Huang Ho River (People's Republic of China) 99

Huldra field (offshore Norway) 20, 62

Humboldt Slide (offshore California) 111, **357**

hydrate (gas) 1, 2, 4, 6, 45, 46, 47, 49, 51, 52, 67, 68, 69, 70, 76–7, 81, 82, 84, 86, 87, 91, 92, 94, 97, 101, 102, 103, 105, 109, 111, 114, 118, 119, 120, 121, 122, 124, 125, 126–7, 139, 147, 163, 174, 175, 176, 177, **178–88**, 197, 200, 201, 236, 237, 238, 239, 240, 244, 245, 247, 252, 262, 263, 264, 276, 295, 311, 323, 333, 334, 335, 344, 345, 346, 347,

348, 349, 350, 351, 352, 353, 354,
   355, 356, 357, 358, 359, 360, **361**,
   362, **367–8**, **370–1**, 372, 373, 374,
   375–6, 377–8, 384
 climate change role 345, 346, 347, 348,
   350, 351, 352, 353, 354
 resource 101, 127, 355, 374, **375–6**
 sampled 49, 68, 69, 70, 77, 82, 84, 86,
   87, 101, 102, 109, 111, 125, 126,
   180, 181, 182, 184, 185, 201, 340,
   376
 seabed 125, 181, 187–8, 244, 276, 340
 stability zone (GHSZ) 68, 82, 119, 128,
   180, 181, 182, 183, 184, 186, 188,
   288, 340, 350, 352, 358, 361, 377
Hydrate Ridge (offshore Oregon) **109–10**,
   180, 182–5, 252, 263–4, 297, 332,
   334, 336, 337, 338, 340, 366, 376
hydraulic 224, 301, 316
 connection 46, 215, 242, 366, 368
 fracturing 209, 222, 223
 gradient 218
 head 66, 200, 244, 259
 pumping 176, 177, 234
 theory 256, 257
hydrocarbon-derived authigenic
   carbonate (HCDAC) 307
hydrocarbon-related diagenetic zones
   (HRDZ) 94, 382
hydrocarbon 2, 4, 6, 7, 8, 25, 27, 36, 40,
   51, 53, 68, 84, 87, 90, 93, 94, 95,
   99, 103, 105, 106, 113, 114, 116,
   117, 118, 122, 123, 124, 125, 126,
   132, 140, 151, 152, 153, 154, 155,
   157, 158, 159, 172, 183, 184, 189,
   202, 203, 217, 219, 220, 228, 237,
   240, 244, 247, 250, 251, 252, 253,
   256, 257, 259, 264, 266, 271, 274,
   275, 276, 279, 295, 299, 301, 302,
   304, 306, 307, 321, 322, 335, 348,
   355, 366, 367, 371, 375, 376, 381,
   383 (see also coal, methane, ethane,
   etc.)
 abiogenic 157, 160
 aliphatic 259
 aromatic 131, 155, 159, 220, 275
 gas 21, 27, 40, 51, 53, 68, 84, 99, 102,
   103, 104, 105, 106, 109, 114, 116,
   118, 123, 124, 125, 149, 150, 155,
   157, 158, 159, 160, 162, 171, 180,
   186, 202, 207, 208, 217, 257, 264,
   295, 301, 316, 325, 335, 348, 380,
   381 (see also methane, ethane, etc.)
 oil 2, 36, 56, 59, 62, 69, 73, 86, 88, 89,
   90, 115, 116, 122, 123, 124, 125,
   131, 132, 134, 155–7, 158, 160,
   161, 162, 170, 192, 195, 197, 198,

200, 201, 202, 207, 215, 217, 219,
   220, 224, 225, 240, 241, 260, 264,
   275, 288, 306, 307, 309, 317, 335,
   355, 362, 375, 376, 378, 279, 382,
   383, 384 (further details listed under
   oil)
 mantle-derived hydrocarbons 159, 160
hydrofracturing 145, 222, 237
hydrogen (H) 96, 102, 103, 143, 144, 145,
   147, 148, 150, 152, 153, 156, 158,
   159, 179, 180, 266, 288, 303, 314,
   325, 326, 327
 deuterium (^{2}H) 84, 162, 381
 hydrogen:carbon ratio (H:C) 155
 hydrogen:deuterium ratio 381
hydrogencarbonate (HCO$_3^-$; also known
   as bicarbonate) 264, 282, 293, 297,
   311
hydrogen sulphide (H$_2$S) 15, 72, 75, 81,
   92, 99, 100, 109, 113, 114, 120,
   125, 132, 144, 145, 148, 160, 180,
   252, 257, 260, 262, 263, 264, 266,
   271, 281, 283, 295, 298, 301, 313,
   318, 326, 327, 335, 366, 383
 as a drilling hazard **366**
hydrogen-oxidising species 93
hydrolysis 152, 160
hydroplastic/plastic deformation 51, 122,
   207, **225–6**, 240, 245, 246
hydrostatic pressure (P_{hydro}) 90, 101, 148,
   176, 178, **214–5**, 216, 221, 222,
   225, 226, 233, 235, 236, 237, 241,
   318, 338, 350, 351, 352, 364, 366,
   370
hydrothermal 1, 45, 72, 75, 76, 86, 87, 93,
   95, 96, 97, 98, 99, 100, 103, 109,
   110, 118, 138, 139, 140, 142, 143,
   145–9, 157–9, 161, 181, 218, 237,
   240, 248, 255, 259, 260–2, 263,
   269, 277, 279, 290, **314–8**, 319,
   320, 321, 323, 331, 333, 345, 347,
   348, 352, 369, 384, 385
 chimney 3, 6, 72, 95–6, 97, 103, 118,
   142, 262, 274, 299, 309, 314, 315,
   316, 317, 324
 fauna 2, 3, 72, 87, 93, 95, 96, 99, 100,
   103, 110, 140, 142, 248, 251, 255,
   256, 259, **260–2**, 263, **265–6**, 268,
   269–71, 274, 275, 276, 277, 279,
   283, 285, 286, 287, 299, 324, 327,
   359
 fluid 3, 4, 69, 76, 87, 96, 98, 100, 103,
   110, 118, 140, 142, 144, 146–50,
   158, 190, 214, 216, 237, 261, 262,
   285, 299, 309, 311, 314, 315, 316,
   318, 323–6, 338, 341, 351, 374
 intrusion-related systems 140, 144, 237

minerals/mineralization 3, 6, 76, 95,
   96, 97, 98, 99, 103, 118, 140, 145,
   149, 298, 307–8, 309, 311, **314–8**,
   319, 320, 321, 355, **374**, 384
 ocean spreading 45, 87, 93, 109, 110,
   118, 138, 139–40, 145, 148, 208,
   262, 277, 290, 314, 315, 316, 317,
   318, 321, 324, 326, 328, 329, 374,
   385
 petroleum generation 87, 98, 118, **158**,
   325
 plume 76, 77, 96, 98, 110, 145, 149,
   158, 268, 285, **323–9**, 331, 350
 reaction zone 146, **147–9**
 recharge zone 142, 146–7
 salt 87, **317–8**
 seamount 142, 145, 262, 278, 324, 325,
   326, 329
 sediment-capped 118, 138, **149**, 158,
   315–6, 325
 serpentinite-associated 72, 142–3, 144,
   149–50, 158, 162, 288, 326
 shallow water 76, 96, 100, 261–2, 277,
   283, 325, 326, 328, 339
 volcanic 75, 95, 96, 98, 99, 100, 103,
   110, 140, 148, 262, 278, 290, 319,
   324, 325, 326, 328
hydrothermal locations
 Ambitle Island (Papua New Guinea)
   259
 Axial Seamount (Juan de Fuca Ridge)
   110–11, 142, 262, 329
 Cape Banza (Lake Tanganyika) 309
 Chamorro (west Pacific Ocean) 99
 Champagne Pool (Taupo Volcanic
   Zone) 311
 Cleft Segment (Juan de Fuca Ridge)
   324
 Co-Axial Segment (Juan de Fuca
   Ridge) 324
 Conical Seamount (west Pacific Ocean)
   98, 149, 150, 198
 DESMOS (Eastern Manus Basin,
   offshore Papua New Guinea) 96
 East Pacific Rise (E Pacific Ocean) 1,
   117, 141, 142, 158–9, 268, 274,
   277, 317, 324, 327, 343
 Edmond (Indian Ocean) 93
 Endeavour Segment (Juan de Fuca
   Ridge) 315, 317, 329
 Escanaba Trough (Juan de Fuca Ridge)
   138, 158, 316, 317
 Explorer Ridge (Juan de Fuca Plate)
   278
 Frolikha Bay (Lake Baikal) 87
 Galapagos Rift (E Pacific Ocean) 4,
   117, 251, **260–1**, 271

hydrothermal locations (*cont.*)
  Grimsey Island (Iceland) 316
  Guaymas Basin (Gulf of California)
    118–19, 158, 264, 266, 278, 299,
    317
  Hancock Seamount (Hawaiin Islands)
    142
  HF2 (Manus Basin, offshore Papua
    New Guinea) 96
  HF3 (Manus Basin, offshore Papua
    New Guinea) 96
  Hook Ridge (Bransfield Strait,
    Antarctica) 118, 148
  Kagoshima Bay (Japan) 100, 261, 269,
    283
  Kairei (Indian Ocean) 93, 261
  Kermadec Arc (SW Pacific Ocean) 98,
    140, 326, 328
  Kos Island (Aegean Sea) 76
  Kraternaya Bight (Kuril Islands) 261,
    283
  Lesbos Island (Aegean Sea) 76
  Logatchev (Mid-Atlantic Ridge) 149,
    317, 385
  Loihi (off Hawaii) 141, 142, 148, 262,
    324, 325
  Lost City (Atlantic Ocean) 72, 142,
    149, 150, 288, 313, 326
  Lucky Strike (Mid-Atlantic Ridge)
    256, 385
  Luhanga (Lake Tanganyika) 309
  MacDonald Seamount (Pacific Ocean)
    142
  Mariana Arc (southwest Pacific Ocean)
    98, 99, 140, 313, 314, 324, 326,
    328
  Marsili Seamount (Tyrrhenian Sea)
    142, 262
  Matupi Harbour (Papua New Guinea)
    96, 283, 325, 328
  Methana Island (Aegean Sea)
    76
  Middle Valley (Juan de Fuca Ridge)
    138, 315, 316, 317
  Milos Island (Aegean Sea) 76, 77, 262,
    283
  New Ireland Spreading Centre (west
    Pacific) 139
  Nisiros Island (Aegean Sea)
  North Fiji (Fiji Islands) 268
  PACMANUS (Eastern Manus Basin,
    offshore Papua New Guinea) 96,
    97, 148
  Palos Verdes Peninsula (California)
    262, 283
  Panarea Island (Aeolian Islands, Italy)
    325

Piip Volcano (NW Pacific Ocean) **103**,
  142, 158, 262, 316
Rainbow (Mid-Atlantic Ridge) 149,
  376, 385
Red Volcano (E Pacific Ocean) 142, 262
Roman Ruins (PACMANUS) 96
Santorini Island (Aegean Sea) 76, 326
Satanic Mills (PACMANUS) 96
Snowcap (PACMANUS) 96
Susu Knolls (eastern Manus Basin,
  offshore Papua New Guinea) 96
TAG (Mid-Atlantic Ridge) 315, 374
Vienna Woods (Manus Basin, offshore
  Papua New Guinea) 96, 286
Whale Island (New Zealand) 283
White Island (New Zealand) 98
Woodlark Spreading Centre (west
  Pacific) 139
Worm Garden (Manus Basin, offshore
  Papua New Guinea) 96
hydroxyarchaeol 264
hyperbolic reflections 15, 17, 62, 192

Ibérico Mud Volcano (Gulf of Cadiz) 296
Ibiza (Spain) **69**
ice
  furrow/gouge/ploughmark/scour (by
    ice, iceberg, or ice sheet) 11, 15,
    18, 45–6, 47, 51, 60, 104, 105, 130,
    131, 132, 178, 192, 238, 239, 247
  rafting **246–7**
  sea ice 192, 348, 351, 252, 353
ice age (glacial period) 59, 115, 120, 134,
  135, 136, 192, 236, 237, 256, 307,
  334, 340, 350, 351, 352 (*see also*
  Pleistocene and Weichselian)
ice streamthroughs 86
ice worm(*Hesiocaeca methanicola*
  (polychaete) 275
iceberg 11, 15, 45, 47, 130, 131, 142, 237,
  238, 247
  furrow/gouge/ploughmark/scour (by
    iceberg, ice, or ice sheet) 11, 15,
    18, 45–6, 47, 51, 60, 104, 105, 130,
    131, 132, 178, 192, 238, 239, 247
iceberg pit **130**, 247
Iceland 72, 158, 209, 316, 328, 329, 374,
  385
igneous 4, 93, 102, 138, 145, 148, 157,
  160, 178, 198, 209, 222, 311, 326,
  329, 348
  activity 72, 137, 140–2, 144, 145, 158,
    159, 324, 326
  volcano 197, 209
ikaite ($CaCO_3.6H_2O$; mineral) **311**, 312
Ikka Fjord (Greenland) 311, 312
illite (clay-like mica mineral) 97, 150

ilmenite ($FeTiO_3$; mineral) 321
India **92–3**, 342, 375, 393
Indian Ocean 91, 93, 141, 256, 261, 277,
  319, 321, 326, 342, 343, 376
Indonesia 94, 172, 175, 259
Indus Fan (offshore Pakistan) 91–2
Inferno Crater (New Zealand) 310
infragravity wave 235
infrared 185, 329, 330, 379
interglacial 115, 307, 350, 351
Intra-plate settings 4, 134, 135, 137,
  140–2, 148, 325
intra-sedimentary doming 17, 73, **169**,
  172, 174, 186, 242
IODP (Integrated Ocean Drilling
  Program) 183–4
  EPSP (Environmental Protection and
    Safety Panel) 183–4, 185
Ionian Sea 75
Iran 189, 199
Ireland 62, 65, 302, 303, 306, 385
Irian Jaya (Indonesia) 202
Irish Sea 59, 60, 61, 168, 284, 286, 386
iron (Fe) 69, 70, 87, 139, 148, 150, 153,
  260, 262, 263, 272, 298, 300, 306,
  314, 315, 317, 318, 319, 338, 374
  oxidation 266
island arc 138, 139, **140**, 148, 321, 325,
  326
  East Caribbean 140
  Izu-Ogasawara (Bonin) 371
  Kermadec Arc 98, 140, 326, 328
  Mariana 98, 99, 140, 313, 314, 324, 326,
    328
  Sanghihe 321
isocubanite ($CuFe_2S_3$; mineral) 315, 316
isopod (crustacean: order Isopoda)
  Synidotea angulata 251
isostasy 59, 291
isotope (*see under* carbon, chlorine,
  hydrogen, radon, strontium)
Israel 79
Italy 196, 198, 202, 289, 298, 304, 325,
  339, 374
Ivanhoe field (UK North Sea) 9
Ixtoc blowout (Mexican Gulf of Mexico)
  123, 275
Izu Peninsula (Japan) 100, 234
Izu-Ogasawara (Bonin) Island Arc 371

J. Storm II (oil rig) 364
Jabiru field (Australian Timor Sea) 95
Jabuka Trough (Adriatic Sea) **73–4**
jack-up rigs 363, 364
Jaco Scarp (offshore Costa Rica) 119, 142
Jaco Seamount (offshore Costa Rica) 142
Jago (submarine) 43, 83, 249, 265

Japan 99–101, 118, 205, 211, 229, 230, 233, 234, 237, 261, 269, 283, 330, 343, 371, 275, 376
Japan Sea 99, 272
Japan Trench 100, 253, 278
Jasper Seamount (east Pacific Ocean) 142
jellybabies (jelly-like chimneys) 265, 300
jellyfish (phylumCnidaria) 285
  Cyanea sp. 285
Johnson Sea Link (submarine) 123, 254
Joillet field (US Gulf of Mexico) 124
joints (discontinuities in rock/sediment) 90, 148, 150, 182, 219
Juan de Fuca hydrothermal plume 285
Juan de Fuca Plate (offshore Oregon) 109
Juan de Fuca Ridge (east Pacific Ocean) 5, 6, 109, 110, 138, 158, 262, 277, 315, 316, 317, 324, 329
  Axial Seamount 110–11, 142, 262, 329
  Cleft Segment 324
  Co-Axial Segment 324
  Explorer Ridge 278
  Dellwood Knolls 138
  Middle Valley 138, 315, 316, 317
juvenile water 150
juveniles (organisms) 105, 249, 284
Jurassic (Mesozoic) 43, 50, 51, 52, 59, 107, 135, 159, 197, 213, 289, 298, 303

Kagoshima Bay (Japan) 100, 261, 269, 283
Kairei vent field (Indian Ocean) 93, 261
Kamchatka Peninsula (Russian Pacific coast) 101, 103, 158, 202
kaolinite ($Al_2Si_2O_5(OH)_4$; clay mineral) 301, 316
Kaoping Canyon (offshore Taiwan) 94
Kastela, Gulf of (Croatia) 72, 329
Karlsefni Trough (Labrador Shelf, Canada) 130
Karmt Shoals (Australian Timor Sea) 94, 304
Karst 309
Karwar (India) 92
Kattegat 52–6, 250, 275, 281, 282, 290, 291, 292, 297, 299, 301, 347, 386
Kavala, Gulf of (Greece) 75
Kazakov Mud Volcano (Sorokin Trough, Black Sea) 84
Kerch Peninsula (Ukraine) 83, 203, 205
Kermadec Arc (southwest Pacific Ocean) 98, 140, 326, 328
kerogen (insoluble amorphous organic remains) 154, 155, 156, 160, 162, 220
Kerynites River (Greece) 232
keyhole 220, 221, 224

Khamamdag Mud Volcano (Azerbaijan) 202
Kick 'em Jenny Submarine Volcano (offshore Grenada) 120, 140, 372
Kidd Mud Volcano (Gulf of Cadiz) 67
Kimmeridge Clay (Jurassic) 43, 44, 59, 212, 213
kinorhyncha (phylum of spiny crown worms) 122
Klamath Delta (California) 111, 112, 230, 357
Kleppe Senior Formation 9, 10, 13, 32, 37
Kodiak Shelf (offshore Alaska) 103
kolobovnik (Lake Baikal ice features) 86
Korea 99
Kornev Mud Volcano (central Black Sea) 83
Kos Island (Aegean Sea) 76
Kraternaya Bight (Kuril Islands) 261, 283
krill (euphausiid crustaceans; order Euphausiacea) 38, 46, 248, 250, 275
Krishna River (India) 93
Kristin field (offshore Norway) 50, 53, 256
Kuril Trench 100
Kuril Islands (Japan/Russia) 102, 103, 261, 283
Kuroko-type mineral deposits 374
Kutch, Gulf of (India) 92

La Goleta seep (offshore California) 375
La Goleta field (offshore California) 375
Labrador (Canada) 129–30
  Current 130
  Shelf 129–30
Læsø (Denmark) 291
LaHave Basin (Scotian Shelf) 7, 8
LaHave Clay 7, 8, 10
lakes 189, 191, 235, 284, 348–9, 357
  brine lakes 80–1, 335 (*see also* Nadir Brine lake)
  tar lakes (Trinidad) 120–1
Lake Baikal (Russia) 86–7, 182, 273, 285, 348–9
Lake Michigan (USA) 129
Lake Superior (Canada/USA) 129, 130, 212, 254–5
Lake Tanganyika (East Africa) 309
Lake Van (Turkey) 310–11
Lambay Deep (Irish Sea) 60
Lamellibrachia sp. (siboglinid tubeworms) 118, 251, 252, 270, 271, 278
Laptev Sea (Russian arctic) 311, 348
Larson Seamounts (E Pacific) 142
last glacial maximum (LGM) 135, 250, 352

Latakia (Syria) 2
Laurentian Channel (Canada) 130
Laurentian Fan (offshore E. Canada) 131, 252, 269, 278, 355
lava 102, 138, 142, 148, 202, 209, 261, 324, 328–9
Lebanon 330
lecithin (chemical additive for drilling mud) 367
Leipzig Mud Volcano (Mediterranean Ridge) 76
Leptonemella aphanothecae (nematode) 250
Lesbos Island (Aegean Sea) 76
Leucothrix sp. (bacterium) 253
Levinsenia gracilis (polychaete) 249
LGM (last glacial maximum) 135, 250, 352
lignin (constituent of woody material) 152, 154, 155
lignite (brown coal) 59, 136, 156, 172
limpet (gastropods; family Acmaeidae) 110, 113, 119, 255, 261, 262, 274
  Lottia limatula 274
  patellacean 113
ling (*Molva molva*; fish) 39, 249, 280
lipid (water-insoluble organic compound) 152, 154, 155, 158
liquefaction 100, 104, 106, 112, 206, 211, 226, 229, 230, 234, 237, 240, 356, 357, 360
lithification 216, 222, 240, 309
Little Bahama Bank 313
Little Salmon Fault (offshore California) 112
lobster (crustaceans; family Nephropidae) 72, 274 (*see also* squat lobster)
  common (*Homarus vulgaris*) 250
  Norwegian (*Nephrops norvegicus*) 36
Loch Eribol (UK) 60
Loch Etive (UK) 60
Logachev Mounds (offshore Ireland) 302
Logatchev hydrothermal vents (Mid-Atlantic Ridge) 149, 317, 385
Logging-while-drilling (LWD) 185
Loihi Seamount (offshore Hawaii) 141, 142, 148, 262, 324, 325
Lokbatan Mud Volcano (Azerbaijan) 203
Lokbatan-type mud volcano 202–4, 205, 225, 226, 240, 246, 333, 336, 347, 352
Lophelia pertusa (cold-water coral; stony coral) 51, 53, 64, 65, 69, 255–8, 302

Lost City (serpentinite-related vents, Atlantic Ocean) 72, 142, 149, 150, 288, 313, 326
Lottia limatula (limpet) 274
Louiseville guyot chain (sw Pacific Ocean) 142
Louisiana (USA) 123, 251, 258
   shelf/slope 123, 251, 258, 304, 381
lucinid bivalves (Lucinidae; clams) 70, 125, 244, 271, **272–3**, 278, 279, 289, 305
   Lucinoma sp. 249, 272
   Myrtea sp. 272
Lucinoma sp. (lucinid bivalve) 249, 272
Lucky Strike hydrothermal vents (Mid-Atlantic Ridge) 256, 385
Luhanga hydrothermal field (Lake Tanganyika) 309
Luzon Arc 93
LWD (logging-while-drilling) 185
Lyngenfjord (Norway) 49–50, 51, 52
Lyopsetta exilis (slender sole; fish) 251

maar 208, 209
   Figge Maar (German North Sea) 335, 364
MacDonald Seamount (Pacific Ocean) 142
Machar field (UK North Sea) 20, 59, 60
macrofauna 69, 72, 87, 91, 96, 99, 100, 112, 122, 125, 193, 249, 250, 251, 256, 261, 262, 269, 273, **274–5**, 276, 280, 281, 282, 289, 349
Macrouridae sp. (rattail or grenadier; fish) 261
Madrepora sp. (cold-water coral; stony coral) 64, 69, 256
mafic
   minerals (e.g. olivine and pyroxene) 158
   rock (e.g. basalt) 146, 160
Magellan Mounds (offshore Ireland) 64, 302–3
Magellan Strait (Chile) 118
Mageneitas River (Greece) 231
magma/magmatic 76, 98, 125, 137, 142, **144–5**, 148, 149, 150, 157, 158, 159, 160, 209, 237, 316, 324, 325, 326
   andesitic (intermediate) 148
   basaltic (basic) 137, 138, 147, 148
   phreatomagmatic 209
   rhyolitic (acidic) 148
magnesium hydroxide (*see* Brucite)
magnetic
   anomaly 102
   survey 132

magnetite (Fe₃O₄; mineral) 149
Mahanadi River (India) 93
Maidstone Mud Volcano (Mediterranean Ridge) 77, 245
Maikopian Clay (Oligocene–Lower Miocene) 83
Maine, Gulf of (USA) **128–9**, 190, 193
Makarov Bank (Azerbaijani Caspian Sea) 203, 336, 337
Makran Accretionary Wedge (offshore Pakistan) 91, 92, 182, 187, 346
Makran Coast (Pakistan) **91–2**, 199, 204
Malacca Strait (Indonesia/Malaysia) 236
Malan Island (Pakistan) 91, 204
Malaysia 93, 370
maldanid polychaetes 253
Malibu Point (California) **117**, 230
Mallik well (Canada) 376
Malyshev Mud Volcano (central Black Sea) 83
Mandovi River (India) 342
Mangalore (India) 92
manganese (Mn) 70, 87, 96, 139, 148, 158, 268, 319, 323, 325, 329, 338
   ferromanganese nodules 290, **319–21**
   Mn and Fe-Mn oxide/hydroxide 99, 153, 314, 321, 319
   oxidation 266
mantle 98, 111, 137, 141, 145, 150, 157, 158, 159, 160, 198, 303, 320, 321
   mantle-derived fluids/volatiles 158, 159, 160, 214, 310, 322
   mantle-derived (juvenile) helium 157, 158, 159, 325
   mantle-derived hydrocarbons 159, 160
   mantle-derived methane 111, 157, **158–9**, 160, 288
   mantle-derived peridotite 72, 143, 148, 157, 214, 288, 313
Manus Basin (Papua New Guinea) 95, **96**, 140, 145, 148, 286
Maorithyas sp. (Thyasirid bivalve) 97
Marble Canyon (British Columbia) 309–10
Marenzelleria viridis (polychaete) 254
Mariana Arc (SW Pacific Ocean) 98, 99, 140, 313, 314, 324, 326, 328
Mariana Trench (SW Pacific Ocean) 98, 99, 142
Marine Protected Area (MPA) 303, 385
Marion Lake (Australia) 308
Mars **247**
Mars field (US Gulf of Mexico) 124
Marsili Seamount (offshore Italy) 142, 262
Martinique (Eastern Caribbean) 140

Matupi Harbour (Papua New Guinea) 96, 283, 325, 328
MBES (multi-beam echo sounder) 2, 10, 11, 12, 14, 16, 19, 22, 40, 41, 50, 52, 61, 64, 68, 71, 80, 82, 83, 373, 381
MDAC (methane-derived authigenic carbonate) 6, 15, 27, 31, 32, 33, 34, 35, 36, 37, 39, 40, 43, 44, 50, 52, 54, 56, 62, 65, 69, 77, 79, 84, 88, 91, 95, 97, 101, 103, 109, 111, 112, 113, 114, 118, 122, 127, 132, 145, 161, 175, 177, 193, 195, 197, 200, 237, 241, 249, 258, 264, 265, 273, 274, 277, 281, 283, **290–307**, 310, 320, 322, 331, 349, 369–70, 383, 386
Measurement-while-drilling (MWD) 369
Mediterranean Sea 2, 67, 69, 75, 76, 142, 246, 256, 294, 326, 329
   Eastern Mediterranean Sea **74–81**, 272, 295
Mediterranean Outflow Water (MOW) 67
Mediterranean Ridge 75, **76–9**, 182, 202, 205, 246, 264
meiofauna 96, 99, 122, 273, **275**, 276
Mekong Delta (Vietnam) 93
mercury (Hg) 98, 262
Merluccius capensis (cape hake; fish) 383
Mersey Estuary (UK) 60
Mesozoic 11, 45, 50, 51, 52, 94, 105, 123
   Cretaceous 7, 50, 51, 81, 107, 113, 126, 132, 140, 159, 212, 256, 289, 303
   Jurassic 43, 50, 51, 52, 59, 107, 135, 159, 197, 213, 289, 298, 303
   Triassic 46, 67, 68, 107, 123, 159, 239
Messinian (Miocene) 80
Messolongi Lagoon (Greece) 75
metal 6, 87, 145, 148, 149, 314, 315, 319, 320, 321, 355, 375, 384
metal sulphides 6, 97, 118, 148, 149, 263, 314–6, 317, 374, 384
   chalcopyrite (CuFeS₂; mineral) 149, 314, 315
   pyrite (FeS₂; mineral) 68, 103, 126, 262, 263, 264, 291, **298**, 300, 306, 314, 316, 318
   pyrrhotite (FeS; mineral) 118, 314, 315, 316
   sphalerite (ZnS; mineral) 149, 314
metal-bearing fluids 87, 96, 208, 262, 314, 315, 316, 317, 318, 327, 374
metalliferous deposits 6, 87, 97, 149, 290, 314, 315, 316, 317, 327, **374**
metagabbro (metamorphosed gabbro) 98

meteor 192
meteoric water 150, 190, 202, 214, 294, 295, 305
meteorite 150, 157, 159, 247
Methana Island (Aegean Sea) 76
methane (CH$_4$) 2, 4, 6, 52, 81, 101, 125, 127, 144, 145, 153–7, 219, 220, 252, 287, 288, 304, 321, 322, 355, 363, 372
    abiogenic (inorganic) 87, 150, 151, 157–60, 161, 162, 190, 303, 325
    atmospheric 6, 58, 85, 103, 127, 161, 187, 285, 288, 323, **345–9**, 350–3
    biogenic (derived from organic matter) 4, 151, 160, 162
    carbon isotope ratio ($^{13}\delta$C) of 40, 57, 91, 104, 161, 162, 250, 263, 264, 271, 278, 281, 282, 291, 295, 295, 316, 335, 343
    density of 163, 216, 219
    fossil **347–8**
    flux through the seabed 49, 58, 72, 81, 85, 122, 126, 132, 181–2, 235, 252, 263, 270, 307, 313, 335, **337–8**, 344, 345, 347, 349, 353, 354, 373
    flux to the atmosphere 6, 58, 81, 86, 103, 116, 338, 340, 342–4, **345–7**, 348, 349, 352, 354, 365, 374
    fractionation of 41, 123, 162
    generation of 4, 126, 134, 135, 136, 151, **153–60**, 162, 178, 193, 216, 263, 330, 337, 341, 342, 343, 348, 350
    hydrate (*see under* hydrate)
    hydrothermal 96, 98, 99, 100, 103, 110, 142, 143, 144, 145, 150, 158, 162, 285, 314, 323, 324, 325, 326, 341, 345, 347, 350, 351, 352,
    oxidation/utilisation 56, 87, 126, 153, 154, 186, 193, 263, 266, 269, 270, 271, 272, 280, 282, 284, 285, 286, 290, 300, 307, 311, 327, 331, 337, 338, 341, 342, 343, 345, 349, 354, 384
    mantle-derived (primordial) 111, 157, **158–9**, 160, 288
    microbial (generated by microbes) 41, 56, 57, 67, 87, 91, 93, 101, 102, 103, 106, 109, 114, 124, 125, 134, 135, 136, 151, **153–4**, 156, 160, 161, 162, 178, 193, 206, 214, 238, 240, 246, 263, 311, 335, 342, 347, 348, 351, 353
    seep 6, 21, 25, 44, 52, 58, 66, 67, 70, 73, 82, 85, 92, 101, 102, 103, 106, 107, 109, 112, 114, 115, 116, 117, 118, 120, 122, 135, 151, 182, 193, 195, 244, 250, 251, 257, 259, 263, 264,

266, 272, 273, 275, 278, 280, 281, 284, 285, 288, 289, 290, 291, 293, 298, 299, 300, 305, 311, 331, 332, 334, 335, 338, 341, 345, 347, 348, 349, 350, 352, 366, 374, 375, 383
    sensor (METS) 114, 332
    thermogenic 40, 41, 57, 59, 94, 103, 104, 114, 121, 125, 149, 151, **154–7**, 161, 162, 214, 263, 303, 311, 316, 325, 335, 349, 351, 362
methane in pore water/sediment 15, 27, 37, 40, 44, 49, 50, 53, 56, 57, 58, 59, 65, 84, 87, 90, 92, 93, 99, 102, 103, 104, 105, 108, 113, 114, 115, 120, 121, 123, 126, 128, 131, 135, 139, 151, 153, 154, 168, 178, 181, 183, 186, 193, 196, 206, 214, 216, 217, 218, 219, 220, 221, 223, 240, 241, 246, 252, 253, 256, 264, 271, 279, 280, 291, 293, 299, 301, 305, 313, 337, 343, 344, 381
methane in the water column 25, 44, 46, 49, 56, 58, 70, 76, 79, 81, 85, 92, 98, 100, 101, 102, 103, 104, 107, 109, 115, 118, 119, 145, 159, 163, 182, 193, 195, 256, 257, 272, 284, 285, 286, 316, 326, 327, 332, 334, **338–45**, 346–9, 352, 353, 354, 380, 382
methane from mud volcanoes 47, 49, 78, 79, 81, 84, 92, 97, 122, 181–2, 202, 214, 252, 335, 338, 341, 344, 347, 348, 353
methane-derived authigenic carbonate (MDAC) 6, 15, 27, 31, 32, 33, 34, 35, 36, 37, 39, 40, 43, 44, 50, 52, 54, 56, 62, 65, 69, 77, 79, 84, 88, 91, 95, 97, 101, 103, 109, 111, 112, 113, 114, 118, 122, 127, 132, 145, 161, 175, 177, 193, 195, 197, 200, 237, 241, 249, 258, 264, 265, 273, 274, 277, 281, 283, **290–307**, 310, 320, 322, 331, 349, 369–70, 383, 386
methanotrophic (methane-oxidising) 107, 122, 250, 251, 253, 263–5, 272, 273, 274, 277, 278, 282, 283, 285, 343, 350
methane-oxidising (methanotrophic) bacteria (*see also* individual genera) 56, 87, 102, 162, 250, 252, 261, 263, 269, 270, 272, 281, 282, 342
methane-rich brine 78, 79, 87, 208, 359
methane-rich plume 46, 49, 70, 92, 98, 107, 109, 145, 159, 182, 326, 327, 332, 334, 338, 346

methanogenesis 4, 126, 134, 135, 136, 151, **153–60**, 162, 178, 193, 216, 263, 330, 337, 341, 342, 343, 348, 350
methanogen 153, 154, 263, 264, 287, 342, 343
methanol (CH$_4$O) 152, 153, 368, 371
Methanosaeta sp. (methanogenic Archaea) 300
Methanothrix sp. (now *Methanosaeta*) 300
methanotrophic (methane-oxidising) 107, 122, 250, 251, 253, 263–5, 272, 273, 274, 277, 278, 282, 283, 285, 343, 350
methanotrophic bacteria (methane-oxidising; *see also* individual genera – listed under bacteria) 56, 87, 102, 250, 252, 261, 269, 270, 272, 281, 282, 342
METI Well (gas hydrate prospecting well) 101
methylamine (CH$_3$NH$_2$) 153
Methylococcus capsulatus (bacterium) 383
Metridium senile (sea anemone) 249
METS methane sensor 114, 332
Mexico, Gulf of 1, 115, **122–6**, 157, 159, 176, 182, 183, 188, 202, 204, 205, 218, 244, 245, 252, 253, 258, 264, 271, 272, 273, 274–5, 276, 278, 285, 286, 295, 299, 304, 307, 330, 362, 364, 367, 368, 369, 376, 378, 379, 380
micrite/micritic 293, 294, 298, 301, 302
microbes 4, 6, 151, 154, 155, 178, 257, 261, **262–9**, 271, 273, 274, 275, 277, 280, 282, 285, 297, 298, 300, 301, 307, 309, 339, 350, 374, 383
    archaea 4, 96, 153, 263–5, 288, 297, 300
    bacteria 4, 56, 76, 78, 87, 92, 93, 96, 102, 158, 250, 251, 252, 261, 262, 264, 266–74, 275, 277, 282, 285, 288, 289, 297, 301, 309, 324, 383
    methanogen 288
    methanotroph 285
    thiotrophs 102, 114, 250, 253, 263, 264, 266, 270, 272, 274, 278, 281, 282, 283 (*see also individual species*: Beggiatoa sp., Thioploca sp., Thiothrix sp., Thiovulum sp.
microbial carbonate 290
microbialite **308–11**, 312
microfauna 249, 250, **262–9**
micro-plates (*listed under* plates)
Mid-Atlantic Ridge 1, **72**, 137, 142, 146, 158, 256, 274, 277, 313, 314, 316, 326, 328, 344, 385
mid-ocean ridge basalt (MORB) 148, 325

Middle Valley (Juan de Fuca Ridge) 138, 315, 316, 317
Midgard field (offshore Norway) 50
migration 4, 8, 41, 44, 60, 93, 159, 163, 182, 196, 189, 197, 212, 214, 216, 240, 241, 244, 245, 263, 304, 339, 352, 360, 383, 384
    capillary 221, 223, 228, 236
    fluid 44, 49, 62, 68, 70, 79, 82, 94, 119, 121, 124, 140, 174, 175, 176, 181, 183, 206, 207, 213, 214, 215, **216–26**, 233, 237, 239, 240, 245, 247, 360
    gas 9, 22, 27, 60, 73, 94, 162, 163, 165, 170, 176, 182, 183, 186, 216, 219, 223, 225, 240, 241, 245, 311
    tertiary 217
    pathway 10, 52, 60, 64, 65, 93, 94, 95, 97, 100, 115, 119, 123, 126, 139, 145, 158, 214, 217, 220, 223, 224, 225, 228, 237, 238, 239, 240, 241, 243, 244, 247, 291, 301, 302, 305, 306, 322, 334, 335, 347, 358, 365, 370, 378
    petroleum 56, 140, 159, 219, 220, 241
    primary 157, 217
    secondary 217
Mikkel field (offshore Norway) 50, 364
Milano Mud Volcano (Mediterranean Ridge) 77, 79, 205
Milford Haven Mud Volcano (Mediterranean Ridge) 76
Miocene (Tertiary) 65, 68, 83, 87, 111, 115, 120, 123, 136, 207, 208, 211, 289, 298, 304
Milos Island (Aegean Sea) 76, 77, 262, 283
minerals/mineralization (*see under individual minerals*)
    hydrothermal-related 3, 6, 76, 95, 96, 97, 98, 99, 103, 118, 140, 145, 149, 298, 307–8, 309, 311, **314–18**, 319, 320, 321, 355, **374**, 384
Mir (submarine) 103, 316
Mississippi (State, USA) 123
    Canyon (US Gulf of Mexico) 244
    Delta (Louisiana) 106, 123, 124, 135, 198, 206, 207, 236, 357, 364
    Fan (US Gulf of Mexico) 124
    River (USA) 207, 235, 236, 251, 368
Mittelgrund (Eckernförde Bay, Germany) 57, 58
Moho (the Mohorovičič discontinuity) 159
mollusc (*see* bivalve, chiton, and gastropod)

Molva molva (ling: fish) 39, 249, 280
monazite (Ce, La, Nd, Th) PO_4; mineral) 301
Monferrato (Italy) 289, 304
Mongolia 86
Monilifera (former sub-phylum of pogonophores; now siboglinid tubeworms) 269
Mono Lake (California) 311
monosaccharide 152
monsoon 91, 92, 343
Monte Norte seamount (west Mediterranean Sea) 69
Monterey Bay (California) **112–14**, 140, 244, 269, 272, 273, 278, 295, 299, 340, 372
Monterey Bay Fault (offshore California) 113, 140
Monterey Canyon (offshore California) 112–14, 140
Monterey Fan (offshore California) 269
Monterey Formation 113, 115
MORB (mid-ocean ridge basalt) 148, 325
Morecambe Bay (UK) 229
Morocco 67, 306
Moscow Mud Volcano (Mediterranean Ridge) 76, 77, 245
Moscow State University Mud Volcano (MSU; central Black Sea) 83
mound
    anhydrite **316–17**
    carbonate (*see* carbonate mound)
    Darwin **64–6**, 384, 385
Mount Manon Mud Volcano (offshore Barbados) 121
Møre Basin (offshore Norway) 140, 237
MPA (Marine Protected Area) 303, 385
Mt Mazama (volcano: Oregon) 268
Mt Crushmore (Monterey Canyon, offshore California) 113
Mt Sakurajima (volcano; Japan) 100
Muck (Island; UK) 60
mud breccia (or olistostrome) 77, 84, 86, 97, 98, 196, 197, 200, 202–4, 214, 225
mud diapir (includes clay and shale diapirs) 4, 51, 52, 54, 55, 67–9, 76–7, 84, 91–2, 93, 94, 97, 101, 112, 119, 120, 121, 122, 172, 183, **195–8**, 200–1, 202, 205, 207–8, 209, 212, 213, 214, 215, 216, 218, 223, 228, 230, 245, 296, 360 (*see also* shallow mud diapir)
mud lump 124, 186, 206, 236
mud mound 124, 196, 200 (*see also* carbonate mud mound)
mud pie 77, 121, 122, 196

mud ridge 77, 196
mud volcano – general (*see also* shallow mud volcano) 2, 4, 6, 45, 51, 60, 70, 73, , 81, 86, 87, 91, 92, 93, 97, 101, 116, 117, 119, 120, 123, 124, 125, 139, 140, 143, 181, 189, 190, 191, **195–205**, 209, 213, 214, 225, 226, 228, **232**, 237, 239, **245–6**, 252–3, 259, 275, 279, 280, 299, 323, 331, 333, **335–6**, 337, 338, 341, 345, 346, 347, 348, 352, 355, 356, 367, 369, 376
    activity types **202–5**
    asphalt 116, 122, 125
    eruption 122, 196, 198, 200, **202–5**, 225, 232, 234, 333, 336, 337, 347, 352, 355
    gryphon 79, **198**, 199, 200, 201, 202, 280, 356
    salse 79, 191, **198**, 199, 200, 210, 202, 280
    serpentinite 98–9
mud volcano areas/belts/groups
    Azerbaijan/Southern Caspian Basin **85–6**, 207, 234, 245
    Barbados accretionary wedge **121–2**, 245, 252, 273
    Central Black Sea **83–4**
    Deep Portuguese Margin field (Gulf of Cadiz) 68, 69
    Gualdalquivir Diapiric Ridge (Gulf of Cadiz) 68
    Gulf of Cadiz (E. Atlantic Ocean) 67–9
    Mediterranean Ridge (E. Mediterranean Sea) 76–9
    Nile Delta **80–1**, 279
    Sorokin trough (Black Sea) 83, **84**, 85, 95, 245
    Spanish-Moroccan field (Gulf of Cadiz) 67, 69
    Taiwan **93–4**
    TASYO field (Gulf of Cadiz) 68
mud volcanoes
    Almazán (Gulf of Cadiz) 68
    Anastasya (Gulf of Cadiz) 68
    Atalante (offshore Barbados)
    Aveiro (Gulf of Cadiz) 68
    Bakhar (Azerbaijan) 200, 201, 205
    Baraza (Gulf of Cadiz) 67, 68
    Bolshoi (Lake Baikal) 87
    Bonjardim (Gulf of Cadiz) 68, 69
    Buzdag (Azerbaijani Caspian Sea) 86
    Chandragup (Pakistan) 92, 199
    Chatham Island (offshore Trinidad) **122**, 204
    Chikishlyar-type **204–5**, 240, 246, 336
    Chirag (Azerbaijani Caspian Sea) 201

Cibeles (Gulf of Cadiz) 68
Complex Mound (US Gulf of Mexico) 205
Cyclops (offshore Barbados) 121
Dashgil (Azerbaijan) 198, 205
Dvurechenskii (DMV; Sorokin Trough, Black Sea) 84
El Pilar (offshore Barbados) 122
Elm (Azerbaijani Caspian Sea) 86
Faro (Gulf of Cadiz) 68
Gades (Gulf of Cadiz) 68
Gelendzhik (Mediterranean Ridge) 76, 202, 245
Gharniarigh-Tapeh (Iran) 199
Ginsburg (Gulf of Cadiz) 67, 68
Håkon Mosby (HMMV; offshore Norway) 47, 49, 50, 181, 197, 200, 236, 245, 253, 254, 338, 344
Hingol Island (Makran Coast, Pakistan) 91
Ibérico (Gulf of Cadiz) 296
Kazakov (Sorokin Trough, Black Sea) 84
Khamamdag (Azerbaijan) 202
Kidd (Gulf of Cadiz) 67
Kornev (central Black Sea) 83
Leipzig (Mediterranean Ridge) 76
Lokbatan (Azerbaijan) 203
Lokbatan-type 202–4, 205, 225, 226, 240, 246, 333, 336, 347, 352
Maidstone (Mediterranean Ridge) 77, 245
Makarov Bank (Azerbaijani Caspian Sea) 203, 336, 337
Malan Island (Makran Coast, Pakistan) 91, 204
Malyshev (central Black Sea) 83
Milano (Mediterranean Ridge) 77, 79, 205
Milford Haven (Mediterranean Ridge) 76
Moscow (Mediterranean Ridge) 76, 77, 245
Moscow State University (MSU; central Black Sea) 83
Mount Manon (offshore Barbados) 121
Napoli (Mediterranean Ridge) 77, 78, 79, 205, 294
Olenin (Gulf of Cadiz) 68
Pipoca (Gulf of Cadiz) 68
Rabat (Gulf of Cadiz) 67, 68
Ribeiro (Gulf of Cadiz) 68
Schugin-type 205, 246
St Petersburg (Gulf of Cadiz) 68
Student (Gulf of Cadiz) 67
Stvor (Mediterranean Ridge) 77
Tarsis (Gulf of Cadiz) 68

TASYO (Gulf of Cadiz) 68
Toronto (Mediterranean Ridge) 77
Tredmar (central Black Sea) 84
TTR (Gulf of Cadiz) 68
Volcano A (offshore Barbados) 121
Volcano F (offshore Barbados) 121
Volcano J (offshore Barbados) 121
Yuma (Gulf of Cadiz) 67, 68
Yuzhmorgeologiya (central Black Sea) 83
multibeam echo sounder (MBES) 2, 10, 11, 12, 14, 16, 19, 22, 40, 41, 50, 52, 61, 64, 68, 71, 80, 82, 83, 373, 381
Mumbai (India) 92
Munida sarsi (squat lobster) 38
Murray Ridge (Indian Ocean) 91
muscovite ($KAl_2(AlSi_3O_{10})(F,OH)_2$; clay mineral) 310
mussels (see also *mytilid bivalves*) 70, 79, 97, 99, 105, 125, 251, 252, 255, 271, 272, 278, 282, 283, 306, 383
mussel farming 67, 383
MWD (measurement-while-drilling) 369
Myojin-Sho submarine Volcano (offshore Japan) 371
Myrtea sp. (lucinid bivalve) 272
mysid (opossum shrimps) 251
mytilid bivalves (Mytilidae; mussels) 119, 261, 269, 271, **272**, 278, 279, 287
Bathymodiolus sp. 97, 99, 252, 272
Mytilus edulis 282
Mytilus edulis (common mussel; mytlid bivalve) 282
Myxine glutinosa (hagfish) 249

Nadir Brine Lake (Mediterranean Ridge) 78, 294
Nam Con Son sedimentary basin (offshore Vietnam) 93
Nameless Seamount (Gulf of Cadiz) 321
Namibia 70, 72, 313, 383
Nankai Trough (offshore Japan) 100, 101, 253, 278, 299, 376
Napoli Mud Volcano (Mediterranean Ridge) 77, 78, 79, 205, 294
napthenes (C_nH_{2n}; also known as *cycloalkanes*) 156, 158
Nares Abyssal Plain (Atlantic Ocean) 173
Natura 2000 (EU network of nature conservation sites) 385–6
Nautile (submarine) 78, 100, 119, 122
Navarin Basin (offshore Alaska) 103, **104–5**
Navarinsky Canyon (offshore Alaska) 105
Nazca Plate (E Pacific Ocean) 119

nekton 33, 36, 248, 273
nematode (Phylum *Nematoda*) 87, 96, 99, 122, 152, 249, 250, 273, 274, 275, 281, 282, 283
Astonomena southwardorum 249
Leptonemella aphanothecae 250
Neogene (Tertiary) 76, 97, 113, 175
Neogastropod 255
nephelometer 70
Nephrops norvegicus (Norwegian lobster) 36
Nephtys sp. (polychaete) 281
Neptunea sp. (gastropod) 112
neptunian dykes 306
Netherlands 9, 60, 192, 385
neural networks 170, 178
New Britain Island (Papua New Guinea) **95–7**, 325
New Britain Trench 95
New Ireland Spreading Centre (west Pacific Ocean) 139
New Jersey (USA) 151, 359, 361
New Jersey continental slope (offshore USA) 361
New Millennium Observatory Network (NeMO; Axial Seamount) 111
New York Bight (USA) 246
New Zealand **97–8**, 188, 199, 203, 283, 299, 310, 325, 374
Newfoundland (Canada) 132, 355
Newfoundland Grand Banks 129, 131, 238, 374, 383
nickel (Ni) 150, 319
Ni-Fe alloys 150
Nicoya Slump (offshore Costa Rica) 119
Niger Delta (Nigeria) **69–70**, 135, 174
Niger Fan (offshore Nigeria) **69–70**
Nigeria 174, 176, 183, 202, 209, 246
Nikolaiika Bay (Greece) 232
Nile Delta (Egypt) **79–81**, 135
Nile Fan (offshore Egypt) **79–81**, 301, 307
Nisiros Island (Aegean Sea) 76
Niue Island (atoll; south Pacific Ocean) 319
Njord field (offshore Norway) 50
nodule
ferromanganese 290, **319–21**
noise (sound) 142, 204, 225, 232, 260, 333
Nome (Alaska) 104
Nordic Basin (offshore Norway) 253
Norne field (offshore Norway) 50
North American Plate 103, 106, 109, 111, 113
North Carolina (USA) 126, 234, 284
North Fiji hydrothermal vents (Fiji Islands) 268

North Pennine Orefield (UK) 315, 374
North Sea **8–44**, 59, 60, 61, 62, 63, 64, 65,
    73, 106, 135, 136, 159, 170 171,
    172, 174, 189, 190, 192, 193, 194,
    195, 206, 207, 210, 211, 212, 217,
    218, 224, 228, 232, 234, 238, 240,
    241, 245, **248–50**, 264, 272, 273,
    274, 275, 278, 280, 281, 284, 285,
    286, **290–1**, 3–1, 307, 333, 334,
    335, 337, 342, 344, 352, 365, 368,
    369, 372, 373, 374, 381, 383,
    384
  Fan 358
Norton Sound (Alaska) 103–4
Norway 21, **49–50**, 51, 53, 54, 140, 145,
    183, 186, 207, 218, 237, 255, 256,
    269, 306, 356, 364, 366, 370, 383,
    385, 386
Norway haddock (*Sebastes vivaparus*; fish)
    32, 249
Norwegian deep-water corals **255–8**, 305,
    386
Norwegian continental shelf 256, 386
Norwegian continental slope 264
Norwegian lobster (*Nephrops norvegicus*)
    36
Norwegian margin 350, 357
Norwegian North Sea 16, 333, 369,
    384
  Block NO1/9 – Tommeliten
    (Norwegian North Sea) 3, 20,
    25–9, 59, 60, 191, 248, 264, 280,
    290, 291, 301, 302, 332, 333
  Block NO24/9 – The Holene
    (Norwegian North Sea) 14, 20,
    33–5, 36, 248, 250
  Block NO25/7 (Norwegian North Sea)
    14, 15, 20, **30–3**, 34, 35, 206, 248
  Block NO26/8 (Norwegian North Sea)
    20, **36**, 37
  Block NO26/9 (Norwegian North Sea)
    20, **36**, 38, 248
Norwegian Sea **50–2**, 65, 178, 181, 206,
    207, 238, 245, 357, 365, 368
Norwegian Trench (Norwegian North
    Sea) 9, 10, 11, 13, 15, 16, 17, 19,
    36, 37, 38, 52, 54, 60, 238, 243,
    248, 250, 381
Nova Scotia (Canada) 7–8, 130, 248
Nowlin Knolls (Gulf of Mexico) 124
NR-1 (submarine) 123
Numulites sp. (foraminifera) 305
nutrient 1, 6 105, 120, 235, 254, 255, 257,
    259, 261, 275, 276, 278, 284–6,
    287, 309, 326, 330

OBS (ocean bottom seismometer) 332

Obturata (vestimentifera; former
    sub-phylum of pogonophores;
    now siboglinid tubeworms) 269
Obsidian Dome Volcano (California) 98
ocean
  Arctic 207, 352, 376
  Atlantic 47, 67, 93, 118, 141, 146, 149,
    255, 256, 259, 261, 272, 277, 278,
    319, 326, 342, 343, 352, 355, 376,
    385
  Indian 91, 93, 141, 256, 261, 277, 319,
    321, 326, 342, 343, 376
  Pacific 93, 98, 102, 117, 118, 138, 140,
    141, 142, 198, 256, 261, 270, 272,
    277, 278, 279, 319, 321, 325, 326,
    343, 352, 376
ocean bottom seismometer (OBS) 332
Ocean Drilling Program (see *ODP*)
ocean methane paradox 343
ocean spreading 1, 4, 45, 72, 87, 95, 117,
    118, **137–8**, 141, 148, 158, 160,
    208, 218, 239, 261, 262, 269, 277,
    290, 324, 325, 326, 327, 352
octocoral (*Anthozoa* subclass Alcyonaria)
    261, 305
  sea pen (order Pennatulacea) 27
    (*individual genera listed under* sea
    pen)
octopus (*Octopoda*)
  Graneledon sp. 261
Oder Bight (German Baltic Sea) 343
Oder, river (Germany) 343
ODP (Ocean Drilling Program): 77, 96,
    98, 119, 151, **183–6**, 215, 288, 317,
    334, 366, 367
  Leg 141 (Chile Triple Junction) 183
  Leg 146 (Hydrate Ridge) 109, 180, 184,
    185–6, 340, 366, 367
  Leg 160 (Mediterranean Ridge) 205
  Leg 164 (Blake Ridge) 126, 127, 184,
    186
  Leg 168 (Endeavour Ridge) 146
  Leg 169 (Middle Valley) 315
  Leg 170 (Costa Rica margin) 118
  Leg 193 (Manus Basin) 95, 145, 148
  Leg 204 (Hydrate Ridge) 109, 184,
    185
  Leg 206 (Cocos Plate) 141
  Site 395 (Leg 109, Atlantic Ocean) 146
  Site 684 (Leg 112, Peru Margin) 299
  Site 808 (Leg 131, Nankai Trough) 101
  Site 889 (Leg 146, Cascadia Margin)
    340
  Site 892 (Leg 146, Cascadia Margin)
    109, 184, 185, 252
  Site 995 (Leg 164, Blake Ridge) 186
  Site 996 (Leg 164, Blake Ridge) 127

offshore basin screening 379
offshore site investigation/survey 8, 21,
    25, 41, 59, 75, 90, 97, 108, 172,
    208, 224, 225, 362, 363, 366, 372,
    382
oil 2, 36, 56, 59, 62, 69, 73, 86, 88, 89, 90,
    115, 116, 122, 123, 124, 125, 131,
    132, 134, 155–7, 158, 160, 161,
    162, 170, 192, 195, 197, 198, 200,
    201, 202, 207, 215, 217, 219, 220,
    224, 225, 240, 241, 260, 264, 275,
    288, 306, 307, 309, 317, 335, 355,
    362, 375, 376, 378, 279, 382, 383,
    384
  exploration/prospecting (*listed under*
    petroleum exploration)
  field (*listed under* petroleum fields)
  industry 1, 6, 9, 44, 59, 62, 115, 123,
    157, 192, 234, 243, 251, 347, 348,
    349, 355, 366, 378
  pollution 1, 4, 115, 123, 251, 362,
    375
  seep 1, 85, 87, 88, 93, 115–17, 123, 124,
    125, 131–2, 158, 190, 197, 200,
    218, 219, 241, 251, 253, 275, 307,
    331, 339, 355, 364, 374–5, 378–83
  slick 1, 85, 93, 115, 123, 124, 125, 131,
    218, 331, 335, 378–80
oily bubble 339, 379, 380
Okhotsk Sea **101–3**, 174, 177, 182, 272,
    280, 284, 299, 340, 348, 352
Oklahoma (USA) 306
Olenin Mud Volcano (Gulf of Cadiz) 68
Oligobrachia sp. (siboglinid tubeworms)
    253
Oligocene (Tertiary) 51, 65, 83, 121, 126,
    136, 211, 298
oligochaete (subclass of annelids; phylum
    Annelida) 87, 250
Olistostrome (mud breccia) 77, 84, 86, 97,
    98, 196, 197, 200, 202–4, 214, 225
olivine (mineral group – *see also* fayalite
    and forsterite) 150, 158
Oman 158
Oman anyssal plain 92
opisthosoma (of siboglinid tubeworms)
    270–1
opossum shrimps (mysids) 251
Orange River/Fan (offshore
    Namibia/South Africa) 70
Ordovician (Palaeozoic) 8, 289
ore 6, 87, 96, 149, 314, 315, 316, **374**,
    375
Oregon (USA) 109, 111, 118, 184, 251,
    268, 278, 281, 299, 334, 342, 359
Oregon (Cascadia) Subduction Zone 109,
    111, 181, 184–6, 252, 263, 366–7

organic 4, 45, 56, 72, **151–3**, 154, 157,
    158, 217, 253, 263, 270, 275, 277,
    278, 297, 304, 337
  carbon 59, 84, 92, 96, 104, 118, 120,
    128, 136, 144, 154, 156, 160, 161,
    186, 282, 286, 330, 350
  matter 4, 58, 66, 67, 86, 93, 96, 106,
    108, 115, 120, 122, 126, 128, 132,
    135, 136, 144, 147, 149, 151–3,
    154, 156, 207, 230, 251, 253, 261,
    263, 269, 271, 273, 274, 275, 283,
    284, 285, 291, 293, 297, 316, 342,
    343, 344, 349
organic-rich sediment 59, 60, 72, 81, 92,
    104, 113, 117, 120, 136, 152, 159,
    197, 206, 214, 246, 293, 311, 347,
    353, 360
Orinoco Delta (Venezuela) 118
Ormen Lange field (offshore Norway)
    186, 358
Ortona (Italy) 74
Oseberg field (Norway North Sea) 62
Oseberg kitchen 36
Oslo (Norway) 9, 385
OSPAR convention 385
ostracod (crustaceans) 87
Otago (New Zealand) 97
otoliths (fish ear bones) 250–1
Ovacik springs (Turkey) 329
overpressure (P_{over}) 67, 92, 127, 148, 163,
    172, 176, 196, 209, **215–16**, 218,
    221, 222, 223, 224, 225, 228, 233,
    236, 238, 240, 241, 243, 244, 246,
    247, 315, 316, 317, 332, 353, 360,
    362, 363, 366, 368 (*see also*
    Equations 7.6, 7.12b, and 7.17)
oxic/anoxic (redox) boundary 81, 83, 252,
    298, 300, 301, 323 (*see also*
    oxylimnion)
oxygen (O) 96, 103, 107, 152, 153, 154,
    158, 241, 255, 266, 268, 271, 273,
    275, 276, 280, 284, 293, 297, 298,
    301, 306, 308, 311, 338, 343,
    383
  BOD (biological oxygen demand) 152
  isotopes 88, 109, 114, 149, **295**, 305
  minimum zone (OMZ) 92, 273, 287
  O:C ratio 155
Oxylimnion (oxic–anoxic boundary) 301
    (*see also* oxic/anoxic boundary)
Øygarden Fault Zone (offshore Norway)
    232

Pacific Ocean 93, 98, 102, 117, 118, 138,
    140, 141, 142, 198, 256, 261, 270,
    272, 277, 278, 279, 319, 321, 325,
    326, 343, 352, 376

Pacific Plate 97, 99, 100, 103, 106, 109,
    113
PACMANUS (hydrothermal site) 96, 97,
    148
Pagarus sp. (hermit crab) 249, 250
pagoda structures **172–4**, 175
Paita (Peru) 119, 279, 299
palaemonid shrimps 100
Palaeocene (Tertiary) 51, 81, 136, 140,
    145, 207, 212, 237, 247, 256, 289,
    353
Palaeozoic (*see also* Devonian, Ordovician,
    Permian) 66, 128, 308
Paleohori Bay (Greece) 76
Palos Verdes Peninsula (California) 262,
    283
Panama 146
Panama City gas seeps area (US Gulf of
    Mexico) 124
Panarea Island (Italy) 325
Pandalus jordani (shrimp) 251
Pangea (super-continent) 123
Panoche Hills (California) 289
Papua New Guinea 95, 283, 286, 359
parametric echo sounder 44, 84, 332, 333
Paramphinome jeffreysii (polychaete) 249
Paramushir Island (Kuril Islands, Russian
    Pacific) 101, 102
Parrita Scarps (offshore Costa Rica) 119,
    142
particulate matter 70, 131, 234, 246, 273,
    338
patellacean limpets 113
pathogen 383
Patras, Gulf of (Greece) 75, 231–2
Pavillion Lake (British Columbia)
    309–10, 312
peat 16, 59, 66, 81, 104, 135, 156, 169, 347
Pechora Sea (Russian Arctic) 207, 208,
    238, 312
PeeDee Belemnite (PDB) 162
Peleponnesos (Greece) 75, 230
Pennatula phosphorea (sea pen) 249
Penobscott Bay (Maine) 128
pentane (C_5H_{12}; hydrocarbon gas) 27, 40,
    44, 118, 125, 259, 295
percolation theory 220
peridotite (ultramafic rock type typically
    derived from the mantle) 149, 150,
    157, 198, 314
  mantle-derived peridotite 72, 143, 148,
    157, 214, 288, 313
  serpentinization of 98, 144, 149, 313,
    326
permafrost 24, 86, 181, 207, 208, 237,
    238, 240, 334, 348, 351, 352,
    353

permeability (K) 145, 150, 157, 174, 178,
    181, 182, 209, 212, 215, **217**, 218,
    220, 223, 224, 233, 236, 237, 238,
    243, 316, 364
Permian (Palaeozoic) 306
Persian Gulf (*see* Arabian Gulf)
Perth (Western Australia) 254
Peru **119–20**, 138, 299
Peru Accretionary Wedge 139
Peru Trench 119, 269, 278, 279
Perviata (former subphylum of
    pogonophores; now siboglinid
    tubeworms) 269
petroleum 1, 3, 4, 7, 36, 43, 59, 81, 90, 94,
    97, 98, 107, 111, 114, 115, 118,
    120, 123, 124, 131, 312, 134, 135,
    151–62, 202, 212, 214, 217, 220,
    221, 223, 240, 258, 259, 260, 278,
    294, 299, 303, 306, 331, 340, 349,
    355, 368, 374, 375, 376, 378, 381,
    382, 383, 384
  abiogenic 157–62
  drilling (exploration/production) 1, 8,
    37, 73, 75, 107, 115, 156, 178, 223,
    260, 335, 355, 362–8, 372, 381,
    384
  exploration/prospecting (oil and gas)
    6, 16, 41, 62, 70, 93, 107, 123, 134,
    157, 160, 178, 331, 332, 335, 355,
    362–3, 374, **376**, **378–83**
  fields (general) 8, 9, 50, 52, 59, 88, 157,
    158, 210
  generation 73, 97, 98, 136, **154–6**,
    157–8, 159, 217
  hydrothermal-associated 87, 98, 118,
    158, 325
  industry/companies 1, 18, 86, 93, 115,
    132, 136, 156, 157, 163, 349, 355,
    383
  migration 56, 90, **156–7**, 159, 190,
    217
  seep 1, 2, 4, 90, 111, 115, 120, 190, 248,
    251, 253, 304, 323, **331–3**, 383
petroleum fields (includes gas and oil
    fields) 8, 9, 50, 52, 59, 88, 157,
    158, 210
  Alba (UK North Sea) 210
  Auger (US Gulf of Mexico) 124
  Azeri (Azerbaijani Caspian Sea) 86
  Åsgard (offshore Norway) 50, 51
  Bach Ho (Vietnam) 157
  Balder (Norwegian North Sea) 211
  Balmoral (UK North Sea) 9
  Berri (Saudi Arabian Arabian Gulf) 88,
    89, 90
  Beryl (UK North Sea) 232
  Bonaccia (Italian Adriatic Sea) 73

petroleum fields (*cont.*)
Britannia (UK North Sea) 9
Bullwinkle (US Gulf of Mexico) 124
Cantarell (Mexican Gulf of Mexico) 399
Cement (Oklahoma) 306–7
Challi (Australian Timor Sea) 95
Chirag (Azerbaijani Caspian Sea) 86
Cognac (US Gulf of Mexico) 381–2
Connemara (offshore Ireland) 62, 303
Cooper (US Gulf of Mexico) 124
Cornea (Australian Timor Sea) 95
Dos Cuadros (offshore California) 116, 364
Draugen (offshore Norway) 50
Ekofisk (Norwegian North Sea) 25, 384
Emma (Italian Adriatic Sea) 73
Estrella (offshore Equatorial Guinea) 382–3
Forth-Harding (UK North Sea) 210
Forties (UK North Sea) 8, 9, 15, 22, 372
Gryphon (UK North Sea) 210, 211
Gullfaks (Norwegian North Sea) 9, 20, 36–7, 38, 171, 215, 363, 370
Guneshli (Azerbaijani Caspian Sea) 86
Heidrun (offshore Norway) 53, 383
Huldra field (offshore Norway) 20, 62
Ivanhoe (UK North Sea) 9
Jabiru (Australian Timor Sea) 95
Joillet (US Gulf of Mexico) 124
Kristin (offshore Norway) 50, 53, 256
La Goleta (offshore California) 375
Machar (UK North Sea) 20, 59, 60
Mars (US Gulf of Mexico) 124
Midgard (offshore Norway) 50
Mikkel (offshore Norway) 50, 364
Njord (offshore Norway) 50
Norne (offshore Norway) 50
Ormen Lange (offshore Norway) 186, 358
Oseberg (Norwegian North Sea) 62
Piper (UK North Sea) 9
Popeye (US Gulf of Mexico) 124
Ram-Powell (US Gulf of Mexico) 124
Safaniya (Saudi Arabian Arabian Gulf) 90
Schwedeneck (German Baltic Sea) 56, 57
Skua (Australian Timor Sea) 95
Snorre (Norwegian North Sea) 9, 36
Snøhvit (Norwegian Barents Sea) 45, 47
Statfjord (Norwegian North Sea) 36, 170
Tartan (UK North Sea) 9

Tommeliten (Norwegian North Sea) 20, 25, 27 (*see also* Tommeliten seeps)
Troll (Norwegian North Sea) 9, 10, 11, 20, 370
Vancouver (US Gulf of Mexico) 124
Veslefrikk (Norwegian North Sea) 9, 20, 62, 307
Vest (Italian Adriatic Sea) 73
Zuluf (Saudi Arabian Arabian Gulf) 88, 89
pH (acidity/alkalinity) 96, 100, 148
Phakettia sp. (sponge) 46
pharmaceutical uses 383
phase separation 146–9, 314, 317
Philippine Sea Plate 100
Phoenician 79
Phoenix Micro-plate 118
phosphate 148, 284, 285, 283, 297, 301, 311, **318–9**, 321, 329, 330
phosphate-rich waters 262, 319
phosphorous (P) 262, 284
phosphorite 318
photosynthesis 4, 151, 152, 261, 269, 277, 281, 283, 286, 287, 304, 343, 349
phyllosilicate (minerals) 301
phytoplankton 151, 152, 155, 281, 282, 343
Piip Submarine Volcano (NW Pacific Ocean) **103**, 142, 158, 262, 316
pile anchor 370
piled structures 370
piles
gas drainage 370
pillow lava 261
pinger 25, 27, 30, 62, 168, 169, 174, 332
pipeline 9, 10, 16, 22, 60, 81, 90, 192, 207, 243, 250, 256, 362, 367, 369, 370, **371**, 386
Haltenpipe (Norwegian Sea) 16, 256, 383
Piper field (UK North Sea) 9
Pipoca Mud Volcano (Gulf of Cadiz) 68
Pisces IV (submarine) 131, 132
Pisces VII (submarine) 101
Pisces XI (submarine) 101
piston core 8, 25, 27, 108, 115, 120, 125, 131, 381, 382
Pitcairn 141
Pitch Lake (Trinidad) 122
plankton 27, 33, 87, 105, 248, 257, 276, 279, 284
bacterio- 251
phyto- 151, 152, 155, 281, 282, 343
zoo- 152, 155, 250, 283, 284, 285, 343, 349

plants (flora) 28, 75, 135, 151, 152, 156, 248, 250, 251, 258, 306, 309, 330, 385, 386
plastic/hydroplastic deformation 51, 122, 207, **225–6**, 240, 245, 246
plate boundary/margin 4, 67, 91, 97, 118, 138, 139, 140, 145
plate convergence/subduction 1, 75, 76, 91, 94, 97, 99, 108, 119, 137, **138–9**, 140, 144, 148, 157, 158, 181, 197, 198, 215, 252–3, 325, 332, 374, 376, 377
plate divergence/ocean spreading 1, 4, 45, 72, 87, 95, 117, 118, **137–8**, 141, 145, 148, 158, 160, 208, 218, 239, 261, 262, 269, 277, 290, 324, 325, 326, 327, 352
transform 113, 134, 137, **140**
plate tectonics 1, 4, 134, **136–43**
plate tectonic setting
back-arc 7, 75, 76, 95, 97, 98, 101, 102, **140**, 145, 148, 324, 326, 374
fore-arc 97, 98, 101, 105, 111, **139–40**, 148, 289, 313, 314, 326
intra-plate 4, 134, 135, 137, **140–2**, 148, 325
island arc 138, 139, **140**, 148, 321, 325, 326 (*see entries under individual island arcs*: East Caribbean, Izu-Ogasawara (Bonin), Kermadec, Mariana and Sanghihe)
plates (and micro-plates)
Aegean 74, 74
African 67, 68, 74
Anatolian 74
Antarctic 118
Arabian 91
Australian 94, 97
Caribbean 118, 121
Cocos 118, 140, 142
Eurasian 67, 74, 91, 94, 100
Explorer 109
Gorda 109, 111
Juan de Fuca 109
Nazca 119
North American 103, 106, 109, 111, 113
Pacific 97, 99, 100, 103, 106, 109, 113
Philippine Sea 100
Phoenix 118
Scotia 118
Solomon Sea 95
South American 118, 119, 121
South Shetland 118
Platform Holly (offshore California) 375

Pleistocene (Quaternary) 10, 59, 93, 94, 104, 123, 129, 135, 151, 192, 205, **209**, 211, 231, 237, 357, 368

plumbing 4, 140, 198, 200, 224, 241, 280, 289, 301, 305, 315, 334, 384

plume

  acoustic 24–5, 27, 43, 44, 54, 61, 65, 67, 77, 89, 90, 97, 102, 103, 104, 105, 118, 132, 142, 193, 194, 195, 231, 285, **331–2**, 333, 340, 372, 379

  bubble 1, 42, 43, 51, 66, 102, 104, 109, 123, 125, 194, 195, 220, 231, 242, 280, 282, 316, 331–2, 333, 334, 336, 339, 364

  event (mega) 285, **324–5**, 328, 329, 333, 334, 347, 352

  flare 82, 84, 101, 102, 103

  freshwater 132, 329, 343

  hydrothermal 76, 77, 96, 98, 110, 145, 149, 158, 268, 285, **323–9**, 331, 350

  methane-rich 46, 49, 70, 92, 98, 107, 109, 144, 159, 182, 326, 327, 332, 334, 338, 346

  sediment 22, 23, 62, 193, 194, 228, 231, 242, 254

  seep 42, 43, 44, 61, 70, 89, 90, 97, 118, 206, 244, 286, 331, 332, 333, 338, 344, 345, 350, 356, 382

  sub-seabed 60, 62, 108, 109, 127, 168

  (*see also* gas chimney)

Plym Estuary (UK) 60

Pobiti Kamani (Bulgaria) 305

pockmark 1, 2, 3, **7–44**, 45–7, 49–50, 51, 52, 53, 54, 56, 57, 58, 60, 62, 65, 66, 67, 68, 69, 70, 71, 73, 74, 75–6, 79, 80, 81, 82, 88–91, 92, 93, 94, 97, 99, 103, 104, 105, 106, 108–9, 111–12, 114–15, 118, 119, 120, 124, 126, 127, 130, 131, 136, 139, 145, 169, 172, 174, 177, 182, 183, 189, **190–5**, 205, 206, 207, 209, 210, 213, 214, 223, 224, 225, 226, 228, 230, 231, 232, 233, 234, 237, 238, 239, 240–5, 246, 247, **248–51**, 252, 256, 257, 273, 274, 275, 280, 281, **290–1**, 299, 303, 331, 335, 346, 351, 357, 358, 359, 361, 363, **369**, 370, 372–3, 386

  activity 9, 13, 14, 22–4, 42, 43, 44, 73, 75–6, 88, 89, 90, 91, 112, 115, 128–9, 193, **194–5**, 231, 232, 234, 243, 250–1, 334, 369

  buried **15**, 17, 24, 73, 74, 238, 243

  density 8, 10–11, 18, 19, 24, 33, 43, 88, 97, 104, 195, 231, 238

fossil **247**

formation 8, 9, 24, 25, 45, 58, 67, 69, 75, 90, 104, 112, 115, 129, 131, 191, 192, 193, 207, 209, 226, 228, 230, 232, 238–9, **242–4**, 245, 246, 247, 302, 338, 365, 369, 372–3

  giant **13**, 41–4, 46–7, 48, 49, 70, 191, 226, 238, 249, 333, 338

  groundwater-associated 56–8, 76, 129, 193, 214, **244**

  size 8, 10–11, 19, 27, 30, 33, 36, 37, 40, 43, 45, 46, 51, 52, 53, 56, 58, 60, 67, 69, 70, 71, 73, 75, 79, 81, 88, 90, 97, 104, 106, 114, 119, 120, 126, 1229, 130, 189, 190, 231, 256, 372, 373

  unit pockmark 13, 15, 129, 30, 36, 37, 39, 43, 190, 191, 242, 256

pockmark areas

  Big Sur (offshore California) **114–15**, 193, 194

  Eckernförde Bay (Germany) **56–8**, 153, 164, 165, 166, 169, 178, 193, 194, 244, 246, 263, 264, 330, 348

  Gulf of Maine (USA) **128–9**, 190, 191, 193

  Norwegian Trench (Norwegian North Sea) 9, **36–41**, 191, 238, 248, 250, 363

  Scotian Shelf (offshore Nova Scotia, Canada) **7–8**, 9, 15, 129–30, 190, 248

  Witch Ground Basin (UK North Sea) 9, 10, 11, 12, 13, 14, 15, **18–25**, 41–4, 60, 62, 189, 191, 193, 195, 228, 232, 234, **238–9**, 247, 249, 281, 285, 334, 353, 372–3, 386

pockmarks

  Braemar (UK16/3, North Sea) 386

  Challenger (UK15/25, North Sea) 40, 43, 44

  Scanner (UK15/25, North Sea) 20, 40, 41, 42, 43, 44, 189, 190, 195, 249, 285, 386

  Scotia (UK15/25, North Sea) 40, 43, 44

  Witch's Hole (UK21/4, North Sea) 20, 21, 22, **24–5**, **372–3**

pogonophore (former designation of siboglinid tubeworms) 269

Pohang (Korea) 99

Point Arena Basin (offshore California) 359

Point Conception (California) 115, 116

Point Fermin (California) 116

Point St George (California) 111

poiseuille flow 218, 225

pollution 1, 2, 4, 123, 235, 251, 259, 275, 375, 379, 385

polychaetes (bristle worms; class Polychaeta, of phylum Annelida) 261 (*see also* serpulid)

  Capitella sp. 251, 281

  Hesiocaeca methanicola (the 'ice worm') 275

  Hetermastus filiformis 281

  Levinsenia gracilis 249

  Maldanid 253

  Marenzelleria viridis 254

  Nephtys sp. 281

  Paramphinome jeffreysii 249

  Spiophanes kroyeri 249

  Sternaspis sp. 281

polygonal faults 129, **211–2**, 213, 258

Polymastia sp. (sponge) 46

polymer 152, 154

polynya (ice-free area) 132, 329, 348, 348

polysaccharide (complex carbohydrate) 152

Popcorn Ridge (east Pacific Ocean) 142

Popeye field (US Gulf of Mexico) 124

Porcupine Basin (northeast Atlantic Ocean) 62, 64, 65, 302, 303

pore fluid 4, 90, 111, 113, 121, 122, 123, 127, 145, 150, 154, 167, 172, 206, 207, 212, 214–16, 223, 225, 226, 229, 233, 237, 240, 241, 245, 284, 299, 362, 363

  pressure (P_{fluid}) 72, 73, 245, 252, 2267, 171, 172, 173, 175, 176, 200, 206, 210, **214–15**, 221, 223, 225, 228, 232, 233, 237, 242, 245, 308, 356, 357, 359, 360–1, 364, 367, 369, 384

pore space/volume 39, 86, 101, 109, 147, 150, 154, 156, 163, 164, 180, 181, 182, 185, 214–15, 217, 220, 221, 260, 291, 320

Pore throat 215, 220, 221, 233, 239, 244, 339

pore water 49, 51, 57, 65, 67, 73, 99, 100, 103, 107, 109, 113, 114, 115, 121, 123, 126, 127, 131, 129, 149, **150–1**, 152, 154, 163, 164, 165, 166, 175, 176, 177, 178, 180, 181, 182, 185, 186, 214–15, 221, 228, 237, 241, 244, 253, 254, 256, 257, 263, 282, 290, 291, 293, 294, 295, 297, 301, 302, 306, 310, 311, 315, 320, 331, 337, 343, 344, 363, 366, 368, 371

  pressure (P_{water}) 100, 104, 175, 221, 242

Port Clarence (Alaska) 104

Portugal 67, 296, 385

Posidonia (sea grass) 75
Posolskiy Fault (Lake Baikal) 87
Poverty Bay (New Zealand) **97**
Pre-Cambrian 311
predator 251, 254, 273, 274, 275, 277, 279, 282, 283
pressure
  atmospheric(P_{atmos}) 72, 214–15, 216, 235, 236, 240
  capillary (P_{cap}) 219, 220, **221**, 223, 239, 241 (*see also* Equation 7.11)
  fluid (P_{fluid}) 72, 73, 245, 252, 2267, 171, 172, 173, 175, 176, 200, 206, 210, **214–15**, 221, 223, 225, 228, 232, 233, 237, 242, 245, 308, 356, 357, 359, 360–1, 364, 367, 369, 384
  fracture (P_{frac}) 222, 363 (*see also* Equation 7.14)
  gas (P_{gas}) 97, 203, 209, 214, **221**, 222, 301, 302, 363, 366, 371 (*see also* Equation 7.12a)
  hydrostatic (P_{hydro}) 90, 101, 148, 176, 178, **214–15**, 216, 221, 222, 225, 226, 233, 235, 236, 237, 241, 318, 338, 350, 351, 352, 364, 366, 370
  over (P_{over}) 67, 92, 127, 148, 163, 172, 176, 196, 209, **215–16**, 218, 221, 222, 223, 224, 225, 228, 233, 236, 238, 240, 241, 243, 244, 246, 247, 315, 316, 317, 332, 353, 360, 362, 363, 366, 368 (*see also* Equations 7.6, 7.12b, and 7.17)
  pore water (P_{water}) 100, 104, 175, **221**, 242 (*see also* Equation 7.10)
Pressure-while-drilling (PWD) 369
primary producer 93, 257, 277, 286
primordial (mantle-derived) methane 111, 157, **158–9**, 160, 288
Prinos Bay (Greece) **75**
Prince Patrick Island (Arctic Canada) 289
propane (C_3H_{8+}; hydrocarbon gas) 27, 40, 44, 84, 85, 99, 102, 103, 105, 109, 115, 116, 123, 131, 158, 161, 180, 202, 281, 295, 301, 380, 381
Pual Ridge (offshore Papua New Guinea) 96
Puerto Rico Trench 253
pulldown (seismic) 27, 169, **170**, 174, 177
pullup (seismic) 170, 175, 176, 177
pumice 98
Purisma Formation 113
Puysegur Ridge (offshore New Zealand) 97
PWD (Pressure-while-drilling) 369
pycnocline 341

pyrite (FeS_2; mineral) 68, 103, 126, 262, 263, 264, 291, **298**, 300, 306, 314, 316, 318
pyrrhotite (FeS; mineral) 118, 314, 315, 316
pyruvate (carboxylate anion of pyruvic acid) 152

quartz (SiO_2; mineral) 69, 223, 292, 301, 315, 317
Quaternary 19, 31, 45, 51, 52, 59, 60, 66, 80, 81, 94, 97, 104, 105, 107, 114, 121, 159, 299, 323, 350, 353, 354, 365 (*see also* Flandrian, Holocene (Recent), Pleistocene, Weichselian (Devensian))
Queen Charlotte Sound (British Columbia) **107**
Quepos Slide (offshore Costa Rica) 119

Rabat Mud Volcano (Gulf of Cadiz) 67, 68
radar 74, 334, 379
  European Radar Satellite 253
  synthetic aperture radar (SAR) 94, 379
radioactive 59, 145, 264
radiolaria /radiolarian ooze 321
radon (^{222}Rn) 139, 22, 320, 329
Rainbow hydrothermal field (Mid-Atlantic Ridge) 149, 376, 385
Ram-Powell field (US Gulf of Mexico) 124
Rande Strait (Galicia, Spain) 267
Ratnagiri (India) 92
rattail (or grenadier; deepwater fish; family Macrouridae) 261
ray (fish: superorder Batoidea) 78, 251
reaction zone (hydrothermal circulation systems) 146, **147–9**
Recent (or Holocene; Quaternary) 7, 45, 56, 57, 59, 72, 76, 79, 81, 88, 93, 99, 104, 106, 111, 115, 128, 129, 135, 161, 193, 229, 231, 289, 294, 299, 349, 354, 357
recharge zone (hydrothermal circulation systems) 142, **146–7**
Red Sea **87–8**, 158, **207–8**, 374
Red Volcano (seamount; E Pacific Ocean) 142, 262
Redfield composition 153
Redondo Beach (California) 111
redox (oxic/anoxic) boundary 81, 83, 252, 298, 300, 301, 323
reef 94, 116, 124, 136, 255, 258, 259, 260, 290, **294**, 304, 319, 320, 344, 381, 384, 385

algal 73, 94
'Bubbling Reefs' (Danish Kattegat) 55,56, **250**, **291**, 301, 386
carbonate mound/reef 60, 61, 64, 68, 79, 94–5, 175, 248, 259, 290, **294**, **302–4**, 305, 306, 307, 344, 381
coral 141, **255–60**, 305, 307, 320, 343–4, 383, 386
  deep-water (cold-water) 49, 53, 64, 66, 69, 120, **255–8**, 290, 302, 305, 383, 384, 385
  Great Barrier Reef (Australia) 259
  Haltenpipe Reef Cluster (HRC; offshore Norway) 53, **256**
  tropical 141, 255, 258, **259–60**
remotely operated vehicle (ROV) 2, 9, 10, 13, 27, 28, 29, 30–3, 34, 35, 36, 39, 40, 41, 43, 44, 46, 57, 79, 83, 85, 90, 107, 112, 114, 115, 182, 206, 280, 285, 300, 301, 332, 336, 372, 373
resistivity (electrical) 9, 99, 185, 186, 369
Rhepoxynius sp. (amphipod) 251
Rhine Graben (Germany) 159
Rhine, river (The Netherlands, Germany) 135
Rhodes (Greece) 209, 211
rhyodacite 98, 148
ria (drowned valley) 60, **135**
  Baixas (Galicia, Spain) **66–7**, 135
Ribeiro Mud Volcano (Gulf of Cadiz) 68
Ridge
  Blake (offshore South Carolina) **126–37**, 136, 181, 182, 183, 272, 288, 329, 338, 373
  Blake Outer (offshore South Carolina) 184, 185
  East Pacific Rise 1, 117, 141, 142, 158–9, 268, 274, 277, 317, 324, 327, 343
  Endeavour (east Pacific Ocean) 146
  Gorda (east Pacific) 138, 277, 316, 324
  Gualdalquivir Diapiric Ridge (Gulf of Cadiz) 68
  Hook Ridge (offshore Antarctica) 118, 148
  Hydrate (east Pacific Ocean) **109–10**, 180, 182–5, 252, 263–4, 297, 332, 334, 336, 337, 338, 340, 366, 376
  Juan de Fuca (east Pacific Ocean) 5, 6, 109, 110, 138, 158, 262, 277, 2315, 316, 317, 324, 329
  Mediterranean (Mediterranean Sea) 75, **76–9**, 182, 202, 205, 246, 264
  Mid-Atlantic (Atlantic Ocean) 1, **72**, 137, 142, 146, 158, 256, 274, 277, 313, 314, 316, 326, 328, 344, 385

Murray (Arabian Sea) 91
Popcorn (east Pacific) 142
Pual (offshore Papua New Guinea) 96
Puysegur (offshore New Zealand) 97
Smooth (Monterey Canyon) 113
Southern Ritchie (offshore New Zealand) 188
Sula (Norwegian Sea) 256, 383, 385
Wyville-Thomson (northeast Atlantic Ocean) 64, 65
Riftia sp. (siboglinid tubeworm) 271
Rimicaris sp. (shrimp) 261, 274
ring-shaped depression 68, **129**
river 67, 70, 134–5, 136, 150, 236, 246, 254, 326, 329, 330, 341–2, 343, 352, 353, 356, 360
    Alsek (Alaska) 107, 206
    Amazon (Brazil) **120**
    Danube (eastern Europe) 81, 342
    Eel (California) 111, 112, 383
    Elbe (northwest Europe) 342
    Fraser (British Columbia) 107, 108, 357
    Godavari (India) 93
    Huang Ho (People's Republic of China) 99
    Kerynites (Greece) 232
    Klamath (California) 111
    Krishna (India) 93
    Mageneitas (Greece) 231
    Mahanadi (India) 93
    Mandovi (India) 342
    Mississippi (southern USA) 207, 235, 236, 251, 368
    Oder (central Europe) 343
    Orange (Namibia/South Africa) 70
    Rhine (The Netherlands, Germany) 135
    Selenga (Mongolia/Russia) 86
    Supsa (Georgia) 85
    Susquehanna (eastern USA) 127
    Thames (UK) 59, 60, 135
    Tyne (UK) 342
    Volga (Russia) 85
    Yangtze (People's Republic of China) 99, 361
    Yesilirmak (Turkey) 81
    Yukon (Alaska) 104
Robert Bank Failure Complex (Fraser Delta, British Columbia) 357
Rockall Trough (northwest Atlantic Ocean) 62, 64, 65, 275, 302, 303, 378
rockfish (*Sebastes elongatus*, fish) 251
Roman 1, 2, 325, 374

Roman Ruins vent site (PACMANUS, offshore Papua New Guinea) 96
Romania 202
ROPOS (ROV) 107
Roseway Banks (offshore Nova Scotia) 7, 8
Roseway Basin (offshore Nova Scotia) 7, 8
roundnose grenadier (*Coryphaenoides rupestris*; fish) 384
ROV (remotely operated vehicle) 2, 9, 10, 13, 27, 28, 29, 30–3, 34, 35, 36, 39, 40, 41, 43, 44, 46, 57, 79, 83, 85, 90, 107, 112, 114, 115, 182, 206, 280, 285, 300, 301, 332, 336, 372, 373
Rovinj seeps (offshore Croatia) 72, 73
Russia/Russian 47, 81, 83, 86, 101, 103, 157, 202, 238, 265, 316
Ryuku Trench (offshore Japan) 101, 205

Sable Island Bank (offshore Nova Scotia) 131
SAC (Special Areas for Conservation) 385
Sackett Bank (US Gulf of Mexico) 258
Safaniya field (Saudi Arabian Arabian Gulf) 90
safety 183–4, 340, 361, 365, 366, 371
Sagami Bay (Japan) **99–100**, 278
Saganu Trough (W Pacific Ocean) 253
Sahara 305–6, 384
Sahul Shoals (Australian Timor Sea) 94
Sakhalin Island (Russian Pacific coast) 101–2, 103
salina **308**, 310
salmon fodder 383
salse (mud volcano feature) 79, 191, **198**, 199, 200, 210, 202, 280
salt (NaCl) 68, 125, 127, 134, 147, 182, 216, 223, 240, 258, 307, 317–18, 329, 342
    deposit 123, 208, 299, **317–18**
    hydrothermal 87, **317–18**
    marsh 78, 298
    mine 317
    movement (diapirism etc.) 80, 123, 124, 172, 223, 317–18, 359
    structure (diapir, dome, nappe, ridge, stock etc.) 3, 25, 27, 28, 56, 57, 59, 60, 68, 80, 123, 124, 125, 126, 127, 178, 182, 197, 208, 218, 223–4, 225, 241, 245, 252, 258, 289, 304, 307, 317–18, 358–9, 360
    tectonics 122, 124, 359
Blake Ridge (offshore South Carolina) **127**, 358

Damon Mound (Texas) 307
Machar field (UK North Sea) 20, 59, 60
Messinian 80
Schwedeneck field (German Baltic Sea) 56
Tommeliten field (Norwegian North Sea) 3, 25, 26–7
San Andreas Fault (California) 113, 140
San Clemente Basin (offshore California) 299
San Clemente Fault (offshore California) 299
San Fernando (California) 117, 230
San Gregorio Fault (offshore California) 113, 140
San Simón (Spain) 66, 266, 267
sand intrusion **209–11**, 240, 368
sand volcano 65, 196, **211–12**, 229–30, 240
sandwave 10, 60–2, 64, 65, 357
Sanghihe Island Arc (west Pacific Ocean) 321
Santa Barbara (California) 2, 115, 375
    Basin (California) 111, 273
    Channel (California) 2, **115–17**, 194, 234–5, 241, 251, 281, 333, 334, 336, 375, 383–4
Santa Cruz Mountains (California) 113
Santa Monica Basin (California) 116
Santorini Island (Aegean Sea) 76, 326
sapropel 156
SAR (synthetic aperture radar) 94, 379
Satanic Mills vents (PACMANUS, offshore Papua New Guinea) 96
Sawu Sea (Indonesia) **94**
Scanner pockmark (UK15/25, North Sea) 20, 40, 41, 42, 43, 44, 189, 190, 195, 249, 285, 386
scanning electron micrographs (SEMs) 164, 292, 296
scavenger 273, 274, 277, 279, 282, 283, 339
Schugin-type mud volcano **205**, 246
Schwedeneck field (German Baltic Sea) 56, 57
SCI (Sites of Conservation Interest) 386
Sclerolinum sp. (siboglinid tubeworms) 253
Scotia pockmark (UK15/25, North Sea) 40, 43, 44
Scotia Sea (southeast Atlantic Ocean) 118
Scotia Plate 118
Scotian Shelf (east Canada) 3, **7–8**, 9, 10, 129, 130, 248

Scotland (UK)/Scottish 60, 62, 238, 266, 305, 384
Scott Inlet (Baffin Island, Canada) **131–2**
Scott Trough (offshore Baffin Island, Canada) 132
SEA (UK's Strategic Environmental Assessment programme) 12, 14, 22, 40, 41, 42, 195, **386**
Sea
    Adriatic **72–4**, 172, 174, 206–7, 275, 329, 334
    Aegean 75, 262, 283, 325, 345
    Alboran 69
    Arabian 92, 342, 343, 346
    Baltic **56–9**, 284, 342, 343, 352, 386
    Barents 13, **45–7**, 48, 49, 50, 52, 192, 238, 239, 273, 334, 352, 353
    Beaufort 103, 348
    Bering **103–5**, 142, 192, 352
    Black 2, **81–5**, 1182, 183, 203, 245, 265, 297, 299, 300, 302, 342, 347
    Caribbean **120–2**, 140, 218, 368, 372
    Caspian **85–6**, 140, 159, 198, 201, 203, 337, 356, 368, 380
    Dead 1, 2, 311, 313
    East China 99
    Greenland 357
    Ionian 75
    Irish 59, 60, 61, 168, 284, 286, 386
    Japan 88, 272
    Laptev 311, 348
    Mediterranean 2, 67, 69, **74–81**, 142, 246, 256, 272, 294, 295, 326, 329
    North **8–44**, 59, 60, 61, 62, 63, 64, 65, 73, 106, 135, 136, 159, 170 171, 172, 174, 189, 190, 192, 193, 194, 195, 206, 207, 210, 211, 212, 217, 218, 224, 228, 232, 234, 238, 240, 241, 245, **248–50**, 264, 272, 273, 274, 275, 278, 280, 281, 284, 285, 286, **290–1**, 3–1, 307, 333, 334, 335, 337, 342, 344, 352, 365, 368, 369, 372, 373, 374, 381, 383, 384
    Norwegian **50–2**, 65, 178, 181, 206, 207, 238, 245, 357, 365, 368
    Okhotsk **101–3**, 174, 177, 182, 272, 280, 284, 299, 340, 348, 352
    Pechora 207, 208, 238, 312
    Red **87–8**, 158, **207–8**, 374
    Sawu **94**
    Scotia 118
    South China **93–4**, 192, 343
    Timor **94–5**, 172, 175, 303, 381
    Tyrrhenian 142, 262
    Yellow 99

sea anemone (order Actinaria) 27, 36, 39, 100, 249, 261, 262
    Bolocera tuediae 249
    Metridium senile 249
    Urticina feline 249
sea cucumber (holothurian; Class Holothuroidea) 100, 255, 259, 262, 283, 284
    Benthodytes sp. 261
sea ice 192, 348, 351, 252, 353
sea level 33, 76, 103, 104, 108, 113, 122, 134, 139, 141, 148, 198, 204, 209, 214, 215, 319, 338, 356, 373
    change (rise/fall) 68, 128, 134, 135, 136, 188, **236**, 284, 304, 350, 351, 352
    eustatic (global) 134, 236, 350, 351
    former 59, 68, 70, 85, 134, 135, 236, 304, 307
    set-up 104
sea pen (order Pennatulacea) 27
    Pennatula phosphorea 249
    Virgularia mirabilis 249
sea surface 6, 15, 27, 58, 72, 74, 76, 82, 85, 87, 93, 98, 100, 102, 112, 116, 117, 123, 125, 131, 140, 142, 180, 204, 207, 214, 215, 222, 230, 234, 235, 240, 252, 253, 284, 285, 304, 329, 331, 332, 333, 334, 335, 339, 344, 345, 348, 349, 354, 364, 365, **378–9**, 382, 383
    bubbles at sea surface 58, 72, 76, 82, 98, 100, 102, 112, 118, 131, 142, 230, 285, 329, 334, 339, 348, 364, 365, 383
sea urchin (class Echinoidea) 72, 249, 261, 274, 279
    Astropecten irregularis 261
seabed dome/doming 15, 19, 25, 27, 30, 31, 49–50, 60, 62, 66, 67, 97, 109, 120, 124, 127, 142, 173, 175, **206**, 209, 223–4, 226, 228, 240, 242, 243
seabed furrow 126
seabed permafrost 24, 238, 348, 351, 352, 353
seamount 98, 100, 119, **141–2**, 145, **262**, 278, 313, 314, 326, 328, 347, 352, 385
    phosphate **318–19**
    serpentinite **142–3**
    subducting 142
    Axial (Juan de Fuca Ridge) **110–11**, 142, 262, 329
    Bounty (south Pacific Ocean) 142

Chamorro (west Pacific Ocean) 99
Conical (west Pacific Ocean) 98, 149, 150, 198
Eratosthenes (east Mediterranean Sea) **79**, 80
Franklin (offshore Papua New Guinea) 95–6
Hancock (Hawaiian Islands) 142
Jaco (offshore Costa Rica) 142
Jasper (E Pacific Ocean) 142
Larson (E Pacific Ocean) 142
Loihi (Hawaiian Islands) 141, 142, 148, 262, 324, 325
MacDonald (Pacific Ocean) 142
Marsili (Tyrrhenian Sea) 142, 262
Monte Norte (west Mediterranean Sea) 69
Nameless (Gulf of Cadiz) 321
Piip Submarine Volcano (northwest Pacific Ocean)
Red Volcano (east Pacific Ocean) 142, 262
Teahitia (south Pacific Ocean) 142
seawater 1, 4, 9, 25, 46, 49, 76, 78, 887, 96, 99, 100, 101, 103, 107, 116, 121, 122, 123, 124, 125, 126, 132, 142, 145, 146, 147, 148, 149, 150, 152, 153, 154, 161, 16, 166, 167, 168, 181, 187, 188, 208, 209, 215, 231, 255, 257, 259, 261, 263, 266, 269, 272, 275, 276, 279, 282, 291, 293, 204, 295, 297, 299, 300, 301, 308, 312, 314, 315, 317, 318, 319, 320, 321, 323, 324, 325, 329, 330, 335, 338, 340, 342, 344, 346, 352, 353, 361, 381
seawater-rock interaction 76, 317
Sebastes elongates (rockfish; fish) 251
Sebastes vivaparus (Norway haddock; fish) 32, 249
sediment compressibility 107, 140, 216, 223, 233, 235, 370
sedimentary diatreme **209**, 225–6, 240, 263
seep/seepage 1–6, 7, 45, 88–91, 113–14, 119, 120, 121, 122, 123, 135, 139, 142, 145, 151, 189, 190, 195, 200, 206, 209, 213, 214, 217, 218, 223, 224, 225, 228, 234–5, 237, 230, 240–1, 242, 244, 245, 246, 247, 248–69, 271–89, 290–1, 293–5, 297, 298, 299, 300, 302, 303, 304, 310, 311, 312, 313, 317, 318, 320, 321, 313, **331–41**, 344, 345, 346, 347, 348, 349, 350, 352, 355, 360, 366, 374, 375, 376, 378–83, 384, 386

asphalt /tar
brine 68, 252, 294–5, 317
fossil **288–9**, 298, 304, 305
gas 1–6, 9–10, 25, 27–8, 40, 43–4, 45,
    46, 51, 52, 53–6, 58, 59, 60, 61, 62,
    63, 64, 65, 66, 67, 68, 70, 72, 73,
    74, 76, 77, 79, 80, 81, 82, 83, 84,
    85, 86, 87, 91, 92, 93, 94, 97, 98,
    99, 102, 103, 104, 105, 106, 107,
    108, 109, 110, 111, 112, 115–17,
    118, 123, 134, 135, 126, 127, 131,
    163, 168, 174, 182, 189, 190, 193,
    194, 197, 200, 206, 207, 208, 219,
    220, 228, 241, 250, 251, 258, 275,
    282, 291, 301, 302, 317, 323,
    331–41, 344, 347, 348, 349, 355,
    364, 373, 374–5, 383
groundwater (submarine groundwater
    discharge: SGD) 1, 2, 72, 75, 76,
    88, 100, 113, 125–6, 128, 133, 134,
    145, **150**, 151, 158, 190, 193,
    217–18, 244, 246, 248, 251, 253–5,
    257, 259, 261, 263, 266, 276,
    308–10, 311, 312, 318, 323,
    329–31, 374
oil 1–6, 7, 85, 87, 93, 115–17, 123, 124,
    125, 131–2, 158, 190 219, 241,
    251, 275, 307, 331, 339, 355, 364,
    374–5, 378–83
petroleum 1, 2, 4, 90, 111, 115, 120,
    190, 248, 251, 253, 304, 323,
    331–3, 383
plume 42, 43, 44, 61, 70, 89, 90, 97,
    118, 206, 244, 286, 331, 332, 333,
    338, 344, 345, 350, 356, 382
self-sealing **301–2**
seep detection **331–3, 378–81**
seep-related fauna 3, 27, 28, 29, 31, 33,
    34, 35, 36, 44, 46, 87, 91, 97, 99,
    100, 102, 107, 112, 113, 118, 119,
    121, 122, 124, 125, 131, 193,
    248–53, 262–83, 300, 336, 349
sediment (soil) compression 107, 140,
    216, 223, 233, 235, 370
sediment plume 22, 23, 62, 193, 194, 228,
    231, 242, 254
sediment (soil) shear strength (S_u) 106,
    167, **215**, 216, 226, 236, 240, 245,
    357, 359, 360, 370 (see also
    Equations 7.4 and 7.13)
seep plume 42, 43, 44, 61, 70, 89, 90, 97,
    118, 206, 244, 286, 331, 332, 333,
    338, 344, 345, 350, 356, 382
seep tent 115, 234, 336, 339, **374–5**
seiche 104
seismic 2, 7, 9, 10, 11, 13, 15, 24, 27, 30,
    36, 40, 41, 46, 51, 55, 60, 65, 66,

69, 72, 73, 74, 75, 77, 81, 85, 86,
    88, 90, 94, 95, 101, 102, 104, 105,
    106, 108, 109, 112, 114, 119, 120,
    124, 127, 132, 137, 141, 153, 163,
    165, 167–78, 183, 184, 185, 186,
    187, 190, 191, 192, 197, 200, 201,
    206, 211, 232, 237, 257, 302, 303,
    304, 334, 357, 358, 359, 364, 366,
    367, 368, 381, 382
evidence of gas **167–78** (see also acoustic
    turbidity, acoustic plume, bright
    spot, enhanced reflection etc.)
seismic (acoustic) velocity (v_p) 9, 27,
    99, 109, 166, 167, 169, 170, 172,
    175, 177, 183, 185, 186, 187
vertical seismic profile (VSP) 109, 185
seismic system
    3D (three-dimensional) 70, 170, 172,
    173, **178**, 201, 209, 211, 212, 247,
    346, 359, 366, 368, 381, 382
    3.5 kHz 9, 25, 54, 62, 63, 67, 105, 111,
    123, 1229, 167, 169, 174, 332
    airgun 43, 105, 167, 184
    boomer 11, 17, 19, 22, 23, 24, 40, 43,
    51, 54, 90, 104, 168, 169
    chirp 53, 99, 169
    digital boomer 42
    pinger 25, 27, 30, 62, 168, 169, 174,
    332
    sparker 38, 43, 104, 105, 108, 172,
    208
    sub-bottom profiler 9, 58, 61, 71, 82,
    99, 118, 129, 224, 231, 256, 331
seismic while drilling (SWD) 369
seismicity/seismic activity 76, 103, 106,
    223, 232, 332, 351, 356, 360 (see
    also earthquake)
Selenga Delta (Lake Baikal) 86
Selenga, river (Mongolia/Russia) 86
Senja Fracture Zone (offshore Norway)
    49
SEM (scanning electron micrograph)
    164, 292, 296
Serendipity Gas Seep Area (US Gulf of
    Mexico) 123, 258
serpentinite **98–9**, 143, 144, **149–50**, 1198,
    240, 245, **313–14**, 326
    mud volcano/seamount 98–9
    Chamorro (west Pacific Ocean) 99
    Conical Seamount (west Pacific
    Ocean) 98, 149, 150, 198
    Logatchev (Mid-Atlantic Ridge)
    149, 317, 385
    Lost City (Atlantic Ocean) 72, 142,
    149, 150, 288, 313, 326
    Rainbow (Mid-Atlantic Ridge) 149,
    376, 385

serpentinization 72, 98, 143, 144, 149,
    150, 158, 162, 214, 288, 303, 313,
    326
serpulid (family Serpulidae; polychaetes)
    120, 204, 261, 289
sessile fauna 27, 33, 248, 301
SGD (submarine groundwater discharge)
    1, 2, 72, 75, 76, 88, 100, 113,
    125–6, 128, 133, 134, 145, **150**,
    151, 158, 190, 193, 217–18, 244,
    246, 248, 251, 253–5, 257, 259,
    261, 263, 266, 276, 308–10, 311,
    312, 318, 323, **329–31**, 374
SGD-related fauna **253–5**, 259, 263,
    266, 276, 330
shallow gas (gassy/gas charged sediment)
    17, 27, 28, 38, 50, 52, 53, 57, 58,
    60, 66, 67, 70, 72, 73, 77, 79, 81,
    82, 84, 91, 92, 93, 108, 112, 120,
    124, 127, 128, 131, 165, 167, 169,
    170, 176, 183, 204, 207, 230, 231,
    233, 234, 235, 256, 302, 337, 357,
    362, 364, 370, 373, 382
    as a geohazard 73, 108, 123, 178, 184,
    225, 238, 275, 355, **362–6**, 368,
    369, 372, 384
shallow mud diapir/volcano 73–4, 172,
    174, **206–7**, 208, 236–7, **246**
shallow water flow (SWF) **368–9**
Shane seep (Santa Barbara Channel,
    California) 194, 333, 334
Shariki Village (Japan) 229
Shark Bay (Western Australia) 308
shear failure **222**, 223, 233, 236, 360 (see
    also Equation 7.13)
shear strength (S_u; of sediment (soil))
    106, 167, **215**, 216, 226, 236, 240,
    245, 357, 359, 360, 370 (see also
    Equations 7.4 and 7.13)
Shelikof Strait (Alaska) **105–6**, 140
shell accumulation 52, 275
Shetland Islands (UK) 31, 62, 232
Shimizu Harbour (Japan) 233
Ship Shoal (licence block, US Gulf of
    Mexico) 124
shipwreck 9, 25, 79, 372, 373
shock crystallisation 317
shrimp (crustaceans; infraorder Caridea)
    27, 31, 36, 38, 39, 46, 70, 78, 89,
    93, 96, 105, 192, 248, 249, 250,
    251, 255, 261, 274, 275, 277, 279,
    281
    Alvinocaris sp. 274
    Callianassa subterranean 192
    Crangon communis 251
    mysid (opossum shrimp) 251
    palaemonid 100

shrimp (*cont.*)
  Pandalus jordani 251
  Rimicaris sp. 261, 274
Siberia (Russia) 86, 311, 348, 349
siboglinid tubeworms (family
    Siboglinidae in the phylum
    Annelida)
  Escarpia sp. 70
  Lamellibrachia sp. 118, 251, 252, 270,
    271, 278
  Oligobrachia sp. 253
  Sclerolinum sp. 253
  Siboglinum sp. 269, 270, 280
  Riftia sp. 271
Siboglinum sp. (siboglinid tubeworm) 269,
    270, 280
Sicily (Italy) 142, 262 372
side-scan sonar 7, 8, 9, 10, 11, 13, 14, 15,
    19, 21, 23, 25, 27, 30, 31, 40, 41,
    42, 43, 47, 50, 52, 55, 58, 61, 64,
    66, 73, 74, 76, 80, 81, 82, 88, 90,
    91, 97, 104, 105, 111, 118, 121,
    126, 128, 129, 130, 141, 168, 194,
    195, 200, 231, 256, 331, 358, 372,
    379, 381
signal starvation (seismic) 95, 171, 173,
    375, **176**, 197
Sigsbee Escarpment (Gulf of Mexico) 123
silicate 99, 118, 314, 316
silicic dome volcanism **98**
Siljan Ring (Sweden) 159
silver (Ag) 97, 98, 310
siphonophore (order Siphonophora;
    hydrozoans) 261, 283
site investigation/survey 8, 21, 25, 41, 59,
    75, 90, 97, 108, 172, 208, 224, 225,
    362, 363, 366, 372, 382
Sites of Conservation Interest (SCI) 386
Skagerrak 9, 10, **52**, 53, 54, 55, 269, 270,
    280, 290, 373
Skandi Inspector (ROV support ship) 372
skate (family Rajidae; fish) 261
Skempton's B coefficient 233
skin friction 370
Skjoldryggen (offshore Norway) 358
Skua field (Australian Timor Sea) 95
slender sole (*Lyopsetta exilis*; fish) 251
slide (submarine slope failure)
  Bear Island Fan (offshore Norway) 47,
    50
  Cabos Blanco (offshore Costa Rica) 119
  Cape Hatteras (offshore South
    Carolina) 126
  Cape Fear (offshore South Carolina)
    126, **127**, **358–9**
  Cape Lookout (offshore South
    Carolina) 358–9

Finneidfjord (Norway) **357–8**
Humboldt (offshore California) 111,
    357
Nicoya Slump (offshore Costa Rica)
    119
Quepos (offshore Costa Rica) 119
Roberts Bank Failure Complex
    (offshore British Columbia) 357
Storegga (offshore Norway) 51, 174,
    176, 186, 350, **357–8**, 359
slope failure/instability 6, 8, 67, 69, 80,
    86, 92, 106, 108, 111, 112, 119,
    126, 129, 131, 136, 151, 187, 202,
    223, 236, 240, 246, 299, 350,
    355–62, 364, 366, 373, 384
slump scar 68, 115, 127
slumping 8, 13, 14, 36, 65, 67, 71, 82, 97,
    106, 118, 119, 202, 206, 209, 238,
    243, 340, 356, 358, 359, 368
smectite (hydrous aluminium
    phyllosilicates; clay minerals) 97,
    14, 150, 212, 214
smoker
  black 96, 103, 274, 314, 315, 324
  white 96, 103, 314, 315, 324
Smooth Ridge (Monterey Bay) 113
SMOW (Standard Mean Ocean Water)
    162, 295
SMTZ (sulphate-methane transition
    zone) **153**, 263, 280, 301
sniffer (underwater hydrocarbon
    detector) 8, 21, 24, 94, 115, 116,
    332, **380**, 381
Snorre field (Norwegian North Sea) 9, 36
Snowcap (hydrothermal site;
    PACMANUS, offshore Papua
    New Guinea) 96
Snøhvit field (Norwegian Barents Sea) 45,
    47
soft corals (subclass Alcyonacea) 46, 255,
    262
Sohm Abyssal Plain (northeast Atlantic
    Ocean) 131, 355
Solar System 157
Solemya sp. (solemyid bivalve) 272,
    299
solemyid bivalves (Solemyidae; clams)
    100, 113, 119, 271, **272**, 278, 289
  Acharax sp. 91, 272
  Solemya sp. 272, 299
Solomon Sea Plate (southwest Pacific
    Ocean) 95
Sonne (research ship) 106
Sorokin Trough (Black Sea) 83, 84, 85,
    245
sound (noise) 142, 204, 225, 232, 260, 333
  (*see also* acoustic)

source rock 43, 81, 93, 123, **154**, 156, 157,
    158, 159, 162, 202, 212, 217, 225,
    303, 376, 381
South Africa 70, 140, 141, 145, 237
South America 118, **119–20**, 138, 186
South American Plate 118, 119, 121
South Carolina (USA) 126, 330
South China Sea **93–4**, 192, 343
South Fladen Pockmark Study Area
    (SFPSA; UK North Sea) 9, 11,
    14, **18–25**, 189, 194, 228, 232, 372,
    373
South Marsh Island (licence block, US
    Gulf of Mexico) 124
South Pass Area (Mississippi Delta) 364
South Shetland Micro-plate (south
    Atlantic Ocean) 118
South Timbalier (licence block, US Gulf
    of Mexico) 124
Southern Caspian Basin (Caspian Sea)
    85–6, 197, 237, 245
Southern Ritchie Ridge (offshore New
    Zealand) 188
Space Shuttle 379
Spain 66–7, 135, 266, 267, 332, 383, 385
Spanish–Moroccan mud volcano field 67,
    69
sparker 38, 43, 104, 105, 108, 172, 208
Special Areas for Conservation (SAC) 385
sphalerite (ZnS; mineral) 149, 314
Spiophanes kroyeri (polychaete) 249
Spirochaeta (bacteria) 96
sponges (phylum Porifera) 33, 46, 72, 78,
    87, 119, 122, 250, 252, 258, 261,
    262, 273, 274, 282, 305
  Chonelasma sp. 46
  Cladorhiza sp. 46, 122, 273
  Geodia sp. 46, 274
  Phakettia sp. 46
  Polymastia sp. 46
  Stelletta sp. 274
spring 134, 244, 308
  groundwater/freshwater 2, 75, 88, 132,
    190, 217, 218, 254, 329, 330
  hot/geothermal 2, 111, 145, 158, 248,
    309, 311, 319
spring sapping 183, **244–5**
squat lobster (decapod crustaceans;
    families Galatheidae and
    Chirostylidae) 253, 261
  Munida sarsi 38
SRBs (sulphate-reducing bacteria) 69,
    153, 261, 263–5, 295, 297, 342
St Lawrence, Gulf of (east Canada) 130
St Lawrence Island (Bering Sea) 105
St Lucia (Eastern Caribbean) 140
St Matthew Island (Alaska) 104

St Petersburg Mud Volcano (Gulf of
    Cadiz) 68
St Vincent (Eastern Caribbean) 140
Standard Mean Ocean Water (SMOW)
    162, 295
Starch (complex carbohydrate) 152
starfish (class Asteroidea) 27, 39, 274
Statfjord field (Norwegian North Sea) 36,
    170
steam (H_2O) 145, 209, 237, 240, 372
Stelletta sp. (sponge) 274
Sternaspis sp. (polychaete) 281
Stillwater igneous complex (Montana)
    145
Stockholm Archipelago **58–9**, 157, 193,
    195, 196
stockwork 145, 148
stony coral (subclass Scleractinia) 262
    Lophelia pertusa 51, 53, 64, 65, 69,
        256–8, 302
    Madrepora oculata 64, 69, 256
Storegga Slide (offshore Norway) 51, 174,
    176, 186, 350, **357–8**, 359
storm 72, 91, 100, 104, 232, 234, 236, 243,
    335, 341, 343, 345, 348, 360
storm surge 104
storm wave 104, 206, **233–4**, 235
strait
    Bransfield (Antarctica) 118, 148
    Florida (Cuba – USA) 311, 312
    Georgia (British Columbia) 107
    Gibraltar (Mediterranean Sea) 67, 69,
        385
    Hecate (British Columbia) 107
    Hormuz (Arabian Gulf) **91–2**, 346
    Magellan (Chile) 118
    Malacca (Indonesia / Malaysia)
        236
    Rande (Galicia, Spain) 267
    Shelikof (Alaska) **105–6**, 140
Strategic Environmental Assessment
    (UK's SEA programme) 12, 14,
    22, 40, 41, 42, 195, **386**
streamthroughs (bubble-formed holes in
    ice cover) 86
stress 4, 207, **214–15**, 216, 217, 226, 229,
    232, 234, 237, 238, 239, 240, 245,
    350, 360, 361, 369, 370, 384
    dominant 214
    effective (σ') 151, **215**, 222, 226, 233,
        234, 236, 350, 384 (*see also*
        Equation 7.3)
    horizontal (σ_h) 214, 222, 223, 236,
        241
    shear 229, 359
    stress (σ) 215, 216, 226
    tectonic 210, 233, 240

vertical (σ_v) **214**, 222, 236, 246, 359,
    360, 384 (*see also* Equation 7.1)
yield 225
stromatolite (microbial formations) 290,
    302, **308–11**
    Exuma Island (eastern Bahama Bank)
        308
Strömma lineament (Sweden) 58
strontium (Sr) 76, 119, 259, 318, 325
Student Mud Volcano (Gulf of Cadiz)
    67
Stvor Mud Volcano (Mediterranean
    Ridge) 77
sub-bottom profiler 9, 58, 61, 71, 82, 99,
    118, 129, 224, 231, 256, 331
subduction (plate convergence) 1, 75, 76,
    91, 94, 97, 99, 108, 119, 137,
    138–9, 140, 144, 148, 157, 158,
    181, 197, 198, 215, 252–3, 325,
    332, 374, 376, 377 (*see also*
    associated accretionary wedges
    and trenches)
    of seamounts 142
subduction (plate convergence) zone
    Aleutian (Alaska) **106–7**, 266, 267, 269,
        297, 307, 311
    Cascadia/Oregon (NE Pacific Ocean)
        109, 111, 181, 184–6, 252, 263,
        366–7
    Peru (SW Pacific Ocean) 119–20
submarine groundwater discharge (SGD;
    includes groundwater springs etc.)
    1, 2, 72, 75, 76, 88, 100, 113,
    125–6, 128, 133, 134, 145, **150**,
    151, 158, 190, 193, 217–18, 244,
    246, 248, 251, 253–5, 257, 259,
    261, 263, 266, 276, 308–10, 311,
    312, 318, 323, **329–31**, 374
    Bahrain 2, 88, 330
    Blake Ridge (offshore South Carolina)
        329
    Cambridge Fjord (Baffin Island,
        Canada) 132, 329, 349
    Cape Henlopen (Delaware, USA)
        253–4
    Chekka (Lebanon) 330
    Eckernförde Bay (Germany) **56–8**, 153,
        164, 165, 166, 169, 178, 193, 194,
        244, 246, 263, 264, 330, 348
    Elaiona Bay (Greece) 76, 244
    Florida Escarpment seeps (US Gulf of
        Mexico) **125–6**, 218, 244, 251,
        255, 266, 318, 329, 330
    Florida Keys (Florida, USA) 259
    Great Bay Estuary (New Hampshire,
        USA) 330
    Gulf of Kastela (Croatia) 72, 329

Ikka Fjord (Greenland) 311
Lake Superior (Canada/USA) **129**,
    130, 212, 254–5
Latakia (Syria) 2
Ovacik springs (Turkey) 329
Perth (Western Australia) 254
Turkey Point (Florida, USA) 330
Yuigahama Beach (Japan) 100
succinate ($HOOC\text{-}CH_2\text{-}CH_2\text{-}COOH$; a
    dicarboxylic acid) 158
suction anchor 371
Suez, Gulf of 1
sugars (represented by $C_6H_{12}O_6$) 152
Sula Ridge/Reef (offshore Norway) 256,
    383, 385
sulphate (SO_4^{2-}) 56, 96, 99, 103, 120,
    148, 163, 193, 252, 263, 264, 266,
    280, 281, 293, 294, 295, 207, 298,
    299, 300, 301, 330, 331, 337, 342
sulphate–methane transition zone
    (SMTZ) **153**, 263, 280, 301
sulphate-reducing bacteria (SRBs) 69,
    153, 261, 263–5, 295, 297, 342
sulphide 6, 93, 113, 118, 126, 148–9, 249,
    251, 252, 255, 263, 266, 268, 271,
    272, 273, 275, 280, 281, 282, 287,
    298, 307, 313, 315–16, 317, 318,
    331, 371, 374
    deposits 6, 87, 97, 149, 290, 314, 315,
        316, 317, 327, **374**
    hydrogen sulphide (H_2S) 15, 72, 75,
        81, 92, 99, 100, 109, 113, 114, 120,
        125, 132, 144, 145, 148, 160, 180,
        252, 257, 260, 262, 263, 264, 266,
        271, 281, 283, 295, 298, 301, 313,
        318, 326, 327, 335, 366, 383
    metal sulphides 6, 97, 118, 148, 149,
        263, 314–16, 317, 374, 384 (*see
        also* chalcopyrite, pyrite,
        pyrrhotite, and sphalerite)
sulphur (S) 7, 72, 87, 103, 122, 142, 160,
    202, 252, 253, 263, 269, 278, 282,
    289, 298, 307, 313, 325, 326, 329
    isotope ($\delta^{34}S$) 298, 299
sulphur-oxidising (thiotrophic) microbes
    102, 114, 250, 253, 263, 266, 272,
    281, 282
    Beggiatoa sp. (sulphur-oxidising
        bacterium) 35, 91, 96, 102, 113,
        125, 132, 193, 200, 249, 250, 251,
        252, 258, 263, 264, 266, 267, 273,
        274, 275, 281, 282, 289, 298, 299,
        301
    Thioploca sp. (sulphur-oxidising
        bacteria) 87, 91, 92, 266
    Thiothrix sp. (sulphur-oxidising
        bacteria) 253, 266

sulphur-oxidising (*cont.*)
  Thiovulum sp. (sulphur-oxidising
    bacteria) 96
Sumba Basin (Indonesia) **172**, 175, 259
Sun/sunlight 1, 4, 152, 204, 255, 349,
  350, 379
supercritical fluids 141, 146, 148, 316,
  317
supercritical water 146, 147–9, 237, 314,
  317, 318
Supersaturation 178, 221, 263, 290, 293,
  295
superthermophile (microbes) 265
Supsa River and canyon (Georgia, eastern
  Black Sea) 85
Sur Basin (offshore California) **114–15**
  pockmark field 114–15, 193, 194
suspension feeder 46, 272, 273, 274
Susquehanna River (eastern USA) 127
Susu Knolls vent site (offshore Papua
  New Guinea) 96
Svalbard (Norway) 268, 289
SWD (seismic-while-drilling) 369
Sweden 58–9, 159, 385
SWF (shallow water flow) **368–9**
Switzerland 385
symbionts 112, 122, 249, 255, 261,
  269–73, 277, 278, 279, 281, 289
  endosymbiont 114, 122, 193, 249, 253,
    269, 270, 271, 272, 278, 282, 298
  ectosymbiont 273
symbiosis/symbiotic 53, 69, 102, 250,
  262, 263, 264, **269–73**, 275, 277,
  283, 289, 304, 305, 350
Synidotea angulata (isopod) 251
synthetic aperture radar (SAR) 94, 379
Syria 2, 158

Table Bluff Anticline (offshore
  California) 112
TAG hydrothermal site (Mid-Atlantic
  Ridge) 315, 374
Tainan Basin (offshore Taiwan) 94
Tainan (Republic of China) **93–4**, 196,
  245, 375
talc ($H_2Mg_3(SiO_3)_4$ or $Mg_3Si_4O_{10}(OH)_2$;
  mineral) 316, 317
Taman Peninsula (Russian Black Sea
  coast) 83
Tamar Estuary (UK) 60
Tamaulipas (Mexico) 123
Tamu childressi (mytilid bivalve) 252
Tanakura Fault (Japan) 99
tar/asphalt 1, 2, 115, 116, 117, 120, 122,
  123, 124, 125, 155, 331, 334
  mounds/mud volcanoes 116, 122, 125
Tarapur (India) 92

Taranaki Basin (offshore New Zealand)
  93, 94, 104, 105, 107, 115, 123,
  129, 131, 136, 159, 207, 210, 211,
  212, 213, 241, 303, 307, 382
Tarsis Mud Volcano (Gulf of Cadiz) 68
Tartan field (UK North Sea) 9
TASYO mud volcano (Gulf of Cadiz) 68
TASYO mud volcano field (Gulf of
  Cadiz) 68
Taupo volcanic zone (New Zealand) **97–8**,
  310, 325
Teahitia Seamount (S. Pacific Ocean) 142
tectonic 58, 67, 86, 99, 107, 111, 119, 122,
  139, 181, 183, 186, 188, 190, 209,
  210, 216, 218, 225, 233, 240, 288,
  304, 314, 351, 368 (*see also* plate
  tectonics)
  compression 67, 68, 93, 94, 97, 111,
    113, 122, 140, 150, 181, 197, 214,
    215, 232, 245
  salt 122, 124, 359
Tees Estuary (UK) 60
television (TV) 55, 372
temperature 4, 96, 114, 153, 154, 162,
  163, 179, 180, 185, 219, 263, 265,
  271, 280, 287, 288, 290, 293, 295,
  297, 311, 315, 318, 321, 350, 352,
  367, 368, 371, 384
  atmosphere 179, 203, 221, 350, 351
  gradient 47, 49, 98, 121, 145, 180, 208,
    261, 322, 328, 352
  mud volcano 49, 84, 121, 204
  STP (standard temperature and
    pressure) 43, 87, 101, 126, 155,
    187
sub-seabed (rock/sediment) 87, 102,
  121, 138, 145, 146, 147, 148, 149,
  150, 154, 155, 156, 157, 158, 159,
  178, 179, 180, 182, 185, 186, 193,
  209, 237, 255, 256, 268, 276, 288,
  299, 316, 318, 321, 322, 328, 352,
  359
submarine groundwater discharge
  (SGD) 126, 318, 330
vent fluid 6, 76, 87, 96, 98, 103, 110,
  118, 149, 200, 201, 205, 309, 314,
  315, 316, 317, 321, 314
water 4, 65, 73, 96, 98, 100, 103, 107,
  110, 111, 125, 128, 146, 149, 178,
  179, 180, 181, 187, 188, 231, 233,
  246, 252, 253, 255, 261, 262, 283,
  285, 312, 329, 334, 338, 341, 343,
  348, 350, 352
tensile strength (*T*) 166, 215, 216, **222**
tension-leg structure 370
tepee 294, 308, 309, 310
termite 153, 345
Tertiary (*see also* Eocene, Miocene,
  Oligocene, Palaeocene, Pliocene)
  7, 10, 16, 22, 43, 44, 50, 59, 60, 66,

Tethys 67
Texas (USA) 123, 258, 275, 307, 359
Thailand 93
Thames River/Estuary (UK) 59, 60, 135
Thasos Island (Greece) 75
Thermocline 285, 341
Thermococcus sp. (Archaea) 96
thermogenic (derived by the
  thermocatalytic cracking of
  organic matter) 4, 18, 27, 40, 41,
  44, 45, 57, 58, 59, 67, 68, 70, 73,
  75, 84, 87, 90, 93, 94, 97, 98, 103,
  104, 105, 114, 118, 121, 123, 124,
  125, 130, 131, 135, 136, 149, 151,
  154–7, 158, 161, 162, 186, 190,
  202, 214, 217, 247, 258, 263, 300,
  311, 325, 335, 351, 360, 362, 381,
  382
thermophilic microbes 96
Thioploca sp. (sulphur-oxidising bacteria)
  87, 91, 92, 266
Thiothrix sp. (sulphur-oxidising bacteria)
  253, 266
thiotroph (sulphur-oxidising microbe)
  102, 114, 250, 253, 263, 266, 272,
  281, 282
  Beggiatoa sp. (sulphur-oxidising
    bacterium) 35, 91, 96, 102, 113,
    125, 132, 193, 200, 249, 250, 251,
    252, 258, 263, 264, 266, 267, 273,
    274, 275, 281, 282, 289, 298, 299,
    301
  Thioploca sp. (sulphur-oxidising
    bacteria) 87, 91, 92, 266
  Thiothrix sp. (sulphur-oxidising
    bacteria) 253, 266
  Thiovulum sp. (sulphur-oxidising
    bacteria) 96
thorium (Th) 59, 101, 145
thrombolite (microbial formations) 302,
  311
thrust 76, 97, 99, 112, 122, 123, 181, 253,
  306, 308
Thyasira sp. (Thyasirid bivalve) 249, 278,
  280, 281
thyasirid bivalves (Thyasiridae; clams) 69,
  97, 102, 271, **272**, 281
  Conchocele bisecta 102, 272
  Maorithyas sp. 97
  Thyasira sp. 249, 278, 280, 281
Tiburon (ROV) 114
tide/tidal 8, 11, 19, 59, 60, 66, 90, 100,
  108, 122, 126, 167, 176, 194, 215,
  226, 228, 229, **233–5**, 240, 242,

246, 247, 261, 262, 266, 267, 274,
    284, 294, 307, 308, 313, 332, 336,
    354, 360, 373, 375, 380
tidal pumping 233, 259, 294
Tierra del Fuego (Argentina/Chile) 118
till (glacially-derived sediment) 7, 45, 47,
    57
time-slicing (seismic process) 178, 209,
    212
Timor (Indonesia) 94, 172, 175
Timor Sea 94–5, 172, 175, 303, 381
Timor Trough 94
Tinrio Basin (Sea of Okhotsk) 102
Tiree (UK) 62
Tjeldbergodden processing plant
    (Norway) 383
TOC (total organic carbon) 59, 154, 282
Tohoku District (Japan) 229
Tommeliten (petroleum field and seeps,
    Norwegian North Sea) 3, 20,
    25–9, 59, 60, 191, 248, 264, 280,
    290, 291, 301, 302, 332, 333
Tonga Trench (SW Pacific Ocean) 142
Tori Shima (Japan) 237
Toronto Mud Volcano (Mediterranean
    Ridge) 77
Torry Bay (Firth of Forth, UK) 59, 266
torsk (*Brosme brosme*; fish) 32, 33, 39, 281
total organic carbon (TOC) 59, 154, 282
transform fault 95, 141, 277, 303
transform plate boundaries 113, 134, 137,
    140
trawling 13, 15, 68, 112, 251, 290, 340,
    372, 384, 385
Tredmar Mud Volcano (Black Sea) 84
trench
    Aleutian 266, 269, 307
    Barbados 252
    Bonin 100
    Hellenic 76, 79, 262
    Japan 100, 253, 278
    Kuril 100
    Mariana 98, 99, 142
    Nankai Trough 100, 101, 253, 278, 299,
        376
    New Britain 95
    Norwegian 9, 10, 11, 13, 15, 16, 17, 19,
        36, 37, 38, 52, 54, 60, 238, 243,
        248, 250, 381
    Peru 119, 269, 278, 279
    Puerto Rico 253
    Ryuku 101, 205
    Tonga 142
trenching (of cables or pipelines) 62,
    369–70
Triassic 46, 67, 68, 107, 123, 159, 239
trilobite (Palaeozoic fossil) 289

Trinidad 120, 122, 199, 204
Tristan da Cunha 141
trochid archeogastropod 255
Troll field (Norwegian North Sea) 9, 10,
    11, 20, 370
trophosome 271
Tsilin Valley (Taiwan) 375
tsunami 189, 235, 237, 355, 357, **359**, 384
TTR Mud Volcano (Gulf of Cadiz) 68
tube anemone (Anthozoa; order
    Ceriantharia)
    Cerianthus lloydii 249
tubeworm – *see* Siboglinid tubeworm
Tubeworm City (Monterey Canyon) 113
tubicolous polychaete 122
turbidite 87, 299, 315, 316, 376
turbidity current 131, 136, 205, 340, 355
Turkey 79, 81, 158, 310–11, 329
Turkey Point (Florida) 330
turrid gastropod (family Turridae) 253,
    255
TV (television) 55, 372
TV-guided equipment 106, 107, 119, 280,
    321, 332, 336
Twenty-eight Fathom Bank (US Gulf of
    Mexico) 258
Tyne Estuary (UK) 342
Tyrrhenian Sea 142, 262

UK (United Kingdom) 181, 294, 357,
    366, 379, 385, 386
UK15/25 (North Sea) 13, 20, 40, **41–4**,
    59, 189, 191, 195, **238–9**, 249, 281,
    285, 291, 334, 353, 386
UK16/3 (North Sea) 386
UKCS (United Kingdom continental
    shelf) 60
UK-sector, North Sea 9, 12, 14, 17, 18,
    20, 60, 63, 218, 247
Ukraine 83, 84, 203, 205
Ulleung Basin (Sea of Japan) 99
Ultramafic rock (e.g. peridotite) 144, 148,
    150, 160, 288
UN Convention on the Law of the Sea
    (UNCLOS) 384, 385
UN World Summit on Sustainable
    Development (WSSD) 384, 385
UNCLOS (UN Convention on the Law
    of the Sea) 384, 385
underground blowout 362, **363**, 364, 368,
    384
underwater rock drill 132
Underwater Swamps (Black Sea) **84–5**
undrained shear strength (S_u) 106, 167,
    215, 216, 226, 236, 240, 245, 357,
    359, 360, 370 (*see also* Equations
    7.4 and 7.13)

unit pockmark 13, 15, 129, 30, 36, 37, 39,
    43, 190, 191, 242, 256
United Kingdom (see *UK*)
United States (see *USA*)
upwelling 49, 122, 182, 308, 319
    mantle 141, 145
    seawater 72, 132, 242, 294, 320, 329,
        341, 343, 383
    upwelling flow **339**
uranium (U) 59, 145, 321
    uranium:thorium (U:Th) dating 101
Ur-Frisia (Pleistocene; southern North
    Sea) 135
Urticina feline (sea anemone) 249
US Atlantic Margin **126–7**, 358
USA (United States of America) 57, 95,
    111, **126–7**, 128, 136, 141, 194,
    278, 248, 329, 330, 361, 365, 375,
    382, 386
    Alabama 123
    Alaska **103–6**, 139, 238, 335, 356
    California 1, 98, **111–17**, 234–5, 241,
        251, 273, 275, 278, 281, 333, 336,
        375
    Delaware 254–5
    Florida 123, 124, 255, 278, 330
    Hawaii 141, 142, 148, 262, 319, 325
    Louisiana 123, 251, 258, 304, 381
    New Jersey 151, 359, 361
    North Carolina 126, 234, 284, 358
    Oklahoma 306
    Oregon 109, 111, 118, 184, 251, 268,
        278, 281, 299, 334, 342, 359
    South Carolina 126, 330, 358
    Texas 123, 258, 275, 307, 359
USSR 98, 157

vagrant 273, 274, 283
VAMP (velocity amplitude feature) 186,
    187, 377
Vancouver field (US Gulf of Mexico) 124
Vancouver Island (British Columbia) 107,
    174, 184
Vaucheria dichterma (alga) 58
velocity amplitude feature (VAMP) 186,
    187, 377
Vema Dome (offshore Norway) 51, 207
veneer boundstone 308
Venezuela 118
Venice lagoon (Italy) 73
Vent SamPler (VESP) 107
vent/venting 2, 4, 45, 144, 189, 190,
    277–80, 283, 288, 290, 323, 332,
    336, 337, 344, 345, 347, 352, 355,
    363, 365, 372, 383
    black smoker 96, 103, 274, 314, 315,
        324

vent/venting (*cont.*)
  cold (seep, mud volcano etc.) 15, 27, 68,
    73, 77, 78, 89, 91, 101, 102, 103,
    106, 107, 109, 112, 114, 116, 118,
    119, 120, 121, 122, 132, 182, 183,
    193, 195, 196, 197, 198, 200, 202,
    204, 209, 224, 228, 230, 231, 234,
    244, 251, 252, 253, 263, 277, 278,
    279, 280, 284, 285, 299, 301, 323,
    332, 336, 3337, 338, 340, 345, 347,
    356, 359, 386
  fauna 72, 87, 93, 95, 96, 100, 140,
    260–2, 263, 265–6, 268–79, 283,
    359
  hot (hydrothermal) 1, 2, 4, 6, 45, 72, 76,
    86, 87, 93, 95, 96, 98, 99, 100, 103,
    109, 110, 118, 140, 142, 144, 145,
    146, 148, 149, 158, 159, 218, 237,
    248, 251, 252, 255, 259, **260–2**,
    266, 268, 269, 270, 271, 272, 274,
    275, 275, 277, 279, 285, 286, 287,
    288, 290, 299, 313, **314–17**, 318,
    319, 320, 321, **323–4**, 325, 326,
    328, 329, 338, 339, 341, 344, 345,
    347, 350, 352, 369, 374, 384, 385
  white smoker 96, 103, 314, 315, 324
Vermilion (licence block, US Gulf of
    Mexico) 124
vertical seismic profile (VSP) 109, 185
vertical stress (σ_v) **214**, 222, 236, 246, 359,
    360, 384 (*see also* Equation 7.1)
vesicle 141, 142, 144
vesicomyid bivalves (Vesicomyidae; clams)
    100, 107, 112, 113, 114, 119, 120,
    122, 125, 252, 261, 269, **271–2**,
    278, 279, 283
  Calyptogena sp. 70, 91, 97, 99, 100, 103,
    112, 113, 118, 120, 122, 252, 253,
    262, 271, 278, 299
Veslefrikk field (Norwegian North Sea) 9,
    20, 62, 307
VESP (Vent SamPler) 107
Vest field (Italian Adriatic Sea) 73
vestimentiferan (former designation of
    siboglinid tubeworms) 269
video 31, 43, 46, 49, 85, 104, 107, 109,
    112, 182, 231, 249, 250, 274, 280,
    333, 334
Vienna Woods hydrothermal vents
    (offshore Papua New Guinea) 96,
    286
Vietnam **93**, 157
Viking Graben (North Sea) 36, 218,
    232
Virgin Islands (US) 294
Virgularia mirabilis (slender sea pen) 249

viscosity 132, 147, 198, 220, 225, 226,
    227
volcanic 98, 148, 202, 209, 317, 319, 320,
    321, 328, 3229, 348, 356
volcanic activity/eruption 69, 95, 96, 97,
    98, 100, 108, 110, 111, 139, 141,
    142, 144, 148, 158, 208, 209, 234,
    237, 268, 288, 324, 328, 345, 360,
    371, 372
  caldera 96, 100, 110, 111, 268, 283, 326,
    329
  diatreme **208–9**
  fluid 96, **144–5**, 158, 290, 326
  rock 96, 118, 141, 142, 145, 261, 319
  Taupo volcanic zone (New Zealand)
    97–8, 310, 325
volcanism 138, 145, 148, 160
  arc 75, 76, **140**, 262, 326, 352
  intra-plate 137, **140–2**
  silicic dome **98**
  spreading centre 138
volcano (*see also* mud volcano and sand
    volcano) 138, 189, 197, 209, 321,
    325, 328
  submarine 4, 98, 103, 110, 120, 140,
    141, 142, 148, 278, 316, 326, 347,
    371, 372
  Axial Seamount (Juan de Fuca Ridge)
    110–11, 142, 262, 329
  Kick'em Jenny (offshore Grenada) 120,
    140, 372
  Mt Mazama (volcano: Oregon)
    268
  Mt Sakurajima (Japan) 100
  Myojin-Sho (offshore Japan) 371
  Obsidian Dome (California) 98
  Piip (northwest Pacific Ocean) **103**,
    142, 158, 262, 316
  White Island (New Zealand) 98
  Volcano A Mud Volcano (offshore
    Barbados) 121
  Volcano F Mud Volcano (offshore
    Barbados) 121
  Volcano J Mud Volcano (offshore
    Barbados) 121
Volga River (Russia) 85
Vøring Basin/Plateau 51, 140, **207**, 237,
    297
VSP (vertical seismic profile) 109,
    185
Vulcan Basin (Australian Timor Sea)
    94
Vulkanolog (research ship) 101, 103

Walrus 105
Walvis Bay (Namibia) 72

water
  connate **150**, 202, 214
  density (ρ_{water}) 215, 219, 240, 285,
    320, 338, 340, 341, 366,
    372
  groundwater 4, 57, 150, 159, 190, 214,
    229, 237–8, 294, 305, 308, 322,
    347, 369
  juvenile 150
  pore water 49, 51, 57, 65, 67, 73, 99,
    100, 103, 107, 109, 113, 114, 115,
    121, 123, 126, 127, 131, 129, 149,
    150–1 152, 154, 163, 164, 165,
    166, 175, 176, 177, 178, 180, 181,
    182, 185, 186, 214–15, 221, 228,
    237, 241, 244, 253, 254, 256, 257,
    263, 282, 290, 291, 293, 294, 295,
    297, 301, 302, 306, 310, 311, 315,
    320, 331, 337, 343, 344, 363, 366,
    368, 371
  seawater 1, 4, 9, 25, 46, 49, 76, 78, 887,
    96, 99, 100, 101, 103, 107, 116,
    121, 122, 123, 124, 125, 126, 132,
    142, 145, 146, 147, 148, 149, 150,
    152, 153, 154, 161, 166, 167, 168,
    181, 187, 188, 208, 209, 215, 231,
    255, 257, 259, 261, 263, 266, 269,
    272, 275, 276, 279, 282, 291, 293,
    204, 295, 297, 299, 300, 301, 308,
    312, 314, 315, 317, 318, 319, 320,
    321, 323, 324, 325, 329, 330, 335,
    338, 340, 342, 344, 346, 352, 353,
    361, 381
water column
  acoustic target 9, 25, 27, 42, 43, 60, 63,
    66, 67, 75, 77, 84, 88, 111, 194,
    195, 231, **331–2**, 333, 334, 338,
    372, 373
  plume
    acoustic 24–5, 27, 43, 44, 54, 61, 65,
      67, 77, 89, 90, 97, 102, 103, 104,
      105, 118, 132, 142, 193, 194, 195,
      231, 285, **331–2**, 333, 340, 372,
      379
    hydrothermal 76, 77, 96, 98, 110,
      145, 149, 158, 268, 285, **323–9**,
      331, 350
    event (mega) 285, **324–5**, 328, 329,
      333, 334, 347, 352
wave **233–6**
  earthquake 228
  infragravity 235
  internal 235
  sandwave 10, 60–2, 64, 65, 357
  sea surface 231, **233–5**, 334, 343, 356,
    379, 380

sediment 120, 127
storm 104, 206, **233–4**, 235
wave length (acoustic; λ) **166**, 167 (*see* Equation 6.1)
wave loading 106, 237, 360
wave pumping 100, 233, 294
wax 152
Weardale Granite (northern England) 315
Weichselian (most-recent Quaternary glaciation; Wisconsin in North America) 45, 54, 59, 128, 256
West Africa 69–72, 136, 172–4, 244, 272, 368
West Cameron (licence block, US Gulf of Mexico) 124
West Flower Garden Bank (Gulf of Mexico) 258
West Vanguard (oil rig) 178, 364, 365
Westshore Terminal (Fraser Delta, British Columbia) 108
wetlands 75, 135, 342, 345, 347
Whakatane Graben (New Zealand) 98
whale 105, 192, 271, 272, 278, 333
Whale Island (New Zealand) 283
whelk (neogastropod) 261
White Island (New Zealand) 98
white smoker 96, 103, 314, 315, 324
whiting 72, **311–13**
wipe-out (acoustic blanking) 123, **172**

Witch Ground Basin 9, 10, 11, 12, 13, 14, 15, 18, 60, 62, 193, 195, 234, 247, 381
Witch Ground Formation 10, 15, 17, 18, 21, 24
Witch's Hole (UK North Sea) 20, 21, 22, **24–5, 372–3**
wolf fish (*Anarhychus lupus*) 249
Woodlark Basin (offshore Papua New Guinea) 95
Woodlark Spreading Centre (offshore Papua New Guinea) 139
World Wildlife Fund for Nature (WWF) 385, 386
Worm Garden (offshore Papua New Guinea) 96
WSSD (UN World Summit on Sustainable Development) 384, 385
Wyville–Thomson Ridge (NE Atlantic Ocean) 64, 65

xenophyophore (protozoans) 65
X-ray 66, 120, 122
X-ray computer-aided tomography (CT scan) 57, 164, 166, 169

Yampi Shelf (Australian Timor Sea) 94, 382

Yangtse River (Peoples' Republic of China) 99, 361
Yaquina Basin (offshore Peru) 120
Yellow Sea 99
Yesilirmak River (Turkey) 81
Yinggehai Basin (offshore People's Republic of China/Taiwan – Peoples' Republic of China) 93
Yucatan Peninsula (Mexico) 123
Yuigahama Beach (Japan) 100
Yukon River (Alaska) 104
Yuma Mud Volcano (Gulf of Cadiz) 67, 68
Yuzhmorgeologiya Mud Volcano (central Black Sea) 83

Zapiola Drift (Argentine Basin, south Atlantic Ocean) 120
Zeepipe pipeline (North Sea) 9, 10, 60
Zhemchung Canyon (Bering Sea) 105
zinc (Zn) 87, 97, 148, 149, 314, 319, 325, 374
Zoarces viviparous (eelpout; fish) 253
Zonenshayn flares (Sea of Okhotsk) 102
Zooplankton 152, 155, 250, 283, 284, 285, 343, 349
zooxanthellae (algal endosymbionts) 255
Zuluf field (Saudi Arabian Arabian Gulf) 88, 89